GENE THERAPY

ADVANCES IN
PHARMACOLOGY

VOLUME 40

GENE THERAPY

Edited by

J. Thomas August

Department of Pharmacology
Johns Hopkins University
Baltimore, Maryland

ADVANCES IN
PHARMACOLOGY

VOLUME 40

ACADEMIC PRESS

San Diego London Boston New York Sydney Tokyo Toronto

This book is printed on acid-free paper. 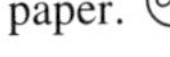

Academic Press
a division of Harcourt Brace & Company
525 B Street, Suite 1900, San Diego, California 92101-4495, USA
http://www.apnet.com

Academic Press Limited
24-28 Oval Road, London NW1 7DX, UK
http://www.hbuk.co.uk/ap/

International Standard Book Number: 0-12-032941-7

PRINTED IN THE UNITED STATES OF AMERICA
97 98 99 00 01 02 BB 9 8 7 6 5 4 3 2 1

Contents

Targeted Tumor Cytotoxicity Mediated by Intracellular Single-Chain Anti-oncogene Antibodies

David T. Curiel

In Vivo Gene Therapy with Adeno-Associated Virus Vectors for Cystic Fibrosis

Terence R. Flotte and Barrie J. Carter

Engineering Herpes Simplex Virus Vectors for Human Gene Therapy

Joseph C. Glorioso, William F. Goins, Martin C. Schmidt, Thomas Oligino,
David M. Krisky, Peggy C. Marconi, James D. Cavalcoli, Ramesh Ramakrishnan,
P. Luigi Poliani, and David J. Fink

Human Adenovirus Vectors for Gene Transfer into Mammalian Cells

Mary M. Hitt, Christina L. Addison, and Frank L. Graham

Anti-oncogene Ribozymes for Cancer Gene Therapy

Akira Irie, Hiroshi Kijima, Tsukasa Ohkawa, David Y. Bouffard, Toshiya Suzuki, Lisa D. Curcio, Per Sonne Holm, Alex Sassani, and Kevin J. Scanlon

Cytokine Gene Transduction in the Immunotherapy of Cancer

Giorgio Parmiani, Mario P. Colombo, Cecilia Melani, and Flavio Arienti

Gene Therapy Approaches to Enhance Antitumor Immunity

Daniel L. Shawler, Habib Fakhrai, Charles Van Beveren, Dan Mercola, Daniel P. Gold, Richard M. Bartholomew, Ivor Royston, and Robert E. Sobol

Modified Steroid Receptors and Steroid-Inducible Promoters as Genetic Switches for Gene Therapy

John H. White

Strategies for Approaching Retinoblastoma Tumor Suppressor Gene Therapy

Hong-Ji Xu

Immunoliposomes for Cancer Treatment

John W. Park, Keelung Hong, Dmitri B. Kirpotin, Demetrios Papahadjopoulos, and Christopher C. Benz

Antisense Inhibition of Virus Infections

R. E. Kilkuskie and A. K. Field

Contributors

Numbers in parentheses indicate the pages on which the authors' contributions begin.

Christina L. Addison (137) Department of Biology, McMaster University, Hamilton, Ontario, Canada L8S 4K1

Flavio Arienti (259) Gene Therapy Program, Division of Experimental Oncology D, Istituto Nazionale Tumori, 20133 Milan, Italy

Richard M. Bartholomew (309) The Immune Response Corporation, Carlsbad, California 92008

Christopher C. Benz (399) Department of Medicine, Division of Hematology-Oncology, University of California, San Francisco, San Francisco, California 94143

David Y. Bouffard (207) Department of Cancer Research, Berlex Biosciences, Richmond, California 94804

Barrie J. Carter (85) Research and Development, Targeted Genetics Corporation, Seattle, Washington 98101

James D. Cavalcoli (103) Department of Molecular Genetics and Biochemistry, University of Pittsburgh School of Medicine, Pittsburgh, Pennsylvania 15261

Mario P. Colombo (259) Gene Therapy Program, Division of Experimental Oncology D, Istituto Nazionale Tumori, 20133 Milan, Italy

Stanley T. Crooke (1) Isis Pharmaceuticals, Inc., Carlsbad, California 92008

Lisa D. Curcio (207) Department of General and Oncologic Surgery, City of Hope National Medical Center, Duarte, California 91010

David T. Curiel (51) Gene Therapy Program, University of Alabama at Birmingham, Birmingham, Alabama 35294

Habib Fakhrai (309) Sidney Kimmel Cancer Center, San Diego, California 92121

A. K. Field (437) Hybridon, Inc., Cambridge, Massachusetts 02139

David J. Fink (103) Departments of Neurology and of Molecular Genetics and Biochemistry, and VA Medical Center, University of Pittsburgh School of Medicine, Pittsburgh, Pennsylvania 15261

Terence R. Flotte (85) Gene Therapy Center and the Departments of Pediatrics and Molecular Genetics and Microbiology, University of Florida, Gainesville, Florida 32610

Joseph C. Glorioso (103) Department of Molecular Genetics and Biochemistry, University of Pittsburgh School of Medicine, Pittsburgh, Pennsylvania 15261

William F. Goins (103) Department of Molecular Genetics and Biochemistry, University of Pittsburgh School of Medicine, Pittsburgh, Pennsylvania 15261

Daniel P. Gold (309) Sidney Kimmel Cancer Center, San Diego, California 92121

Frank L. Graham (137) Departments of Biology and Pathology, McMaster University, Hamilton, Ontario, Canada L8S 4K1

Mary M. Hitt (137) Department of Biology, McMaster University, Hamilton, Ontario, Canada L8S 4K1

Per Sonne Holm (207) Section of Biochemical Pharmacology, Department of Medical Oncology, City of Hope National Medical Center, Duarte, California 91010*

Keelung Hong (399) Department of Cellular and Molecular Pharmacology, University of California, San Francisco, San Francisco, California 94143

Akira Irie (207) Department of Cancer Research, Berlex Biosciences, Richmond, California 94804

Hiroshi Kijima (207) Department of Cancer Research, Berlex Biosciences, Richmond, California 94804†

R. E. Kilkuskie (437) Hybridon, Inc., Cambridge, Massachusetts 02139

Dmitri B. Kirpotin (399) Department of Cellular and Molecular Pharmacology, University of California, San Francisco, San Francisco, California 94143

David M. Krisky (103) Department of Molecular Genetics and Biochemistry, University of Pittsburgh School of Medicine, Pittsburgh, Pennsylvania 15261

---

* *Current address:* Institut fur Pathologie, Charité, Humboldt-Universitat, D-10117 Berlin, Germany

† *Current address:* Department of Pathology, Tokai University School of Medicine, Bohseidai, Isehara, Kanawaga, Japan 269-11

Peggy C. Marconi (103) Department of Molecular Genetics and Biochemistry, University of Pittsburgh School of Medicine, Pittsburgh, Pennsylvania 15261

Cecilia Melani (259) Gene Therapy Program, Division of Experimental Oncology D, Istituto Nazionale Tumori, 20133 Milan, Italy

Dan Mercola (309) Sidney Kimmel Cancer Center, San Diego, California 92121

Tsukasa Ohkawa (207) Department of Cancer Research, Berlex Biosciences, Richmond, California 94804

Thomas Oligino (103) Department of Molecular Genetics and Biochemistry, University of Pittsburgh School of Medicine, Pittsburgh, Pennsylvania 15261

Demetrios Papahadjopoulos (399) Department of Cellular and Molecular Pharmacology, University of California, San Francisco, San Francisco, California 94143

John W. Park (399) Department of Medicine, Division of Hematology-Oncology, University of California, San Francisco, San Francisco, California 94143

Giorgio Parmiani (259) Gene Therapy Program, Division of Experimental Oncology D, Istituto Nazionale Tumori, 20133 Milan, Italy

P. Luigi Poliani (103) Department of Neurology, University of Pittsburgh School of Medicine, Pittsburgh, Pennsylvania 15261

Ramesh Ramakrishnan (103) Department of Molecular Genetics and Biochemistry, University of Pittsburgh School of Medicine, Pittsburgh, Pennsylvania 15261

Ivor Royston (309) Sidney Kimmel Cancer Center, San Diego, California 92121

Alex Sassani (207) Department of Cancer Research, Berlex Biosciences, Richmond, California 94804

Kevin J. Scanlon (207) Department of Cancer Research, Berlex Biosciences, Richmond, California 94804

Martin C. Schmidt (103) Department of Molecular Genetics and Biochemistry, University of Pittsburgh School of Medicine, Pittsburgh, Pennsylvania 15261

Daniel L. Shawler (309) Sidney Kimmel Cancer Center, San Diego, California 92121

Robert E. Sobol (309) Sidney Kimmel Cancer Center, San Diego, California 92121

Toshiya Suzuki (207) Department of Cancer Research, Berlex Biosciences, Richmond, California 94804

Charles Van Beveren (309)   Sidney Kimmel Cancer Center, San Diego, California 92121

John H. White (339)   Department of Physiology, McGill University, Montreal, Quebec, Canada H3G 1Y6

Hong-Ji Xu (369)   Department of Molecular Oncology, Division of Medicine, The University of Texas M. D. Anderson Cancer Center, Houston, Texas 77030

Stanley T. Crooke

Isis Pharmaceuticals, Inc.
Carlsbad, California 92008

Advances in Understanding the Pharmacological Properties of Antisense Oligonucleotides

I. Introduction

Interest in developing antisense technology and in exploiting it for therapeutic purposes has become intense. Although progress has been gratifyingly rapid, the technology remains in its infancy and the questions that remain to be answered still outnumber the questions for which there are answers. Appropriately, considerable debate continues about the breadth of the utility of the approach and about the type of data required to prove that a drug works through an antisense mechanism.

The objectives of this chapter are to provide a summary of progress, to assess the status of the technology, to place the technology in the pharmacological context in which it is best understood, and to deal with some of the controversies with regard to the technology and the interpretation of experiments.

Advances in Pharmacology, Volume 40

II. History

Clearly, the antisense concept derives from an understanding of nucleic acid structure and function and depends on Watson–Crick hybridization (Watson and Crick, 1953). Thus, arguably, the demonstration that nucleic acid hybridization is feasible (Gillespie and Spiegelman, 1965) and the advances in *in situ* hybridization and diagnostic probe technology (Thompson and Gillespie, 1990) lay the most basic elements of the foundation supporting the antisense concept.

However, the first clear enunciation of the concept of exploiting antisense oligonucleotides as therapeutic agents was in the work of Zamecnik and Stephenson (1978). In their publication, these authors reported the synthesis of an oligodeoxyribonucleotide 13 nucleotides long that was complementary to a sequence in the Rous sarcoma virus genome. They suggested that this oligonucleotide could be stabilized by 3′- and 5′-terminal modifications and showed evidence of antiviral activity. More important, they discussed possible sites for binding in RNA and mechanisms of action of oligonucleotides.

Although less precisely focused on the therapeutic potential of antisense oligonucleotides, the work of Miller and Ts'o and their collaborators during the same period helped establish the foundation for antisense research and reestablish an interest in phosphate backbone modifications as approaches to improve the properties of oligonucleotides (Ts'o *et al.*, 1983; Barrett *et al.*, 1974; Miller, 1989). Their focus on ethyl phosphotriester-modified oligonucleotides as a potential medicinal chemical solution to pharmacokinetic limitations of oligonucleotides presaged much of the medicinal chemistry to be performed on oligonucleotides.

Despite the observations of Miller and Ts'o and Zamecnik and colleagues, interest in antisense research was quite limited until the late 1980s, when advances in several areas provided technical solutions to a number of impediments. As antisense drug design requires an understanding of the sequence of the RNA target, the explosive growth in availability of viral and human genomic sequences provided the information from which "receptor sequences" could be selected. The development of methods for synthesis of research quantities of oligonucleotide drugs then supported antisense experiments with both phosphodiester and modified oligonucleotides (Caruthers, 1985; Alvarado-Urbina *et al.*, 1981). The inception of the third key component (medicinal chemistry) forming the foundation of oligonucleotide therapeutics, in fact, is the synthesis in 1969 of phosphorothioate poly(rI)–poly(rC) as a means of stabilizing the polynucleotide (DeClercq *et al.*, 1969). Subsequently, Miller and Ts'o initiated studies on the neutral phosphate analogs, methylphosphonates (Ts'o *et al.*, 1983), and groups at the National Institutes of Health, the Food and Drug Administration, and the Worcester Foundation investigated phosphorothioate oligonucleotides (Marcus-Sekura *et al.*, 1987; Matsukura *et al.*, 1987; Agrawal *et al.*, 1988; Goodchild *et al.*, 1988; Sarin *et al.*, 1988). With these advances forming the foundation for oligonucleotide therapeutics and the initial studies suggesting *in vitro* activities against a number of viral and mammalian targets (Agrawal *et al.*, 1988; Gao *et al.*, 1989; Smith *et al.*, 1986; Agris *et al.*, 1986; Heikkila *et al.*, 1987; Wickström *et al.*, 1989), interest in oligonucleotide therapeutics intensified.

III. Proof of Mechanism

A. Factors That May Influence Experimental Interpretations

Clearly, the ultimate biological effect of an oligonucleotide will be influenced by the local concentration of the oligonucleotide at the target RNA, the concentration of the RNA, the rates of synthesis and degradation of the RNA, the type of terminating mechanism, and the rates of the events that result in termination of the activity of the RNA. At present, we understand essentially nothing about the interplay of these factors.

1. Oligonucleotide Purity

Currently, phosphorothioate oligonucleotides can be prepared consistently and with excellent purity (S. T. Crooke and Lebleu, 1993). However, this has been the case for only the past 3 to 4 years. Prior to that time, synthetic methods were evolving and analytical methods were inadequate. In fact, our laboratory reported that different synthetic and purification procedures resulted in oligonucleotides that varied in cellular toxicity (R. M. Crooke, 1991) and that potency varied from batch to batch. Although these are no longer synthetic problems that phosphorothioates, they undoubtedly complicated earlier studies. More important, with each new analog class, new synthetic, purification, and analytical challenges are encountered.

2. Oligonucleotide Structure

Antisense oligonucleotides are designed to be single stranded. We now understand that certain sequences (e.g., stretches of guanosine residues) are prone to adopt more complex structures (Wyatt *et al.*, 1994). The potential to form secondary and tertiary structures also varies as a function of the chemical class. For example, higher affinity 2′-modified oligonucleotides have a greater tendency to self-hybridize, resulting in more stable oligonucleotide duplexes than would be expected on the basis of rules derived from oligonucleotides (S. M. Freier, unpublished results).

3. RNA Structure

RNA is structured. The structure of the RNA has a profound influence on the affinity of the oligonucleotide and on the rate of binding of the oligonucleotide to its RNA target (Freier, 1993; Ecker, 1993). Moreover, RNA structure produces asymmetrical binding sites that then result in divergent affinity constants depending on the position of oligonucleotide in that structure (Lima *et al.*, 1992; Ecker *et al.*, 1992; Ecker, 1993). This, in turn, influences the optimal length of an oligonucleotide needed to achieve maximal affinity. We understand little about how RNA structure and RNA–protein interactions influence antisense drug action.

4. Variations in in Vitro Cellular Uptake and Distribution

Studies in several laboratories have clearly demonstrated that cells in tissue culture may take up phosphorothioate oligonucleotides via an active process, and that the uptake of these oligonucleotides is highly variable, depending on many conditions (R. M. Crooke, 1991; S. T. Crooke *et al.*, 1994). Cell type has a dramatic effect on total uptake, kinetics of uptake, and pattern of subcellular distribution. At present, there is no unifying hypothesis to explain these differences. Tissue culture conditions, such as the type of medium, the degree of confluence, and the presence of serum, can all have enormous effects on uptake (S. T. Crooke *et al.*, 1994). Oligonucleotide chemical class obviously influences the characteristics of uptake as well as the mechanism of uptake. Within the phosphorothioate class of oligonucleotides, uptake varies as a function of length, but not linearly (R. M. Crooke, 1991). Uptake varies as a function of sequence, and stability in cells is also influenced by sequence (S. T. Crooke *et al.*, 1994; R. M. Crooke *et al.*, 1995).

Given the foregoing, it is obvious that conclusions about *in vitro* uptake must be carefully made and generalizations are virtually impossible. Thus, before an oligonucleotide could be said to be inactive *in vitro,* it should be studied in several cell lines. Furthermore, while it may be absolutely correct that receptor-mediated endocytosis is a mechanism of uptake of phosphorothioate oligonucleotides (Loke *et al.*, 1989), it is obvious that a generalization that all phosphorothioates are taken up by all cells *in vitro* primarily by receptor-mediated endocytosis is simply unwarranted.

Finally, extrapolations from *in vitro* uptake studies to predictions about *in vivo* pharmacokinetic behavior are entirely inappropriate; in fact, there are now several lines of evidence in animals and humans that, even after careful consideration of all *in vitro* uptake data, one cannot predict *in vivo* pharmacokinetics of the compounds (Cossum *et al.*, 1993, 1994; S. T. Crooke *et al.*, 1994; Sands *et al.*, 1994).

5. Binding to and Effects of Binding to Protein and Other Non-Nucleic-Acid Targets

Phosphorothioate oligonucleotides tend to bind to many proteins, and those interactions are influenced by many factors. The effects of binding can influence cell uptake, distribution, metabolism, and excretion. They may induce nonantisense effects that can be mistakenly interpreted as antisense or complicate the identification of an antisense mechanism. By inhibiting RNase H, protein binding may inhibit the antisense activity of some oligonucleotides. Finally, binding to proteins can certainly have toxicological consequences.

In addition to proteins, oligonucleotides may interact with other biological molecules, such as lipids or carbohydrates, and such interactions, like those with proteins, will be influenced by the chemical class of oligonucleotide studied. Unfortunately, essentially no data bearing on such interactions are currently available.

An especially complicated experimental situation is encountered in many *in vitro* antiviral assays. In these assays, high concentrations of drugs, viruses, and cells are often coincubated. The sensitivity of each virus to nonantisense effects of oligonucleotides varies depending on the nature of the virion proteins and the characteristics of the oligonucleotides (Cowsert, 1993; Azad *et al.*, 1993). This has resulted in considerable confusion. In particular for human immunodeficiency virus (HIV), herpes simplex virus, cytomegaloviruses, and influenza virus, the nonantisense effects have been so dominant that identifying oligonucleotides that work via an antisense mechanism has been difficult. Given the artificial character of such assays, it is difficult to know whether nonantisense mechanisms would be as dominant *in vivo* or result in antiviral activity.

6. Terminating Mechanisms

It has been amply demonstrated that oligonucleotides may employ several terminating mechanisms. The dominant terminating mechanism is influenced by RNA receptor site, oligonucleotide chemical class, cell type, and probably many other factors (for review, see S. T. Crooke *et al.*, 1994). Obviously, as variations in terminating mechanism may result in significant changes in antisense potency and studies have shown significant variations from cell type to cell type *in vitro,* it is essential that the terminating mechanism be well understood. Unfortunately, at present, our understanding of terminating mechanisms remains rudimentary.

7. Effects of Control Oligonucleotides

A number of types of control oligonucleotides, including randomized oligonucleotides, have been used. Unfortunately, we know little to nothing about the potential biological effects of such "controls"; the more complicated a biological system and test, the more likely that "control" oligonucleotides may have activities that complicate interpretations. Thus, when a control oligonucleotide displays a surprising activity, the mechanism of that activity should be explored carefully before concluding that the effects of the control oligonucleotide prove that the activity of the putative antisense oligonucleotide is not due to an antisense mechanism.

8. *Kinetics of Effects*

Many rate constants may affect the activities of antisense oligonucleotides (e.g., the rate of synthesis and degradation of the target RNA and its protein; the rates of uptake into cells; the rates of distribution, extrusion, and metabolism of an oligonucleotide in cells; and similar pharmacokinetic considerations in animals). Despite this, relatively few time courses have been reported, and *in vitro* studies that range from a few hours to several days have been reported. In animals, we have a growing body of information on pharmacokinetics, but in most studies reported to date, the doses and schedules were chosen arbitrarily; again, little information on duration of effect and onset of action has been presented.

Clearly, more careful kinetic studies are required, and rational *in vitro* and *in vivo* dose schedules must be developed.

B. Recommendations: Positive Demonstration of Antisense Mechanism and Specificity

Until more is understood about how antisense drugs work, it is essential to positively demonstrate effects consistent with an antisense mechanism. For RNase H-activating oligonucleotides, Northern blot analysis showing selective loss of the target RNA is the best choice, and many laboratories are publishing reports of such activities *in vitro* and *in vivo* (Chiang *et al.*, 1991; Dean and McKay, 1994; Skorski *et al.*, 1994; Hijiya *et al.*, 1994). Ideally, a demonstration that closely related isotypes are unaffected should be included. In brief, then, for proof of mechanism, the following steps are recommended:

- Perform careful dose–response curves *in vitro*, using several cell lines and methods of *in vitro* delivery.
- Correlate the rank order potency *in vivo* with that observed *in vitro* after thorough dose–response curves are generated *in vivo*.
- Perform careful "gene walks" for all RNA species and oligonucleotide chemical classes.
- Perform careful time courses before drawing conclusions about potency.
- Directly demonstrate the proposed mechanism of action by measuring the target RNA and/or protein.
- Evaluate specificity and therapeutic indices via studies on closely related isotypes and with appropriate toxicological studies.
- Perform sufficient pharmacokinetics to define rational dosing schedules for pharmacological studies.
- When control oligonucleotides display surprising activities, determine the mechanisms involved.

IV. Molecular Mechanisms of Antisense Drugs

A. Occupancy Only-Mediated Mechanisms

Classic competitive antagonists are thought to alter biological activities because they bind to receptors, preventing natural agonists from binding, then inducing normal biological processes. Binding of oligonucleotides to specific sequences may inhibit the interaction of the RNA with proteins, other nucleic acids, or other factors required for essential steps in the intermediary metabolism of the RNA or its utilization by the cell.

1. Inhibition of Splicing

A key step in the intermediary metabolism of most mRNA molecules is the excision of introns. These "splicing" reactions are sequence specific and require the concerted action of spliceosomes. Consequently, oligonucleotides that bind to sequences required for splicing may prevent binding of necessary factors or physically prevent the required cleavage reactions. This, then, would result in inhibition of the production of the mature mRNA. Although there are several examples of oligonucleotides directed to splice junctions, none of the studies present data showing inhibition of RNA processing, accumulation of splicing intermediates, or a reduction in mature mRNA. Nor are there published data in which the structure of the RNA at the splice junction was probed and the oligonucleotides demonstrated to hybridize to the sequences for which they were designed (McManaway *et al.*, 1990; Kulka *et al.*, 1989; Zamecnik *et al.*, 1986; Smith *et al.*, 1985b). Activities have been reported for anti-c-*myc* and antiviral oligonucleotides with phosphodiester, methyl phosphonate, and phosphorothioate backbones. An oligonucleotide has been reported to induce alternative splicing in a cell-free splicing system, and in that system RNA analyses confirmed the putative mechanism (Dominski and Kole, 1993).

In our laboratory, we have attempted to characterize the factors that determine whether splicing inhibition is effected by an antisense drug (Hodges and Crooke, 1995). To this end, a number of luciferase-reporter plasmids containing various introns were constructed and transfected into HeLa cells. The effects of antisense drugs designed to bind to various sites were then characterized. The effects of RNase H-competent oligonucleotides were compared with those of oligonucleotides that do not serve as RNase H substrates. The major conclusions from this study were, first, that most of the splicing inhibition reported in earlier studies was probably due to nonspecific effects. Second, less effectively spliced introns are better targets than those with strong consensus splicing signals. Third, the 3′ splice site and branch point are usually the best sites to which to target to the oligonucleotide to inhibit splicing. Fourth, RNase H-competent oligonucleotides are usually more potent than even higher affinity oligonucleotides that inhibit by occupancy only.

2. Translational Arrest

A mechanism for which many oligonucleotides have been designed is to arrest translation of targeted protein by binding to the translation initiation codon. The positioning of the initiation codon within the area of complementarity of the oligonucleotide and the length of oligonucleotide used have varied considerably. Again, unfortunately, only in relatively few studies have the oligonucleotides in fact been shown to bind to the sites for which they were designed, and data that directly support translation arrest as the mechanism have been lacking.

Target RNA species that have been reported to be inhibited by a translational arrest mechanism include HIV (Agrawal *et al.*, 1988), vesicular stomatitis virus (VSV) (Lemaitre *et al.*, 1987), N-*myc* (Rosolen *et al.*, 1990), and a number of normal cellular genes (Vasanthakumar and Ahmed, 1989; Sburlati *et al.*, 1991; Zheng *et al.*, 1989; Maier *et al.*, 1990).

In our laboratories, we have shown that a significant number of targets may be inhibited by binding to translation initiation codons. For example, ISIS 1082 hybridizes to the AUG codon for the UL13 gene of herpesvirus types 1 and 2. RNase H studies confirmed that it binds selectively in this area. *In vitro* protein synthesis studies confirmed that it inhibited the synthesis of the UL13 protein, and studies in HeLa cells showed that it inhibited the growth of herpes type 1 and type 2 with a 50% inhibitory concentration (IC_{50}) of 200–400 nM by translation arrest (Mirabelli *et al.*, 1991). Similarly, ISIS 1753, a 30-mer phosphorothioate complementary to the translation initiation codon and surrounding sequences of the E2 gene of bovine papillomavirus, was highly effective and its activity was shown to be due to translation arrest. ISIS 2105, a 20-mer phosphorothioate complementary to the same region in human papillomavirus, was shown to be a potent inhibitor. Compounds complementary to the translation initiation codon of the E2 gene were the most potent of the more than 50 compounds studied that were complementary to various other regions in the RNA (Cowsert *et al.*, 1993). We have shown inhibition of translation of a number of other mRNA species by compounds designed to bind to the translation codon as well.

In conclusion, translation arrest represents an important mechanism of action for antisense drugs. A number of examples purporting to employ this mechanism have been reported, and studies on several compounds have provided data that unambiguously demonstrate that this mechanism can result in potent antisense drugs. However, little is understood about the precise events that lead to translation arrest.

3. Disruption of Necessary RNA Structure

RNA adopts a variety of three-dimensional structures induced by intramolecular hybridization, the most common of which is the stem–loop. These structures play crucial roles in a variety of functions. They are used to provide additional stability for RNA and as recognition motifs for a number of proteins, nucleic acids, and ribonucleoproteins that participate in the intermediate metabolism and activities of RNA species. Thus, given the potential general activity of the mechanism, it is surprising that occupancy-based disruption of RNA has not been more extensively exploited.

As an example, we designed a series of oligonucleotides that bind to the important stem–loop present in all RNA species in HIV, the *trans*-activator response (TAR) element. We synthesized a number of oligonucleotides designed to disrupt the TAR element and showed that several did indeed bind to the TAR element, disrupt the structure, and inhibit TAR-mediated production of a reporter gene (Vickers *et al.*, 1991). Furthermore, general rules useful in disrupting stem–loop structures were developed as well (Ecker *et al.*, 1992).

Although designed to induce relatively nonspecific cytotoxic effects, two other examples are noteworthy. Oligonucleotides designed to bind to a 17-nucleotide loop in *Xenopus* 28S RNA required for ribosome stability and protein synthesis inhibited protein synthesis when injected into *Xenopus* oocytes (Saxena and Ackerman, 1990). Similarly, oligonucleotides designed to bind to highly conserved sequences in 5.8S RNA inhibited protein synthesis in rabbit reticulocyte and wheat germ systems (Walker *et al.*, 1990).

B. Occupancy-Activated Destabilization

RNA molecules regulate their own metabolism. A number of structural features of RNA are known to influence stability, various processing events, subcellular distribution, and transport. It is likely that, as RNA intermediary metabolism is better understood, many other regulatory features and mechanisms will be identified.

I. 5 Capping

A key early step in RNA processing is 5′ capping (Fig. 1). This stabilizes pre-mRNA and is important for the stability of mature mRNA. It also is important in binding to the nuclear matrix and transport of mRNA out of the nucleus. As the structure of the cap is unique and understood, it presents an interesting target.

Several oligonucleotides that bind near the cap site have been shown to be active, presumably by inhibiting the binding of proteins required to cap the RNA. For example, the synthesis of simian virus 40 (SV40) T antigen was reported to be most sensitive to an oligonucleotide linked to polylysine and targeted to the 5′ cap site of RNA (Westerman *et al.*, 1989). However, again, in no published study has this putative mechanism been rigorously demonstrated. In fact, in no published study have the oligonucleotides been shown to bind to the sequences for which they were designed.

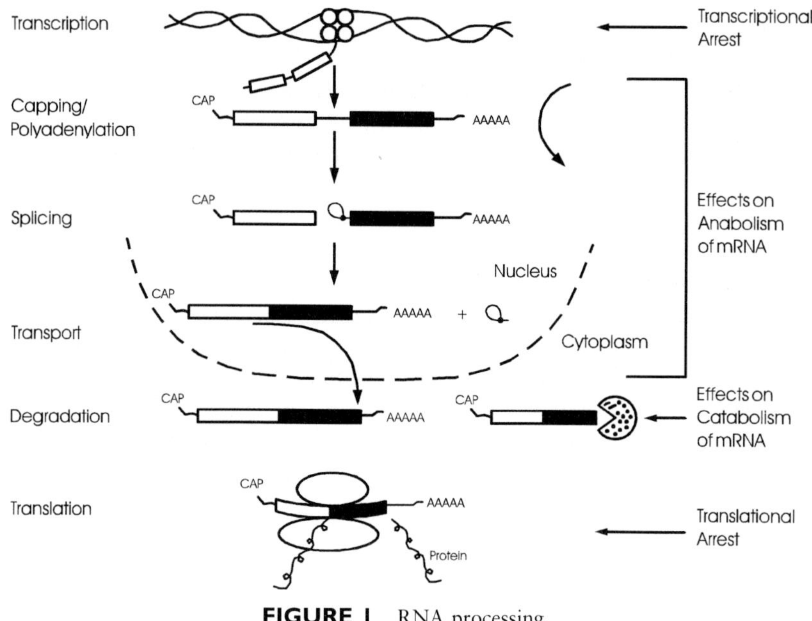

FIGURE I   RNA processing.

In our laboratory, we have designed oligonucleotides to bind to 5′ cap structures and reagents to specifically cleave the unique 5′ cap structure (Baker, 1993). These studies demonstrate the 5′ cap-targeted oligonucleotides were capable of inhibiting the binding of the translation initiation factor eIF-4α (Baker *et al.*, 1992).

2. Inhibition of 3 Polyadenylation

In the 3′-untranslated region of pre-mRNA molecules are sequences that result in the posttranscriptional addition of long (hundreds of nucleotides) tracts of polyadenylate. Polyadenylation stabilizes mRNA and may play other roles in the intermediate metabolism of RNA species. Theoretically, interactions in the 3′-terminal region of pre-mRNA could inhibit polyadenylation and destabilize the RNA species. Although there are a number of oligonucleotides that interact in the 3′-untranslated region and display antisense activities (Chiang *et al.*, 1991), to date no study has reported evidence of alterations in polyadenylation.

C. Other Mechanisms

In addition to 5′ capping and 3′ adenylation, there are clearly other sequences in the 5′- and 3′-untranslated regions of mRNA that affect the stability of the molecules. Again, there are a number of antisense drugs that may work by these mechanisms.

Zamecnik and Stephenson (1978) reported that a 13-mer targeted to untranslated 3′- and 5′-terminal sequences in Rous sarcoma viruses was active. Oligonucleotides conjugated to an acridine derivative and targeted to a 3′-terminal sequence in type A influenza viruses were reported to be active (Zerial *et al.*, 1987; Thuong *et al.*, 1989; Helene and Toulme, 1989), against several RNA targets, and studies in our laboratories have shown that sequences in the 3′-untranslated region of RNA molecules are often the most sensitive. For example, ISIS 1939 is a 20-mer phosphorothioate that binds to and appears to disrupt a predicted stem–loop structure in the 3′-untranslated region of the mRNA for the intercellular adhesion molecule (ICAM), a potent antisense inhibitor. However, inasmuch as a 2′-methoxy analog of ISIS 1939 was much less active, it is likely that, in addition to destabilization to cellular nucleolytic activity, activation of RNase H (see the following section) is also involved in the activity of ISIS 1939 (Chiang *et al.*, 1991).

D. Activation of RNase H

RNase H is a ubiquitous enzyme that degrades the RNA strand of an RNA–DNA duplex. It has been identified in organisms as diverse as viruses and human cells (for review, see Crouch and Dirksen, 1985). At least two classes of RNase H have been identified in eukaryotic cells. Multiple enzymes with RNase H activity have been observed in prokaryotes (Crouch and Dirksen, 1985).

Although RNase H is involved in DNA replication, it may play other roles in the cell and is found in the cytoplasm as well as the nucleus (Crum *et al.*, 1988). However, the concentration of the enzyme in the nucleus is thought to be greater, and some of the enzyme found in cytoplasmic preparations may be due to nuclear leakage.

RNase H activity is quite variable in cells. It is absent or minimal in rabbit reticulocytes (Haeuptle *et al.*, 1986) but present in wheat germ extracts (Crouch and Dirksen, 1985). In HL-60 cells, for example, the level of activity in undifferentiated cells is greatest, relatively high in dimethyl sulfoxide (DMSO)- and vitamin D-differentiated cells, and much lower in phorbol myristate acetate (PMA)-differentiated cells (G. D. Hoke, unpublished data).

The precise recognition elements for RNase H are not known. However, it has been shown that tetramer-length oligonucleotides with DNA-like properties can activate RNase H (Doris-Keller, 1979). Changes in the sugar moiety influence RNase H activation, as sugar modifications that result in RNA-like oligonucleotides (e.g., 2′-fluoro or 2′-methoxy) do not appear to serve as substrates for RNase H (Kawasaki *et al.*, 1993; Sproat *et al.*, 1989). Alterations in the orientation of the sugar relative to the base can also affect RNase H activation, as α-oligonucleotides are unable to induce RNase H or may require parallel annealing (Morvan *et al.*, 1991; Gagnor *et al.*, 1989).

In addition, backbone modifications influence the ability of oligonucleotides to activate RNase H. Methyl phosphonates do not activate RNase H (Maher *et al.,* 1989; Miller, 1989). In contrast, phosphorothioates are excellent substrates (Mirabelli *et al.,* 1991; Cazenave *et al.,* 1989; G. D. Hoke, unpublished data). In addition, chimeric molecules have been studied as oligonucleotides that bind to RNA and activate RNase H (Quartin *et al.,* 1989; Furdon *et al.,* 1989). For example, oligonucleotides composed of wings of 2′-methoxy phosphonates and a five-base gap of deoxyoligonucleotides bind to their target RNA and activate RNase H (Quartin *et al.,* 1989; Furdon *et al.,* 1989). Furthermore, a single ribonucleotide in a sequence of deoxyribonucleotides was shown to be sufficient to serve as a substrate for RNase H when bound to its complementary deoxyoligonucleotide (Eder and Walder, 1991).

That it is possible to take advantage of chimeric oligonucleotides designed to activate RNase H, have greater affinity for their RNA receptors, and enhance specificity has also been demonstrated (Monia *et al.,* 1993; Giles and Tidd, 1992). RNase H-mediated cleavage of target transcript was much more selective when deoxyoligonucleotides composed of methyl phosphonate deoxyoligonucleotide wings and phosphodiester gaps were compared to full phosphodiester oligonucleotides (Giles and Tidd, 1992).

Despite the information about RNase H and the demonstration that many oligonucleotides may activate RNase H in lysate and purified enzyme assays (Walder and Walder, 1988; Minshull and Hunt, 1986; Gagnor *et al.,* 1987), relatively little is yet known about the role of structural features in RNA targets in activating RNase H. In fact, direct proof that RNase H activation is, in fact, the mechanism of action of oligonucleotides in cells is to a large extent lacking.

Studies in our laboratories provide additional, albeit indirect, insights into these questions. ISIS 1939 is a 20-mer phosphorothioate complementary to a sequence in the 3′-untranslated region of ICAM-1 RNA (Chiang *et al.,* 1991). It inhibits ICAM production in human umbilical vein endothelial cells, and northern blots demonstrate that ICAM-1 mRNA is rapidly degraded. A 2′-methoxy analog of ISIS 1939 displays higher affinity for the RNA than the phosphorothioate, is stable in cells, but inhibits ICAM-1 protein production much less potently than ISIS 1939. It is likely that ISIS 1939 destabilizes the RNA and activates RNase H. In contrast, ISIS 1570, an 18-mer phosphorothioate that is complementary to the translation initiation codon of the ICAM-1 message, inhibited production of the protein but caused no degradation of the RNA. Thus, two oligonucleotides that are capable of activating RNase H had different effects depending on the site in the mRNA at which they bound (Chiang *et al.,* 1991).

A more direct demonstration that RNase H is likely a key factor in the activity of many antisense oligonucleotide was provided by studies in which a reverse-ligation polymerase chain reaction (PCR) was used to identify cleavage products from *bcr-abl* mRNA in cells treated with phosphorothioate oligonucleotides (Giles *et al.*, 1995).

Given the emerging role of chimeric oligonucleotides with modifications in the 3′ and 5′ wings designed to enhance affinity for the target RNA and nuclease stability and a DNA-type gap to serve as a substrate for RNase H, studies focused on understanding the effects of various modifications on the efficiency of the enzyme(s) are also of considerable importance. In one such study on *Escherichia coli* RNase H, we have reported that the enzyme displays minimal sequence specificity and is processive. When a chimeric oligonucleotide with 2′-modified sugars in the wings was hybridized to the RNA, the initial site of cleavage was the nucleotide adjacent to the methoxy–deoxy junction closest to the 3′ end of the RNA substrate. The initial rate of cleavage increased as the size of the DNA gap increased, and the efficiency of the enzyme was considerably less against an RNA target duplexed with a chimeric antisense oligonucleotide than a full DNA-type oligonucleotide (S. T. Crooke *et al.*, 1995).

V. Characteristics of Phosphorothioate Oligodeoxynucleotides

A. Introduction

Of the first-generation oligonucleotide analogs, the class that has resulted in the broadest range of activities and about which the most is known is the phosphorothioate class. Phosphorothioate oligonucleotides were first synthesized in 1969, when a poly(rI)–poly(rC) phosphorothioate was synthesized (DeClerq *et al.*, 1969). Their modification clearly achieves the objective of increased nuclease stability. In this class of oligonucleotides, one of the oxygen atoms in the phosphate group is replaced with a sulfur. The resulting compound is negatively charged, as is a chiral phosphodiester, but much more resistant to nucleases (Cohen, 1993).

B. Hybridization

The hybridization of phosphorothioate oligonucleotides to DNA and RNA has been thoroughly characterized (S. T. Crooke, 1992, 1993; see S. T. Crooke and Lebleu, 1993, for reviews). The T_m of a phosphorothioate oligodeoxynucleotide for RNA is approximately 0.5°C less per nucleotide than that for a corresponding phosphodiester oligodeoxynucleotide. This reduction in T_m per nucleotide is virtually independent of the number of phosphorothioate units substituted for phosphodiesters. However, sequence context has some influence, as the ΔT_m can vary from -0.3 to 1.0°C, depending on sequence. Compared with RNA and RNA duplex formation, a phosphorothioate oligodeoxynucleotide has a T_m approximately $-2.2°C$ lower per unit (Freier, 1993). This means that, to be effective *in vitro*, phosphorothioate oligodeoxynucleotides must typically be 17–20 nucleotides in length (Monia *et al.*, 1992, 1993) and that invasion of double-stranded regions in RNA is difficult (Vickers *et al.*, 1992; Lima *et al.*, 1992).

Association rates of phosphorothioate oligodeoxynucleotide to unstructured RNA targets are typically 10^6–10^7 M^{-1} sec^{-1}, independent of oligonucleotide length or sequence (Lima *et al.*, 1992; Freier, 1993). Association rates to structured RNA targets can vary from 10^2 to 10^8 M^{-1} sec^{-1}, depending on the structure of the RNA, site of binding in the structure, and other factors (Freier, 1993). Said another way, association rates for oligonucleotides that display acceptable affinity constants are sufficient to support biological activity at therapeutically achievable concentrations.

The specificity of hybridization of phosphorothioate oligonucleotides is, in general, slightly greater than that of phosphodiester analogs. For example, a T–C mismatch results in a 7.7 or 12.8°C reduction in T_m, respectively, for a phosphodiester or phosphorothioate oligodeoxynucleotide 18 nucleotides in length with the mismatch centered (Freier, 1993). Thus, from this perspective, the phosphorothioate modification is quite attractive.

C. Interactions with Proteins

Phosphorothioate oligonucleotides bind to proteins. The interactions with proteins can be divided into nonspecific, sequence-specific, and structure-specific binding events, each of which may have different characteristics and effects. Nonspecific binding to a wide variety of proteins has been demonstrated. An example of this type of binding is the interaction of phosphorothioate oligonucleotides with serum albumin. The affinity of such interactions is low. The K_d for albumin is approximately 200 μM (R. M. Crooke *et al.*, 1996), thus in a similar range with aspirin or penicillin (Joos and Hall, 1969). Furthermore, in this study, no competition between phosphorothioate oligonucleotides and several drugs that bind to bovine serum albumin was observed. In this study, binding and competition were determined in an assay in which electrospray mass spectrometry was used. In contrast, in a study in which an equilibrium dissociation constant was derived from an assay using albumin loaded on a CH-Sephadex column, the K_m ranged from 1 to 5 $\times$ 10^{-5} M for bovine serum albumin and from 2 to 3 $\times$ 10^{-4} M for human serum albumin. Moreover, warfarin and indomethacin were reported to compete for binding to serum albumin (Srinivasan *et al.*, 1995). Clearly, much more work is required before definitive conclusions can be drawn.

Phosphorothioate oligonucleotides can interact with nucleic acid-binding proteins, such as transcription factors, and single-strand nucleic acid-binding proteins. However, little is known about these binding events.

In addition, it has been reported that phosphorothioates bind to an 80-kDa membrane protein that was suggested to be involved in cellular uptake processes (Loke *et al.,* 1989). However, again, little is known about the affinities, sequence, or structure specificities of these putative interactions.

Phosphorothioates interact with nucleases and DNA polymerases. These compounds are slowly metabolized by both endonucleases and exonucleases (S. T. Crooke, 1992) and inhibit these enzymes (S. T. Crooke *et al.,* 1995). The inhibition of these enzymes appears to be competitive, and this may account for some early data suggesting that phosphorothioates are almost infinitely stable to nucleases. In these studies, the oligonucleotide-to-enzyme ratio was high, and thus the enzyme was inhibited. Phosphorothioates also bind to RNase H when in an RNA–DNA duplex, and the duplex serves as a substrate for RNase H (Gao *et al.,* 1991). At higher concentrations, presumably by binding as a single strand to RNase H, phosphorothioates inhibit the enzyme (Gao *et al.,* 1991; S. T. Crooke *et al.,* 1995). Again, the oligonucleotides appear to be competitive antagonists for the DNA–RNA substrate.

Phosphorothioates have been shown to be competitive inhibitors of DNA polymerases α and β with respect to the DNA template, and noncompetitive inhibitors of DNA polymerases γ and δ (Gao *et al.,* 1991). Despite this inhibition, several studies have suggested that phosphorothioates might serve as primers for polymerases and be extended (Stein and Cheng, 1993; Agrawal *et al.,* 1991; S. T. Crooke *et al.,* 1995). In our laboratories, we have shown extensions of 2–3 nucleotides only. At present, a full explanation as to why no longer extensions are observed is not available.

Phosphorothioate oligonucleotides have been reported to be competitive inhibitors for HIV reverse transcriptase (RT) (Majumdar *et al.,* 1989) and inhibit RT-associated RNase H activity (Cheng *et al.,* 1991). They have been reported to bind to the cell surface protein, CD4 (Stein *et al.,* 1991), and to protein kinase C. Various viral polymerases have also been shown to be inhibited by phosphorothioates (for review, see Stein and Cheng, 1993). In addition, we have shown potent, non-sequence-specific inhibition of RNA splicing by phosphorothioates (Hodges and Crooke, 1995).

Like other oligonucleotides, phosphorothioates can adopt a variety of secondary structures. As a general rule, self-complementary oligonucleotides are avoided, if possible, to avoid duplex formation between oligonucleotides. However, other structures that are less well understood can also form. For example, oligonucleotides containing runs of guanosines can form tetrameric structures called G quartets, and these appear to interact with a number of proteins with relatively greater affinity than unstructured oligonucleotides (Wyatt *et al.,* 1994).

In conclusion, phosphorothioate oligonucleotides may interact with a wide range of proteins via several types of mechanisms. These interactions may influence the pharmacokinetic, pharmacologic, and toxicologic properties of these molecules. They may also complicate studies on the mechanism of action of these drugs.

D. Pharmacokinetic Properties

To study the pharmacokinetics of phosphorothioate oligonucleotides, a variety of labeling techniques have been used. In some cases, 3′- or 5′-^{32}P end-labeled or fluorescently labeled oligonucleotides have been used in *in vitro* or *in vivo* studies. These are probably less satisfactory than internally labeled compounds because terminal phosphates are rapidly removed by phosphatases, and fluorescently labeled oligonucleotides have physicochemical properties that differ from those of the unmodified oligonucleotides. Consequently, either uniformly ^{35}S-labeled or base-labeled phosphorothioates are preferable for pharmacokinetic studies. In our laboratories, a tritium exchange method that labels a slowly exchanging proton at the C-8 position in purines was developed and proved to be useful (Graham *et al.*, 1993). A method that added radioactive methyl groups via *S*-adenosylmethionine has also been successfully used (Sands *et al.*, 1994). Finally, advances in extraction, separation, and detection methods have resulted in methods that provide excellent pharmacokinetic analyses without radiolabeling (S. T. Crooke *et al.*, 1996).

1. Nuclease Stability

The principal metabolic pathway for oligonucleotides is cleavage via endonucleases and exonucleases. Phosphorothioate oligonucleotides, while quite stable to various nucleases (Hoke *et al.*, 1991; Wickström, 1986; Campbell *et al.*, 1990), are competitive inhibitors of nucleases (Gao *et al.*, 1991; S. T. Crooke *et al.*, 1995). Consequently, the stability of phosphorothioate oligonucleotides to nucleases is probably less than initially thought, as high concentrations (that inhibited nucleases) of oligonucleotides were employed in the early studies. Similarly, phosphorothioate oligonucleotides are degraded slowly by cells in tissue culture, with a half-life of 12–24 hr (Hoke *et al.*, 1991; S. T. Crooke *et al.*, 1995), and are slowly metabolized in animals (Cossum *et al.*, 1993). The pattern of metabolites suggests primarily exonuclease activity with perhaps modest contributions by endonucleases. However, a number of lines of evidence suggest that, in many cells and tissues, endonucleases play an important role in the metabolism of oligonucleotides. For example, 3′- and 5′-modified oligonucleotides with phosphodiester backbones have been shown to be relatively rapidly degraded in cells and after administration to animals (Sands *et al.*, 1995; Miyao *et al.*, 1995). Thus, strategies in which oligonucleotides are modified at only the 3′ and 5′ terminus as a means of enhancing stability have not proved to be successful.

2. In Vitro Cellular Uptake

Phosphorothioate oligonucleotides are taken up by a wide range of cells *in vitro* (R. M. Crooke, 1991, 1993a; R. M. Crooke *et al.*, 1995; Neckers,

1993; Gao *et al.*, 1991). In fact, uptake of phosphorothioate oligonucleotides into a prokaryote, *Vibrio parahaemolyticus*, has been reported (Chrissey *et al.*, 1993), as has uptake into *Schistosoma mansoni* (Tao *et al.*, 1995). Uptake is time and temperature dependent. It is also influenced by cell type, cell culture conditions, media, and sequence and length of the oligonucleotide (R. M. Crooke *et al.*, 1995). No obvious correlation between the lineage of cells, whether the cells are transformed or virally infected, and uptake has been identified (R. M. Crooke *et al.*, 1995). Nor are the factors that result in differences in uptake of different sequences of oligonucleotide understood. Although several studies have suggested that receptor-mediated endocytosis may be a significant mechanism of cellular uptake (Loke *et al.*, 1989), the data are not yet compelling enough to conclude that receptor-mediated endocytosis accounts for a significant portion of the uptake in most cells.

Numerous studies have shown that phosphorothioate oligonucleotides distribute broadly in most cells, once taken up (R. M. Crooke, 1993a; R. M. Crooke *et al.*, 1995). Again, however, significant differences in subcellular distribution between various types of cells have been noted.

Cationic lipids and other approaches have been used to enhance uptake of phosphorothioate oligonucleotides in cells that take up little oligonucleotide *in vitro* (Bennett *et al.*, 1992, 1993; Quattrone *et al.*, 1994). Again, however, there are substantial variations from cell type to cell type. Other approaches to enhance intracellular uptake *in vitro* have included streptolysin D treatment of cells and the use of dextran sulfate (Giles *et al.*, 1995) and other liposome formulations (Wang *et al.*, 1995), as well as physical means such as microinjections (for review, see S. T. Crooke, 1995a).

3. In Vivo Pharmacokinetics

Phosphorothioate oligonucleotides bind to serum albumin and α_2-macroglobulin. The apparent affinity for albumin is quite low (200–400 μM) (S. T. Crooke *et al.*, 1996; Srinivasan *et al.*, 1995) and comparable to the low-affinity binding observed for a number of drugs (e.g. aspirin, penicillin) (Joos and Hall, 1969). Serum protein binding, therefore, provides a repository for these drugs and prevents rapid renal excretion. As serum protein binding is saturable, at higher doses, intact oligomer may be found in urine (Agrawal *et al.*, 1991; Iversen, 1991). Studies in our laboratory suggest that, in rats, oligonucleotides administered intravenously at doses of 15 to 20 mg/kg saturate the serum protein binding capacity (J. Leeds, unpublished data).

Phosphorothioate oligonucleotides are rapidly and extensively absorbed after parenteral administration. For example, in rats, after an intradermal dose of 3.6 mg of ^{14}C-labeled ISIS 2105 (a 20-mer phosphorothioate) per kilogram, approximately 70% of the dose was absorbed within 4 hr and total systemic bioavailability was in excess of 90% (Cossum *et al.*, 1994). After intradermal injection in humans, absorption of ISIS 2105 was similar to that observed in rats (S. T. Crooke *et al.*, 1994). Subcutaneous administration to rats and monkeys results in somewhat lower bioavailability and greater distribution to lymph, as would be expected (J. Leeds, unpublished observations).

Distribution of phosphorothioate oligonucleotides from blood after absorption or intravenous administration is extremely rapid. We have reported distribution half-lives of less than 1 hr (Cossum *et al.*, 1993, 1994), and similar data have been reported by others (Agrawal *et al.*, 1991; Iversen, 1991). Blood and plasma clearance is multiexponential, with a terminal elimination half-life from 40 to 60 hr in all species except humans. In humans, the terminal elimination half-life may be somewhat longer (S. T. Crooke *et al.*, 1994).

Phosphorothioates distribute broadly to all peripheral tissues. Liver, kidney, bone marrow, skeletal muscle, and skin accumulate the highest percentage of a dose, but other tissues display small quantities of drug (Cossum *et al.*, 1993, 1994). No evidence of significant penetration of the blood–brain barrier has been reported. The rates of incorporation and clearance from tissues vary as a function of the organ studied, with liver accumulating drug most rapidly (20% of a dose within 1–2 hr) and other tissues accumulating drug more slowly. Similarly, elimination of drug is more rapid from liver than from any other tissue (e.g., terminal half-life from liver, 62 hr; from renal medulla, 156 hr). The distribution into the kidney has been studied more extensively, and drug has been shown to be present in Bowman's capsule, the proximal convoluted tubule, the brush border membrane, and within renal tubular epithelial cells (Rappaport *et al.*, 1995). The data suggested that the oligonucleotides are filtered by the glomerulus and then reabsorbed by the proximal convoluted tubule epithelial cells. Moreover, it was suggested that reabsorption might be mediated by interactions with specific proteins in the brush border membranes.

At relatively low doses, clearance of phosphorothioate oligonucleotides is due primarily to metabolism (Iversen, 1991; Cossum *et al.*, 1993, 1994). Metabolism, mediated by exonucleases and endonucleases, results in shorter oligonucleotides and, ultimately, nucleosides that are degraded by normal metabolic pathways. Although no direct evidence of base excision or modification has been reported, these are theoretical possibilities that may occur. In one study, a higher molecular weight radioactive material was observed in urine, but not fully characterized (Agrawal *et al.*, 1991). Clearly, the potential for conjugation reactions and extension of oligonucleotides via these drugs serving as primers for polymerases must be explored in more detail. In a thorough study, 20-nucleotide phosphodiester and phosphorothioate oligonucleotides were administered intravenously at a dose of 6 mg/kg to mice. The oligonucleotides were internally labeled with C^3H_3 by methylation of an internal deoxycytidine residue using *Hha*I methylase and S-[^{3}H]adenosylmethionine (Sands *et al.*, 1994). The observations for the phosphorothioate oligonucleotide were entirely consistent with those made in our studies. In addition, autoradiographic analyses showed drug in renal cortical cells (Sands *et al.*, 1994).

One study of prolonged infusions of a phosphorothioate oligonucleotide to human beings has been reported (Bayever *et al.*, 1993). In this study, five patients with leukemia were given 10-day intravenous infusions at a dosage of 0.05 mg/kg/hr. Elimination half-lives reportedly varied from 5.9 to 14.7 days. Urinary recovery of radioactivity was reported to be 30–60% of the total dose, with 30% of the radioactivity being intact drug. Metabolites in urine included both higher and lower molecular weight compounds. In contrast, when GEM-91 (a 25-mer phosphorothioate oligodeoxynucleotide) was administered to human subjects as a 2-hr intravenous (i.v.) infusion at a dose of 0.1 mg/kg, a peak plasma concentration of 295.8 ng/ml was observed at the cessation of the infusion. Plasma clearance of total radioactivity was biexponential with initial and terminal elimination half-lives of 0.18 and 26.71 hr, respectively. However, degradation was extensive and intact drug pharmacokinetic models were not presented. Nearly 50% of the administered radioactivity was recovered in urine, but most of the radioactivity represented degradates. In fact, no intact drug was found in the urine at any time (R. Zhang *et al.*, 1995a).

In a more recent study (Glover *et al.*, 1997) in which the level of intact drug was carefully evaluated by the use of capillary gel electrophoresis, the pharmacokinetics of ISIS 2302, a 20-mer phosphorothioate oligodeoxynucleotide, after a 2-hr infusion were determined. Doses from 0.06 to 2.0 mg/kg were studied, and the peak plasma concentrations were shown to increase linearly with dose, with the 2-mg/kg dose resulting in peak plasma concentrations of intact drug of approximately 9.5 μg/ml. Clearance from plasma, however, was dose dependent, with the 2-mg/kg dose having a clearance of 1.28 ml min^{-1} kg^{-1}, while that of 0.5 mg/kg was 2.07 ml min^{-1} kg^{-1}. Essentially, no intact drug was found in urine.

Clearly, the two most recent studies differ from the initial report in several respects, but the most likely explanation is related to the evolution of assay methods not to the difference between compounds. Overall, the behavior of phosphorothioates in the plasma of human subjects appears to be similar to that in other species.

We have also performed oral bioavailability experiments in rodents treated with an H2 antagonist to avoid acid-mediated depurination or precipitation. In these studies, limited (<5%) bioavailability was observed (S. T. Crooke, unpublished observations). However, it seems likely that the principal limiting factor in the oral bioavailability of phosphorothioates may be degradation in the gut rather than absorption. Studies using everted rat jejunum sacs demonstrated passive transport across the intestinal epithelium (Hughes *et al.*, 1995). Furthermore, studies using more stable 2'-methoxy phosphorothioate oligonucleotides showed a significant increase in oral bioavailability that appeared to be associated with the improved stability of the analogs (Agrawal *et al.*, 1995).

In summary, pharmacokinetics studies of several phosphorothioates demonstrate that they are well absorbed from parenteral sites, distribute broadly to all peripheral tissues, do not cross the blood–brain barrier, and are eliminated primarily by slow metabolism. In short, systemic dosing once a day or every other day should be feasible. Although the similarities between oligonucleotides of different sequences are far greater than the differences, additional studies are required before it can be determined whether there are subtle effects of sequence on the pharmacokinetic profile of this class of drugs.

E. Pharmacological Properties

I. Molecular Pharmacology

Antisense oligonucleotides are designed to bind to RNA targets via Watson–Crick hybridization. As RNA can adopt a variety of secondary structures via Watson–Crick hybridization, one useful way to think of antisense oligonucleotides is as competitive antagonists for self-complementary regions of the target RNA. Obviously, the creation of oligonucleotides with the highest affinity per nucleotide unit is pharmacologically important, and a comparison of the affinity of the oligonucleotide with that of a complementary RNA oligonucleotide is the most sensible comparison. In this context, phosphorothioate oligodeoxynucleotides are relatively competitively disadvantaged, as the affinity per nucleotide unit of oligomer is less than that of RNA ($>-2°C$ T_m per unit) (Cook, 1993). This results in a requirement of at least 15–17 nucleotides in order to have sufficient affinity to produce biological activity (Monia *et al.*, 1992).

Although multiple mechanisms by which an oligonucleotide may terminate the activity of an RNA species to which it binds are possible, examples of biological activity have been reported for only three oligonucleotides. Antisense oligonucleotides have been reported to inhibit RNA splicing (Kulka *et al.*, 1989) and translation (Agrawal *et al.*, 1988) of mRNA and to induce degradation of RNA by RNase H (Chiang *et al.*, 1991). Without question, the mechanism that has resulted in the most potent compounds and is best understood is RNase H activation. To serve as a substrate for RNase H, a duplex between RNA and a "DNA-like" oligonucleotide is required. Specifically, a sugar moiety in the oligonucleotide that induces a duplex conformation equivalent to that of a DNA–RNA duplex and a charged phosphate are required (for review, see Mirabelli and Crooke, 1993). Thus, phosphorothioate oligodeoxynucleotides are expected to induce RNase H-mediated cleavage of the RNA when bound. As discussed in Section VI, many chemical approaches that enhance the affinity of an oligonucleotide for RNA result in duplexes that are no longer substrates for RNase H.

Selection of sites at which optimal antisense activity may be induced in an RNA molecule is complex; it depends on the terminating mechanism and is influenced by the chemical class of the oligonucleotide. Each RNA appears to display unique patterns of sites of sensitivity. Within the phosphorothioate oligodeoxynucleotide chemical class, studies in our laboratory have shown that antisense activity can vary from undetectable to 100% by shifting an oligonucleotide by just a few bases in the RNA target (S. T. Crooke, 1992; Chiang *et al.*, 1991; Bennett and Crooke, 1994). Although significant progress has been made in developing general rules that help define potentially optimal sites in RNA species, to a large extent this remains an empirical process that must be performed for each RNA target and every new chemical class of oligonucleotides.

Phosphorothioates have also been shown to have effects inconsistent with the antisense mechanism for which they were designed. Some of these effects are due to sequence- and structure-specific, as well as nonspecific, interactions with proteins. These effects are particularly prominent in *in vitro* tests for antiviral activity, as often high concentrations of cells, viruses, and oligonucleotides are coincubated (Azad *et al.*, 1993; Wagner *et al.*, 1993). Human immunodeficiency virus (HIV) is particularly problematic, as many oligonucleotides bind to the gp120 protein of the virus (Wyatt *et al.*, 1994). However, the potential for confusion arising from the misinterpretation of an activity as being due to an antisense mechanism when, in fact, it is due to nonantisense effects is certainly not limited to antiviral or just *in vitro* tests (Barton and Lemoine, 1995; Burgess *et al.*, 1995; Hertl *et al.*, 1995). Again, these data simply urge caution and argue for careful dose–response curves, direct analyses of target protein or RNA, and inclusion of appropriate controls before conclusions are drawn concerning the mechanisms of action of oligonucleotide-based drugs. In addition to protein interactions, other factors, such as overrepresented sequences of RNA and unusual structures that may be adopted by oligonucleotides, can contribute to unexpected results (Wyatt *et al.*, 1994).

Given the variability in cellular uptake of oligonucleotides, the variability in potency as a function of binding site in an RNA target and potential nonantisense activities of oligonucleotides, careful evaluation of dose–response curves and clear demonstration of the antisense mechanism are required before conclusions can be drawn from *in vitro* experiments. Nevertheless, numerous well-controlled studies have been reported in which antisense activity was conclusively demonstrated. As many of these studies have been reviewed previously (S. T. Crooke, 1992, 1993; S. T. Crooke and Lebleu, 1993; S. T. Crooke, 1995a; Nagel *et al.*, 1993; Stein and Cheng, 1993), suffice it to say that antisense effects of phosphorothioate oligodeoxynucleotides against a variety of targets are well documented.

2. In Vivo Pharmacological Activities

A relatively large number of reports of *in vivo* activities of phosphorothioate oligonucleotides have now appeared, documenting activities after both local and systemic administration (Table I) (for review, see S. T. Crooke, 1995b). However, for only a few of these reports have sufficient studies been performed to warrant relatively firm conclusions concerning the mechanism of action. Consequently, this chapter reviews in some detail only a few reports that provide sufficient data to support a relatively firm conclusion with regard to mechanism of action. Local effects have been reported for phosphorothioate and methyl phosphonate oligonucleotides. A phosphorothioate oligonucleotide designed to inhibit c-Myb production and applied locally was shown to inhibit intimal accumulation in the rat carotid artery (Simons *et al.*, 1992). In this study, a Northern blot showed a significant reduction in c-*myb* RNA in animals treated with the antisense compound, but no effect in animals treated with a control oligonucleotide. In one study, the effects of the oligonucleotide were suggested to be due to a nonantisense mechanism (Burgess *et al.*, 1995). However, only one dose level was studied, so much remains to be done before definitive conclusions are possible.

Similar effects were reported for phosphorothioate oligodeoxynucleotides designed to inhibit cyclin-dependent kinases (CDC-2 and CDK-2). Again, the antisense oligonucleotide inhibited intimal thickening and cyclin-dependent kinase activity, while a control oligonucleotide had no effect (Abe *et al.*, 1994). In addition, local administration of a phosphorothioate oligonucleotide designed to inhibit N-*myc* resulted in reduction in N-*myc* expression and slower growth of a subcutaneously transplanted human tumor in nude mice (Whitesell *et al.*, 1991).

Antisense oligonucleotides administered intraventricularly have been reported to induce a variety of effects in the central nervous system. Intraventricular injection of antisense oligonucleotides to neuropeptide Y-Y1 receptors reduced the density of the receptors and resulted in behavioral signs of anxiety (Wahlestedt *et al.*, 1993). Similarly, an antisense oligonucleotide designed to bind to NMDA-R1 receptor channel RNA inhibited the synthesis of these channels and reduced the volume of focal ischemia produced by occlusion of the middle cerebral artery in rats (Wahlestedt *et al.*, 1993).

In a series of well-controlled studies, antisense oligonucleotides administered intraventricularly selectively inhibited dopamine type 2 receptor expression, dopamine type 2 receptor RNA levels, and behavioral effects in animals with chemical lesions. Controls included randomized oligonucleotides and the observation that no effects were observed on dopamine type 1 receptor or RNA levels (Weiss *et al.*, 1993; Zhou *et al.*, 1994; Qin *et al.*, 1995). This laboratory also reported the selective reduction of dopamine type 1 receptor and RNA levels with the appropriate oligonucleotide (Zhang *et al.*, 1994).

Similar observations were reported in studies on AT-1 angiotensin receptors and tryptophan hydroxylase. In studies in rats, direct observations of AT-1 and AT-2 receptor densities in various sites in the brain after administration of different doses of phosphorothioate antisense, sense, and scrambled oligonucleotides were reported (Ambuhl *et al.*, 1995). Again, in rats, intraventricular administration of phosphorothioate antisense oligonucleotide resulted in a decrease in tryptophan hydroxylase levels in the brain, while a scrambled control did not (McCarthy *et al.*, 1995).

Injection of antisense oligonucleotides to synaptosomal-associated protein 25 into the vitreous body of rat embryos reduced the expression of the protein and inhibited neurite elongation by rat cortical neurons (Osen-Sand *et al.*, 1993).

In addition to local and regional effects of antisense oligonucleotides, a growing number of well-controlled studies have demonstrated systemic effects of phosphorothioate oligodeoxynucleotides. Expression of interleukin 1 in mice was inhibited by systemic administration of antisense oligonucleotides (Burch and Mahan, 1991). Oligonucleotides to the NF-κB p65 subunit administered intraperitoneally at 40 mg/kg every 3 days slowed tumor growth in mice transgenic for the human T cell leukemia viruses (Kitajima *et al.*, 1992). Similar results with other antisense oligonucleotides were shown in another *in vivo* tumor model after either prolonged subcutaneous infusion or intermittent subcutaneous injection (Higgins *et al.*, 1993).

Several reports further extend the studies of phosphorothioate oligonucleotides as antitumor agents in mice. In one study, a phosphorothioate oligonucleotide directed to inhibition of the *bcr-abl* oncogene was administered at a dose of 1 mg/day for 9 days intravenously to immunodeficient mice injected with human leukemic cells. The drug was shown to inhibit the development of leukemic colonies in the mice and to selectively reduce *bcr-abl* RNA levels in peripheral blood lymphocytes, spleen, bone marrow, liver, lungs, and brain (Skorski *et al.*, 1994). However, it is possible that the effects on the RNA levels were secondary to effects on the growth of various cell types. In the second study, a phosphorothioate oligonucleotide antisense to the protooncogene *myb* inhibited the growth of human melanoma in mice. Again, *myb* mRNA levels appeared to be selectively reduced (Hijiya *et al.*, 1994).

A number of studies from our laboratories that directly examined target RNA levels, target protein levels, and pharmacological effects using a wide range of control oligonucleotides and examination of the effects on closely related isotypes have been completed. Single and chronic daily administration of a phosphorothioate oligonucleotide designed to inhibit mouse protein kinase C α (PKC-α) selectively inhibited expression of PKC-α RNA in mouse liver without effects on any other isotype. The effects lasted at least 24 hr after a dose, and a clear dose–response curve was observed with a dose of 10 to 15 mg/kg intraperitoneally, reducing PKC-α RNA levels in liver by 50% 24 hr after a dose (Dean and McKay, 1994).

TABLE I   Reported Activity of Antisense Oligonucleotides in Animal Models

Target	Route	Species	Ref.
Cardiovascular models			
c-*myb*	Topical	Rat	Simons *et al.* (1992)
cdc2 kinase	Topical	Rat	Morishita *et al.* (1993)
PCNA	Topical	Rat	Morishita *et al.* (1993)
cdc2 kinase	Topical	Rat	Abe *et al.* (1994)
CDK2	Topical	Rat	Abe *et al.* (1994)
Cyclin B1	Topical	Rat	Morishita *et al.* (1994)
PCNA	Topical	Rat	Simons *et al.* (1994)
Angiotensin 1 receptor	Intracerebral	Rat	Gyurko *et al.* (1993)
Angiotensinogen	Intracerebral	Rat	Phillips *et al.* (1994)
c-*fos*	Intracerebral	Rat	Suzuki *et al.* (1994)
Inflammatory models			
Type 1 IL-1 receptor	Intradermal	Mouse	Burch and Mahan (1991)
ICAM-1	Intravenous	Mouse	Stepkowski *et al.* (1994)
Cancer models			
N-*myc*	Subcutaneous	Mouse	Whitesell *et al.* (1991)
NF-κB p65	Intraperitoneal	Mouse	Kitajima *et al.* (1992)
c-*myb*	Subcutaneous	Mouse	Ratajczak *et al.* (1992)
p120 nucleolar antigen	Intraperitoneal	Mouse	Perlaky *et al.* (1993)
NK-κB p65	Subcutaneous	Mouse	Higgins *et al.* (1993)
Protein kinase C-α	Intraperitoneal	Mouse	Dean and McKay (1994)
c-*myb*	Subcutaneous	Mouse	Hijya *et al.* (1994)
Ha-*ras*	Intratumor	Mouse	Schwab *et al.* (1994)
BCR-ABL	Intravenous	Mouse	Skorski *et al.* (1994)
PTHrP	Intraventricular	Rat	Akino *et al.* (1996)
Neurological models			
c-*fos*	Intracerebral	Rat	Chiasson *et al.* (1992)
SNAP-25	Intracerebral	Chicken	Osen-Sand *et al.* (1993)
Kinesin heavy chain	Intravitreal	Rabbit	Amaratunga *et al.* (1993)
Arginine vasopressin	Intracerebral	Rat	Flanagan *et al.* (1993)
c-*fos*	Intracerebral	Rat	Heilig *et al.* (1993)
Progesterone receptor	Intracerebral	Rat	Pollio *et al.* (1993)
Dopamine D_2 receptor	Intracerebral	Rat	Zhang and Creese (1993)
Y-Y1 receptor	Intracerebral	Rat	Wahlestadt *et al.* (1993)
Neuropeptide Y	Intracerebral	Rat	Akabayashi *et al.* (1994)
κ opioid receptor	Intracerebral	Rat	Adams *et al.* (1994)
IGF-1	Intracerebral	Rat	Castro-Alamancos and Torres-Aleman (1994)
κ-opioid receptor	Intracerebral	Rat	Adams *et al.* (1994)
c-*fos*	Intraspinal	Rat	Gillardon *et al.* (1994)
c-*fos*	Intracerebral	Rat	Hooper *et al.* (1994)
c-*fos*	Intraspinal	Rat	Woodburn *et al.* (1994)
NMDA receptor	Intracerebral	Rat	Kindy (1994)
CREB	Intracerebral	Rat	Konradi *et al.* (1994)
δ opioid receptor	Intracerebral	Mice	Lai *et al.* (1994)
Progesterone receptor	Intracerebral	Rat	Mani *et al.* (1994)
GAD65	Intracerebral	Rat	McCarthy *et al.* (1994)
GAD67	Intracerebral	Rat	McCarthy *et al.* (1994)

continues

TABLE I (*Continued*)

Target	Route	Species	Ref.
AT1-angiotensin receptor	Intracerebral	Rat	Sakai *et al.* (1995)
Tryptophan hydroxylase	Intracerebral	Mouse	McCarthy *et al.* (1995)
AT1-angiotensin receptor	Intracerebral	Rat	Ambuhl *et al.* (1995)
CRH$_1$-corticotropin-releasing hormone receptor	Intracerebral	Rat	Liebsch *et al.* (1995)
δ opioid receptor	Intracerebral	Rat	Cha *et al.* (1995)
δ opioid receptor	Intracerebral	Mouse	Mizoguchi *et al.* (1995)
Oxytocin	Intracerebral	Rat	
Oxytocin	Intracerebral	Rat	Neumann *et al.* (1994)
Substance P receptor	Intracerebral	Rat	Ogo *et al.* (1994)
Tyrosine hydroxylase	Intracerebral	Rat	Skutella *et al.* (1994)
c-*jun*	Intracerebral	Rat	Tischmeyer *et al.* (1994)
D$_1$ dopamine receptor	Intracerebral	Mouse	Zhang *et al.* (1994)
D$_2$ dopamine receptor	Intracerebral	Mouse	Zhou *et al.* (1994)
D$_2$ dopamine receptor	Intracerebral	Mouse	Weiss *et al.* (1993)
D$_2$ dopamine receptor	Intracerebral	Mouse	Qin *et al.* (1995)
Viral models			
HSV-1		Mouse	Kulka *et al.* (1989)
Tick-borne encephalitis		Mouse	Vlassov (1989)
Duck hepatitis virus	Intravenous	Duck	Offensperger *et al.* (1993)

A phosphorothioate oligonucleotide designed to inhibit human PKC-α expression selectively inhibited expression of PKC-α RNA and PKC-α protein in human tumor cell lines implanted subcutaneously in nude mice after intravenous administration (Dean *et al.*, 1996). In these studies, effects on RNA and protein levels were highly specific and observed at dosages lower than 6 mg/kg/day and antitumor effects were detected at dosages as low as 0.6 mg/kg/day. A large number of control oligonucleotides failed to show activity.

In a similar series of studies, Monia *et al.* (1996a,b) demonstrated highly specific loss of human c-*raf* kinase RNA in human tumor xenografts and antitumor activity that correlated with the loss of RNA. Moreover, a series of control oligonucleotides with one to seven mismatches showed decreasing potency *in vitro* and precisely the same rank-order potencies *in vivo*.

Finally, a single injection of a phosphorothioate oligonucleotide designed to inhibit cAMP-dependent protein kinase type 1 was reported to reduce RNA and protein levels selectively in human tumor xenografts and to reduce tumor growth (Nesterova and Cho-Chung, 1995).

Thus, there is a growing body of evidence that phosphorothioate oligonucleotides can induce potent systemic and local effects *in vivo*. More important, there are now a number of studies with sufficient controls and direct observation of target RNA and protein levels to suggest highly specific effects that are difficult to explain by any mechanism other than antisense. As would be expected, the potency of these effects varies depending on the target, the organ, and the end point measured as well as the route of administration and the time after a dose when the effect is measured.

In conclusion, although it is of obvious importance to interpret *in vivo* activity data cautiously, and it is clearly necessary to include a range of controls and to evaluate effects on target RNA and protein levels and control RNA and protein levels directly, it is difficult to argue with the conclusion today that some effects have been observed in animals that are most likely primarily due to an antisense mechanism.

In addition, in studies on patients with cytomegalovirus-induced retinitis, local injections of ISIS 2922 have resulted in impressive efficacy, although it is obviously impossible to prove that the mechanism of action is antisense in these studies (Hutcherson *et al.*, 1995).

F. Toxicological Properties

I. In Vitro

In our laboratory, we have evaluated the toxicities of scores of phosphorothioate oligodeoxynucleotides in a significant number of cell lines in tissue culture. As a general rule, no significant cytotoxicity is induced at concentrations below 100 μM oligonucleotide. In addition, with a few exceptions, no significant effect on macromolecular synthesis is observed at concentrations below 100 μM (R. M. Crooke, 1993b; S. T. Crooke, 1993).

Polynucleotides and other polyanions have been shown to cause release of cytokines (Colby, 1971). Also, bacterial DNA species have been reported to be mitogenic for lymphocytes *in vitro* (Messina *et al.*, 1991). Furthermore, oligodeoxynucleotides (30–45 nucleotides in length) were reported to induce interferons and enhance natural killer cell activity (Kuramoto *et al.*, 1992). In the latter study, the oligonucleotides that displayed natural killer (NK) cell-stimulating activity contained specific palindromic sequences and tended to be guanosine rich. Collectively, these observations indicate that nucleic acids may have broad immunostimulatory activity.

It has been shown that phosphorothioate oligonucleotides stimulate B lymphocyte proliferation in a mouse splenocyte preparation (analogous to bacterial DNA) (Psietsky and Reich, 1993), and the response may underlie the observations of lymphoid hyperplasia in the spleen and lymph nodes of rodents caused by repeated administration of these compounds (see Section V,F,3). We also have evidence of enhanced cytokine release by immunocompetent cells when exposed to phosphorothioates *in vitro* (S. T. Crooke *et al.*, 1996). In this study, both human keratinocytes and an *in vitro* model of human skin released interleukin 1α (IL-1α) when treated with 250 mM to 1 mM of phosphorothioate oligonucleotides. The effects seemed to be dependent on the phosphorothioate backbone and independent of sequence or 2′ modification. In a study in which murine B lymphocytes were treated with phosphodiester oligonucleotides, B cell activation was induced by oligonucleotides with unmethylated CpG dinucleotides (Krieg *et al.*, 1995). This has been extrapolated to suggest that the CpG motif may be required for immune stimulation of oligonucleotide analogs such as phosphorothioates. This clearly is not the case with regard to release of IL-1α from keratinocytes (S. T. Crooke *et al.*, 1996). Nor is it the case with regard to *in vivo* immune stimulation (see Section V,F,3).

2. Genotoxicity

As with any new chemical class of therapeutic agents, concerns about genotoxicity cannot be dismissed because little *in vitro* testing has been performed and no data from long-term studies of oligonucleotides are available. Clearly, given the limitations in our understanding about the basic mechanisms that might be involved, empirical data must be generated. We have performed mutagenicity studies on two phosphorothioate oligonucleotides, ISIS 2105 and ISIS 2922, and found them to be nonmutagenic at all concentrations studied (S. T. Crooke *et al.*, 1994).

Two mechanisms of genotoxicity that may be unique to oligonucleotides have been considered. One possibility is that an oligonucleotide analog could be integrated into the genome and produce mutagenic events. Although integration of an oligonucleotide into the genome is conceivable, it is likely to be extremely rare. For most viruses, viral DNA integration is itself a rare event and, of course, viruses have evolved specialized enzyme-mediated mechanisms to achieve integration. Moreover, preliminary studies in our laboratory have shown that phosphorothioate oligodeoxynucleotides are generally poor substrates for DNA polymerases, and it is unlikely that enzymes such as integrases, gyrases, and topoisomerases (that have obligate DNA cleavage as intermediate steps in their enzymatic processes) will accept these compounds as substrates. Consequently, it would seem that the risk of genotoxicity due to genomic integration is no greater and probably less than that of other potential mechanisms, for example, alteration of the activity of growth factors, cytokine release, nonspecific effects on membranes that might trigger arachidonic acid release, or inappropriate intracellular signaling. Presumably, new analogs that deviate significantly more from natural DNA would be even less likely to be integrated.

A second concern that has been raised about possible genotoxicity is the risk that oligonucleotides might be degraded to toxic or carcinogenic metabolites. However, metabolism of phosphorothioate oligodeoxynucleotides by base excision would release normal bases, which presumably would be nongenotoxic. Similarly, oxidation of the phosphorothioate backbone to the natural phosphodiester structure would also yield nonmutagenic (and probably nontoxic) metabolites. Finally, it is possible that phosphorothioate bonds could be hydrolyzed slowly, releasing nucleoside phosphorothioates that presumably would be rapidly oxidized to natural (nontoxic) nucleoside phosphates. However, oligonucleotides with modified bases, backbones, or both may pose different risks.

3. *In Vivo*

The acute 50% lethal dose (LD_{50}) in mice of all phosphorothioate oligonucleotides tested to date is in excess of 500 mg/kg (D. L. Kornbrust, unpublished observations). In rodents, we have had the opportunity to evaluate the acute and chronic toxicities of multiple phosphorothioate oligonucleotides administered by multiple routes (Henry *et al.*, 1997a,b). The consistent dose-limiting toxicity was immune stimulation manifested by lymphoid hyperplasia, splenomegaly, and a multiorgan monocellular infiltrate. These effects occurred only with chronic administration at doses >20 mg/kg and were dose dependent. Liver and kidney were the organs most prominently affected by monocellular infiltrates. All of these effects appeared to be reversible, and chronic intradermal administration appeared to be the most toxic route, probably because of high local concentrations of the drugs resulting in local cytokine release and initiation of a cytokine cascade. There were no obvious effects of sequence. At doses of 100 mg/kg and greater, minor increases in liver enzyme levels and mild thrombocytopenia were also observed.

In monkeys, however, the toxicological profile of phosphorothioate oligonucleotides is quite different. The most prominent dose-limiting side effect is sporadic reduction in blood pressure associated with bradycardia. When these events are observed, they are often associated with activation of C-5 complement; they are related to dose and to peak plasma concentration. This appears to be related to the activation of the alternative pathway (Henry *et al.*, 1997d). All phosphorothioate oligonucleotides tested to date appear to induce these effects although there may be slight variations in potency as a function of sequence, length, or both (Cornish *et al.*, 1993; Galbraith *et al.*, 1994; Henry *et al.*, 1997c).

A second prominent toxicological effect in the monkey is the prolongation of activated partial thromboplastin time. At higher doses, evidence of clotting abnormalities is observed. Again, these effects are dose and peak plasma concentration dependent (Henry *et al.*, 1997c; Galbraith *et al.*, 1994). Although no evidence of sequence dependence has been observed, there appears to be a linear correlation between number of phosphorothioate linkages and potency between 18 and 25 nucleotides (P. Nicklin, unpublished observations). The mechanisms responsible for these effects are likely to be complex, but preliminary data suggest that direct interactions with thrombin may be at least partially responsible for the effects observed (Henry *et al.*, 1997e).

In humans, again the toxicological profile differs. When ISIS 2922 is administered intravitreally to patients with cytomegalovirus retinitis, the most common adverse event is an anterior chamber inflammation easily managed with steroids. A relatively rare and dose-related adverse event is the occurrence of morphological changes in the retina associated with loss in peripheral vision (Hutcherson *et al.*, 1995).

ISIS 2105, a 20-mer phosphorothioate designed to inhibit the replication of human papillomaviruses that cause genital warts, is administered intradermally at doses as high as 3 mg per wart weekly for 3 weeks; essentially no toxicities have been observed, and there is, remarkably, a complete absence of local inflammation (L. Grillone, unpublished results).

Administration every other day of 2-hr intravenous infusions of ISIS 2302 at doses as high as 2 mg/kg resulted in no significant toxicities, no evidence of immune stimulation, and no hypotension. A slight subclinical increase in activated partial thromboplastin time (APTT) was observed at the 2-mg/kg dose (Glover *et al.*, 1996).

G. Therapeutic Index

In Fig. 2, an attempt to put the toxicities and their dose–response relationships in a therapeutic context is shown. This is particularly important because considerable confusion has arisen concerning the potential utility of phosphorothioate oligonucleotides for selected therapeutic purposes deriving from unsophisticated interpretation of toxicological data. As can be

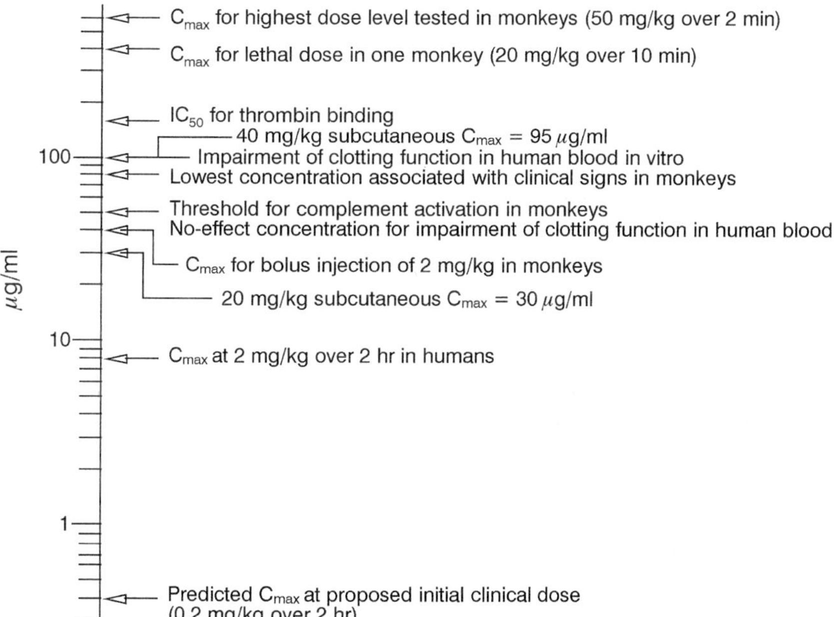

FIGURE 2 Plasma concentrations of ISIS 2302 at which various activities are observed. These concentrations are determined by extracting plasma and analyzing by capillary gel electrophoresis and represent intact ISIS 2302.

readily seen, the immune stimulation induced by these compounds appears to be particularly prominent in rodents and is unlikely to be dose limiting in humans. Nor have we, to date, observed hypotensive events in human subjects. Thus, this toxicity appears to occur at lower doses in monkeys than in humans and certainly is not dose limiting in humans.

On the basis of our experience to date, we believe that the dose-limiting toxicity in humans will consist of clotting abnormalities, and this will be associated with peak plasma concentrations well in excess of 10 μg/ml. In animals, pharmacological activities have been observed with i.v. bolus doses from 0.006 to 10–15 mg/kg, depending on the target, the end point, the organ studied, and the time after a dose when the effect is measured. Thus, it would appear that phosphorothioate oligonucleotides have a therapeutic index that supports their evaluation for a number of therapeutic indications.

H. Conclusions

Phosphorothioate oligonucleotides have perhaps outperformed many expectations. They display attractive parenteral pharmacokinetic properties. They have produced potent systemic effects in a number of animal models, and in many experiments the antisense mechanism has been directly demonstrated as the hoped-for selectivity. Furthermore, these compounds appear to display satisfactory therapeutic indices for many indications.

Nevertheless, phosphorothioates clearly have significant limits (Table II). Pharmacodynamically, they have a relatively low affinity per nucleotide unit. This means that longer oligonucleotides are required for biological activity and that invasion of many RNA structures may not be possible. At higher concentrations, these compounds inhibit RNase H as well. Thus, the higher end of the pharmacologic dose–response curve is lost. Pharmacokinetically, phosphorothioates do not cross the blood–brain barrier, are not

TABLE II Phosphorothioate
Oligonucleotides: Limits

Pharmacodynamic
   Low affinity per nucleotide unit
   Inhibition of RNase H at high concentrations
Pharmacokinetic
   Limited bioavailability
   Limited blood–brain barrier penetration
   Dose-dependent pharmacokinetics
   Possible drug–drug interactions
Toxicologic
   Release of cytokines
   Complement-associated effects on blood pressure?
   Clotting effects

significantly bioavailable orally, and may display dose-dependent pharmaco-kinetics. Toxicologically, clearly the release of cytokines, activation of complement, and interference with clotting will pose dose limits if they are encountered in the clinical setting.

As several clinical trials are in progress with phosphorothioates and others will be initiated shortly, we shall soon have more definitive information about the activities, toxicities, and value of this class of antisense drugs in human beings.

VI. The Medicinal Chemistry of Oligonucleotides ________________

A. Introduction

The core of any rational drug discovery program is medicinal chemistry. Although the synthesis of modified nucleic acids has been a subject of interest for some time, the intense focus on the medicinal chemistry of oligonucleotides dates perhaps to no more than 5 years prior to the publication of this chapter. Consequently, the scope of medicinal chemistry has expanded enormously, but the biological data to support conclusions about synthetic strategies are only beginning to emerge.

Modifications in the base, sugar, and phosphate moieties of oligonucleotides have been reported. The subjects of medicinal chemical programs include approaches to create enhanced affinity and more selective affinity for RNA or duplex structures; to enhance the ability to cleave nucleic acid targets; to enhance nuclease stability, cellular uptake, and distribution; and to improve *in vivo* tissue distribution, metabolism, and clearance.

B. Heterocycle Modifications

1. Pyrimidine Modifications

A relatively large number of modified pyrimidines have been synthesized and are now incorporated into oligonucleotides and evaluated. The principal sites of modification are C-2, C-4, C-5, and C-6. These and other nucleoside analogs have been thoroughly reviewed (Sanghvi, 1993). Consequently, a brief summary of the analogs that displayed interesting properties is incorporated here.

Inasmuch as the C-2 position is involved in Watson–Crick hybridization, C-2-modified pyrimidine-containing oligonucleotides have shown unattractive hybridization properties. An oligonucleotide containing 2-thiothymidine was found to hybridize well to DNA and, in fact, even better to RNA (ΔT_m 1.5°C modification) (Swayze *et al.*, unpublished results).

In contrast, several modifications in the 4-position that have interesting properties have been reported. 4-Thiopyrimidines have been incorporated into oligonucleotides with no significant negative effect on hybridization (Nikiforov and Connolly, 1991). A bicyclic and an N^4-methoxy analog of cytosine were shown to hybridize with both purine bases in DNA with T_m values approximately equal to those of natural base pairs (Lin and Brown, 1989). In addition, a fluorescent base has been incorporated into oligonucleotides and shown to enhance DNA–DNA duplex stability (Inoue and Ohtsuka, 1985).

A large number of modifications at the C-5 position have also been reported, including halogenated nucleosides. Although the stability of duplexes may be enhanced by incorporating 5-halogenated nucleosides, the occasional mispairing with G and the potential that the oligonucleotide might degrade and release toxic nucleoside analogs cause concern (see Sanghvi, 1993, for review).

Furthermore, oligonucleotides containing 5-propynylpyrimidine modifications have been shown to enhance the duplex stability (ΔT_m 1.6°C per modification; Wagner et al., 1993) and support the RNase H activity. The 5-heteroarylpyrimidines were also shown to influence the stability of duplexes (ΔT_m 1.7°C per modification; Gutierrez et al., 1994). A more dramatic influence was reported for the tricyclic 2′-deoxycytidine analogs, exhibiting an enhancement of 2–5°C per modification, depending on the positioning of the modified bases (Lin et al., 1995). It is believed that the enhanced binding properties of these analogs are due to extended stacking and increased hydrophobic interactions.

In general, as expected, modifications in the C-6 position of pyrimidines are highly duplex destabilizing (Sanghvi et al., 1993). Oligonucleotides containing 6-azapyrimidines have been shown to reduce the T_m by 1–2°C per modification, but to enhance the nuclease stability of oligonucleotides and to support RNase H-induced degradation of RNA targets (Sanghvi, 1993).

2. Purine Modifications

Although numerous purine analogs have been synthesized, when incorporated into oligonucleotides they usually have resulted in destabilization of duplexes. However, there are a few exceptions in which a purine modification has had a stabilizing effect. A brief summary of some of these analogs follows.

Generally, N-1 modifications of the purine moiety have resulted in destabilization of the duplex (Hagenberg et al., 1973). Similarly, C-2 modifications have usually resulted in destabilization. However, 2-6-diaminopurine has been reported to enhance hybridization by approximately 1°C per modification when paired with T (Sproat et al., 1991). Of the 3-position-substituted bases reported to date, only the 3-deazaadenosine analog has been shown to have no negative effect on hybridization.

Modifications at the C-6 and C-7 positions have likewise resulted in only a few interesting bases from the point of view of hybridization. Inosine has been shown to have little effect on duplex stability, but, because it can pair and stack with all four normal DNA bases, it behaves as a universal base and creates an ambiguous position in an oligonucleotide (Martin *et al.*, 1985). Incorporation of 7-deazainosine into oligonucleotides was destabilizing, and this was considered to be due to its relatively hydrophobic nature (Santa Lucia *et al.*, 1991). 7-Deazaguanine was similarly destabilizing, but when 8-aza-7-deazaguanine was incorporated into oligonucleotides, it enhanced hybridizations (Seela *et al.*, 1989). Thus, on occasion, introduction of more than one modification in a nucleobase may compensate for destabilizing effects of some modifications. Interestingly, a 7-iodo-7-deazaguanine residue has been incorporated into oligonucleotides and shown to enhance the binding affinity dramatically (ΔT_m 10.0°C per modification compared to 7-deazaguanine; Seela *et al.*, 1995). The increase in T_m value was attributed to (1) the hydrophobic nature of the modification, (2) increased stacking interaction, and (3) favorable pK_a of the base.

In contrast, some C-8-substituted bases have yielded improved nuclease resistance when incorporated in oligonucleotides (Sanghvi, 1993) but seem to be somewhat destabilizing.

3. Oligonucleotide Conjugates

Although conjugation of various functionalities to oligonucleotides has been reported to achieve a number of important objectives, the data supporting some of the claims are limited and generalizations are not possible on the basis of the data now available.

a. Nuclease Stability   Numerous 3′ modifications have been reported to enhance the stability of oligonucleotides in serum (Manoharan, 1993). Both neutral and charged substituents have been reported to stabilize oligonucleotides in serum and, as a general rule, the stability of a conjugated oligonucleotide tends to be greater as bulkier substituents are added. Inasmuch as the principal nuclease in serum is a 3′-exonuclease, it is not surprising that 5′ modifications have resulted in significantly less stabilization. Internal modifications of base, sugar, and backbone have also been reported to enhance nuclease stability at or near the modified nucleoside (Manoharan, 1993). Oligonucleotides modified with Thionotriester (adamantyl, cholesteryl, and others) have shown improved nuclease stability, cellular association, and binding affinity (Z. Zhang *et al.*, 1995).

The demonstration that modifications may induce nuclease stability sufficient to enhance activity in cells in tissue culture and in animals has proved to be much more complicated because of the presence of 5′-exonucleases and endonucleases. In our laboratory, 3′ modifications and internal point modifications have not provided sufficient nuclease stability to demonstrate pharmacological activity in cells (Hoke *et al.*, 1991). In fact, even a 5-nucleotide-long phosphodiester gap in the middle of a phosphorothioate oligonucleotide resulted in sufficient loss of nuclease resistance to cause complete loss of pharmacological activity (Monia *et al.*, 1992).

In mice, neither a 5′-cholesterol nor a 5′-C-18 amine conjugate altered the metabolic rate of a phosphorothioate oligodeoxynucleotide in liver, kidney, or plasma (S. T. Crooke *et al.*, 1996). Furthermore, blocking the 3′ and 5′ termini of a phosphodiester oligonucleotide did not markedly enhance the nuclease stability of the parent compound in mice (Sands *et al.*, 1995). However, 3′ modification of a phosphorothioate oligonucleotide was reported to enhance its stability in mice relative to the parent phosphorothioate (Temsamani *et al.*, 1993). Moreover, a phosphorothioate oligonucleotide with a 3′ hairpin loop was reported to be more stable in rats than its parent (R. Zhang *et al.*, 1995b). Thus, 3′ modifications may enhance the stability of the relatively stable phosphorothioates sufficiently to be of value.

b. Enhanced Cellular Uptake   Although oligonucleotides have been shown to be taken up by a number of cell lines in tissue culture, with perhaps the most compelling data relating to phosphorothioate oligonucleotides, a clear objective has been to improve cellular uptake of oligonucleotides (R. M. Crooke, 1991; S. T. Crooke *et al.*, 1994). Inasmuch as the mechanisms of cellular uptake of oligonucleotides are still poorly understood, the medicinal chemistry approaches have been largely empirical and are based on many unproved assumptions.

Because phosphodiester and phosphorothioate oligonucleotides are water soluble, the conjugation of lipophilic substituents to enhance membrane permeability has been a subject of considerable interest. Unfortunately, studies in this area have not been systematic and, at present, there is little information about the changes in physiochemical properties of oligonucleotides actually affected by specific lipid conjugates. Phospholipids, cholesterol and cholesterol derivatives, cholic acid, and simple alkyl chains have been conjugated to oligonucleotides at various sites in the oligonucleotide. The effects of these modifications on cellular uptake have been assessed by means of fluorescent, or radiolabeled, oligonucleotides or by measurement of pharmacological activities. From the perspective of medicinal chemistry, few systematic studies have been performed. The activities of oligonucleotides conjugated to short alkyl chains, adamantine, daunomycin, fluorescein, cholesterol, and porphyrin were compared in one study (Boutorine *et al.*, 1991). A cholesterol modification was reported to be more effective at enhancing uptake than the other substituents. It also seems likely that the effects of various conjugates on cellular uptake may be affected by the cell type and target studied. For example, we have studied cholic acid conjugates of phosphorothioate deoxyoligonucleotides or phosphorothioate 2′-methoxyoligonucleotides and observed enhanced activity against HIV and no effect on the activity of ICAM-directed oligonucleotides.

In addition, polycationic substitutions and various groups designed to bind to cellular carrier systems have been synthesized. Although many compounds have been synthesized (see Manoharan, 1993, for review), the data reported to date are insufficient to draw firm conclusions about the value of such approaches or structure activity relationships.

c. RNA-Cleaving Groups   Oligonucleotide conjugates have been reported to act as artificial ribonucleases, albeit at low efficiencies (for review, see DeMesmaeker *et al.*, 1995). Conjugation of chemically reactive groups such as alkylating agents, photoinduced azides, porphyrin, and psoralen have been used extensively to effect a cross-linking of oligonucleotide and the target RNA. In principle, this treatment may lead to translation arrest. In addition, lanthanides and complexes thereof have been reported to cleave RNA via a hydrolytic pathway. A novel europium complex was covalently linked to an oligonucleotide and shown to cleave 88% of the complementary RNA at physiological pH (Hall *et al.*, 1994).

d. In Vivo Effects   To date, relatively few *in vivo* studies have been reported. The properties of 5′-cholesterol and 5′-C-18 amine conjugates of a 20-mer phosphorothioate oligodeoxynucleotide have been determined in mice. Both compounds increased the fraction of an i.v. bolus dose found in the liver. The cholesterol conjugate, in fact, resulted in more than 80% of the dose accumulating in the liver. Neither conjugate enhanced stability in plasma, liver, or kidney (S. T. Crooke *et al.*, 1996). Interestingly, the only significant change in the toxicity profile was a slight increase in effects on serum transaminases and histopathological changes indicative of slight liver toxicity associated with the cholesterol conjugate (Henry *et al.*, 1997f). A 5′-cholesterol phosphorothioate conjugate has also been reported to have a longer elimination half-life, to be more potent, and to induce greater liver toxicity in rats (Desjardins *et al.*, 1995).

4. Sugar Modifications

The focus of second-generation oligonucleotide modifications has centered on the sugar moiety. In oligonucleotides, the pentofuranose sugar ring connects the nucleobase to the phosphate and also positions the nucleobases for effective stacking. A symposium series has been published on the carbohydrate modifications in antisense research (Sanghvi and Cook, 1994) and covers this topic in great detail. Therefore, the content of the following discussion is restricted to a summary of the main events in this area.

A growing number of oligonucleotides in which the pentofuranose ring is modified or replaced have been reported (Breslauer *et al.*, 1986). Uniform modifications at the 2′-position have been shown to enhance hybridization to RNA and, in some cases, to enhance nuclease resistance)Breslauer *et al.*, 1986). Chimeric oligonucleotides containing 2′-deoxyoligonucleotide gaps with 2′-modified wings have been shown to be more potent than parent molecules (Monia *et al.,* 1993).

Other sugar modifications include α-oligonucleotides, carbocyclic oligonucleotides, and hexapyranosyloligonucleotides (Breslauer *et al.,* 1986). Of these, α-oligonucleotides have been most extensively studied. They hybridize in parallel fashion to single-stranded DNA and RNA and are nuclease resistant. However, they have been reported to be oligonucleotides designed to inhibit Ha-*ras* expression. All these oligonucleotides support RNase H and, as can be seen, a direct correlation between affinity and potency exists.

A growing number of oligonucleotides in which the C-2′-position of the sugar ring is modified have been reported (Manoharan, 1993; DeMesmaeker *et al.,* 1995). These modifications include lipophilic alkyl groups, intercalators, amphipathic amino-alkyl tethers, positively charged polyamines, highly electronegative fluoro or fluoro alkyl moities, and sterically bulky methylthio derivatives. The beneficial effects of a C-2′ substitution on the antisense oligonucleotide cellular uptake, nuclease resistance, and binding affinity have been well documented in the literature. In addition, excellent review articles have appeared on the synthesis and properties of C-2′-modified oligonucleotides (Lamond and Sproat, 1993; Sproat and Lamond, 1993; Parmentier *et al.,* 1994; DeMesmaeker *et al.,* 1995).

Other modifications of the sugar moiety have also been studied, including other sites as well as more substantial modifications. However, much less is known about the antisense effects of these modifications (for review, see S. T. Crooke, 1995).

2′-Methoxy-substituted phosphorothioate oligonucleotides have been reported to be more stable in mice than their parent compounds and to display enhanced oral bioavailability (R. Zhang *et al.,* 1995b; Agrawal *et al.,* 1995). The analogs displayed tissue distribution similar to that of the parent phosphorothioate.

Similarly, we have compared the pharmacokinetics of 2′-propoxy-modified phosphodiester and phosphorothioate deoxynucleotides (S. T. Crooke *et al.,* 1996). As expected, the 2′-propoxy modification increased lipophilicity and nuclease resistance. In fact, in mice the 2′-propoxy phosphorothioate was too stable in liver or kidney for measurement of an elimination half-life.

Interestingly, the 2′-propoxy phosphodiester was much less stable than the parent phosphorothioate in all organs except the kidney, in which the 2′-propoxy phosphodiester was remarkably stable. The 2′-propoxy phosphodiester did not bind to albumin significantly, while the affinity of the phosphorothioate for albumin was enhanced. The only difference in toxicity between the analogs was a slight increase in renal toxicity associated with the 2′-propoxy phosphodiester analog (Henry *et al.,* 1997f).

Incorporation of the 2′-methoxyethyoxy group into oligonucleotides increased the T_m by 1.1°C per modification when hybridized to the complement RNA. In a similar manner, several other 2′-O-alkoxy modifications have been reported to enhance the affinity (Martin, 1995). The increase in affinity with these modifications was attributed to (1) the favorable gauche effect of side chain and (2) additional solvation of the alkoxy substituent in water.

More substantial carbohydrate modifications have also been studied. Hexose-containing oligonucleotides were created and found to have low affinity for RNA (Pitsch *et al.*, 1995). Also, the 4′-oxygen has been replaced with sulfur. Although a single substitution of a 4′-thio-modified nucleoside resulted in destabilization of a duplex, incorporation of two 4′-thio-modified nucleosides increased the affinity of the duplex (Bellon *et al.*, 1994). Finally, bicyclic sugars have been synthesized with the hope that preorganization into more rigid structures would enhance hybridization. Several of these modifications have been reported to enhance hybridization (for review, see Sanghvi and Cook, 1994).

5. Backbone Modifications

Substantial progress in creating new backbones for oligonucleotides that replace the phosphate or the sugar–phosphate unit has been made. The objectives of these programs are to improve hybridization by removing the negative charge, enhance stability, and potentially improve pharmacokinetics.

For a review of the backbone modifications reported to date, please see S. T. Crooke (1995) and Sangvhi and Cook, (1994). Suffice it to say that numerous modifications have been made that replace phosphate, retain hybridization, alter charge, and enhance stability. Since these modifications are now being evaluated *in vitro* and *in vivo*, a preliminary assessment should be possible shortly.

Replacement of the entire sugar–phosphate unit has also been accomplished, and the oligonucleotides produced have displayed interesting characteristics. Peptide nucleic acid (PNA) oligonucleotides have been shown to bind to single-stranded DNA and RNA with extraordinary affinity and high sequence specificity. They have been shown to be able to invade some double-stranded nucleic acid structures. PNA oligonucleotides can form triple-stranded structures with DNA or RNA.

PNA oligonucleotides were shown to be able to act as antisense and transcriptional inhibitors when microinjected in cells (Hanvey *et al.*, 1992). PNA oligonucleotides appear to be quite stable in nucleases and peptidases as well.

In summary, then, in the past 5 years, enormous advances in the medicinal chemistry of oligonucleotides have been reported. Modifications at nearly every position in oligonucleotides have been attempted, and numerous potentially interesting analogs have been identified. Although it is far too early to determine which of the modifications may be most useful for particular purposes, it is clear that a wealth of new chemicals is available for systematic evaluation and that these studies should provide important insights into the structure activity relationships (SAR) of oligonucleotide analogs.

VII. Conclusions

Although many more questions about antisense remain to be answered than are answered, progress has continued to be gratifying. Clearly, as more is learned, we will be in a position to perform progressively more sophisticated studies and to understand more of the factors that determine whether an oligonucleotide actually works via an antisense mechanism. We should also have the opportunity to learn a great deal more about this class of drugs as additional studies are completed in human beings.

Acknowledgments

The author thanks Colleen Matzinger for excellent typographic and administrative assistance and Drs. Frank Bennett and Yogesh Sanghvi for review and for constructive comments about the manuscript.

References

Abe, J., Zhou, W., Taguchi, J., *et al.* (1994). Suppression of neointimal smooth muscle cell accumulation *in vivo* by antisense CDC2 and CDK2 oligonucleotides in rat carotid artery. *Biochem. Biophys. Res. Commun.* **198,** 16.

Adams, J. U., Chen, X. H., deRiel, J. K., *et al.* (1994). *In vivo* treatment with antisense oligodeoxynucleotide to kappa-opioid receptors inhibited kappa-agonist-induced analgesia in rats. *Regul. Pept.* **54,** 1.

Agrawal, S., Goodchild, J., Civeira, M. P., *et al.* (1988). Oligodeoxynucleoside phosphoramidates and phosphorothioates as inhibitors of human immunodeficiency virus. *Proc. Natl. Acad. Sci. U.S.A.* **85,** 7079.

Agrawal, S., Temsamani, J., and Tang, J. Y. (1991). Pharmacokinetics, biodistribution and stability of oligodeoxynucleotide phosphorothioates in mice. *Proc. Natl. Acad. Sci. U.S.A.* **88,** 7595.

Agrawal, S., Zhang, X., Lu, Z., *et al.* (1995). Absorption, tissue distribution and *in vivo* stability in rats of a hybrid antisense oligonucleotide following oral administration. *Biochem. Pharmacol.* **50,** 571.

Agris, C. H., Blake, K. R., Miller, P. S., *et al.* (1986). Inhibition of vesicular stomatitis virus protein synthesis and infection by sequence-specific oligodeoxyribonuclease methylphosphonates. *Biochemistry* **25,** 6268.

Akabayashi, A., Wahlestedt, C., Alexander, J. T., *et al.* (1994). Specific inhibition of endogenous neuropeptide Y synthesis in arcuate nucleus by antisense oligonucleotides suppresses feeding behavior and insulin secretion. *Mol. Brain Res.* **21,** 55.

Akino, K., Ohtsuru, A., Yano, H., *et al.* (1996). Antisense inhibition of parathyroid hormone-related peptide gene expression reduces malignant pituitary tumor progression and metastases in the rat. *Cancer Res.* **56,** 77.

Alvarado-Urbina, G., Sathe, G. M., Liu, W. C., *et al.* (1981). Automated synthesis of gene fragments. *Science* **214**, 270.

Amaratunga, A., Morin, P. J., Kosik, K. S., *et al.* (1993). Inhibition of kinesin synthesis and rapid anterograde axonal transport *in vivo* by an antisense oligonucleotide. *J. Biol. Chem.* **268**, 17427.

Ambuhl, P., Gyurko, R., and Phillips, M. I. (1995). A decrease in angiotensin receptor binding in rat brain nuclei by antisense oligonucleotides to the angiotensin AT_1 receptor. *Regul. Pept.* **59**, 171.

Azad, R. F., Driver, V. B., Tanaka, K., *et al.* (1993). Antiviral activity of a phosphorothioate oligonucleotide complementary to RNA of the human cytomegalovirus major immediate-early region. *Antimicrob. Agents Chemothe.* **37**, 1945.

Baker, B., (1993). Decapitation of a $5'$ capped oligoribonucleotide by ortho-phenanthroline:Cu(II). *J. Am. Chem. Soc.* **115**, 3378.

Baker, B. F., Miraglia, L., and Hagedorn, C. H. (1992). Modulation of eukaryotic initiation factor-4E binding to $5'$ capped oligoribonucleotides by modified antisense oligonucleotides. *J. Biol. Chem.* **267**, 11495.

Barrett, J. C., Miller, P. S., and Ts'o, P. O. P. (1974). Inhibitory effect of a complex formation with oligodeoxyribonucleotide ethyl phosphotriesters on transfer ribonucleic acid aminoacylation. *Biochemistry* **13**, 4897.

Barton, C. M., and Lemoine, N. R. (1995). Antisense oligonucleotides directed against p53 have antiproliferative effects unrelated to effects on p53 expression. *Br. J. Cancer* **71**, 429.

Bayever, E., Iversen, P. L., Bishop, M. R., *et al.* (1993). Systemic administration of a phosphorothioate oligonucleotide with a sequence complementary to p53 for acute myelogenous leukemia and myelodysplastic syndrome: Initial results of a phase 1 trial. *Antisense Res. Dev.* **3**, 383.

Bellon, L., Leydier, C., Barascut, J.-L., *et al.* (1994). 4'-Thio RNA: A novel class of sugar-modified β-RNA. *In* "Carbohydrate Modifications in Antisense Research" (Y. S. Sanghvi and P. D. Cook, eds.), p. 68. American Chemical Society, Washington, D.C.

Bennett, C. F., and Crooke, S. T. (1996). Oligonucleotide-based inhibitors of cytokine expression and function. *In* "Therapeutic Modulation of Cytokines," a volume in "Pharmacology and Toxicology: Basic and Clinical Aspects" (B. Henderson and M. W. Bodmer, eds.), pp. 171–193. CRC Press, Boca Raton, Florida.

Bennett, C. F., Chiang, M.-Y., Chan, H. *et al.* (1992). Cationic lipids enhance cellular uptake and activity of phosphorothioate antisense oligonucleotides. *Mol. Pharmacol.* **41**, 1023.

Bennett, C. F., Chiang, M.-Y., Chan, H., *et al.* (1993). Use of cationic lipids to enhance the biological activity of antisense oligonucleotides. *J. Liposome Res.* **3**, 85.

Boutorine A., Huet, C., Saison, T., *et al.* (1991). Cell penetration studies of oligonucleotides derivatized with cholesterol and porphyrins. *Conf. Nucleic Acid Ther.*, p. 60 (abstr.).

Breslauer, K. J., Frank, R., Blocker, H., *et al.* (1986). Predicting DNA duplex stability from base sequence. *Proc. Natl. Acad. Sci. U.S.A.* **83**, 8746.

Burch, R. M., and Mahan, L. C. (1991). Oligonucleotides antisense to the interleukin 1 receptor mRNA block the effects of interleukin 1 in cultured murine and human fibroblasts and in mice. *J. Clin. Invest.* **88**, 1190.

Burgess, T. L., Fisher, E. F., Ross, S. L., *et al.* (1995). The antiproliferative activity of c-*myb* and c-*myc* antisense oligonucleotides in smooth muscle cells is caused by a nonantisense mechanism. *Proc. Natl. Acad. Sci. U.S.A.* **92**, 4051.

Campbell, J. M., Bacon, T. A., and Wickström, E. (1990). Oligodeoxynucleoside phosphorothioate stability in subcellular extracts, culture media, sera and cerebrospinal fluid. *J. Biochem. Biophys. Methods* **20**, 259.

Caruthers, M. H. (1985). Gene synthesis machines: DNA chemistry and its uses. *Science* **230**, 281.

Castro-Alamancos, M. A., and Torres-Aleman, I. (1994). Learning of the conditioned eyeblink response is impaired by an antisense insulin-like growth factor I oligonucleotide. *Proc. Natl. Acad. Sci. U.S.A.* **91**, 10203.

Cazenave, C., Stein, C. A., Loreau, N., *et al.* (1989). Comparative inhibition of rabbit globin mRNA translation by modified antisense oligodeoxynucleotides. *Nucleic Acids Res.* **17**, 4255.

Cha, X. Y., Xu, H., Ni, Q., *et al.* (1995). Opioid peptide receptor studies. 4. Antisense oligodeoxynucleotide to the δ-opiod receptor delineates opiod receptor subtypes. *Regul. Pept.* **59**, 247.

Cheng, Y.-C., Gao, W., and Han, F. (1991). Against human immunodeficiency virus and herpes viruses. *Nucleosides Nucleotides* **10**, 155.

Chiang, M. Y., Chan, H., Zounes, M. A., *et al.* (1991). Antisense oligonucleotides inhibit ICAM-1 expression by two distinct mechanisms. *J. Biol. Chem.* **266**, 18162.

Chiasson, B. J., Hooper, M. L., Murphy, P. R., *et al.* (1992). Antisense oligonucleotide eliminates *in vivo* expression of c-*fos* in mammalian brain. *Eur. J. Pharmacol.* **227**, 451.

Chrissey, L. A., Walz, S. E., Pazirandeh, M., *et al.* (1993). Internalization of oligodeoxyribonucleotides by *Vibrio parahaemolyticus*. *Antisense Res. Dev.* **3**, 367.

Cohen, J. S. (1993). Phosphorothioate oligodeoxynucleotides. *In* "Antisense Research and Applications" (S. T. Crooke and B. Lebleu, eds.), p. 205. CRC Press, Boca Raton, Florida.

Colby, C. (1971). The induction of interferon by natural and synthetic polynucleotides. *Prog. Nucleic Acid Res. Mol. Biol.* **11**, 1.

Cook, P. D. (1993). Medicinal chemistry strategies for antisense research. *In* "Antisense Research and Applications" (S. T. Crooke and B. Lebleu, eds.), p. 303. CRC Press, Boca Raton, Florida.

Cornish, K. G., Iversen, P., Smith, L., *et al.* (1993). Cardiovascular effects of a phosphorothioate oligonucleotide with sequence antisense to p53 in the conscious rhesus monkey. *Pharmacol. Commun.* **3**, 239.

Cossum, P. A., Sasmor, H., Dellinger, D., *et al.* (1993). Disposition of the ^{14}C-labeled phosphorothioate oligonucleotide ISIS 2105 after intravenous administration to rats. *J. Pharmacol. Exp. Ther.* **267**, 1181.

Cossum, P. A., Truong, L., Owens, S. R., *et al.* (1994). Pharmacokinetics of a ^{14}C labeled phosphorothioate oligonucleotide, ISIS 2105, after intradermal administration to rats. *J. Pharmacol. Exp. Ther.* **269**, 89.

Cowsert, L. M. (1993). Antiviral activities of oligonucleotides. *In* "Antisense Research and Applications" (S. T. Crooke and B. Lebleu, eds.), p. 521. CRC Press, Boca Raton, Florida.

Cowsert, L. M., Fox, M. C., Zon, G., *et al.* (1993). *In vitro* evaluation of phosphorothioate oligonucleotides targeted to the E_2 mRNA of papillomavirus: Potential treatment of genital warts. *Antimicrob. Agents Chemother.* **37**, 171.

Crooke, R. M. (1991). *In vitro* toxicology and pharmacokinetics of antisense oligonucleotides. *Anti-Cancer Drug Des.* **6**, 609.

Crooke, R. M. (1993a). Cellular uptake, distribution and metabolism of phosphorothioate, phosphorodiester and methylphosphonate oligonucleotides. *In* "Antisense Research and Applications" (S. T. Crooke and B. Lebleu, eds.), p. 427. CRC Press, Boca Raton, Florida.

Crooke, R. M. (1993b). *In vitro* and *in vivo* toxicology of first generation analogs. *In* "Antisense Research and Applications" (S. T. Crooke and B. Lebleu, eds.), p. 471. CRC Press, Boca Raton, Florida.

Crooke, R. M., Graham, M. J., Cooke, M. E., *et al.* (1995). *In vitro* pharmacokinetics of phosphorothioate antisense oligonucleotides. *J. Pharmacol. Toxicol.* **275**, 462.

Crooke, R. M., Crooke, S. T., Graham, M. J., *et al.* (1996). Effect of antisense oligonucleotides on cytokine release from human keratinocytes in an *in vitro* model of skin. *Toxicol. Appl. Pharmacol.* **140**, 85–93.

Crooke, S. T. (1992). Therapeutic applications of oligonucleotides. *Annu. Rev. Pharmacol. Toxicol.* **32**, 329.

Crooke, S. T. (1993). Progress toward oligonucleotide therapeutics: Pharmacodynamic properties. *FASEB J.* **7**, 533.

Crooke, S. T. (1995a). Oligonucleotide therapeutics. *In* "Burger's Medicinal Chemistry and Drug Discovery" (M. E. Wolff, ed.), 5th Ed., Vol. 1, p. 863. Wiley, New York.

Crooke, S. T. (1995b). "Therapeutic Applications of Oligonucleotides." R. G. Landes Co., Austin, Texas.

Crooke, S. T., and Lebleu, B., eds. (1993). "Antisense Research and Applications." CRC Press, Boca Raton, Florida.

Crooke, S. T., Grillone, L. R., Tendolkar, A., *et al.* (1994). A pharmacokinetic evaluation of ^{14}C labeled afovirsen sodium in genital wart patients. *J. Pharmacol. Exp. Ther.* **56,** 641.

Crooke, S. T., Lemonidis, K. M., Neilson, L., *et al.* (1995). Kinetic characteristics of *E. coli* RNase H1: Cleavage of various antisense oligonucleotide–RNA duplexes. *Biochem. J.* **312,** 599.

Crooke, S. T., Graham, M. J., Zuckerman, J. E., *et al.* (1996). Pharmacokinetic properties of several novel oligonucleotide analogs in mice. *J. Pharmacol. Exp. Ther.* **277**(2), 923–937.

Crouch, R. J., and Dirksen, M.-L. (1985). Ribonucleases H. *In* "Nucleases" (S. M. Linn and R. J. Roberts, eds.), p. 211. Cold Spring Harbor Laboratory Press, Cold Spring Harbor, New York.

Crum, C., Johnson, J. D., Nelson, A., *et al.* (1988). Complementary oligodeoxynucleotide mediated inhibition of tobacco mosaic virus RNA translation *in vitro. Nucleic Acid Res.* **16,** 4569.

Dean, N. M., and McKay, R. (1994). Inhibition of PKC-α expression in mice after systemic administration of phosphorothioate antisense oligonucleotides. *Proc. Natl. Acad. Sci. U.S.A.* **91,** 11762.

Dean, N. M., McKay, R., Miraglia, L., *et al.* (1996). Inhibition of growth of human tumor cell lines in nude mice by an antisense oligonucleotide inhibitor of PKC-α expression. *Cancer Res.* **56**(15), 3499–3507.

DeClercq, E., Eckstein, F., and Merigan, T. C. (1969). Interferon induction increased through chemical modification of a synthetic polyribonucleotide. *Science* **165,** 1137.

DeMesmaeker, A., Haner, R., Martin, P., *et al.* (1995). Antisense oligonucleotides. *Acc. Chem. Res.* **28,** 366.

Desjardins, J., Mata, J., Brown, T., *et al.* (1995). Cholesteryl-conjugated phosphorothioate oligodeoxynucleotides modulate CYP2B1 expression *in vivo. J. Drug Target.* **2,** 477.

Dominski, Z., and Kole, R. (1993). Restoration of correct splicing in thalassemic pre-mRNA by antisense oligonucleotides. *Proc. Natl. Acad. Sci. U.S.A.* **90,** 673.

Doris-Keller, H. (1979). Site specific enzymatic cleavage of RNA. *Nucleic Acids Res.* **7,** 179.

Ecker, D. J. (1993). Strategies for invasion of RNA secondary structure. *In* "Antisense Research and Applications" (S. T. Crooke and B. Lebleu, eds.), p. 387. CRC Press, Boca Raton, Florida.

Ecker, D. J., Vickers, T. A., Bruice, T. W., *et al.* (1992). Pseudo-half knot formation with RNA. *Science* **257,** 958.

Eder, P. S., and Walder, J. A. (1991). Ribonuclease H from K562 human erythroleukemia cells. *J. Biol. Chem.* **206,** 6472.

Flanagan, L. M., McCarthy, M. M., Brooks, P. J., Pfaff, D. W., and McEwen, B. S. (1993). Arginine vasopressin levels after daily infusions of antisense oligonucleotides into the supraoptic nucleus. *Ann. N.Y. Acad. Sci.* **689,** 520.

Freier, S. M. (1993). Hybridization: Considerations affecting antisense drugs. *In* "Antisense Research and Applications" (S. T. Crooke and B. Lebleu, eds.), p. 67. CRC Press, Boca Raton, Florida.

Furdon, P., Dominski, Z., and Kole, R. (1989). RNase H cleavage of RNA hybridized to oligonucleotides containing methylphosphonate, phosphorothioate and phosphodiester bonds. *Nucleic Acids Res.* **17,** 9193.

Gagnor, C., Bertrand, J., Thenet, S., *et al.* (1987). α-DNA. VI: Comparative study of α- and β-anomeric oligodeoxyribonucleotides in hybridization to mRNA and in cell free translation inhibition. *Nucleic Acids Res.* **15,** 10419.

Gagnor, C., Rayner, B., Leonetti, J.-P., *et al.* (1989). α-DNA. IX. Parallel annealing of α-anomeric oligodeoxyribonucleotides to natural mRNA is required for interference in RNase H mediated hydrolysis and reverse transcription. *Nucleic Acids Res.* **17**, 5107.

Galbraith, W. M., Hobson, W. C., Giclas, P. C., *et al.* (1994). Complement activation and hemodynamic changes following intravenous administration of phosphorothioate oligonucleotides in the monkey. *Antisense Res. Dev.* **4**, 201.

Gao, W., Stein, C. A., Cohen, J. S., *et al.* (1989). Effect of phosphorothioate homo-oligodeoxynucleotides on herpes simplex virus type 2-induced DNA polymerase. *J. Biol. Chem.* **264**, 11521.

Gao, W.-Y., Han, F.-S., Storm, C., *et al.* (1991). Phosphorothioate oligonucleotides are inhibitors of human DNA polymerases and RNase H: Implications for antisense technology. *Mol. Pharmacol.* **41**, 223.

Giles, R. V., and Tidd, D. M. (1992). Increased specificity for antisense oligodeoxynucleotide targeting of RNA cleavage of RNase H using chimeric methylphosphorodiester structures. *Nucleic Acids Res.* **20**, 763.

Giles, R. V., Spiller, D. G., and Tidd, D. M. (1995). Detection of ribonuclease H-generated mRNA fragments in human leukemia cells following reversible membrane permeabilization in the presence of antisense oligodeoxynucleotides. *Antisense Res. Dev.* **5**, 23.

Gillardon, F., Beck, H., Uhlmann, E., *et al.* (1994). Inhibition of c-*fos* protein expression in rat spinal cord by antisense oligodeoxynucleotide superfusion. *Eur. J. Neurosci.* **6**, 880.

Gillespie, D., and Spiegelman, S. (1965). A quantitative assay for DNA–RNA hybrids with DNA immobilized on a membrane. *J. Mol. Biol.* **12**, 829–842.

Glover, J. M., Leeds, J. M., Mant, T. G. K., *et al.* (1997). Phase I safety and pharmacokinetic profile of an ICAM-1 antisense oligonucleotide (ISIS 2302). *J. Pharmacol. Exp. Ther.* In press.

Goodchild, J., Agrawal, S., Civeira, M. P., *et al.* (1988). Inhibition of human immunodeficiency virus replication by antisense oligonucleotides. *Proc. Natl. Acad. Sci. U.S.A.* **85**, 5507.

Graham, M. J., Freier, S. M., Crooke, R. M., *et al.* (1993). Tritium labeling of antisense oligonucleotides by exchange with tritiated water. *Nucleic Acids Res.* **21**, 3737.

Gutierrez, A. J., Terhorst, T. J., Matteucci, M. D., *et al.* (1994). 5-Heteroaryl-2′-deoxyuridine analogs. Synthesis and incorporation into high-affinity oligonucleotides. *J. Am. Chem. Soc.* **116**, 5540.

Gyurko, F., Wielbo, D., and Phillips, M. I. (1993). Antisense inhibition of AT_1 receptor mRNA and angiotensinogen mRNA in the brain of spontaneously hypertensive rats reduces hypertension of neurogenic origin. *Regul. Pept.* **49**, 167.

Haeuptle, M. T., Frank, R., and Dobberstein, B. (1986). Translation arrest by oligodeoxynucleotides complementary to mRNA coding sequences yields polypeptides of predetermined length. *Nucleic Acids Res.* **14**, 1427.

Hagenberg, L., Gassen, H. G., and Matthaei, H. (1973). Synthesis and coding properties of poly(c^1A), poly(c^3A), poly(c^7A) and poly(h^6A). *Biochem. Biophys. Res. Commun.* **50**, 1104.

Hall, J., Husken, D., Pieles, U., *et al.* (1994). Efficient sequence-specific cleavage of RNA using novel europium complexes conjugated to oligonucleotides. *Chem. Biol.* **1**, 185.

Hanvey, J. C., Peffer, N. J., Bisi, J. E., *et al.* (1992). Antisense and antigene properties of peptide nucleic acids. *Science* **258**, 1481.

Heikkila, R., Schwab, G., Wickström, E., *et al.* (1987). A c-*myc* antisense oligonucleotide inhibits entry into S phase but not progress from G0 to G1. *Nature (London)* **328**, 445.

Heilig, M., Engel, J. A., and Soderpalm, B. (1993). c-*fos* antisense in the nucleus accumbens blocks the locomotor stimulant action of cocaine. *Eur. J. Pharmacol.* **236**, 339.

Helene, C., and Toulme, J. J. (1989). Control of gene expression by oligonucleotides covalently linked to intercalating agents and nucleic acid-cleaving reagents. *In* "Oligodeoxynucleotides: Antisense Inhibitors of Gene Expression" (J. S. Cohen, ed.), p. 137. CRC Press, Boca Raton, Florida.

Henry, S. P., Orr, J. L., Bruner, R. H., *et al.* (1997a). Comparison of the toxicities profile of ISIS 1082 and ISIS 2105, phosphorothioate oligonucleotides, following subacute intradermal administration in Sprague-Dawley rats. *Toxicology* 116(1–3), 77–88.

Henry, S. P., Taylor, J., Midgley, L., *et al.* (1997b). Evaluation of the toxicity profile of ISIS 2302, a phosphorothioate oligonucleotide. 1. 4-week study in CD-1 mice. *Antisense Nucleic Drug Dev.* In press.

Henry, S. P., Bolte, H., Auletta, C., *et al.* (1997c). Evaluation of the toxicity of ISIS 2302, a phosphorothioate oligonucleotide. 2. 4-week study in cynomolgus monkeys. *Toxicology* In press.

Henry, S. P., Giclas, P., Leeds, J., *et al.* (1997d). Activation of the alternative pathway of complement by a phosphorothioate oligonucleotide: Potential mechanism of action. *J. Pharmacol. Exp. Ther.* In press.

Henry, S. P., Novotny, W., Leeds, J., *et al.* (1997e). Inhibition of clotting parameters by a phosphorothioate oligonucleotide. In preparation.

Henry, S. P., Zuckerman, J. E., Rojko, J., *et al.* (1997f). Toxicologic properties of several novel oligonucleotide analogs in mice. *Anti-cancer Drug Des.* 12(1), 1–14.

Hertl, M., Neckers, L. M., and Katz, S. I. (1995). Inhibition of interferon-γ-induced intercellular adhesion molecule-1 expression on human keratinocytes by phosphorothioate antisense oligodeoxynucleotides is the consequence of antisense-specific and antisense-non-specific effects. *J. Invest. Dermatol.* 104, 813.

Higgins, K. A., Perez, J. R., Coleman, T. A., *et al.* (1993). Antisense inhibition of the p65 subunit of NF-κB blocks tumorigenicity and causes tumor regression. *Proc. Natl. Acad. Sci. U.S.A.* 90, 9901.

Hijiya, N., Zhang, J., Ratajczak, M. Z., *et al.* (1994). Biological and therapeutic significance of MYB expression in human melanoma. *Proc. Natl. Acad. Sci. U.S.A.* 91, 4499.

Hodges, D., and Crooke, S. T. (1995). Inhibition of wild-type and mutated luciferase-adenovirus pre-mRNAs by antisense oligonucleotides. *Mol. Pharmacol.* 48, 905.

Hoke, G. D., Draper, K., Freier, S. M., *et al.* (1991). Effects of phosphorothioate capping on antisense oligonucleotide stability, hybridization and antiviral efficacy versus herpes simplex virus infection. *Nucleic Acids Res.* 19, 5743.

Hooper, M. L., Chiasson, B. J., and Robertson, H. A. (1994). Infusion into the brain of an antisense oligonucleotide to the immediate-early gene c-*fos* suppresses production of *fos* and produces a behavioral effect. *Neuroscience* 63, 917.

Hughes, J. A., Avrutskaya, A. V., Brouwer, K. L. R., *et al.* (1995). Radiolabeling of methylphosphonate and phosphorothioate oligonucleotides and evaluation of their transport in everted rat jejunum sacs. *Pharmacol. Res.* 12, 817.

Hutcherson, S. L., Palestine, A. G., Cantrill, H. L., *et al.* (1995). Antisense oligonucleotide safety and efficacy for CMV retinitis in AIDS patients. *Int. Conf. Antimicrob. Agents Chemother., 35th*, p. 204.

Inoue, H., and Ohtsuka, E. (1985). Synthesis and hybridization of dodecadeoxyribonucleotides containing a fluorescent pyridopyrimidine deoxynucleoside. *Nucleic Acids Res.* 13, 7119.

Iversen, P. (1991). *In vivo* studies with phosphorothioate oligonucleotides: Pharamcokinetic prologue. *Anti-Cancer Drug Des.* 6, 531.

Joos, R. W., and Hall, W. H. (1969). Determination of binding constants of serum albumin for penicillin. *J. Pharmacol. Exp. Ther.* 166, 113.

Kawasaki, A. M., Casper, M. D., Freier, S. M., *et al.* (1993). Uniformly modified 2′-deoxy-2′-fluoro phosphorothioate oligonucleotides as nuclease resistant antisense compounds with high affinity and specificity for RNA targets. *J. Med. Chem.* 36, 831.

Kindy, M. S. (1994). NMDA receptor inhibition using antisense oligonucleotides prevents delayed neuronal death in gerbil hippocampus following cerebral ischemia. *Neurosci. Res. Commun.* 14, 175.

Kitajima, I., Shinohara, T., Bilakovics, J., *et al.* (1992). Ablation of transplanted HTLV-1 tax-transformed tumors in mice by antisense inhibition of NF-κB. *Science* 258, 1792.

Konradi, C., Cole, R. L., Heckers, S., *et al.* (1994). Amphetamine regulates gene expression in rat striatum via transcription factor CREB. *J. Neurosci.* **14,** 5623.

Krieg, A. M., Yi, A.-K., Matson, S., *et al.* (1995). CpG motifs in bacterial DNA trigger direct B-cell activation. *Nature (London)* **374,** 546.

Kulka, M., Smith, C., Aurelian, L., *et al.* (1989). Site specificity of the inhibitory effects of oligo(nucleoside methylphosphonates) complementary to the acceptor splice junction of herpes simplex virus type 1 immediately early mRNA. *Proc. Natl. Acad. Sci. U.S.A.* **86,** 6868.

Kuramoto, E., Yano, O., Kimura, Y., *et al.* (1992). Oligonucleotide sequences required for natural killer cell activation. *Jpn. J. Cancer Res.* **83,** 1128.

Lai, J., Bilsky, E. J., Rothman, R. B., *et al.* (1994). Treatment with antisense oligodeoxynucleotide to the opioid-δ receptor selectively inhibits δ_2-agonist antinociception. *NeuroReport* **5,** 1049.

Lamond, A. I., and Sproat, B. S. (1993). Antisense oligonucleotides made of 2'-O-alkyl RNA: Their properties and applications in RNA biochemistry. *FEBS Lett.* **325,** 123.

Lemaitre, M., Bayard, B., and Lebleu, B. (1987). Specific antiviral activity of a poly(l-lysine)-conjugated oligodeoxyribonucleotide sequence complementary to vesicular stomatitis virus N-protein mRNA initiation site. *Proc. Natl. Acad. Sci. U.S.A.* **84,** 648.

Liebsch, G., Landgraf, R., Gerstberger, R., *et al.* (1995). Chronic infusion of a CRH_1 receptor antisense oligodeoxynucleotide into the central nucleus of the amygdala reduced anxiety-related behavior in socially defeated rats. *Regul. Pept.* **59,** 229.

Lima, W. F., Monia, B. P., Ecker, D. J., *et al.* (1992). Implication of RNA structure on antisense oligonucleotide hybridization kinetics. *Biochemistry* **31,** 12055.

Lin, K. T. P., and Brown, D. M. (1989). Synthesis and duplex stability of oligonucleotides containing cytosine–thymine analogues. *Nucleic Acids Res.* **17,** 10373.

Lin, K.-Y., Jone, R. J., and Matteucci, M. (1995). Tricyclic-2'-deoxycytidine analogs: Syntheses and incorporation into oligodeoxynucleotides which have enhanced binding to complementary RNA. *J. Am. Chem. Soc.* **117,** 3873.

Loke, S. L., Stein, C. A., Zhang, X. H., *et al.* (1989). Characterization of oligonucleotide transport into living cells. *Proc. Natl. Acad. Sci. U.S.A.* **86,** 3474.

Maher, J. L., III, Wold, B., and Dervan, P. G. (1989). Inhibition of DNA binding properties by oligonucleotide-directed triple helix formation. *Science* **245,** 725.

Maier, J. A. M., Voulalas, P., Roeder, D., *et al.* (1990). Extension of the life-span of human endothelial cells by an interleukin-1α antisense oligomer. *Science* **249,** 1570.

Majumdar, C., Stein, C. A., Cohen, J. S., *et al.* (1989). Stepwise mechanism of HIV reverse transcriptase: Primer function of phosphorothioate oligodeoxynucleotide. *Biochemistry* **28,** 1340.

Mani, S. K., Blaustein, J. D., Allen, J. M. C., *et al.* (1994). Inhibition of rat sexual behavior by antisense oligonucleotides to the progesterone receptor. *Endocrinology (Baltimore)* **135,** 278.

Manoharan, M. (1993). Designer antisense oligonucleotides: Conjugation chemistry and functionality placement. *In* "Antisense Research and Therapeutics" (S. T. Crooke and B. Lebleu, eds.), p. 303. CRC Press, Boca Raton, Florida.

Marcus-Sekura, C. J., Woerner, A. M., Shinozuka, K., *et al.* (1987). Comparative inhibition of chloramphenicol acetyltransferase gene expression by antisense oligonucleotide analogues having alkyl phosphotriester, methylphosphonate and phosphorothioate linkages. *Nucleic Acids Res.* **15,** 5749.

Martin, F. H., Castro, M. M., Aboul-ela, F., *et al.* (1985). Base pairing involving deoxyinosine: Implications for probe design. *Nucleic Acids Res.* **13,** 8927.

Martin, P. (1995). New access to 2'-O-alkylated ribonucleosides and properties of 2'-O-alkylated oligoribonucleotides. *Helv. Chim. Acta* **78,** 486.

Matsukura, M., Shinozuka, K., Zon, G., *et al.* (1987). Phosphorothioate analogs of oligodeoxyribonucleotides: Inhibitors of replication and cytopathic effects of human immunodeficiency virus. *Proc. Natl. Acad. Sci. U.S.A.* **84,** 7706.

McCarthy, M. M., Masters, D. B., Rimvall, K., *et al.* (1994). Intracerebral administration of antisense oligodeoxynucleotides to GAD_{65} and GAD_{67} mRNAs modulate reproductive behavior in the female rat. *Brain Res.* **636,** 209.

McCarthy, M. M., Nielsen, D. A., and Goldman, D. (1995). Antisense oligonucleotide inhibition of tryptophan hydroxylase activity in mouse brain. *Regul. Pept.* **59,** 163.

McManaway, M. E., Neckers, L. M., Loke, S. L., *et al.* (1990). Tumour-specific inhibition of lymphoma growth by an antisense oligodeoxynucleotide. *Lancet* **335,** 808.

Messina, J. P., Gilkeson, G. S., and Pisetsky, D. S. (1991). Stimulation of *in vitro* murine lymphocyte proliferation by bacterial DNA. *J. Immunol.* **147,** 1759.

Miller, P. S. (1989). Non-ionic antisense oligonucleotides. *In* "Oligodeoxynucleotides: Antisense Inhibitors of Gene Expression" (J. S. Cohen, ed.), p. 79. CRC Press, Boca Raton, Florida.

Minshull, J., and Hunt, T. (1986). The use of single-stranded DNA and RNase H to promote quantitative hybrid arrest of translation of mRNA/DNA hybrids in reticulocyte lysate cell-free translations. *Nucleic Acids Res.* **14,** 6433.

Mirabelli, C. K., and Crooke, S. T. (1993). Antisense oligonucleotides in the context of modern molecular drug discovery and development. *In* "Antisense Research and Applications" (S. T. Crooke and B. Lebleu, eds.), p. 7. CRC Press, Boca Raton, Florida.

Mirabelli, C. K., Bennett, C. F., Anderson, K., *et al.* (1991). *In vitro* and *in vivo* pharmacologic activities of antisense oligonucleotides. *Anti-Cancer Drug Des.* **6,** 647.

Miyao, T., Takakura, Y., Akiyama, T., *et al.* (1995). Stability and pharmacokinetic characteristics of oligonucleotides modified at terminal linkages in mice. *Antisense Res. Dev.* **5,** 115.

Mizoguchi, H., Narita, M., Nagase, H., *et al.* (1995). Antisense oligodeoxynucleotide to a δ-opioid receptor blocks the antinociception induced by cold water swimming. *Regul. Pept.* **59,** 255.

Monia, B. P., Johnston, J. F., Ecker, D. J., *et al.* (1992). Selective inhibition of mutant Ha-*ras* mRNA expression by antisense oligonucleotides. *J. Biol. Chem.* **267,** 19954.

Monia, B. P., Lesnik, E. A., Gonzalez, C., *et al.* (1993). Evaluation of 2′-modified oligonucleotides containing 2′-deoxy gaps as antisense inhibitors of gene expression. *J. Biol. Chem.* **268.**

Monia, B. P., Johnston, J. F., Geiger, T., *et al.* (1996a). Antitumor activity of a phosphorothioate antisense oligodeoxynucleotide targeted against c-*raf* kinase. *Nat. Med.* **2**(6), 668–675.

Monia, B. P., Johnston, J. F., Sasmor, H., *et al.* (1996b). Nuclease resistance and antisense activity of modified oligonucleotides targeted to Ha-*ras. J. Biol. Chem.* **271**(24), 14533–14540.

Morishita, R., Gibbons, G. H., Ellison, K. E., *et al.* (1993). Single intraluminal delivery of antisense cdc2 kinase and proliferating-cell nuclear antigen oligonucleotides results in chronic inhibition of neointimal hyperplasia. *Proc. Natl. Acad. Sci. U.S.A.* **90,** 8474.

Morishita, R., Gibbons, G. H., Kaneda, Y., *et al.* (1994). Pharmacokinetics of antisense oligodeoxyribonucleotides (cyclin B_1 and CDC 2 kinase) in the vessel wall *in vivo:* Enhanced therapeutic utility for restenosis by HVJ-liposome delivery. *Gene* **149,** 13.

Morvan, F., Rayner, B., and Imbach, J.-L. (1991). α-Oligonucleotides: A unique class of modified chimeric nucleic acids. *Anti-Cancer Drug Des.* **6,** 521.

Nagel, K. M., Holstad, S. G., and Isenberg, K. E. (1993). Oligonucleotide pharmacotherapy: An antigene strategy. *Pharmacotherapy* **13,** 177.

Neckers, L. M. (1993). Cellular internalization of oligodeoxynucleotides. *In* "Antisense Research and Applications" (S. T. Crooke and B. Lebleu, eds.), p. 451. CRC Press, Boca Raton, Florida.

Nesterova, M., and Cho-Chung, Y. S. (1995). A single injection protein kinase α-directed antisense treatment to inhibit tumor growth. *Nat. Med.* **1,** 528.

Neumann, I., Porter, D. W. F., Landgraf, R., *et al.* (1994). Rapid effect on suckling of an oxytocin antisense oligonucleotide administered into rat supraoptc nucleus. *Bull. Am. Phys. Soc.,* p. R852.

Nikiforov, T. T., and Connolly, B. A. (1991). The synthesis of oligodeoxynucleotides containing 4-thiothymidine residues. *Tetrahedron Lett.* **32**, 3851.

Offensperger, W.-B., Offensperger, S., Walter, E., *et al.* (1993). *In vivo* inhibition of duck hepatitis B virus replication and gene expression by phosphorothioate modified antisense oligodeoxynucleotides. *EMBO J.* **12**, 1257.

Ogo, H., Hirai, Y., Miki, S., *et al.* (1994). Modulation of substance P/neurokinin-1 receptor in human astrocytoma cells by antisense oligodeoxynucleotides. *Gen. Pharmacol.* **25**, 1131.

Osen-Sand, A., Catsicas, M., Staple, J. K., *et al.* (1993). Inhibition of axonal growth by SNAP-25 antisense oligonucleotides *in vitro* and *in vivo. Nature (London)* **364**, 445.

Parmentier, G., Schmitt, G., Dolle, F., *et al.* (1994). A convergent synthesis of 2'-O-methyl uridine. *Tetrahedron* **50**, 5361.

Perlaky, L., Saijo, Y., Busch, R. K., *et al.* (1993). Growth inhibition of human tumor cell lines by antisense oligonucleotides designed to inhibit p120 expression. *Anti-Cancer Drug Des.* **8**, 3.

Phillips, M. I., Wielbo, D., and Gyurko, R. (1994). Antisense inhibition of hypertension: A new strategy for renin–angiotensin candidate genes. *Kidney Int.* **46**, 1554.

Pitsch, S., Krishnamurthy, R., Bolli, M., *et al.* (1995). Pyranosyl-RNA ("p-RNA"): Base-pairing selectivity and potential to replicate. *Helv. Chim. Acta* **78**, 1621.

Pollio, G., Xue, P., Zanisi, M., *et al.* (1993). Antisense oligonucleotide blocks progesterone-induced lordosis behavior in ovariectomized rats. *Mol. Brain Res.* **19**, 135.

Psietsky, D. S., and Reich, C. F. (1993). Stimulation of murine lymphocyte proliferation by a phosphorothioate oligonucleotide with antisense activity for herpes simplex virus. *Life Sci.* **54**, 101.

Qin, Z.-H., Zhou, L.-W., Zhang, S.-P., *et al.* (1995). D_2 dopamine receptor antisense oligodeoxynucleotide inhibits the synthesis of a functional pool of D_2 dopamine receptors. *Mol. Pharmacol.* **48**, 730.

Quartin, R., Brakel, C., and Wetmur, J. (1989). Number and distribution of methylphosphonate linkages in oligodeoxynucleotides affect exo- and endonuclease sensitivity and ability to form RNase H substrates. *Nucleic Acids Res.* **17**, 7253.

Quattrone, A., Papucci, L., Schiavone, *et al.* (1994). Intracellular enhancement of intact antisense oligonucleotide steady-state levels by cationic lipids. *Anti-Cancer Drug Des.* **9**, 549.

Rappaport, J., Hanss, B., Kopp, J. B., *et al.* (1995). Transport of phosphorothioate oligonucleotides in kidney: Implications for molecular therapy. *Kidney Int.* **47**, 1462.

Ratajczak, M. Z., Kant, J. A., Luger, S. M., *et al.* (1992). *In vivo* treatment of human leukemia in a *scid* mouse model with c-*myb* antisense oligodeoxynucleotides. *Proc. Natl. Acad. Sci. U.S.A.* **89**, 11823.

Rosolen, A., Whitesell, L., Olegalo, M., *et al.* (1990). Antisense inhibition of single copy N-myc expression results in decreased cell growth without reduction of c-myc protein in a neuroepithelioma cell line. *Cancer Res.* **50**, 6316.

Sakai, R. R., Ma, L. Y., He, P. F., *et al.* (1995). Intracerebroventricular administration of angiotensin type 1 (AT_1) receptor antisense oligonucleotide attenuates thirst in the rat. *Regul. Pept.* **59**, 183.

Sands, H., Gorey-Feret, L. J., Cocuzza, A. J., *et al.* (1994). Biodistribution and metabolism of internally ³H-labeled oligonucleotides. I. Comparion of a phosphodiester and a phosphorothioate. *Mol. Pharmacol.* **45**, 932.

Sands, H., Gorey-Feret, L. J., Ho, S. P., *et al.* (1995). Biodistribution and metabolism of internally ³H-labeled oligonucleotides. II. 3',5'-Blocked oligonucleotides. *Mol. Pharmacol.* **47**, 636.

Sanghvi, Y. S. (1993). Heterocyclic base modifications in nucleic acids and their applications in antisense oligonucleotides. *In* "Antisense Research and Applications" (S. T. Crooke and B. Lebleu, eds.), p. 273. CRC Press, Boca Raton, Florida.

Sanghvi, Y. S., and Cook, P. D., eds. (1994). "Carbohydrate Modifications in Antisense Research." American Chemical Society, Washington, D.C.

Sanghvi, Y. S., Hoke, G. D., Freier, S. M., *et al.* (1993). Antisense oligonucleotides: Synthesis and biological evaluation of oligodeoxynucleotides containing modified pyrimidines. *Nucleic Acids Res.* **21**, 3197.

Santa Lucia, J., Kierzek, R., and Turner, D. H. (1991). Functional group substitutions as probes of hydrogen bonding between GA mismatches in RNA internal loops. *J. Am. Chem. Soc.* **113**, 4314.

Sarin, P. S., Agrawal, S., and Civeira, M. P. (1988). Inhibition of acquired immunodeficiency syndrome virus by oligodeoxynucleoside methylphosphonates. *Proc. Natl. Acad. Sci. U.S.A.* **85**, 7448.

Saxena, S. K., and Ackerman, E. J. (1990). Microinjected oligonucleotides complementary to the α-sarcin loop of 28 S RNA abolish protein synthesis in *Xenopus* oocytes. *J. Biol. Chem.* **265**, 3263.

Sburlati, A. R., Manrow, R. E., and Berger, S. L. (1991). Prothymosin α antisense oligomers inhibit myeloma cell division. *Proc. Natl. Acad. Sci. U.S.A.* **88**, 253.

Schwab, G., Chavany, C., Duroux, I., *et al.* (1994). Antisense oligonucleotides adsorbed to polyalkylcyanoacrylate nanoparticles specifically inhibit mutated Ha-*ras* mediated cell proliferation and tumorigenicity in nude mice. *Proc. Natl. Acad. Sci. U.S.A.* **91**, 10460.

Seela, F., Kaiser, K., and Bindig, U. (1989). 2′-Deoxy-β-D-ribofuranosides of N^6-methylated 7-dazaadenine and 8-aza-7-deazaadenine: Solid phase synthesis of oligodeoxyribonucleotides and properties of self-complementary duplexes. *Helv. Chim. Acta* **72**, 868.

Seela, F., Ramzaeva, N., and Chen, Y. (1995). Oligonucleotide duplex stability controlled by the 7-substituents of 7-deazaguanine bases. *Bioorg. Med. Chem. Lett.* **5**, 3049.

Simons, M., Edelman, E. R., DeKeyser, J.-L., *et al.* (1992). Antisense c-*myb* oligonucleotides inhibit intimal arterial smooth muscle cell accumulation *in vivo*. *Nature (London)* **359**, 67.

Simons, M., Edelman, E. R., and Rosenberg, R. D. (1994). Antisense proliferating cell nuclear antigen oligonucleotides inhibit intimal hyperplasia in a rat carotid artery injury model. *J. Clin. Invest.* **9**, 2351.

Skorski, T., Nieborowska-Skorska, M., Nicolaides, N. C., *et al.* (1994). Suppression of Ph[1] leukemia cell growth in mice by BCR-ABL antisense oligodeoxynucleotide. *Proc. Natl. Acad. Sci. U.S.A.* **91**, 4504.

Skutella, T., Probst, J. C., Jirikowski, G. F., *et al.* (1994). Ventral tegmental area (VTA) injections of tyrosine hydroxylase phosphorothioate antisense oligonucleotide suppress operant behavior in rats. *Neurosci. Lett.* **167**, 55.

Smith, C. C., Aurelian, L., Reddy, M. P., *et al.* (1986). Antiviral effect of an oligo(nucleoside methylphosphonate) complementary to the splice junction of herpes simplex virus type 1 immediate early pre-mRNAs 4 and 5. *Proc. Natl. Acad. Sci. U.S.A.* **83**, 2787.

Sproat, B. S., and Lamond, A. I. (1993). 2′-O-Alkyloligoribonucleotides. *In* "Antisense Research and Applications" (S. T. Crooke and B. Lebleu, eds.), p. 351. CRC Press, Boca Raton, Florida.

Sproat, B. S., Lamond, A. L., Beijer, B., *et al.* (1989). Highly efficient chemical synthesis of 2′-O-methyloligoribonucleotides and tetrabiotinylated derivatives; novel probes that are resistant to degradation by RNA or DNA specific nucleases. *Nucleic Acids Res.* **17**, 3373.

Sproat, B. S., Iribarren, A. M., *et al.* (1991). New synthetic routes to synthons suitable for 2′-O-allyloligoribonucleotide assembly. *Nucleic Acids Res.* **19**, 733.

Srinivasan, S. K., Tewary, H. K., and Iversen, P. L. (1995). Characterization of binding sites, extent of binding and drug interactions of oligonucleotides with albumin. *Antisense Res. Dev.* **5**, 131.

Stein, C. A., and Cheng, Y.-C. (1993). Antisense oligonucleotides as therapeutic agents. Is the bullet really magic? *Science* **261**, 1004.

Stein, C. A., and Cohen, J. S. (1989). Phosphorothioate oligodeoxynucleotide analogues. In "Oligodeoxynucleotides: Antisense Inhibitors of Gene Expression" (J. S. Cohen, ed.), p. 97. CRC Press, Boca Raton, FL.

Stein, C. A., Neckers, L. M., Nair, B. C., *et al.* (1991). Phosphorothioate oligodeoxycytidine interferes with binding of HIV-1 with gp120 to CD4. *J. AIDS* **4,** 686.

Stepkowski, S. M., Tu, Y., Condon, T. P., *et al.* (1994). Blocking of heart allograft rejection by intercellular adhesion molecule-1 antisense oligonucleotides alone or in combination with other immunosuppressive modalities. *J. Immunol.* **10,** 5336.

Suzuki, S., Pilowsky, P., Minson, J., *et al.* (1994). c-*fos* antisense in rostral ventral medulla reduces arterial blood pressure. *J. Physiol. (London)* R1418.

Tao, L.-F., Marz, K. A., Wongwit, W., *et al.* (1995). Uptake, intracellular distribution, and stability of oligodeoxynucleotide phosphorothioate by *Schistosoma mansoni. Antisense Res. Dev.* **5,** 123.

Temsamani, J., Tang, J.-Y., Padmapriya, A., *et al.* (1993). Pharmacokinetics, biodistribution and stability of capped oligodeoxynucleotide phosphorothioates in mice. *Antisense Res. Dev.* **3,** 277.

Thompson, J. D., and Gillespie, D. (1990). Current concepts in quantitative molecular hybridization. *Clin. Biochem.* **23,** 261–266.

Thuong, N. T., Asseline, U., and Monteney-Garestier, T. (1989). Oligodeoxynucleotides covalently linked to intercalating and reactive substances: Synthesis, characterization and physicochemical studies. *In* "Oligodeoxynucleotides: Antisense Inhibitors of Gene Expression" (J. S. Cohen, ed.), p. 25. CRC Press, Boca Raton, Florida.

Tischmeyer, W., Grimm, R., Schicknick, H., *et al.* (1994). Sequence-specific impairment of learning by c-*jun* antisense oligonucleotides. *NeuroReport* **5,** 1501.

Ts'o, P. O. P., Miller, P. S., and Greene, J. J. (1983). Nucleic acid analogs with targeted delivery at chemotherapeutic agents. *In* "Development of Target-Oriented Anticancer Drugs" (Y. C. Cheng, B. Gox, and M. Minkoff, eds.), p. 189. Raven Press, New York.

Vasanthakumar, G., and Ahmed, N. K. (1989). Modulation of drug resistance in a daunorubicin resistant subline with oligonucleoside methylphosphonates. *Cancer Commun.* **1,** 225.

Vickers, T., Baker B. F., Cook, P. D., *et al.* (1991). Inhibition of HIV-LTR gene expression by oligonucleotides targeted to the TAR element. *Nucleic Acids Res.* **19,** 3359.

Vlassov, V. V. (1989). Inhibition of tick-borne viral encephalitis expression using covalently linked oligonucleotide analogs. *Meet., Oligodeoxynucleotides Antisense Inhibitors Gene Express.: Ther. Implications,* Rockville, Maryland, 1989, 15.

Wagner, R. W., Matteucci, M. D., Lewis, J. G., *et al.* (1993). Antisense gene inhibition by oligonucleotides containing C-5 propyne pyrimidines. *Science* **260,** 1510.

Wahlestedt, C., Pich, E. M., Koob, G. F., *et al.* (1993). Modulation of anxiety and neuropeptide Y-Y1 receptors by antisense oligodeoxynucleotides. *Science* **259,** 528.

Walder, R. Y., and Walder, J. A. (1988). Role of RNase H in hybrid-arrested translation by antisense oligonucleotides. *Proc. Natl. Acad. Sci. U.S.A.* **85,** 5011.

Walker, K., Elela, S. A., and Nazar, R. N. (1990). Inhibition of protein synthesis by anti-5.8S rRNA oligodeoxyribonucleotides. *J. Biol. Chem.* **265,** 2428.

Wang, S., Lee, R. J., Cauchon, G., *et al.* (1995). Delivery of antisense oligodeoxyribonucleotides against the human epidermal growth factor receptor into cultured KB cells with liposomes conjugated to folate via polyethylene glycol. *Proc. Natl. Acad. Sci. U.S.A.* **92,** 3318.

Watson, J. D., and Crick, F. H. C. (1953). Molecular structure of nucleic acids: A structure for deoxyribose nucleic acid. *Nature (London)* **171,** 737.

Weiss, B., Zhou, L.-W., Zhang, S.-P., *et al.* (1993). Antisense oligodeoxynucleotide inhibits D_2 dopamine receptor-mediated behavior and D_2 messenger RNA. *Neuroscience* **55,** 607.

Westerman, P., Gross, B., and Hoinkis, G. (1989). Inhibition of expression of SV40 virus large T-antigen by antisense oligodeoxyribonucleotides. *Biomed. Biochem. Acta* **48,** 85.

Whitesell, L., Rosolen, A., and Neckers, L. M. (1991). *In vivo* modulation of N-myc expression by continuous perfusion with an antisense oligonucleotide. *Antisense Res. Dev.* **1,** 343.

Wickström, E. (1986). Oligodeoxynucleotide stability in subcellular extracts and culture media. *J. Biochem. Biophys. Methods* **13,** 97.

Wickström, E. L., Bacon, T. A., Gonzalez, A., *et al.* (1989). Anti-c-*myc* DNA increases differentiation and decreases colony formation by HL-60 cells *in vitro*. *Cell Dev. Biol.* **25**, 297.

Woodburn, V. L., Hunter, J. C., Durieux, C., Poat, J. A., and Hughes, J. (1994). The effect of c-*fos* antisense in the formalin-paw test. *Regul. Pept.* **54**, 327.

Wyatt, J. R., Vickers, T. A., Roberson, J. L., *et al.* (1994). Combinatorially selected guanosine-quartet structure is a potent inhibitor of human immunodeficiency virus envelope-mediated cell fusion. *Proc. Natl. Acad. Sci. U.S.A.* **91**, 1356.

Zamecnik, P. C., and Stephenson, M. L. (1978). Inhibition of Rous sarcoma virus replication and cell transformation by a specific oligodeoxynucleotide. *Proc. Natl. Acad. Sci. U.S.A.* **75**, 280.

Zamecnik, P. C., Goodchild, J., Taguchi, Y., *et al.* (1986). Inhibition of replication and expression of human T-cell lymphotropic virus type III in cultured cells by exogenous synthetic oligonucleotides complementary to viral RNA. *Proc. Natl. Acad. Sci. U.S.A.* **83**, 4143.

Zerial, A., Thuong, N. T., and Helene, C. (1987). Selective inhibition of the cytopathic effect of type A influenza viruses by oligodeoxynucleotides covalently linked to an intercalating agent. *Nucleic Acids Res.* **15**, 9909.

Zhang, M., and Creese, I. (1993). Antisense oligodeoxynucleotide reduces brain dopamine D_2 receptors: Behavioral correlates. *Neurosci. Lett.* **161**, 223, 4947.

Zhang, R., Yan, J., Shahinian, H., *et al.* (1995a). Pharmacokinetics of an anti-human immunodeficiency virus antisense oligodeoxynucleotide phosphorothioate (GEM 91) in HIV-infected subjects. *Clin. Pharmacol. Ther.* **58**, 44.

Zhang, R., Lu, Z., Zhang, X., *et al.* (1995b). *In vivo* stability and disposition of a self-stabilized oligodeoxynucleotide phosphorothioate in rats. *Clin. Chem. (Winston-Salem, N.C.)* **41**, 836.

Zhang, S.-P., Zhou, L.-W., and Weiss, B. (1994). Oligodeoxynucleotide antisense to the D_1 dopamine receptor mRNA inhibits D_1 dopamine receptor-mediated behaviors in normal mice and in mice lesioned with 6-hydroxydopamine. *J. Pharmacol. Exp. Ther.* **271**, 1462.

Zhang, Z., Tang, J. X., and Tang, J. Y. (1995). Syntheses and properties of novel thiono triester modified antisense oligodeoxynucleotide phosphorothioates. *Bioorg. Med. Chem. Lett.* **5**, 1735.

Zheng, H., Sahai, B. M., Kilgannon, P., *et al.* (1989). Specific inhibition of cell-surface T-cell receptor expression by antisense oligodeoxynucleotides and its effect on the production of an antigen-specific regulatory T-cell factor. *Proc. Natl. Acad. Sci. U.S.A.* **86**, 3758.

Zhou, L.-W., Zhang, S.-P., Qin, Z.-H., *et al.* (1994). *In vivo* administration of an oligodeoxynucleotide antisense to the D_2 dopamine receptor messenger RNA inhibits D_2 dopamine receptor-mediated behavior and the expression of D_2 dopamine receptors in mouse striatum. *J. Pharmacol. Exp. Ther.* **268**, 1015.

David T. Curiel

Gene Therapy Program
University of Alabama at Birmingham
Birmingham, Alabama 35294

Targeted Tumor Cytotoxicity Mediated by Intracellular Single-Chain Anti-oncogene Antibodies

There is increasing recognition that cancer results from a series of accumulated, acquired genetic lesions. To a larger and larger extent, the genetic lesions associated with malignant transformation and progression are being identified (1). The recognition of the molecular basis of carcinogenesis makes it rational to consider genetic approaches to therapy. In this regard, a number of strategies have been developed to accomplish cancer gene therapy (2–8). These approaches include: (a) mutation compensation, (b) molecular chemotherapy, and (c) genetic immunopotentiation. For mutation compensation, gene therapy techniques are designed to rectify the molecular lesions in the cell having undergone malignant transformation. For molecular chemotherapy, methods have been developed to achieve selective delivery or expression of a toxin gene in cancer cells to achieve their eradication. Genetic immunopotentiation strategies attempt to achieve active immunization against tumor-associated antigens by gene transfer methods. Whereas the biology of each malignant disease target will likely dictate the approach taken, the

Advances in Pharmacology, Volume 40

overwhelming majority of human clinical gene therapy trials to date have been based on the genetic immunopotentiation approach. For most tumor types, however, the absence of clear clinical evidence of antitumor effect has suggested the need for alternative approaches to achieve positive results. In addition the lack of clear efficacy of any of the human studies to date argue for the development of additional, novel approaches for anticancer gene therapy.

The genetic lesions etiologic of malignant transformation may be thought of as a critical compilation of two general types: aberrant expression of "dominant" oncogenes or loss of expression of "tumor suppressor" genes. Gene therapy strategies have been proposed to achieve correction of each of these lesions. For approaching the loss of function of a tumor suppressor gene, the logical intervention is replacement of the deficient function with a wild-type tumor suppressor gene counterpart. This strategy has been shown to allow phenotypic correction *in vitro*. For instance, Vogelstein *et al.* have demonstrated that delivery of the wild-type p53 gene to transformed p53-deficient colonic carcinoma cells can abrogate the malignant phenotype (9). In addition, similar studies carried out with other tumor suppressor genes and other tumor targets have further demonstrated the potential therapeutic effects achievable with reestablishment of wild-type tumor suppressor gene function (10–12). This concept has also been demonstrated in *in vivo* models. Roth *et al.* have shown that delivery of the wild-type p53 gene via recombinant retrovirus or recombinant adenovirus by direct *in vivo* delivery can have a therapeutic effect in a murine model employing human lung cancer xenografts (13). Similar findings have been noted in the context of replacement of the p16 tumor suppressor gene (14). Importantly, it has been shown that despite the presence of multiple genetic lesions, the targeted rectification of only one of these is, in many instances, sufficient to revert the neoplastic phenotype (10,11). This work has established the rationale for human clinical gene therapy trials designed to achieve mutation compensation in epithelial carcinomas of multiple sites including lung, liver, and the head and neck.

For dominant oncogenes, aberrant expression of the corresponding gene product elicits the associated neoplastic transformation. In this context, molecular therapeutic interventions are designed to ablate expression of the dominant oncogene. The most universally employed method to achieve this is the utilization of antisense molecules (DNA or RNA oligonucleotides) (15–18). These molecules are designed to specifically target sequences to achieve blockade of the encoded genetic informational flow. Approaches have included the use of triplex DNA to achieve functional ablation of transcriptional activation through blockade of transcription factor-binding sites. This has been used in *in vitro* model systems for targeting the c-myc (19), ras (20), and erbB-2 oncogenes (21). Targeting has also been achieved at levels of gene expression distal to transcription. Specific antisense binding to transcribed RNA sequences may interrupt the flow of genetic information through several mechanisms including degradation, impaired transport, and translational arrest. These interventions may be accomplished by simple antisense oligonucleotides, as well as by antisense molecules that possess catalytic activity to accomplish cleavage of target sense sequences (20,22,23). A variety of experimental models have demonstrated the utility of the antisense approach as an anticancer therapeutic. Importantly, several studies have shown the ability to selectively ablate a dominant oncogene with reversion of the malignant phenotype (19,20,23). In selected instances, the *in vivo* demonstration of this effect could also be accomplished by direct, *in vivo* delivery of the antisense molecules (24,25). Thus, the antisense approach offers the potential to achieve targeted disruption of specific genes in anticancer therapy models.

Despite the potentially novel therapeutic strategies offered by the antisense approach, this methodology has in practice been associated with severe limitations. These practical constraints have limited wide employment of this technology in human gene therapy anticancer protocols. In this regard, there do not exist universal rules dictating the efficacy of a given antisense oligonucleotide for achieving specific gene inhibition (15–17). Thus, despite the utility of antisense inhibition, there are a great many cancer-related genes that have resisted attempts to achieve their antisense ablation. In addition, delivery of antisense molecules has been highly problematic (15–17). It is often difficult to achieve effective sustained intracellular levels of the antisense molecules sufficient for a therapeutic effect. To circumvent this problem, a number of design modifications have been developed to enhance their stability (15–17,26). In addition, a number of vector approaches have been explored for effective cellular delivery (25,27,28). Despite these maneuvers, the overriding limitations to the employment of this therapeutic modality remain the idiosyncratic efficacy of specific antisense for a given target gene and the suboptimal delivery of antisense molecules.

As an alternative, the employment of intracellular single-chain antibodies has been explored as a means to achieve targeted knockout of a cellular product. In this regard, techniques have been developed to allow the derivation of recombinant molecules that possess antigen-binding specificities expropriated from immunoglobins. These single-chain immunoglobin (sFv) molecules retain the antigen-binding specificity of the immunoglobin from which they were derived; however, they lack other functional domains characterizing the parent molecule. The basis of constructing sFvs has been established. Pastan *et al.* have developed methods to derive cDNAs that encode the variable regions of specific immunoglobins (29,30). Specifically, a single-chain antibody (sFv) gene is derived that contains the coding sequences for variable regions from the heavy chain (V_H) and the light chain (V_L) of the immunoglobin separated by a short linker (L) of hydrophobic amino acids. The resultant recombinant molecule, when expressed in prokaryotic systems, is a single-chain antibody (sFv) that retains the antigen recognition and binding profile of the parent. The development of recombinant immunotoxins employing sFv moieties achieves cell-specific binding of the toxin to the exterior of the target cell, allowing receptor-mediated endocytosis to accomplish toxin internalization. A variety of strategies employing the recombinant sFv-directed immunotoxins have been developed by a number of investigators (29–34). In addition, it has been shown that sFv molecules may be expressed intracellularly in eukaryotic cells by gene transfer of sFv cDNAs. The encoded sFv may be expressed in the target cell and localized to specific, targeted subcellular compartments by appropriate signal molecules (35). Importantly, these intracellular sFvs may recognize and bind antigen within the target cell. Marasco *et al.* have shown that intracellular antibodies against HIV can abrogate production of progeny virion in human immunodeficiency virus (HIV)-infected cells (36,37). Thus, intracellular sFvs serve as unique vehicles to achieve intracellular knockout of specific gene products.

Targets for the intracellular antibody knockout method have included viral antigens in the context of HIV infection. In addition, partial phenotypic reversion has been noted with anti-Ras single-chain antibodies (38). We also wish to explore the utility of this approach in targeting transforming oncoproteins. In this regard, erbB-2 is a 185-kDa transmembrane protein kinase receptor with extensive homology to the epithelial growth factor receptor family (39). Several lines of evidence suggest that aberrant expression of the erbB-2 gene may play an important role in neoplastic transformation and progression. Specifically, ectopic expression of erbB-2 has been shown to be capable of transforming rodent fibroblasts *in vitro* (40). In addition, transgenic mice carrying either normal or mutant erbB-2 develop a variety of tumors, including neoplasms of mammary origin (41). Importantly, it has been shown that amplification and/or overexpression of the erbB-2 gene occurs in a variety of human epithelial carcinomas, including malignancies of the ovary, breast, gastrointestinal tract, salivary gland, and lung (42). In the instance of breast or ovarian carcinoma, a direct correlation has been noted between overrexpression of erbB-2 and aggressive tumor growth with reduced overall patient survival (43,44). As erbB-2 overexpression may be a key event in malignant transformation and progression, strategies to ablate its expression would be rational as a therapeutic modality. This fact has led to the development of therapeutic strategies to target tumor cells exhibiting increased surface levels of erbB-2. Specifically, monoclonal antibodies (MAbs) have been developed that exhibit high-affinity binding to the extracellular domains of the erbB-2 protein (45,46). A number of studies have demonstrated that a subset of these MAbs can elicit growth inhibition of erbB-2-overexpressing tumor cells, both *in vitro* and *in vivo* (47,48). On the basis of these observations, clinical trials in humans have been undertaken that exploit the direct antiproliferative effect of anti-erbB-2 MAbs (47). Antibody-based tumor targeting has also been utilized with radiolabeled anti-erbB-2 MAbs (49). In addition, antitumor therapies directed at erbB-2 have been developed for targeted immunotoxins (50). These experimental strategies have employed recombinant fusion proteins consisting of various bacterial toxins selectively targeted to tumors by virtue of single-chain anti-erbB-2 antibody moieties. Clinical trials to date, however, have not duplicated the impressive results obtained with *in vitro* or *in vivo* model systems.

The conceptual basis of the above strategies is dependent on exploiting cell surface erbB-2 expression as a marker of malignant cells. In addition, other methods have been proposed to target erbB-2-overexpressing tumor cells by means of directly modulating levels of the oncoprotein. These approaches have included antisense strategies targeted to the transcriptional and posttranscriptional levels of gene expression. In the former instance, triplex-forming oligonucleotides have been shown to be capable of binding the erbB-2 promoter region to inhibit transcription of the erbB-2 gene (21). In addition, antisense oligonucleotides targeted to the erbB-2 transcript have accomplished phenotypic alterations in erbB-2-overexpressing tumor cells including downregulation of cell surface expression and inhibition of proliferation (51,52). Thus, the conceptual basis of tumor targeting based on modulation of erbB-2 level has been established.

As a method to effect targeted oncoprotein ablation, we have developed a novel strategy: construction of a gene encoding a single-chain immunoglobin (sFv) directed against the specific oncoprotein, erbB-2, to effect selected oncoprotein knockout. We hypothesized that if an anti-erbB-2 sFv were localized to the endoplasmic reticulum (ER) of SKOV3 cells (an ovarian carcinoma cell line that overexpresses erbB-2), the nascent erbB-2 protein would be entrapped within the ER of the cells, and therefore unable to achieve its normal cell surface localization. It was further hypothesized that this intracellular entrapment would prevent erbB-2, a transmembrane tyrosine kinase receptor, from interacting with its ligand, thus abrogating one autocrine growth factor loop thought to be involved in malignant transformation in erbB-2-overexpressing cell lines. To prevent maturational processing of the nascent erbB-2 protein during synthesis, a gene construct was thus designed that encoded an ER-directed form of the anti-erbB-2 sFv (pGT21) (Fig. 1). As a control, a similar anti-erbB-2 sFv was designed that lacked a signal sequence dictating its localization to the ER (pGT20). These sFv constructs were cloned into the eukaryotic expression vector pcDNA3, which directs high-level gene expression from the cytomegalovirus (CMV) early intermediate promoter/enhancer. For this analysis, the plasmid DNAs pcDNA3, pGT20, and pGT21 were transfected into the erbB-2-overexpressing ovarian carcinoma cell line SKOV3, using the adenovirus–polylysine (AdpL) method developed by our group (53). Preliminary experiments had demonstrated that the adenovirus–polylysine–DNA complexes containing a β-galactosidase reporter gene (pCMVβ) effected detectable levels of reporter

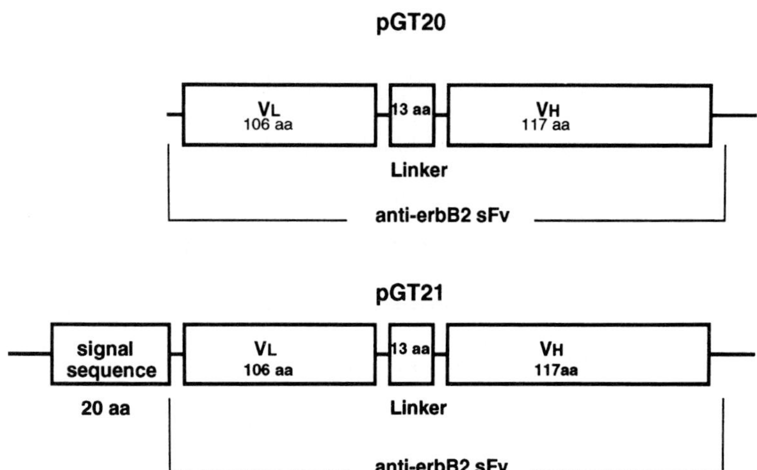

FIGURE I   Construction of anti-erbB-2 sFv expression plasmids. Gene constructs encoding anti-erbB-2 sFvs were derived from the plasmid e23sFv by PCR methods. The sFv gene constructs were cloned into the eukaryotic expression vector pcDNA3. The plasmid pGT20 was predicted to express a non-ER form (cytosolic) of the sFv. The plasmid pGT21 was predicted to express an ER form of the sFv. [Reproduced with permission from Ref. 73, from *Gene Therapy*.]

gene expression in >95% of targeted cells. At various times after transfection, the cells were evaluated for cell surface expression of erbB-2 using an anti-human erbB-2 monoclonal antibody. Cells transfected with the irrelevant plasmid DNA, pcDNA3, exhibited high levels of cell surface erbB-2, as would be expected (Fig. 2). In addition, SKOV3 cells transfected with the non-ER (cytosolic) form of the anti-erbB-2 sFv (pGT20) exhibited levels of cell surface erbB-2 similar to the control. In marked contrast, SKOV3 cells transfected with pGT21, which encodes an ER form of the anti-erbB-2 sFv, demonstrated marked downregulation of cell surface erbB-2 expression. This downregulation appeared to be time dependent with cell surface erbB-2 levels progressively declining from 48 to 96 hr posttransfection. At 96 hr posttransfection, fewer than 10% of the pGT21-transfected cells exhibited detectable levels of cell surface erbB-2 protein. The cells otherwise appeared morphologically indistinguishable from the control groups.

To determine whether modulation of cell surface expression of erbB-2 levels effected cellular proliferation in the SKOV3 cells, the various gene constructs were transfected using the AdpL vector as before. For this analysis, immunohistochemistry for the proliferation-associated antigen Ki-67 was employed (54) (Fig. 3). Transfection of cells with the control plasmid pcDNA3 resulted in the immunohistochemical detection of active cellular proliferation as indicated by intense nuclear staining. In addition, transfection with the non-ER form of the anti-erbB-2 sFv did not result in any net change in cell proliferation by this assay. In marked contrast, transfection

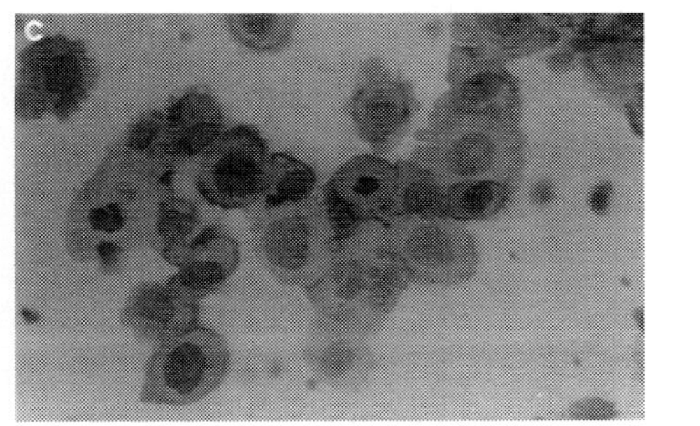

FIGURE 2 Effect of intracellular anti-erbB-2 sFv on cell surface expression of erbB-2 protein. The human ovarian carcinoma cell line SKOV3 was transfected by the AdpL method with the described plasmid constructs and analyzed for cell surface erbB-2 at 96 hr posttransfection, using an anti-human erbB-2 monoclonal antibody. Original magnification: ×400. (A) Transfection with control plasmid pcDNA3. (B) Transfection with non-ER form of anti-erbB-2 sFv plasmid pGT20. (C) Transfection with ER form of anti-erbB-2 sFv plasmid pGT21. [Reproduced with permission from Ref. 73, from *Gene Therapy*.]

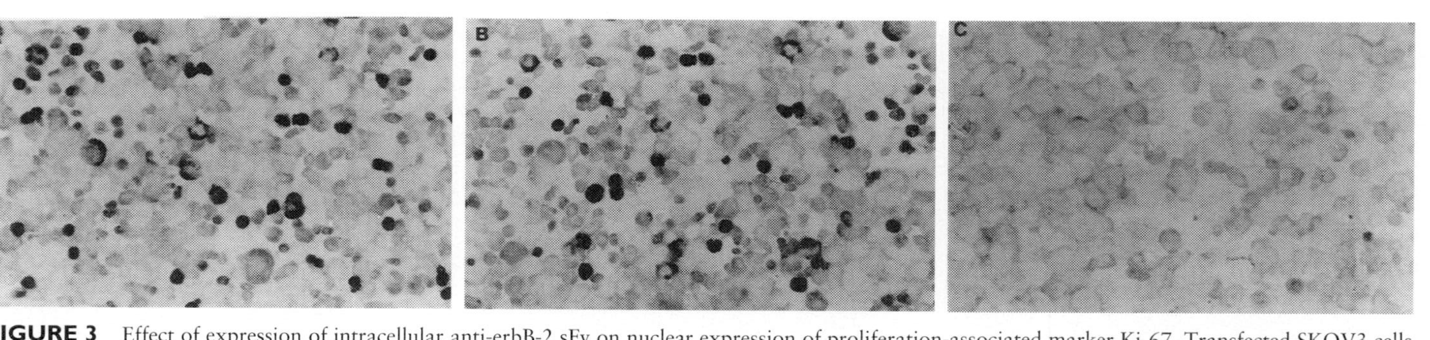

FIGURE 3　Effect of expression of intracellular anti-erbB-2 sFv on nuclear expression of proliferation-associated marker Ki-67. Transfected SKOV3 cells were analyzed for Ki-67 expression 96 hr posttransfection, using a mouse monoclonal antibody. Original magnification: ×200. (A) Control plasmid pcDNA3. (B) Non-ER form of anti-erbB-2 sFv plasmid, pGT20. (C) ER form of anti-erbB-2 sFv plasmid, pGT21. [Reproduced with permission from Ref. 73, from *Gene Therapy*.]

of the erbB-2-overexpressing cell line SKOV3 with the ER form of the anti-erb-B-2 sFv resulted in a dramatic inhibition of cellular proliferation. Because the ER-directed anti-erbB-2 sFv exhibited a prominent antiproliferative effect, it was hypothesized that it might also exhibit a tumoricidal effect in cells stably modified to express this gene construct. As the plasmids pcDNA3, pGT20, and pGT21 contained neomycin selectable markers, they were used to derive stable clones. In a preliminary experiment, the various plasmid constructs were used to derive G418-resistant clones in HeLa cells, a human cervical carcinoma cell line not characterized by overexpression of erbB-2. After selection, the number of clones derived from transfection with pGT20 and pGT21 was not significantly different (Table I). Further, the number of clones did not differ after transfection with the control plasmid pcDNA3. A similar analysis was then carried out with the erbB-2-overexpressing tumor line SKOV3 as the target. In this study, the number of clones derived with pGT20 did not differ from the number derived with the control plasmid pcDNA3 (Table I). Transfection with pGT21, however, resulted in a dramatic reduction in the number of stable clones derived ($p < 0.001$). Thus, it appeared that the expression of the ER form of the anti-erbB-2 sFv was incompatible with long-term viability of stably transfected SKOV3 cells. Further, this effect appeared specific for erbB-2-overexpressing cells, as this

TABLE I Derivation of Stable Colonies after Transfection with Anti-erbB2 sFv Expression Plasmids[a]

	G418-resistant colonies[b]	
	---	---
Cell line	*Anti-erbB2 sFv (non-ER form, pGT20)*	*Anti-erbB2 sFv (ER form, pGT21)*
SKOV3	36	5
	28	5
	23	3
	26	3
	27	3
SW626	21	18
	24	16
	21	16
	28	21
	20	19
HeLa	68	77
	84	83
	91	93
	77	69
	88	89

[a] Reproduced with permission from Ref. 73, from *Gene Therapy*.
[b] G418-resistant colonies were counted after staining with crystal violet at day 21 post-transfection.

differential clone survival was not noted in the HeLa cells. Thus, as would be predicted, non-erbB-2-expressing tumor cells did not appear to be affected by this specific anti-erbB-2 intervention.

As the expression of the anti-erbB-2 sFv appeared incompatible with the derivation of stable clones from SKOV3, we sought to determine whether the effect of expression of the single-chain antibody might be cytocidal. To test this concept, plasmid DNAs that encoded either the cytosolic form or the ER form of the anti-erbB-2 sFv, as well as the control plasmid pcDNA3, were delivered to SKOV3 cells. Transfected cells were evaluated for growth rates by analysis of the time-dependent increase in cell number. Cells transfected with the irrelevant plasmid DNA, pcDNA3, as well as the cytosolic pGT20 showed a time-dependent increase in cell number (Fig. 4). In marked contrast, transfection with the ER form of the anti-erbB-2 sFv, pGT21, resulted in a significant inhibition of cell growth. Analysis of cell growth kinetics suggested that intracellular expression of the ER form of the anti-erbB-2 sFv was cytocidal to erbB-2-overexpressing tumor cells, and not

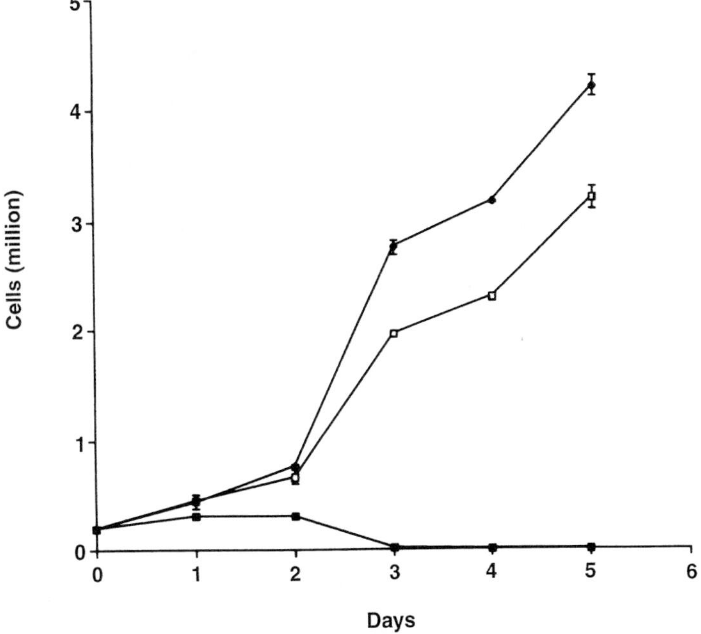

FIGURE 4 Growth rate measurements of SKOV3 cells transfected with anti-erbB-2 sFv-encoding plasmids. The erbB-2-overexpressing ovarian carcinoma cell line SKOV3 was transfected with a control plasmid (pcDNA3; □), a plasmid encoding a cytosolic form of the anti-erbB-2 sFv gene (pGT20; ◆), or a plasmid encoding an endoplasmic reticulum form of the anti-erbB-2 sFv gene (pGT21; ■). Cell numbers were counted in triplicate by trypan blue exclusion of viable cells at the indicated times posttransfection. [From J. Deshane, J. Grim, S. Loechel, G. P. Siegal, R. D. Alvarez, and D. T. Curiel: Intracellular antibody against erbB-2 mediates targeted tumor cell eradication by apoptosis. *Cancer Gene Therapy* 3:2, 89–98, 1996. Reprinted by permission of Appleton & Lange, Inc.]

cytostatic, as originally hypothesized. To more accurately determine the precise biologic effect of anti-erbB-2 sFv expression in the erbB-2-overex-pressing tumor target, the XTT assay was employed as a direct determination of cell viability by quantifying a specific cellular enzymatic reaction (55). For this analysis, the ovarian carcinoma line SKOV3 was transfected, as before, with the plasmid DNAs pcDNA3, pGT20, and pGT21. Transfection with pGT21 resulted in a time-dependent decrease in cell viability, with a >95% decrement in the number of viable cells by 72 hr posttransfection (Fig. 5). Transfection with the control plasmids pcDNA3 and pGT20, however, did not exert any significant effect on cell viability. As an additional control, the ER form of the anti-erbB-2 sFv had no observable effect on the non-erbB-2-expressing human cervical carcinoma line HeLa. We have also transduced non-erbB-2-expressing human cell lines from a variety of different tissues including bladder, liver, mesothelium, and kidney, and have not noted any significant cytotoxicity. These results confirmed our previous study demonstrating that the ER form of the sFv uniquely elicits phenotypic alterations in erbB-2-expressing cells.

The foregoing studies are consistent with the concept that entrapment of erbB-2 within the ER of erbB-2-overexpressing tumor cells elicits a selective cytotoxicity. To further delineate the mechanistic basis of this effect, studies were carried out to determine if programmed cell death, i.e., apoptosis, was occurring. As before, the plasmid DNA constructs pcDNA3, pGT20, and pGT21 were delivered to the erbB-2-overexpressing SKOV3 cells and the non-erbB-2-expressing tumor cell line HeLa. At specific time points post-transfection, cells were harvested and evaluated for evidence of nuclear DNA fragmentation, a hallmark of programmed cell death (56). In the HeLa cells, transfection with the various constructs did not demonstrate any evidence of apoptotic cellular events as determined by morphologic appearance or alterations in DNA as measured by gel electrophoresis (Fig. 6A). Transfec-tion of the SKOV3 cells with the control plasmid pcDNA3 or the cytosolic anti-erbB-2 sFv pGT20 similarly did not elicit any evidence of cellular apop-tosis. When the SKOV3 cells were transfected with the ER form of the anti-erbB-2 sFv, however, marked changes in chromosomal DNA were noted. These changes were first detected at 47 hr posttransfection and revealed on a 2% agarose gel as a characteristic 200-bp apoptotic ladder (Fig. 6B). As independent confirmation, the presence of apoptotic nuclei was evaluated employing differential nuclear uptake of DNA-binding dyes (56). In this analysis, SKOV3 cells transfected with the plasmid DNA pGT21 showed intense nuclear staining characteristic of cellular apoptosis. These alterations were not seen in cells transfected with the control plasmids pcDNA3 and pGT20 (Fig. 7A–C).

Quantitative analysis demonstrated that >90% of the transfected SKOV3 cells exhibited apoptotic nuclear changes, whereas cells transfected with pcDNA3 and pGT20 did not exhibit levels of apoptosis different from

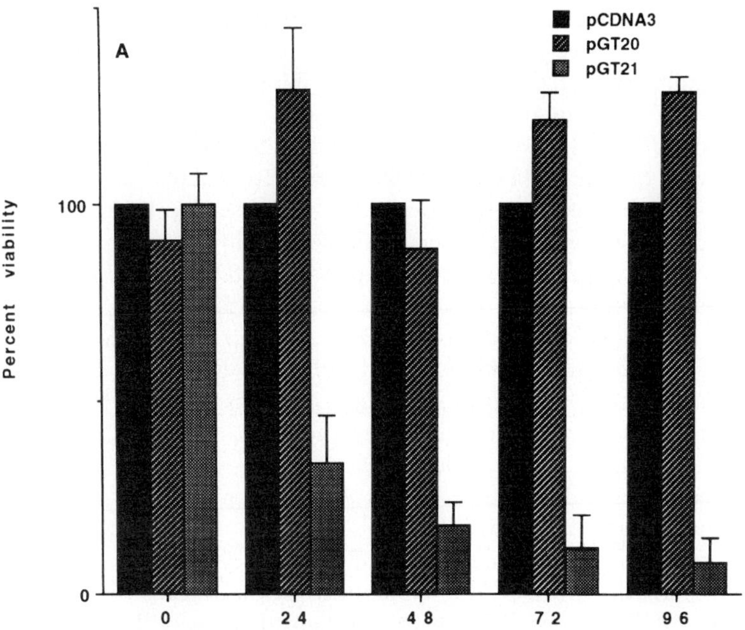
A
pCDNA3
pGT20
pGT21
Percent viability
100
0
0
2 4
4 8
7 2
9 6
Time(hrs)

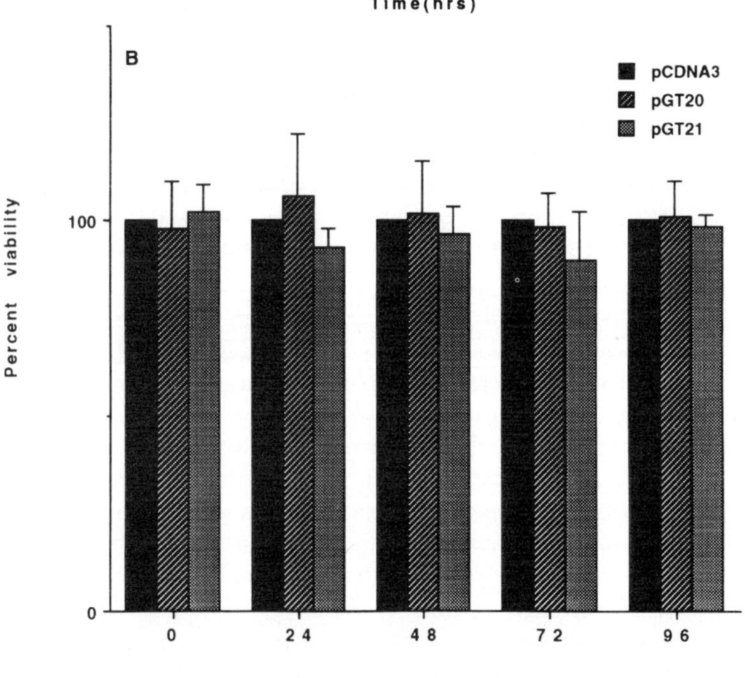
B
pCDNA3
pGT20
pGT21
Percent viability
100
0
0
2 4
4 8
7 2
9 6
Time(hrs)

A. HeLa

B. SKOV3

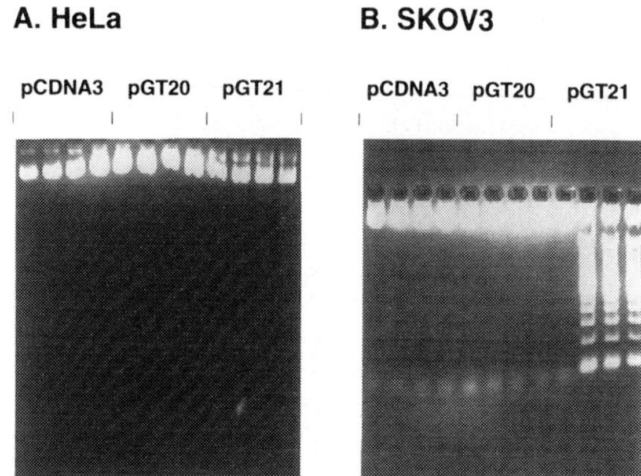

FIGURE 6 Determination of apoptotic DNA fragmentation induced by ER anti-erbB-2 sFv. Tumor cells were transfected with the plasmids pcDNA3, pGT20, and pGT21. At the indicated time points posttransfection, cells were harvested and chromosomal DNA analyzed by gel electrophoresis. (A) Transfection of the non-erbB-2-overexpressing human cervical cell line HeLa. (B) Transfection of the erbB-2-overexpressing human ovarian carcinoma cell line SKOV3. [From J. Deshane, J. Grim, S. Loechel, G. P. Siegal, R. D. Alvarez, and D. T. Curiel: Intracellular antibody against erbB-2 mediates targeted tumor cell eradication by apoptosis. *Cancer Gene Therapy* 3:2, 89–98, 1996. Reprinted by permission of Appleton & Lange, Inc.]

untrasfected controls. Thus, the basis of the cytocidal effect of the ER-directed anti-erbB-2 sFv in the erbB-2-overexpressing cells was the induction of apoptosis. In the context of dominant oncogene-induced tumorigenesis, downregulation of overexpressed immortalizing growth factor receptors may induce cellular apoptosis (57,58). This suggests that the abrogation of the immortalizing stimulus allows cells to reengage the previously overridden apoptotic program. Alternatively, ablation of dominant oncogene function may result in proliferative arrest, without induction of programmed cell death (59,60). The precise mechanism distinguishing these alternate responses to oncogene ablation is not presently clear. It is interesting to note that erbB-2 downregulation mediated by antisense oligonucleotides induces proliferative arrest, but not apoptosis in erbB-2-overexpressing tumor targets (51,52). In contrast, we have induced apoptosis by virtue of an alternate mechanism of erbB-2 downregulation. This suggests that erbB-2 downregulation, per se, is not inductive of apoptosis.

FIGURE 5 Effect of expression of intracellular anti-erbB-2 sFv genes on tumor cell viability in (A) the erbB-2-overexpressing human ovarian carcinoma cell line SKOV3, and in (B) the non-erbB-2-expressing cervical carcinoma cell line HeLa. Tumor cell targets were transfected with the plasmids pcDNA3, pGT20, and pGT21. At the indicated times posttransfection, cell viability was determined employing the XTT assay. Assays were performed 12 times at each time point. [From J. Deshane, J. Grim, S. Loechel, G. P. Siegal, R. D. Alvarez, and D. T. Curiel: Intracellular antibody against erbB-2 mediates targeted tumor cell eradication by apoptosis. *Cancer Gene Therapy* 3:2, 89–98, 1996. Reprinted by permission of Appleton & Lange, Inc.]

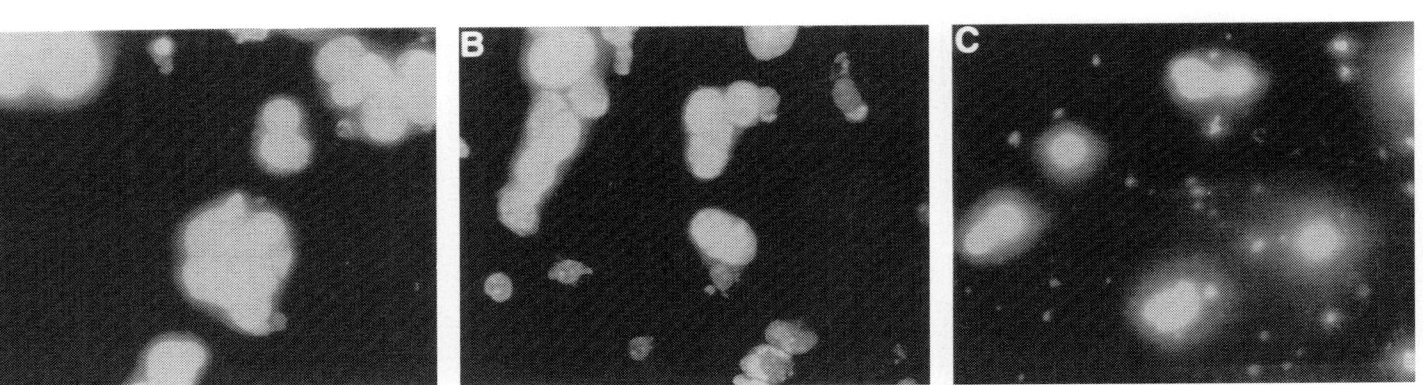

FIGURE 7 Determination of apoptotic nuclear staining induced by ER anti-erbB-2 sFv. Tumor cell targets were transfected with the plasmids pcDNA3, pGT20, and pGT21. At 24 hr posttransfection, cells were harvested and nuclear uptake of fluorescent DNA-binding dyes was determined. (A) SKOV3 cells transfected with pcDNA3. (B) SKOV3 cells transfected with pGT20. (C) SKOV3 cells transfected with pGT21. Original magnification: ×400. [From J. Deshane, J. Grim, S. Loechel, G. P. Siegal, R. D. Alvarez, and D. T. Curiel: Intracellular antibody against erbB-2 mediates targeted tumor cell eradication by apoptosis. *Cancer Gene Therapy* 3:2, 89–98, 1996. Reprinted by permission of Appleton & Lange, Inc.]

To determine the basis whereby the anti-erbB-2 sFv induced apoptosis, we attempted to reproduce this phenomenon in a different system. For this analysis, ectopic localization of erbB-2, in non-erbB-2-transformed tumor cells, was accomplished by cotransfection of HeLa cells with wild-type human erbB-2 cDNA and the cDNA for the ER form of the anti-erB-2 sFv. Transfection of the non-erbB-2-expressing HeLa cell line with the erbB-2 cDNA did not result in any change in cell viability, identical to that observed employing the irrelevant plasmid DNA control pcDNA3. In contrast, cotransfection of the erbB-2 cDNA with the anti-erbB-2 sFv construct caused a marked cytocidal effect (Fig. 8). This cytotoxicity could also be shown to be on the basis of induction of apoptosis as was observed in SKOV3 cells transfected with the anti-erbB-2 sFv (data not shown). Thus, even where erbB-2 does not contribute to the transformed phenotype, ectopic localization of erbB-2 within the ER still induced apoptosis. Consistent with this concept, the cytocidal effect of ER entrapment of erbB-2 could be reversed by overexpression of bcl-2, a gene that encodes a mitochondrial protein that functions to promote cell survival through interference with the apoptosis program. In this regard overexpression of the bcl-2 gene has been shown to revert apoptotic cell death induced by a variety of stimuli (61–64). Whereas the ER form of the anti-erbB-2 sFv-induced apoptotic cell death in the erbB-2-overexpressing ovarian carcinoma cell line SKOV3, this effect was abrogated by cotransduction of these cells with the bcl-2 gene (Fig. 9). These findings corroborate the concept that the ectopic localization of the

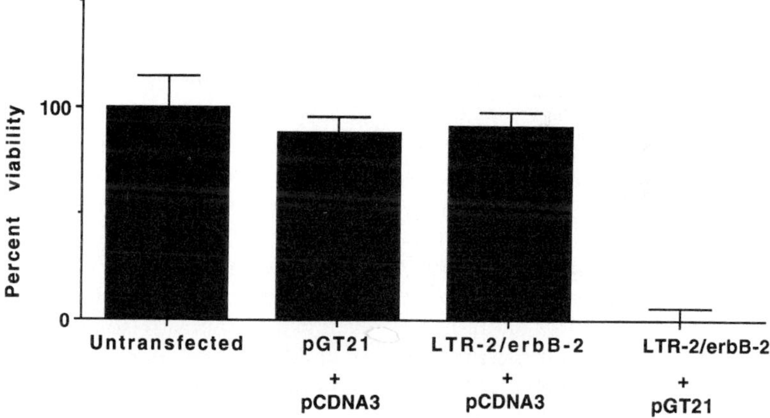

FIGURE 8 Effect of coexpression of erbB-2 and the anti-erbB-2 sFv on HeLa cell viability. The non-erbB-2-expressing human tumor cell line HeLa was transfected with plasmids encoding the ER form of the anti-erbB-2 sFv (pGT21) and/or the human erbB-2 expression vector LTR-2/erbB-2. At 96 hr posttransfection, cell viability was determined employing the XTT assay. The mean of eight assays is shown. [From J. Deshane, J. Grim, S. Loechel, G. P. Siegal, R. D. Alvarez, and D. T. Curiel: Intracellular antibody against erbB-2 mediates targeted tumor cell eradication by apoptosis. *Cancer Gene Therapy* 3:2, 89–98, 1996. Reprinted by permission of Appleton & Lange, Inc.]

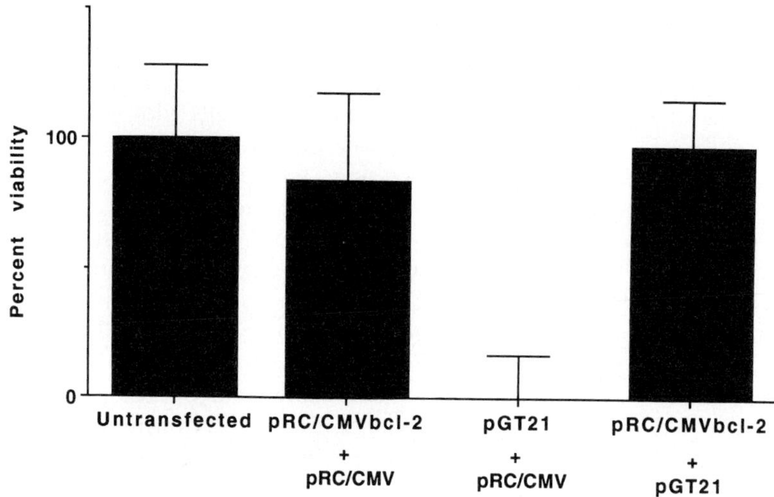

FIGURE 9   Effect of bcl-2 expression on anti-erbB-2 sFv-mediated apoptosis. The erbB-2-overexpressing ovarian carcinoma cell line SKOV3 was cotransfected with the plasmids pGT21, pRC/CMV control plasmid, and pRC/CMV bcl-2 encoding the wild-type bcl-2 cDNA. Cells were evaluated for viability at 72 hr posttransfection, employing the XTT assay. The mean of eight assays is shown. [From J. Deshane, J. Grim, S. Loechel, G. P. Siegal, R. D. Alvarez, and D. T. Curiel: Intracellular antibody against erbB-2 mediates targeted tumor cell eradication by apoptosis. *Cancer Gene Therapy* 3:2, 89–98, 1996. Reprinted by permission of Appleton & Lange, Inc.]

erbB-2 oncoprotein specifically induces apoptosis. Further, this indicates that the induced apoptotic pathway is analogous to described mechanisms converging through the bcl-2 proto-oncogene. These data suggested that abrogation of a transforming oncogene is not necessarily the basis for apoptosis induction. The fact that mislocalization of erbB-2, in a non-erbB-2-transformed cell, caused apoptosis suggests that mislocalization may be a general means to induce cellular apoptosis. It is thus reasonable to speculate that other endogenous or xenotropic oncogenes, when abrogated in this manner, may trigger apoptotic cell death.

We then sought to examine the effects of the anti-erbB-2 sFv in human tumor material isolated from a patient with primary ovarian carcinoma of epithelial origin. For this analysis, we developed methods to isolate primary ovarian tumor cells that maintain their viability and proliferation capacity *in vitro* for approximately 7–10 days. In addition, the amount of cell surface erbB-2 in these tumor explants had been rapidly determined employing a sensitive enzyme-linked immunosorbent assay (ELISA). To establish the biologic effects of intracellular single-chain antibody knockout of erbB-2 in these primary ovarian carcinoma cells, the various anti-erbB-2 sFv constructs were delivered to cells employing the AdpL vector system followed by the XTT assay for determination of cell viability. Control experiments employing a LacZ reporter gene had demonstrated that >95% of the isolated human primary ovarian carcinoma cells could be transduced in this manner.

The human ovarian carcinoma cell line SKOV3 was employed as an additional control for these experiments. In this analysis, the ER form of the anti-erbB-2 sFv exhibited a specific cytotoxic effect in the human primary tumor cells at 96 hr posttransfection. Interestingly, the magnitude of the sFv-mediated cell killing observed in the primary tumor material was as great as that observed in the erbB-2-overexpressing cell line SKOV3 (Fig. 10). These findings strongly suggest that ovarian cancer cell lines represent appropriate models of the operative mechanisms utilized in actual patient tumor cells. These results thus exclude the possibility that the observed sFv-mediated cytotoxicity represents only an *in vitro* phenomenon. We have demonstrated here the utility of selective cytotoxicity in fresh tumor material derived from humans.

We next undertook to determine if human ovarian cancer cells could be selectively killed in a murine model of malignant ascites. For these studies, we engrafted athymic nude mice with the erbB-2-overexpressing human ovarian carcinoma line SKOV3. This model allows for the development of malignant ascites and peritoneal implants of neoplastic cells in a manner that parallels the human disease. For gene therapy to be of practical utility in human ovarian carcinoma, vector strategies must be capable of accomplishing direct, *in situ* delivery of heterologous genes to tumor *in vivo*. We

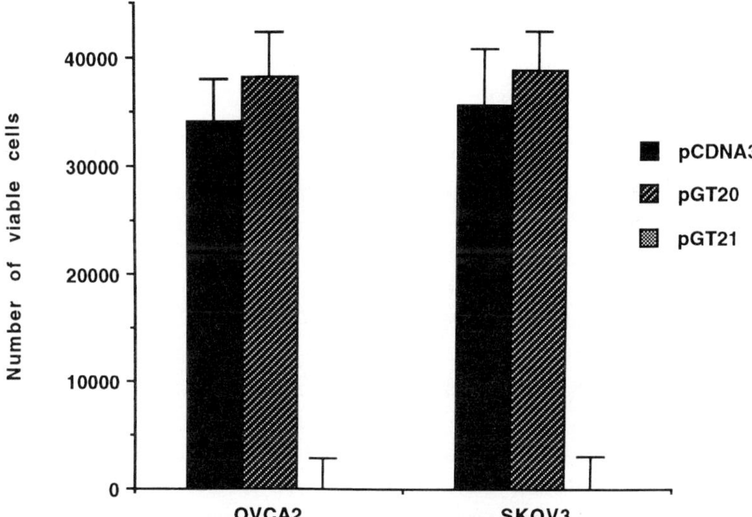

FIGURE 10   Effect of expression of the anti-erbB-2 sFv gene on human ovarian tumor cell viability. The erbB-2-expressing human primary ovarian carcinoma cells isolated from malignant ascites were transfected with pcDNA3, pGT20, or pGT21. ErbB-2-overexpressing ovarian carcinoma cells (SKOV3) were used as additional controls. Cells were assayed for viability by the XTT assay at 96 hr posttransfection. This experiment was replicated 10 times. Data represent mean ± SEM. [Reproduced with permission from Ref. 72.]

thus undertook a preliminary analysis to determine which of a series of vector systems could accomplish efficient *in situ* transduction of the mobile tumor cells found in ovarian carcinoma malignant ascites fluid. We chose for this analysis vector systems that had been previously reported to be capable of achieving a reasonable level of *in vivo* gene transfer. Athymic nude mice (BALB/c) were transplanted intraperitoneally with 1×10^7 SKOV3 cells. After 48 hr, vectors were administered by the intraperitoneal route to accomplish delivery of an *Escherichia coli* β-galactosidase reporter gene construct (LacZ) to target the mobile neoplastic cells. Evaluated vector systems included adenovirus–polylysine–DNA complexes (AdpL), liposomes (DOTAP), and a recombinant adenovirus encoding (lacZ (AdCMVLacZ). Forty-eight hours after vector administration, mobile tumor cells were harvested by peritoneal lavage and analyzed for expression of the LacZ reporter gene. This was accomplished by a fluorescence-activated double-sorting procedure (Fig. 11). In this analysis, the highest level of gene transfer was accomplished with the recombinant adenovirus: the transduction frequency achieved with this vector was >80%. These initial studies do not imply that the adenovirus will ultimately be the optimal vector for *in vivo* use in human ovarian carcinoma. In this regard, issues related to vector safety, toxicity, immunogenicity, and efficacy in the context of more advanced disease will need to be considered. This vector does, however, give us the means to ask additional questions at present, related to the potential efficacy of the anti-erbB-2 sFv approach as a gene therapy strategy in these model systems.

As the recombinant adenovirus proved of utility for *in situ* transduction of mobile neoplastic cells *in vivo* was asked whether it was possible to accomplish anti-erbB-2 sFv-mediated selective toxicity in this setting. We therefore constructed a recombinant adenovirus encoding the ER form of the anti-erbB-2 sFv (Ad21), using standard methods of homologous recombination. The resultant recombinant virus is E1A/B deleted and, thus, replication incompetent. Preliminary studies confirmed the structural integrity of the recombinant adenovirus genome. To establish that the anti-erbB-2 sFv gene functioned in this vector configuration, *in vitro* analysis was carried out employing the SKOV3 cells as the target. Cells were analyzed for viability employing the XTT assay. In this analysis, it could be seen that the anti-erb-2 sFv-encoding adenovirus accomplished the same selective cytotoxicity in the erbB-2-overexpressing targets as observed with AdpL-mediated delivery (Fig. 12). Notably, the adenovirus encoding an irrelevant gene (LacZ) had no effects on cell viability, even when delivered at an identical multiplicity of infection (MOI). Thus, a replication-defective adenovirus encoding the anti-erbB-2 sFv has been constructed that retains the capacity to express an ER-anti-erbB-2 sFv. This vector can achieve selective cytotoxicity based on the encoded sFv in human ovarian carcinoma cell lines, *in vitro*.

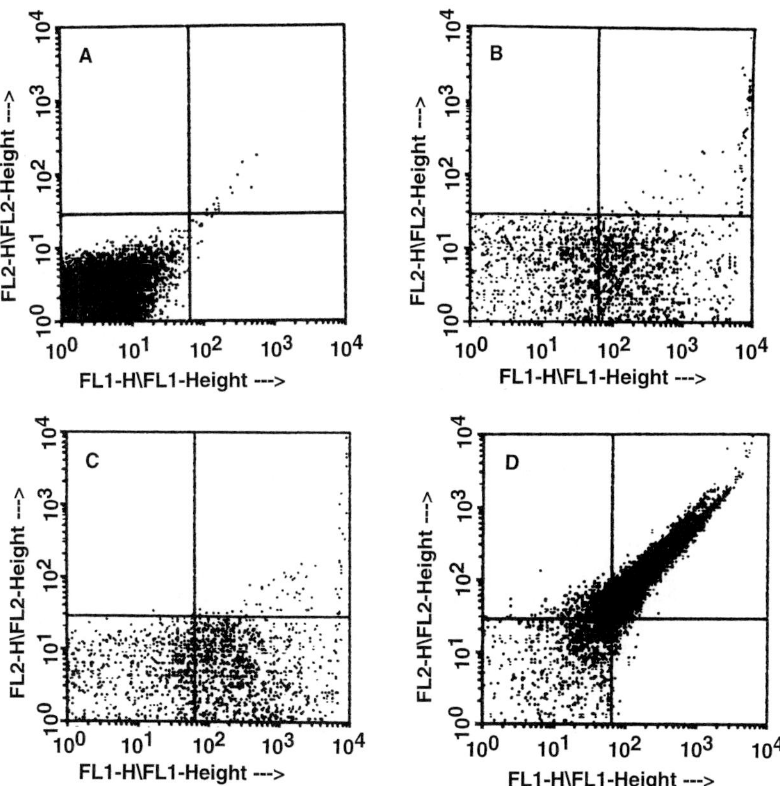

FIGURE I I   The efficacy of various vectors in accomplishing *in vivo* gene delivery. Intraperitoneally transplanted SKOV3 cells were challenged with different vector systems delivering
the LacZ reporter gene. Peritoneal lavage contents were subjected to FACS analysis for LacZ
expression in erbB-2-overexpressing tumor cells. Analyzed vectors included (A) control,
(B) liposomes (DOTAP), (C) AdpL complexes, and (D) recombinant adenoviruses. [Reproduced
from Ref. 75, from the *Journal of Clinical Investigation*, 1995, **96**, 2980–2989, by copyright
permission of The American Society for Clinical Investigation.]

To determine the feasibility of employing the adenoviral vector for *in situ*
tumor cell killing via anti-erbB-2 sFv gene delivery, we undertook treatment
experiments employing an orthotopic murine model. As before, SKOV3
cells were xenotransplanted into SCID mice. Forty-eight hours after engraftment with SKOV3 cells, the SCID mice were challenged intraperitoneally
with the E1A/B-deleted recombinant adenovirus encoding the anti-erbB-2
sFv (Ad21) or an E1A/B-deleted recombinant adenovirus encoding the irrelevant reporter gene LacZ (AdCMVLacZ). Ninety-six hours after treatment,
the animals underwent peritoneal lavage for analysis of harvested mobile
tumor cells. Cells were analyzed for cell viability employing the XTT assay.
It could be seen that the number of viable cells was dramatically decreased

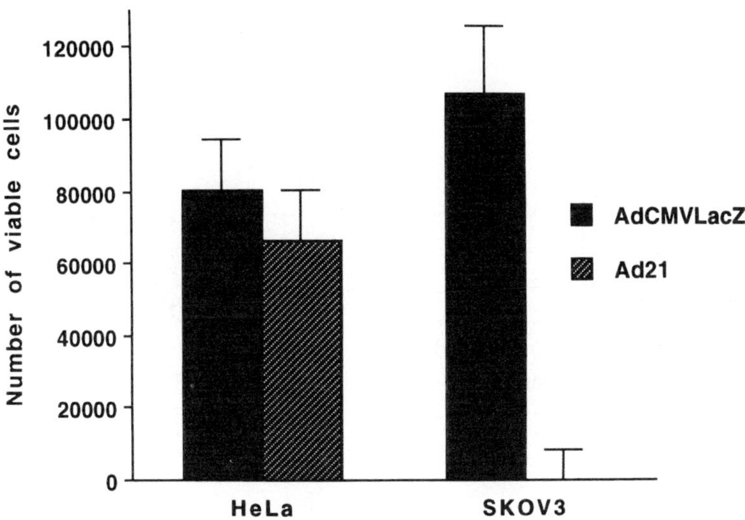

FIGURE 12   Effect of expression of a recombinant adenovirus-encoded anti-erbB-2 sFv gene on viability of erbB-2-overexpressing human ovarian tumor cells. SKOV3 cells and HeLa cells were infected with recombinant adenovirus encoding the cytosolic or ER-directed anti-erbB-2 sFvs. As a control, cells were also infected with an adenovirus encoding the irrelevant reporter gene LacZ (AdCMVLacZ). The XTT assayed was employed to determine cell viability 96 hr postinfection. This experiment was replicated 10 times. Data represent the mean ± SEM. [Reproduced from Ref. 75, from the *Journal of Clinical Investigation, 1995, **96**, 2980–2989*, by copyright permission of The American Society for Clinical Investigation.]

in the Ad21 group compared to the AdCMVLacZ group (Fig. 13). This cytotoxicity appeared to be specifically associated with the anti-erbB-2 sFv-encoding adenovirus. Analysis of the mechanism of cell death demonstrated that the Ad21 virus induced cellular apoptosis (data not shown). Thus, the recombinant adenovirus encoding the anti-erbB-2 sFv accomplished a specific cytotoxicity in mobile neoplastic cells in an orthotopic murine model of human ovarian cancer.

As outlined above, intracellular expression of anti-erbB-2 sFv accomplishes downregulation of cell surface erbB-2 in ovarian cancer cell lines. Furthermore, we demonstrated that an endoplasmic reticulum-directed form of the anti-erbB-2 sFv was uniquely capable of achieving this effect, presumably on the basis of entrapment of nascent erbB-2 in the ER of transduced tumor cells. We next sought to determine if similar effects would be achieved in breast cancer cells, based on the hypothesis that similar results might be obtained in this neoplastic target where erbB-2 also may play a prominent transforming role (41–44). For this analysis, the first cellular target was the erbB-2-overexpressing human breast carcinoma cell line MDA-MB-361. Gene transfer to these cells was accomplished by delivery of either a control adenovirus, AdCMVLacZ, or the ER-directed anti-erbB-2 sFv, Ad21. Trans-

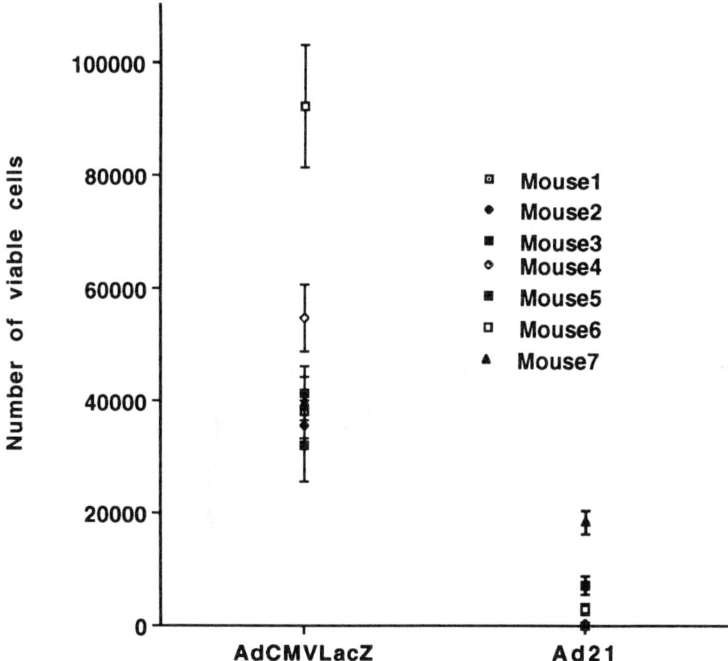

FIGURE 13   Specific cytotoxicity of an anti-erbB-2 sFv-encoding adenovirus against human ovarian cancer targets. In a murine model of malignant ascites, SCID mice were injected intraperitoneally with 10^7 SKOV3 cells. After 48 hr, animals were challenged intraperitoneally with 10^9 plaque-forming units of Ad21 or AdCMVLacZ. After 96 hr, mobile cells were harvested by peritoneal lavage and analyzed for survival by the XTT assay. [Reproduced from Ref. 75, from the *Journal of Clinical Investigation*, 1995, **96**, 2980–2989, by copyright permission of The American Society for Clinical Investigation.]

duced cells were evaluated 24 hr after viral infection for cell surface erbB-2 expression by immunohistochemistry. In this analysis, the adenovirus encoding the ER-directed anti-erbB-2 sFv accomplished a significant reduction in the amount of detectable cell surface erbB-2 (data not shown). In contrast, transduction with the control virus AdCMVLacZ did not result in any detectable downregulation of cell surface erbB-2 levels when compared to untransfected cells. Thus, intracellular expression of the ER-directed anti-erbB-2 sFv mediates downregulation of cell surface erbB-2 in a human breast cancer cell line overexpressing erbB-2. We also sought to determine if the sFv-mediated erbB-2 downregulation achieved in breast adenocarcinoma cells would likewise induce specific tumor cell killing. For this analysis, a panel of human breast cancer cell lines was infected with either Ad21 or AdCMVLacZ recombinant adenovirus. In this study, the various breast cancer cell lines differed appreciably with respect to sFv-mediated cytotoxic-

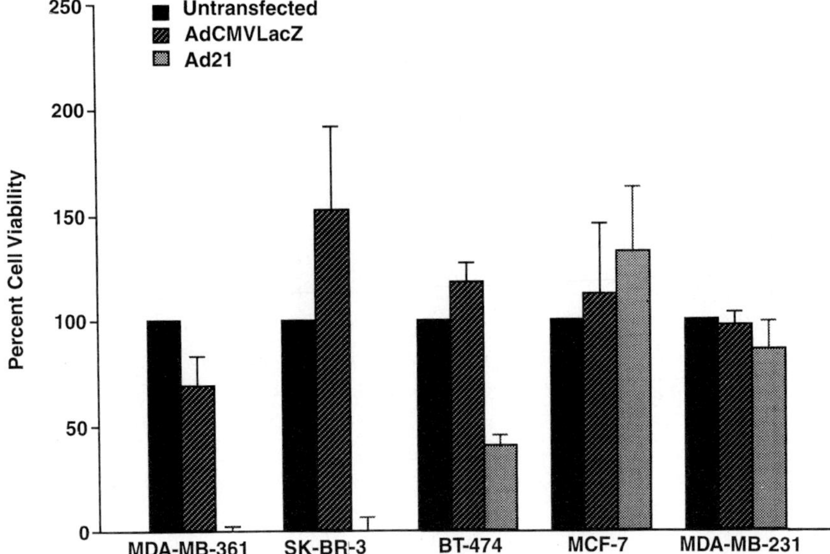

FIGURE 14   Anti-erbB-2 sFV-mediated cytotoxicity in human breast cancer cell lines. The β-galactosidase-encoding adenovirus, AdCMVLacZ, and Ad21, which encodes an ER-directed anti-erbB-2 sFv, were employed to evaluate cell viability by an MTS assay at 72 hr postinfection. Experiments were repeated three times. Results are reported as mean ± SEM.

ity (Fig. 14). None of the lines infected with the control virus AdCMVLacZ exhibited any cytotoxicity after infection. However, three of the breast cancer cell lines, SK-BR-3, MDA-MB-361, and BT-474, showed marked reduction in cell survival at 96 hr when infected with the anti-erbB-2-encoding virus, Ad21. In contrast, the cell lines MCF-7 and MDA-MB-231 were relatively resistant to the cytotoxic effect on the sFv. This latter result could have reflected inadequate transduction frequency in these instances, and thus may not have revealed a true resistance to sFv-mediated antitumor effects. To exclude this possibility, transduction efficiencies were determined employing the reporter gene encoding adenovirus, AdCMVLacZ. Following viral infection, fluorescence-activated cell sorting (FACS) analyses for β-galactosidase-positive cells were undertaken. In this study, both cell lines exhibited a >95% transduction frequency (data not shown). Therefore the differential cytotoxicity noted in this study appeared to reflect differences in breast cancer cell sensitivity to anti-erbB-2-mediated cytotoxicity and not differences in transducibility. Thus, while it could be shown that the anti-erbB-2 sFv approach caused cytotoxicity in human breast cancer cell lines, the effect was not uniform with respect to this tissue type.

Whereas erbB-2 overexpression has been noted in up to 30% of human breast cancer cells within this group, levels may be extremely variable. We hypothesized that the differences in anti-erbB-2 sFv sensitivity might reflect differences in the erbB-2 levels of target cells. To evaluate this possibility, we determined the levels of cellular erbB-2 in the tumor cell targets by a sensitive ELISA. Whereas previous studies examined this parameter in other cell types at the mRNA level of gene expression, few carried out this characterization with respect fo erbB-2 protein expression (64–66). In this study, the breast cancer cells assayed differed appreciably with respect to cellular erbB-2 levels (Fig. 15) Significantly, a correlation was noted to exist between cellular erbB-2 levels and susceptibility to sFv-mediated cytotoxicity. Specifically, the cell lines that appeared resistant to the anti-erbB-2 possessed the lowest levels of cellular erbB-2. In contrast, the cell lines that exhibited susceptibility to the anti-erbB-2 sFv overexpressed erbB-2. Thus, this analysis suggests that the endogenous level of cellular erbB-2 predicts tumor cell sensitivity to anti-erbB-2 sFv-mediated killing in the context of breast carcinoma.

Because of the association of erbB-2 overexpression with a subset of lung cancers (67), we also undertook studies to determine if this interventional approach could also accomplish targeted tumor cell killing in the context of lung cancer. To evaluate the toxicity of our anti-erbB-2 sFv independent

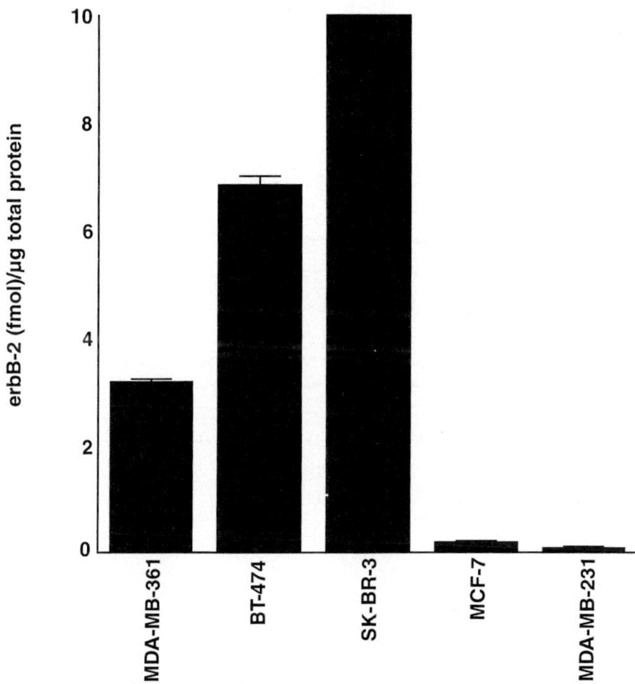

FIGURE 15 Levels of erbB-2 protein in human breast cancer cell lines. A panel of human breast cancer cell lines was evaluated for cellular levels of the erbB-2 protein by ELISA. Relative erbB-2 levels were quantified against a standard curve. Determinants were repeated three times with results expressed as mean ± SEM.

of transfection efficiency, we attempted to derive stable clones from several human lung cancer cell lines. A differential in stable clone derivation is a common assay used to indicate cytotoxicity due to expression of particular genetic constructs. These experiments have the added advantage of correcting for transfection efficiency, which can complicate interpretation of results using transient expression systems. The plasmid pGT21, encoding the ER form of the sFv, and the control plasmid pcDNA3 were used in these experiments. As shown in Table II, all erbB-2-positive cell lines showed a significant differential in stable clone derivation with decreased clones noted with pGT21 transfection. This result indicates that expression of the erbB-2 sFv construct significantly reduces the number of viable stable clones. Of note, the lung cancer cell line NCI-H520, which does not express erbB-2, showed no difference in the number of stable clones derived fro either construct. These results indicate that the expression of the anti-erbB-2 sFv in erbB-2-positive lung cancer cell lines is relatively incompatible with outgrowth of tumor clones. Also, erbB-2 status is correlated with sensitivity to the sFv, as was noted in the context of carcinoma of the breast. After verifying high-level transfection of the lung cancer cell line A549 with this vector system, we next analyzed the direct effects of the anti-erbB-2 sFv on cell viability. The A549 cells were transfected with either pcDNA3, pGT20, or pGT21 via the AdpL vector system, and then analyzed 5 days posttransfection for viable cell number (Fig. 16). SKOV3 and HeLa cells were used as positive and negative controls, respectively (data not shown). In the case of A549 cells and the SKOV3 control, there was an observed decrease in the number of viable cells such that greater than 95% of tumor cells had been killed at 5 days posttransfection. This cytotoxic effect was noted only in the context of the ER-targeted anti-erbB-2 sFv construct pGT21. In marked contrast, HeLa cells showed no significant differences in viable cell number when

TABLE II   Effect of Anti-erb-2 sFv on Derivation of Stable Clones[a,b]

	G418-resistant colonies	
Human lung cancer cell line	*Plasmid DNA control (pCDNA3)*	*Anti-erbB2 sFv ER form (pGT21)*
NCI-H520	27 ± 5.0	22.4 ± 4.1
H1299	4 ± 1.6	1 ± 0.82
H358	17.25 ± 5.9	5 ± 1.2
SKLU	10 ± 4.2	0.25 ± 0.5

[a] Reproduced with permission from Ref. 76, from the *American Journal of Respiratory Cell and Molecular Biology.*

[b] Cell lines were transfected with either the pcDNA3 or pGT21 construct, using DOTAP. At 48 hr, G418 was added to the medium. At 21 days posttransfection, colony numbers were evaluated. Results are from five separate experiments and are expressed as a mean ± SEM.

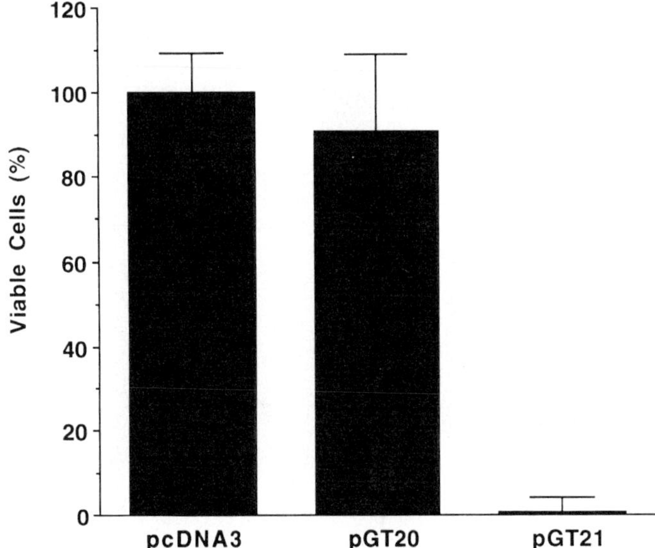

FIGURE 16   Effect of the anti-erbB-2 sFv on cell viability in the cell line A549. A549 cells were transfected via the AdpL system with either the pcDNA3, pGT20, or pGT21 construct and evaluated at 5 days posttransfection, using an MTS assay of cell viability. Percent cell viability is expressed as a percentage of cells transfected with pcDNA3 control plasmid DNA. Results from six experiments are expressed as mean ± SEM. [Reproduced with permission from Ref. 76, from the *American Journal of Respiratory Cell and Molecular Biology.*]

transfected with the vector control and either of the sFv constructs. Thus, the anti-erbB-2 sFv induces selective cytotoxicity in erbB-2-expressing lung cancer cells.

Given the capacity of the sFv approach to modulate erbB-2 levels, we hypothesized that this approach might also provide a means to enhance tumor cell chemosensitivity. In this regard, several studies have demonstrated that the sensitivity to chemotherapeutic agents can be enhanced by cytostatic monoclonal antibodies directed against the erbB-2 protein in breast and ovarian carcinoma cell lines (68–70). Thus, we hypothesized that erbB-2 downmodulation accomplished via intracellular expression of an anti-erbB-2 sFv would enhance the sensitivity of erbB-2-overexpressing cells to CDDP (cisplatin). As a proof of concept, a human ovarian carcinoma cell line overexpressing erbB-2 (SKOV3) was transiently transfected with a plasmid encoding the anti-erbB-2 sFv, pGT21, using the AdpL method. CDDP (2 μg/ml) was then added and cell viability was assayed at 72 hr (Fig. 17). Untransfected SKOV3 cells and cells transfected with the control plasmid, pcDNA3, with and without CDDP served as controls. Transient transfection with pGT21 induced enhanced sensitivity to CDDP expressed as a percentage of untreated SKOV3 cells: CDDP, 40.3% (±8.9%); pGT21, 6.3% (±2.2%);

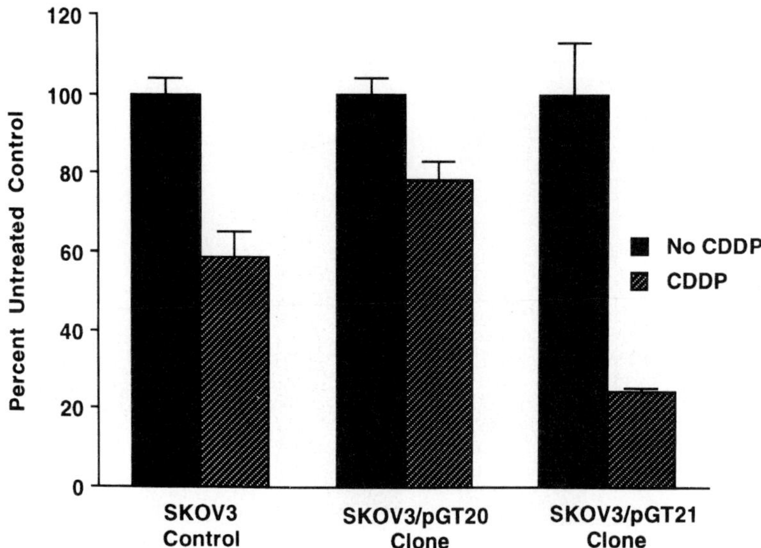

FIGURE 17   Sensitivity of anti-erbB-2 sFv-expressing SKOV3 clones to CDDP. SKOV3/pGT21 clones expressing the ER form of the anti-erbB-2 sFv demonstrate enhanced chemosensitivity to CDDP. SKOV3 cells, SKOV3/pGT20 clones, and SKOV3/pGT21 clones were treated with CDDP (2 μg/ml) and incubated for 72 hr. Cell viability was then measured using an MTS assay. Experiments were performed in triplicate and the results depict a representative experiment ± SEM.

pGT21/CDDP, 2.8% (±0.48%). The difference between the mean percentage of viable cells treated with pGT21 and pGT21/CDDP was calculated to be statistically significant ($p < 0.01$). Cells transiently transfected with the control plasmid (pcDNA3) did not demonstrate enhanced sensitivity to CDDP, expressed as a percentage of untreated SKOV3: pcDNA3, 85.2% (±11.0%); pcDNA3/CDDP, 50.1% (±5.0%). These results demonstrate enhanced cytotoxicity of SKOV3 cells to CDDP after transfection with an anti-erbB-2 sFv.

We next sought to evaluate this concept employing cells stably modified to express the anti-erbB-2 sFv. For this analysis, SKOV3 cells were transfected with neomycin-containing plasmids constructs encoding a cytosolic form of anti-erbB-2 sFv (pGT20) or an ER form of anti-erbB-2 sFv (pGT21), and the stable clonal populations isolated and expanded. Previous studies have shown that phenotypic alterations induced by the anti-erbB-2 sFv are specific to the ER form of the sFv, thus SKOV3/pGT20 clones served as negative controls. Initial analysis confirmed the phenotype of the derived clones with respect to erbB-2 expression. A specific ELISA demonstrated that the SKOV3/pGT21 clone had reduced levels of cellular erbB-2 compared to both the parental SKOV3 cells and the SKOV3/pGT20 clone (Fig. 18). Control studies confirmed that SKOV3 cells overexpressed erbB-2 and that

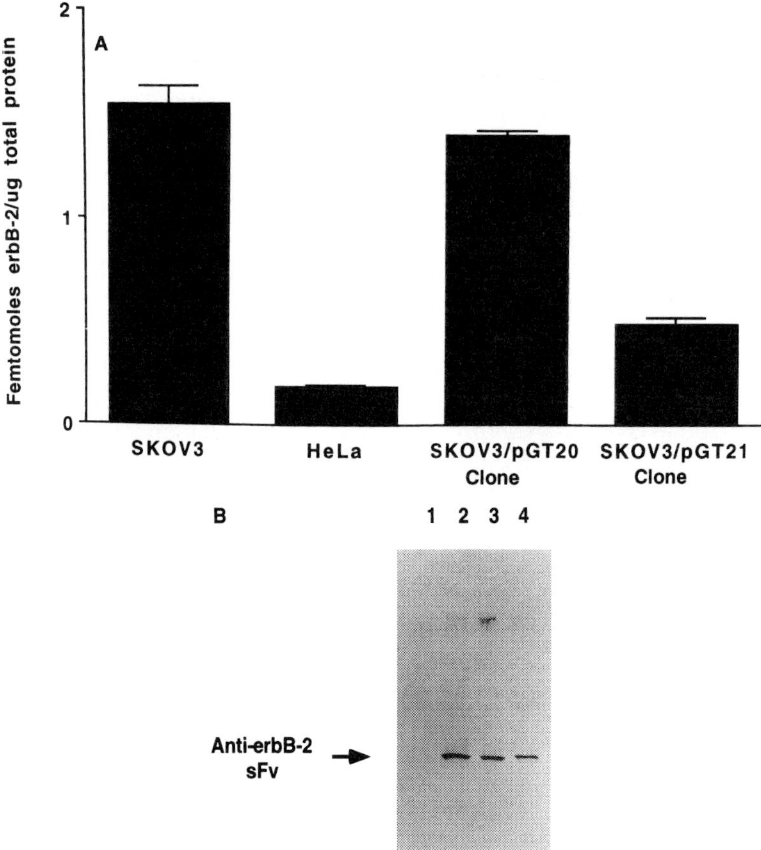

FIGURE 18 Characterization of anti-erbB-2 sFv-expressing SKOV3 clones. (A) Determination of cell surface erbB-2 protein expression in stable clones by an ELISA. Relative erbB-2 levels were extrapolated from a standard curve. SKOV3 cells are a positive control while HeLa, an erbB-2-negative human cervical cell line, served as a negative control. Results are expressed as the mean $\pm$ SEM. (B) Determination of the presence of anti-erbB-2 sFv in stable clones by Western blot. Cells were solubilized and the samples were electrophoresed on a 4–20% gradient gel and transferred to a PVDF membrane by electroblotting. The sFv was detected in the blot by probing with a polyclonal rabbit anti-sFv antibody. Cell lysates: Lane 1, untransfected SKOV3 cells; lane 2, SKOV3 cells transfected with pGT21; lane 3, SKOV3/pGT20 clone; lane 4, SKOV3/pGT21 clone.

the SKOV3/pGT20 clone did not differ from the parental control in this respect. In addition, both clones demonstrated constitutive expression of their respective anti-erbB-2 sFv. Thus, stable cell clones were derived that produced distinct phenotypes with respect to erbB-2 downmodulation. Importantly, stable expression of the cytosolic form of the anti-erbB-2 sFv was not associated with altered erbB-2 levels in derived clones, while stable expression of the ER-directed anti-erbB-2 sFv was associated with decreased erbB-2 levels in the derived clone. The described clones were then evaluated for chemosensitivity to CDDP. For this analysis SKOV3/pGT21 clones and SKOV3/pGT20 clones were treated with CDDP (2 μg/ml). Parental SKOV3 cells were used as controls. Cell viability was assayed after a 72-hr incubation. Enhanced sensitivity to CDDP was specific to clones stably expressing pGT21, the ER form of anti-erbB-2 sFv, and expressed as a percentage of the untreated cells (Fig. 19): SKOV3, 58.85% ($\pm$6.15%); SKOV3/pGT20 clone, 78.4% ($\pm$4.65%); SKOV3/pGT21 clone, 24.3% ($\pm$1.20%). The findings in this model system are consistent with the results obtained with transient transfection of SKOV3 cells. Thus, the distinctly different sensitivity to CDDP among the clonal cell populations confirms that sFv-mediated erbB-2 downmodulation is capable of altering tumor cell chemoresistance.

Our findings are in accord with other efforts to enhance cytotoxicity to cisplatin in erbB-2-overexpressing tumor cells using monoclonal antibodies. In this regard, Hancock *et al.* demonstrated enhanced sensitivity to cisplatin in breast and ovarian tumor cell lines using a monoclonal antibody to erbB-2 (68,69). Pietras and Paik have also demonstrated a synergistic effect of monoclonal antibodies to erbB-2 and chemotherapeutic agents (70). In addition, Arteaga *et al.* observed enhanced sensitivity of SKBR3 cells to cisplatin after treatment with a monoclonal antibody to erbB-2. These authors demonstrated that activation of the erbB-2 protein kinase by an agonistic monoclonal antibody is associated with increased cisplatin/DNA intrastrand adduct formation and a delayed rate of adduct decay (69). These studies suggest that antibody-mediated erbB-2 downmodulation with resultant alterations in DNA repair mechanisms is a key component of the enhanced tumor cell chemosensitivity. In contrast, the mechanism of sFv-mediated chemosensitivity may be dependent on other pathways. In this regard, Graus-Porta *et al.*, using an ER-directed form of an anti-erbB-2 sFv, demonstrated suppression of cell surface protein expression that was associated with phenotypic alterations including impairment of the mitogen-activated protein kinase pathway, induction of growth by neu differentiation factor, and induction of c-fos expression (71). These authors also noted impairment of epidermal growth factor (EGF)-mediated protein kinase activity. Thus, the anti-erbB-2 sFv-mediated enhanced chemosensitivity and associated induction of apoptosis may occur by an alternative mechanism that remains to be fully elucidated.

We have developed a novel method of gene product ablation that has been employed to knock out a selected oncogene. The results from our preliminary data have demonstrated consequences from the standpoint of the development of a novel therapeutic modality. To achieve selective gene product ablation, an intracellular single-chain immunoglobulin was expressed in target cells. The intracellularly expressed sFv prevented the normal maturation of the transmembrane tyrosine kinase receptor erbB-2, which likely occurred on the basis of ER entrapment of the nascent protein during

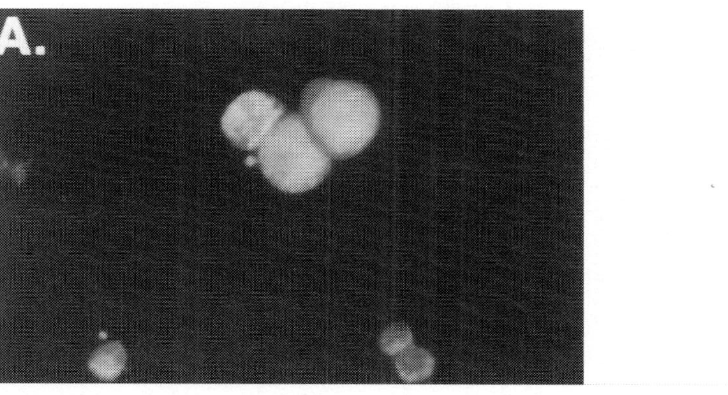

FIGURE 19 Induction of apoptosis by CDDP in anti-erbB-2 sFv-expressing SKOV3 clones. The presence of apoptotic nuclear staining in SKOV3/pGT21 clones treated with CDDP (2 μg/ml) was determined by nuclear uptake of fluorescent DNA-binding dye. (A) Untreated SKOV3/pGT21 clones. (B) SKOV3/pGT21 clones treated with CDDP.

biosynthesis. The intracellular expression of an sFv has been described by Marasco *et al.* (36,37). In their study, the cellular expression of an anti-HIV sFv resulted in abrogation of the HIV infectious cycle with a functional reduction of viral production. We have also accomplished intracellular single-chain immunoglobin-mediated knockout utilizing gene transfer of an sFv construct. In this instance, however, the consequence was ablation of transforming oncogene, as opposed to a xenogeneic viral gene product. A consequence of this novel methodology is the definition of a potential therapeutic modality for the achievement of selective killing of specific target cells. In the context of anticancer therapeutics, it has been proposed that selective abrogation of specific target gene products can revert them from the malignant phenotype. The overexpression of dominant oncogenes has been shown to be a critical determinant of neoplastic transformation and progression. Targeted disruption of selected oncogenes may accomplish reversion from the malignant phenotype or initiation of cell death. The utilization of intracellular single-chain immunoglobins represents another strategy for these purposes. This methodology offers certain potential advantages over previous genetic techniques for therapeutic gene ablation. Many MAbs have been developed against a variety of cancer-related gene products. It would thus be logical to convert these reagents to sFvs that would possess the potential for therapeutic utility. In this schema, the ablation construct might be designed with *a priori* knowledge of its specific recognition of its cellular target. As in the example reported here, expression of intracellular immunoglobins may possess the potential to achieve a highly selective effect on specific gene products with the end result being cell killing. In addition, the employment of DNA-based methods for delivery is implicit in the sFv strategy. This allows the reagents to be delivered employing a variety of high efficiency vehicles. In contrast, effective delivery of antisense constructs has represented a major limitation to their practical employment to date. Thus, the intracellular antibody strategy may offer significant practical advantages over antisense methods as a means to achieve selective tumor cytotoxicity based on targeted knockout of oncogene or oncogene products.

Acknowledgments

The author acknowledges the expert editorial assistance of Connie Howton. In addition, the author is indebted to Gene P. Siegal, M.D., Ronald D. Alvarez, M.D., Jessy Deshane, Jon Grim, and Mack Barnes, M.D.

References

1. Sikora, K. (1993). Gene therapy for cancer. *Trends Biotechnol.* **11**, 197–201.
2. Dorudi, S., Northover, J. M. A., Vile, R. G. (1993). Gene transfer therapy in cancer. *Br. J. Surg.* **80**, 566–572.

3. Freeman, S. M., and Zwiebel, J. A. (1993). Gene therapy of cancer. *Cancer Invest.* **11**, 676–688.

4. Vanchieri, C. (1993). Opportunities "opening up" for gene therapy. *J. Natl. Cancer Inst.* **85**, 90–91.

5. Lemoine, N., and Sikora, K. (1993). Interventional genetics and cancer treatment. *Br. Med. J.* **306**, 665–666.

6. Gutierrez, A. A., Lemoine, N. R., and Sikora, K. (1992). Gene therapy for cancer. *Lancet* **339**, 715–721.

7. Karp, J. E., and Broder, S. (1994). New directions in molecular medicine. *Cancer Res.* **54**, 653–665.

8. Rosenberg, S. A., Anderson, W. F., Blaese, M., Hwu, P., Yannelli, J. R., Yang, J. C., Topalian, S. L., Schwartzentruber, D. J., Weber, J. S., Ettinghausen, S. E., Parkinson, D. N., and White, D. E. (1993). The development of gene therapy for the treatment of cancer. *Ann. Surg.* **218**, 455–464.

9. Baker, S. J., Markowitz, S., Fearon, E. R., Willson, J. K. V., and Volgelstein, B. (1990). Suppression of human colorectal carcinoma cell growth by wild-type p53. *Science* **249**, 912–915.

10. Huang, H.-J. S., Yee, J.-K., Shew, J.-Y., Chen, P.-L., Bookstein, R., Friedmann, T., Lee, E. Y., and Lee, W.-H. (1988). Suppression of neoplastic phenotype by replacement of the RB gene in human cancer cells. *Science* **242**, 1563–1566.

11. Bookstein, R., Shew, J.-Y., Chen, P.-L., Scully, P., and Lee, W.-H. (1990). Suppression of tumorigenicity of human prostate carcinoma cells by replacing a mutated RB gene. *Science* **247**, 712–716.

12. Fujiwara, T., Grimm, E. A., Mukhopadhyay, T., Cai, D. W., Owen-Schaub, L. B., and Roth, J. A. (1993). A retroviral wild-type p53 expression vector penetrates human lung cancer spheroids and inhibits growth by inducing apoptosis. *Cancer Res.* **53**, 4129–4133.

13. Zhan, W.-W., Fang, X., Mazur, W., French, B. A., Georges, R. N., and Roth, J. A. (1994). High-efficiency gene transfer and high-level expression of wild-type p53 in human lung cancer cells mediated by recombinant adenovirus. *Cancer Gene Ther.* **1**, 5–13.

14. Fueyo, J., Gomez-Manzano, C., Yung, W. K. A., Clayman, G. L., Liu, T.-J., Bruner, J., Levin, V. A., and Kyritisis, A. P. (1996). Adenovirus-mediated p16/CDKN2 gene transfer induces growth arrest and modifies the transformed phenotype of glioma cells. *Oncogene* **12**, 103–110.

15. Krol, A. V., Mol, J. N. M., and Stuitje, A. R. (1988). Modulation of eukaryotic gene expression by complement RNA or DNA sequences. *BioTechniques* **6**, 958–976.

16. Helene, C., and Toulme, J.-J. (1990). Specific regulation of gene expression by antisense, sense and antigene nucleic acids. *Biochim. Biophys. Acta* **104**, 99–125.

17. Zon, G. (1988). Oligonucleotide as potential chemotherapeutic agents. *Pharm. Res.* **5**, 539–549.

18. Stein, C. A., and Cheng, Y.-C. (1993). Antisense oligonucleotides as therapeutic agents—Is the bullet really magic? *Science* **261**, 1004–1012.

19. Wickström, E. L., Bacon, T. A., Gonzalez, A., Freeman, D. L., Lyman, G. H., and Wickström, E. (1988). Human promyelocytic leukemia HL-60 cell proliferation and c-Myc protein expression are inhibited by antisense pentadecadeoxynucleotide targeted against c-*myc* mRNA. *Proc. Natl. Acad. Sci. U.S.A.* **85**, 1028–1032.

20. Kashani-Sabet, M., Funato, T., Florenes, V. A., Fodstad, O., and Scanlon, K. J. (1994). Suppression of a neoplastic phenotype *in vivo* by an anti-*ras* ribozyme. *Cancer Res.* **54**, 900–902.

21. Ebbinghaus, S. W., Gee, J. E., Rodu, B., Mayfield, C. A., Sanders, G., and Miller, D. M. (1993). Triplex formation inhibits HER-2/neu transcription *in vivo*. *J. Clin. Invest.* **92**, 2433–2439.

22. Haseloff, J., and Gerlach, W. L. (1988). Simple RNA enzymes with new and highly specific endoribonuclease activities. *Nature* (*London*) **334**, 585–591.

23. Kashani-Sabet, M., Funato, T., Tone, T., Jiao, L., Wang, W., and Yoshida, E. (1992). Reversal of the malignant phenotype by an anti-*ras* ribosome. *Antisense Res. Dev.* **2,** 3–15.

24. Ratahczak, M. Z., Kant, J. A., Luger, S. M., Hijiya, N., Zhang, J., Zon, G., and Gerwirtz, A. M. (1992). *In vivo* treatment of human leukemia in a scid mouse model with c-*myb* antisense oligonucleotides. *Proc. Natl. Acad. Sci. U.S.A.* **89,** 11823–11827.

25. Georges, R. N., Mukhopadhyay, T., Zhang, Y., Yen, N., and Roth, J. A. (1993). Prevention of orthotopic human lung cancer growth by intratreacheal instillation of a retroviral antisense K-*ras* construct. *Cancer Res.* **53,** 1743–1746.

26. Iverson, P. L., Zhu, S., Meyer, A., and Zon, G. (1992). Cellular uptake and subcellular distribution of phosphorothioate into cultured cells. *Antisense Res. Dev.* **2,** 211–222.

27. Citro, G., Perrotti, D., Cucco, C., D'Agnano, I., Sacchi, A., Zupi, G., and Calabretta, B. (1992). Inhibition of leukemia cell proliferation by receptor-mediated uptake of c-*myb* antisense oligodeoxynucleotides. *Proc. Natl. Acad. Sci. U.S.A.* **89,** 7031–7035.

28. Bergan, R., Connell, Y., Fahmy, B., and Neckers, L. (1993). Electroporation enhances c-*myc* antisense oligodeoxynucleotide efficacy. *Nucleic Acids Res.* **21,** 3567–3573.

29. Theuer, C. P., and Pastan, I. (1993). Immunotoxins and recombinant toxins in the treatment of solid carcinomas. *Am. J. Surg.* **166,** 284–288.

30. Brinkmann, U., Pai, L. H., FitzGerald, D. J., Willingham, M., and Pastan, I. (1991). B3-(Fv)-PE38KDEL, a single chain immunotoxin that causes complete regression of a human carcinoma in mice. *Proc. Natl. Acad. Sci. U.S.A.* **88,** 8616–8620.

31. Colcher, D., Bird, R., Roselli, M., Hardmann, K. M., Johnson, S., Pope, S., Dodd, S. W., Pantoliano, M. W., Milenic, D. E., and Schlom, J. (1990). *In vivo* tumor targeting of a recombinant single-chain antigen-binding protein. *J. Natl. Cancer Inst.* **82,** 1191–1197.

32. Wawrzynczak, E. J. (1992). Rational design of immunotoxins: Current progress and future prospects. *Anti-Cancer Drug Des.* **7,** 427–441.

33. Mykebust, A. T., Godal, A., and Fodstad, O. (1994). Targeted therapy with immunotoxins in a nude rat model for leptomeningeal growth of human small cell cancer. *Cancer Res.* **54,** 2146–2150.

34. Friedman, P. N., Chance, D. F., Trail, P. A., and Siegall, C. B. (1993). Antitumor activity of the single-chain immunotoxin BR96 sFv-PE40 against established breast and lung tumor xenografts. *J. Immunol.* **150,** 3054–3061.

35. Biocca, S., Pierandrei-Amaldi, P., and Cattaneo, A. (1993). Intracellular expression of anti-p21^{ras} single chain Fv fragments inhibits meiotic maturation of *Xenopus* oocytes. *Biochem. Biophys. Res. Commun.* **197,** 422–427.

36. Marasco, W. A., Haseltine, W. A., and Chen, S.-Y. (1993). Design, intracellular expression, and activity of a human anti-human immunodeficiency virus type 1 gp120 single-chain antibody. *Proc. Natl. Acad. Sci. U.S.A.* **90,** 7889–7893.

37. Chen, S.-Y., Bagley, J., and Marasco, W. A. (1994). Intracellular antibodies as a new class of therapeutic molecules for gene therapy. *Hum. Gene Ther.* **5,** 595–601.

38. Werge, T. M., Biocca, S., and Cattaneo, A. (1990). Cloning and intracellular expression of a monoclonal antibody to the p21^{ras} protein. *FEBS Lett.* **274,** 193–198.

39. Yarden, Y., and Ullrich, A. (1988). Growth factor receptor tyrosine kinases. *Annu. Rev. Biochem.* **57,** 443–478.

40. Hudziak, R. M., Schlessinger, J., and Ulrich, A. (1987). Increased expression of the putative growth factor receptor p185HER2 causes transformation and tumorigenesis of NIH 3T3 cells. *Proc. Natl. Acad. Sci. U.S.A.* **84,** 7159–7163.

41. Muller, W. J., Sinn, E., Pattengale, P. K., Wallace, R., and Leder, P. (1988). Single-step induction of mammary adenocarcinoma in transgenic mice bearing the activated c-*neu* oncogene. *Cell* **54,** 105–115.

42. Slamon, D. J., Godolphin, W., Jones, L. A., Holt, J. A., Wong, S. G., and Keith, D. E. (1989). Studies of the HER-2/neu proto-oncogene in the human breast and ovarian cancer. *Science* **244,** 707–712.

43. Hynes, N. E. (1993). Amplification and overexpression of the erbB-2 gene in human tumors: Its involvement in tumor development, significance as a prognostic factor, and potential as a target for cancer therapy. *Cancer Biol.* **4,** 19–26.

44. Giovanella, B. C., Vardeman, D. M., Williams, L. J., Taylor, D. J., Deipolyi, P. D., Greef, P. J., Stehlin, J. S., Ullrich, A., Cailleau, R., Slamon, D. J., and Gary, H. E. (1991). Heterotransplantation of human breast carcinomas in nude mice. Correlation between successful heterotransplants, poor prognosis and amplification of HER-2/Neu oncogene. *Int. J. Cancer* **47,** 66–71.

45. Drebin, J. A., Link, V. C., and Greene, M. I. (1988). Monoclonal antibodies specific for the *neu* oncogene product directly mediate anti-tumor effects *in vivo*. *Oncogene* **2,** 387–394.

46. Fendley, B. M., Winget, M., Hudziak, R. M., Lipari, M. T., Naperi, M. A., and Ullrich, A. (1990). Characterization of murine monoclonal antibodies reactive to either the human epidermal growth factor receptor or HER2/neu gene product. *Cancer Res.* **50,** 1550–1558.

47. Carter, P., Presta, L., Gorman, C. M., Ridgeway, J. B. B., Henner, D., Wong, W. L. T., Rowland, A. M., Kotts, C., Carver, M. E., and Shepard, H. M. (1992). Humanization of an anti-p185HER2 antibody for human cancer therapy. *Proc. Natl. Acad. Sci. U.S.A.* **89,** 4285–4289.

48. Hurwitz, E., Stancovski, I., Sela, M., and Yarden, Y. (1995). Suppression and promotion of tumor growth by monoclonal antibodies to ErbB-2 differentially correlate with cellular uptake. *Proc. Natl. Acad. Sci. U.S.A.* **92,** 3353–3357.

49. De Santes, K., Slamon, D., Anderson, S. K., Shepard, M., Fendly, B., Maneval, D., and Press, O. (1992). Radiolabelled antibody targeting of the HER-2/neu oncoprotein. *Cancer Res.* **52,** 1916–1923.

50. Batra, J. K., Kazpryzyk, P. G., Bird, R. E., Pastan, I., and King, C. R. (1992). Recombinant anti-erbB-2 immunotoxins containing *Pseudomonas* exotoxin. *Proc. Natl. Acad. Sci. U.S.A.* **89,** 5867–5871.

51. Bertram, J., Killian, M., Brysch, W., Schlingensiepen, K.-H., and Kneba, M. (1994). Reduction of erbB-2 gene product in mammary carcinoma cell lines by erbB2 mRNA-specific and tyrosine kinase consensus phosphorothioate antisense oligonucleotides. *Biochem. Biophys. Res. Commun.* **200,** 661–667.

52. Brysch, W., Magal, E., Louis, J.-C., Knust, M., Klinger, I., and Schlinggensiepen, R. (1994). Inhibition of p185c-erbB-2 proto-oncogene expression by antisense oligodeoxynucleotides down-regulates p185-associated tyrosine-kinase activity and strongly inhibits mammary tumor-cell proliferation. *Cancer Gene Ther.* **1,** 99–105.

53. Curiel, D. T., Wagner, E., Cotten, M., Birnsteil, M. L., Li, C.-M., Loechel, S., Agarwal, S., and Hu, P.-C. (1992). High efficiency gene transfer mediated by adenovirus coupled to DNA–polylysine complexes via an antibody bridge. *Hum. Gene Ther.* **3,** 147–154.

54. Gerdes, J., Lemke, H., Baisch, H., Wacker, H. H., Schwab, U., and Stein, H. (1984). Cell cycle analysis of a cell proliferation-associated human nuclear antigen defined by the monoclonal antibody Ki-67. *J. Immunol.* **133,** 1710–1715.

55. Hinshaw, V. S., Olsen, C. W., Dybhahl-Sissoko, N., and Evans, D. (1994). Apoptosis: A mechanism of cell killing by influenza A and B viruses. *J. Virol.* **68,** 3667–3673.

56. Duke, R. C., and Cohen, J. J. (1992). Morphological and biochemical assays of apoptosis. *Curr. Protocols, Suppl.* **3.**

57. Vaux, D. L. (1993). Toward an understanding of the molecular mechanisms of physiological cell death. *Proc. Natl. Acad. Sci. U.S.A.* **90,** 786–789.

58. Williams, G. T. (1991). Programmed cell death: Apoptosis and oncogenesis. *Cell (Cambridge, Mass.)* **65,** 1097–1098.

59. Thompson, C. B. (1995). Apoptosis in the pathogenesis and treatment of disease. *Science* **267,** 1456–1462.

60. Steller, H. (1995). Mechanisms and genes of cellular suicide. *Science* **267,** 1445–1449.

61. Carson, W. E., Haldar S., Baiocchi, R. A., Croce, C. M., and Caligiuri, M. A. (1994). The c-kit ligand suppresses apoptosis of human natural killer cells through the upregulation of bcl-2. *Proc. Natl. Acad. Sci. U.S.A.* **91,** 7553–7557.

62. Miyashita, T., and Reed, J. C. (1992). Bcl-2 gene transfer increases relative resistance of S49.1 and WEHI7.2 lymphoid cells to cell death and DNA fragmentation induced by glucocorticoids and multiple chemotherapeutic drugs. *Cancer Res.* **52,** 5407–5411.

63. Miyashita, T., Reed, J. C. (1993). Bcl-2 oncoprotein blocks chemotherapy-induced apoptosis in a human leukemia cell line. *Blood* **81,** 151–157.

64. Kraus, M. H., Popescu, N. C., Amsbaugh, S. C., and Kin, C. R. (1987). Overexpression of the EGF receptor-related proto-oncogene erbB-2 in human mammary tumor cell lines by different molecular mechanisms. *EMBO J.* **6,** 605–610.

65. Hynes, N. E., Gerber, H. A., Saurer, S., and Groner, B. (1989). Overexpression of the c-erbB-2 protein in human breast tumor lines. *J. Cell. Biochem.* **39,** 167–173.

66. Iglehart, J. D., Kraus, M. H., Langton, B. C., Huper, G., Kerns, B. J., and Marks, J. R. (1990). Increased erbB-2 gene copies and expression in multiple stages of breast cancer. *Cancer Res.* **50,** 6701–6707.

67. Kern, J. A., Schwartz, D. A., Nordberg, J. E., Weiner, D. B., Greene, M. I., Torney, L., and Robinson, R. A. (1990). p185neu expression in human lung adenocarcinomas predicts shortened survival. *Cancer Res.* **50**(16), 5184–5187.

68. Hancock, M. C., Langton, B. C., Chan, T., Toy, P., Monahan, J. J., Mischak, R. P., and Shawver, L. K. (1991). A monoclonal antibody against the c-erbB-2 protein enhances the cytotoxicity of *cis*-diamminedichloroplatinum against human breast and ovarian tumor cell lines. *Cancer Res.* **51,** 4575–4580.

69. Arteaga, C. L., Winnier, A. R., Poirier, M. C., Lopez-Larraza, D. M., Shawver, L. K., Hurd, S. D., and Stewart, S. J. (1994). p185c-erbB-2 singaling enhances cisplatin-induced cytotoxicity in human breast carcinoma cells: Association between oncogenic receptor tyrosine kinase and drug-induced DNA repair. *Cancer Res.* **54,** 3758–3765.

70. Pietras, R. J., Fendly, B. M., Chazin, V. R., Pegram, M. D., Howell, S. B., and Slamon, D. J. (1994). Antibody to HER-2/neu receptor blocks DNA repair after cisplatin in human breast and ovarian cancer cells. *Oncogene* **9,** 1829–1838.

71. Graus-Porta, D., Beerli, R. R., and Hynes, N. E. (1995). Single-chain antibody-mediated intracellular retention of erbB-2 impairs neu differentiation factor and epidermal growth factor signaling. *Mol. Cell. Biol.* **15,** 1182–1191.

72. Deshane, J., Cabrera, G., Grim, J. E., Siegal, G. P., Pike, J., and Alvarez, R. D. (1995). Targeted eradication of ovarian cancer mediated by intracellular expression of anti-erbB-2 single chain antibody. *Gyn. Onc.* **59,** 8–14.

73. Deshane, J., Loechel, F., Conry, R. M., Siegal, G. P., King, C. R., and Curiel, D. T. (1994). Intracellular single-chain antibody directed against erbB-2 down-regulates cell surface erbB-2 and exhibits a selective anti-proliferative effect in erbB-2 overexpressing cancer cells lines. *Gene Ther.* **1,** 332–337.

74. Deshane, J., Grim, J., Loechel, S., Siegal, G., Alvarez, R. D., and Curiel, D. T. (1996). Intracellular antibody against erbB-2 mediates targeted tumor cell eradication by apoptosis. *Cancer Gene Ther.* **3,** 89–98.

75. Deshane, J., Siegal, G. P., Alvarez, R. D., Wang, M. H., Feng, M., Cabrera, G., Liu, T., Kay, M., and Curiel, D. T. (1995). Targeted tumor killing via an intracellular antibody against erbB-2. *J. Clin. Invest.* **96,** 2980–2989.

76. Grim, J., Deshane, J., Feng, M., Lieber, A., Kay, M., and Curiel, D. T. (1996). erbB-2 knock-out employing an intracellular single-chain antibody (sF) accomplishes specific toxicity in erbB-2 expressing lung cancer cell. *Am. J. Respir. Cell Mol. Biol.* **15,** 348–354.

Terence R. Flotte*
Barrie J. Carter†

*Gene Therapy Center
University of Florida
Gainesville, Florida 32610

†Targeted Genetics Corporation
Seattle, Washington 98101

In Vivo Gene Therapy with Adeno-Associated Virus Vectors for Cystic Fibrosis

I. The Problem of Gene Therapy for Cystic Fibrosis Lung Disease

A. Cystic Fibrosis Lung Disease

Cystic fibrosis (CF) is a common autosomal recessive disorder that was recognized clinically as a triad of chronic obstructive pulmonary disease, exocrine pancreatic insufficiency, and abnormally high sweat electrolyte concentrations (di Sant' Agnese *et al.*, 1953). Patients with this disorder suffer from intestinal malabsorption with subsequent malnutrition and from chronic suppurative bronchitis and bronchiectasis due to infection with *Staphylococcus aureus*, *Pseudomonas aeruginosa*, and *Haemophilus influenzae* (Boat *et al.*, 1989). Without modern treatments, the life expectancy was less than 10 years in the 1950s and early 1960s.

The severe disease sequelae combined with an incidence of 1 in 2750 live births make CF a major health problem in the United States (Boat *et*

Advances in Pharmacology, Volume 40

al., 1989). Current therapies are based on the following three principles: (1) pancreatic enzyme replacement and other nutritional support, (2) airway clearance with physical therapy and associated techniques, and (3) aggressive antibiotic therapy directed against *Staphylococcus* and *Pseudomonas*. These therapies have resulted in an increase in median life expectancy to nearly 30 years (Fitzsimmons, 1994). Despite the remarkable improvement, pulmonary disease has remained the principal cause of morbidity and mortality. None of the standard therapies has been able to prevent the development and progression of lung disease in these patients.

Efforts have focused on better defining the basic defect in CF, so as to devise more specific therapies. Quinton and Bijman (1983) discovered that the unifying physiological defect in organs affected in CF was a defect in cAMP-activated chloride secretion. The discovery of the gene responsible for CF, by positional cloning (Riordan *et al.*, 1989), has led to a much more detailed understanding of the functions of the defective protein, which is called the cystic fibrosis transmembrane conductance regulator (CFTR).

B. The Cystic Fibrosis Transmembrane Conductance Regulator

The CFTR is a member of the ATP-binding cassette (ABC) superfamily of membrane transporters. Its deduced structure consists of 1480 amino acid residues, including (sequentially from the amino terminus) 6 membrane-spanning domains, a nucleotide-binding fold, a regulatory (R) domain, another 6 membrane-spanning domains, and a second nucleotide-binding domain (Welsh and Smith, 1993) (Fig. 1). This structure is encoded by 4.4 kb of coding sequences within a 6.5-kb mRNA transcribed from exons spread over 250 kb of genomic DNA. The CFTR acts as an ATP-dependent, cAMP-activated apical chloride channel in the respiratory and gastrointestinal epithelium and in the duct of the sweat gland, with a linear current–voltage

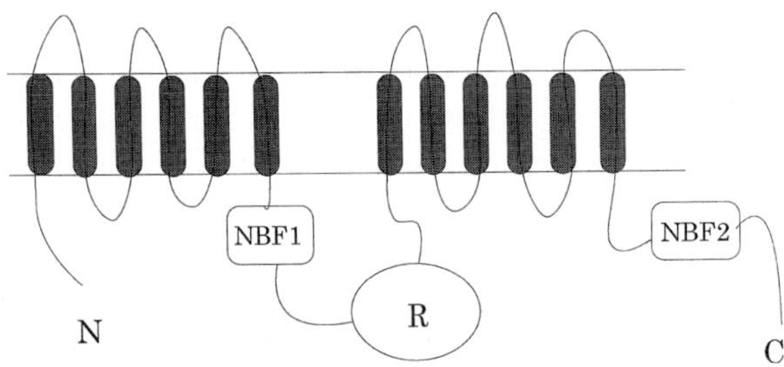

FIGURE I   Structure of CFTR. The putative structure of the CFTR molecule includes from the amino (N) terminus: six membrane-spanning regions (shaded ovals), a first nucleotide-binding fold (NBF1), a regulatory domain (R), six more membrane-spanning regions, a second nucleotide-binding fold (NBF2), and the carboxyl terminus (C). (See text for details.)

relationship, and a conductance of approximately 10 picosiemens (pS). The CFTR also regulates other channels in the membrane, including the amiloride-sensitive epithelial sodium channel and the outwardly rectifying chloride channel (ORCC). The ORCC current–voltage relationship is characterized by outward rectification and a conductance of 40 pS (Egan *et al.*, 1992). The CFTR may also play a role in endosome turnover (Bradbury *et al.*, 1992), regulation of endosomal pH (Lukacs *et al.*, 1992), and in the sialylation of membrane GM_1 gangliosides (Imundo *et al.*, 1994).

C. Obstacles to Gene Therapy

The discovery of the CFTR has also enabled complementation of the CF defect in cultured cells (Drumm *et al.*, 1990) and, theoretically, *in vivo*. In this setting, the CFTR cDNA may be viewed as a pharmacologic agent that must be delivered to a specific intracellular compartment, the target cell nucleus, in order to have its effects (Afione *et al.*, 1995). While complementation of defective chloride transport has been accomplished *in vitro* using a variety of viral and nonviral vectors carrying the CFTR gene, phenotypic correction of CF lung disease is remarkably more difficult. This difficulty reflects the diverse potential target cells and tissues and the limited number of *in vivo* gene transfer systems currently available for the task. Several issues must be addressed before gene therapy can have a therapeutic impact.

First, the target cells for CF gene therapy in the lungs have not been clearly defined. Studies of endogenous CFTR expression reveal levels of expression in the surface epithelium that are surprisingly low considering that CFTR appears to be primarily responsible for regulating the hydration and electrolyte composition of the epithelial lining fluid. Foci of increased expression have been found in serous cells in submucosal glands (Engelhardt *et al.*, 1992), but these gland structures are found only in the larger airways, while clinical disease begins in the smaller peripheral airways. Therefore, it remains unclear whether the primary cell targets should be the surface epithelial cells, the submucosal gland cells, or both types of cells.

A second issue is that none of the potential target cells is likely to be suitable for *ex vivo* manipulation and reimplantation. The typical retroviral approaches to gene transfer would include *ex vivo* culture, stimulation of cell division, transduction, and expansion of the complemented cell population. Gene therapy for CF, in contrast, will require that cells be transduced *in vivo*, maintaining their native tissue architecture and organization. The proliferation rate of the target cells under these conditions is likely to be slow, which may create additional challenges for some vector systems (Miller *et al.*, 1990; Russell *et al.*, 1994).

The third issue is that the airways of CF patients become inflamed and infected at a very early age. Studies indicate that, even prior to onset of apparent clinical symptoms, most patients have increased levels of proinflammatory cytokines [interleukin 6 (IL-6), IL-8, and tumor necrosis factor α (TNF-α)] and infection with *S. aureus* and *P. aeruginosa* (Konstan *et al.*, 1994; Kronberg *et al.*, 1993). This could create two potential difficulties for CF gene therapy. There may be by-products of inflammation, such as thick purulent sputum, which could impair the efficacy of gene transfer vectors by blocking access to binding sites on the target cell. The preexisting inflammation in CF lungs may prime them for inflammatory and immune reactions to viral vectors, thus limiting the safety or efficacy of this approach. The latter effect may be particularly problematic for adenovirus (Ad)-based vectors, which tend to be proinflammatory in animal models (Simon *et al.*, 1993; Yang *et al.*, 1994).

The fourth major issue for CF gene therapy is the challenge of achieving long-lasting expression. Since CF lung disease progresses gradually over many years, it is likely that successful gene therapy will require CFTR expression over many years. This again could present a problem with Ad vectors and with liposome-based vectors, both of which tend to be transient in their expression (Rosenfeld *et al.*, 1991, 1992; Engelhardt *et al.*, 1993; Zabner *et al.*, 1993; Crystal *et al.*, 1994; Caplen *et al.*, 1995). The alternative approach of frequent, repeat delivery may also have significant difficulties due to enhanced immune or inflammatory responses.

II. Adeno-Associated Virus as a Potential Vector System for Cystic Fibrosis Gene Therapy

The challenge of CF gene therapy has served as a catalyst for the clinical development of a number of gene transfer systems including adenovirus and the nonpathogenic human parovovirus, adeno-associated virus (AAV) serotype 2. A number of aspects of AAV make it a potentially useful agent for gene therapy of CF lung disease (Carter, 1992). These include its tropism for cells in the respiratory tract, its nonpathogenic nature, and its tendency for long-term persistence.

A. Adeno-Associated Virus Biology

Adeno-associated virus was first isolated as a contaminant of adenoviruses isolated from cell cultures. It was later found to possess characteristics similar to other parvoviruses, including a structure consisting of a 4.7-kb single-stranded DNA (ssDNA) genome encapsidated in a 20-nm nonenveloped icosahedral virion (reviewed in Berns, 1990). There are several different serotypes including avian, canine, simian, and human strains (Blacklow, 1985). AAV1, -2, and -3 are all common human isolates. Adeno-associated virus shedding has been found in individuals infected with Ad, but it has not been associated with disease, and has not been found to significantly alter concomitant Ad infection (Blacklow *et al.*, 1971). The prevalence of AAV seropositivity in the adult population has been noted to be as high as 85%.

Adeno-associated virus serotype 2 has been cloned (Samulski *et al.*, 1982; Laughlin *et al.*, 1983) and sequenced (Srivastava *et al.*, 1983), and its life cycle has been studied in detail (Carter, 1990; Berns, 1990; Carter *et al.*, 1990). The AAV genome (Fig. 2) consists of two genes: *rep*, which encodes functions required for replication, and *cap*, which encodes structural proteins for the virion (Fig. 2). The *rep* gene is transcribed from two transcription promoters: the p5 promoter and the internal p19 promoter. Each mRNA is present in both spliced and unspliced forms to allow for production of four Rep proteins designated Rep78, Rep68, Rep52, and Rep40 (Mendelson *et al.*, 1986). The p5 Rep proteins have DNA-binding, helicase, and nicking functions required for resolution of AAV termini (McCarty *et al.*, 1994a,b). These proteins also have transcriptional regulatory functions (Beaton *et al.*, 1989), and may play a role in suppression of tumorigenesis from coinfecting oncogenic viruses. Rep may also play a role in site-specific AAV integration (see below).

The AAB *cap* gene is transcribed from the p40 promoter. Splicing of the p40 transcript utilizes either one of two different splice acceptor sites.

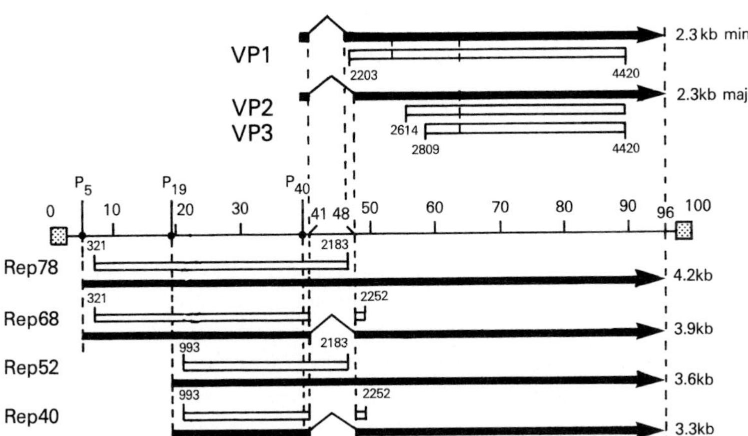

FIGURE 2 Organization of the AAV2 genome. The genome is shown as a single bar with a 100-map unit scale (1 map unit = 1% of genome size, approximately 47 bp). The ITRs are shown as stippled boxes. The transcription promoters (p_5, p_{19}, p_{40}) are shown as solid circles. The polyadenylation signal is at map position 96. RNA transcripts from AAV promoters are shown as heavy arrows with the introns indicated by carets. Protein-coding regions are shown as open boxes and the numbers indicate the locations of initiation and termination codons. The three capsid proteins are VP1, VP2, and VP3; the four Rep proteins are Rep78, Rep68, Rep52, and Rep40. [Redrawn with permission from Smuda and Carter, *Virology*, 1991.]

One version of these two p40 mRNAs can initiate translation from either of two start codons, so that a total of three different structural proteins are produced: VP1, VP2, and VP3. VP1, VP2 and VP3 have overlapping peptide sequences but differ only in the amount of amino-terminal sequence.

Flanking the two AAV genes are the inverted terminal repeats (ITRs), which serve as origins for DNA replication and as packaging signals in a productive life cycle. The ITRs have also been shown to have enhancer effects (Beaton *et al.*, 1989) and, when isolated from downstream sequences, to possess intrinsic transcriptional promoter activity (Flotte *et al.*, 1993b).

Productive infection with AAV generally requires a helper virus, such as adenovirus, vaccinia, or herpes simplex virus. Low-level replication can be supported by ultraviolet (UV) irradiation or by treating cells with genotoxic drugs. In the absence of helper virus infection, wild-type AAV persists in a latent state (Cheung *et al.*, 1980; Hoggan *et al.*, 1973), which most often includes integration of viral DNA into a specific region of chromosome 19, the AAVS1 site (Kotin *et al.*, 1990, 1991, 1992; Samulski *et al.*, 1991; Samulski, 1993). If cells with integrated wild-type AAV are later infected with helper virus, the AAV DNA will be rescued, i.e., excised, replicated, and packaged, as in a productive infection. Rep78 and Rep68 proteins are capable of binding to both the AAV ITR (McCarty *et al.*, 1994a,b) and the AAVS1 site (Weitzmann *et al.*, 1994), and so may play a role in directing site-specific integration (Giraud *et al.*, 1994).

B. Adeno-Associated Virus-Based Vectors

The first recombinant AAV vectors were constructed by deleting portions of the AAV coding sequence and substituting the transgene of interest (Fig. 3). (Tratschin *et al.*, 1984, 1985; Hermonat and Muzyczka, 1984). If the vector sequence inserted between the ITRs is within the packaging limit of approximately 4.7 kb, and *rep* and *cap* are expressed in adenovirus-infected cells, then the vector sequences can be packaged into infectious particles. This is now generally accomplished by deleting both *rep* and *cap* from the vector to allow for maximal insert size and to decrease the frequency of recombination events between the vector and the complementing *rep/cap* DNA, which could result in the production of wild-type AAV (Flotte *et al.*, 1995a). Samulski *et al.* (1989) produced *rep/cap*-deleted vectors that could be complemented by the packaging plasmid pAAV/Ad, whose sequences did not overlap with the vector. This system produced wild-type-free stocks of recombinant AAV vectors expressing the *neo* gene, which were relatively free of wild-type AAV, and that could be produced at high titers sufficient to transduce up to 70% of a target cell population. These findings, along with the nonpathogenic nature of AAV, and the natural tropism for the respiratory tract, helped to establish the rationale for pursu-

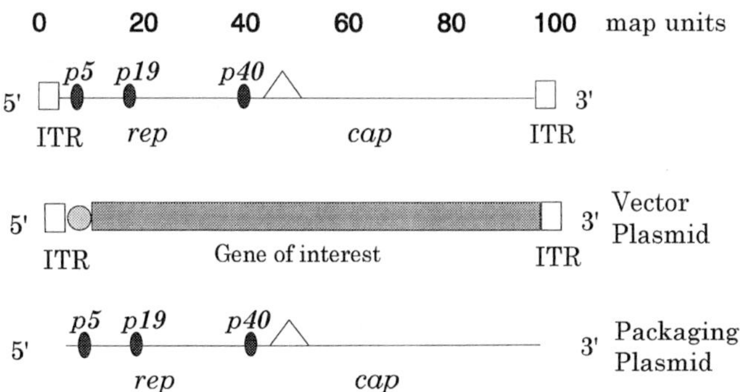

FIGURE 3   Structure of AAV-based vectors. *Top*: The AAV2 genome with a map unit scale is shown at the top, with the ITRs (open boxes) and the transcription promoters (p$_5$, p$_{19}$, p$_{40}$) as dark shaded circles. *Middle*: Vector plasmids are constructed by inserting the transgene of interest (lightly shaded bar) and a suitable promoter (lightly shaded circle) between the ITRs, which serve as replication origins and packaging signals. *Bottom*: For the vector DNA to be packaged into infectious AAV particles, it must be cotransfected with a packaging plasmid that lacks ITRs into adenovirus-infected cells.

ing studies of the capacity of AAV vectors to transduce airway epithelial cells and to complement the CF defect *in vitro* and *in vivo* (Flotte *et al.*, 1995b).

III. Adeno-Associated Virus Vector Transduction of Cystic Fibrosis Bronchial Epithelial Cells and Complementation of the Cystic Fibrosis Defect

A. Adeno-Associated Virus Vectors in Airway Epithelial Cells

In addition to the general challenges of CF gene therapy outlined previously, an additional obstacle to AAV-mediated CF gene therapy was posed by the size of the CFTR coding sequence (4.44 kb) relative to the packaging limit of wild-type AAV (4.7 kb). This made it necessary to select the smallest possible promoter that would be active in the target cell population. Initial studies to address this point examined the transcriptional activity of the AAV p5 promoter, which forms a convenient 263-nucleotide cassette when included with the left-hand ITR (Flotte *et al.*, 1992). The p5 promoter contains a number of upstream regulatory elements, including a bifunctional YY1 element, which can be upregulated by the AdE1 a gene product and downregulated by AAV Rep and by nuclear factors present in some cells.

To determine the activity of p5 in cell types affected in CF, p5–chloramphenicol acetyltransferase (CAT) and p5*neo* constructs were transfected into the CF bronchial epithelial cell line (IB3-1) and into the pancreatic adenocarcinoma cell line (CFPAC-l), using cationic liposomes. The CAT activity and geneticin sulfate (G418)-resistant (neo^+) colony formation were 5- to 10-fold higher than that observed in the same cell lines transfected with the pSV2CAT and pSV2neo constructs, in which the simian virus (SV40) early promoter was used to express the transgenes. Furthermore, p5 constructs that retained an upstream cAMP response element (CRE) had a level of CAT expression induced by twofold in the presence of 10 μM forskolin, while the CRE-deleted versions of p5 showed no induction.

Packaged AAVp5*neo* stocks were then tested in the IB3-1 cell line at multiplicities of infection ranging from 0 to 850 particles per cell. The expression of *neo* was assessed by G418 selection. In these experiments, a dose-related increase in the percentage of cells transduced was observed up to a maximum of 60 to 70%. This finding indicated that IB3-1 cells were permissive for AAV vector transduction, and that it was likely that a high proportion of cells could be transduced at a high multiplicity of infection.

B. Adeno-Associated Virus Vectors to Complement the Cystic Fibrosis Defect

Similar studies were then performed with AAV vectors expressing the CFTR cDNA from the AAV p5 promoter (Flotte *et al.*, 1993b). These studies indicated that liposome-mediated transfection with an AAV–p5–CFTR construct resulted in expression of CFTR mRNA. Complemented cells demonstrated a cAMP-activated component of chloride efflux on a radioisotope ($^{36}Cl^-$) tracer assay, a pattern characteristic of functional CFTR expression, while the parental IB3-1 cells showed no evidence of cAMP-mediated activation.

Unexpectedly, a construct in which only the AAV ITR was present upstream from the CFTR cDNA also showed evidence of CFTR protein expression and function. This indicated that the ITR alone was capable of functioning as a transcription promoter. Examination of the ITR sequence indicated that a consensus RNA initiator (*inr*) site was present in the "d" sequence of the ITR, and that several binding sites for the transcription factor Sp1 were present upstream. The presence of promoter activity in the ITR was confirmed using an AAV–ITR–CAT construct, which demonstrated promoter activity at a level approximately 40% of that seen with AAV–p5–CAT when assayed in an adenovirus-transformed human embryonic kidney cell line (293 cells).

Both AAV–p5–CFTR and AAV–ITR–CFTR constructs were successfully packaged into recombinant virions (Flotte *et al.*, 1993b). Packaged constructs were used to transduce IB3-1 cells in the absence of selective pressure. Immunologic analysis of IB3 cell clones picked randomly by serial dilution indicated that measurable levels of CFTR protein expression were present, along with functional correction of cAMP-mediated $^{36}Cl^-$-mediated efflux. The pooled, unselected population of transduced cells showed a similar pattern, with somewhat lower levels of protein expression and functional correction.

A more detailed characterization of the electrophysiologic properties of parental and complemented cells was performed using excised patch-clamp analysis (Egan *et al.*, 1992) and whole-cell current analysis (Schwiebert *et al.*, 1994). In both settings, complemented cells demonstrated restored cAMP-mediated regulation of both the small, linear chloride conductance characteristic of recombinant CFTR expression and the ORCC, which had previously been associated with the CF defect (Fig. 4). These studies suggested that the levels of CFTR expression produced with AAV–CFTR vectors may be sufficient to result in physiologic correction of the CF defect.

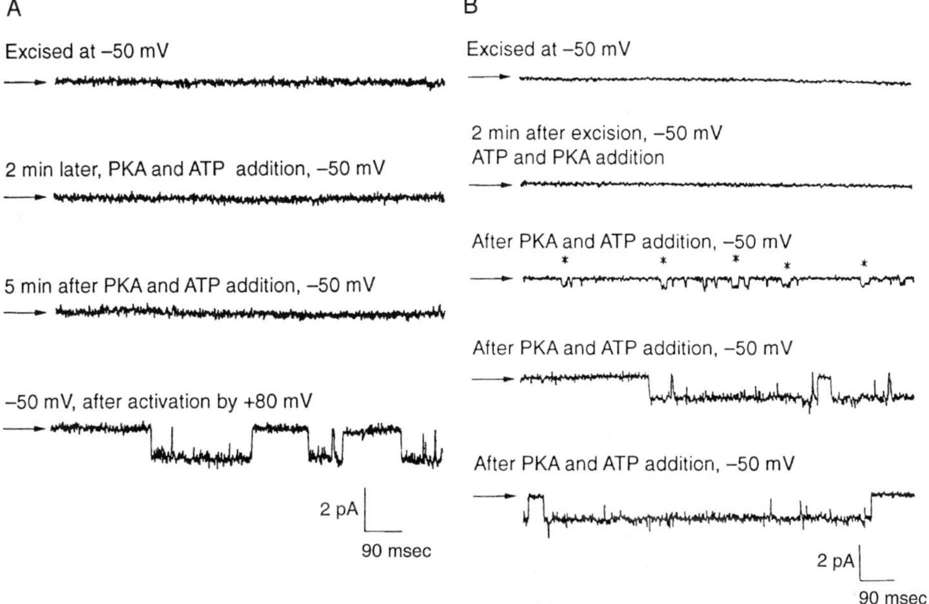

FIGURE 4　　Restoration of CFTR chloride channel activity and regulation of the outwardly rectifying chloride channel (ORCC) in CF bronchial epithelial cells complemented with AAV–CFTR. Excised patch-clamp recordings from the parental IB3-1 cell line (A) indicate the absence of any basal chloride channel activity or any channel opening in response to protein kinase A (PKA) and ATP addition. The fact that an ORCC is present in the patch is confirmed by channel opening in the bottom tracing, which was activated by a depolarizing (+80-mV) pulse. In (B), a tracing from complemented cells shows both the presence of PKA/ATP-activated CFTR channels [small deflections marked by asterisks (*) in third tracing] and increased activation of the ORCC (larger deflections in the fourth and fifth tracings). [Redrawn with permission from *Nature*, Egan *et al.* (1992). Defective regulation of outwardly rectifying Cl⁻ channels by protein kinase A corrected by insertion of CFTR. **358**, 581–584. Copyright 1992 MacMillan Magazines Limited.]

IV. *In Vivo* Gene Transfer with AAV–CFTR in the Bronchial Epithelium

A. The New Zealand White Rabbit Model

To determine whether AAV vectors would be functional in the *in vivo* setting, studies were performed in which doses of 1×10^{10} total particles of packaged AAV–ITR–CFTR were instilled through a fiber-optic bronchoscope into the right lower lobe (RLL) of New Zealand White rabbits (Flotte *et al.*, 1993a). Animals were sacrificed at timed intervals ranging from 3 days to 6 months after vector instillation. Vector DNA transfer was assessed by DNA *in situ* PCR amplifcation. RNA expression was determined by reverse transcriptase-PCR (RT-PCR) (Fig. 5), and protein expression was determined by immunoblotting and immunohistochemistry with antibodies directed against the R domain of the CFTR and against a unique vector-specific N-terminal epitope. These studies indicated that vector DNA had been transferred to approximately 50% of bronchial epithelial cells near the infusion site and that vector RNA and protein expression could be

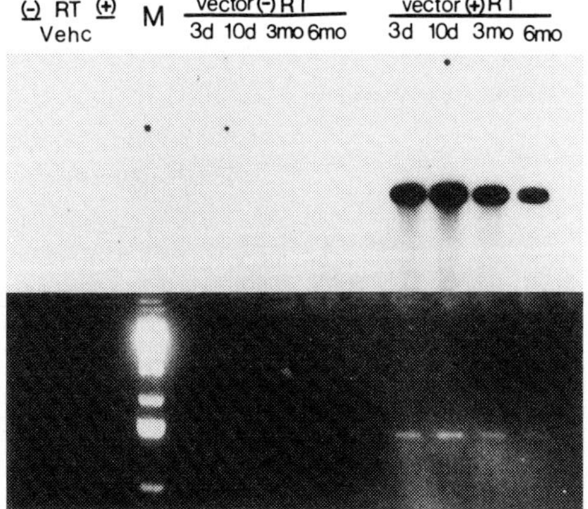

FIGURE 5   RT-PCR indicates CFTR mRNA expression is present in rabbit lungs for 6 months after endobronchial administration of AAV–CFTR. RT-PCR was performed on DNase-treated RNA extracts from lungs harvested at 3 days, 10 days, 3 months, and 6 months from rabbits treated with 1×10^{10} total particles of AAV–CFTR via fiber-optic bronchoscopy. PCR products were analyzed by ethidium bromide-stained agarose gel electrophoresis (*bottom*) and Southern blot analysis (*top*) with an internal CFTR-specific probe. The presence of signal corresponding to the expected 900-bp fragment (see comparison with 1-kb marker, lane M) indicates that vector RNA expression is present. The absence of signal in samples treated without reverse transcriptase (−RT) indicates that signals are due to RNA expression rather than vector DNA contamination. The absence of signal with the vehicle control (Vehc) lung RNA extract indicates the specificity of the human CFTR primers and probes. [Reproduced with permission from Flotte *et al.*, 1993a.]

detected for 6 months after vector instillation. There was no apparent toxicity from vector administration as judged by histopathology.

A more recent study of AAV–CFTR gene transfer in the rabbit airway has defined the dose-response relationship over a range of vector doses from 1×10^6 to 1×10^{10} total particles (Flotte *et al.*, 1995b). In that study, the cell-type specificity of AAV–CFTR gene transfer was determined by colocalization with cell surface markers specific for basal cells and ciliated cells within the bronchial epithelium. The threshold for detection of gene transfer to ciliated cells was approximately 1×10^8 total particles, while that for basal cells was 1×10^{10}. This difference may reflect a physical barrier to access to the basal cell population, since they reside deeper within the epithelium and are not in contact with the airway surface. The transduction of basal cells is potentially important since basal cells represent the progenitor cell population for the trachea and bronchi. Overall, these experiments provided important information about the safety, bioactivity, and dose response of AAV–CFTR in the airway.

B. A Nonhuman Primate Model

Another study sought to define the dose–response relationship for safety and biological activity of AAV–CFTR and its duration of expression in a nonhuman primate species, the rhesus macaque (Conrad *et al.*, 1996). In the macaque, AAV–CFTR bioactivity was examined using a DNA *in situ* PCR to detect DNA transfer and RT-PCR and RNase protection assays for vector RNA expression. These studies indicated that vector gene transfer and expression were detectable at doses of 10^8 or greater particles per animal, when delivered to a single lung segment by fiber-optic bronchoscopy. Gene transfer was stable over the 6-month study period.

Safety studies in this model included "clinically" meaningful variables such as pulmonary function testing (PFTs), chest X rays, and arterial blood gases. Bronchoalveolar lavage (BAL) studies were used to quantitate proinflammatory cytokines and cell counts as indicators of airway inflammation. Finally, histopathologic analyses were performed. All of these studies were normal. Interestingly, some vector DNA was detectable outside the lungs in several of the animals. Additional studies (T. R. Flotte and B. J. Carter, unpublished data) have been performed that indicate that some vector DNA can be detected in peripheral blood within the first 4 hr after vector administration, which may account for the observed dissemination of vector DNA. It is important to note that vector DNA levels in extrapulmonary organs, such as liver or kidney, were low and that there was no evidence of pathology at any of these sites. There were no detectable vector DNA sequences in the gonads.

The rhesus monkey has also been used as an *in vivo* model of AAV vector interaction with wild-type AAV and adenovirus (Afione *et al.*, 1996).

For these studies a host-range mutant (Ad2HR405) of a human adenovirus, capable of replicating efficiently in monkey cells, was used (Brough *et al.*, 1985). The interactions between AAV–CFTR, wild-type AAV2, and Ad2HR405 in the respiratory tract were studied in three different protocols. In the first protocol, animals were infected with AAV–CFTR in the right lower lobe (RLL) bronchus of the lung and 10 weeks later had wild-type AAV2 and Ad2HR405 infection established in the nose. In this case, wild-type AAV2 and Ad infection were established, but were limited to the upper respiratory tract, and the recombinant was not shed. In the second protocol, wild-type AAV2 infection was done first in the RLL, followed several weeks later by AAV–CFTR inoculation, and then several weeks after by nasal Ad infection. In this case, AAV–CFTR was shed at 3 days from the nose and at 6 days from the lung, but it was at low levels. In the third protocol, wild-type AAV2 infection was performed in the nose, followed by AAV–CFTR intranasally and then by Ad infection intranasally. Interestingly, AAV–CFTR could not be liberated from the nose, and the nasal fluid was found to have a marked neutralization effect on both wild-type AAV and AAV–CFTR *in vitro*. Overall, these data suggest that rescue of recombinant AAV–CFTR sequences is possible, but not likely to occur at a high frequency.

V. Effects of Cell Proliferation on Expression from Adeno-Associated Virus Vectors

One important issue regarding AAV is gene transfer agent for clinical CF gene therapy relates to the relative levels of mRNA and protein expression achievable from AAV vectors in primary cells or in slowly dividing cell populations. One study (Halbert *et al.*, 1995) of AAV gene transfer in respiratory epithelial cells found a substantial difference between the gene transfer efficiency in primary nasal polyp cells as compared with the immortalized IB3-1 cell line. In a related study performed in quiescent primary human skin fibroblasts, entry of ssDNA from AAV vectors was found to occur without substantial conversion to double-stranded DNA (dsDNA) (Russell *et al.*, 1994). Expression levels were also low in those cells. If cells were switched from nonproliferating conditions into a state of rapid turnover, a significant proportion of the cells would then be "recruited" to express the transgene. In a related study, Alexander *et al.* (1994) found that DNA-damaging agents such as hydroxyureas were able to enhance transduction of nondividing cells.

These findings are consistent with those of another study (Flotte *et al.*, 1994), which compared CF bronchial epithelial cells under conditions of either rapid or slow proliferation. In those studies, the dose–response curve for vector reporter gene expression showed that expression was less efficient in slowly proliferating cells. Interestingly, conversion of input vector DNA to double-stranded DNA episomal forms did eventually occur in the slowly proliferating bronchial epithelial cell population, but vector DNA integration could not be detected in these cells. This suggests that vector DNA can exist in multiple forms (single-stranded episomal, double-stranded integrated) in transduced cells, and that the cell proliferation rate could affect the relative abundance of these forms in certain cell types (Kearns *et al.,* 1996).

Examination of cells transduced *in vivo* in rhesus monkeys have shown similar conversion to dsDNA episomal forms (Afione *et al.,* 1996). Additional studies of these cells using Southern blot and fluorescent *in situ* hybridization (FISH) analysis have not shown evidence of vector DNA integration. Both vector DNA presence and vector RNA expression were found to persist in the lungs of these monkeys for 6-months, suggesting that these episomal dsDNA forms are capable of persistence and expression over a period of time.

VI. Clinical Trials of AAV–CFTR Gene Transfer in Humans —

At present two clinical trials of AAV–CFTR gene transfer have been approved by the National Institutes of Health (NIH, Bethesda, MD) Recombinant DNA Advisory Committee. The first of these trials will consist of combined intranasal and single-lobe endobronchial administration (via bronchoscopy) in a phase I dose-escalation protocol designed to test the safety and bioactivity of AAV–CFTR. Nasal brushings will be used to test for the presence of vector DNA and RNA in the nasal epithelium, and nasal transepithelial potential difference measurements will be used to examine CFTR function. The primary safety outcomes will relate to the status of pulmonary inflammation (as judged by BAL fluid examination, pulmonary function tests, and other clinical variables). The second study involves direct instillation of AAV–CFTR vector into the maxillary sinuses of CF patients. This study will allow for detection of bioactivity and safety of vector delivered in the relatively isolated environment of a paranasal sinus. One additional feature of the study is the potential to directly measure indicators of inflammation in the sinus lavage fluid and by computed tomography (CT) scanning, which could give an indication of a clinical improvement in sinus disease.

Several additional studies will be required to move from localized administration to more generalized delivery to the entire lower respiratory tract. The eventual goal of these studies will be to find a vector construct, which at a specific dose and given by a specific route of administration, will be safe and have long-term effects on CFTR function beginning at an age prior to onset of irreversible lung damage.

References

Afione, S. A., Conrad, C. K., Adams, R., Reynolds, T. C., Guggino, W. B., Carter, B. J., and Flotte, T. R. (1996). *In vivo* model of adeno-associated virus vector persistence and rescue. *J. Virol.* **70,** 3235–3241.

Afione, S. A., Conrad, C. K., and Flotte, T. R. (1995). Gene therapy vectors as drug delivery systems. *Clin. Pharmacokinet.* **28**, 181–189.

Alexander, I. E., Russell, D. W., and Miller, A. D. (1994). DNA damaging agents greatly increase the transduction of non-dividing cell by adeno-associated virus vectors. *J. Virol.* **68**, 8282–8287.

Beaton, A., Palumbo, P., and Berns, K. I. (1989). Expression from the adeno-associated virus p5 and p19 promoters is negatively regulated *in trans* by the *rep* protein. *J. Virol.* **63**, 4450–4454.

Berns, K. I. (1990). Parvoviridae and their replication. *In* "Virology" (B. N. Fields, D. M. Knipe, R. M. Chanock, J. L. Melnick, M. S. Hirsch, T. P. Monath, and B. Roizman, eds.), pp. 1743–1764. Raven Press, New York.

Blacklow, N. R. (1985). Adeno-associated viruses of humans. *In* "Parvoviruses and Human Disease" (J. Pattison, ed.), pp. 165–174. CRC Press, Boca Raton, Florida.

Blacklow, N. R., Hoggan, M. D., Kapikian, A. Z., Austin, J. B., and Rowe, W. P. (1971). Epidemiology of adeno-associated virus infection in a nursery population. *Am. J. Epidemiol.* **94**, 359–366.

Boat, T. F., Welsh, M. J., and Beaudet, A. L. (1989). Cystic fibrosis. *In* "The Metabolic Basis of Inherited Disease" (C. L. Scriver, A. L. Beaudet, W. S. Sly, and D. Valle, eds.), 6th Ed., pp. 2649–2680. McGraw-Hill, New York.

Bradbury, N. A., Jilling, T., Berta, G., Sorscher, E. J., Bridges, R. J., and Kirk, K. L. (1992). Regulation of plasma membrane recycling by CFTR. *Science* **256**, 530–532.

Brough, D. E., Rice, S. A., Sell, S., and Klessig, D. F. (1985). Restricted changes in the adenovirus DNA-binding protein that lead to extended host range or temperature-sensitive phenotypes. *J. Virol.* **55**, 206–212.

Caplen, N. J., Alton, E. W. F. W., Middleton, P. G., Dorin, J. R., Stevenson, B. J., Gao, X., Durham, S. R., Jeffery, P. K., Hodson, M. D., Coutelle, C., Huang, L., Porteous, D. J., Williamson, R., and Geddes, D. M. (1995). Liposome-mediated CFTR gene transfer to the nasal epithelium of patients with cystic fibrosis. *Nat. Med.* **1**, 39–46.

Carter, B. J. (1990). The growth cycle of adeno-associated virus. *In* "Handbook of Parvoviruses" (P. Tjissen, ed.), Vol. 1, pp. 155–168. CRC Press, Boca Raton, Florida.

Carter, B. J. (1992). Adeno-associated virus vectors. *Curr. Opin. Biotechnol.* **3**, 533–539.

Carter, B. J., Marcus-Sekura, C. J., Laughlin, C. A., and Ketner, G. (1983). Properties of an adenovirus type 2 mutant Ad2d1807 having a deletion near the right-hand genome terminus: Failure to help AAV replication. *Virology* **126**, 505–516.

Carter, B. J., Mendelson, E., and Trempe, J. P. (1990). AAV DNA replication, integration, and genetics. *In* "Handbook of Parvoviruses" (P. Tjissen, ed.), Vol. 1, pp. 169–226. CRC Press, Boca Raton, Florida.

Cheung, A. M. K., Hoggan, M. D., Hauswirth, W. W., and Berns, D. I. (1980). Integration of the adeno-associated virus genome into cellular DNA in latently infected human Detroit 6 cells. *J. Virol.* **33**, 739–748.

Conrad, C. K., Allen, S., Afione, S. A., Reynolds, T. C., Barrazza-Ortiz, X., Fee-Maki, M., Adams, R., Askin, F., Carter, B. J., Guggino, W. B., and Flotte, T. R. (1996). Safety of single dose administration of an adeno-associated virus (AAV)–CFTR vector in the primate lung. *Gene Therapy* **3**, 658–668.

Crystal, R. G., McElvaney, N. G., Rosenfeld, M. A., Chu, C.-S., Mastrangeli, A. A., Jogn, G. H., Brody, S. L., Jaffe, H. A., Eissa, N. T., and Danel, C. (1994). Administration of an adenovirus containing the human CFTR cDNA to the repiratory tract of individuals with cystic fibrosis. *Nat. Genet.* **8**, 42–51.

di Sant'Agnese, P. A., Darling, R. C., Perea, G. A., *et al.* (1953). Abnormal electrolyte composition of sweat in cystic fibrosis of the pancreas. Clinical significance and relationship to disease. *Pediatrics* **12**, 549–563.

Drumm, M. L., Pope, H. A., Cliff, W. H., Rommens, J. M., Sheila, S. A., Tsui, L.-C., Collins, F. S., Frizzell, R. A., and Wilson, J. M. (1990). Correction of the cystic fibrosis defect *in vitro* by retrovirus-mediated gene transfer. *Cell (Cambridge, Mass.)* **62**, 1227–1233.

Egan, M., Flotte, T., Afione, S., Solow, R., Zeitlin, P. L., Carter, B. J., and Guggino, W. B. (1992). Correction of defective PKA regulation of outwardly rectifying chloride channels by insertion of cystic fibrosis transmembrane conductance regulator into CF airway epithelial cells. *Nature (London)* **358**, 581–584.

Engelhardt, J. F., Yankaskas, J. R., Ernst, S. A., Yang, Y., Marino, C. R., Boucher, R. C., Cohn, J. A., and Wilson, J. M. (1992). Submucosal glands are the predominant site of CFTR expression in the human bronchus. *Nat. Genet.* **2**, 240–248.

Engelhardt, J. F., Simon, R. H., Yang, Y., Zepeda, M., Weber-Pendelton, S., Doranz, B., Grossman, M., and Wilson, J. M. (1993). Adenovirus-mediated transfer of the CFTR gene to lung of nonhuman primates: Biological efficacy study. *Hum. Gene Ther.* **4**, 759–769.

Fitzsimmons, S. C. (1994). "CF Foundation Patient Registry," 1993 Annu. Data Rep.

Flotte, T. R., Solow, R., Owens, R. A., Afione, S., Zeitlin, P. L., and Carter, B. J. (1992). Gene expression from adeno-associated virus vectors in airway epithelial cells. *Am. J. Respir. Cell Mol. Biol.* **7**, 349–356.

Flotte, T. R., Afione, S. A., Conrad, C. K., McGrath, S. A., Solow, R., Oka, H., Zeitlin, P. L., Guggino, W. B., and Carter, B. J. (1993a). Stable *in vivo* expression of the cystic fibrosis transmembrane conductance regulator with an adeno-associated virus vector. *Proc. Natl. Acad. Sci. U.S.A.* **90**, 10613–10617.

Flotte, T. R., Afione, S. A., Solow, R., Drumm, M. L., Markakis, D., Guggino, W. B., Zeitlin, P. L., and Carter, B. J. (1993b). Expression of the cystic fibrosis transmembrane conductance regulator from a novel adeno-associated virus promoter. *J. Biol. Chem.* **268**, 3781–3790.

Flotte, T. R., Afione, S. A., and Zeitlin, P. L. (1994). Adeno-associated virus vector gene expression occurs in non-dividing cells in the absence of vector DNA integration. *Am. J. Respir. Cell. Mol. Biol.* **11**, 517–521,

Flotte, T. R., Barraza-Ortiz, X., Solow, R., Afione, S. A., Carter, B. J., and Guggino, W. B. (1995a). An improved system for packaging recombinant adeno-associated virus vectors capable of *in vivo* transduction. *Gene Ther.* **2**, 29–37.

Flotte, T. R., Conrad, C., Reynolds, T., Afione, S., Adams, R., Allen, S., Guggino, W. B., and Carter, B. J. (1995b). Preclinical evaluation of AAV vectors expressing the human CFTR cDNA. *J. Cell. Biochem.* **21A**, 364.

Giraud, C., Winocour, E., and Berns, K. I. (1994). Site-specific integration by adeno-associated virus is directed by a cellular DNA sequence. *Proc. Natl. Acad. Sci. U.S.A.* **91**, 10039–10043.

Halbert, C. L., Alexander, I. E., Wolgamot, G. M., and Miller, A. D. (1995). Adeno-associated virus vectors transduce primary cells much less efficiently than immortalized cells. *J. Virol.* **69**, 1473–1479.

Hermonat, P. L., and Muzyczka, N. (1984). Use of adeno-associated virus as a mammalian DNA cloning vector: Transduction of neomycin resistance into mammalian tissue culture cells. *Proc. Natl. Acad. Sci. U.S.A.* **81**, 6466–6470.

Hoggan, M. D., Thomas, G. F., and Johnson, F. B. (1973). Continuous carriage of adeno-associated virus genome in cell culture in the absence of helper adenovirus. *In Possible Episomes Eukaryotes, Proc. Lepetit Colloq. 4th,* Cocoyac, Mexcio, *1972,* pp. 243–253.

Imundo, L., Barasch, J., Prince, A., and Al-Awqati, Q. (1995). Cystic fibrosis epithelial cells have a receptor for pathogenic bacteria on their apical surface. *Proc. Natl. Acad. Sci. USA* **92**, 3019–3023.

Kearns, W. G., Afione, S. A., Fulmer, S. B., Caruso, J., Flotte, T. R., and Cutting, G. R. (1996). Recombinant adeno-associated virus (AAV) vectors do not integrate in a site-specific fashion but persist in episomal form. *Gene Therapy* **3**, 748–755.

Konstan, M. W., Hilliard, K. A., Norvell, T. M., and Berger, M. (1994). Bronchoalveolar lavage findings in cystic fibrosis patients with stable, clinically mild lung disease suggest ongoing infection and inflammation. *Am. J. Respir. Crit. Care Med.* **150**, 448–454.

Kotin, R. M., Siniscalco, M., Samulski, R. J., Zhu, X., Hunter, L., Laughlin, C. A., McLaughlin, S., Muzyczka, N., Rocchi, M., and Berns, K. I. (1990). Site-specific integration by adeno-associated virus. *Proc. Natl. Acad. Sci. U.S.A.* **87,** 2210–2215.

Kotin, R. M., Menninger, J. C., Ward, D. C., and Berns, K. I. (1991). Mapping and direct visualization of a region-specific viral DNA integration site on chromosome 19q13-qter. *Genomics* **10,** 831–834.

Kotin, R. M., Linden, R. M., and Berns, K. I. (1992). Characterization of a preferred site on chromosome 19q for integration of adeno-associated virus DNA by nonhomologous recombination. *EMBO J.* **11,** 5071–5076.

Kronberg, G., Hansen, M. B., Svenson, M., Fomsgaard, A., Hoiby, N., and Bendtzen, K. (1993). Cytokines in sputum and serum from patients with cystic fibrosis and chronic *Pseudomonas aeruginosa* infection as markers of destructive inflammation in the lungs. *Pediatr. Pulmonol.* **15,** 292–297.

Laughlin, C. A., Tratschin, J.-D., Coon, H., and Carter, B. J. (1983). Cloning of infectious adeno-associated virus genomes in bacterial plasmids. *Gene* **23,** 65–73.

Lukacs, G. L., Chang, X. B., Kartner, N., Rotstein, O. D., Riordan, J. R., and Grinstein, S. (1992). The cystic fibrosis transmembrane regulator is present and functional in endosomes. Role as a determinant of endosomal pH. *J. Biol. Chem.* **267,** 14568–14572.

McCarty, D. M., Pereira, D. J., Zolotkhin, I., Zhou, X., Ryan, J. H., and Muzyczka, N. (1994a). Identification of linear DNA sequences that specifically bind the adeno-associated virus Rep protein. *J. Virol.* **68,** 4988–4997.

McCarty, D. M., Ryan, J. H., Zolotkhin, S., Zhou, X., Muzyczka, N. (1994b). Interaction of the adeno-associated virus Rep protein with a sequence within the A palindrome of the viral terminal repeat. *J. Virol.* **68,** 4998–5006.

Mendelson, E., Trempe, J. P., and Carter, B. J. (1986). Identification of the trans-acting rep proteins of adeno-associated virus using antibodies to a synthetic polypeptide. *J. Virol.* **60,** 823–832.

Miller, D. G., Adam, M. A., and Miller, A. D. (1990). Gene transfer by retrovirus vectors occurs only in cells that are actively replicating at the time of infection. *Mol. Cell. Biol.* **10,** 4239–4242.

Quinton, P. M., and Bijman, J. (1983). Higher bioelectric potentials due to decreased chloride absorption in the sweat glands of patients with cystic fibrosis. *N. Engl. J. Med.* **308,** 1185–1189.

Riordan, J. R., Rommens, J. M., Kerem, B. S., Alon, N., Rozmahel, R., Grzelczak, Z., Zielenski, J., Lok, S., Plavsic, N., Chou, J.-L., Drumm, M. L., Iannuzzi, M. C., Collons, F. S., and Tsui, L.-C. (1989). Identification of the cystic fibrosis gene: Cloning and characterization of the complementary DNA. *Science* **245,** 1066–1073.

Rosenfeld, M. A., Siegfried, W., Yoshimura, K., Yoneyama, K., Fukayama, M., Stier, L. E., Paakko, P. K., Gilardi, P., Stratford-Perricaudet, L. D., Perricaudet, M., Jaalat, S., Pavirani, A., Lecocq, J.-P., and Crystal, R. G. (1991). Adenovirus-mediated transfer a recombinant alpha 1-antitrypsin gene to the lung epithelium *in vivo. Science* **252,** 431–434.

Rosenfeld, M. A., Yoshimura, K., Trapnell, B. C., Yoneyama, K., Rosenthal, E. R., Dalemans, W., Fukayama, M., Bargon, J., Stier, L. E., Stratford-Perricaudet, L., Perricaudet, M., Guggino, W. B., Lecocq, J.-P., and Crystal, R. G. (1992). *In vivo* transfer of the human cystic fibrosis transmembrane conductance regulator gene to the airway epithelium. *Cell (Cambridge, Mass.)* **68,** 143–155.

Russell, D. W., Miller, A. D., and Alexander, I. E. (1994). Adeno-associated virus vectors preferentially transduce cells in S phase. *Proc. Natl. Acad. Sci. U.S.A.* **91,** 8915–8919.

Samulski, R. J. (1993). Adeno-associated virus: Integration at a specific chromosomal locus. *Curr. Opin. Biotechnol.* **3,** 74–80.

Samulski, R. J., Berns, K. I., Tan, M., and Muzyczka, N. (1982). Cloning of adeno-associated virus into pBR322: Rescue of intact virus from the recombinant plasmid in human cells. *Proc. Natl. Acad. Sci. U.S.A.* **79,** 2077–2081.

Samulski, R. J., Chang, L.-S., and Shenk, T. (1989). Helper-free stocks of recombinant adeno-associated viruses: Normal integration does not require viral gene expression. *J. Virol.* **63**, 3822–3828.

Samulski, R. J., Zhu, X., Xiao, X., Brook, J. D., Housman, D. E., Epstein, N., and Hunter, L. A. (1991). Targeted integration of adeno-associated virus (AAV) into human chromosome 19. *EMBO. J.* **10**, 3941–3950.

Schwiebert, E., Flotte, T. R., Cutting, G., and Guggino, W. B. (1994). CFTR and outwardly rectifying chloride channels contribute to whole cell chloride currents in normal airway epithelial cells are defectively regulated in cystic fibrosis. *Am. J. Physiol.* **266**, (*Cell Physiol.* **35**), C1464–C1477.

Simon, R. H., Engelhardt, J. F., Yang, Y., Zepeda, M., Weber-Pendleton, S., Grossman, M., and Wilson, J. M. (1993). Adenovirus-mediated transfer of the CFTR gene to lung of nonhuman primates: Toxicity study. *Hum. Gene Ther.* **4**, 771–780.

Smuda, J. W., and Carter, B. J. (1991). Adeno-associated viruses having nonsense mutations in the capsid genes: Growth in mammalian cells containing an inducible amber suppressor. *Virology* **184**, 310–318.

Srivastava, A., Lusby, E. W., and Berns, K. I. (1983). Nucleotide sequence and organization of the adeno-associated virus 2 genome. *J. Virol.* **45**, 555–564.

Tratschin, J.-D., West, M. H. P., Sandbank, R., and Carter, B. J. (1984). A human parvovirus, adeno-associated virus, as a eukaryotic vector: Transient expression and encapsidation of the prokaryotic gene for chloramphenicol acetyltransferase. *Mol. Cell. Biol.* **4**, 2072–2087.

Tratschin, J.-D., Miller, I. L., Smith, M. G., and Carter, B. J. (1985). Adeno-associated virus vector for high-frequency integration, expression and rescue of genes in mammalian cells. *Mol. Cell. Biol.* **5**, 3251–3260.

Weitzmann, M. D., Kyöstiö, S. R. M., Kotin, R. M., and Owens, R. A. (1994). Adeno-associated virus (AAV) Rep proteins mediate complex formation between AAV DNA and its integration site in human DNA. *Proc. Natl. Acad. Sci. U.S.A.* **91**, 5808–5812.

Welsh, M. J., and Smith, A. E. (1993). Molecular mechanisms of CFTR chloride channel dysfunction in cystic fibrosis. *Cell (Cambridge, Mass.)* **73**, 1251–1254.

Yang, Y., Nunes, F. A., Berensci, K., Furht, E. F., Gönzcö, E., and Wilson, J. M. (1994). Cellular immunity to viral antigens limits E1-deleted adenoviruses for gene therapy. *Proc. Natl. Acad. Sci. U.S.A.* **91**, 4407–4411.

Zabner, L., Couture, L. A., Gregory, R. J., Graham, S. M., Smith, A. E., and Welsh, M. J. (1993). Adenovirus-mediated gene transfer transiently corrects the chloride transport defect in nasal epithelia of patients with cystic fibrosis. *Cell (Cambridge, Mass.)* **75**, 207–216.

Joseph C. Glorioso*
William F. Goins*
Martin C. Schmidt*
Thomas Oligino*
David M. Krisky*
Peggy C. Marconi*
James D. Cavalcoli*
Ramesh Ramakrishnan*
P. Luigi Poliani†
David. J. Fink*,†

*Department of Molecular Genetics and Biochemistry
University of Pittsburgh School of Medicine
Pittsburgh, Pennsylvania 15261

†Department of Neurology
University of Pittsburgh School of Medicine
Pittsburgh, Pennsylvania 15261

Engineering Herpes Simplex Virus Vectors for Human Gene Therapy

I. Introduction

The rapid and expanding knowledge of the molecular genetic and biochemical basis of human disease coupled with major advances in technologies related to gene manipulation has made possible a new era in genetic medicine. Genetic medicine includes new techniques for molecular diagnosis that allow the identification of presymptomatic individuals, the prediction of risk susceptibility, and the opportunity to treat established or developing disease by gene therapy.

Gene therapy is a burgeoning field still in its early stages. The basic principle is that a gene vector can be used to introduce and locally express a therapeutic product to ameliorate a disease process. In the few years since the first gene therapy protocol was initiated in 1990 at the National Institutes of Health (NIH, Bethesda, MD) by French Anderson, Mike Blaese, and Steve Rosenberg, a large number of human trials to treat a variety of different

Advances in Pharmacology, Volume 40

diseases by gene transfer have been undertaken around the world. Despite the interest and enthusiasm generated by the promise of gene therapy, there are still major hurdles to overcome to make this form of therapy practical and broadly useful to patients. Indeed, real success in the treatment of any disease by gene transfer has yet to be realized.

The central impediment to successful gene therapy is the development of effective gene delivery systems for specific tissues. The gene must be introduced into the appropriate cell type, transducing a sufficient number of cells to have a therapeutic effect. In addition to the problem of delivery, therapeutic efficacy will require expression of the transgene to an appropriate level and for a specific duration. This feature requires that promoter systems be devised that are not repressed and can respond to appropriate regulatory signals. The most efficient delivery systems for *in vivo* gene therapy employ recombinant viral vectors, which are by nature both pathogenic and immunogenic. Often expression of the transgene itself may induce an immune response in the context of the vector, even in the face of preexisting tolerance to the transgene product. This problem necessitates engineering the viral delivery system in a manner to render the vector less cytotoxic and to prevent immune-mediated destruction of transduced cells or inactivation of the therapeutic protein. Finally, issues related to vector safety and manufacture are substantial in themselves, pressing the bounds of pharmaceutical manufacturing technology and regulation. What have we learned so far, and what are the next steps required to develop effective gene therapy?

Gene delivery in human clinical trials has been carried out using two basic strategies. *Ex vivo* gene transfer employs recombinant retroviral vectors to insert the transgene into the host chromosomes of dividing cells explanted from the patient; these cells are then surgically reintroduced to deliver the transgene product. Experimental applications of *ex vivo* therapy include the delivery of the transgene to bone marrow stem cells, differentiated macrophages and lymphocytes derived from peripheral blood, fibroblasts derived from skin biopsy, hepatocytes, synoviocytes, or tumor cells derived from accessible malignancies. *Ex vivo* strategies have a number of inherent advantages: (1) the target cells can be grown as a largely homogeneous population, infected under controlled conditions in which the transduced cells can be cloned or enriched by growth in the presence of a selection medium; (2) the transduced cells can be pretested for transgene expression prior to transplantation; and (3) the potential for the development of antiviral immunity is eliminated. Despite these advantages, *ex vivo* gene therapy is limited to applications in which cell transplantation is feasible, and is relatively expensive and cumbersome since each treatment requires that cells be obtained from the individual patient, cultured, and transduced prior to reimplantation.

In vivo approaches, in which a vector is employed to deliver the therapeutic gene directly into cells in the host, represents a second strategy with obvious advantages. However, the development of an effective vector for gene delivery *in vivo* has proven to be more complex than the development of vectors for *ex vivo* transduction. Vectors for *in vivo* gene transfer must be capable of transducing a large number of nondividing cells *in situ*. The vectors must be constructed so as to be noncytotoxic, and in addition must induce minimal inflammation or antiviral immunity, which could limit expression or repeat dosing with the vector. The vector also should be stable, free of recombinant wild-type virus, and ideally constructed so as not to produce any viral gene products after infection *in vivo*. Theoretically, targeting of these vectors to specific cell types may be achieved by virus attachment to its cognate receptor, or the engineering of recombinant vectors that bind to specific cellular receptors. It should also be possible to achieve tissue-specific expression through the use of selected cellular promoter and/or enhancer elements. These are demanding criteria that are not currently met by any vector for *in vivo* gene delivery, but rather define the goals and challenges for which creative solutions may eventually be found.

Many viruses that infect humans have evolved through natural selection to possess features that may be exploited for vector design. Detailed understanding of the molecular mechanisms through which viruses achieve these ends can be used to disarm these viruses by deleting genes that contribute to viral pathogenicity while at the same time taking advantage of the natural biology of the specific virus system to carry out stable gene transfer to the desired tissue. Lentiviruses, herpesviruses, parvoviruses, adeno-associated viruses, coxsackie viruses, and the hepatitis viruses all replicate in host tissue before reaching their ultimate target, where they may persist for the life of the individual without triggering elimination by the host immune response. These viruses frequently persist in a particular cell type and vary widely in their associated pathogenesis within the specific host. Many of these viruses also continue to express genes that allow them to persist, in contrast to lytic viruses, which cause diseases but do not persist in the body or establish latency, and that therefore may be less amenable to gene transfer applications where long-term gene persistence and gene expression are required. The challenge is to engineer viruses that cannot replicate but nonetheless are capable of reaching the target tissue without causing disease or reactivating at a later time with subsequent spread to other individuals.

The natural biology of herpes simplex virus type 1 (HSV-1) makes it attractive as a gene delivery vehicle for the central and peripheral nervous systems, and experimental studies suggest it may be useful for the direct delivery of transgenes to other tissues as well. HSV-1 naturally replicates in skin or mucosal membranes to high titers without producing viremia (Fig. 1). These surface tissues are the natural portal of entry into the body where infection is initiated by direct contact. Replication in epithelial cells precedes transmission to the nervous system, although in animal studies mutant viruses can be delivered by injection into the projection field of

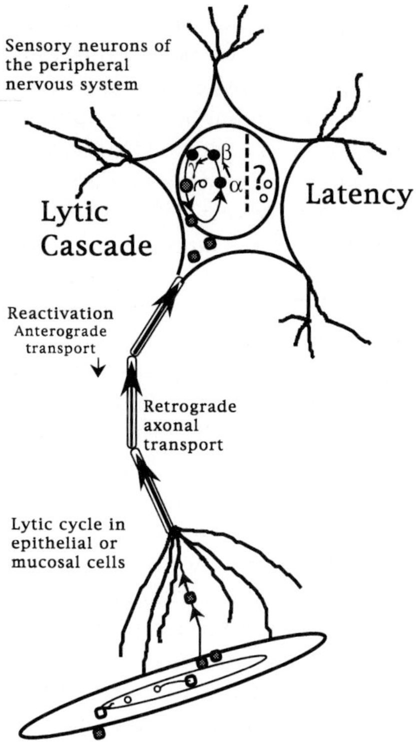

FIGURE I   HSV-1 life cycle. Schematic representation of the HSV-1 life cycle in neurons following retrograde transport of the viral nucleocapsid to the nerve cell body, wherein the virus can either replicate or enter latency.

peripheral neurons, or by direct injection into the brain (Bak *et al.*, 1977; Fink *et al.*, 1992; McFarland *et al.*, 1986). Following epithelial infection, the virus invades the nervous system through the axon terminals of sensory ganglia and by retrograde transport ultimately reaches the cytoskeletal framework of the nerve cell body in a manner that protects the genome from degradation by enzymes in the cytoplasm. The virus DNA is then injected through a modified penton (Newcomb and Brown, 1994) into the nucleus through a nuclear pore, along with several viral proteins that assist activation of the viral gene expression program. Despite the fact that HSV-1 can express its replication functions in neurons, the replication cycle is characteristically aborted in favor of the establishment of latency, a state in which the viral genome is almost entirely quiescent except for the expression of latency-associated transcripts (LATs) in the neuronal cell nucleus (Croen *et al.*, 1987; Deatly *et al.*, 1988; Gordon *et al.*, 1988; Rock *et al.*, 1987; Spivack and Fraser, 1987; Stevens *et al.*, 1987). The major LAT is a 2-kb RNA molecule that is a noncapped, nonpolyadenylated intron spliced from a large 8.3-kb poly(A)$^+$ RNA containing 16 potential open reading frames (ORFs) for which a protein product has not been found in latently infected neurons (Devi-Rao *et al.*, 1991; Dobson *et al.*, 1989; Doerig *et al.*, 1991; Farrell *et al.*, 1991). There are also less abundant LATs expressed that may be spliced variants of the 2-kb LAT (Spivack and Fraser, 1988; Spivack *et al.*, 1991; Wagner *et al.*, 1988). The LATs are expressed in virtually all latently infected neuronal cells. A fraction of latently infected neurons express LATs detectable by *in situ* hybridization (Coen *et al.*, 1989; Jacobson *et al.*, 1989; Ramakrishnan *et al.*, 1994a; Sauer *et al.*, 1987; Sawtell and Thompson, 1992), but all latent genomes express low levels of LATs detectable by the more sensitive *in situ* reverse transcriptase-polymerase chain reaction (RT-PCR) methods (Ramakrishnan *et al.*, 1994b). The presence of LATs is not required for the establishment or maintenance of viral latency (Fareed and Spivack, 1994; Ho and Mocarski, 1989; Javier *et al.*, 1988; Perng *et al.*, 1996; Sedarati *et al.*, 1989; Steiner *et al.*, 1989), although in some animal models there are reports of effects on reactivation kinetics and abundance of reactivated virus (Bloom *et al.*, 1996; Hill *et al.*, 1990; Leib *et al.*, 1991). Thus the LAT gene can be replaced by a foreign gene and the promoter sequences responsible for LAT expression used in vector systems to drive expression of the foreign gene during viral latency. The question is, can the pathogenesis of this virus be tamed, yet still retain the features of latency with expression of a therapeutic gene rather than the LATs? Indications are that the answer is yes. Moreover, there are some surprising new possibilities for engineering HSV vectors for gene transfer that are described below.

In this chapter, aspects of the structure and molecular biology of HSV-1 relevant to engineering HSV vectors for the nervous system and other tissues are discussed and some findings from our laboratories are described as examples of our attempts to exploit this virus for human gene therapy applications. In particular, problems related to vector behavior and design are described, including methods for the removal of cytotoxic genes and the production of complementing cell lines; methods to reduce vector immunogenicity, alter the host range, and provide added space for incorporation of transgenes; and finally strategies to express transgenes in different tissue types.

II. Structure of the Herpes Simplex Virus Particle, Its Genome Organization, and Its Lytic Replication Cycle

A. Virus Structure

The infectious virus particle is 110 nm in diameter (Roizman and Furlong, 1974); it is composed of an icosahedral nucleocapsid surrounded by a protein matrix, the tegument, which in turn is surrounded by a glycolipid-containing envelope (Fig. 2). The trilaminar membrane envelope is acquired during budding from the inner lamellae of the nuclear membrane and the particle passes through the Golgi in vesicles through a process probably involving membrane exchange (Campadelli *et al.*, 1993; Johnson and Spear, 1982). The envelope contains at least 11 glycoproteins that play an essential role in the adsorption and penetration of the host cell (Spear, 1993). The adsorption process involves initial binding to heparan sulfate moieties on the cell surface, primarily by glycoproteins B (gB) and C (gC) (Herold *et al.*, 1991; Shieh and Spear, 1994; Wudunn and Spear, 1989), followed by an apparently higher affinity binding to a second receptor recognized by glycoprotein D (gD) (Johnson *et al.*, 1990; Ligas and Johnson, 1988). The virus penetrates the cell by fusion of the virus envelope with the cell surface membrane and requires the presence of gD, gB, and the gH/gL complex (Cai *et al.*, 1987; Desai *et al.*, 1988; Hutchinson *et al.*, 1992; Ligas and Johnson, 1988; Roop *et al.*, 1993). Fusion may be mediated by gB although it is still not certain (Cai *et al.*, 1988; Navarro *et al.*, 1992). The deenveloped particle enters the cytoplasm, where it is guided to the nucleus. The viral DNA subsequently enters the nucleus to begin the productive replication cycle.

B. Genome Structure and Organization of Genes

The viral genome is packaged as a toroidal-shaped structure condensed by spermine and spermidine molecules (Furlong *et al.*, 1972; Gibson and Roizman, 1971). The genome is a linear double-stranded DNA molecule 152 kb in length and contains two unique segments (U_L and U_S), each flanked by inverted repeat (IR) components (Fig. 2). Of the 81 known genes, 38 are essential for production of infectious virus particles in cell culture while the remaining 43 genes are not essential for replication *in vitro* but contribute to the virus life cycle *in vivo*. Nonessential genes can be individually deleted from the viral genome without preventing virus replication under permissive tissue culture conditions used for culturing virus. Some deletion mutants grow less vigorously than wild-type virus and the removal of multiple genes can significantly impair replication. In general, the accessory genes contribute to the virus host range, increase pathogenesis, help the virus-infected cell elude immune surveillance, increase virus growth in nondividing cells such as neurons, and assist in the establishment, maintenance, or reactivation from latency.

C. The Virus Lytic Cycle

Following entry into the cell nucleus, the viral genome circularizes and the cascade of viral gene expression (Honess and Roizman, 1974)

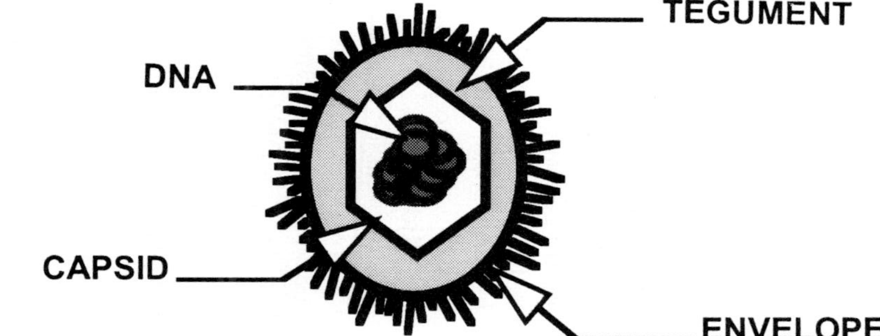

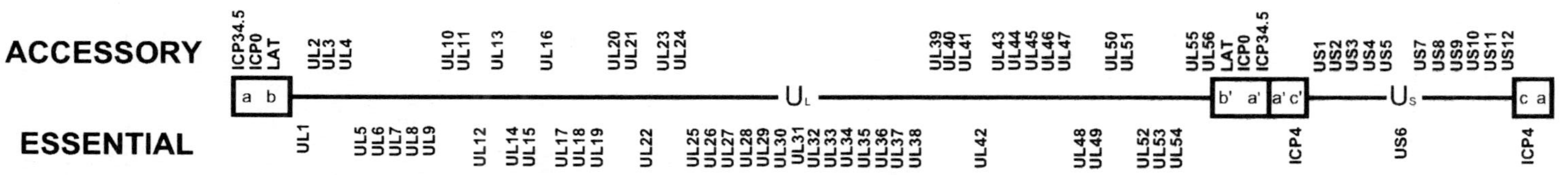

FIGURE 2 HSV-1 particle structure and genome organization. (A) Schematic illustration of the HSV virion, showing the capsid, tegument, and glycoprotein-containing lipid envelope. (B) Schematic representation of the HSV genome, showing the unique long (U_L) and unique short (U_S) segments, each bounded by inverted repeat (IR) elements. The location of the essential genes that are required for viral replication *in vitro*, and the nonessential or accessory genes, which may be deleted without affecting replication *in vitro*, are shown.

is initiated by binding of a viral tegument protein, VP16, in combination with two cellular transcription factors Oct 1 and HCF, to TAATGARAT enhancer sequences located in the promoters of the five immediate early (IE) genes (Gerster and Roeder, 1988; Katan *et al.*, 1990; Kristie and Sharp, 1993; O'Hare and Goding, 1988; Preston *et al.*, 1988; Werstuck and Capone, 1993; Wilson *et al.*, 1993). These genes encode infected cell proteins (ICPs) 0, 4, 22, 27, and 47 named according to their molecular sizes in order of appearance in sodium dodecyl sulfate (SDS) polyacrylamide gels of infected cell viral proteins. A sixth gene, ICP6, is expressed both as an IE and early (E) function since its promoter contains the VP16 responsive element and is also transactivated by ICP0 (Desai *et al.*, 1993). The IE gene products ICP4 and 27 are essential (DeLuca and Schaffer, 1985; Sacks *et al.*, 1985), while the other IE gene products are accessory functions. ICP0, 4, and 27 enhance expression of E and late (L) genes (DeLuca and Schaffer, 1985; Dixon and Schaffer, 1980; Preston, 1979; Sacks *et al.*, 1985; Watson and Clements, 1980). In addition, late gene expression requires viral DNA synthesis (Holland *et al.*, 1980; Mavromara-Nazos and Roizman, 1987). The IE genes are the only viral genes that can be expressed in the absence of viral protein synthesis. In addition to transcriptional regulation functions, ICP27 affects the splicing, polyadenylation and stability of mRNA (Brown *et al.*, 1995; McGregor *et al.*, 1996; Sandri-Goldin and Hibbard, 1996; Sandri-Goldin *et al.*, 1995; Smith *et al.*, 1992), ICP22 may aid in the usurping of cellular RNA polymerase by phosphorylation (Rice *et al.*, 1994), and ICP47 inhibits MHC class I antigen presentation (discussed below) (Hill *et al.*, 1995; York *et al.*, 1994). The E genes are expressed in response to IE gene induction and are largely products that carry out viral DNA synthesis. Nine of these gene products, including the viral DNA polymerase and origin-binding protein, are essential for viral genome replication (Challberg, 1986). The genome is thought to be replicated by a rolling circle mechanism forming head-to-tail concatemers (Jacob *et al.*, 1979; Skaliter *et al.*, 1996). During DNA replication the U_L and U_S components can invert by homologous recombination events involving the inverted repeat to form four possible isomers, all of which appear to be infectious (Davison and Wilkie, 1983; Jacob *et al.*, 1979; Mocarski and Roizman, 1982). The L genes encode mainly structural proteins that assemble into capsids in a well-ordered manner with head full packaging of the viral genome and stabilization of the nucleocapsid (Deiss *et al.*, 1986; Frenkel *et al.*, 1976). The tegument assembles around the mature capsids prior to budding through the nuclear membrane, where the virus acquires its envelope. The lytic cycle almost always results in cell lysis, with the possible exception of replication in sensory neurons during the establishment of latency or during viral reactivation.

III. Herpes Simplex Virus Cytotoxicity and the Complementation of Deletion Mutants

Because HSV has a well-described lytic cycle in which the viral genes are expressed in a highly ordered cascade (Honess and Roizman, 1974), it is possible to nearly eliminate early and completely prevent viral late gene expression by deletion of just one major immediate early gene, ICP4, whose function is essential for expression of all later viral genes (DeLuca and Schaffer, 1985; Dixon and Schaffer, 1980; Preston, 1979; Watson and Clements, 1980). Following deletion of both ICP4 and ICP27, E and L gene expression cannot be detected by standard methods (Samaniego *et al.*, 1995). Because ICP4 is an essential viral gene, the virus cannot replicate unless this missing function is supplied, which can be accomplished by engineering a cell line to express the ICP4 protein under the control of its cognate IE gene promoter integrated into the cellular genome in a location where the gene is transcriptionally silent. On infection of these cells with mutant virus, VP16 in the viral tegument activates expression of the ICP4 gene embedded in the cellular genome, thus providing the essential function required for viral replication to the virus in trans (Fig. 3). In this same manner, deletion of other essential viral genes can also be complemented.

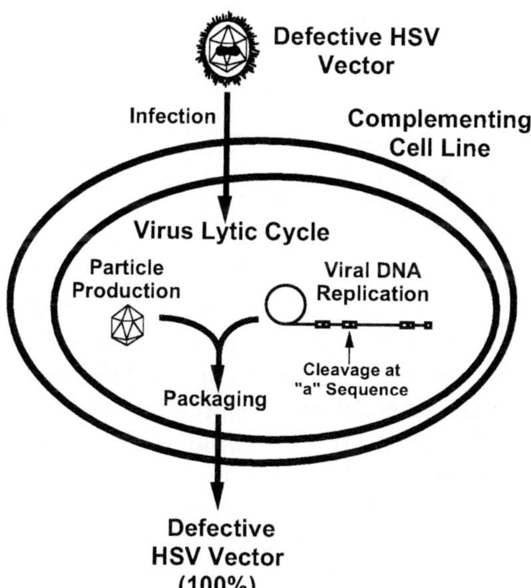

FIGURE 3 Replication-defective HSV-1 vectors. Production of replication-defective full-length HSV-based vectors is carried out in cell lines that are engineered to provide the deleted essential gene(s) in trans. These resulting progeny vectors are incapable of replicating in cells *in vivo* because of the missing essential gene(s).

Mutants deleted for ICP4 have been engineered to contain an *Escherichia coli lacZ* (β-galactosidase) reporter gene whose transcription is controlled by the human cytomegalovirus (HCMV) IE promoter (Stinski and Roehr, 1985; Thomsen *et al.*, 1984), one of the strongest promoters known and one that functions in a broad range of cell types (Furth *et al.*, 1991; Koedood *et al.*, 1995; Schmidt *et al.*, 1990). This promoter can express *lacZ* in the absence of E and L viral gene expression and thus is functionally similar to an IE gene. On infection of noncomplementing cells, the virus expresses *lacZ*, the other IE genes (ICP0, 22, 27, and 47), and ICP6, the large subunit of the viral ribonucleotide reductase. The cell is efficiently transduced for β-galactosidase (β-Gal) production as detected by 5-bromo-4-chloro-3-indolyl-β-D-galactopyranoside (X-Gal) staining (Fig. 4, see color plate); a single plaque-forming unit (PFU) of virus is able to produce enough β-Gal for detection. However, at multiplicities of infection as low as 0.3 the cells are eventually killed by the virus even though viral E and L genes are not expressed. This observation, in addition to reports that ultraviolet (UV) irradiation of wild-type virus nearly eliminates viral toxicity (Johnson *et al.*, 1992; Leiden *et al.*, 1980) and transduction of cells with single IE genes can kill cells (Johnson *et al.*, 1992), suggests that the remaining IE gene products may be responsible for viral toxicity in the absence of the expression of viral lytic genes.

There are a variety of viral genes whose products might contribute to cell death even in the absence of viral replication, ranging from functions that destabilize cellular RNA (Kwong *et al.*, 1988; Oroskar and Read, 1989; Read and Frenkel, 1983) to those that have more global effects including the alteration of cellular transcription (DeLuca *et al.*, 1985; Everett and Maul, 1994; Everett, 1987; Maul and Everett, 1994; Russell *et al.*, 1987) and the degradation of cellular DNA (Johnson *et al.*, 1992). Cellular genes are activated on infection and the production of stress proteins could contribute to cell death. To use HSV vectors broadly for gene transfer, it will be necessary to remove those viral genes (or block their expression) that cause cell necrosis and disrupt normal host cell metabolism and function.

This ICP4-deleted virus with the *lacZ* reporter gene driven by the HCMV promoter has been tested for the ability to express β-Gal in the central nervous system following stereotactic inoculation of rat hippocampus. Four important observations were derived from these studies: (1) the virus is less toxic for neurons in brain than the cell culture studies would predict; (2) the virus expresses IE genes and β-Gal but expression is transient, disappearing within a few days; (3) the viral latency transcript (LAT) can be detected long after IE gene and β-Gal expression had been shut off; and (4) both viral DNA and LAT expression can be detected for up to 1 year after infection, and appear to persist in a stable manner. These results demonstrate that the viral lytic cycle is not required for the establishment of latency and, in contrast to the latency promoter, a strong foreign gene promoter was shut off with the same kinetics as viral lytic cycle gene promoters. We have

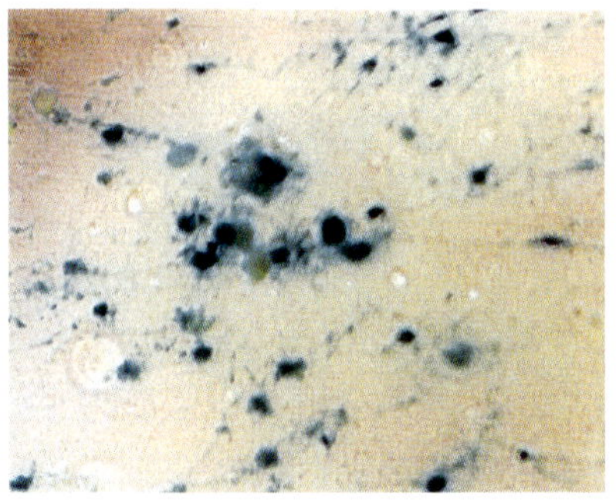
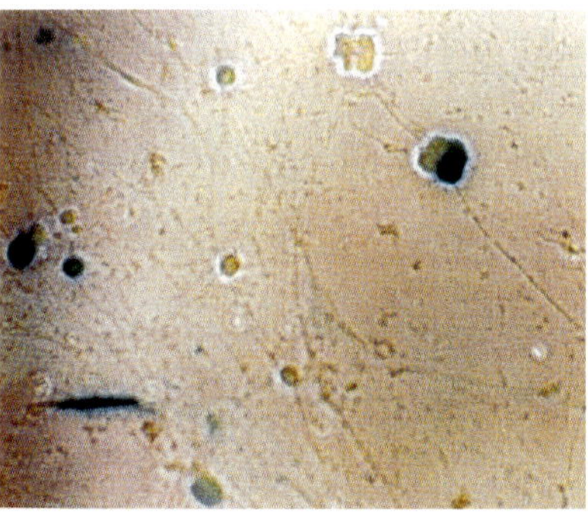

FIGURE 4 Transgene expression by replication-defective HSV-1 vectors in DRG neurons in culture. Cultures of DRG neurons were infected with replication-defective HSV-1 vectors containing deletions of single (ICP4⁻ : : HCMV IEp-*lacZ*) or multiple (ICP4⁻/22⁻/27⁻ : : HCMV IEp-*lacZ*) IE genes and stained with X-Gal to detect the presence of the transgene (*lacZ*) at 24 hr postinfection.

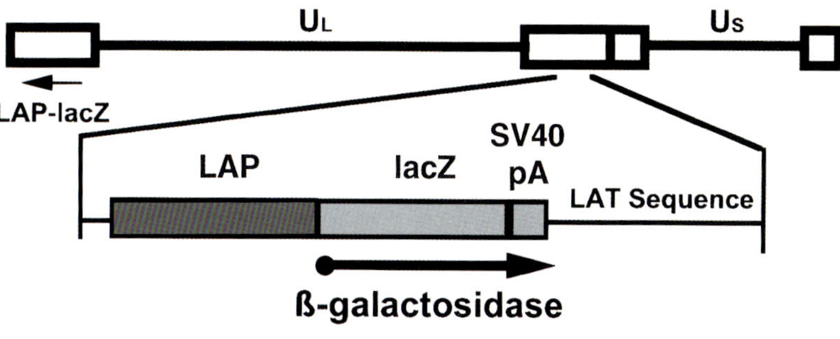

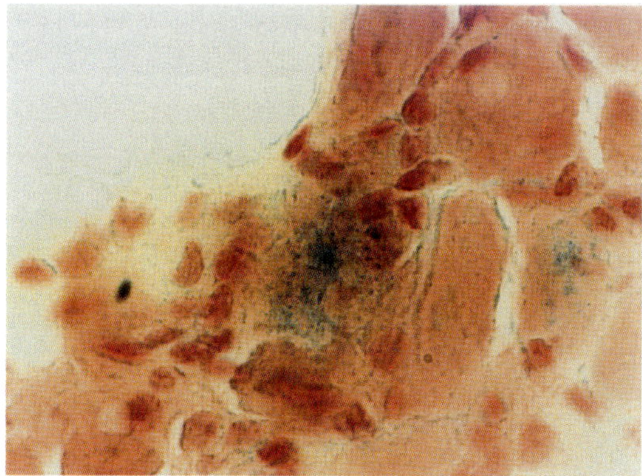

FIGURE 7 HSV-1 LAP-driven transgene expression in latently infected mouse peripheral nerves. Mice were infected by topical corneal scarification with the VPO vector (KOS, LAT⁻ : : LAP-*lacZ*) in which the *lacZ* reporter gene was inserted into the LAT intron at +42 of the stable 2-kb LAT. The vector expressed the transgene as evidenced by the presence of X-gal-positive punctate blue-staining neurons at 56 days postinfection, a time consistent with latent infection.

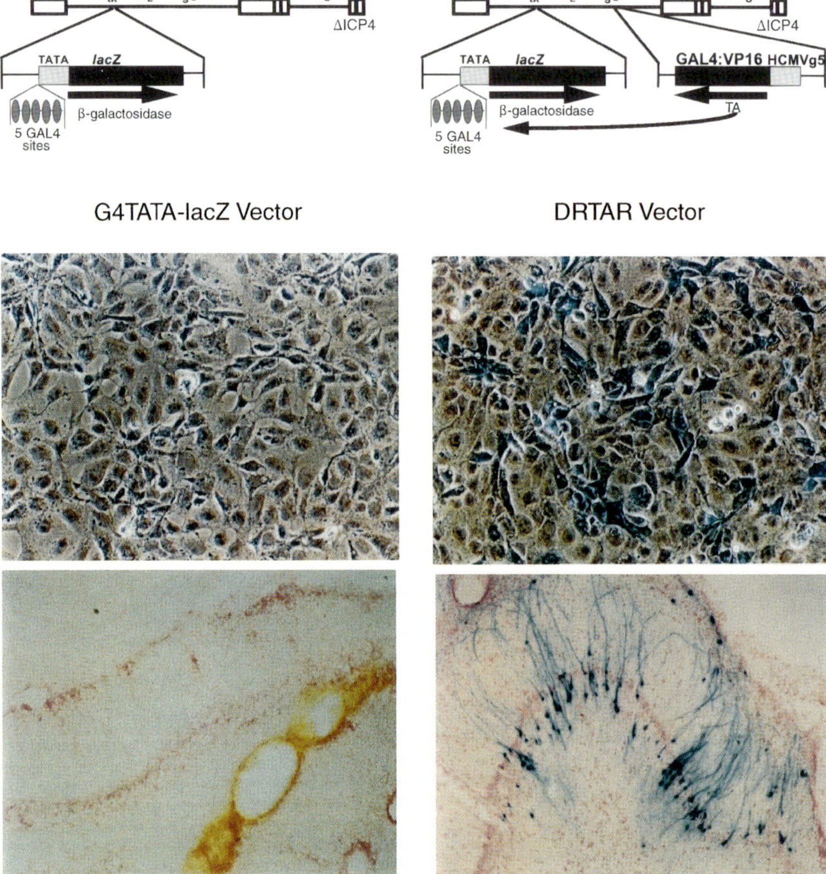

FIGURE 8 Use of the Gal4 : VP16 transactivator system to increase transgene expression *in vitro* and *in vivo*. Replication-defective recombinant HSV-1 vectors were constructed to contain either a minimal promoter with five tandem Gal4-binding sites driving the *lacZ* reporter gene (G4TATA-*lacZ*) or this reporter cassette and a cassette that expresses the Gal4 : VP16 transactivator (TA) from the strong HCMV IE gene promoter (DRTAR). Transgene expression from both vectors was examined *in vitro* in Vero cells in culture or *in vivo* following stereotatic inoculation of rat hippocampus. The Gal4 : VP16 TA was able to boost dramatically the level of transgene expression from the Gal4-sensitive promoter both *in vitro* and *in vivo*.

tested a wide variety of other viral and cellular promoters for persistence of gene expression in the background of the ICP4 mutant and thus far all promoters tested appear to be effectively shut off, even mammalian promoters, which are neuron specific, driving constitutive expression of their natural product in CNS neurons. The persistence of LAT expression during viral latency has led to considerable effort toward determining the molecular basis of LAT promoter activity, with the goal of exploiting this promoter to express genes during latency (see below).

The phenomenon of promoter shutoff is intriguing since other viral vectors such as adenovirus and adeno-associated virus containing the HCMV promoter to drive *lacZ* expression in rat brain do not appear to show immediate shutoff (Akli *et al.*, 1993; Davidson *et al.*, 1993; Kaplitt *et al.*, 1994b; Le Gal La Salle *et al.*, 1993), suggesting that the HCMV promoter is capable of remaining active in brain. In addition, the HCMV promoter has been used to express transgenes in brain neurons in transgenic animals as part of the neuronal cell genome (Furth *et al.*, 1991; Koedood *et al.*, 1995; Schmidt *et al.*, 1990). Moreover, HSV-packaged plasmid vectors (amplicons) (described below) have been reported to express reporter genes for several weeks using an HSV IE promoter that is shut off quickly as part of the defective viral genome (During *et al.*, 1994; Geller and Breakefield, 1988; Geller and Freese, 1990). Several possible explanations could account for the observed differences in promoter function in defective HSV viruses versus other gene delivery vehicles. These might include (1) effects on promoter activity that act in cis since the reporter gene cassette is recombined into the defective HSV genome and (2) activities that act in trans and likely include the activation of functions that prevent viral protein synthesis such as interferon α or viral gene products that directly or indirectly inactivate promoters embedded in the viral genome. The latter seems less attractive on the one hand since replication-defective viruses establish "latency" as determined by the expression of the LATs, so that latency would appear to depend on a "passive" mechanism from the standpoint of the virus. Perhaps neuronal cells contain an active repressor mechanism or simply do not support viral replication. On the other hand, other herpesviruses (e.g., Epstein–Barr virus) are known to control latency actively and it might be considered counterintuitive to assume that HSV is not also capable of controlling the establishment of latency, especially given its other sophisticated methods for interacting with the host. The establishment of "latency" by defective viruses could be misleading since the normal pathway to latency would be obviated. Moreover, why would promoters that contain neuronal enhancers and normally function constitutively in brain neurons be silenced if silencing was simply a passive process? Finally, this vector can express *lacZ* for weeks in cells other than neurons such as muscle fibers (described below) and thus the shutoff phenomenon may be limited to neurons. This interesting problem has led to the development of an array of multiple IE gene mutants for analysis of the promoter shutoff phenomenon, with rather surprising results.

Figure 5 shows a list of IE gene mutants available as of this writing along with cell lines used to complement the essential IE genes encoding ICP4 and ICP27. These mutants are difficult to isolate since some IE genes (ICP4 and 0) are diploid in the genome, and two others share the same promoter (ICP22 and 47) located in the inverted repeat sequences surrounding the U_S component of the genome. While only two of these genes are essential, ICP22 and 0 also might be complemented from the cellular genome in order to achieve high-titer virus production. While each individual gene has been deleted and most double-mutant combinations are now available, we have yet to isolate a mutant with all the IE genes removed from a single genome.

The multiple gene knockouts have proved useful for several reasons. First, they have provided space for the introduction of multiple foreign genes

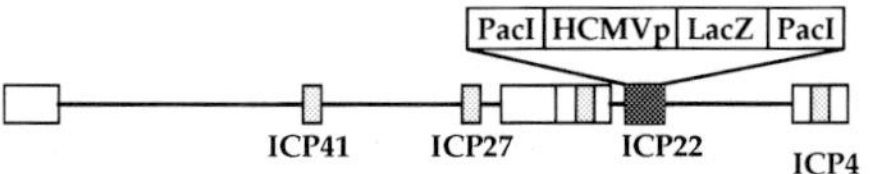

Virus Isolate	Test via Growth on Complementing Cell Lines				Southern Blot Analysis		Genes Deleted
	7B (4/27)	E5 (4)	N23-25 (27)	Vero	ICP4TK	UL41	
NEDLS10	+	+	+	+	–	–	22⁻,41⁻
NEDLS30	+	+	+	+	–	–	22⁻,41⁻
NEDLS34	+	–	+	–	–	+	22⁻,27⁻
NEDLS39	+	+	+	+	–	–	22⁻,41⁻
NEDLS40	+	–	–	–	+	+	4⁻,22⁻,27⁻
NEDLS60	+	–	–	–	+	+	4⁻,22⁻,27⁻
NEDLS64	+	–	–	–	+	+	4⁻,22⁻,27⁻
NEDLS67	+	–	–	–	+	+	4⁻,22⁻,27⁻
NEDLS71	+	–	–	–	+	–	4⁻,22⁻,27⁻,41⁻
NEDLS73	+	–	–	–	+	+	4⁻,22⁻,27⁻
NEDLS74	+	–	–	–	+	+	4⁻,22⁻,27⁻
NEDLS75	+	–	–	–	+	+	4⁻,22⁻,27⁻
NEDLS78	+	+	–	–	–	–	4⁻,22⁻,41⁻

FIGURE 5   List of replication-defective HSV-1 vectors containing deletions in HSV-1 genes involved in cytotoxicity. *Top:* A general schematic showing the location of the various HSV-1 genes involved in cytotoxicity as well as the site for introduction of a reporter gene cassette. *Bottom:* The various replication-defective mutants were tested for growth on cell lines that complement ICP4 (E5), ICP27 (N23-25), or both ICP4 and 27 (7B), or in noncomplementing Vero cells. Southern blot analysis was employed to confirm the presence of either a transgene cassette (ICP4-TK) or the nonessential virion host shutoff (vhs) gene UL41.

into the viral genome. Second, these multiple gene deletion mutants are highly reduced in cytotoxicity. For example, many cell lines can be infected with multiply deleted mutants at multiplicities of infection approaching 10 with the majority of cells surviving and capable of subsequent growth and division. This has broadened the utility of these vectors for gene transfer to other cell types where the virus does not establish latency and would otherwise be cytotoxic. Here the viral genomes can persist in a state that mimicks latency although the LATs are not expressed, consistent with the LAT promoter responsiveness to neuronal signals. Third, the deletion of the IE genes has demonstrated some interesting effects on foreign promoter function. For example, on infection of cells, a mutant virus lacking all of the immediate early genes except ICP0 and ICP47 is able to express *lacZ* under control of the HCMV IE promoter for weeks in cell culture and in primary neuronal cell cultures (Fig. 4). However, on injection of this virus into rat hippocampus, *lacZ* gene expression was weak compared to expression in the background of the single ICP4 deletion mutant. In contrast to the ICP4 mutant, expression continued and was not immediately shut off in the multiple deletion mutant, and the ICP0 gene product continued to be expressed for several weeks. As expected, E and L genes, like LATs, were not expressed. The multiply deleted viral mutant did not display the properties of true latent infection but appeared to be arrested in a prelatent state. These findings suggest that the other IE gene and/or other expressed viral products are in some way contributing to promoter shutoff and perhaps to the establishment of latency. It remains to be determined whether the promoter activity of the IE genes and the HCMV promoter is dependent on the expression of ICP0. These findings raise the interesting possibility that foreign promoters and the HSV-1 IE promoters themselves can be released from shutoff in the absence of the IE gene products. We are in the process of testing neuron-specific promoters for sustained activity in the background of these mutants. Success in these experiments would solve both the problem of viral toxicity and promoter shutoff, thereby greatly improving the utility of these viruses for CNS applications.

IV. Herpes Simplex Virus Amplicon Vectors

An alternative approach to the use of genomic HSV vectors is the construction of amplicons, often referred to in the literature as "defective HSV vectors" (Geller and Breakefield, 1988; Geller and Freese, 1990). Amplicons are plasmid vectors that contain both an *E. coli* and HSV origin of replication and HSV packaging signals, that together enable plasmid propagation in *E. coli* and amplicon DNA synthesis in mammalian cells on cotransfection with defective HSV. The concatemerized amplicons and defective HSV genomes are subsequently packaged into HSV particles for delivery to neurons (Fig. 6). Amplicons were discovered initially in studies of defective interfering HSV particles that arise spontaneously in populations of replicating virus (Spaete and Frenkel, 1982). Unfortunately, the ratio of amplicon to helper virus-containing particles is difficult to control. To favorably increase the amplicon : helper virus ratio the preparation must be passaged repeatedly, with undesired opportunities for recombination between the helper virus and the complementing gene used to support the growth of the helper virus. In one published long-term study of amplicon-mediated gene transfer, a number of animals died as a result of recombinant wild-type virus in the amplicon preparation (During *et al.*, 1994). A number of strategies are currently being explored in an attempt to circumvent packaging of helper virus.

Despite the problem of wild-type recombination, amplicons have been used in a number of experimental models to transfer genes to neurons *in vitro* (Battleman *et al.*, 1993; Casaccia-Bonnefil *et al.*, 1993; Geller *et al.*, 1993; Geschwind *et al.*, 1994; Ho *et al.*, 1993) and *in vivo* (During *et al.*, 1994; Kaplitt *et al.*, 1994a). Amplicons have been used (1) to express nerve growth factor (NGF) in sympathetic ganglia, with the biological effect of blocking axotomy-induced loss of adrenergic phenotype, (2) to express tyrosine hydroxylase (TH) in 6-hydroxydopamine-lesioned rat striatum, with the biological effect of blocking apomorphine-induced rotational behavior, (3) to express a glucose transporter (GT) gene in hippocampal neurons to reduce kainate-induced seizure damage in the CA3 cell field of hippocampus, and (4) to express bcl-2 in hippocampus with a reduction in the size of the ischemic penumbra after experimental cerebral ischemia. With the exception of the TH experiments, transgene expression in each of these cases was transient. The expression of transgene message but not protein was demonstrated in the short-term experiments; however, detection of the transgene even by RT-PCR in the long-term experiments was inconsistent. Other investigators report that amplicons that express a transgene using the HSV IE ICP22/47 promoter initially peak (presumably due to VP16 transactivation) and then diminish to low levels within 30 days. It may be possible, however, to use cellular promoters in the amplicon vectors to express transgenes and indeed there are several reports of this use (During *et al.*, 1994; Kaplitt *et al.*, 1994a). Our current view is that our multiply deleted defective viruses and amplicons will merge in a functional sense with respect to their ability to express transgenes long term, unimpeded by viral functions. Defective viruses should prove to be more useful inasmuch as their production will be less complicated, it will be easier to verify their purity, and they can contain more space for foreign DNA insertion into the vector genome.

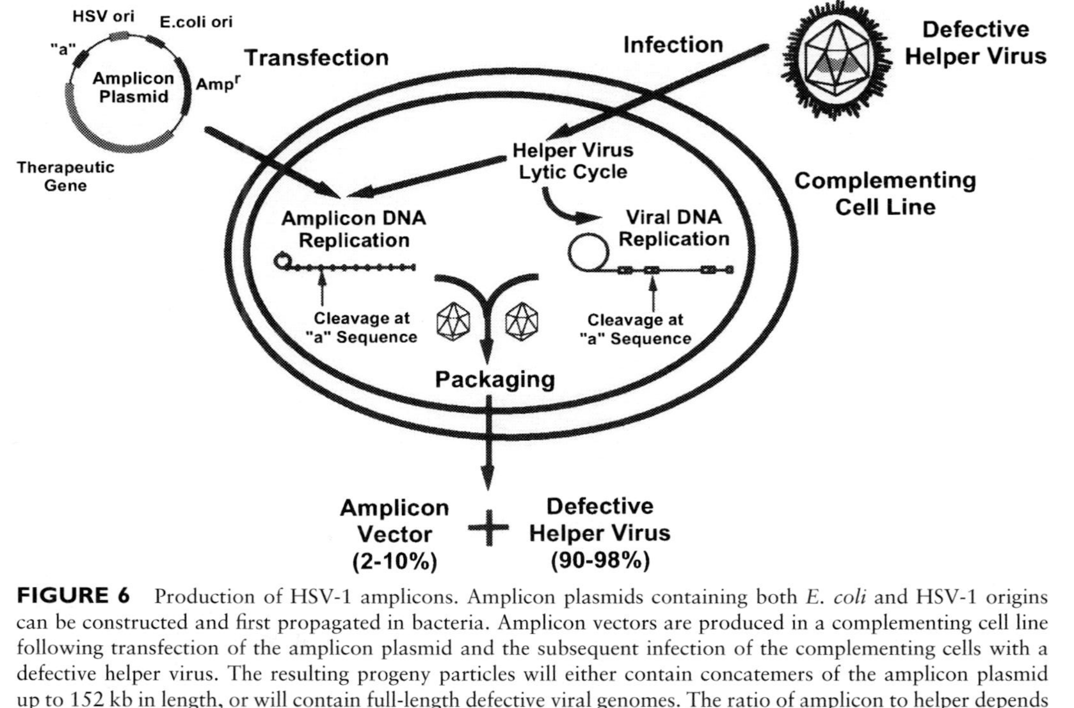

FIGURE 6 Production of HSV-1 amplicons. Amplicon plasmids containing both *E. coli* and HSV-1 origins can be constructed and first propagated in bacteria. Amplicon vectors are produced in a complementing cell line following transfection of the amplicon plasmid and the subsequent infection of the complementing cells with a defective helper virus. The resulting progeny particles will either contain concatemers of the amplicon plasmid up to 152 kb in length, or will contain full-length defective viral genomes. The ratio of amplicon to helper depends on transfection efficiency and can be increased by multiple passage on the complementing cell line, which can thereby increase the likelihood of generating wild-type recombinants.

V. Herpes Simplex Virus Immunology and Antigenic Stealthing

Herpesviruses are among those that have developed specific strategies for escaping immune surveillance. Herpes simplex virus, for example, carries an IE gene that encodes ICP47, a protein that specifically inhibits MHC class I-mediated antigen presentation and consequently recognition by HSV-specific ctytotoxic T lymphocytes (CTLs) that recognize the virus-infected cell through its class I-specific T cell receptor (York *et al.*, 1994). ICP47 interferes with the formation of antigen class I complexes in the endoplasmic reticulum (ER). This results in the degradation of the class I glycoproteins, which normally leave the ER, pass through the Golgi, and appear on the cell surface as a peptide antigen class I complex associated with β_2-microglobulin. ICP47 carries out its function by associating with the transporter associated with antigen presentation (TAP) (Hill *et al.*, 1995). It is of significant interest that Epstein–Barr virus (EBV) through its EBNA-1 latency gene product, human cytomegalovirus (HCMV) through its UL18 product, and adenovirus (AV) through its 19K E3 gene product all block class I antigen presentation by interfering with class I antigen association at different stages in this pathway. In addition to this mechanism EBV expresses an altered form of the cytokine interleukin 10 (i.e., vIL-10) (Hsu *et al.*, 1990; Vieira *et al.*, 1991) that can also alter the immune response locally by interfering with T cell expansion due to T_H cell activity (deWaal-Malefyt *et al.*, 1991; Moore *et al.*, 1990). These genes, in addition to viral products that interfere with apoptosis such as the HSV $\gamma34.5$ gene product (Chou and Roizman, 1992, 1994; He *et al.*, 1996), suggest that the combination of immune interference and prevention of programmed cell death are powerful viral strategies to prevent destruction of the infected cell, particularly those cells in which the virus will eventually persist. It might well be imagined that for applications involving HSV vectors in which the immune system will need to be eluded and transgene products continuously expressed, that a combination of antigen stealthing genes introduced into the virus would be effective in allowing the vector to become established in the host tissue following direct gene transfer. Moreover, several of the HSV glycoproteins can help avoid complement inactivation and reduce antibody-dependent complement-mediated virus neutralization and antibody-dependent cell-mediated killing (ADCC) of HSV-infected cells. Glycoprotein C is well described for its ability to bind the C_3b component of complement and interfere with the alternate pathway for complement activation (Hung *et al.*, 1992, 1994). Glycoproteins I and E together can bind the Fc component of IgG, thus functioning as an Fc receptor (Johnson *et al.*, 1988). Expression of gI and gE on infected cell surface membranes can bind normal IgG and reduce ADCC (Adler *et al.*, 1978; Lehner *et al.*, 1975).

VI. Alteration of the Viral Host Range and Vector Targeting

The precise cascade of molecular events resulting in HSV-1 entry into susceptible host cells is still poorly understood despite years of study by a large number of laboratories. The mechanism by which HSV-1 accomplishes infection has been suggested to be a two-step process (Herold *et al.*, 1994). First, the virus binds to the cell surface through weak interactions of at least two viral glycoproteins (gB and gC) with heparan sulfate (HS), a common component of mammalian cell surface (Shieh and Spear, 1994; Wudunn and Spear, 1989). It is through this nonspecific adsorption that the virus is held in close proximity to the cell membrane, providing an opportunity for interaction with a low-density receptor resulting in tighter virus binding to the cell surface (Johnson *et al.*, 1994; Ligas and Johnson, 1988). The fusion at neutral pH of the two membranes is then triggered and the capsid delivered into the cytoplasm of the host cell.

Of the 11 HSV-1 glycoproteins so far identified, only 4 have been demonstrated to be essential for penetration. These include glycoproteins B, D, H, and L (gB, gD, gH, and gL) (Cai *et al.*, 1988; Desai *et al.*, 1988; Hutchinson *et al.*, 1992; Ligas and Johnson, 1988). If gD has a role restricted to adsorption, then gB and/or the gH/gL complex promote entry as fusion factors. The rest of the viral glycoproteins appear to be dispensable for virus infection with the possible exception of gK, which does not appear to be incorporated into infectious virus particles (Hutchinson and Johnson, 1995). However, it has been demonstrated that infectivity was decreased 10-fold when viruses were deleted of glycoprotein C (gC), a dispensable glycoprotein (Herold *et al.*, 1991).

We have found that removal of the HS-binding domain from HSV-1 gB can prevent gB attachent to HS on cells without interfering with its ability to mediate virus penetration. A double mutant deleted for gC in combination with the gB HS-binding domain mutation is highly impaired in its ability to attach to cells but forms normal plaques. It is this virus background that can be used for incorporation of a novel binding ligand using a recombinant gC gene. It should be possible thus to avoid the natural first step in virus attachment to HS and redirect the initial attachment phase to a ligand-specific receptor.

So far, most attempts to modify target cell specificity have been performed on retrovirus through the construction of pseudotype particles in which a virus is enveloped by the membrane of a second virus. The host cell specificity is then determined by the virus providing the envelope protein. For example, the G protein of vesicular stomatitis virus (VSV) was able to interact with the nucleocapsid of Moloney murine leukemia virus (MoMLV) in the formation of an MoMLV(VSV) pseudotype conferring to the pseudotype particle a broader cell host range than normally found with MoMLV

(Emi *et al.*, 1991). In this instance the targeted host cell is limited due to the specificity defined by available viral envelope proteins. However, the targeting of MoMLV virus has been performed using an engineered fusion protein capable of inducing viral infection of a cell expressing a selected receptor (Kasahara *et al.*, 1994). In these experiments, a retroviral vector (MoMLV) packaged with a chimeric envelope glycoprotein, consisting of the hormonal protein erythropoietin (EPO) fused to the N-terminal portion of the ectropic virus Env protein, infects mouse fibroblasts transformed to express 10,000 EPO receptors per cell more efficiently than control fibroblasts. Moreover, this ligand-mediated targeting virus infects human cells that express EPO receptors that were not normally infected by the wild-type virus. The principle of ligand-mediated targeting has been established by these experiments. Work in progress is attempting to use the EPO ligand to target HSV binding to EPO receptor-bearing cells. Success in these experiments would provide interesting new opportunities for HSV vector targeting *in vivo*.

VII. The Herpes Simplex Virus Latency-Associated Transcript Promoter System and Its Utility in the Expression of Foreign Genes

A promising method to express foreign genes in the nervous system could involve exploiting the natural viral latency promoter system. The transcriptional control of LAT expression, however, is complex. Two latency active promoters, LAP1 and LAP2, have been identified. LAP1, containing a TATA box and basal transcriptional regulatory sequences, lies approximately 700–1300 bp upstream of the 5′ end of the major 2-kb species of LAT2 (Dobson *et al.*, 1989). A second promoter element, LAP2, located between LAP1 and the 5′ terminus of the 2-kb LAT, is highly GC-rich, sharing sequence homology with many eukaryotic housekeeping and protooncogene gene promoters (Goins *et al.*, 1994). Both LAP1 and LAP2 are active in transient transfection assays, with the activity of LAP1 being substantially higher than that of LAP2 (Goins *et al.*, 1994). However, the LAP1 promoter element alone is unable to drive reporter gene expression during latency in neurons of the trigeminal ganglion when the LAP1-*lacZ* cassette is placed in an ectopic (gC) locus in the viral genome (Lokensgard *et al.*, 1994), or when *lacZ* was inserted downstream of LAP1 in its native locus (Margolis *et al.*, 1993). In contrast, we have found that LAP2 sequences are capable of driving reporter gene expression from the natural LAT locus (Chen *et al.*, 1996) or an ectopic (gC) locus (Goins *et al.*, 1994) in the PNS neurons of the trigeminal ganglion (Fig. 7, see color plate). In the central nervous system, long-term expression of the reporter gene RNA can be detected by RT-PCR after inoculation into hippocampus. These results suggest that LAP2 sequences must contain an element that can function as a weak promoter in an isolated situation (Goins *et al.*, 1994), and in the native locus may serve to alter chromatin structure or recruit other factors to influence LAP1 activity.

The level of expression driven by LAP2 is, however, very low, and so will likely require some other manipulation in order to produce potentially therapeutic levels of gene products. One approach we have employed is the use of a constitutive transcriptional transactivator, consisting of the yeast Ga14 DNA-binding domain fused to the HSV-1 VP16 acidic transactivation domain (Carey *et al.*, 1990; Chasman *et al.*, 1989; Sadowski *et al.*, 1988). We have already demonstrated that Ga14 : VP16 expressed from the viral genome under the control of the HCMV immediate early promoter is capable of activating a minimal promoter containing five Ga14-binding sites juxtaposed to the adenovirus E1b minimal TATA box placed either in the cellular genome, in another vector used for coinfection, or with both the Ga14 : VP16 transactivator and the Ga14-sensitive TATA box promoter in the same vector (Oligino *et al.*, 1996). This system was capable of boosting transgene expression in cell culture as well as *in vivo* in rat CNS (Fig. 8, see color plate). An adaptation of this system with LAP2 driving the expression of Ga14 : VP16 may be useful in enhancing the level of transgene expression from latent viral genomes.

VIII. Application of Herpes Simplex Virus Vectors

A. Neurodegenerative Disease

Herpes simplex virus-based gene transfer vectors hold promise for the treatment of a number of diseases; however, on the basis of the natural biology of the virus, neurodegenerative diseases of the central or peripheral nervous system are particularly attractive candidates for treatment. Among the large number of such diseases, relatively common diseases including Parkinson's, Alzheimer's, and Huntington's chorea in the central nervous system, and amyotrophic lateral sclerosis and peripheral sensory neuropathies in the peripheral nervous system, serve as paradigms for the various treatment options using HSV vector technology.

Parkinson's disease (PD) is characterized by degeneration of dopaminergic neurons of the substantia nigra, resulting in dopamine deficiency in the terminal field of those neurons in the striatum. The mainstay of current pharmacotherapy for PD is the administration of levodopa or L-dopa (L-dihydroxyphenylalanine), which is converted to dopamine by remaining dopaminergic terminals in the striatum, or the use of direct dopamine receptor agonists. However, despite some therapy with these agents (and the addition of anticholinergic drugs) the disease progresses and patients become unresponsive to continued drug therapy, leading ultimately to immobility and death. One gene therapy approach to PD is to use a vector to transfer the tyrosine hydroxylase (TH) gene to neurons of the striatum, in order to provide for local production of dopamine from otherwise nondopaminergic cells. This type of therapy may well require regulation of dopamine production, since in contrast to the rigidity and bradykinesia resulting from dopamine deficiency, excess dopamine in the striatum causes unwanted adventitious movements. A second approach would be to rescue the dopaminergic neurons of the substantia nigra from degeneration through the delivery of a protective "trophic" factor. Studies have shown that in experimental animal models of PD glial-derived neurotrophic factor (GDNF) can protect nigral neurons from degeneration caused by 6-hydroxydopamine or by the neurotoxin 1-methyl-4-phenyl-1,2,3,6-tetrahydropyridine (MPTP) (Hoffer *et al.*, 1995; Kearns and Gash, 1995; Tomac *et al.*, 1995). If the degeneration of nigral neurons in the human were analogous to these models, local continuous delivery of GDNF, either to the region of the substantia nigra or to the terminals in the striatum, might prevent disease progression. On the basis of an understanding of trophic factors, it is likely that production of the tropic factor would not require exquisite regulation, but that constitutive low-level production would suffice. Because GDNF may have trophic effects in other cell types as well, local production would be preferable to systemic or even intraventricular administration, if either of those routes of administration were possible.

Alzheimer's disease (AD) is a widespread neurodegenerative condition of brain that affects many cell types, but is particularly severe for the cholinergic neurons of the nucleus basalis. Thus no single neurotransmitter replacement therapy is likely to suffice, and it seems unlikely that a single trophic factor could rescue the many cell types involved. The hallmark of AD is the accumulation of extracellular amyloid in senile plaques in many regions of brain, and there is increasing evidence that the amyloid peptide may be pathogentic in this disease process. Therefore a gene product that prevented the accumulation of amyloid, either by correct amyloid processing within cells, or by breaking down extracellular amyloid, might prevent the development of the disease process. Identification of potential candidate genes involved in these processes awaits further elucidation of the biology of amyloid processing.

Huntington's disease (HD) is a dominantly inherited disease causing progressive dementia, choreiform movement, and ultimately death, characterized by loss of cells in the caudate and putamen. The cell loss is a direct result of the presence of expanded CAG repeats in the gene for the Huntington protein, located on chromosome 4, although the mechanism of cell death resulting from the abnormal protein product is not known. Unlike the delivery of TH or trophic factors where expression in only a small number of cells would be required to release the product (dopamine or GDNF) into the local extracellular environment, a vector to treat HD would be effective only in cells into which the vector was delivered, and would therefore require global cellular delivery.

Amyotrophic lateral sclerosis (ALS) is a degenerative disease that specifically affects motor neurons of the brain and spinal cord, resulting in complete paralysis, sparing sensation and cognition. In some dominantly inherited cases of ALS, mutations in the gene encoding superoxide dismutase (SOD) have been identified, although the mechanism by which this mutation leads to decrease is not known. In the *wobbler* mouse animal models of ALS, administration of the trophic factor CNTF (ciliary neurotrophic factor) prevents neuronal degeneration; however, in human trials, the amount of CNTF that could be administered systemically was limited by the cytokine-like side effects of the peptide; in addition, the half-life of the administered drug was short, and no therapeutic effect could be demonstrated at the doses employed. Delivery of the gene product directly into the motor neurons would circumvent this problem and thereby allow the therapy to be successful.

B. Herpes Simplex Virus Vector Gene Delivery to Muscle and Gene Therapy for Muscular Dystrophy

Muscle has long been recognized as an excellent target for gene delivery for vaccines, production of soluble products, and for treatment of neuromuscular disease. We have explored the use of HSV vectors for gene transfer to muscle with the eventual aim of using this viral vector to deliver genes for treatment of muscular dystrophy including the Duchenne type, one of the most prevalent heritable human diseases. Duchenne muscular dystrophy (DMD) is a devastating muscle-wasting syndrome characterized by a lack of dystrophin expression at the sarcolemma of muscle fibers (Arahata *et al.*, 1988; Hoffman *et al.*, 1987; Zubryzcka-Gaarn *et al.*, 1988). This protein appears to function in the maintenance of muscle membrane integrity. Dystrophin is one of the largest known human genes and mutations arise at high frequency, making it one of the most common genetic diseases (affecting 1 in 2000 males). There is no treatment and affected children usually die in their late teens of heart or respiratory failure.

Two different approaches have been considered for restoration of dystrophin to dystrophic muscle: myoblast transplantation and gene therapy. Myoblast transplantation (MT) consists of implantation of normal myoblast precursors (satellite cells) into diseased muscle to create a reservoir of myoblasts capable of dystrophin expression (Huard *et al.*, 1994; Karpati and Acsadi, 1994; Morgan *et al.*, 1993). Myoblast transplantation in both animals and human trials has not been successful, primarily owing to transplantation rejection and difficulty in delivery. Gene therapy is also subject to difficulties in delivery, requiring a systemic approach involving intravenous inoculation of vector.

A number of vectors have been tried for gene delivery to muscle including naked DNA, retroviruses, and adenoviruses. Naked DNA proved to be inefficient although stable gene delivery was possible (Acsadi *et al.*, 1991; Danko *et al.*, 1993) and retroviruses have not been found to infect muscle fibers (Dunckley *et al.*, 1992; Salvatori *et al.*, 1993). Moreover, the dystrophin cDNA is large, amounting to 14 kb, and thus standard adenoviral (AV) vectors are unable to accommodate the full-length coding sequence. Completely "gutted" AV vectors have been reported that can accommodate the dystrophin gene; however, these vectors are contaminated with helper virus that engender immune rejection of the vector-delivered transgene. Moreover, AV vectors infect mature muscle fibers poorly (and myoblasts preferentially) (Acsadi *et al.*, 1994; Quantin *et al.*, 1992; Ragot *et al.*, 1993; Vincent *et al.*, 1993), the mature fibers having a low density of the AV receptor. Unfortunately, myoblasts begin to disappear as the disease progresses, requiring gene delivery to muscle fibers. Adenovirus is also highly immunogenic, perhaps because of the high doses of vector required for infection of muscle, which will make repeat dosing difficult; indeed, repeat dosing of AV has not been generally possible owing to the production of neutralizing antibodies on first administration of vector (Yang *et al.*, 1994).

Herpes simplex virus vectors solve some of these problems since it can easily accommodate the full-length dystrophin cDNA plus tissue-specific regulatory sequences. Herpes simplex virus also infects muscle with much greater efficiency, requiring only 1% as much virus to achieve the same level of transduction of myoblasts and myotubes (Huard *et al.*, 1995), and HSV infects both types of muscle tissue equally both *in vitro* and *in vivo* in mice. However, both HSV and AV infect mature muscle poorly. Herpes simplex virus is significantly impeded by the muscle basal lamina, which acts as a physical barrier to infection (Huard *et al.*, 1996). Moreover, HSV induces inflammation in muscle, which ultimately leads to a loss of transduced fibers. We have yet to test the multiply deleted vectors described above *in vivo* but they infect muscle cells *in vitro* efficiently and express the transgene vigorously using the HCMV IE promoter. Moreover, our first-generation vectors are able to achieve long-term expression in severe combined immunodeficiency (SCID) animals, indicating that unlike the brain, long-term expression in muscle may not be a problem. Current experiments are aimed at developing methods for temporarily opening the basal laminum for HSV infection and in developing HSV vectors that express antigenic stealthing genes, which may make vector-infected cells less likely to be rejected by the immune response.

C. Cancer

After many decades of intensive research designed to understand the molecular basis of human cancer in hopes of identifying potential targets for the development of anticancer drugs, our most effective chemotherapeutic agents are still based largely on inhibiting the growth of tumor cells by targeting DNA replication. Antimetabolites have not proved to be very effective and single drugs can vary substantially in efficacy. Currently, the combination of early diagnosis and surgery have the best outcome. However, not all tumors are operable and often tumors have metastasized before they become recognized. The most common human malignancies such as colon, breast, and lung cancers can be effectively treated in the early stages, whereas later in the progression of the disease it is far more difficult. Some cancers are almost uniformly fatal, such as glioblastoma multiforme. The combination of surgical debulking, chemo- and radiation therapies, and in some cases immunotherapy has extended life considerably. Resistance to cancer drugs generally leaves patients without hope and thus new types of interventions are required.

Gene therapy for treatment of cancer may offer a treatment alternative to patients without other options and if used early and wisely in others, the outcome could be encouraging. There are a number of considerations in applying gene therapy to cancer, which include the selection of the appropriate therapeutic gene(s), the specific effect or mechanism, the selection of target tissue, which may include the surounding stromal tissue as well as the tumor itself, and finally issues related to the vector and method of delivery. The overriding problem in cancer gene therapy is the fact that metastatic cancer is a systemic disease and thus even if gene transfer were effective in destroying a tumor locally, this is not enough and may only delay the inevitable. What features does HSV have that may make it useful in cancer gene therapy and which types of tumors may be particularly good targets?

Herpes simplex virus may be well suited for treatment of glioblastoma and other primary brain tumors or tumors that arise as metastases from other non-CNS tissues. Gliomas, for example, often produce large masses in the brain with the tumor invading the normal surrounding brain tissue, making the complete surgical resection of these tumors impossible. Moreover, the blood–brain barrier is intact, making access to infiltrating tumor cells inaccessible by systemic delivery, unless the blood–brain barrier is disrupted (using mannitol, for example). Even using this approach, the architecture of newly synthesized blood vessels within the tumor is irregular and blood flow uneven. Nevertheless, it may be possible to use HSV vectors that are compromised in their ability to replicate in normal nondividing neurons while retaining their ability to replicate in the tumor cells. The use of conditional replication-competent viruses could in theory allow for spread in tumor tissue without damaging normal brain, thus increasing the effectiveness. Such mutants include those lacking the viral thymidine kinase (Boviatsis *et al.*, 1994; Kosz-Vnenchak *et al.*, 1990; Markert *et al.*, 1993; Martuza *et al.*, 1991), the ribonucleotide reductase (Mineta *et al.*, 1994; Yamada *et al.*, 1991), a protein kinase (Fink *et al.*, 1992), or a gene (γ34.5) required for growth specifically in neurons (Chamber *et al.*, 1995; MacLean *et al.*, 1991;

Whitley *et al.*, 1993). Deleting these genes in combination creates viruses that are highly compromised in their ability to replicate in and kill neuronal cells yet retain the ability to replicate in and kill tumor cells. Highly compromised viruses grow poorly in tumor cells, however, and although multiple deletions increase safety, efficacy is compromised. If direct killing of tumor cells is coupled with gene expression that can activate anticancer prodrugs *in situ*, perhaps the effectiveness of these anticancer vectors would be enhanced. Suitable prodrugs include the natural HSV-1 thymidine kinase coupled with ganciclovir therapy and cytosine deaminase coupled with 5-fluorocytosine administration. Both activated metabolites can kill neighboring cells through a bystander effect. That is, phosphorylated ganciclovir can be taken up by neighboring cells across gap junctions while 5-fluorouracil is released from the cell and can be taken up by distal cells through the process of diffusion. The combination of delivery of the virus to tumor through a systemic mechanism along with viral replication locally, spread to neighboring cells, and bystander killing should enhance the chances for broad destruction of glioma cells.

There are also safety concerns in using these vectors related to toxicity for endothelial cells, normal glial cells, and microglia; the virus could potentially enter the meningeal fluid, where it could cause meningitis and destruction of appendymal cells lining the ventricles of the brain. These potential problems may be greatly reduced if the vector contains ligands that will target it to tumor cells or at least block infection of neuronal cells. Specific cell lysis may also be possible if an essential viral gene such as ICP4 were transcriptionally regulated by a glial-specific promoter, further ensuring that the virus would replicate only in tumor tissue.

Despite these novel uses of HSV vectors, it would appear likely that some glial cells might escape these killing mechanisms and thus it is essential that additional surveillance mechanisms be invoked to ensure removal of the tumor. There is a considerable amount of interest in using cytokine genes, costimulator molecules, tumor antigens, and recruitment molecules to enhance the immune response to the tumor. The development of antitumor immunity could circumvent the need for replication-competent vectors since tumor-specific cytotoxic T lymphocytes constantly move through the brain parenchyma searching for target cells. There is a growing body of literature to suggest that local expression of cytokines can enhance CTL activation at least in animal model systems and these bear testing in human brain cancer. Herpes simplex virus offers the potential for combinational gene therapy in this regard since multiple immunomodulatory genes can be recombined into the virus and tested individually and in combination.

IX. Summary and Future Directions

Herpes simplex virus has considerable potential as a gene vector for the nervous system and other tissues. Certain aspects of the virus biology, if better understood, should be helpful in improving the utility of the virus as a vector. For example, a great deal must be learned about the ability of the virus to express genes during latency. In particular, the exact cis–trans interactions that control latency gene expression are not yet determined. This will no doubt be accomplished in the near future as substantial numbers of latency promoter mutants altered in sites recognized by DNA-binding factors are now available. These reagents will provide the needed genetic material for dissection of their contribution to promoter function. Insight into latency promoter cis-functional elements should allow us to enhance promoter action by introducing additional elements that respond to factors present in neurons. In addition to promoter elements, we should be able to take advantage of natural signal transduction pathways in neurons to activate and regulate vector-bearing transgenes. The state of the viral genome will also no doubt play a role in gene expression during latency. More information on chromatin structure, DNA elements that affect methylation, and nucleosome interaction with the latency viral DNA will help to explain promoter shutoff and thus reveal methods to maintain promoter function during latency. We should be able to eliminate all viral gene expression in multiple IE gene mutants, freeing the transgene promoter to respond to cellular signals and increase tissue specificity of the vector. Such vectors will be safe for manufacture, nearly devoid of toxicity for normal cells and tissue, and far less immunogenic, affording the opportunity for vector repeat dosing and stable gene transfer. We should be able to devise systems to regulate transgenes using drugs that activate signals that trigger promoter function. Finally, HSV vectors may be required that have targeting ligands for more effective use *in vivo*. This will require the modification of the viral envelope in a manner to prevent binding to the normal viral receptors and redirect virus attachment to specific cell types, using targeting ligands engineered into the vector envelope.

Cancer will likely be the first target application for HSV vectors. Here vector toxicity and immunity are less of an issue. Indeed, we are in the process of constructing vectors with multiple cytokine genes and drug activation genes for use as anticancer vectors and cancer vaccines. However, the best use of HSV will likely involve applications in the nervous system. As we become more knowledgeable about the virus biology it would appear possible to engineer HSV vectors for treatment of currently intractable neurodegenerative diseases. Success in these attempts will realize the true potential of this vector system for human gene therapy.

References

Acsadi, G., Dickson, G., Love, D., Jani, A., Walsh, F., Gurusinghe, A., Wolff, J., and Davies, K. (1991). Human dystrophin expression in mdx mice after intramuscular injection of DNA constructs. *Nature (London)* **352**, 815–818.

Acsadi, G., Jani, A., Massie, B., Simoneau, M., Holland, P., Blaschuk, K., and Karpati, G. (1994). A differential efficiency of adenovirus-mediated *in vivo* gene transfer into skeletal muscle cells of different maturity. *Hum. Mol. Genet.* **3**, 579–584.

Adler, R., Glorioso, J. C., Cossman, J., and Levine, M. (1978). Possible role of Fc receptors on cells infected and transformed by herpesvirus: Escape from immune cytolysis. *Infect. Immun.* **21**, 442–447.

Akli, S., Cailland, C., Vigne, E., Stratford-Perricaudet, L., Poenaru, L., Perricaudet, M., and Peschanski, M. (1993). Transfer of a foreign gene into the brain using adenovirus vectors. *Nat. Genet.* **3**, 224–228.

Arahata, K., Ishiura, S., Ishiguro, T., Tsukahara, T., Suhara, Y., Eguchi, C., Ishiara, T., Nonaka, I., Ozawa, E., and Sugita, H. (1988). Immunostaining of skeletal and cardiac muscle surface membrane with antibody against Duchenne muscular dystrophy peptide. *Nature (London)* **333**, 861–863.

Bak, I. J., Markhan, C. H., and Cook, M. L. (1977). Intra-axonal transport of herpes simplex virus in the rat central nervous system. *Brain Res.* **136**, 415–429.

Battleman, D., Geller, A., and Chao, M. (1993). HSV-1 vector-mediated gene transfer of the human nerve growth factor receptor p75hNGFR defines high-affinity NGF binding. *J. Neurosci.* **13**, 941–951.

Bloom, D., Hill, J., Devi-Rao, G., Wagner, E., Feldman, L., and Stevens, J. (1996). A 348-base-pair region in the latency-associated transcript facilitates herpes simplex virus type 1 reactivation. *J. Virol.* **70**, 2449–2459.

Boviatsis, E., Chase, M., Wei, M., Tamiya, T., Hurford, R., Kowall, N., Tepper, R., Breakefield, X., and Chiocca, E. (1994). Gene transfer into experimental brain tumors mediated by adenovirus, herpes simplex virus, and retrovirus vectors. *Hum. Gene Ther.* **5**, 183–191.

Brown, C. R., Nakamura, M. S., Mosca, J. D., Hayward, G. S., Straus, S. T., and Perera, L. P. (1995). Herpes simplex virus trans-regulatory protein ICP27 stabilizes and binds to 3′ ends of labile mRNA. *J. Virol.* **69**, 7187–7195.

Cai, W., Person, S., Warner, S., Zhou, J., and Glorioso, J. (1987). Linker-insertion nonsense and restriction-site deletion mutations of the gB glycoprotein gene of herpes simplex virus type 1. *J. Virol.* **61**, 714–721.

Cai, W., Gu, B., and Person, S. (1988). Role of glycoprotein B of herpes simplex virus type 1 in viral entry and cell fusion. *J. Virol.* **62**, 2596–2604.

Campadelli, G., Brandimarti, R., Lazzaro, C. D., Ward, P., Roizman, B., and Torrisi, M. (1993). Fragmentation and dispersal of Golgi proteins and redistribution of glycoproteins and glycolipids processed through Golgi following infection with herpes simplex virus 1. *Proc. Natl. Acad. Sci. U.S.A.* **90**, 2798–2802.

Carey, M., Leatherwood, J., and Ptashne, M. (1990). A potent GAL4 derivative activates transcription at a distance *in vitro*. *Science* **247**, 710–712.

Casaccia-Bonnefil, P., Benedikz, E., Shen, H., Stelzer, A., Edelstein, D., Geschwind, M., Brownlee, M., Federoff, H. J., and Bergold, P. J. (1993). Localized gene transfer into organotypic hippocampal slice cultures and acute hippocampal slices. *J. Neurosci. Methods* **50**, 341–351.

Challberg, M. (1986). A method of identifying the viral genes required for herpesvirus DNA replication. *Proc. Natl. Acad. Sci. U.S.A.* **83**, 9094–9098.

Chamber, R., Gillespie, G. Y., Soroceanu, L., Andreansky, S., Chatterjee, S., Chou, J., Roizman, B., and Whitley, R. (1995). Comparison of genetically engineered herpes simplex viruses for the treatment of brain tumors in a scid mouse model of human malignant glioma. *Proc. Natl. Acad. Sci. U.S.A.* **92**, 1411–1415.

Chasman, D. I., Leatherwood, M., Carey, M., Ptashne, M., and Kornberg, R. D. (1989). Activation of yeast polymerase II transcription by herpesvirus VP16 and GAL4 derivative *in vitro*. *Mol. Cell. Biol.* **9**, 4746–4749.

Chen, X., Schmidt, M. C., Goins, W. F., Hendricks, R. L., Fink, D., and Glorioso, J. C. (1996). HSV-1 Vector-mediated *lacZ* gene transfer to the nervous system: Long term gene expression using the viral latency-active promoter. *Gene Ther.* (submitted for publication).

Chou, J., and Roizman, B. (1992). The γ 34.5 gene of herpes simplex virus 1 precludes neuroblastoma cells from triggering total shutoff protein synthesis characteristic of programmed cell death in neuronal cells. *Proc. Natl. Acad. Sci. U.S.A.* **89**, 3266–3270.

Chou, J., and Roizman, B. (1994). Herpes simplex virus 1 γ_1 34.5 gene function which blocks the host response to infection maps in the homologous domain of the genes expressed during growth arrest and DNA damage. *Proc. Natl. Acad. Sci. U.S.A.* **91**, 5247–5251.

Coen, D. M., Kosz-Venchak, M., Jacobson, J. G., Leib, D. A., Bogard, C. L., Schaffer, P. A., Tyler, K. L., and Knipe, D. M. (1989). Thymidine kinase-negative herpes simplex virus mutants establish latency in mouse trigeminal ganglia but do not reactivate. *Proc. Natl. Acad. Sci. U.S.A.* **86**, 4736–4740.

Croen, K. D., Ostrove, J. M., Dragovic, L. J., Smialek, J. E., and Straus, S. E. (1987). Latent herpes simplex virus in human trigeminal ganglia. Detection of an immediate early gene "anti-sense" transcript by *in situ* hybridization. *N. Engl. J. Med.* **317**, 1427–1432.

Danko, I., Fritz, J., Latendresse, J., Herweijer, H., Schultz, E., and Wolff, J. (1993). Dystrophin expression improves myofiber survival in mdx muscles following intramuscular plasmid DNA injection. *Hum. Mol. Genet.* **2**, 2055–2061.

Davidson, G., Allen, E., Kozarsky, K. F., Wilson, J., and Roessler, B. (1993). A model system for *in vivo* gene transfer into the central nervous system using an adenoviral vector. *Nat. Genet.* **3**, 219–223.

Davison, A., and Wilkie, N. (1983). Inversion of the two segments of the herpes simplex virus genome in intertypic recombinants. *J. Gen. Virol.* **64**, 1–18.

Deatly, A. M., Spivack, J. G., Lavi, E., O'Boyle, D., and Fraser, N. W. (1988). Latent herpes simplex virus type 1 transcripts in peripheral and central nervous system tissues of mice map to similar regions of the viral genome. *J. Virol.* **62**, 749–756.

Deiss, L., Chou, J., and Frenkel, N. (1986). Functional domain within the "a" sequence involved in the cleavage-packaging of herpes simplex virus DNA. *J. Virol.* **57**, 605–618.

DeLuca, N. A., and Schaffer, P. A. (1985). Activation of immediate-early, early, and late promoters by temperature-sensitive and wild-type forms of herpes simplex virus type 1 protein ICP4. *Mol. Cell. Biol.* **5**, 1997–2008.

DeLuca, N. A., McCarthy, A. M., and Schaffer, P. A. (1985). Isolation and characterization of deletion mutants of herpes simplex virus type 1 in the gene encoding immediate-early regulatory protein ICP4. *J. Virol.* **56**, 558–570.

Desai, P., Schaffer, P., and Minson, A. (1988). Excretion of noninfectious virus particles lacking glycoprotein H by a temperature-sensitive mutant of herpes-simplex virus type 1: Evidence that gH is essential for virion infectivity. *J. Gen. Virol.* **69**, 1147–1156.

Desai, P., Ramakrishnan, R., Lin, Z. W., Ozak, B., Glorioso, J. C., and Levine, M. (1993). The RR1 gene of herpes simplex virus type 1 is uniquely transactivated by ICP0 during infection. *J. Virol.* **67**, 6125–6135.

Devi-Rao, G. B., Goddart, S. A., Hecht, L. M., Rochford, R., Rice, M. K., and Wagner, E. K. (1991). Relationship between polyadenylated and nonpolyadenylated HSV type 1 latency-associated transcripts. *J. Gen. Virol.* **65**, 2179–2190.

deWaal-Malefyt, R., Haanen, J., and Spits, H. (1991). Interleukin 10 (IL-10) and viral IL-10 strongly reduce antigen-specific human T cell proliferation by diminishing the antigen-presenting capacity of monocytes via down-regulation of class II major histocompatibility complex expression. *J. Exp. Med.* **174**, 915–924.

Dixon, R. A. F., and Schaffer, P. A. (1980). Fine-structuring mapping and functional analysis of temperature-sensitive mutants in the gene encoding the herpes simplex virus type 1 immediate early protein VP175. *J. Virol.* **36**, 189–203.

Dobson, A. T., Sederati, F., Devi Rao, G., Flanagan, W. M., Farrell, M. J., Stevens, J. G., Wagner, E. K., and Feldman, L. T. (1989). Identification of the latency-associated transcript promoter by expression of rabbit β-globin mRNA in the mouse sensory nerve ganglia latently infected with a recombinant herpes simplex virus. *J. Virol.* **63**, 3844–3851.

Doerig, C., Pizer, L. I., and Wilcox, C. L. (1991). An antigen encoded by the latency-associated transcripts in neuronal cell cultures latently infected with herpes simplex virus type 1. *J. Virol.* **65**, 2724–2727.

Dunckley, M., Love, D., Davies, K., Walsh, F., Morris, G., and Discon, G. (1992). Retroviral-mediated transfer of a dystrophin minigene into mdx mouse myoblasts *in vitro*. *FEBS Lett.* **2**, 128–134.

During, M., Naegele, J., O'Malley, K., and Geller, A. (1994). Long-term behavioral recovery in parkinsonian rats by an HSV vector expressing tyrosine hydroxylase. *Science* **266**, 1399–1403.

Emi, N., Friedmann, T., and Yee, J. (1991). Pseudotype formation of murine leukemia virus with the G protein of vesicular stomatitis virus. *J. Virol.* **65**, 1202–1207.

Everett, R., and Maul, G. (1994). HSV-1 IE protein Vmw110 causes redistribution of PML. *EMBO J.* **13**, 5062–5069.

Everett, R. D. (1987). The regulation of transcription of viral and cellular genes by herpesvirus immediate-early gene products. *Anticancer Res.* **7**, 589–604.

Fareed, M., and Spivack, J. (1994). Two open reading frames (ORF1 and ORF2) within the 2.0-kilobase latency-associated transcript of herpes simplex virus type 1 are not essential for reactivation from latency. *J. Virol.* **68**, 8071–8081.

Farrell, M. J., Dobson, A. T., and Feldman, L. T. (1991). Herpes simplex virus latency-associated transcript is a stable intron. *Proc. Natl. Acad. Sci. U.S.A.* **88**, 790–794.

Fink, D. J., Sternberg, L. R., Weber, P. C., Mata, M., Goins, W. F., and Glorioso, J. C. (1992). *In vivo* expression of β-galactoside in hippocampal neurons by HSV-mediated gene transfer. *Hum. Gene Ther.* **3**, 11–19.

Frenkel, N., Locker, H., Batterson, W., Hayward, G., and Roizman, B. (1976). Anatomy of herpes simplex DNA. VI. Defective DNA originates from the S component. *J. Virol.* **20**, 527–531.

Furlong, D., Swift, H., and Roizman, B. (1972). Arrangement of herpesvirus deoxyribonucleic acid in the core. *J. Virol.* **10**, 1071–1074.

Furth, P., Hennighausen, L., Baker, C., Beatty, B., and Woychick, R. (1991). The variability in activity of the universally expressed human cytomegalovirus immediate early gene 1 enhancer/promoter in transgenic mice. *Nucleic Acids Res.* **19**, 6205–6208.

Geller, A., and Breakefield, X. (1988). A defective HSV-1 vector expresses *Escherichia coli* β-galactosidase in cultured peripheral neurons. *Science* **241**, 1667–1669.

Geller, A., and Freese, A. (1990). Infection of cultured central nervous system neurons with a defective herpes simplex virus 1 vector results in stable expression of *Escherichia coli* β-galactosidase. *Proc. Natl. Acad. Sci. U.S.A.* **87**, 1149–1153.

Geller, A., During, M., Haycock, J., Freese, A., and Neve, R. (1993). Long-term increases in neurotransmitter release from neuronal cells expressing a constitutively active adenylate cyclase from a herpes simplex virus type 1 vector. *Proc. Natl. Acad. Sci. U.S.A.* **90**, 7603–7607.

Gerster, T., and Roeder, R. (1988). A herpesvirus trans-activating protein interacts with transcription factor OTF-1 and other cellular proteins. *Proc. Natl. Acad. Sci. U.S.A.* **85**, 6347–6351.

Geschwind, M., Kessler, J., Geller, A., and Federoff, H. (1994). Transfer of the nerve growth factor gene into cell lines and cultured neurons using a defective herpes simplex virus vector. Transfer to the NGF gene into cells by a HSV-1 vector. *Brain Res.* **24**, 327–335.

Gibson, W., and Roizman, B. (1971). Compartmentalization of spermine and spermidine in the herpes simplex virion. *Proc. Natl. Acad. Sci. U.S.A.* **68**, 2818–2821.

Goins, W. F., Sternberg, L. R., Croen, K. D., Krause, P. R., Hendricks, R. L., Fink, D. J., Straus, S. E., Levine, M., and Glorioso, J. C. (1994). A novel latency-active promoter is contained within the herpes simplex virus type 1 U_L flanking repeats. *J. Virol.* **68**, 2239–2252.

Gordon, Y. J., Johnson, B., Romanonski, E., and Araullo-Cruz, T. (1988). RNA complementary to herpes simplex virus type 1 ICP0 gene demonstrated in neurons of human trigeminal ganglia. *J. Virol.* **62**, 1832–1835.

He, B., Chou, J., Liebermann, D. A., Hoffman, B., and Roizman, B. (1996). The carboxyl terminus of the murine MyoD116 gene substitutes for the corresponding domain of the γ_1 34.5 gene of herpes simplex virus to preclude the premature shutoff of total protein synthesis in infected human cells. *J. Virol.* **70**, 84–90.

Herold, B., Wudunn, D., Soltys, N., and Spear, P. (1991). Glycoprotein C of herpes simplex virus type 1 plays a principal role in the adsorption of virus to cells and in infectivity. *J. Virol.* **65**, 1090–1098.

Herold, B., Visalli, R., Susmarski, N., Brandt, C., and Spear, P. (1994). Glycoprotein C-independent binding of herpes simplex virus to cells requires cell surface herparan sulfate and glycoprotein B. *J. Gen. Virol.* **74**, 1211–1222.

Hill, A., Jugovic, P., York, I., Russ, G., Bennink, J., Yewdell, J., Ploegh, H., and Johnson, D. (1995). Herpes simplex virus turns off the TAP to evade host immunity. *Nature (London)* **375**, 411–415.

Hill, J. M., Sedarati, F., Javier, R. T., Wagner, E. K., and Stevens, J. G. (1990). Herpes simplex virus latent phase transcription facilitiates *in vivo* reactivation. *Virology* **174**, 117–125.

Ho, D., Mocarski, E., and Sapolsky, R. (1993). Altering central nervous system physiology with a defective herpes simplex virus vector expressing the glucose transporter gene. *Proc. Natl. Acad. Sci. U.S.A.* **90**, 3655–3659.

Ho, D. Y., and Mocarski, E. S. (1989). Herpes simplex virus latent RNA (LAT) is not required for latent infection in the mouse. *Proc. Natl. Acad. Sci. U.S.A.* **86**, 7596–7600.

Hoffer, B., A., Bowenkamp, K., Huettl, P., Hudson, J., Martin, D., Lin, L.-F., and Gerhard, G. (1995). GDNF reverses toxin-induced injury to midbrain dopaminergic neurons *in vivo*. *Neurosci. Lett.* **182**, 107–111.

Hoffman, E., Brown, R., and Kunkel, L. (1987). Dystrophin: The protein product of the Duchenne muscular dystrophy locus. *Cell (Cambridge, Mass.)* **51**, 919–928.

Holland, L. E., Anderson, K. P., Shipman, C., and Wagner, E. K. (1980). Viral DNA synthesis is required for efficient expression of specific herpes simplex virus type 1 mRNA. *Virology* **101**, 10–24.

Honess, R., and Roizman, B. (1974). Regulation of herpes simplex virus macromolecular synthesis. I. Cascade regulation of the synthesis of three groups of viral proteins. *J. Virol.* **14**, 8–19.

Hsu, D., deWaal Malefyt, R., and Fiorentino, D. (1990). Expression of interleukin-10 activity by Epstein–Barr virus protein BCRF1. *Science* **250**, 830–832.

Huard, J., Acsadi, G., Jani, A., Massie, B., and Karpati, G. (1994). Gene transfer into skeletal muscles by isogenic myoblasts. *Hum. Gene Ther.* **5**, 949–958.

Huard, J., Lochmuller, H., Acsadi, G., Jani, A., Holland, P., Guérin, C., Massie, B., and Karpati, G. (1995). Differential short-term transduction efficiency of adult versus newborn mouse tissues by adenoviral recombinants. *Exp. Med. Pathol.* **62**, 131–143.

Huard, J., Akkaraju, G., Watkins, S. C., Pike-Cavalcoli, M., and Glorioso, J. C. (1997). LacZ gene transfer to skeletal muscle using a replication-defective herpes simplex virus type 1 mutant vector. *Hum. Gene Ther.* **8**, 439–452.

Hung, S., Srinivasan, S., Friedman, H., Eisenberg, R., and Cohen, G. (1992). Structural basis of C3b binding by glycoprotein C of herpes simplex virus. *J. Virol.* **66**, 4013–4027.

Hung, S., Peng, C., Kostavasili, I., Friedman, H., Lambrid, J., Eisenberg, R., and Cohen, G. (1994). The interaction of glycoprotein C of herpes simplex virus types 1 and 2 with the alternative complement pathway. *Virology* **203**, 299–312.

Hutchinson, L., and Johnson, D. (1995). Herpes simplex virus glycoprotein K promotes egress of virus particles. *J. Virol.* **69**, 5401–5413.

Hutchinson, L., Browne, H., Wargent, V., Davis-Poynter, N., Primorac, S., Goldsmith, K., Minson, A., and Johnson, D. (1992). A novel herpes simplex virus glycoprotein, gL, forms a complex with glycoprotein H (gH) and affects normal folding and surface expression of gH. *J. Virol.* **6,** 2240–2250.

Jacob, R., Morse, L., and Roizman, B. (1979). Anatomy of herpes simplex virus DNA. XII. Accumulation of head-to-tail concatemers in nuclei of infected cells and their role in the generation of the four isomeric arrangements of viral DNA. *J. Virol.* **29,** 448–457.

Jacobson, J., Leib, D., Goldstein, D., Bogard, C., Schaffer, P., Weller, S., and Cohen, D. (1989). A herpes simplex virus ribonucleotide reductase deletion mutant is defective for productive, acute, and reactivatible infections for mice and for replication in mouse cells. *Virology* **173,** 276–283.

Javier, R. T., Stevens, J. G., Dissette, V. B., and Wagner, E. K. (1988). A herpes simplex virus transcript abundant in latently infected neurons is dispensible for establishment of the latent state. *Virology* **166,** 254–257.

Johnson, A. C., Jinno, Y., and Merlino, G. T. (1988). Modulation of epidermal growth factor receptor proto-oncogene transcription by a promoter site sensitive to S1 nuclease. *Mol. Cell. Biol.* **8,** 4174–4184.

Johnson, D., and Spear, P. (1982). Monensin inhibits the processing of herpes simplex virus glycoproteins, their transport to the cell surface, and the egress of virions from infected cells. *J. Virol.* **43,** 1102–1112.

Johnson, D., Burke, R., and Gregory, T. (1990). Soluble forms of herpes simplex virus glycoprotein D bind to limited number of cell surface receptors and inhibit virus entry into cells. *J. Virol.* **64,** 2569–2576.

Johnson, P., Miyanohara, A., Levine, F., Cahill, T., and Friedmann, T. (1992). Cytotoxicity of a replication-defective mutant herpes simplex virus type 1. *J. Virol.* **66,** 2952–2965.

Johnson, P., Wang, M., and Friedman, T. (1994). Improved cell survival by the reduction of immediate-early gene expression in replication-defective mutants of herpes simplex virus type 1 but not by mutation of the virion host shutoff function. *J. Virol.* **68,** 6347–6362.

Kaplitt, M., Kwong, A., Kleopoulos, S., Mobbs, C., Rabkin, S., and Pfaff, D. (1994a). Preproenkephalin promoter yields region-specific and long-term expression in adult brain after direct *in vivo* gene transfer via a defective herpes simplex viral vector. *Proc. Natl. Acad. Sci. U.S.A.* **91,** 8979–8983.

Kaplitt, M., Leone, P., Samulski, R., Xiao, X., Pfaff, D., O'Malley, K., and During, M. (1994b). Long-term gene expression and phenotype correction using adeno-associated virus vectors in the mammalian brain. *Nat. Genet.* **8,** 148–154.

Karpati, G., and Acsadi, G. (1994). The principles of gene therapy in Duchenne muscular dystrophy. *Clin. Invest. Med.* **17,** 531–541.

Kasahara, N., Dozy, M., and Kan, W. W. (1994). Tissue-specific targeting of retroviral vectors through ligand–receptor interactions. *Science* **266,** 1373–1376.

Katan, M., Haigh, A. Verrijzer, C., van der Vliet, P., and O'Hare, P. (1990). Characterization of a cellular factor which interacts functionally with Oct-1 in the assembly of a multicomponent transcription complex. *Nucleic Acids Res.* **18,** 6871–6880.

Kearns, C., and Gash, D. (1995). GDNF protects nigral dopamine neurons against 6-hydroxydopamine *in vivo*. *Brain Res.* **672,** 104–111.

Koedood, M., Fichtel, A., Meier, P., and Mitchell, P. (1995). Human cytomegalovirus (HCMV) immediate-early enhancer/promoter specificity during embryogenesis defines target tissues of congenital HCMV infection. *J. Virol.* **69,** 2194–2207.

Kosz-Vnenchak, M., Coen, D., and Knipe, D. (1990). Restricted expression of herpes simplex virus lytic genes during establishment of latent infection by thymidine kinase-negative mutant viruses. *J. Virol.* **64,** 5396–5402.

Kristie, T., and Sharp, P. (1993). Purification of the cellular C1 factor required for the stable recognition of the Oct-1 homeodomain by herpes simplex virus α-trans-induction factor (VP16). *J. Biol. Chem.* **268,** 6525–6534.

Kwong, A. D., Kruper, J. A., and Frenkel, N. (1988). Herpes simplex virus virion host shutoff function. *J. Virol.* **62,** 912–921.

Le Gal La Salle, G., Robert, J., Berrard, S., Ridoux, V., Stratford-Perricaudet, L., Perricaudet, M., and Mallet, J. (1993). An adenovirus vector for gene transfer into neurons and glia in the brain. *Science* **259**, 988–990.

Lehner, T., Wilton, J., and Shillitoe, E. (1975). Immunological basis for latency, recurrences, and putative oncogenicity of herpes simplex virus. *Lancet* **2**, 60–62.

Leib, D. A., Nadeau, K. C., Rundle, S. A., and Schaffer, P. A. (1991). Promoter of the latency-associated transcripts of herpes simplex virus type 1 contains a functional cAMP-response element: Role of the latency-associated transcripts and cAMP in reactivation of viral latency. *Proc. Natl. Acad. Sci. U.S.A.* **88**, 48–52.

Leiden, J., Frenkel, N., and Rapp, F. (1980). Identification of the herpes simplex virus DNA sequences present in six herpes simplex virus thymidine kinase-transformed mouse cell lines. *J. Virol.* **33**, 272–285.

Ligas, M., and Johnson, D. (1988). A herpes simplex virus mutant in which glycoprotein D sequences are replaced by β-galactosidase sequences binds to but is unable to penetrate into cells. *J. Virol.* **62**, 1486–1494.

Lokensgard, J. R., Bloom, D. C., Dobson, A. T., and Feldman, L. T. (1994). Long-term promoter activity during herpes simplex virus latency. *J. Virol.* **68**, 7148–7158.

MacLean, A., ul-Fareed, M., Robertson, L., Harland, J., and Brown, S. (1991). Herpes simplex virus type 1 deletion variants 1714 and 1716 pinpoint neurovirulence-related sequences in Glasgow strain 17+ between immediate early gene 1 and the "a" sequence. *J. Gen. Virol.* **72**, 631–639.

Margolis, T. P., Bloom, D. C., Dobson, A. T., Feldman, L. T., and Stevens, J. G. (1993). Decreased reporter gene expression during latent infection with HSV LAT promoter constructs. *Virology* **197**, 585–592.

Markert, J., Malick, A., Coen, D., and Martuza, R. (1993). Reduction of elimination of encephalitis in experimental glioma therapy model with attenuated herpes simplex mutants that retain susceptibility to acyclovir. *Neurosurgery* **32**, 597–603.

Martuza, R., Malick, A., Markert, J., Ruffner, K., and Coen, D. (1991). Experimental therapy of human glioma by means of a genetically engineered virus mutant. *Science* **252**, 854–856.

Maul, G., and Everett, R. (1994). The nuclear location of PML, a cellular member of the C3HC4 zinc-binding domain protein family, is rearranged during herpes simplex virus infection by the C3HC4 viral protein ICP0. *J. Gen. Virol.* **75**, 1223–1233.

Mavromara-Nazos, P., and Roizman, B. (1967). Activation of herpes simplex virus 1 γ_2 genes by viral DNA replication. *Virology* **161**, 593–598.

McFarland, D. J., Sikora, E., and Hotchkin, J. (1986). The production of focal herpes encephalitis in mice by stereotaxic inoculation of virus. Anatomical and behavioral effects. *J. Neurol. Sci.* **72**, 307–318.

McGregor, F., Phelan, A., Dunlop, J., and Clements, J. (1996). Regulation of herpes simplex virus poly(A) site usage and the action of immediate-early protein IE63 in the early–late switch. *J. Virol.* **70**, 1931–1940.

Mineta, T., Rabkin, S., and Martuza, R. (1994). Treatment of malignant gliomas using ganciclovir-hypersensitive, ribonucleotide reductase-deficient herpes simplex viral mutant. *Cancer Res.* **54**, 3963–3966.

Mocarski, E., and Roizman, B. (1982). Structure and role of the herpes simplex virus DNA termini in inversion, circularization and generation of virion DNA. *Cell (Cambridge, Mass.)* **31**, 89–97.

Moore, K., Vieira, P., Fiorentino, D., Trounstine, M., Khan, T., and Mosmann, T. (1990). Homology of cytokine synthesis inhibitory factor (IL-10) to the Epstein–Barr virus gene BCRFI. *Science* **248**, 1230–1234.

Morgan, J., Pagel, C., Sherrat, T., and Partridge, T. (1993). Long-term persistence and migration of myogenic cells injected into preirradiated muscles of mdx mice. *J. Neurol. Sci.* **115**, 191–200.

Navarro, D., Paz, P., and Pereira, L. (1992). Domains of herpes simplex virus 1 glycoprotein B that function in virus penetration, cell-to-cell spread, and cell fusion. *Virology* **186,** 99–112.

Newcomb, W., and Brown, J. (1994). Induced extrusion of DNA from the capsid of herpes simplex virus type 1. *J. Virol.* **68,** 433–440.

O'Hare, P., and Goding, C. (1988). Herpes simplex virus regulatory elements and the immunoglobulin octamer domain bind a common factor and are both targets for virion transactivation. *Cell (Cambridge, Mass.)* **52,** 435–445.

Oligino, T., Poliani, P. L., Marconi, P., Bender, M. A., Schmidt, M. C., Fink, D. J., and Glorioso, J. C. (1996). *In vivo* transgene activation from an HSV-based gene vector by GAL4 : VP16. *Gene Ther.* **3,** 892–899.

Oroskar, A., and Read, G. (1989). Control of mRNA stability by the virion host shutoff function of herpes simplex virus. *J. Virol.* **63,** 1897–1906.

Perng, G.-C., Chokephaibulkit, K., Thompson, R., Sawtell, N., Slanina, S., Ghiasi, H., Nesburn, A., and Wechsler, S. (1996). The region of the herpes simplex virus type 1 LAT gene that is colinear with the ICP34.5 gene is not involved in spontaneous reactivation. *J. Virol.* **70,** 282–291.

Preston, C. (1979). Control of herpes simplex virus type 1 mRNA synthesis in cells infected with wild-type virus or the temperature-sensitive mutant tsK. *J. Virol.* **29,** 275–284.

Preston, C., Frame, M., and Campbell, M. (1988). A complex formed between cell components and an HSV structural polypeptide binds to a viral immediate early gene regulatory DNA sequence. *Cell (Cambridge, Mass.)* **52,** 425–434.

Quantin, B., Perricaudet, L., Tajbakhsh, S., and Mandel, J.-L. (1992). Adenovirus as an expression vector in muscle cells *in vivo. Proc. Natl. Acad. Sci. U.S.A.* **89,** 2581–2584.

Ragot, T., Vincent, N., Chafey, P., Gilgenkrantz, H., Couton, D., Cartaud, J., Briand, P., Kaplan, J.-C., Perricaudet, M., and Kahn, A. (1993). Efficient adenovirus-mediated transfer of a human minidystrophin gene to skeletal muscle of mdx mice. *Nature (London)* **324,** 647–650.

Ramakrishnan, R., Fink, D. J., Guihua, J., Desai, P., Glorioso, J. C., and Levine, M. (1994a). Competitive quantitative polymerase chain reaction (PCR) analysis of herpes simplex virus type 1 DNA and LAT RNA in latently infected cells of the rat brain. *J. Virol.* **68,** 1864–1870.

Ramakrishnan, R., Levine, M., and Fink, D. (1994b). PCR-based analysis of herpes simplex virus type 1 latency in the rat trigeminal ganglion established with a ribonucleotide reductase-deficient mutant. *J. Virol.* **68,** 7083–7091.

Read, G. S., and Frenkel, N. (1983). Herpes simplex virus mutants defective in the virion-associated shutoff of host polypeptide synthesis and exhibiting abnormal synthesis of α (immediate early) viral polypeptides. *J. Virol.* **46,** 498–512.

Rice, S., Long, M., Lam, V., and Spencer, C. (1994). RNA polymerase II is aberrantly phosphorylated and localized to viral replication compartments following herpes simplex virus infection. *J. Virol.* **68,** 988–1001.

Rock, D. L., Nesburn, A. B., Ghiasi, H., Ong, J., Lewis, T. L., Lokensgard, J. R., and Wechsler, S. (1987). Detection of latency-related viral RNAs in trigeminal ganglia of rabbits latently infected with herpes simplex virus type 1. *J. Virol.* **61,** 3820–3826.

Roizman, B., and Furlong, D. (1974). The replication of herpesviruses. *In* "Comprehensive Virology" (H. Fraenkel-Conrat and R. R. Wagner, eds.), pp. 229–403. Plenum, New York.

Roop, C., Hutchinson, L., and Johnson, D. (1993). A mutant herpes simplex virus type 1 unable to express glycoprotein L cannot enter cells, and its particles lack glycoprotein H. *J. Virol.* **67,** 2285–2297.

Russell, J., Stow, E., Stow, N., and Preston, C. (1987). Abnormal forms of the herpes simplex virus immediate early polypeptide Vmw175 induce the cellular stress response. *J. Gen. Virol.* **68,** 2397–2406.

Sacks, W., Greene, C., Aschman, D., and Schaffer, P. (1985). Herpes simplex virus type 1 ICP27 is essential regulatory protein. *J. Virol.* **55**, 796–805.

Sadowski, I., Ma, J., Triezenberg, S., and Ptashne, M. (1988). GAL4/VP16 is an unusually potent transcriptional activator. *Nature (London)* **335**, 563–564.

Salvatori, G., Ferrari, G., Messogiorno, A., Servidei, A., Coletta, M., Tonali, P., Giavazzi, R., Cossu, G., and Mavilio, F. (1993). Retroviral vector-mediated gene transfer into human primary myogenic cells leads to expression in muscle fibers *in vivo*. *Hum. Gene Ther.* **4**, 713–723.

Samaniego, L., Webb, A., and DeLuca, N. (1995). Functional interaction between herpes simplex virus immediate-early proteins during infection: Gene expression as a consequence of ICP27 and different domains of ICP4. *J. Virol.* **69**, 5705–5715.

Sandri-Goldin, R., and Hibbard, M. (1996). The herpes simplex virus type 1 regulatory protein ICP27 coimmunoprecipitates with anti-sm antiserum, and the C terminus appears to be required for this interaction. *J. Virol.* **70**, 108–118.

Sandri-Goldin, R., Hibbard, M., and Hardwicke, M. (1995). The C-terminal repressor region of herpes simplex virus type 1 ICP27 is required for the redistribution of small nuclear ribonucleoprotein particles and splicing factor SC25; however, these alterations are not sufficient to inhibit host cell splicing. *J. Virol.* **69**, 6063–6076.

Sauer, B., Whealy, M., Robbins, A., and Enquist, L. (1987). Site-specific insertion of DNA into a pseudorabies virus vector. *Proc. Natl. Acad. Sci. U.S.A.* **84**, 3160–3167.

Sawtell, N. M., and Thompson, R. L. (1992). Herpes simplex virus type 1 latency-associated transcription unit promotes anatomical site-dependent establishment and reactivation from latency. *J. Virol.* **66**, 2157–2169.

Schmidt, E., Christoph, G., Zeller, R., and Leder, P. (1990). The cytomegalovirus enhancer: A pan-active control element in transgenic mice. *Mol. Cell. Biol.* **10**, 4406–4411.

Sedarati, F., Izumi, K. M., Wagner, E. K., and Stevens, J. G. (1989). Herpes simplex virus type 1 latency-associated transcript plays no role in establishment or maintenance of a latent infection in murine sensory neurons. *J. Virol.* **63**, 4455–4458.

Shieh, M.-T., and Spear, P. (1994). Herpes virus-induced cell fusion that is dependent on cell surface heparan sulfate on soluble heparin. *J. Virol.* **68**, 1224–1228.

Skaliter, R., Makhov, A., Griffith, J., and Lehman, I. (1996). Rolling circle DNA replication by extracts of herpes simplex virus type 1-infected human cells. *J. Virol.* **70**, 1132–1136.

Smith, I. L., Hardwicke, M. A., and Sandri-Goldin, R. M. (1992). Evidence that the herpes simplex virus immediate early protein ICP27 acts posttranscriptionally during infection to regulate gene expression. *Virology* **186**, 74–86.

Spaete, R., and Frenkel, N. (1982). The herpes simplex virus amplicon: A new eucaryotic defective-virus cloning amplifying vector. *Cell (Cambridge, Mass.)* **30**, 295–304.

Spear, P. (1993). Membrane fusion induced by herpes simplex virus. *In* "Viral Fusion Mechanisms" (J. Bentz, ed.), pp. 201–232. CRC Press, Boca Raton, Florida.

Spivack, J., and Fraser, N. (1988). Expression of herpes simplex virus type 1 latency-associated transcripts in trigeminal ganglia of mice during acute infection and reactivation of latent infection. *J. Virol.* **62**, 1479–1485.

Spivack, J. G., and Fraser, N. W. (1987). Detection of herpes simplex virus type 1 transcripts during latent infection in mice. *J. Virol.* **61**, 3841–3847.

Spivack, J. G., Woods, G. M., and Fraser, N. W. (1991). Identification of a novel latency-specific splice donor signal within HSV type 1 2.0-kilobase latency-associated transcript (LAT): Translation inhibition of LAT open reading frames by the intron within the 2.0-kilobase LAT. *J. Virol.* **65**, 6800–6810.

Steiner, I., Spivack, J. G., Lirette, R. P., Brown, S. M., MacLean, A. R., Subak-Sharpe, J., and Fraser, N. W. (1989). Herpes simplex virus type 1 latency-associated transcripts are evidently not essential for latent infection. *EMBO J.* **8**, 505–511.

Stevens, J. G., Wagner, E. K., Devi-Rao, G. B., Cook, M. L., and Feldman, L. T. (1987). RNA complementary to a herpesvirus α gene mRNA is prominent in latently infected neurons. *Science* **255**, 1056–1059.

Stinski, M. F., and Roehr, T. J. (1985). Activation of the major immediate early gene of human cytomegalovirus by cis-acting elements in the promoter-regulatory sequence and by virus-specific trans-acting components. *J. Virol.* **55**, 431–441.

Thomsen, D. R., Sternberg, R. M., Goins, W. F., and Stinski, M. F. (1984). Promoter regulatory region of the major immediate early gene of human cytomegalovirus. *Proc. Natl. Acad. Sci. U.S.A.* **81**, 659–663.

Tomac, A., Linquist, E., Lin, L.-F., Ogren, S., Young, D., Hoffer, B., and Olson, L. (1995). Protection and repair of the nigrostriatal dopaminergic system by GDNF *in vivo*. *Nature (London)* **373**, 335–339.

Vieira, P., Waal-Malefty, R. D., and Dang, M. (1991). Isolation and expression of human cytokine synthesis inhibitory factor cDNA clones: Homology to Epstein–Barr virus open reading frame BCRFI. *Proc. Natl. Acad. Sci. U.S.A.* **88**, 1172–1176.

Vincent, N., Ragot, T., Gilgenkrantz, H., Couton, D., Chafey, P., Grégoire, A., Briand, P., Kaplan, J.-C., Kahn, A., and Perricaudet, M. (1993). Long-term correction of mouse dystrophic degeneration by adenovirus-mediated transfer of a minidystrophin gene. *Nat. Genet.* **5**, 130–134.

Wagner, E. K., Devi-Rao, G., Feldman, L. T., Dobson, A. T., Zhang, Y., Elanagan, W. F., and Stevens, T. (1988). Physical characterization of the herpes simplex virus latency-associated transcript in neurons. *J. Virol.* **63**, 1194–2002.

Watson, R., and Clements, J. (1980). A herpes simplex virus type 1 function continuously required for early and late virus RNA synthesis. *Nature (London)* **285**, 329–330.

Werstuck, G., and Capone, J. (1993). An unusual cellular factor potentiates protein–DNA complex assembly Oct-1 and Vmw65. *J. Biol. Chem.* **268**, 1272–1278.

Whitley, R., Kern, E., Chatterjee, S., Chou, J., and Roizman, B. (1993). Replication of establishment of latency, and induced reactivation of herpes simplex virus γ_1 34.5 deletion mutants in rodent models. *J. Clin. Invest.* **91**, 2837–2843.

Wilson, A., LaMarco, K., Peterson, M., and Herr, W. (1993). The VP16 accessory protein HCF is a family of polypeptides processed from a large precursor protein. *Cell (Cambridge, Mass.)* **74**, 115–125.

Wudunn, D., and Spear, P. (1989). Initial interaction of herpes simplex virus with cells is binding to heparan sulfate. *J. Virol.* **63**, 52–58.

Yamada, Y., Kimura, H., Morishima, T., Daikoku, T., Maeno, K., and Nishiyama, K. (1991). The pathogenicity of ribonucleotide reductase-null mutants of herpes simplex virus type 1 in mice. *J. Infect. Dis.* **164**, 1091–1097.

Yang, Y., Nunes, F., Berencis, K., Gonczol, E., Engelhardt, J., and Wilson, J. (1994). Inactivation of E2a in recombinant adenoviruses improves the prospect for gene therapy in cystic fibrosis. *Nat. Genet.* **7**, 362–369.

York, I., Roo, C., Andrews, D., Riddell, S., Graham, F., and Johnson, D. (1994). A cytosolic herpes simplex virus protein inhibits antigen presentation to CD8+ T lymphocytes. *Cell (Cambridge Mass.)* **77**, 525–535.

Zubrzycka-Gaarn, E. E., Bulman, D. E., Karpati, G., Burghes, A. H., Belfall, B., Klamut, H. J., Talbot, J., Hodges, R. S., Ray, P. N., and Worton, R. G. (1988). The Duchenne muscular dystrophy gene is localized in the sarcolemma of human skeletal muscle. *Nature (London)* **333**, 466–469.

Mary M. Hitt*
Christina L. Addison*
Frank L. Graham*,†

*Department of Biology
McMaster University
Hamilton, Ontario, Canada L8S 4K1

†Department of Pathology
McMaster University
Hamilton, Ontario, Canada L8S 4K1

Human Adenovirus Vectors for Gene Transfer into Mammalian Cells

Adenovirus (Ad) vectors are probably the most efficient means currently available for delivering foreign genes into mammalian cells both *in vivo* and in cell culture. The adenoviruses have been well characterized in the four decades since their first isolation, and these investigations have revealed several features of Ad biology that contribute to the effectiveness of the virus as a gene transfer vector. Numerous studies have shown that a wide variety of cell types and tissues of many different species can be infected by Ad. In addition, both dividing and nondividing cells can be infected at high efficiencies. The infection is rapid and requires no exposure to toxic or harmful substances, in contrast to some methods for introducing foreign DNA into cells. Furthermore, the genome is relatively easy to manipulate using standard molecular biological techniques, facilitating construction of recombinants. Finally, recombinant vectors can be easily produced and purified on a large scale, yielding viral stocks with titers up to 10^{12} plaque-forming units (PFU)/ml.

Advances in Pharmacology, Volume 40

Because the Ad vector system holds such great potential for expression of foreign DNA, particularly in gene therapy regimens, it has been the subject of intense research. A number of excellent reviews (Graham and Prevec, 1992; Berkner, 1992; Trapnell, 1993; Becker *et al.*, 1994; Bramson *et al.*, 1995) are available that describe many of these investigations. It is the goal of this chapter in part to highlight some of these landmark studies, but mainly to focus on advances made in Ad vector design, as well as to describe the current status of Ad vector applications for expression analysis and in gene therapy.

I. The Structure and Lytic Cycle of Adenovirus

More than 100 different serotypes of Ad have been identified, about half of which are derived from humans (Hierholzer *et al.*, 1988). Among the earliest findings from Ad studies was the observation that certain Ad serotypes were oncogenic in newborn rodents, the first such observation with a human virus (reviewed by Branton *et al.*, 1985). This prompted intensive research into the molecular biology and genetics of Ads, an activity that continues unabated. The human Ad serotypes 2, 5, 7, and 12 have been studied most extensively, but all of the known human serotypes generally share similar structural and biological features (reviewed in Ginsberg, 1984). The virus has a double-stranded linear DNA genome of about 30–40 kilobase pairs (kb) with inverted terminal repeats (ITRs) at either end. The genome is packaged within the viral capsid to form a nonenveloped virion with a diameter of approximately 140 nm. The icosahedral capsid is composed predominantly of virally encoded hexon, penton base, and fiber proteins. A three-dimensional structure has been proposed for Ad on the basis of crystallographic and cryoelectron microscopic data (Stewart *et al.*, 1991, 1993).

Viral entry into the target cell occurs in two separable phases: attachment and internalization. First, the fiber protein projecting from the virion binds with high affinity to an as yet unidentified receptor on the surface of the cell (Philipson *et al.*, 1968; Defer *et al.*, 1990). Competition studies indicate that at least some of the different serotypes utilize widespread but distinct attachment receptors (Defer *et al.*, 1990; Stevenson *et al.*, 1995). Next, internalization of the virus occurs by binding of penton base to α_v integrins (the cellular receptors for the extracellular matrix protein vitronectin) (Wickham *et al.*, 1993) followed by receptor-mediated endocytosis. Once inside the cell, the endosomal membrane is lysed in a process mediated by the penton base, releasing the contents of the endosome to the cytoplasm (Seth *et al.*, 1984; Curiel *et al.*, 1991). Interestingly, even defective Ad particles will facilitate the uptake of nonviral macromolecules into host cells by receptor-mediated endocytosis (Defer *et al.*, 1990; Seth *et al.*, 1994), a property that has been exploited to introduce foreign DNA into mammalian cells (Yoshimura *et al.*, 1993; Michael and Curiel, 1994; Cotten, 1995).

Transcription, replication, and packaging of Ad DNA occur within the nucleus of the infected cell, with the first two processes utilizing host cell functions in addition to viral proteins. Expression of the Ad genome can be divided roughly into two phases; early, which occurs prior to DNA replication, and late, which occurs after the onset of DNA replication. The genes that are expressed early after infection (within the first 6–8 hr) are located in four regions of the genome (Fig. 1): early region 1 (E1), which is composed of the two transcription units E1A and E1B, E2, E3, and E4 (reviewed by Nevins, 1987). Multiple mRNAs are produced from each of these regions. E1A proteins, which are among the first viral products expressed after infection, are involved in transcriptional regulation of the virus and are required for replication. E1B products are important for the early-to-late transition by influencing viral and cellular mRNA metabolism and host protein shutoff. In addition, E1B proteins counteract E1A-induced apoptosis (White *et al.*, 1991). The E2 region encodes proteins required for replication of viral DNA: the 72-kDa DNA-binding protein encoded at approximately 60 to 70 map units (mu), and the DNA polymerase and terminal protein precursor (pTP), encoded between 10 and 25 mu.

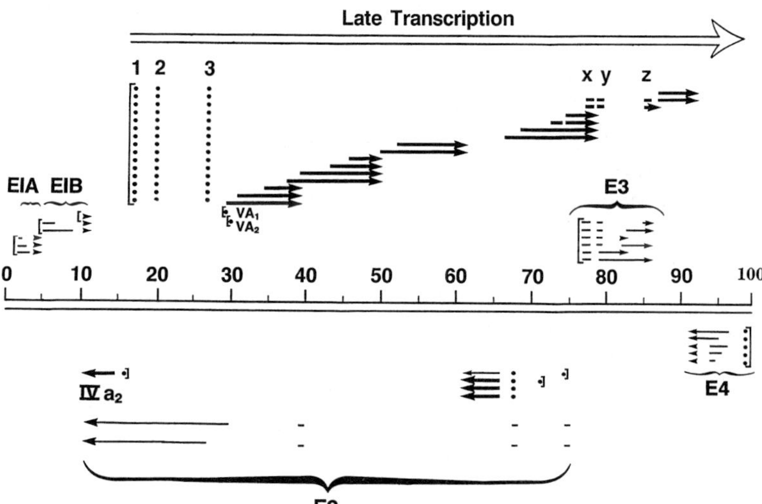

FIGURE 1   Transcription map of Ad5. The genome length is given as 100 map units (mu) with each map unit equal to 360 bp. Early transcripts are indicated by light lines and late transcripts by heavy lines, with the arrowhead at the 3′ end of the message. The primary transcript initiating at the MLP (shown as an open arrow) is processed by alternative splicing and polyadenylation into the five families of late transcripts. The tripartite leader of these late messages is indicated by 1, 2, and 3 at mu 16.5, 19.5, and 26.5, respectively.

The E3 region is not required for Ad replication in cultured cells, but E3-encoded proteins play an important role *in vivo* by modulating the host immune response to virally infected cells (reviewed by Wold and Gooding, 1989). At least eight different infected cell proteins map to this region, two of which, the 19- and 14.7-kDa proteins, have been studied in great detail. The 19-kDa protein binds to major histocompatibility complex (MHC) class I molecules and prevents their transport to the cell surface, thus reducing recognition and subsequent lysis of infected cells by cytotoxic T lymphocytes (CTLs). The 14.7-kDa protein protects infected cells by preventing cytolysis by tumor necrosis factor α.

The E4 region is less well characterized than the other early regions of Ad. Products of E4 facilitate DNA replication, enhance late viral gene expression, and decrease host protein synthesis. At least seven open reading frames (ORFs) map to E4, although all of these except ORF6 may be dispensible for growth of the virus in cultured cells (Bridge and Ketner, 1989). The 34-kDa ORF6 gene product complexes with the 55-kDa E1B product and it seems probable that this complex functions in host protein shutoff (Sarnow *et al.*, 1984; Babiss *et al.*, 1985; Pilder *et al.*, 1986; Cutt *et al.*, 1987).

Viral DNA replication, which begins at about 6–8 hr postinfection in permissive cells, requires both viral-encoded and cellular factors (reviewed in Ginsberg, 1984). The ITRs are the only sequences required in *cis* for Ad DNA replication (Hay, 1985). DNA synthesis is initiated by a protein-priming step involving covalent linkage of dCMP to pTP. The 3′-OH of the pTP-dCMP then serves as a primer to initiate DNA synthesis at either 5′ end of the genome, which continues through to the 3′ end of the template by a strand displacement mechanism. The ITR sequences found at the termini of the displaced parental strand can anneal to form a panhandle structure, which can also function as a template for initiation of DNA synthesis (Daniell, 1976; Lippe and Graham, 1989).

Following the onset of viral DNA replication, the late regions of the genome are expressed. This is accompanied by inhibition of both host mRNA transport and host protein synthesis, such that nearly all the protein synthesized late in infection is viral. Most of the late transcripts initiate from the major late promoter (MLP) and are differentially spliced to encode almost all of the structural virion proteins (reviewed in Ginsberg, 1984). These MLP-directed mRNAs, the L1 to L5 families of transcripts, all derive from the same primary transcript that terminates at a site near the right end of the genome, and contain the well-characterized tripartite leader sequence at their 5′ ends. Members within each family share a common cleavage and polyadenylation [poly(A)] site, but differ in the intron splicing that joins the tripartite leader to the main body of the message. The structural proteins penton base, hexon, and fiber are encoded by L2, L3, and L5 transcripts, respectively.

In addition to the MLP, three other late promoters are also active at this time (those of pIX and pIVa2, and the late E2 promoter). pIX is important for heat stability of the virus (Colby and Shenk, 1981) and for efficient packaging of full-length genomes (Ghosh-Choudhury *et al.*, 1987). Other predominantly late transcripts include the RNA polymerase III-transcribed VA RNAs, which are required for the translation of late viral mRNAs (Thimmapaya *et al.*,1982). By 30–40 hr postinfection, virus production is complete, resulting in approximately 1000 PFU, and roughly 10 to 100 times as many viral particles, per host cell.

II. Recombinant Adenovirus Vector Construction

A. First-Generation Adenovirus Vectors

The human Ad serotypes 2 and 5 form the backbone of what are now called the first-generation Ad vectors. There are several reasons for selecting these two types. They are by far the best characterized of all the serotypes, they replicate to higher yields than most other human Ad serotypes, and the complete DNA sequences of both Ad2 and Ad5, which are 95% homologous, are available (Roberts *et al.*, 1984; Chroboczek *et al.*, 1992). Although none of the Ad serotypes has ever been associated with human malignancies, Ad2 and Ad5 have added safety for animal and human studies in that they belong to subgroup C of Ad, a subclass which is nononcogenic in rodents (deletion of the E1 region would render even the highly oncogenic group A viruses nontumorigenic). Recombinant vectors can be produced with several different modifications of the Ad2 or Ad5 backbone, each with differing utility depending on the application.

1. Replication-Defective versus Nondefective Vectors

Because genome sizes up to 105% of wild type can be packaged (Bett *et al.*, 1993), Ad2 and Ad5 can accept foreign DNA inserts of up to 1.8 to 2 kb without any concomitant deletion of viral sequences. To accommodate larger inserts, Ads have been modified by deletion of the E1 and/or E3 regions to produce the first-generation vectors. E1-deleted viruses are replication defective unless E1 is supplied in *trans*, and such viruses are generally propagated in 293 cells, a human embryonic kidney cell line transformed by the E1 region of Ad5 (Graham *et al.*, 1977). The left ITR [nucleotides (nt) 1–103 of Ad5] and the viral packaging signal (nt 194–358) (Hearing *et al.*, 1987) are required in *cis* for viral replication, and the gene encoding protein IX (nt 3525–4088) is required for packaging of full-length viral DNA. Therefore up to 3.2 kb of E1 can be deleted without affecting growth of the virus in 293 cells. Because replication of the virus leads to death of the infected cell as well as dissemination of the recombinant Ad, gene therapy protocols and many *in vitro* expression studies require the use of the highly attenuated E1-deleted Ad vectors.

Viruses with only E3 deleted (the maximum deletion reported to date is 3.1 kb; Bett *et al.*, 1994), on the other hand, are replication proficient. Because the copy number of the virus increases dramatically during the course of infection, a higher level of recombinant gene expression theoretically can be obtained with $E1^+$ vectors. Therefore $E1^+E3^-$ vectors may be the most suitable for large-scale *in vitro* recombinant protein production and for live Ad vaccines provided the expression cassettes are not greater than about 5 kb. In at least one investigation, however, the same level of transgene expression was obtained for both defective and nondefective vectors (Levrero *et al.*, 1991), although these results may have been influenced by the site of insertion (E1) as well as the replication ability of the viruses. In practice, most Ad vector backbones are both E1 and E3 deleted.

2. Sites for Insertion of Foreign DNA

Most foreign sequences have been inserted into first-generation Ad vectors in place of either E1 or E3. In E1 replacement vectors, foreign genes are essentially always inserted as expression cassettes under the control of heterologous promoters. However, because the E1A enhancer overlaps the packaging signal, it is present in most E1-deleted vectors and may exert some influence over genes inserted in this region. Translocating the packaging signal to the right end of the viral genome (Hearing and Shenk, 1983) would alleviate any problems associated with this influence. Foreign DNA can be inserted in either the leftward or rightward orientation (with respect to the Ad genome; see Fig. 1), although inserts oriented rightward in E1 frequently express at a higher level than those inserted leftward (Hitt *et al.*, 1995). Some genes, however, cannot be rescued in the rightward orientation (M. M. Hitt, C. L. Addison, and F. L. Graham, unpublished observations), possibly due to aberrant transcripts extending into Ad sequences downstream of the expression cassette that are required for virus viability, or because levels of transgene expression are potentially too high to permit viral replication. To avoid aberrant transcription, it is desirable, and sometimes necessary, to include a heterologous poly(A) signal 3' of rightward-oriented genes to ensure efficient transcriptional termination.

In defective (i.e., E1-deleted) vectors, foreign DNA inserted in either orientation in E3 has been expressed (Bramson *et al.*, 1996a). In contrast, in nondefective $E1^+$ vectors, it is often observed that only DNA inserted in the rightward orientation in E3 is expressed at high levels (Johnson *et al.*, 1988; Schneider *et al.*, 1989; Mittal *et al.*, 1993; Yarosh *et al.*, 1996). In such cases, the inserted gene lacked a strong promoter and expression was largely due to transcription initiating in viral sequences upstream of the insert, either at the E3 promoter or the MLP and continuing through the inserted sequence.

In addition to the E1 and E3 replacement vectors, replication-proficient recombinant vectors have been generated that have foreign DNA inserted in the right end of the Ad genome between the E4 promoter and the right ITR (Saito *et al.*, 1985; Mason *et al.*, 1990). One of these vectors, containing a rightward-directed insert under the control of a strong promoter, expressed high levels of transgene product (Mason *et al.*, 1990). It may also be possible to insert foreign DNA in the E2 region between the DNA polymerase and the pTP genes (Munz and Young, 1987) without disrupting virus viability.

3. Vector Systems and Recombinant Virus Construction

There are several strategies commonly used to generate first-generation recombinant Ad vectors, and protocols detailing these strategies are available (Hitt *et al.*, 1995; Spector and Samaniego, 1995; Gerard and Meidell, 1995). All involve cloning the gene of interest, usually with exogenous promoter and poly(A) sequences, into a plasmid such that the insert is flanked by Ad sequences homologous to the region of the viral genome targeted for DNA insertion. DNA from this shuttle plasmid is then rescued into infectious viral DNA either by *in vitro* ligation followed by transfection or more usually by *in vivo* homologous recombination in cotransfected 293 cells, as described below.

Sequences can be rescued into E1 of Ad5 by insertion into a "shuttle" plasmid that contains the left 16% of the viral genome from which most of E1 has been deleted, but which retains all or most of the left ITR, the packaging signal, and pIX coding sequence. Shuttle plasmids have been constructed that contain multicloning sites flanked by strong promoters and poly(A) signals for optimal expression in Ad vectors (Hitt *et al.*, 1995; Gerard and Meidell, 1995). If convenient restriction sites are available in the vector backbone, this shuttle can be ligated *in vitro* to a fragment of viral DNA and used to transfect E1-complementing 293 cells to generate infectious virions. Owing to the large size of the virus genome, however, appropriate unique restriction sites are not often available, and most protocols rely on *in vivo* homologous recombination. In this case, parental Ad5 DNA (the strain dl309; Jones and Shenk, 1979) can be cleaved with two or more restriction enzymes (e.g., *Cla*I and *Xba*I) that recognize sites in the left terminus, thus separating the left ITR and packaging signal from the rest of the genome and minimizing the possibility of religation of the parental fragments. The large fragment of restricted viral DNA, preferably purified by agarose gel electrophoresis, can then be cotransfected with the shuttle plasmid into 293 cells (Fig. 2A). As little as 1 kb of overlap between the plasmid and the viral DNA is sufficient for recombination (Berkner and Sharp, 1983). Infectivity of Ad DNA is increased by one or two orders of magnitude if the terminal protein is not removed from the viral DNA during purification (Sharp *et al.*, 1976). Although the efficiency of vector isolation is high with this system, it is sometimes complicated by the preferential

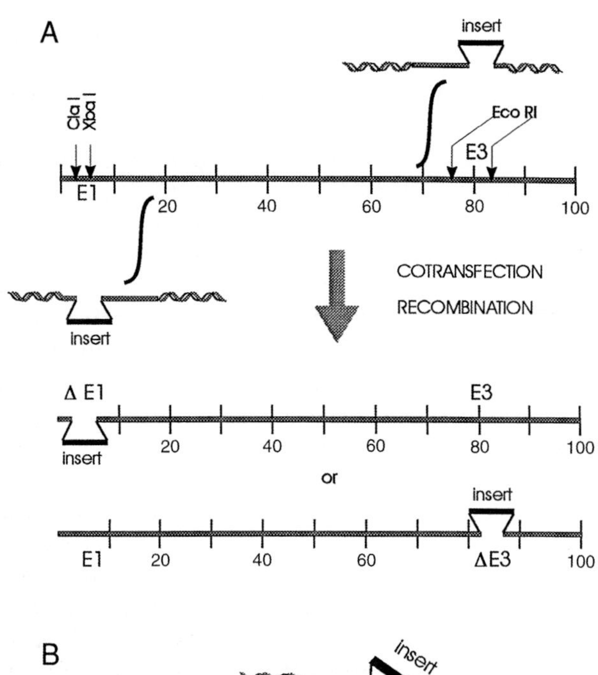

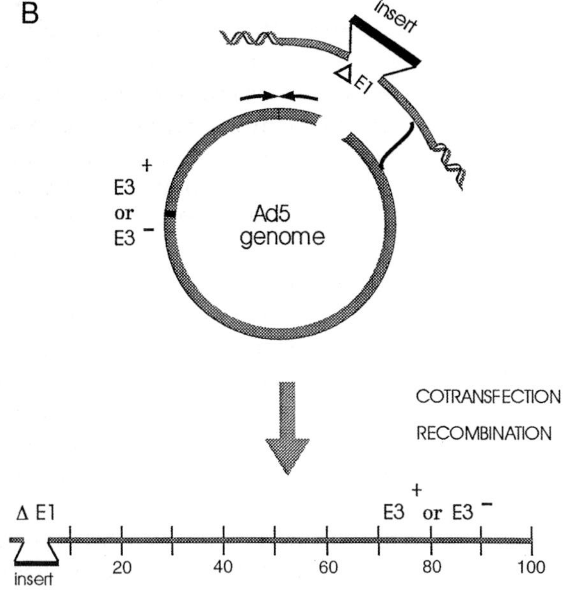

FIGURE 2   Strategies for rescue of foreign DNA sequences into first-generation Ad vectors. (A) Rescue by *in vivo* recombination between the linear viral genome and an E1 (shown below the genome map) or an E3 (shown above the genome map) shuttle plasmid carrying the foreign DNA sequence. Restriction sites suitable for digesting the linear viral DNA prior to cotransfection with the shuttle plasmid are shown as thin black arrows. Note that the *Xba*I site at the left end is unique only in the dl309 mutant of Ad. (B) Rescue by *in vivo* recombination

rescue of parental virus, which frequently has a growth advantage over the recombinant. To facilitate recombinant virus rescue, two strategies have been proposed in which the parental viruses carry selectable markers. A conditionally lethal parental virus has been constructed that contains the herpes simplex virus type 1 (HSV-1) thymidine kinase gene in the E1 region (Imler *et al.*, 1995a). Recombinants derived from this vector by E1 replacement can replicate in the presence of ganciclovir, whereas the parental virus cannot. Another vector backbone has been generated in which the gene encoding β-galactosidase is inserted in E1 (Schaack *et al.*, 1995a). Plaques formed by this parental virus can be easily distinguished from recombinants by their blue color after staining with 5-bromo-4-chloro-3-indolyl-β-D-galactopyranoside (X-Gal). A possible problem with both of these strategies is that viruses that have deleted all or part of the marker genes will have the same phenotype as vectors containing the desired insert.

As an alternative strategy, the shuttle plasmid can be recombined *in vivo* with a plasmid containing a full-length modified Ad genome (see Fig. 2B). Plasmids carrying a circular form of Ad5 with the ITRs joined head to tail, and a bacterial origin of replication and selectable marker inserted in either E1 or E3, are infectious in 293 cells (Graham, 1984; Ghosh-Choudhury *et al.*, 1986). Two-plasmid recombination systems have been devised that reduce and in some cases virtually eliminate rescue of all viruses other than the desired recombinant. In one strategy, the Ad5 genomic plasmid pJM17 has an insert of bacterial sequences in E1 that exceeds the packaging constraints of the viral capsid (McGrory *et al.*, 1988). A second system utilizes genomic plasmids from which the packaging signal has been deleted (Bett *et al.*, 1994). In both systems, following cotransfection of 293 cells with the shuttle plasmid carrying the transgene and the Ad genomic plasmid, infectious virus is produced by homologous recombination between the two plasmids yielding the desired recombinant.

Similar methods can be used to rescue foreign DNA into the E3 region (or between the start of E4 and the right ITR) of Ad. In these cases, the shuttle plasmid should contain, flanking the transgene, Ad sequences from the right end of the genome. This shuttle can be cotransfected either with viral DNA that has been cleaved at the right end (e.g., by *Eco*RI for Ad5-based vectors), or with Ad5 genomic plasmids (Ghosh-Choudhury *et al.*, 1986; Johnson *et al.*, 1988; Mittal *et al.*, 1993). Alternatively, the transgene can be cloned directly into E3 of a genomic plasmid lacking the packaging

---

between a circular Ad genome plasmid and an E1 shuttle plasmid carrying the foreign DNA insert. The ITRs joined head to tail in the genome plasmid are indicated by the thick black arrows. The resulting E1- and E3-replacement vectors are shown below the broad shaded arrows in (A) and (B). Adenovirus sequences are indicated by shaded bars, foreign DNA inserts are indicated by solid black bars, and plasmid sequences are indicated by helical lines.

signal, and recombinant virus generated by cotransfecting with an E1-type shuttle plasmid or with a plasmid containing an intact E1 region (Bett *et al.*, 1994; Bramson *et al.*, 1996a).

Adenovirus serotype 2 viral DNA has been cloned as a yeast artificial chromosome (YAC) (Ketner *et al.*, 1994). This Ad YAC can be genetically manipulated in yeast by homologous recombination, producing recombinant DNA that is infectious when introduced into 293 cells. Because manipulation of the Ad genome in yeast is more flexible than manipulations *in vitro,* this method for generating recombinant Ad may be useful, particularly for the construction of complex vectors with multiple inserts, or by laboratories familiar with the handling of YACs.

4. Problems with First-Generation Adenovirus Vectors

In the past 15 years, recombinant Ad vectors have proved their utility for foreign gene expression in mammalian cells and have shown potential as recombinant viral vaccines, but a number of problems have been encountered using these first-generation Ad vectors for gene therapy. Even with E1-deleted vectors, a basal level of viral gene expression is observed *in vitro* (Rich *et al.*, 1993; Nevins, 1981; Spergel *et al.*, 1992) and *in vivo* (Engelhardt *et al.*, 1993a; Y. Yang *et al.*, 1994a). Although this level of expression is generally insufficient for virus production (Rich *et al.*, 1993), it is more than adequate in some cases to induce a host immune response against Ad proteins *in vivo* (Y. Yang *et al.*, 1994a). In most animal studies using first-generation vectors, the host immune system not only clears the Ad-infected, transgene-expressing cells (Y. Yang *et al.*, 1994a,b, 1995a, 1996; Yang and Wilson, 1995), but it also can prevent expression of the transgene after a second administration of the recombinant virus (Dai *et al.*, 1995; Y. Yang *et al.*, 1995a; Gilgenkrantz *et al.*, 1995). Although the expression of viral proteins in infected cells is likely to be the main inducer of an immmune response against those cells, the virions in the original inoculum used for infection (McCoy *et al.*, 1995) as well as the transgene itself (Tripathy *et al.*, 1996) have been shown to play a role in immune stimulation. For Ad gene therapy to be successful, it will be necessary to prevent induction of the immune response, by blocking expression of the immunogenic viral proteins and/or blocking the ability of the host to mount an immune response.

Another problem with first-generation recombinant Ad vectors is the potential for inadvertantly generating replication-competent Ad (RCA). The presence of even low levels of RCA might be a serious problem for gene therapy under certain conditions. The level of late gene expression in cells infected with nondefective viruses can be quite high, as discussed in Section I, resulting in a strong host immune response against the virus and Ad-infected cells. Furthermore, in humans, RCA can potentially act as a complementing helper for replication of the recombinant virus, thus mobilizing the transgene within the treated individual and possibly even between individuals. The increase in transgene copy number resulting from replication of the recombinant Ad could produce inappropriately high levels of transgene expression, which might be detrimental.

Because RCA can be generated by recombination between the recombinant virus and the E1 sequences in the 293 cell line in which the virus is propagated, recombinant vector preparations should be routinely tested for contaminating RCA. This is particularly important if the recombinant grows poorly, since it will be quickly outgrown by RCA during serial passage. Lochmuller *et al.* (1994) have compared a number of RCA detection methods including growth in noncomplementing human cell lines, Southern blot analysis, and polymerase chain reaction (PCR) amplification, with the latter being the most sensitive as well as the least time consuming. The ultimate solution to the problem of contaminating RCA is to develop cell lines for recombinant virus propagation that complement the replication defect but contain Ad E1 sequences in a form that prevents recombination, and/or to introduce additional mutations in the vector that preclude its ability to replicate in the host even if recombination with wild-type E1 sequences has occurred. Such modifications of Ad vector systems, as well as those that are designed to reduce the immune response (IR) against Ad in the absence of RCA, are designated as second-generation vector systems in this chapter.

B. Construction of Second-Generation Adenovirus Vectors and Propagating Cell Lines

1. Second-Generation Vectors That Can Be Propagated in 293 Cells

Several vectors have been designed that reduce the potential for RCA generated by recombination with wild-type E1 sequences. An Ad2 vector has been constructed that contains only ORF6 in E4 (Armentano *et al.*, 1995) and can be propagated in 293 cells. However, in cotton rats, which are permissive for human Ad production, replication of $E1^{+}$ derivatives of this vector is reduced and delayed compared to wild-type Ad, suggesting that any RCA generated during propagation of the virus would be attenuated in their ability to replicate *in vivo*. It is not clear whether this modification of the E4 region, in the absence of E1, affects the basal level of viral gene expression observed *in vivo* with first-generation vectors. However, this modification is not sufficient to prevent the IR generated by long-term repeat administration *in vivo* (Kaplan *et al.*, 1996).

To address the problem of host IR against Ad-infected cells, Engelhardt *et al.* (1994a) have constructed a vector that carries a temperature sensitive (ts) mutation in the E2A region encoding the 72-kDa DNA-binding protein. This mutation results in reduced levels of late gene expression in cells infected at 37°C. In addition, recombination between this vector and wild-type E1 sequences would generate virus that is impaired in late gene expression and viral replication. When compared to first-generation vectors *in vivo,* the tsE2A vector promoted prolonged transgene expression and reduced CTL infiltration on administration to the mouse liver or to cotton rat airway epithelium. However, expression was transient and accompanied by lymphocyte infiltration in the alveoli of *in vivo* ts-E2A vector-transduced cotton rat lungs (Engelhardt *et al.,* 1994a,b).

Wild-type Ad can complement the replication defect of gene therapy vectors administered to the lung, potentially mobilizing the recombinant virus. To reduce the possibility of dissemination, an E1-replacement vector has been constructed carrying a mutation that reduces the packaging efficiency of the recombinant (Imler *et al.,* 1995b). In a mixed infection, DNA with a wild-type packaging signal is preferentially packaged, leading to the desired dilution of the recombinant. Care would be necessary in producing this virus in 293 cells, however, to avoid inadvertent production of RCA that would have a significant growth advantage over the vector.

2. Alternative Cell Lines for Virus Propagation

The E1-complementing cell line 293 (Graham *et al.,* 1977), which carries approximately 4 kb from the left end of the Ad5 genome, has been used for more than two decades for the rescue and propagation of human Ad mutants with defects in E1. These cells express E1A and E1B gene products, but not the late protein IX, even though pIX sequences are within E1B. Attempts have been made to generate new Ad5 E1-complementing human cell lines, but not all have been successful, and some do not stably express E1 (Grodzicker and Klessig, 1980; Babiss *et al.,* 1983). A new complementing cell line designated 911 has been produced by transfecting human embryonic retinoblasts with a plasmid containing the Ad5 E1 region (nt 79–5789) (Fallaux *et al.,* 1996). 911 cells share many of the properties of 293 cells in that both are easy to transfect, both support a high level of recombination and virus production, and both are suitable for use in plaque assays. Because of homology between the Ad sequences in 911 cells and those in most recombinant Ad vectors, RCA could potentially be generated during recombinant virus production, but none has yet been detected (Fallaux *et al.,* 1996).

E1-complementing cell lines have also been generated by transfecting human lung carcinoma A549 cells with expression cassettes containing Ad5 E1 from nt 505 to 4034 (Imler *et al.,* 1996a). As this does not include the left ITR and packaging signal and lacks overlap with the left end sequences of most Ad5 E1-replacement vectors, homologous recombination between sequences in the cell lines and recombinant vectors cannot occur to give rise to RCA. Two lines, pTG6564-17 and pTG6559-5, expressed the E1A proteins and the E1B 19-kDa protein (but not the 55-kDa E1B protein or pIX) for at least 3 months in culture, and supported E1-deleted virus replication, although at a 10-fold reduced level compared to 293 cells. The ability of these lines to support virus rescue following cotransfection was not reported.

3. Alternative Complementation Systems

Several approaches have been taken in designing recombinant Ad vectors that allow minimal Ad gene expression in infected cells (to reduce host IR) and that have a minimal chance of generating RCA during propagation. Two types of E1 plus E4 complementing cell lines have been produced by introducing inducible Ad5 E4 sequences into the 293 cell line. Inducible expression of E4 is necessary since E4 ORF6 and the 55-kDa E1B products are thought to interact to shut off host protein synthesis (Sarnow *et al.*, 1984; Bridge and Ketner, 1990). The IGRP2 line carries the distal ORF6+7 fragment of the E4 region under the control of the dexamethasone-inducible mouse mammary tumor virus (MMTV) promoter (Yeh *et al.*, 1996). Although the level of induced E4 expression was low, expression of ORF6 and the spliced ORF6/7 products was detected by reverse transcription-based PCR (RT-PCR). More importantly, E1- and E4-deleted virus could be rescued and propagated in these cells without producing contaminating RCA, and such viruses should be greatly reduced in late gene expression *in vivo*.

Cell lines that complement E1, E4, and protein IX have been produced by transfecting 293 cells with constructs containing the E4 transcription unit under control of the MMTV promoter and the pIX gene under control of the metallothionein promoter (Krougliak and Graham, 1995). The lines VK10-9 and VK2-20 inducibly express E4 and pIX proteins detectable by immunoprecipitation and at levels sufficient for the rescue and propagation of E1-, E3-, E4-, and pIX-deleted virus. Vector backbones with these deletions would have a cloning capacity of about 11 kb, are predicted to be completely refractory to production of RCA by homologous recombination, and should induce minimal late gene expression in infected cells.

A novel strategy for producing doubly-deleted Ad vectors has been reported (Scaria *et al.*, 1995). The E4-deficient Ad dl1014, which is unable to grow in 293 cells, was conjugated to an expression plasmid containing E4 ORF6 or the entire E4 region in an Ad–polylysine–DNA complex, which was then used to transfect 293 cells. The plasmid-encoded E4 complemented the defective virus, resulting in efficient replication of E4$^-$ Ad. Although some technical problems, such as those limiting large-scale virus production, would first have to be solved, it may be possible to use this approach for generating E1$^-$E4$^-$ virus (or viruses with other mutations in addition to E1) by a similar complementation system, without a need for generating new cell lines.

293 cells have also been modified by transformation with either the Ad5 gene encoding the precursor terminal protein (pTP) (Schaack *et al.*, 1995b) or the Ad2 DNA polymerase gene (Amalfitano *et al.*, 1996). The resulting cell lines express sufficient levels of the Ad proteins to support the growth of ts-pTP or ts-polymerase mutant Ads at the nonpermissive

temperature. Because the coding sequences of the polymerase and pTP genes are on the strand complementary to the major late primary transcript and overlap late genes, it is not clear whether either of these genes could be deleted without interfering with expression of late proteins. Thus these systems might not increase the cloning capacity of the vector, but might allow generation of vectors that are more highly attenuated and cannot produce RCA by recombination.

An Ad vector, Av3nBg, has been constructed with deletions in E1, E2A, and E3 (Gorziglia *et al.*, 1996). This vector can be propagated in a complementing cell line, AE1-E2a derived from A549 cells, which contains the E1 and E2A region genes under the control of the glucocorticoid-responsive GRE5 and MMTV promoters, respectively. The viral titers are equivalent to those of first-generation vectors grown in 293 cells. In contrast to infection with E1-deleted, ts-E2A viruses, which resulted in significant late gene expression and replication at multiplicities of infection (MOIs) of 500 PFU/cell, neither late gene expression nor replication was detectable following infection of noncomplementing cells with Av3nBg at MOIs up to 1500 PFU/cell. This vector system, like the E4- and E2 pTP-complementing systems described above, should prevent generation of RCA during propagation and, *in vivo*, should minimize late gene expression and prevent dissemination of the virus.

4. Helper-Dependent Systems

The two major advantages of helper-dependent Ad vector systems are their large cloning capacity (theoretically up to 35 kb) and the potential to remove any Ad genes whose expression may be detrimental to the desired application of the virus. The main disadvantages of helper-dependent systems are the need to separate the helper and the vector viruses, and technical difficulties in stably maintaining the vector in the presence of competing helper virus.

One of the first systems developed for expression of foreign genes in recombinant Ad-infected cells was a helper-dependent system (Thummel *et al.*, 1982). In this system, sequences downstream of the MLP in Ad were replaced with a DNA fragment encoding a foreign gene and the gene for simian virus type 40 (SV40) T antigen (Tag). The expression of Tag allowed for selection of the desired recombinant in CV-1 monkey cells, which are nonpermissive for growth of Ad in the absence of Tag. Because the Ad sequences replaced by the foreign DNA insert were essential for virus viability, amplification of the recombinant virus required wild type as helper. This approach was utilized for production of SV40 Tag itself (Thummel *et al.*, 1983), HSV thymidine kinase (TK) (Yamada *et al.*, 1985), and polyomavirus tumor antigens (Mansour *et al.*, 1985), as well as for analysis of *cis*-acting regulatory elements of the MLP (Mansour *et al.*, 1986), although most of the virus in the stock preparations was wild type, even after selection.

Attention has turned again to helper-dependent vectors and a novel helper system has been described by Mitani *et al.* (1995a) for the propagation of Ad recombinants in which the region of the Ad5 viral genome encoding L1, L2, VAI, VAII, and pTP was replaced by a selectable marker gene (β-geo, a fusion of *lacZ* and the neomycin resistance gene; Friedrich and Soriano, 1991). This DNA was packaged and propagated in the presence of wild-type Ad2. The helper could be partially separated from the recombinant by CsCl gradient centrifugation, presumably due in part to the difference in helper vs vector genome size (36 vs 33.7 kb, respectively). Nonetheless, the final preparation contained mostly helper virus. The titer of the vector, determined from the number of infected cells staining blue after X-Gal treatment, was at least 4×10^6 transducing particles per milliliter.

This strategy was further developed to propagate a recombinant virus (AdDYSβgal) containing a 28-kb foreign DNA segment encoding the full-length dystrophin cDNA under control of the muscle-specific creatine kinase promoter together with the *lacZ* reporter gene (Kochanek *et al.*, 1996). The Ad5 sequence from nt 1 to 358, which includes the left ITR and the packaging signal, was reiterated on either end of the foreign DNA sequence. Rescue and propagation of the vector was helped by SV5, a mutant Ad5 that was deleted in E1, E3, and part of the packaging signal such that the packaging efficiency of the helper was reduced by 100-fold. After six serial propagations in 293 cells and CsCl equilibrium centrifugation, the final titer of AdDYSβgal was 5×10^8 transducing particles per milliliter, with only 1% helper contamination. This strategy looks promising, although whether this level of purity is sufficient to prevent host IR to Ad-infected cells remains to be tested. In addition, this system could potentially generate RCA by recombination between the helper and the E1 sequences of the 293 cells, which might substantially reduce the purity of the final recombinant preparations.

In a similar manner, helper-dependent vectors were generated containing either *lacZ* or the cystic fibrosis transmembrane conductance regulator (CFTR) cDNA flanked on the left by the Ad5 5′ ITR and packaging signal and on the right by the 3′ ITR (Fisher *et al.*, 1996). After growth of the *lacZ* vector in 293 cells with an E1-replacement helper virus encoding a second marker gene, the preparation was purified by CsCl equilibrium centrifugation. Analysis of gradient fractions revealed that a portion of the virions contained multiple concatameric forms of the recombinant vector DNA, with the peak of *lacZ* transduction activity correlating with a trimeric species of about 15 kb. Serial propagation enhanced the proportion of the 15-kb concatamer relative to the monomeric form, suggesting that the smaller vectors are at a growth disadvantage. The final titer following CsCl gradient centrifugation was about 10^8 transducing particles per milliliter, and the degree of helper contamination, although not specified in the report, appeared to be about 5% from the data presented in Fig. 6 of Fisher *et al.* (1996).

Although quite promising at this initial stage, helper-dependent systems will require significant improvements to permit large-scale production and purification of the vector, particularly with regard to separation from the helper, in order to be useful for gene therapy.

C. Factors That Influence the Specificity, Level, and Duration of Expression in Target Cells

One of the factors that determines the type of cell that can be infected by Ad both *in vivo* and *in vitro* is the presence of α_v integrins, the receptors for virus internalization, on the cell surface (Wickham *et al.*, 1993). Although these integrins are widespread among most cell types, there are some exceptions (Goldman and Wilson, 1995; Acsadi *et al.*, 1994a,b). Furthermore, induction of α_v integrins on human monocytes and T lymphocytes *in vitro* was shown to greatly enhance Ad-mediated gene delivery to these cells (Huang *et al.*, 1995). *In vivo,* the route of administration also determines the type of cell transduced. Intraperitoneal (i.p.) injection of Ad vectors in mice has resulted in efficient transduction of liver, spleen (Mittal *et al.*, 1993), peritoneum, and diaphragm (Huard *et al.*, 1995). Intravenous (i.v.) delivery has been shown to transduce primarily the liver, and to a smaller extent the diaphragm and kidney (Huard *et al.*, 1995). Other tissues and organs have been targeted by direct delivery, e.g., nasal delivery to the airway epithelium, which is discussed in greater detail in Section IV. Even with direct administration to the kidney, however, the targeted cell population differed depending on whether the recombinant virus was perfused into the renal artery or infused through a retrograde catheter (Moullier *et al.*, 1994). A summary of cell types and tissues that have been transduced by recombinant Ad vectors is shown in Table I. Michael *et al.* (1995) have developed a novel strategy for potentially targeting specific cell types by addition of a peptide ligand to the fiber protein to alter its binding specificity. Although this study reported the genetic fusion of fiber to a single specific ligand, this strategy might be applicable to targeting numerous cell types for which peptide ligands have been identified.

The level of expression in cells infected by Ad vectors is influenced to a large extent by the promoter driving expression of the transgene. A comparison of the human β-actin, human cytomegalovirus (HCMV) immediate early, Ad major late, and SV40 early and late promoters has indicated that the HCMV immediate early promoter directs the highest level of expression in the widest variety of human and animal cell types *in vitro* (Xu *et al.*, 1995). However, the HCMV promoter as well as the Rous sarcoma virus (RSV) long terminal repeat (LTR) have shown some degree of cell specificity (Lowenstein *et al.*, 1995). In our hands, the murine CMV immediate early promoter (Dorsch-Hasler *et al.*, 1985) has been the strongest promoter by far for *in vitro* expression in rodent cells, and rivals the HCMV

TABLE I   Cell Types and Tissues Transduced by Adenovirus Vectors

Cell type/tissue	Species	Infection	Ref.
Airway epithelium	Mouse	*In vivo*	Katkin *et al.* (1995); Y. Yang *et al.* (1995a); Zsengeller *et al.* (1995)
	Cotton rat	*In vivo*	Rosenfeld *et al.* (1992); Engelhardt *et al.* (1994b); Katkin *et al.* (1995)
	Sheep (fetal)	*In vivo*	McCray *et al.* (1995); Vincent *et al.* (1995)
	Rhesus monkey	*In vivo*	Bout *et al.* (1994b); St. George *et al.* (1996)
	Baboon	*In vivo*	Simon *et al.* (1993); Engelhardt *et al.* (1993b); Goldman *et al.* (1995a)
	Human	*In vitro, ex vivo*	Dupuit *et al.* (1995); Amin *et al.* (1995); Pilewski *et al.* (1995b); Teramoto *et al.* (1995)
		In xenograft	Engelhardt *et al.* (1993a); Grubb *et al.* (1994)
		In vivo	Reviewed in O'Neal and Beaudet (1994)
	Human (fetal)	*Ex vivo*	Ballard *et al.* (1995)
		In xenograft	Peault *et al.* (1994)
Airway submucosal gland	Human	In xenograft	Pilewski *et al.* (1995a)
Pulmonary endothelium	Rat	*In vivo*	Schachtner *et al.* (1995)
	Sheep	*In vivo*	Lemarchand *et al.* (1994)
	Pig	*In vivo*	Muller *et al.* (1994)
Skeletal muscle	Mouse	*In vivo*	Quantin *et al.* (1992); Acsadi *et al.* (1994b, 1995)
	Human	*In vitro*	Acsadi *et al.* (1994a)
Diaphragm	Mouse	*In vivo*	Petrof *et al.* (1995)
Cardiac muscle	Mouse	*In vivo*	Stratford-Perricaudet *et al.* (1992); Acsadi *et al.* (1996)
	Rat	*In vivo*	Kass-Eisler *et al.* (1993); Guzman *et al.* (1993a)
	Rabbit	*In vivo*	Barr *et al.* (1994)
	Pig	*In vivo*	Muhlhauser *et al.* (1996); French *et al.* (1994b)
Vascular smooth muscle cells	Rat	*In vitro*	Guzman *et al.* (1993b)
	Cow	*In vitro*	March *et al.* (1995)
	Dog, pig	*In vivo, in vitro*	Mazur *et al.* (1994)
	Human	*In vitro*	Mazur *et al.* (1994)
Endothelial cells	Pig, human	*In vitro*	Wrighton *et al.* (1996)
Arteries	Rat	*In vivo*	Guzman *et al.* (1993b); Lee *et al.* (1993); Schulick *et al.* (1995a,b)
	Rabbit	*In vivo*	Steg *et al.* (1994); Landau *et al.* (1995); Feldman *et al.* (1995); Newman *et al.* (1995)

(continues)

TABLE I *(continued)*

Cell type/tissue	Species	Infection	Ref.
	Sheep	*In vivo*	Lemarchand *et al.* (1993); Rome *et al.* (1994)
	Pig	*In vivo*	French *et al.* (1994a)
Aortic transplants	Rabbit	*Ex vivo*	Mehra *et al.* (1996)
Vein grafts	Pig	*Ex vivo*	S. J. Chen *et al.* (1994)
Liver transplants	Rat	*Ex vivo*	Shaked *et al.* (1994); Drazan *et al.* (1995a,b)
	Hamster to rat	*Ex vivo*	Drazan *et al.* (1995d)
Hepatocytes	Rat, mouse, human	*In vitro*	Jaffe *et al.* (1992); Li *et al.* (1993); Morsy *et al.* (1993)
Liver	Mouse	*In vivo*	Stratford-Perricaudet *et al.* (1990); Li *et al.* (1993); Huard *et al.* (1995); Engelhardt *et al.* (1994a)
	Rat	*In vivo*	Jaffe *et al.* (1992); Drazan *et al.* (1995c)
	Rat	*Ex vivo*	Csete *et al.* (1994b)
	Rabbit	*In vivo*	Hughes *et al.* (1996); Kozarsky *et al.* (1994)
	Dog	*In vivo*	Kay *et al.* (1994)
Renal tubular cells	Rat	*In vivo*	Moullier *et al.* (1994)
Bladder	Rat	*In vivo*	Morris *et al.* (1994)
Genitourinary epithelium	Human, mouse	*In vitro*	Bass *et al.* (1995)
	Mouse	*In vivo*	Bass *et al.* (1995)
Skin	Murine	*In vivo, ex vivo*	Setoguchi *et al.* (1994a)
Salivary glands	Human	*In vitro*	Mastrangeli *et al.* (1994)
	Rat	*In vivo*	Mastrangeli *et al.* (1994)
Oral epithelium	Hamster	*In vivo*	Clayman *et al.* (1995)
Synoviocytes	Human	*In vitro*	Roessler *et al.* (1995)
	Rabbits	*In vivo*	Roessler *et al.* (1993)
Neuronal cells	Rat	*In vitro*	Le Gal La Salle *et al.* (1993); Caillaud *et al.* (1993); Shy *et al.* (1995); Shering *et al.* (1996); Lowenstein *et al.* (1995)
		In vivo	Draghia *et al.* (1995)
Brain	Mouse	*In vivo*	Davidson *et al.* (1993)
	Rat	*In vivo*	Bajocchi *et al.* (1993); Akli *et al.* (1993); Byrnes *et al.* (1995); Le Gal La Salle *et al.* (1993)
Sciatic nerve	Rat	*In vivo*	Shy *et al.* (1995)
Astrocytes	Rat	*Ex vivo*	Ridoux *et al.* (1994)
CNS ependymal cells	Rat	*In vivo*	Bajocchi *et al.* (1993)
Neuronal progenitors	Human (fetal)	In xenograft	Sabate *et al.* (1995)
	Rat	*Ex vivo*	Gage *et al.* (1995)

(continues)

TABLE I *(continued)*

Cell type/tissue	Species	Infection	Ref.
Retinal epithelium	Mouse	*In vivo*	Bennett *et al.* (1994); T. Li *et al.* (1994)
Other ocular tissues	Mouse	*In vivo*	Budenz *et al.* (1995)
Pancreatic islets	Human	*In vitro*	Csete *et al.* (1994a)
	Mouse	*Ex vivo*	Csete *et al.* (1995)
Gall bladder	Human	*In vitro*	Maeda *et al.* (1994)
	Sheep	*Ex vivo*	Maeda *et al.* (1994)
Intestinal epithelial stem cells	Mouse	*Ex vivo*	F. C. Campbell, personal communication
Bone marrow	Human	*In vitro*	Mitani *et al.* (1994)
Macrophages	Human	*In vitro*	Haddada *et al.* (1993b)
CD34$^+$ progenitors	Human	*In vitro*	U. Pettersson, personal communication
Monocytes	Human	*In vitro*	U. Pettersson, personal communication
Leukemic hematopoetic cells	Human	*In vitro*	U. Pettersson, personal communication
Mesothelioma cells	Human	In xenograft	Brody *et al.* (1994a); Smythe *et al.* (1994)
Gliosarcoma	Rat	*In vitro, in vivo*	Boviatsis *et al.* (1994)
Lung carcinoma	Mouse	*In vivo*	Tang *et al.* (1994)
	Human	*In vitro, ex vivo*	Tang *et al.* (1994)
Hepatoma	Human	*In vitro, in* xenograft	Arbuthnot *et al.* (1996)

promoter in human cells, yielding about 10 to 100 μg of transgene product per 10^6 infected cells (Addison *et al.*, 1997). *In vivo*, the highest level of transgene expression reported to date is 1 mg of protein per milliliter of plasma following i.v. delivery to the mouse of an Ad vector carrying the transgene under the control of the HCMV promoter (Kolls *et al.*, 1994). For expression in limited cell types, some tissue-specific promoters have been demonstrated to maintain their specificity to some degree in Ad vectors (see Table II) but the level of expression driven by these cellular promoters is usually considerably lower than that driven by the viral promoters described previously.

In vitro, the duration of gene expression is generally high for up to about 1 week in proliferating cells. *In vivo*, a high level of expression can often be maintained for 2 to 4 weeks and considerably longer periods have been reported (Stratford-Perricaudet *et al.*, 1990, 1992; Barr *et al.*, 1995). The duration of expression may be limited both *in vivo* and *in vitro* by loss of the episomally maintained nonreplicating vector genome during cell proliferation and potentially by loss of transgene promoter activity in the transduced cells. In addition, the length of expression *in vivo* is limited by clearance of the infected cell population by the host immune system, although this limitation may be alleviated by improvements in vector design or by modulation of the host immune response.

III. Expression of Foreign DNA Sequences in Adenovirus Vectors

A. Overexpression and Characterization of Recombinant Proteins

A selection of Ad vectors that have been constructed to facilitate characterization of the recombinant protein products is listed in Table IIA. The high level of transgene expression in recombinant Ad-infected cells has been exploited to produce recombinant proteins on a large scale (cf. Garnier *et al.*, 1994). Up to 90 mg of functional recombinant protein has been obtained per liter of culture volume, corresponding to 10–20% of the total cellular protein content, following infection of 293 cells with recombinant Ad vectors. This method of recombinant protein production was superior to production in baculovirus systems in overall yield of soluble functional products (Massie *et al.*, 1995).

Recombinant Ad vector-infected mammalian cells not only express high levels of transgene products, but these recombinant proteins are also post-translationally modified correctly and are biologically active. For this reason, Ad vectors have facilitated *in vitro* characterizations of cellular products that are normally present at relatively low levels in mammalian cells. Adenovirus vectors have been constructed encoding genes for the human vitamin D_3 receptor (Smith *et al.*, 1991), human secreted placental alkaline phosphatase (Doronin *et al.*, 1993), protein tyrosine phosphatase 1C (Garnier *et al.*, 1994), as well as numerous cDNAs used in gene therapy (see Section IV) and used for analysis in cultured cells. Adenovirus vectors have also been instrumental in *in vivo* analyses of low abundance cellular proteins. *In vivo* delivery of cytokine-expressing Ad vectors to the lung has provided an excellent model for investigating the role of various cytokines in lung inflammation (cf. Gauldie *et al.*, 1996; also see Table IIA). In a similar manner, infection of hypothyroid mice infected with Ad vectors containing wild-type or mutant thyroid hormone receptor β genes has provided an animal model for studies on thyroid hormone resistance (Hayashi *et al.*, 1996).

Several genes isolated from yeast, including those encoding the homothallic (HO) endonuclease, the TATA box-binding protein, and the transcription factor TFIIB, have been inserted into Ad vectors in order to investi-

TABLE II Adenovirus Vectors Constructed for Analytical Studies

Transgene	*Ref.*
A. Recombinant protein analysis	
Human vitamin D_3 receptor	Smith *et al.* (1991)
Human secreted placental alkaline phosphatase	Doronin *et al.* (1993)
Protein tyrosine phosphatase 1C	Garnier *et al.* (1994)
Rat and murine interleukin 6	Braciak *et al.* (1993); Xing *et al.* (1994); Richards *et al.* (1995)
Murine GM-CSF[a]	Xing *et al.* (1996a,b)
Murine RANTES[a]	Braciak *et al.* (1996)
Rat macrophage inflammatory protein 2	Gauldie *et al.* (1996)
Porcine transforming growth factor β	Gauldie *et al.* (1996)
VSV glycoprotein	Schneider *et al.* (1989)
HSV thymidine kinase	Yamada *et al.* (1985); Haj-Ahmad and Graham (1986)
HSV ICP0	Zhu *et al.* (1988)
HSV gB	Johnson *et al.* (1988)
HSV gC	Witmer *et al.* (1990)
HSV gE and gI	Hanke *et al.* (1990)
HSV gK	Hutchinson *et al.* (1992, 1993)
HSV ICP47	York *et al.* (1994)
HSV ribonucleotide reductase subunits	Huang *et al.* (1988); Massie *et al.* (1995)
HCMV gH and gL	R. Milne and J. Booth, unpublished results
Polyomavirus T antigens	Mansour *et al.* (1985); Massie *et al.* (1986); Davidson and Hassell (1987)
SV40 large T antigen	Thummel *et al.* (1983)
Yeast HO endonuclease	H. Young, personal communication
Yeast TATA box-binding protein	H. Young, personal communication
Transcription factor TFIIB	E. Falck-Pedersen and H. Young, personal communication
B. Transcriptional analysis	
Human CFTR promoter	Imler *et al.* (1996b)
Skeletal α-actin promoter	Quantin *et al.* (1992)
Mouse β-globin, rat albumin, and mouse Ig promoters	Friedman *et al.* (1986); Babiss *et al.* (1986)
Rat tyrosine hydroxylase promoter	J. J. Robert, personal communication
Chicken acetylcholine receptor promoter	Bessereau *et al.* (1994)
Rabbit surfactant protein A promoter	Alcorn *et al.* (1993)
Thyroid hormone/retinoic acid response element	Hayashi *et al.* (1994)
Complement factor 3 promoter	Varley *et al.* (1995)
Serum amyloid A3 promoter	Varley *et al.* (1995)
HIV LTR and RSV LTR	Rice and Matthews (1988a,b)
Ig heavy chain splicing and poly(A)	Ruether *et al.* (1988)

[a] GM-CSF, Granulocyte-macrophage colony-stimulating factor; RANTES, regulated on activation, normal T expressed and secreted.

gate their activities in mammalian cells (H. Young and E. Falck-Pedersen, personal communication). In addition, many genes derived from viruses other than Ad have been successfully inserted into Ad vectors for characterization of the encoded viral proteins in isolation from other proteins encoded by the parental virus. For example, an Ad vector containing the HSV-1 ICP47 gene (and not a control Ad) induced MHC class I transport abnormalities in infected fibroblasts, confirming results suggested by experiments with HSV mutants, and demonstrating that other HSV-1 polypeptides were not required for this activity (York *et al.*, 1994). Adenovirus vectors have also been useful for the delivery of genes encoding various viral glycoproteins to cultured cells for use as targets in CTL assays. Using this approach, HSV proteins and epitopes on those proteins have been identified that are preferentially recognized by HSV-specific CTLs (Witmer *et al.*, 1990; Hanke *et al.*, 1991).

Recombinant Ad vectors encoding a broad range of microbial proteins have been developed for use as live recombinant vaccines. For a summary of Ad vectors that have been constructed primarily for vaccine use, the reader is referred to the review by Graham and Prevec (1992).

B. Transcriptional Analysis of Heterologous DNA

Because of their ability to infect a wide range of cell types at high efficiencies, recombinant Ads have become powerful tools for the analysis of *cis*-acting elements involved in transcriptional regulation in different tissues or in response to different stimuli (Table IIB). Some of the tissue-specific promoters that have been analyzed following insertion into Ad vectors include the human CFTR promoter (Imler *et al.*, 1996b), the skeletal α-actin promoter (Quantin *et al.*, 1992), and the β-globin, albumin, and immunoglobulin promoters (Friedman *et al.*, 1986; Babiss *et al.*, 1986). In general these promoters were found to retain tissue specificity in the Ad virus background, although expression was sometimes observed in cell lines that did not express the corresponding endogenous gene.

The insertion of inducible promoters into Ad vectors has facilitated mapping of the various inducer-responsive elements in those promoters. A cytokine-responsive element in the rat tyrosine hydroxylase promoter (J. J. Robert, personal communication) as well as the sequence responsible for electrical activity-dependent expression of the chicken acetylcholine receptor gene (Bessereau *et al.*, 1994) and the cyclic AMP responsive elements of rabbit surfactant protein A promoter (Alcorn *et al.*, 1993) have all been mapped using recombinant Ad vectors *in vitro*. In addition, Ad vectors containing a thyroid hormone/retinoic acid-responsive element (Hayashi *et al.*, 1994) and the promoters for two murine acute-phase protein genes (Varley *et al.*, 1995) were found to respond appropriately to their inducers in mice, suggesting that at least some of these *cis*-acting elements may be useful for *in vivo* investigational or therapeutic purposes requiring cell-specific or inducible expression.

Two novel methods have been reported in which expression of sequences carried by one Ad vector was modulated by gene products of a second recombinant Ad vector. In one investigation, expression of a reporter gene under control of a tumor necrosis factor (TNF)-responsive promoter in an Ad vector was induced after coinfection with an Ad vector encoding TNF-α (Hersh *et al.*, 1995). The second method for controlling gene expression utilizes the site-specific Cre recombinase gene in one vector and recognition sites for Cre, *loxP*, in the second vector. In two separate studies, Ad vectors carrying the Cre gene mediated excision of *loxP*-flanked spacer DNA inserted between the promoter and coding sequence of reporter genes in the target vector, resulting in activation of the reporter (Anton and Graham, 1995; Kanegae *et al.*, 1995). The Ad Cre vector has also been shown to induce site-specific excision in cultured cells engineered to contain *loxP* sites in their chromosomal DNA (Wang *et al.*, 1995). An advantage of the Cre system is that constructs containing potentially any promoter, including the highly efficient viral promoters, can be made inducible, which would be useful for numerous applications.

C. Stable Transformation of Mammalian Cells by Adenovirus-Mediated Gene Transfer

Recombinant Ad vectors expressing appropriate genes can induce permanent transformation of cellular phenotypes by nonhomologous integration into the host chromosome. A variety of Ad vectors encoding selectable markers, including adenine phosphoribosyltransferase (APRT) (Konan *et al.*, 1991), dihydrofolate reductase (DHFR) (Ghosh-Choudhury and Graham, 1987), TK (Haj-Ahmad and Graham, 1986), and neomycin resistance (neo) (Van Doren *et al.*, 1984; Karlsson *et al.*, 1985), have been used to establish stably transformed cell lines. The frequency of transformation by an E1-deleted AdTK vector was approximately 2 to 5 TK$^+$ colonies per 10^5 cells (Wilson and Graham, 1989).

In addition, primary human fibroblasts have been transformed by Ad vectors in which the E1 region was replaced by the SV40 Tag gene, generating immortal cell lines (Van Doren and Gluzman, 1984). This technique has been used to establish a cell line from adult human hepatocytes (Ueno *et al.*, 1993) and should be valuable in generating lines from other primary cells that have low proliferative capacities in culture.

Reports indicate that Ad may have potential as a vector for gene targeting by homologous recombination in mammalian cells. A replication-proficient Ad APRT vector (at an MOI of 10 PFU/cell) stably corrected the APRT deficiency in APRT$^-$ CHO cells at a frequency of 1 in 10^5 to 10^6 infected cells, although less than 20% of the transductants resulted from homologous recombination events (Wang and Taylor, 1993). Similar results were obtained with a replication-defective Ad recombinant (Mitani *et al.*, 1995b). Recombination was observed at a somewhat higher frequency (6 to 8 colonies per 10^4 cells) using an Ad carrying an intact *neo* gene as a vector (at an MOI of 400 PFU/cell) and a nuclear papillomavirus plasmid carrying a mutant *neo* gene as a target. In this case at least one copy of the mutant *neo* gene in all of the transductants had been precisely corrected by homologous recombination with the vector, although some evidence of nonhomologous recombination was also detected (Fujita *et al.*, 1995). It is not yet known whether gene targeting to the chromosomes would be as efficient as targeting to this multicopy episomal plasmid.

IV. Gene Therapy with Adenovirus Vectors

Research on Ad vectors for their potential use in gene therapy has grown enormously. A summary of recombinant Ads that have been investigated as possible gene therapy vectors is given in Table III, and many of these are described in detail below. The main advantages of Ad vectors over other systems commonly used for gene therapy are the technical ease of preparing and purifying relatively large quantities of the recombinant vector and the variety of cells that can be infected by these viruses. The fact that quiescent cells can be easily infected by Ad and efficiently express the encoded transgenes makes Ad vectors particularly appealing since administration of the virus directly to the recipient is possible, in contrast to the need with many other vector systems for explantation of tissue and transformation or transduction of cultured cells *in vitro*.

There are, however, some drawbacks to Ad as a gene therapy vector. Because of the high infectivity of the virus, steps must be taken to minimize mobilization of the transgene during and after treatment. In many cases this can be accomplished by using replication-defective Ad vectors. Administration of Ad vectors to the lung might be especially problematic because even E1-deleted viruses might be amplified by complementation with wild-type Ads from latent infections or from the environment since the virus has a natural tropism for lung epithelium. In 20% of the normal individuals tested, Ad2 or Ad5 E1A sequences have been detected by PCR amplification of cellular DNA from the respiratory epithelium (Eissa *et al.*, 1994). Furthermore, E1A proteins also have been detected in human lung tissue (Elliott *et al.*, 1995). As described in Section II,B,1, second-generation vectors have been constructed that may alleviate the problem of dissemination following complementation with endogenous E1 sequences (Imler *et al.*, 1995b).

Particularly important is the effect of the immune system of the recipient on the efficacy of Ad vector therapy. Individuals with previous exposure to Ad2 or Ad5 (the most commonly used vector backbones) may already have sufficiently high levels of neutralizing antibodies to the virus to block initial infection by the vector. Furthermore, as discussed in Section II,A,4, first-generation vectors have been shown to induce potent cellular and humoral antiviral immune responses, resulting in both transient expression of the transgene as well as an inability to readminister the virus successfully. In some applications, for example cancer therapy with Ad vectors encoding cytokines, transient expression is probably preferable and the stimulation of an immune response could be an advantage. However, for gene therapy in which long-term expression or repeated administration of the vector is desired, this host immunity poses a serious problem.

A. Modulation of the Immune Response Induced by Adenovirus Gene Transfer

Several approaches have been adopted to reduce the immune response in animals treated with Ad vectors, allowing repeated administration of the vector and/or prolonged transgene expression. Transient immunosuppression of mice with a single dose of deoxyspergualin at the same time as i.v. administration of an Ad vector effectively blocked formation of anti-Ad neutralizing antibody, permitting readministration of the vector (T. A. G. Smith *et al.*, 1996). Likewise, recombinant interleukin 12 (IL-12) or interferon γ (IFN-γ) administered at the same time as the recombinant Ad to the mouse airway prevented the generation of neutralizing antibody (Y. Yang *et al.*, 1995b). The latter treatment had no effect on the duration of gene expression, however, which was still transient. Vector administration accompanied by daily treatment with the immunosuppressant drug FK506 (Lochmuller *et al.*, 1995; Vilquin *et al.*, 1995) blocked both cellular and humoral responses, permitting long-term transgene expression in adult mouse muscle. This type of therapy, however, would result in generalized immunosuppression of the host, which is not ideal for long-term treatment. A more attractive therapy involves coadministration of recombinant Ad with soluble CTLA4Ig, which blocks the costimulatory signal required for T cell activation by the antigen-presenting cell (Kay *et al.*, 1995). Mice subjected to this single-dose treatment expressed the transgene for over 5 months, and had markedly reduced T cell responses to virally infected cells, compared to controls receiving vector alone. Although the production of antiviral antibodies was reduced, a sufficient level remained to block readministration of recombinant virus. Another possible approach to preventing activation of the host immune system following Ad infection may be to incorporate into the vector backbone additional viral genes that are known to interfere with host recognition of infected cells. Such viral genes might include those encoding the Ad5 E3 19-kDa protein (Wold and Gooding, 1989) and HSV ICP47 (York *et al.*, 1994), described in Sections I and III,A, respectively. To this end, an Ad vector was constructed that contained the

TABLE III Analysis of Potential Adenovirus Therapy Vectors

Disorder	Transgene	Target tissue	Infection	Ref.
Cystic fibrosis	Human CFTR	Cotton rat airway	*In vivo*	Rosenfeld *et al.* (1992); Mastrangeli *et al.* (1993); Zabner *et al.* (1994a); Yei *et al.* (1994a,b)
		Rhesus monkey airway	*In vivo*	Zabner *et al.* (1994a); Bout *et al.* (1994a); Brody *et al.* (1994a); Kaplan *et al.* (1996); Sene *et al.* (1995); St. George *et al.* (1996)
		Baboon airway	*In vivo*	Simon *et al.* (1993); Engelhardt *et al.* (1993b); Goldman *et al.* (1995a)
		CF mouse airway	*In vivo*	Grubb *et al.* (1994); Yang *et al.* (1994c)
		CF human airway epithelium	*In vitro*	Rich *et al.* (1993); Zabner *et al.* (1994b); Rosenfeld *et al.* (1994); Johnson *et al.* (1995); Grubb *et al.* (1994); Mittereder *et al.* (1994)
			In xenograft	Engelhardt *et al.* (1993a); Goldman *et al.* (1995b)
			In vivo	Reviewed by O'Neal and Beaudet (1994); Knowles *et al.* (1995); Hay *et al.* (1995); Crystal *et al.* (1994)
		Mouse liver	*In vivo*	Yang *et al.* (1993)
	Human pancreatic lipase	Sheep gallbladder	*Ex vivo*	Maeda *et al.* (1994)
		Human gall bladder epithelial cells	*In vitro*	Maeda *et al.* (1994)
Respiratory distress syndrome	Human SP-B	Mouse lung epithelial cells	*In vitro*	Yei *et al.* (1994c)
		Cotton rat lung	*In vivo*	Yei *et al.* (1994c)
	Human SP-A, SP-B	Rat airway	*In vivo*	Korst *et al.* (1995)
		Human lung epithelial cells	*In vitro*	Korst *et al.* (1995)
Pulmonary edema	Rat Na$^+$,K$^+$-ATPase β_1 subunit	Rat lung epithelium	*In vitro*	P. Factor, personal communication

Oxidant stress	Human catalase	Human airway	In vitro	Yoo et al. (1994)
		Human endothelial cells	In vitro	Erzurum et al. (1993)
	Human HO-1	Rabbit corneal epithelium	In vitro	Abraham et al. (1995)
		Rabbit eye	In vivo	Abraham et al. (1995)
Arrythmia	Drosophila Shaker potassium channel	Rat cardiac myocytes	In vitro	Johns et al. (1995)
		Rat liver	In vivo	Johns et al. (1995)
Ischemia	hsp70i[a]	Rat cardiac myocytes	In vitro	Mestril et al. (1996)
	Acidic FGF[a]	Mouse tissue	In vivo	Muhlhauser et al. (1995a)
		Human endothelial cells	In vitro	Muhlhauser et al. (1995a)
	VEGF$_{165}$[a]	Mouse tissue	In vivo	Muhlhauser et al. (1995b)
		Human endothelial cells	In vitro	Muhlhauser et al. (1995b)
Muscular dystrophy	Human dystrophin minigenes	mdx mouse muscle	In vitro	Clemens et al. (1995)
			In vivo	Ragot et al. (1993); Vincent et al. (1993); Acsadi et al. (1996)
	Mouse dystrophin (full length)	mdx mouse myoblasts	In vitro	Kochanek et al. (1996)
Restenosis	Hirudin	Rat carotid artery	In vivo	Rade et al. (1996)
	HSV TK	Rat carotid artery	In vivo	Guzman et al. (1994); Chang et al. (1995a)
		Pig carotid artery	In vivo	Ohno et al. (1994)
	p21^{waf1}	Rat carotid artery	In vivo	Chang et al. (1995b)
	IκBα	Human, pig endothelial cells	In vitro	Wrighton et al. (1996)
Graft rejection	vIL-10	Rat liver allografts	In vivo	Drazan et al. (1995b)
	sVCAM	Pig vein grafts	Ex vivo	S. J. Chen et al. (1994)
Inflammation	Human IL-1 receptor antagonist	Rabbit synovium	In vivo	Roessler et al. (1995)
		Human synoviocytes	In vitro	Roessler et al. (1995)
Neuronal damage	sCNTF	Rat astrocytes	In vitro	G. M. Smith et al. (1996)
	NT-3	Rat astrocytes	In vitro	G. M. Smith et al. (1996)
	B50/GAP43	Rat olfactory neurons	In vivo	Verhaagen et al. (1995)

(continues)

TABLE III (continued)

Disorder	Transgene	Target tissue	Infection	Ref.
Parkinson's disease	Tyrosine hydroxylase	Rat brain	*In vivo*	Horellou *et al.* (1994)
Atherosclerosis, hypercholesterolemia	Human LDL receptor	FH mouse liver	*In vivo*	Ishibashi *et al.* (1993)
		FH rabbit liver	*In vivo*	Kozarsky *et al.* (1994)
		FH human hepatocytes	*In vitro*	Kozarsky *et al.* (1993)
	Rat cholesterol 7α-hydroxylase	Hamster liver	*In vivo*	Spady *et al.* (1995)
	Rat apo-B mRNA editing protein	Mouse liver	*In vivo*	Teng *et al.* (1994); Hughes *et al.* (1996)
		Rabbit liver	*In vivo*	Hughes *et al.* (1996)
	Human apo A-I	Mouse liver	*In vivo*	Kopfler *et al.* (1994)
α_1-antitrypsin deficiency	Human AAT	Murine keratinocytes	*Ex vivo, in vivo*	Setoguchi *et al.* (1994a)
		Mouse liver	*In vivo*	Kay *et al.* (1995)
		Cotton rat peritoneum	*In vivo*	Setoguchi *et al.* (1994b)
		Human airway	*In vitro*	Siegfried *et al.* (1995)
		Human endothelium	*In vitro, ex vivo*	Lemarchand *et al.* (1992)
Hemophilia A	Human factor VIII	Mouse liver	*In vivo*	Connelly *et al.* (1995)
		Human hepatocytes	*In vitro*	Connelly *et al.* (1995)
Hemophilia B	Human factor IX	Mouse liver	*In vivo*	Smith *et al.* (1993)
	Canine factor IX	Hemophilia B dog liver	*In vivo*	Kay *et al.* (1994); Fang *et al.* (1995)
		Mouse muscle	*In vivo*	Dai *et al.* (1995)
Endotoxic shock	TNF inhibitor	Mouse liver	*In vivo*	Kolls *et al.* (1994, 1995)
Anemia	Human Epo	Mouse muscle	*In vivo*	Tripathy *et al.* (1994, 1996)
		Cotton rat liver	*In vivo*	Setoguchi *et al.* (1994c)
		Human hepatocytes	*In vitro*	Setoguchi *et al.* (1994c)
	Mouse Epo	Mouse muscle	*In vivo*	Tripathy *et al.* (1996)
	Monkey Epo	Mouse liver	In vivo	Descamps *et al.* (1994)
Thrombocytopenia	HST-1	Mouse liver	*In vivo*	Konishi *et al.* (1995)
Ornithine-δ-aminotransferase deficiency	Human OAT	Human retinal epithelial cells	*In vitro*	Sullivan *et al.* (1996)

Ornithine transcarbamylase deficiency	Rat OTC	Mouse liver	*In vivo*	Stratford-Perricaudet *et al.* (1990)
	Human OTC	Human hepatocytes	*In vitro*	Morsy *et al.* (1993)
Adenosine deaminase deficiency	Human ADA[a]	ADA human bone marrow	*In vitro*	Mitani *et al.* (1994)
Protoporphyria	Human ferrochelatase	Human fibroblasts	*In vitro*	Magness and Brenner (1995)
Cancer	Cytosine deaminase	Human colon carcinoma	*In vitro*, in xenograft	Hirschowitz *et al.* (1995)
	HSV TK	Mouse prostate cancer	*In vivo*	Hall *et al.*(1996); Shaker *et al.* (1996); Eastham *et al.* (1996)
		Rat glioma	*In vivo*	S.-H. Chen *et al.* (1994)
		Rat gliosarcoma	*In vivo*	Perez-Cruet *et al.* (1994); Vincent *et al.* (1996)
		Rat mammary adenocarcinoma	*In vivo*	Colak *et al.* (1995)
		Human prostate cancer	*In vitro*	Eastham *et al.* (1996)
		Human mesothelioma	In xenograft	Smythe *et al.* (1995)
		Human hepatocellular carcinoma	*In vitro*, in xenograft	Qian *et al.* (1995a,b)
		Human head and neck cancer cells	In xenograft	O'Malley *et al.* (1995)
	Human p53	Rat liver	*In vivo*	Drazan *et al.* (1994)
		Mouse lung	*In vivo*	W.-W. Zhang *et al.* (1995)
		Murine prostate cancer cells	*In vitro, in vivo*	Eastham *et al.* (1995)
		Human ovarian carcinoma	*In vitro*	Bacchetti and Graham (1993)
		Human breast cancer cells	*In vitro*	Katayose *et al.* (1995a,b)
		Human prostate cancer	*In vitro, ex vivo*	C. Yang *et al.* (1995)
		Human lung carcinoma	*In vitro*	Katayose *et al.* (1995a,b); W.-W. Zhang *et al.* (1995)
			In xenograft	Fujiware *et al.* (1994)
		Human head and neck cancer cells	*In vitro*, in xenograft	Liu *et al.* (1994)

(*continues*)

TABLE III (continued)

Disorder	Transgene	Target tissue	Infection	Ref.
	Human p21[wafl]	Murine renal carcinoma	*In vitro, in vivo*	Z.-Y. Yang *et al.* (1995)
		Murine prostate cancer cells	*In vitro, in vivo*	Eastham *et al.* (1995)
		Murine melanoma	*In vitro*	Z.-Y. Yang *et al.* (1995)
		Human lung carcinoma	*In vitro*	Katayose *et al.* (1995b)
		Human breast cancer cells	*In vitro*	Katayose *et al.* (1995b)
		Human melanoma	*In vitro*	Z.-Y. Yang *et al.* (1995)
	Human p16[ink4]	Human lung cancer cells	*In vitro, in* xenograft	Jin *et al.* (1995)
	C-CAM1[a]	Human prostate cancer cells	*In vitro, in* xenograft	Kleinerman *et al.* (1995)
	Anti-*ras* ribozyme	Human bladder carcinoma	*In vitro, ex vivo*	Feng *et al.* (1995)
	Ad5 E1A	Human ovarian tumors	*In vitro, in vivo*	Y. Zhang *et al.* (1995)
	Human IFN	Hamster melanoma	In xenograft	Zhang *et al.* (1996)
		Human breast cancer cells	In xenograft	Zhang *et al.* (1996)
		Human leukemia	In xenograft	Zhang *et al.* (1996)
	Murine IL-2	Murine mastocytoma	*Ex vivo, in vivo*	Haddada *et al.* (1993a); Cordier *et al.* (1995)
	Human IL-2	Murine mammary adenocarcinoma	*Ex vivo, in vivo*	Addison *et al.* (1995a)
	Murine IL-4	Murine mammary adenocarcinoma	*Ex vivo, in vivo*	Addison *et al.* (1995b)
	Murine IL-12	Murine mammary adenocarcinoma	*In vivo*	Bramson *et al.* (1996b)

[a] hsp 70i, Heat shock protein 70i; FGF, fibroblast growth factor; VEGF, vascular endothelial growth factor; sVCAM, soluble vascular cell adhesion molecule; OTC, ornithine transcarbamylase; C-CAM1, epithelium-specific cell adhesion molecule; ADA, adenosine deaminase.

Ad E3 19-kDa ORF under the control of a constitutive promoter (Lee *et al.*, 1995) and was shown to induce a significantly lower antivector cytotoxic T cell response than other first-generation Ad vectors. All of these studies suggest that immunomodulation may allow long-term Ad gene therapy, but clearly further investigation is required to determine which regimen is most beneficial, and whether these results can be extended to humans.

B. Gene Transfer to the Lung

Much of the research employing gene transfer to the lung has been directed toward developing protocols for the treatment of cystic fibrosis (CF). Cystic fibrosis is one of the most common inherited recessive diseases and is caused by mutations in the CFTR gene that result in decreased Cl^- conductance and increased Na^+ uptake in affected cells. Although this defect affects a number of organs, 95% of the deaths associated with CF are due to respiratory failure and therefore most CF therapies have been directed to the lung (reviewed in Collins, 1992; O'Neal and Beaudet, 1994; Johnson, 1995). Because of its natural tropism for the lung, among other properties, Ad is considered a potentially useful vector for CFTR gene transfer.

The safety and efficacy of Ad vectors for *in vivo* gene transfer to the lung have been studied in several animal models. Although initial studies with cotton rats suggested efficient gene transfer to all major types of airway epithelial cells (Mastrangeli *et al.*, 1993), more recent investigations indicate that differentiated columnar airway epithelium is relatively resistant to Ad-mediated gene expression (Grubb *et al.*, 1994; Dupuit *et al.*, 1995), most likely due to the lack of $\alpha_v\beta_5$ cell surface receptors required for virus internalization (Goldman and Wilson, 1995). This cell type specificity may explain some of the variability in levels of transgene expression between different studies. Nonetheless, Ad vectors encoding the human CFTR or marker genes efficiently transduce the airway epithelia of cotton rats (Rosenfeld *et al.*, 1992; Zabner *et al.*, 1994a; Yei *et al.*, 1994a,b), rhesus monkeys (Zabner *et al.*, 1994a; Bout *et al.*, 1994a,b; Brody *et al.*, 1994b), and baboons (Simon *et al.*, 1993). In general, transduction was accompanied by increases in inflammation and virus-neutralizing antibodies in a dose-dependent manner and, in one case, a low level of virus shedding (Bout *et al.*, 1994a). Although this degree of immune response might not pose a risk for the recipient, it diminishes the efficacy of Ad gene transfer by limiting the duration of transgene expression and preventing repeated administration of the vector.

A limited number of animal studies have been performed using second-generation Ad vectors (see Section II,B). An Ad vector carrying a temperature-sensitive E2A gene showed prolonged expression in the cotton rat lung compared to first-generation vectors (Engelhardt *et al.*, 1994b). A similar effect was seen by administering that vector to the baboon lung (Goldman *et al.*, 1995a), and to the lungs of a mouse model for CF (Y.

Yang *et al.*, 1994c). Another second-generation Ad vector from which all of the E4 region except ORF6 had been deleted, however, induced both cellular and humoral responses in rhesus monkeys after long-term repeated administration (Kaplan *et al.*, 1996). Although the E2A mutant vector might be more defective than the vector with a modified E4 region, it is not clear whether it would induce the same level of cellular and humoral immune responses if it were analyzed by the more stringent test of repeated administration over long periods.

Human cells transduced *in vitro* with the Ad CFTR vector have been shown to express a functional CFTR gene product. Even at low MOIs (i.e., 1–10 PFU/cell), infection with the recombinant virus corrected the Cl^- transport defect in monolayers of airway epithelial cells derived from CF patients (Rich *et al.*, 1993; Zabner *et al.*, 1994b; Rosenfeld *et al.*, 1994). A separate study using much higher doses of virus (10^4 PFU/cell) found both the Cl^- secretion and Na^+ absorption in CF airway epithelial cells corrected to normal levels following infection with a CFTR recombinant Ad vector (Johnson *et al.*, 1995). CFTR gene transfer has also been investigated in human bronchial xenografts in immunocompromised mice (Englehardt *et al.*, 1993a; Goldman *et al.*, 1995b). *In vivo* infection of CF xenografts with high doses of CFTR recombinant Ads reconstituted Cl^- transport, but correction of Na^+ hyperabsorption was variable (Goldman *et al.*, 1995b), corroborating results obtained with a mouse model for CF (Grubb *et al.*, 1994), and suggesting that the two ion transport mechanisms are differentially regulated. Furthermore, these data indicate that high doses of recombinant virus may be required for the treatment of CF, in contrast to the predictions of earlier studies.

Several trials have been initiated to test first-generation CFTR Ad vectors in patients with cystic fibrosis (reviewed in O'Neal and Beaudet, 1994; Boucher *et al.*, 1994; Wilson, 1994; Welsh *et al.*, 1995; Crystal, 1995a,b). In an early study, doses at as low an MOI as 1 PFU/cell to the nasal epithelia of CF patients transiently corrected the chloride transport defect, with no virus-associated adverse effects (Zabner *et al.*, 1993). Subsequent investigations have reported results ranging from no correction of sodium or chloride ion transport at doses from an MOI of 1 up to 1000 PFU/cell, with significant inflammation at the highest dose (Knowles *et al.*, 1995), to at least partial correction of both transport systems (Hay *et al.*, 1995). Knowles *et al.* (1995) have suggested that the inconsistencies in reported efficacy may be due to variations in the protocols used to assess ion transport abnormalities specific to CF. In an evaluation of the safety of CFTR Ad administration to both the nasal and bronchial epithelia of CF patients, Crystal *et al.* (1994) found transient expression of the transgene consistent with humoral and cellular responses to the vector in some cases, but no long-term adverse effects and no neutralizing antibody against the virus. Because the effectiveness of first-generation Ad vector gene therapy for CF appears to be limited by the host inflammtory response, it will be interesting to see whether the second-generation Ad vectors are more successful.

Adenovirus vectors have also been examined as potential therapeutic agents for respiratory distress syndrome in newborns. This congenital lethal disease is due to a deficiency of surfactant protein B(SP-B) and possibly other surfactant proteins as well, which are normally expressed in the lung epithelium during and after fetal development (Nogee *et al.*, 1993). Preliminary studies have begun to assess the feasibility of administering Ad vectors to fetal lung tissue. Adenovirus vectors have been used successfully to transfer marker genes to human fetal lung *ex vivo* (Ballard *et al.*, 1995), and to the lungs of fetal sheep *in utero* (McCray *et al.*, 1995; Vincent *et al.*, 1995). Expression persisted for nearly 1 month *ex vivo*, but was limited to 2 weeks *in vivo* and was accompanied by an inflammatory response. Although human SP-B and SP-A cDNAs carried by recombinant Ads are efficiently expressed in lung epithelial cells *in vitro* as well as in rat and cotton rat lung *in vivo* (Yei *et al.*, 1994c; Korst *et al.*, 1995), the immune response against these vectors will clearly have to be prevented for long-term gene expression to be realized.

C. Gene Transfer to Skeletal and Cardiac Muscle

Adenovirus vectors have been used successfully for gene transfer to both skeletal and heart muscle cells (reviewed by Acsadi *et al.*, 1995). Intravenous administration of Ad recombinants carrying reporter genes effectively transduced mouse skeletal and cardiac muscles, resulting in reporter gene expression in these nonproliferating cells for up to 1 year (Stratford-Perricaudet *et al.*, 1992). The level of transgene expression in skeletal muscles is determined to a large extent by the maturity of the cells (Acsadi *et al.*, 1994b). Neonatal mice injected intramuscularly with Ad reporter vectors expressed the transgene at high levels in approximately half of the total muscle fibers, whereas expression in adult mice was 20-fold lower. Intermediate levels of reporter gene expression were observed in regenerating adult mouse muscles. This may be due partly to the decreased levels of $\alpha_v\beta_5$ integrins required for Ad internalization on the surface of mature skeletal muscle cells (Acsadi *et al.*, 1994a). In contrast to skeletal muscle, direct injection of both adult and neonatal rat myocardium with Ad reporter constructs resulted in high levels of expression (Kass-Eisler *et al.*, 1993; Guzman *et al.*, 1993a; Acsadi *et al.*, 1995). In larger animals the route of administration was found to influence the level of expression in the heart, with either direct intramyocardial injection (French *et al.*, 1994b; Muhlhauser *et al.*, 1996) or the less invasive catheter-mediated delivery to the coronary artery (Barr *et al.*, 1994) being the most effective.

Muscle cells have been targeted primarily for the therapy of Duchenne muscular dystrophy (DMD). Duchenne muscular dystrophy is a progressively degenerative disease for which there is no known cure. Death usually occurs in the second decade as a result of respiratory or cardiac failure. Duchenne muscular dystrophy is caused by mutations in the gene encoding dystrophin, a 427-kDa cytoskeletal protein, resulting in the absence of functional dystrophin protein (Hoffman *et al.*, 1987). Because the size of the normal dystrophin cDNA (12 kb) exceeds the capacity of first-generation Ad vectors, most MD gene therapy studies have utilized a 6.3-kb truncated cDNA (known as the dystrophin minigene) isolated from a patient with a milder form of this disease, Becker muscular dystrophy (England *et al.*, 1990). Dystrophin minigenes have also been engineered by deleting segments of the normal dystrophin cDNA, but characterization of these constructs is still at a preliminary stage (Clemens *et al.*, 1995).

Injection of a recombinant Ad dystrophin minigene vector into the hind limb of neonatal mdx mice (a mouse model defective in expression of dystrophin) resulted in up to half of the muscle fibers expressing the transgene 1 week after injection, and 10% still expressing after 6 months (Ragot *et al.*, 1993; Vincent *et al.*, 1993). In a similar study, although the initial level of expression was high, the duration of expression was shorter, and limited by a host immune response against the transduced cells (Acsadi *et al.*, 1996). In all of these cases the dystrophin minigene appeared to arrest muscle degeneration to some extent. A novel Ad vector has been reported in which all of the Ad coding sequences have been replaced by an expression cassette composed of a muscle-specific promoter, the full-length murine dystrophin cDNA, and a marker gene (Kochanek *et al.*, 1996). This recombinant vector was shown to express the full-length dystrophin protein in primary mdx mouse myoblasts *in vitro*. Although some hurdles must be cleared (see Section II,A) before Ad can be used as a vector for the gene therapy of MD in humans, this avenue of research may be the most promising of all potential therapies for this disease.

D. Gene Transfer to the Vasculature

Adenovirus vectors have been investigated as agents for gene therapy of vascular disorders, and in particular restenosis, a disorder characterized by abnormal proliferation of vascular smooth muscle cells (VSMCs) within the blood vessel following arterial injury such as that caused by balloon angioplasty. Recombinant Ad vectors are appealing for the treatment of restenosis for several reasons: expression of the transgene can be targeted to the artery during the angioplasty procedure, a high level of transient local expression is both sufficient and desirable, and other therapies have not been successful.

Adenoviruses containing reporter genes were shown to express recombinant protein effectively in human, porcine, canine, rat, and bovine VSMCs *in vitro* (Guzman *et al.*, 1993b; Mazur *et al.*, 1994; March *et al.*, 1995). *In vivo*, the method of vector delivery to uninjured arteries determined the cell type infected: in general, endothelial cells were the primary targets (Guzman *et al.*, 1993b; Lemarchand *et al.*, 1993; Rome *et al.*, 1994; Schulick *et al.*, 1995a), with VSMCs infected only after abrasion of the endothelial cell barrier (Lee *et al.*, 1993; Steg *et al.*, 1994). Although in one study little difference was seen in recombinant protein expression following Ad vector delivery to atherosclerotic, balloon-injured (i.e., restenotic), or normal arteries (French *et al.*, 1994a), others have reported a considerable reduction in the number of cells transduced in the atherosclerotic or injured arteries compared to uninjured arteries (Landau *et al.*, 1995; Feldman *et al.*, 1995). Therefore the most effective gene therapies for restenosis may involve expression of recombinant proteins that would block injury-induced proliferation of the majority of VSMCs, even when only a small proportion are infected. One such vector has been investigated, that encoding the thrombin inhibitor hirudin (Rade *et al.*, 1996). This vector was found to induce hirudin secretion, and to inhibit thrombin activity locally when administered *in vivo* to injured rat carotid arteries, leading to a 35% reduction in VSMC proliferation. Another Ad vector that appears promising for the treatment of restenosis is one encoding the HSV thymidine kinase (TK) gene (Ohno *et al.*, 1994; Guzman *et al.*, 1994; Chang *et al.*, 1995a). Thymidine kinase induces sensitivity to the drug ganciclovir (GCV) in both the TK-expressing cell as well as surrounding cells. This vector, in combination with GCV, was effective in reducing VSMC proliferation following arterial injury. A recombinant Ad encoding the cyclinD/cyclin-dependent kinase inhibitor, p21[wafl], which blocks cell cycle progression, also showed efficacy in a rat model for restenosis (Chang *et al.*, 1995b).

Successful transplantation of vascularized organs or tissue can be impeded by accelerated VSMC proliferation similar to that resulting from arterial injury, as well as by the host immune response. Potentially, the former may be alleviated by Ad vectors such as those described previously. In preliminary investigations to examine this possibility, Ad vectors administered *ex vivo* during the transplantation procedure have been shown to effectively transduce aortic transplants (Mehra *et al.*, 1996), liver transplants (Shaked *et al.*, 1994; Drazan *et al.*, 1995a), and vein grafts (S. J. Chen *et al.*, 1994). To modulate inflammation following transplantation, an Ad vector has been constructed that contains a cDNA encoding an inhibitor of NF-κB, IκBα (Wrighton *et al.*, 1996). Although *in vivo* results have not yet been reported, infection of endothelial cells *in vitro* with this recombinant resulted in inhibition of inflammatory cytokine-induced gene expression, and induction of other antiinflammatory effects. A recombinant Ad vector encoding the cDNA for the Epstein–Barr virus homolog of interleukin 10 (vIL-10), a potent immunosuppressant, induced physiologically relevant levels of vIL-10 *in vivo* in rat liver allografts (Drazan *et al.*, 1995b), suggesting that this vector may be useful in antirejection gene therapies.

Adenovirus-mediated transfer of genes encoding angiogenic factors has been explored as a treatment for ischemic diseases, in which tissues or organs are oxygen deprived owing to decreased blood flow. Vectors encoding human acidic fibroblast growth factor and vascular endothelial growth factor have been generated and shown to induce proliferation of infected endothelial cells *in vitro* and neovascularization *in vivo* (Muhlhauser *et al.*, 1995a,b), suggesting that this approach may be useful therapeutically for ischemic disorders.

E. Gene Transfer to the Brain and Central Nervous System

Because the blood–brain barrier prevents systemic delivery of proteins to the brain, effective gene therapy of brain disorders must target this organ directly. Adenovirus vectors have been shown *in vitro* to transduce a variety of neuronal cells efficiently, including astrocytes, neurons, glia, Schwann cells, brain fibroblasts (Le Gal La Salle *et al.*, 1993; Caillaud *et al.*, 1993; Shy *et al.*, 1995; Shering *et al.*, 1996), and neuronal progenitor cells (Gage *et al.*, 1995). In addition, Ad vectors have been used successfully *in vivo* to deliver transgenes to various regions of the brain by intracerebral injection of virus (Davidson *et al.*, 1993; Bajocchi *et al.*, 1993; Akli *et al.*, 1993; Horellou *et al.*, 1994). In some cases, expression of a *lacZ* reporter gene could be detected for at least 45 days postinoculation (Akli *et al.*, 1993) without cytotoxic effects at doses of less than 10^7 PFU. Inflammation associated with intracerebral administration of Ad vectors (at doses of approximately 10^6 PFU) has been reported although transgene expression nonetheless persisted for 2 months (Byrnes *et al.*, 1995). These studies suggest that Ad vectors may be used successfully for gene transfer to the brain and may be useful for therapies to prevent neurodegenerative disorders. To this end, G. M. Smith *et al.* (1996) transduced primary astrocytes with Ad vectors encoding the neurotrophic factors ciliary neurotrophic factor (sCNTF) or neurotrophin 3 (NT-3), and found that these cells secreted bioactive factors that could support the growth and survival of peripheral and central neuronal populations. Verhaagen *et al.* (1995) used an Ad vector to express the neurotrophic factor B50/GAP43, and successfully transduced mature olfactory neurons *in vivo* following intranasal administration of the vector. Expression of these neurotrophic factors could be therapeutic and promote neuronal regeneration following damage. Other neuronal disorders, such as Parkinson's disease, may also be targets for Ad-mediated gene therapy. Horellou *et al.* (1994) demonstrated in a rat model of Parkinson's disease that intracerebral injection of an Ad vector expressing tyrosine hydroxylase could reduce the severity of the disease, thus indicating that this type of therapy may be applicable clinically. In addition, the transduction of neuronal progenitor cells with Ad vectors followed by implantation of these transduced cells into the brain led to their engraftment and differentiation into mature neurons (Gage *et al.*, 1995). This technique would allow for the potential engraftment of healthy neuronal cells, which, if expressing neurotrophic factors, could prolong their own survival or could be engineered to express therapeutic proteins to combat other neurological disorders. These early results and others are promising and indicate that Ad vectors may be extremely useful for neurological studies (for a review of the use of Ad vectors for neurological disorders, see Lowenstein *et al.*, 1996).

F. Gene Transfer to the Liver

Between 50 and 100% of the cells in primary cultures of rat, mouse, and human hepatocytes can be transduced by Ad vectors *in vitro* (Jaffe *et al.*, 1992; Li *et al.*, 1993; Morsy *et al.*, 1993). *In vivo*, the method of vector administration strongly influences the proportion of hepatocytes that are transduced. Because most of the virus is deposited in the liver following i.v. injection (Huard *et al.*, 1995), this route of delivery has been suitable for some applications. In one of the first reports on the use of Ad *in vivo* as a vector for gene therapy, Stratford-Perricaudet *et al.* (1990) demonstrated transfer and expression of the cDNA encoding ornithine transcarbamylase to mouse liver via i.v. injection. However, higher efficiencies of gene transfer have been obtained by direct viral infusion into the portal vein, resulting in more than 95% transduction of mouse hepatocytes (Li *et al.*, 1993), and up to 30% transduction of rat liver cells (Jaffe *et al.*, 1992; Drazan *et al.*, 1995c), without significant pathology.

Gene therapy vectors have been used to target the liver in the treatment of a number of disorders including atherosclerosis. The risk of developing atherosclerosis, a common cardiovascular disease, is influenced by the balance of serum cholesterol in low-density lipoprotein (LDL) particles *vs* high-density lipoprotein (HDL) particles. High levels of circulating LDLs are associated with increased risk, whereas high levels of HDLs are associated with decreased risk. Because many of the processes that regulate plasma lipoprotein concentrations and cholesterol metabolism occur in the liver (Dietschy *et al.*, 1993), several gene therapies for atherosclerosis and hypercholesterolemia have been directed to this organ. To decrease circulating LDL concentrations, recombinant Ads have been constructed that encode the LDL receptor, which is normally involved in LDL uptake by the liver and is reduced in patients with familial hypercholesterolemia (FH). These vectors have been shown to express the LDL receptor in primary hepatocyte cultures derived from patients with FH (Kozarsky *et al.*, 1993), as well as *in vivo* in mouse (Ishibashi *et al.*, 1993) and rabbit (Kozarsky *et al.*, 1994) models for FH. In the mouse model, an increased LDL level (relative to normal mice) was corrected by i.v. administration of the receptor vector. In the rabbit model, intraportal vein infusion of the receptor vector induced an acute, although transient, reduction in serum cholesterol. A second administration of the recombinant virus to the rabbit was blocked by neutralizing antibody, however, so it is likely that modulation of the immune response against virus and virus-infected cells will be required to realize the full potential of these vectors in the treatment of FH.

Additional Ad vectors have been designed that target other processes involved in determining the levels of LDL and HDL cholesterol in serum. Intravenous administration of a vector encoding rat cholesterol 7α-hydroxylase to hamsters has been shown to decrease plasma LDL cholesterol by 60–75% by converting cholesterol to bile acids (Spady *et al.*, 1995). Teng *et al.* (1994) and Hughes *et al.* (1996) have constructed Ad vectors that encode the catalytic component of the apoB mRNA editing complex of rat. Expression of this gene in mice reduced the formation of both LDLs and lipoprotein(a) by producing a truncated form of apoB that, in the full-length form, is an essential component of these two lipoproteins. In rabbits, however, both the recombinant vector and a control Ad induced elevated LDL concentrations, offsetting the therapeutic value of the transgene (Hughes *et al.*, 1996), and suggesting a need for further investigation. In an alternative approach for modulating lipoprotein balance in the serum, Kopfler *et al.* (1994) have constructed an Ad vector containing the gene encoding human apolipoprotein A-I, which induced an increase in HDL cholesterol following administration to normal mice. All of these investigations highlight the potential of Ad vectors for gene therapy to reduce the risk of cardiovascular disease, and the need to develop further strategies to prolong transgene expression.

A novel type of Ad vector has been reported that contains an expression cassette for a prototype ribozyme, specific for human growth hormone (hGH) mRNA (Lieber and Kay, 1996). Administration of the ribozyme vector to the livers of hGH-expressing transgenic mice results in a reduction of up to 96% in the level of hepatic hGH mRNA. Potentially, Ad ribozyme vectors can be designed and used to treat infectious liver diseases such as viral hepatitis.

G. Expression of Serum Proteins

Several common diseases resulting from deficiencies in serum proteins are caused by single gene mutations and as such are possible candidates for Ad gene therapy. α_1-antitrypsin (AAT), which is normally produced in the liver and delivered to the circulation, is reduced in patients with hereditary progressive emphysema. Adenovirus vectors encoding human AAT have been constructed and used to transfer the gene to a variety of tissues and cell types, including human respiratory epithelium *in vitro* (Siegfried *et al.*, 1995), human umbilical vein endothelial cells *in vitro* and *ex vivo* (Lemarchand *et al.*, 1992), murine keratinocytes *in vitro* and *in vivo* (Setoguchi *et al.*, 1994a), cotton rat peritoneal mesothelium *in vivo* (Setoguchi *et al.*, 1994b), and mouse hepatocytes *in vivo* (Kay *et al.*, 1995). In all animal experiments, significant, although transient, levels of AAT were detected in the serum with the maximum reaching therapeutic levels (700 μg/ml) (Kay *et al.*, 1995).

Hemophilias A and B are caused by deficiencies in the blood coagulation factors VIII and IX, respectively. An Ad vector containing a modified factor VIII cDNA efficiently transduced human hepatocytes *in vitro*, and, following tail vein injection of mice, induced expression of biologically active factor VIII at serum levels exceeding the normal level found in humans (Connelly *et al.*, 1995), indicating the potential therapeutic value of this vector for the treatment of hemophilia A. In a similar manner, factor IX Ad vectors induced therapeutic levels of factor IX in the sera of normal mice following tail vein injection (Smith *et al.*, 1993), and in hemophilia B dogs following portal vein infusion (Kay *et al.*, 1994; Fang *et al.*, 1995). In the latter case, the treatment was highly successful as the defect in blood coagulation in this dog model for hemophilia B was completely, although transiently, corrected.

Tumor necrosis factor (TNF) is a cytokine involved in the pathogenesis of endotoxic shock, cachexia, and some chronic inflammatory diseases (reviewed by Tracey and Cerami, 1993). An Ad vector that encodes a chimeric protein capable of binding and neutralizing TNF and lymphotoxin was administered i.v. to mice and shown to induce high levels of the secreted TNF inhibitors, which effectively protected the mice against endotoxic shock (Kolls *et al.*, 1994, 1995). Thus this vector may have therapeutic potential in instances where abrogation of TNF or lymphotoxin activity is beneficial.

Other Ad vectors have been constructed to facilitate an increase in production of specific blood cell types. Erythropoietin (Epo), a major regulator of erythropoiesis, is currently administered as a recombinant protein to some patients suffering from severe anemia. Adenovirus vectors containing the human or monkey Epo cDNA have been administered to mice (Tripathy *et al.*, 1994, 1996; Descamps *et al.*, 1994) and cotton rats (Setoguchi *et al.*, 1994c) and in all cases there was a significant increase in serum Epo as well as hematocrit levels, indicating an increase in red blood cell mass, with a duration of up to 6 months. The HST-1 gene product, which has been shown to induce an increase in the number of platelets in the blood, has been efficiently expressed in a mouse model for thrombocytopenia following injection with an Ad vector encoding the HST-1 gene (Konishi *et al.*, 1995). This treatment was more effective in reducing thrombocytopenia than any other cytokine treatment reported to date. Both the Epo and HST-1 vectors may be useful to induce a transient elevation in the concentration of specific blood cell types as required for many clinical applications.

H. Gene Transfer to Ocular Tissue

Successful Ad-mediated gene transfer to ocular tissues has been reported by a number of investigators. Injection of Ad reporter vectors into the subretinal space of the mouse eye has resulted in efficient expression in retinal pigment epithelium, and to a lesser extent in photoreceptor cells, for a duration of 1 to 3 months (Bennett *et al.*, 1994; T. Li *et al.*, 1994). Corneal endothelial, trabecular meshwork, lens epithelial, and iris epithelial cells have also been shown to support reporter gene expression following Ad vector administration to mouse ocular tissues *in vivo*, although in this experiment expression was limited to 2 weeks (Budenz *et al.*, 1995). These studies prompted the analysis of Ad vectors for retinal gene delivery of ornithine δ-aminotransferase (OAT), a mitochondrial matrix enzyme, the deficiency of which results primarily in loss of sight. *In vitro* experiments employing a recombinant Ad OAT vector to infect cultured human retinal epithelial cells have demonstrated highly efficient transduction and expression of the transgene (Sullivan *et al.*, 1996). At the highest MOI (150 PFU/cell) some mitochondrial toxicity and morphological changes in the cells were observed, but these side effects could possibly be alleviated by improvements in vector design. Another recombinant Ad vector has been constructed containing the gene for human heme-oxygenase-1 (HO-1), a stress protein that counteracts oxidative injury. Following infection with this vector, HO-1 gene transfer and expression were demonstrated in rabbit corneal epithelial cells *in vitro* and in various rabbit eye tissues *in vivo*, indicating a potential for Ad HO-1 vector treatment to protect against oxidant stress, which could contribute to the alleviation of several eye diseases including cataracts (Abraham *et al.*, 1995).

I. Cancer Gene Therapy

Although most work in cancer gene therapy to date has been carried out using retroviral vectors, Ad vectors may ultimately become the vector of choice for cancer therapy. Analogous to observations that Ad vectors primarily infect cells surrounding the site of injection in the liver (Jaffe *et al.*, 1992), intratumoral administration of recombinant virus has been shown to induce transgene expression localized to the tumor (Tang *et al.*, 1994; Bramson *et al.*, 1996b). Because Ad rarely integrates into the host cell genome, the expression from the vector is transient and therefore one would avoid persistence of the foreign gene. This is of critical importance in immunotherapies and perhaps some antioncogene or tumor suppressor therapies in which persistent expression might produce adverse side effects. The host response against Ad-infected cells should not hinder the use of Ad for gene therapy of cancer since the destruction of infected tumor cells by the immune system would be beneficial. Furthermore, in immunotherapy the virus itself may act an adjuvant, which might aid in eliciting specific antitumor responses and protective immunity. Adenovirus vectors expressing suicide genes, tumor suppressor genes, antioncogene factors, and immunomodulating proteins have all been used *in vivo* as antitumor therapies. Most protocols involve *ex vivo* infection of tumor cells and subsequent implantation *in vivo*, or direct intratumoral injections *in vivo*. Some of these techniques are discussed in more detail below.

I. Suicide Genes

This mode of tumor therapy involves the introduction of a gene encoding an enzyme that converts a nontoxic drug to one that is lethal to cells. The two most commonly used systems introduce the genes for *Escherichia coli* cytosine-deaminase (CD), which confers sensitivity to the prodrug 5-fluorocytosine (5FC), or the HSV TK gene, which confers sensitivity to the drug GCV. An Ad vector expressing CD was shown *in vitro* to suppress the growth of HT29 human colon carcinoma-derived cells in a dose-dependent manner (Hirschowitz *et al.*, 1995) on treatment with 5FC. In addition, on injection into established HT29 tumors in nude mice and subsequent treatment with 5FC, there was a four- to fivefold reduction in tumor size compared to controls, although the tumors continued to grow until the animals became moribund.

The HSV TK gene has been more widely used and has also been shown to be effective in mediating antitumor responses *in vivo*. Although this strategy has long been shown to be effective when delivered via retroviral vectors, S.-H. Chen *et al.* (1994) were among the first to demonstrate the efficacy of this therapy using Ad vectors as the mode of delivery. They injected C6 glioma cells intracerebrally in rats, and 8 days later followed with an intratumoral injection of an Ad vector expressing HSV TK. Animals treated with GCV were found to have a 500-fold reduction in tumor size compared to control animals. Two of 10 animals became tumor free while the majority retained small residual tumors around the injection site. Unfortunately these animals were sacrificed 20 days posttreatment and it is unknown whether there might have been complete regression in the two animals or if they would have developed tumors at a later date. Perez-Cruet *et al.* (1994) also demonstrated significant efficacy using an Ad HSV TK vector with the 9L gliosarcoma model. They found that GCV-treated animals survived for the duration of the experiment (80–120 days) in contrast to the controls, which died by 22 days, although it is unclear whether the surviving animals were free of residual tumors at the time the experiment was terminated. Tumors were completely eradicated in immunocompetent but not in nude mice, suggesting a role for the immune system in the antitumor responses mediated by HSV TK. In similar experiments, Colak *et al.* (1995) injected MATB mammary adenocarcinoma cells into the brains of rats as a metastatic model, and 7 days later injected the tumors with an HSV TK-expressing Ad vector and treated with GCV. They found that by 16 days postinjection all control animals had died, whereas all treated animals remained alive and appeared to have no residual tumors. However, by 27 days postinjection, although no evidence of tumor was present at the injection site, tumors were present at other sites in the brain and the animals eventually succumbed to tumor load. Adenovirus vectors expressing HSV TK have also been shown to be efficacious in animal models for mesothelioma (Smythe *et al.*, 1995), prostate cancer (Hall *et al.*, 1996; Shaker *et al.*, 1996; Eastham *et al.*, 1996), hepatocellular carcinoma (Qian *et al.*, 1995a), and head and neck squamous cell carcinoma (O'Malley *et al.*, 1995).

Although expression of suicide genes has been shown to involve a "bystander effect" in that only a fraction of the cells need express the suicide gene to mediate the destruction of surrounding nontransduced cells, it would appear that this treatment alone may not mediate total tumor eradication. The use of suicide gene delivery in combination with other antitumor approaches may lead to more successful tumor ablation *in vivo*.

2. Tumor Suppressor Genes and Oncogene Inactivation

The study of cell cycle regulation has led to the discovery of a number of proteins that activate (protooncogene products) or inhibit (tumor suppressors) the growth and proliferation of cells. Many of these genes have been found to be mutated in a variety of tumors, and these mutations are responsible for the proliferative capacity and resistance of these cells to commonly used therapies such as γ irradiation and chemotherapy. The reintroduction of wild-type forms of tumor suppressor genes into cells harboring mutated copies or the treatment of cells with antioncogenic factors (such as dominant-negative forms of the oncogenes or ribozymes targeting the oncogenes) are two approaches for the elimination of cells that have gained a higher proliferative capacity.

a. p53   The tumor suppressor p53 plays a role in arresting the cell cycle at the G_1 checkpoint in response to DNA damage and in inducing apoptosis in cells that have extensive radiation damage (reviewed by Hartwell and Kastan, 1994; Hinds and Weinberg, 1994). It is the most commonly mutated gene found in human cancers (Hollstein *et al.*, 1991) and mutations that inactivate p53 are thought to lead to increased genomic instability in cells, allowing for the accumulation of additional mutations (Livingstone *et al.*, 1992). Reintroduction of p53 into cells harboring this mutation should restore the capacity to arrest at a G_1 checkpoint and to undergo apoptosis in response to ionizing radiation. Several groups have constructed Ad vectors expressing p53 under the control of various promoters. Following Ad p53 infection *in vitro*, the biological effects of p53 could be easily detected, including the upregulation of p21[wafl], an overall growth suppression, and an increase in the number of cells undergoing apoptosis in a variety of tumor cell lines carrying p53 mutations (Bacchetti and Graham, 1993; Liu *et al.*, 1994; C. Yang *et al.*, 1995; Katayose *et al.*, 1995a). Furthermore, administration of Adp53 vectors has been shown to be efficacious in several tumor models. In one study it was found that tumor cells infected *ex vivo* with

Adp53 are less tumorigenic *in vivo* than control vector-infected tumor cells (C. Yang *et al.*, 1995). Fujiwara *et al.* (1994) have shown that administration of an Adp53 vector following intratracheal administration of H226Br tumor cells in nude mice significantly inhibited tumor growth with only 25% of the treated mice developing tumors (70–80% of controls develop tumors). Animals that did develop tumors after receiving Adp53 had tumors that were reduced in size compared to control animals. The efficacy of Adp53 treatment has also been demonstrated with established tumors. Liu *et al.* (1994) showed that following instillation of an Adp53 vector into established subcutaneous (s.c.) squamous carcinoma of the head and neck tumor cells (Tu-177 and Tu-138), complete tumor regression occurred in 2 of 7 animals, and in the remaining animals tumor growth was reduced 60-fold relative to control virus-treated animals.

   b. p21^{wafl} and p16^{ink4}   The tumor suppressors p21^{wafl} and p16^{ink4} are members of the family of cyclin-dependent kinase (CDK) inhibitors that bind and inactivate CDKs and thus play a role in the maintenance of the cell cycle checkpoints and progression (Xiong *et al.*, 1993; Harper *et al.*, 1993; Serrano *et al.*, 1995). p53 has been shown to upregulate the production of p21^{wafl} protein in response to DNA damage induced by ionizing radiation, and thus induce a G_1 cell cycle arrest (Y. Li *et al.*, 1994). p16^{ink4} is thought to regulate the kinase responsible for phosphorylation of the retinoblastoma protein, allowing cells to progress through S phase (Serrano *et al.*, 1995). In addition, p16^{ink4} maps to a chromosomal locus that is deleted in many tumor types (Caldas *et al.*, 1994; Cheng *et al.*, 1994; Kamb *et al.*, 1994). These observations and others suggest that these proteins themselves may be tumor suppressors, and as such, candidates for antitumor therapy.

   The effect of p21^{wafl} on tumor growth has been studied using p21^{wafl} recombinant Ad vectors. Following infection of cultured cells [human H-358 lung carcinoma, MDA-MB-231 and MCF7 breast cancer cells (Katayose *et al.*, 1995b), murine B16BL6 melanoma, Renca renal carcinoma (Z.-Y. Yang *et al.*, 1995), or 148-IPA murine prostate cancer cells (Eastham *et al.*, 1995)], it was demonstrated that expression of p21^{wafl} induced a growth arrest at the G_0/G_1 checkpoint without inducing apoptosis. *Ex vivo* infected Renca tumor cells were also found to be less tumorigenic following transplantation *in vivo* (Z.-Y. Yang *et al.*, 1995). In addition, growth suppression of established s.c. prostate tumors could be demonstrated on intratumoral injection of Adp21^{wafl}, with treated animals surviving at least twice as long as controls (Eastham *et al.*, 1995). There has been one report of complete regression of renal carcinoma tumors in mice following intratumoral injection of Adp21^{wafl}, but in this study five repeated daily injections of the vector were given, and the animals were studied for only 40 days posttreatment (Z.-Y. Yang *et al.*, 1995).

Studies on p16^{ink4} have been more limited than those on p21^{waf1}. When an Ad vector encoding this gene was used to infect lung cancer cell lines that do not express this protein it was found that expression of p16^{ink4} blocked the entry of these infected cells into the S phase of the cell cycle and inhibited tumor cell growth (Jin *et al.*, 1995). *In vivo*, only 50% of animals receiving p16^{ink4}-expressing cells formed tumors (versus 88% for controls), and the tumors were reduced in size compared to controls. Direct intratumoral injection of recombinant virus resulted in a twofold reduction in tumor volume compared to controls. Taken together, these results suggest that the reintroduction of wild-type tumor suppressor genes by Ad vector delivery could be effective in the treatment of tumors harboring mutations of these genes.

c. Ras Antagonists   An additional means of eradicating tumor cells is by abrogating the oncogenic properties of the tumor cell. Many tumor cells have been shown to express a form of activated *ras* (H-*ras*) which is involved in cellular signal transduction (Lowy, 1993). Mutational activation of Ras protein leads to aberrant signaling and transformation of cells, and thus elimination of the active form of Ras would result in a growth suppression of these cells. An innovative approach has been developed using an Ad vector-encoded hammerhead ribozyme targeted to the *ras* oncogene (Feng *et al.*, 1995). The EJ human bladder carcinoma cell line that possesses an activated form of *ras* was used as a model system in these studies. On infection with the Ad ribozyme vector, the level of *ras* transcription was specifically reduced, and significant suppression of cell growth occurred with no viable cells being detected by 5 days postinfection. Furthermore, the tumorigenicity of these cells was abrogated since injection into nude mice of EJ cells transduced by the Ad ribozyme vector resulted in no tumor development while animals injected with control virus-infected cells developed tumors rapidly.

d. HER-2/neu Antagonists   The HER-2/*neu* oncogene product is an epidermal growth factor receptor-like protein that is overexpressed in many types of cancers, and this overexpression has been correlated with tumor size, frequency of relapse, and metastases (Slamon *et al.*, 1987). In addition, transgenic mice carrying a mutationally activated form of HER-2/*neu* develop mammary adenocarcinomas that have a high frequency of metastases (Muller *et al.*, 1988). This supports a role for HER-2/*neu* in the progression of tumorigenesis and therefore inhibitors of this molecule are possible candidates for tumor therapy. Yu *et al.* (1995) found that expression of Ad5 E1A proteins could repress transcription of the HER-2/*neu* oncogene, and could suppress tumor growth when stably transfected into tumor cells which overexpress HER-2/*neu*. It was subsequently shown that i.p. injection of a replication-defective Ad vector expressing the E1A proteins, but not the E1B proteins, could prolong the survival of animals bearing established SK-OV3 ovarian tumors which overexpress HER-2/*neu* (Y. Zhang *et al.*, 1995).

These studies indicate that therapeutic effects may be achieved using Ad vectors expressing anti-oncogene factors or tumor suppressor gene products. Although more work in the field of tumor suppressor restoration and onco-gene ablation is required, initial studies suggest that this form of therapy may be potent, particularly when used in combination with other forms of cancer therapy. It should be noted, however, that in all of the tumor therapies previously mentioned, with perhaps the exception of those using certain suicide genes, there is a strong requirement for infection of at least a large fraction of tumor cells since "bystander effects" have not been shown to play a significant role.

3. Immunotherapy Using Adenovirus Vectors

Tumor cell growth can be inhibited by immune effector cells such as lymphokine activated killer (LAK) cells (Lotze *et al.*, 1981; Grimm *et al.*, 1982), tumor-infiltrating lymphocytes (TILs) (Cameron *et al.*, 1990; Lind-gren *et al.*, 1993), CTLs (Fearon *et al.*, 1990), and natural killer (NK) cells (Hackett *et al.*, 1986). One approach to immunotherapy involves the delivery of exogenous cytokine molecules that activate and stimulate the proliferation of antitumor immune effectors. However, because tumor cells themselves may synthesize molecules that inhibit the activity of these effector cells, immunotherapies designed to counteract the effects of immunomodulatory proteins produced by the tumor may also stimulate eradication of the tumor.

To date, Ad vectors have been used to deliver genes for the cytokines interferon (IFN), interleukin 2 (IL-2), IL-4, and IL-12 in tumor models *in vivo*. A replication-proficient Ad vector encoding the human IFN consensus gene was used to treat nude mice carrying tumors derived either from a human breast cancer line (MDA-MB-435) or from a human myelogenous leukemic cell line (K562) (Zhang *et al.*, 1996). When the Ad–IFN recombi-nant was injected 24 hr after tumor cells were injected into the mice, growth of both types of tumors was completely blocked. Interestingly, tumor growth was significantly delayed in the breast cancer model treated with a wild-type Ad control. In a similar manner, repeated injections into established MDA-MB-435 tumors resulted in complete regression with Ad–IFN and a block in further tumor growth with wild-type virus. The ability of wild-type virus to block tumor growth *in vivo* correlated with the ability of the virus to induce cell lysis, presumably by virus production, *in vitro*, suggesting that regression in the human breast cancer xenograft model might have been due to both IFN therapy and virus-induced lysis.

Other investigations on cancer immunotherapy have utilized replication-defective Ad recombinants. Haddada *et al.* (1993a) demonstrated that the use of an Ad vector encoding IL-2 could reduce the tumorigenicity of P815 mastocytoma cells following *in vitro* infection and subsequent injection into mice. In this case only 20% of the mice receiving IL-2-transduced cells developed tumors, and those that remained tumor free were protected from a subsequent challenge. Direct injection of AdIL-2 vectors into established P815 mastocytoma tumors induced tumor regression in 45–70% of animals. These mice remained tumor free for more than 8 months and were also protected from later challenge (Cordier *et al.*, 1995). Another murine cancer model has been developed (Guy *et al.*, 1992), employing primary mammary adenocarcinoma cells derived from transgenic animals carrying the polyoma middle T antigen (PyMidT) under control of the MMTV LTR. The tumor cells (designated PyMidT cells) are explanted from the transgenic animals, cultured briefly *ex vivo,* then injected s.c. into syngeneic mice, forming solid tumors in approximately 15–21 days. In contrast to results with the P815 tumor model, all mice injected with *ex vivo* AdIL-2-infected PyMidT cells developed tumors, although a significant delay in the onset of tumor formation and a prolongation of survival compared to control animals were observed (Addison *et al.*, 1995a). Intratumoral injection of established PyMidT tumors with the AdIL-2 vector, however, was far more effective, with complete tumor regression in half of the mice by 3 to 4 weeks postinjection. These animals were protected from subsequent challenge with tumor cells on the opposite flank and remained tumor free for at least 18 months (Addison *et al.*, 1995a; C. L. Addison and F. L. Graham, unpublished results). In further studies, up to 30% of animals bearing tumors on both the right and left flanks could be cured of both tumors by injection of AdIL-2 at only one of the sites. These results strongly suggest that intratumoral injection of AdIL-2 virus can mediate systemic immunity that is effective on distal tumors and perhaps metastases.

Injection of an Ad vector expressing IL-4 has also been shown to be effective in inducing antitumor responses in the PyMidT model. *Ex vivo* infection of PyMidT cells with an AdIL-4 vector prior to injection into animals induced a considerable delay (greater than 12 weeks) in the onset of tumorigenesis (Addison *et al.*, 1995b). More than half of the animals receiving AdIL-4-transduced cells remained tumor free for up to 18 months postinjection. Direct delivery of AdIL-4 to the established tumor was also effective, with 50% of treated animals undergoing complete tumor regression by 8 to 10 weeks postinfection, and resisting subsequent challenge. The differences between AdIL-2 treatment and AdIL-4 treatment, both in the kinetics of tumor regression (3 to 4 weeks vs 8 to 10 weeks, respectively, for complete regression) and in the type of immune cell infiltrate in the tumors (lymphocytes vs eosinophils, respectively), suggest that the two cytokines are likely to mediate tumor eradication by different mechanisms.

The efficacy of an Ad vector expressing IL-12 (Bramson *et al.*, 1996a) was also examined in the PyMidT model of mammary adenocarcinoma. Injection of AdIL-12 into preexisting tumors resulted in initial regressions in greater than 75% of animals, with 40% of these going on to demonstrate complete and permanent tumor regression and protection from challenge (Bramson *et al.*, 1996b).

Studies demonstrate that Ad vectors may be extremely useful as gene delivery systems in cancer therapies, particularly in immunotherapy where the vector itself may act as an adjuvant. Preliminary studies have shown that Ad vectors expressing cytokines can mediate regression of primary tumors and induce protective systemic immunity. These observations suggest that the use of Ad vectors in immunotherapies may lead to protection of patients from recurrent tumors and metastases, a problem that conventional cancer treatments have difficulty in overcoming. The use of Ad vectors as gene delivery systems readily permits combinatorial therapies, i.e., the use of suicide genes with tumor suppressor or cytokine therapy, as well as combinations of cytokine vectors, and these types of treatments may prove more efficacious than any one used alone. It is hoped that preclinical investigations completed in the near future will result in the development of strategies that prove therapeutic in a clinical setting.

V. Conclusions

The high efficiency of gene transfer mediated by recombinant Ad has prompted the construction of dozens of first-generation Ad vectors for high-level transgene expression in mammalian cells. These vectors have been designed and employed for uses as diverse as recombinant protein production and characterization, transcriptional analysis, establishment of stably transformed cell lines, and gene therapy. Some of the most encouraging applications of Ad vectors in gene therapy, such as the immunotherapy of cancer, require transient high-level expression of the transgene in target cells, which is easily achieved with first-generation Ad vectors. Gene replacement therapies, however, require prolonged expression and/or repeated administration of the vector, both of which have been problematic owing to activation of the host immune system against Ad and Ad-infected cells. These difficulties are being addressed by the development of second-generation vectors and of strategies to modulate the host immune response. These improvements, and no doubt others yet to come, hold promise for the generation of genuinely effective vectors for treatment of genetic disorders and other diseases.

Acknowledgments

We thank the many authors who have allowed us to discuss their results prior to publication, and those who generously sent us recent reprints on Ad vector construction or applications. We also thank Jonathan Bramson for critically reviewing this manuscript. Research in the authors' laboratory has been supported by grants from the Medical Research Council, the

National Cancer Institute of Canada (NCIC), the Natural Sciences and Engineering Research Council, and the National Institutes of Health. F.L.G. is a Terry Fox Research Scientist of the NCIC. C.L.A. is a research student of the NCIC supported with funds provided by the Canadian Cancer Society.

References

Abraham, N. G., da Silva, J. L., Lavrovsky, Y., Stoltz, R. A., Kappas, A., Dunn, M. W., and Schwartzman, M. L. (1995). Adenovirus-mediated heme oxygenase-1 gene transfer into rabbit ocular tissues. *Invest. Ophthalmol. Visual Sci.* **36,** 2202–2210.

Acsadi, G., Jani, A., Huard, J. Blaschuk, K., Massie, B., Holland, P., Lochmuller, H., and Karpati, G. (1994a). Cultured human myoblasts and myotubes show markedly different transducibility by replication-defective adenovirus recombinants. *Gene Ther.* **1,** 338–340.

Acsadi, G., Jani, A., Massie, B., Simoneau, M., Holland, P., Blaschuk, K., and Karpati, G. (1994b). A differential efficiency of adenovirus-mediated *in vivo* gene transfer into skeletal muscle cells of different maturity. *Hum. Mol. Genet.* **3,** 579–584.

Acsadi, G., Massie, B., and Jani, A. (1995). Adenovirus-mediated gene transfer into striated muscle. *J. Mol. Med.* **73,** 165–180.

Acsadi, G., Lochmuller, H., Jani, A., Huard, J., Massie, B., Prescott, S., Simoneau, M., Petrof, B. J., and Karpati, G. (1996). Dystrophin expression in muscles of mdx mice after adenovirus-mediated *in vivo* gene transfer. *Hum. Gene Ther.* **7,** 129–140.

Addison, C. L., Braciak, T., Ralston, R., Muller, W. J., Gauldie, J., and Graham, F. L. (1995a). Intra-tumoral injection of an adenovirus expressing interleukin-2 induces regression and immunity in a murine breast cancer model. *Proc. Natl. Acad. Sci. U.S.A.* **92,** 8522–8526.

Addison, C. L., Gauldie, J., Muller, W. J., and Graham, F. L. (1995b). An adenoviral vector expressing interleukin-4 modulates tumorigenicity and induces regression in a murine breast cancer model. *Int. J. Oncol.* **7,** 1253–1260.

Addison, C. L., Hitt, M. M., Kunsken, D., and Graham, F. L. (1997). Comparison of the human versus murine cytomegalovirus immediate early gene promoters for transgene expression by adenoviral vectors. *J. Gen. Virol.* (in press).

Akli, S., Caillaud, C., Vigne, E., Stratford-Perricaudet, L. D., Poenaru, L., Perricaudet, M., Kahn, A., and Peschanski, M. R. (1993). Transfer of a foreign gene into the brain using adenovirus vectors. *Nat. Genet.* **3,** 224–228.

Alcorn, J. L., Gao, E., Chen, Q., Smith, M. E., Gerard, R. D., and Mendelson, C. R. (1993). Genomic elements involved in transcriptional regulation of the rabbit surfactant protein-A gene. *Mol. Endocrinol.* **7,** 1072–1085.

Amalfitano, A., Begy, C. R., and Chamberlain, J. S. (1996). Improved adenovirus packaging cell lines to support the growth of replication-defective gene-delivery vectors. *Proc. Natl. Acad. Sci. U.S.A.* **93,** 3352–3356.

Amin, R., Wilmott, R., Schwarz, Y., Trapnell, B., and Stark, J. (1995). Replication-deficient adenovirus induces expression of interleukin-8 by airway epithelial cells *in vitro.* *Hum. Gene Ther.* **6,** 145–153.

Anton, M., and Graham, F. L. (1995). Site-specific recombination mediated by an adenovirus vector expressing the Cre recombinase protein: A molecular switch for control of gene expression. *J. Virol.* **69,** 4600–4606.

Arbuthnot, P. B., Bralet, M.-P., LeJossic, C., Dedieu, J.-F., Perricaudet, M., Brechot, C., and Ferry, N. (1996). *In vitro* and *in vivo* hepatoma cell-specific expression of a gene transferred with an adenoviral vector. *Hum. Gene Ther.* **7,** 1503–1514.

Armentano, D., Sookdeo, C. C., Hehir, K. M., Gregory, R. J., St. George, J. A., Prince, G. A., Wadsworth, S. C., and Smith, A. E. (1995). Characterization of an adenovirus gene transfer vector containing an E4 deletion. *Hum. Gene Ther.* **6,** 1343–1353.

Babiss, L. E., Young, C. S. H., Fisher, P. B., and Ginsberg, H. S. (1983). Expression of adenovirus E1A and E1B gene products and the *Escherichia coli* XGPRT gene in KB cells. *J. Virol.* **46**, 454–465.

Babiss, L. E., Ginsberg, H. S., and Darnell, J. E., Jr. (1985). Adenovirus E1B proteins are required for accumulation of late viral mRNA and for effects on cellular mRNA translation and transport. *Mol. Cell. Biol.* **5**, 2552–2558.

Babiss, L. E., Friedman, J. M., and Darnell, J. E., Jr. (1986). Cellular promoters incorporated into the adenovirus genome: Effects of viral regulatory elements on transcription rates and cell specificity of albumin and β-globin promoters. *Mol. Cell. Biol.* **6**, 3798–3806.

Bacchetti, S., and Graham, F. L. (1993). Inhibition of cell proliferation by an adenovirus vector expressing the human wild type p53 protein. *Int. J. Oncol.* **3**, 781–788.

Bajocchi, G., Feldman, S. H., Crystal, R. G., and Mastrangeli, A. (1993). Direct *in vivo* gene transfer to ependymal cells in the central nervous system using recombinant adenovirus vectors. *Nat. Genet.* **3**, 229–234.

Ballard, P. L., Zepeda, M. L., Schwartz, M., Lopez, N., and Wilson, J. M. (1995). Adenovirus-mediated gene transfer to human fetal lung *ex vivo*. *Am. J. Physiol.* **268**, L839–L845.

Barr, D., Tubb, J., Ferguson, D., Scaria, A., Lieber, A., Wilson, C., Perkins, J., and Kay, M. A. (1995). Strain related variations in adenovirally mediated transgene expression from mouse hepatocytes *in vivo*: Comparisons between immunocompetent and immunodeficient inbred strains. *Gene Ther.* **2**, 151–155.

Barr, E., Carroll, J., Kalynych, A. M., Tripathy, S. K., Kozarsky, K., Wilson, J. M., and Leiden, J. M. (1994). Efficient catheter-mediated gene transfer into the heart using replication-defective adenovirus. *Gene Ther.* **1**, 51–58.

Bass, C., Cabrera, G., Elgavish, A., Robert, B., Siegal, G. P., Anderson, S. C., Maneval, D. C., and Curiel, D. T. (1995). Recombinant adenovirus-mediated gene transfer to genitourinary epithelium *in vitro* and *in vivo*. *Cancer Gene Ther.* **2**, 97–104.

Becker, T. C., Noel, R. J., Coats, W. S., Gomez-Foix, A. M., Alam, T., Gerard, R. D., and Newgard, C. B. (1994). Use of recombinant adenovirus for metabolic engineering of mammalian cells. *Methods Cell Biol.* **43**, 161–189.

Bennett, J., Wilson, J., Sun, D., Forbes, B., and Maguire, A. (1994). Adenovirus vector-mediated *in vivo* gene transfer into adult murine retina. *Invest. Ophthalmol. Visual Sci.* **35**, 2535–2542.

Berkner, K. L. (1992). Expression of heterologous sequences in adenoviral vectors. *Curr. Top. Microbiol. Immunol.* **158**, 39–66.

Berkner, K. L., and Sharp, P. A. (1983). Generation of adenovirus by transfection of plasmids. *Nucleic Acids Res.* **11**, 6003–6020.

Bessereau, J. L., Stratford-Perricaudet, L. D., Piette, J., Le Poupon, C., and Changeux, J. P. (1994). *In vivo* and *in vitro* analysis of electrical activity dependent expression of muscle acetylcholine receptor genes using adenovirus. *Proc. Natl. Acad. Sci. U.S.A.* **91**, 1304–1308.

Bett, A. J., Prevec, L., and Graham, F. L. (1993). Packaging capacity and stability of human adenovirus type 5 vectors. *J. Virol.* **67**, 5911–5921.

Bett, A. J., Haddara, W., Prevec, L., and Graham, F. L. (1994). An efficient and flexible system for construction of adenovirus vectors with inserts or deletion in early regions 1 and 3. *Proc. Natl. Acad. Sci. U.S.A.* **91**, 8802–8806.

Boucher, R. C., Knowles, M. R., Johnson, L. G., Olsen, J. C., Pickles, R., Wilson, J. M., Engelhardt, J., Yang, Y., and Grossman, M. (1994). Gene therapy for cystic fibrosis using E1-deleted adenovirus: A phase I trial in the nasal cavity. *Hum. Gene Ther.* **5**, 615–639.

Bout, A., Imler, J.-L., Schultz, H., Perricaudet, M., Zurcher, C., Herbrink, P., Valerio, D., and Pavirani, A. (1994a). *In vivo* adenovirus-mediated transfer of human CFTR cDNA to rhesus monkey airway epithelium: Efficacy, toxicity, and safety. *Gene Ther.* **1**, 385–394.

Bout, A., Perricaudet, M., Baskin, G., Imler, J. L., Scholte, B. J., Pavirani, A., and Valerio, D. (1994b). Lung gene therapy: *In vivo* adenovirus-mediated gene transfer to rhesus monkey airway epithelium. *Hum. Gene Ther.* **5**, 3–10.

Boviatsis, E. J., Chase, M., Wei, M. X., Tamiya, T., Hurford, R. K., Jr., Kowall, N. W., Tepper, R. I., Breakefield, X. O., and Chiocca, E. A. (1994). Gene transfer into experimental brain tumors mediated by adenovirus, herpes simplex virus, and retrovirus vectors. *Hum. Gene Ther.* **5**, 183–191.

Braciak, T. A., Mittal, S. K., Graham, F. L., Richards, C. D., and Gauldie, J. (1993). Construction of recombinant human type 5 adenoviruses expressing rodent IL-6 genes. An approach to investigate *in vivo* cytokine function. *J. Immunol.* **151**, 5145–5153.

Braciak, T. A., Bacon, K., Xing, Z., Torry, D. J., Graham, F. L., Schall, T. J., Richards, C. D., Croitoru, K., and Gauldie, J. (1996). Overexpression of RANTES using a recombinant adenovirus vector induces the tissue directed recruitment of monocytes to the lung. *J. Immunol.* **157**, 5076–5084.

Bramson, J. L., Graham, F. L., and Gauldie, J. (1995). The use of adenoviral vectors for gene therapy and gene transfer *in vivo*. *Curr. Opin. Biotechnol.* **6**, 590–595.

Bramson, J., Hitt, M., Gallichan, W. S., Rosenthal, K. L., Gauldie, J., and Graham, F. L. (1996a). Construction of a double recombinant adenovirus vector expressing a heterodimeric cytokine: *In vitro* and *in vivo* production of biologically active interleukin-12. *Hum. Gene Ther.* **7**, 333–342.

Bramson, J., Hitt, M., Addison, C. L., Muller, W. J., Gauldie, J., and Graham, F. L. (1996b). Direct intratumoral injection of an adenovirus expressing interleukin-12 induces regression and long-lasting immunity that is associated with highly localized expression of interleukin-12. *Hum. Gene Ther.* **7**, 1995–2002.

Branton, P. E., Bayley, S. T., and Graham, F. L. (1985). Transformation by human adenoviruses. *Biochim. Biophys. Acta* **780**, 67–94.

Bridge, E., and Ketner, G. (1989). Redundant control of adenovirus late gene expression by early region 4. *J. Virol.* **63**, 631–638.

Bridge, E., and Ketner, G. (1990). Interaction of adenoviral E4 and E1b products in late gene expression. *Virology* **174**, 345–353.

Brody, S. L., Jaffe, H. A., Han, S. K., Wersto, R. P., and Crystal, R. G. (1994a). Direct *in vivo* gene transfer and expression in malignant cells using adenovirus vectors. *Hum. Gene Ther.* **5**, 437–447.

Brody, S. L., Metzger, M., Danel, C., Rosenfeld, M. A., and Crystal, R. G. (1994b). Acute responses of non-human primates to airway delivery of an adenovirus vector containing the human cystic fibrosis transmembrane conductance regulator cDNA. *Hum. Gene Ther.* **5**, 821–836.

Budenz, D. L., Bennett, J., Alonso, L., and Maguire, A. (1995). *In vivo* gene transfer into murine corneal endothelial and trabecular meshwork cells. *Invest. Ophthalmol. Visual Sci.* **36**, 2211–2215.

Byrnes, A. P., Rusby, J. E., Wood, M. J. A., and Charlton, H. M. (1995). Adenovrus gene transfer causes inflammation in the brain. *Neuroscience* **66**, 1015–1024.

Caillaud, C., Akli, S., Vigne, E., Koulakoff, A., Perricaudet, M., Poenaru, L., Kahn, A., and Berwald-Netter, Y. (1993). Adenoviral vector as a gene delivery system into cultured rat neuronal and glial cells. *Eur. J. Neurosci.* **5**, 1287–1291.

Caldas, C., Hahn, S. A., Dacosta, L. T., Redston, M. S., Schutte, M., Seymour, A. B., Weinstein, C. L., Hruban, R. H., Yeo, C. J., and Kern, S. E. (1994). Frequent somatic mutations and homozygous deletions of the p16 (MTS1) gene in pancreatic adenocarcinoma. *Nat. Genet.* **8**, 27–32.

Cameron, R. B., Spiess, P. J., and Rosenberg, S. A. (1990). Synergistic antitumor activity of tumor-infiltrating lymphocytes, interleukin-2, and local tumor irradiation. *J. Exp. Med.* **171**, 249–263.

Chang, M. W., Ohno, T., Gordon, D., Lu, M. M., Nabel, G. J., Nabel, E. G., and Leiden, J. M. (1995a). Adenovirus-mediated transfer of the herpes simplex virus thymidine kinase gene inhibits vascular smooth muscle cell proliferation and neointima formation following balloon angioplasty of the rat carotid artery. *Mol. Med.* **1**, 172–181.

Chang, M. W., Barr, E., Lu, M. M., Barton, K., and Leiden, J. M. (1995b). Adenovirus-mediated over-expression of the cyclin/cyclin-dependent kinase inhibitor, p21, inhibits vascular smooth muscle cell proliferation and neointima formation in the rat carotid artery model of balloon angioplasty. *J. Clin. Invest.* **96**, 2260–2268.

Chen, S.-H., Shine, H. D., Goodman, J. C., Grossman, R. G., and Woo, S. L. C. (1994). Gene therapy for brain tumors: Regression of experimental gliomas by adenovirus-mediated gene transfer *in vivo*. *Proc. Natl. Acad. Sci. U.S.A.* **91**, 3054–3057.

Chen, S. J., Wilson, J. M., and Muller, D. W. (1994). Adenovirus-mediated gene transfer of soluble vascular cell adhesion molecule to porcine interposition vein grafts. *Circulation* **89**, 1922–1928.

Cheng, J. Q., Jhanwar, S. C., Klein, W. M., Bell, D. W., Lee, W., Altomare, D. A., Nobori, T., Olopade, O. I., Buckler, A. J., and Testa, J. R. (1994). p16 alterations and deletion mapping of 9p21-p22 in malignant mesothelioma. *Cancer Res.* **54**, 5547–5551.

Chroboczek, J., Bieber, F., and Jacrot, B. (1992). The sequence of the genome of adenovirus type 5 and its comparison with the genome of adenovirus type 2. *Virology* **186**, 280–285.

Clayman, G. L., Trapnell, B. C., Mittereder, N., Liu, T. J., Eicher, S., Zhang, S., and Shillitoe, E. J. (1995). Transduction of normal and malignant oral epithelium by an adenovirus vector: The effect of dose and treatment time on transduction efficiency and tissue penetration. *Cancer Gene Ther.* **2**, 105–111.

Clemens, P. R., Krause, T. L., Chan, S., Korb, K. E., Graham, F. L., and Caskey, C. T. (1995). Recombinant truncated dystrophin minigenes: Construction, expression, and adenoviral delivery. *Hum. Gene Ther.* **6**, 1477–1485.

Colak, A., Goodman, J. C., Chen, S.-H., Woo, S. L. C., Grossman, R. G., and Shine, H. D. (1995). Adenovirus-mediated gene therapy in an experimental model of breast cancer metastatic to the brain. *Hum. Gene Ther.* **6**, 1317–1322.

Colby, W. W., and Shenk, T. (1981). Adenovirus type 5 virions can be assembled *in vivo* in the absence of detectable polypeptide IX. *J. Virol.* **39**, 977–980.

Collins, F. S. (1992). Cystic fibrosis: Molecular biology and therapeutic implications. *Science* **256**, 774–779.

Connelly, S., Smith, T. A. G., Dhir, G., Gardner, J. M., Mehaffey, M. G., Zaret, K. S., McClelland, A., and Kaleko, M. (1995). *In vivo* gene delivery and expression of physiological levels of functional human factor VIII in mice. *Hum. Gene Ther.* **6**, 185–193.

Cordier, L., Duffour, M.-T., Sabourin, J.-C., Lee, M. G., Cabannes, J., Ragot, T., Perricaudet, M., and Haddada, H. (1995). Complete recovery of mice from a pre-established tumor by direct intratumoral delivery of an adenovirus vector harboring the murine IL-2 gene. *Gene Ther.* **2**, 16–21.

Cotten, M. (1995). Adenovirus-augmented, receptor-mediated gene delivery and some solutions to the common toxicity problems. *In* "The Molecular Repertoire of Adenoviruses III: Biology and Pathogenesis" (W. Doerfler and P. Boehm, eds.), pp. 283–295. Springer, Berlin.

Crystal, R. G., principal investigator (1995a). Clinical protocol: A phase I study, in cystic fibrosis patients, of the safety, toxicity, and biological efficacy of a single administration of a replication deficient, recombinant adenovirus carrying the cDNA of the normal cystic fibrosis transmembrane conductance regulator gene in the lung. *Hum. Gene Ther.* **6**, 643–666.

Crystal, R. G., principal investigator (1995b). Clinical protocol: Evaluation of repeat administration of a replication deficient, recombinant adenovirus containing the normal cystic fibrosis transmembrane conductance regulator cDNA to the airways of individuals with cystic fibrosis. *Hum. Gene Ther.* **6**, 667–703.

Crystal, R. G., McElvaney, N. G., Rosenfeld, M. A., Chu, C. S., Mastrangeli, A., Hay, J. G., Brody, S. L., Jaffe, H. A., Eissa, N. T., and Danel, C. (1994). Administration of an adenovirus containing the human CFTR cDNA to the respiratory tract of individuals with cystic fibrosis. *Nat. Genet.* **8**, 42–51.

Csete, M. E., Afra, R., Mullen, Y., Drazan, K. E., Benhamou, P. Y., and Shaked, A. (1994a). Adenoviral-mediated gene transfer to pancreatic islets does not alter islet function. *Transplant. Proc.* **26,** 756–757.

Csete, M. E., Drazan, K. E., Van Bree, M., McIntee, D. F., McBride, W. H., Bett, A., Graham, F. L., Busuttil, R. W., Berk, A. J., and Shaked, A. (1994b). Adenovirus-mediated gene transfer in the transplant setting. I. Conditions for expression of transferred genes in cold-preserved hepatocytes. *Transplantation* **57,** 1502–1507.

Csete, M. E., Benhamou, P. Y., Drazan, K. E., Wu, L., McIntee, D. F., Afra, R., Mullen, Y., Busuttil, R. W., and Shaked, A. (1995). Efficient gene transfer to pancreatic islets mediated by adenoviral vectors. *Transplantation* **59,** 263–268.

Curiel, D. T., Agarwal, S., Wagner, E., and Cotten, M. (1991). Adenovirus enhancement of transferrin-polylysine-mediated gene delivery. *Proc. Natl. Acad. Sci. U.S.A.* **88,** 8850–8854.

Cutt, J. R., Shenk, T., and Hearing, P. (1987). Analysis of adenovirus early region 4-encoded polypeptides synthesized in productively infected cells. *J. Virol.* **61,** 543–552.

Dai, Y., Schwarz, E. M., Gu, D., Zhang, W. W., Sarvetnick, N., and Verma, I. M. (1995). Cellular and humoral immune responses to adenoviral vectors containing factor IX gene: Tolerization of factor IX and vector antigens allows for long-term expression. *Proc. Natl. Acad. Sci. U.S.A.* **92,** 1401–1405.

Daniell, E. (1976). Genome structure of incomplete particles of adenovirus. *J. Virol.* **19,** 685–708.

Davidson, B. L., Allen, E. D., Kozarsky, K. F., Wilson, J. M., and Roessler, B. J. (1993). A model system for *in vivo* gene transfer into the central nervous system using an adenoviral vector. *Nat. Genet.* **3,** 219–223.

Davidson, D., and Hassell, J. A. (1987). Overproduction of polyoma virus middle T antigen in mammalian cells through the use of an adenovirus vector. *J. Virol.* **61,** 1226–1239.

Defer, C., Belin, M.-T., Caillet-Boudin, M.-L., and Boulanger, P. (1990). Human adenovirus–host cell interactions: Comparative study with members of subgroups B and C. *J. Virol.* **64,** 3661–3673.

Descamps, V., Blumenfeld, N., Villeval, J. L., Vainchenker, W., Perricaudet, M., and Beuzard, Y. (1994). Erythropoietin gene transfer and expression in adult normal mice: Use of an adenovirus vector. *Hum. Gene Ther.* **5,** 979–985.

Dietschy, J. M., Turley, S. D., and Spady, D. K. (1993). Role of liver in the maintenance of cholesterol and low density lipoprotein homeostasis in different animal species, including man. *J. Lipid Res.* **34,** 1637–1659.

Doronin, K. K., Zakharchuk, A. N., Grinenko, N. F., Yurov, G. K., Krougliak, V. A., and Naroditsky, B. S. (1993). Expression of the gene encoding secreted placental alkaline phosphatase (SEAP) by a nondefective adenovirus vector. *Gene* **126,** 247–250.

Dorsch-Hasler, K., Keil, G. M., Weber, F., Jasin, M., Schaffner, W., and Koszinowski, U. H. (1985). A long and complex enhancer activates transcription of the gene coding for the highly abundant immediate early mRNA in murine cytomegalovirus. *Proc. Natl. Acad. Sci. U.S.A.* **82,** 8325–8329.

Draghia, R., Caillaud, C., Manicom, R., Pavirani, A., Kahn, A., and Poenaru, L. (1995). Gene delivery into the central nervous system by nasal instillation in rats. *Gene Ther.* **2,** 418–423.

Drazan, K. E., Shen, X. D., Csete, M. E., Zhang, W. W., Roth, J. A., Busuttil, R. W., and Shaked, A. (1994). *In vivo* adenoviral-mediated human p53 tumor suppressor gene transfer and expression in rat liver after resection. *Surgery (St. Louis)* **116,** 197–204.

Drazan, K. E., Wu, L., Shen, X.-D., Bullington, D., Jurim, O., Busuttil, R. W., and Shaked, A. (1995a). Adenovirus-mediated gene transfer in the transplant setting. III. Variables affecting gene transfer in liver grafts. *Transplantation* **59,** 670–673.

Drazan, K. E., Wu, L., Olthoff, K. M., Jurim, O., Busuttil, R. W., and Shaked, A. (1995b). Transduction of hepatic allografts achieves local levels of viral IL-10 which suppress alloreactivity *in vitro. J. Surg. Res.* **59,** 219–223.

Drazan, K. E., Csete, M. E., Shen, X.-D., Bullington, D., Cottle, G., Busuttil, R. W., and Shaked, A. (1995c). Hepatic function is preserved following liver-directed, adenovirus-mediated gene transfer. *J. Surg. Res.* **59**, 299–304.

Drazan, K. E., Shen, X.-D., Wu, L., Jurim, O., Busuttil, R. W., and Shaked, A. (1995d). Adenoviral-mediated gene transfer to liver xenografts results in successful expression and protein production. *Transplant. Proc.* **27**, 331–332.

Dupuit, F., Zahm, J.-M., Pierrot, D., Brezillon, S., Bonnet, N., Imler, J.-L., Pavirani, A., and Puchelle, E. (1995). Regenerating cells in human airway surface epithelium represent preferential targets for recombinant adenovirus. *Hum. Gene Ther.* **6**, 1185–1193.

Eastham, J. A., Hall, S. J., Sehgal, I., Wang, J., Timme, T. L., Yang, G., Connell-Crowley, L., Elledge, S. J., Zhang, W.-W., Harper, J. W., and Thompson, T. C. (1995). *In vivo* gene therapy with p53 or p21 adenovirus for prostate cancer. *Cancer Res.* **55**, 5151–5155.

Eastham, J. A., Chen, S.-H., Sehgal, I., Yang, G., Timme, T. L., Hall, S. J., Woo, S. L. C., and Thompson, T. C. (1996). Prostate cancer gene therapy: Herpes simplex virus thymidine kinase gene transduction followed by ganciclovir in mouse and human prostate cancer models. *Hum. Gene Ther.* **7**, 515–523.

Eissa, N. T., Chu, C. S., Danel, C., and Crystal, R. G. (1994). Evaluation of the respiratory epithelium of normals and individuals with cystic fibrosis for the presence of adenovirus E1a sequences relevant to the use of E1a-adenovirus vectors for gene therapy for the respiratory manifestations of cystic fibrosis. *Hum. Gene Ther.* **5**, 1105–1114.

Elliott, W. M., Hayashi, S., and Hogg, J. C. (1995). Immunodetection of adenoviral E1A proteins in human lung tissue. *Am. J. Respir. Cell Mol. Biol.* **12**, 642–648.

Engelhardt, J. F., Yang, Y., Stratford-Perricaudet, L. D., Allen, E. D., Kozarsky, K., Perricaudet, M., Yankaskas, J. R., and Wilson, J. M. (1993a). Direct gene transfer of human CFTR into human bronchial epithelia of xenografts with E1-deleted adenoviruses. *Nat. Genet.* **4**, 27–34.

Engelhardt, J. F., Simon, R. H., Yang, Y., Zepeda, M., Weber-Pendleton, S., Doranz, B., Grossman, M., and Wilson, J. M. (1993b). Adenovirus-mediated transfer of the CFTR gene to lung of nonhuman primates: Biological efficacy study. *Hum. Gene Ther.* **4**, 759–769.

Engelhardt, J. F., Ye, X., Doranz, B., and Wilson, J. M. (1994a). Ablation of E2A in recombinant adenoviruses improves transgene persistence and decreases inflammatory response in mouse liver. *Proc. Natl. Acad. Sci. U.S.A.* **91**, 6196–6200.

Engelhardt, J. F., Litzky, L., and Wilson, J. M. (1994b). Prolonged transgene expression in cotton rat lung with recombinant adenoviruses defective in E2a. *Hum. Gene Ther.* **5**, 1217–1229.

England, S. B., Nicholson, L. V. B., Johnson, M. A., Forrest, S. M., Love, D. R., Zubrzyczka-Gaarn, E. E., Bulman, D. E., Harris, J. B., and Davies, K. (1990). Very mild muscular dystrophy associated with the deletion of 46% of dystrophin. *Nature (London)* **343**, 180–182.

Erzurum, S. C., Lemarchand, P., Rosenfeld, M. A., Yoo, J. H., and Crystal, R. G. (1993). Protection of human endothelial cells from oxidant injury by adenovirus-mediated transfer of the human catalase cDNA. *Nucleic Acids Res.* **21**, 1607–1612.

Fallaux, F. J., Kranenburg, O., Cramer, S. J., Houweling, A., van Ormondt, H., Hoeben, R. C., and van der Eb, A. J. (1996). Characterization of 911: A new helper cell line for the tritration and propagation of early-region-1-deleted adenovrial vectors. *Hum. Gene Ther.* **7**, 215–222.

Fang, B., Eisensmith, R. C., Wang, H., Kay, M. A., Cross, R. E., Landen, C. N., Gordon, G., Bellinger, D. A., Read, M. S., Hu, P. C., Brinkhous, K. M., and Woo, S. L. C. (1995). Gene therapy for hemophilia B: Host immunosuppression prolongs the therapeutic effect of adenovirus-mediated factor IX expression. *Hum. Gene Ther.* **6**, 1039–1044.

Fearon, E. R., Pardoll, D., Itaya, T., Golumbek, P., Levitsky, H. I., Simons, J. W., Karasuyama, H., Vogelstein, B., and Frost, P. (1990). Interleukin-2 production by tumor cells bypasses

T helper function in the generation of an antitumor response. *Cell* (*Cambridge, Mass.*) **60**, 397–403.

Feldman, L. J., Steg, P. G., Zheng, L. P., Chen, D., Kearney, M., McGarr, S. E., Barry, J. J., Dedieu, J. F., Perricaudet, M., and Isner, J. M. (1995). Low-efficiency of percutaneous adenovirus-mediated arterial gene transfer in the atherosclerotic rabbit. *J. Clin. Invest.* **95**, 2662–2671.

Feng, M., Cabrera, G., Deshane, J., Scanlon, K. J., and Curiel, D. T. (1995). Neoplastic reversion accomplished by high efficiency adenoviral-mediated delivery of an anti-*ras* ribozyme. *Cancer Res.* **55**, 2024–2028.

Fisher, K. J., Choi, H., Burda, J., Chen, S.-J., and Wilson, J. M. (1996). Recombinant adenovirus deleted of all viral genes for gene therapy of cystic fibrosis. *Virology* **217**, 11–22.

French, B. A., Mazur, W., Ali, N. M., Geske, R. S., Finnigan, J. P., Rogers, G. P., Roberts, R., and Raizner, A. E. (1994a). Percutaneous transluminal *in vivo* gene transfer by recombinant adenovirus in normal porcine coronary arteries, atherosclerotic arteries, and two models of coronary restenosis. *Circulation* **90**, 2402–2413.

French, B. A., Mazur, W., Geske, R. S., and Bolli, R. (1994b). Direct *in vivo* gene transfer into porcine myocardium using replication-deficient adenoviral vectors. *Circulation* **90**, 2414–2424.

Friedman, J. M., Babiss, L. E., Clayton, D. F., and Darnell, J. E., Jr. (1986). Cellular promoters incorporated into the adenovirus genome: Cell specificity of albumin and immunoglobulin expression. *Mol. Cell. Biol.* **6**, 3791–3797.

Friedrich, G., and Soriano, P. (1991). Promoter traps in embryonic stem cells: A genetic screen to identify and mutate developmental genes in mice. *Genes Dev.* **5**, 1513–1523.

Fujita, A., Sakagami, D., Kanegae, Y., Saito, I., and Kobayashi, I. (1995). Gene targeting with a replication-defective adenovirus vector. *J. Virol.* **69**, 6180–6190.

Fujiwara, T., Grimm, E. A., Mukhopadhyay, T., Zhang, W.-W., Owen-Schaub, L. B., and Roth, J. A. (1994). Induction of chemosensitivity in human lung cancer cells *in vivo* by adenovirus-mediated transfer of the wild-type p53 gene. *Cancer Res.* **54**, 2287–2291.

Gage, F. H., Coates, P. W., Palmer, T. D., Kuhn, H. G., Fisher, L. J., Suhonen, J. O., Peterson, D. A., Suhr, S. T., and Ray, J. (1995). Survival and differentiation of adult neuronal progenitor cells transplanted to the adult brain. *Proc. Natl. Acad. Sci. U.S.A.* **92**, 11879–11883.

Garnier, A., Cote, J., Nadeau, I., Kamen, A., and Massie, B. (1994). Scale-up of the adenovirus expression system for the production of recombinant protein in human 293S cells. *Cytotechnology* **15**, 145–155.

Gauldie, J., Graham, F. L., Xing, Z., Braciak, T., Foley, R., and Sime, P. J. (1996). Adenovirus vector mediated cytokines gene transfer to lung tissue. *Ann. N. Y. Acad. Sci.* **796**, 235–244.

Gerard, R. D., and Meidell, R. S. (1995). Adenovirus vectors. *In* "DNA Cloning: A Practical Approach" (B. D. Hames and D. Glover, eds.), pp. 285–306. Oxford University Press, Oxford, UK.

Ghosh-Choudhury, G., and Graham, F. L. (1987). Stable transfer of a mouse dihydrofolate reductase gene into a deficient cell line using a human adenovirus vector. *Biochem. Biophys. Res. Commun.* **147**, 964–973.

Ghosh-Choudhury, G., Haj-Ahmad, Y., Brinkley, P., Rudy, J., and Graham, F. L. (1986). Human adenovirus cloning vectors based on infectious bacterial plasmids. *Gene* **50**, 161–171.

Ghosh-Choudhury, G., Haj-Ahmad, Y., and Graham, F. L. (1987). Protein IX, a minor component of the human adenovirus capsid, is essential for the packaging of full length genomes. *EMBO J.* **6**, 1733–1739.

Gilgenkrantz, H., Duboc, D., Juillard, V., Couton, D., Pavirani, A., Guillet, J. G., Briand, P., and Kahn, A. (1995). Transient expression of genes transferred *in vivo* into heart using first-generation adenoviral vectors: Role of the immune response. *Hum. Gene Ther.* **6**, 1265–1274.

Ginsberg, H. S., ed. (1984). "The Adenoviruses." Plenum, New York.

Goldman, M. J., and Wilson, J. M. (1995). Expression of $\alpha_v\beta_5$ integrin is necessary for efficient adenovirus-mediated gene transfer in the human airway. *J. Virol.* **69**, 5951–5958.

Goldman, M. J., Litzky, L. A., Engelhardt, J. F., and Wilson, J. M. (1995a). Transfer of the CFTR gene to the lung of nonhuman primates with E1-deleted, E2a-defective recombinant adenoviruses: A preclinical toxicology study. *Hum. Gene Ther.* **6**, 839–851.

Goldman, M. J., Yang, Y., and Wilson, J. M. (1995b). Gene therapy in a xenograft model of cystic fibrosis lung corrects chloride transport more effectively than the sodium defect. *Nat. Genet.* **9**, 126–131.

Gorziglia, M. I., Kadan, M. J., Yei, S., Lim, J., Lee, G. M., Luthra, R., and Trapnell, B. C. (1996). Elimination of both E1 and E2a from adenovirus vectors further improves prospects for *in vivo* human gene therapy. *J. Viol.* **70**, 4173–4178.

Graham, F. L. (1984). Covalently closed circles of human adenovirus DNA are infectious. *EMBO J.* **3**, 2917–2922.

Graham, F. L., and Prevec, L. (1992). Adenovirus-based expression vectors and recombinant vaccines. *In* "Vaccines: New Approaches to Immunological Problems" (R. W. Ellis, ed.), pp. 363–389. Butterworth-Heinemann, Boston.

Graham, F. L., Smiley, J., Russell, W. C., and Nairn, R. (1977). Characteristics of a human cell line transformed by DNA from human adenovirus 5. *J. Gen. Virol.* **36**, 59–72.

Grimm, E. A., Mazumder, A., Zhang, H. Z., and Rosenberg, S. A. (1982). Lymphokine-activated killer cell phenomenon: Lysis of natural killer-resistant fresh solid tumor cells by interleukin-2-activated autologous human peripheral blood lymphocytes. *J. Exp. Med.* **155**, 1823–1841.

Grodzicker, T., and Klessig, D. F. (1980). Expression of unselected adenovirus genes in human cells cotransformed with the HSV-1 TK gene and adenovirus 2 DNA. *Cell (Cambridge, Mass.)* **21**, 453–463.

Grubb, B. R., Pickles, R. J., Ye, H., Yankaskas, J. R., Vick, R. N., Engelhardt, J. F., Wilson, J. M., Johnson, L. G., and Boucher, R. C. (1994). Inefficient gene transfer by adenovirus vector to cystic fibrosis airway epithelia of mice and humans. *Nature (London)* **371**, 802–806.

Guy, C. T., Cardiff, R. D., and Muller, W. J. (1992). Induction of mammary tumors by expression of polyoma middle T oncogene: A transgenic mouse model for metastatic disease. *Mol. Cell. Biol.* **12**, 954–961.

Guzman, R. J., Lemarchand, P., Crystal, R. G., Epstein, S. E., and Finkel, T. (1993a). Efficient gene transfer into myocardium by direct injection of adenovirus vectors. *Circ. Res.* **73**, 1202–1207.

Guzman, R. J., Lemarchand, P., Crystal, R. G., Epstein, S. E., and Finkel, T. (1993b). Efficient and selective adenovirus-mediated gene transfer into the vascular neointima. *Circulation* **88**, 2838–2848.

Guzman, R. J., Hirschowitz, E. A., Brody, S. L., Crystal, R. G., Epstein, S. E., and Finkel, T. (1994). *In vivo* suppression of injury-induced vascular smooth muscle cell accumulation using adenovirus-mediated transfer of the herpes simplex virus thymidine kinase gene. *Proc. Natl. Acad. Sci. U.S.A.* **91**, 10732–10736.

Hackett, J., Jr., Tutt, M., Lipscomb, M., Bennett, M., Koo, G., and Kumar, V. (1986). Origin and differentiation of natural killer cells. II. Functional and morphological studies of purified NK1.1+ cells. *J. Immunol.* **136**, 3124–3131.

Haddada, H., Ragot, T., Cordier, L., Duffour, M. T., and Perricaudet, M. (1993a). Adenoviral interleukin-2 gene transfer into P815 tumor cells abrogates tumorigenicity and induces antitumoral immunity in mice. *Hum. Gene Ther.* **4**, 703–711.

Haddada, H., Lopez, M., Martinache, C., Ragot, T., Abina, M. A., and Perricaudet, M. (1993b). Efficient adenovirus-mediated gene transfer into human blood monocyte-derived macrophages. *Biochem. Biophys. Res. Commun.* **195**, 1174–1183.

Haj-Ahmad, Y., and Graham, F. L. (1986). Development of a helper-independent human adenovirus vector and its use in the transfer of the herpes simplex virus thymidine kinase gene. *J. Virol.* **57**, 267–274.

Hall, J., Mutchnik, S. E., Yang, G., Chen, S.-H., Woo, S. L. C., and Thompson, T. (1996). "An Orthotopic Model for Metastatic Mouse Prostate Cancer: Characterization and Use for Gene Therapy," Annu. Meet. Abstra. American Urological Assoc., Orlando, Florida.

Hanke, T., Graham, F. L., Lulitanond, V., and Johnson, D. C. (1990). Herpes simplex virus IgG Fc receptors induced using recombinant adenovirus vectors expressing glycoproteins E and I. *Virology* **177**, 437–444.

Hanke, T., Graham, F. L., Rosenthal, K. L., and Johnson, D. C. (1991). Identification of an immunodominant cytotoxic T-lymphocyte recognition site in glycoprotein B of herpes simplex virus by using recombinant adenovirus vectors and synthetic peptides. *J. Virol.* **65**, 1177–1186.

Harper, J. W., Adami, G. R., Wei, N., Keyomarsi, K., and Elledge, S. J. (1993). The p21 Cdk-interacting protein Cip1 is a potent inhibitor of G1 cyclin-dependent kinases. *Cell (Cambridge, Mass.)* **75**, 805–816.

Hartwell, L. H., and Kastan, M. B. (1994). Cell cycle control and cancer. *Science* **266**, 1821–1828.

Hay, J. G., McElvaney, N. G., Herena, J., and Crystal, R. G. (1995). Modfication of nasal epithelial potential differences of individuals with cystic fibrosis consequent to local administration of a normal CFTR cDNA adenovirus gene transfer vector. *Hum. Gene Ther.* **6**, 1487–1496.

Hay, R. T. (1985). The origin of adenovirus DNA replication: Minimal DNA sequence requirement *in vivo*. *EMBO J.* **4**, 421–426.

Hayashi, Y., DePaoli, A. M., Burant, C. F., and Refetoff, S. (1994). Expression of a thyroid hormone responsive recombinant gene introduced into adult mice livers by replication-defective adenovirus can be regulated by endogenous thyroid hormone receptor. *J. Biol. Chem.* **269**, 23872–23875.

Hayashi, Y., Mangoura, D., and Refetoff, S. (1996). A mouse model of resistance to thyroid hormone produced by somatic gene transfer of a mutant thyroid hormone receptor. *Mol. Endocrinol.* **10**, 100–106.

Hearing, P., and Shenk, T. (1983). The adenovirus type 5 E1A transcriptional control region contains a duplicated enhancer element. *Cell (Cambridge, Mass.)* **33**, 695–703.

Hearing, P., Samulski, R. J., Wishart, W. L., and Shenk, T. (1987). Identification of a repeated sequence element required for efficient encapsidation of the adenovirus type 5 chromosome. *J. Virol.* **61**, 2555–2558.

Hersh, J., Crystal, R. G., and Bewig, B. (1995). Modulation of gene expression after replication-deficient, recombinant adenovirus-mediated gene transfer by the product of a second adenovirus vector. *Gene Ther.* **2**, 124–131.

Hierholzer, J. C., Wigand, R., Anderson, L. J., Adrian, T., and Gold, J. W. M. (1988). Adenoviruses from patients with AIDS: A plethora of serotypes and a description of five new serotypes of subgenus D (types 43–47). *J. Infect. Dis.* **158**, 804–813.

Hinds, P. W., and Weinberg, R. A. (1994). Tumor suppressor genes. *Curr. Opin. Genet. Dev.* **4**, 135–141.

Hirschowitz, E. A., Ohwada, A., Pascal, W. R., Russi, T. J., and Crystal, R. G. (1995). *In vivo* adenovirus-mediated gene transfer of the *Escherichia coli* cytosine deaminase gene to human colon carcinoma-derived tumors induces chemosensitivity to 5-fluorocytosine. *Hum. Gene Ther.* **6**, 1055–1063.

Hitt, M., Bett, A. J., Addison, C. L., Prevec, L., and Graham, F. L. (1995). Techniques for human adenovirus vector construction and characterization. *Methods Mol. Genet.* **7**, 13–30.

Hoffman, E. P., Brown, R. H. Jr., and Kunkel, L. M. (1987). Dystrophin: The protein product of the Duchenne muscular dystrophy locus. *Cell (Cambridge, Mass.)* **51**, 919–928.

Hollstein, M., Sidransky, D., Vogelstein, B., and Harris, C. C. (1991). p53 mutations in human cancers. *Science* **253**, 49–53.

Horellou, P., Vigne, E., Castel, M. N., Barneoud, P., Colin, P., Perricaudet, M., Delaere, P., and Mallet, J. (1994). Direct intracerebral gene transfer of an adenoviral vector expressing tyrosine hydroxylase in a rat model of Parkinson's disease. *NeuroReport* **6**, 49–53.

Huang, A., Jacobi, G., Haj-Ahmad, Y., and Bacchetti, S. (1988). Expression of the HSV-2 ribonucleotide reductase subunits in adenovirus vectors or stably transformed cells: Restoration of enzymatic activity by reassociation of enzyme subunits in the absence of other HSV proteins. *Virology* **163**, 462–470.

Huang, S., Endo, R. I., and Nemerow, G. R. (1995). Upregulation of integrins $\alpha_v\beta_3$ and $\alpha_v\beta_5$ on human monocytes and T lymphocytes facilitates adenovirus-mediated gene delivery. *J. Virol.* **69**, 2257–2263.

Huard, J., Lochmuller, H., Acsadi, G., Jani, A., Massie, B., and Karpati, G. (1995). The route of administration is a major determinant of the transduction efficiency of rat tissues by adenoviral recombinants. *Gene Ther.* **2**, 107–115.

Hughes, S. D., Rouy, D., Navaratnam, N., Scott, J., and Rubin, E. M. (1996). Gene transfer of cytidine deaminase apoBEC-1 lowers lipoprotein(a) in transgenic mice and induces apolipoprotein B editing in rabbits. *Hum. Gene Ther.* **7**, 39–49.

Hutchinson, L., Goldsmith, K., Snoddy, D., Ghosh, H., Graham, F. L., and Johnson, D. C. (1992). Indentification and characterization of a novel herpes simplex virus glycoprotein, gK, involved in cell fusion. *J. Virol.* **66**, 5603–5609.

Hutchinson, L., Graham, F. L., Cai, W., Debroy, C., Person, S., and Johnson, D. C. (1993). Herpes simplex virus (HSV) glycoproteins B and K inhibit cell fusion induced by HSV syncytial mutants. *Virology* **196**, 514–531.

Imler, J.-L., Chartier, C., Dieterle, A., Dreyer, D., Mehtali, M., and Pavirani, A. (1995a). An efficient procedure to select and recover recombinant adenovirus vectors. *Gene Ther.* **2**, 263–268.

Imler, J.-L., Bout, A., Dreyer, D., Dieterle, A., Schultz, H., Valerio, D., Mehtali, M., and Pavirani, A. (1995b). *Trans*-complementation of E1-deleted adenovirus: A new vector to reduce the possibility of codissemination of wild-type and recombinant adenoviruses. *Hum. Gene Ther.* **6**, 711–721.

Imler, J.-L., Chartier, C., Dreyer, D., Dieterle, A., Sainte-Marie, M., Faure, T., Pavirani, A., and Mehtali, M. (1996a). Novel complementation cell lines derived from human lung carcinoma A549 cells support the growth of E1-deleted adenovirus vectors. *Gene Ther.* **3**, 75–84.

Imler, J.-L., Dupuit, F., Chartier, C., Accart, N., Dieterle, A., Schultz, H., Puchelle, E., and Pavirani, A. (1996b). Targeting cell-specific gene expression with an adenovirus vector containing the *lacZ* gene under the control of the CFTR promoter. *Gene Ther.* **3**, 49–58.

Ishibashi, S., Brown, M. S., Goldstein, J. L., Gerard, R. D., Hammer, R. E., and Herz, J. (1993). Hypercholesterolemia in low density lipoprotein receptor knockout mice and its reversal by adenovirus-mediated gene delivery. *J. Clin. Invest.* **92**, 883–893.

Jaffe, H. A., Danel, C., Longenecker, G., Metzger, M., Setoguchi, Y., Rosenfeld, M. A., Gant, T. W., Thorgeirsson, S. S., Stratford-Perricaudet, L. D., Perricaudet, M., Pavirani, A., Lecocq, J.-P., and Crystal, R. G. (1992). Adenovirus-mediated *in vivo* gene transfer and expression in normal rat liver. *Nat. Genet.* **1**, 372–378.

Jin, X., Nguyen, D., Zhang, W.-W., Kyritsis, A. P., and Roth, J. A. (1995). Cell cycle arrest and inhibition of tumor cell proliferation by the p16[INK4] gene mediated by an adenovirus vector. *Cancer Res.* **55**, 3250–3253.

Johns, D. C., Nuss, H. B., Chiamvimonvat, N., Ramza, B. M., Marban, E., and Lawrence, J. H. (1995). Adenovirus-mediated expression of a voltage-gated potassium channel *in vitro* (rat cardiomyocytes) and *in vivo* (rat liver). *J. Clin. Invest.* **96**, 1152–1158.

Johnson, D. C., Ghosh-Choudhury, G., Smiley, J. R., Fallis, L., and Graham, F. L. (1988). Abundant expression of herpes simplex virus glycoprotein gB using an adenovirus vector. *Virology* **164**, 1–14.

Johnson, L. G. (1995). Gene therapy for cystic fibrosis. *Chest* **107**, 77S–83S.

Johnson, L. G., Boyles, S. E., Wilson, J., and Boucher, R. C. (1995). Normalization of raised sodium absorption and raised calcium-mediated chloride secretion by adenovirus-mediated expression of cystic fibrosis transmembrane conductance regulator in primary human cystic fibrosis airway epithelial cells. *J. Clin. Invest.* **95**, 1377–1382.

Jones, N., and Shenk, T. (1979). Isolation of adenovirus type 5 host range deletion mutants defective for transformation of rat embyro cells. *Cell (Cambridge, Mass.)* **17**, 683–689.

Kamb, A., Gruis, N. A., Weaverfeldhaus, J., Lui, Q. Y., Harshman, K., Tavtigian, S. V., Stockert, E., Day, R. S., Johnson, B. E., and Skolnick, M. H. (1994). A cell cycle regulator potentially involved in genesis of many tumor types. *Science* **264**, 436–440.

Kanegae, Y., Lee, G., Sato, Y., Tanaka, M., Nakai, M., Sakaki, T., Sugano, S., and Saito, I. (1995). Efficient gene activation in mammalian cells by using recombinant adenovirus expressing site-specific Cre recombinase. *Nucleic Acids Res.* **23**, 3816–3821.

Kaplan, J. M., St. George, J. A. Pennington, S. E., Keyes, L. D., Johnson, R. P., Wadsworth, S. C., and Smith, A. E. (1996). Humoral and cellular immune responses of non-human primates to long term repeated lung exposure to Ad2/CFTR-2. *Gene Ther.* **3**, 117–127.

Karlsson, S., Humphries, R. K., Gluzman, Y., and Nienhuis, A. W. (1985). Transfer of genes into hematopoetic cells using recombinant DNA viruses. *Proc. Natl. Acad. Sci. U.S.A.* **82**, 158–162.

Kass-Eisler, A., Falck-Pedersen, E., Alvira, M., Rivera, J., Buttrick, P. M., Wittenberg, B. A., Cipriani, L., and Leinwand, L. A. (1993). Quantitative determination of adenovirus-mediated gene delivery to rat cardiac myocytes *in vitro* and *in vivo*. *Proc. Natl. Acad. Sci. U.S.A.* **90**, 11498–11502.

Katayose, D., Gudas, J., Nguyen, H., Srivastava, S., Cowan, K. H., and Seth, P. (1995a). Cytotoxic effects of adenovirus-mediated wild-type p53 protein expression in normal and tumor mammary epithelial cells. *Clin. Cancer Res.* **1**, 889–897.

Katayose, D., Wersto, R., Cowan, K. H., and Seth, P. (1995b). Effects of a recombinant adenovirus expressing WAF1/Cip1 on cell growth, cell cycle, and apoptosis. *Cell Growth Differ.* **6**, 1207–1212.

Katkin, J. P., Gilbert, B. E., Langston, C., French, K., and Beaudet, A. L. (1995). Aerosol delivery of a beta-galactosidase adenoviral vector to the lungs of rodents. *Hum. Gene Ther.* **6**, 985–995.

Kay, M. A., Landen, C. N., Rothenberg, S. R., Taylor, L. A., Leland, F., Wiehle, S., Fang, B., Bellinger, D., Finegold, M., Thompson, A. R., Read, M., Brinkhous, K. M., and Woo, S. L. C. (1994). *In vivo* hepatic gene therapy: Complete albeit transient correction of factor IX deficiency in hemophilia B dogs. *Proc. Natl. Acad. Sci. U.S.A.* **91**, 2353–2357.

Kay, M. A., Holterman, A.-X., Meuse, L., Gown, A., Ochs, H., Linsley, P., and Wilson, C. B. (1995). Long term hepatic adenovirus-mediated gene expression in mice following CTLA4Ig administration. *Nat. Genet.* **11**, 191–197.

Ketner, G., Spencer, F., Tugendreich, S., Connelly, C., and Hieter, P. (1994). Efficient manipulation of the human adenovirus genome as an infectious yeast artificial chromosome clone. *Proc. Natl. Acad. Sci. U.S.A.* **91**, 6186–6190.

Kleinerman, D. I., Zhang, W.-W., Lin, S.-H., Van, N. T., von Eschenbach, A. C., and Hsieh, J.-T. (1995). Application of a tumor suppressor (C-CAM1)-expressing recombinant adenovirus in androgen-independent human prostate cancer therapy: A preclinical study. *Cancer Res.* **55**, 2831–2836.

Knowles, M. R., Hohneker, K., Zhou, Z., Olsen, J. C., Noah, T. L., Hu, P. C., Leigh, M. W., Engelhardt, J. F., Edwards, L. J., Jones, K. R., Grossman, M., Wilson, J. M., Johnson, L. G., and Boucher, R. C. (1995). A controlled study of adenoviral-vector-mediated gene transfer in the nasal epithelium of patients with cystic fibrosis. *N. Engl. J. Med.* **333**, 823–831.

Kochanek, S., Clemens, P. R., Mitani, K., Chen, H. H., Chan, S., and Caskey, T. (1996). A new adenoviral vector: Replacement of all viral coding sequences with 28 kb of DNA expressing both full-length dystrophin and β-galactosidase. *Proc. Natl. Acad. Sci. U.S.A.* **93**, 5731–5736.

Kolls, J., Peppel, K., Silva, M., and Beutler, B. (1994). Prolonged and effective blockade of tumor necrosis factor activity through adeonovirus-mediated gene transfer. *Proc. Natl. Acad. Sci. U.S.A.* **91**, 215–219.

Kolls, J. K., Lei, D., Nelson, S., Summer, W. R., Greenberg, S., and Beutler, B. (1995). Adenovirus-mediated blockade of tumor necrosis factor in mice protects against endotoxic shock yet impairs pulmonary host defense. *J. Infect. Dis.* **171**, 570–575.

Konan, V., Sahota, A., Graham, F. L., and Taylor, M. W. (1991). Transduction of the CHO *aprt* gene into mouse L cells using an adeno-5/APRT recombinant virus. *Somatic Cell Mol. Genet.* **17**, 359–368.

Konishi, H., Ochiya, T., Sakamoto, H., Tsukamoto, M., Saito, I., Muto, T., Sugimura, T., and Terada, M. (1995). Effective prevention of thrombocytopenia in mice using adenovirus-mediated transfer of HST-1 (FGF-4) gene. *J. Clin. Invest.* **96**, 1125–1130.

Kopfler, W. P., Willard, M., Betz, T., Willard, J. E., Gerard, R. D., and Meidell, R. S. (1994). Adenovirus-mediated transfer of a gene encoding human apolipoprotein A–I into normal mice increases circulating high-density lipoprotein cholesterol. *Circulation* **90**, 1319–1327.

Korst, R. J., Bewig, B., and Crystal, R. G. (1995). *In vitro* and *in vivo* transfer and expression of human surfactant SP-A- and SP-B-associated protein cDNAs mediated by replication-deficient, recombinant adenoviral vectors. *Hum. Gene Ther.* **6**, 277–287.

Kozarsky, K., Grossman, M., and Wilson, J. M. (1993). Adenovirus-mediated correction of the genetic defect in hepatocytes from patients with familial hypercholesterolemia. *Somatic Cell Mol. Genet.* **19**, 449–458.

Kozarsky, K. F., McKinley, D. R., Austin, L. L., Raper, S. E., Stratford-Perricaudet, L. D., and Wilson, J. M. (1994). *In vivo* correction of low density lipoprotein receptor deficiency in the Watanabe heritable hyperlipidemic rabbit with recombinant adenoviruses. *J. Biol. Chem.* **269**, 13695–13702.

Krougliak, V., and Graham, F. L. (1995). Development of cell lines capable of complementing E1, E4, and protein IX defective adenovirus type 5 mutants. *Hum. Gene Ther.* **6**, 1575–1586.

Landau, C., Pirwitz, M. J., Willard, M. A., Gerard, R. D., Meidell, R. S., and Willard, J. E. (1995). Adenoviral mediated gene transfer to atherosclerotic arteries after balloon angioplasty. *Am. Heart J.* **129**, 1051–1057.

Lee, M. G., Abina, M. A., Haddada, H., and Perricaudet, M. (1995). The constitutive expression of the immunomodulatory gp19k protein in E1$^-$, E3$^-$ adenoviral vectors strongly reduces the host cytotoxic T cell response against the vector. *Gene Ther.* **2**, 256–262.

Lee, S. W., Trapnell, B. C., Rade, J. J., Virmani, R., and Dichek, D. A. (1993). *In vivo* adenoviral vector-mediated gene transfer into balloon-injured rat carotid arteries. *Circ. Res.* **73**, 797–807.

Le Gal La Salle, G., Robert, J. J., Berrard, S., Ridoux, V., Stratford-Perricaudet, L. D., Perricaudet, M., and Mallet, J. (1993). An adenovirus vector for gene transfer into neurons and glia in the brain. *Science* **259**, 988–990.

Lemarchand, P., Jaffe, H. A., Danel, C., Cid, M. C., Kleinman, H. K., Stratford-Perricaudet, L., Perricaudet, M., Pavirani, A., Lecocq, J. P., and Crystal, R. G. (1992). Adenovirus-mediated transfer of a recombinant human α1-antitrypsin cDNA to human endothelial cells. *Proc Natl. Acad. Sci. U.S.A.* **89**, 6482–6486.

Lemarchand, P., Jones, M., Yamada, I., and Crystal, R. G. (1993). *In vivo* gene transfer and expression in normal uninjured blood vessels using replication-deficient recombinant adenovirus vectors. *Circ. Res.* **72**, 1132–1138.

Lemarchand, P., Jones, M., Danel, C., Yamada, I., Mastrangeli, A., and Crystal, R. G. (1994). *In vivo* adenovirus-mediated gene transfer to lungs via pulmonary artery. *J. Appl. Physiol.* **76**, 2840–2845.

Levrero, M., Barban, V., Manteca, S., Ballay, A., Balsamo, C., Avantaggiati, M. L., Natoli, G., Skellekens, H., Tiollais, P., and Perricaudet, M. (1991). Defective and nondefective adenovirus vectors for expressing foreign genes *in vitro* and *in vivo*. *Gene* **101**, 195–202.

Li, Q., Kay, M. A., Finegold, M., Stratford-Perricaudet, L. D., and Woo, S. L. C. (1993). Assessment of recombinant adenoviral vectors for hepatic gene therapy. *Hum. Gene Ther.* **4**, 403–409.

Li, T., Adamian, M., Roof, D. J., Berson, E. L., Dryja, T. P., Roessler, B. J., and Davidson, B. L. (1994). *In vivo* transfer of a reporter gene to the retina mediated by an adenoviral vector. *Invest. Ophthalmol. Visual Sci.* **35**, 2543–2549.

Li, Y., Jenkins, C. W., Nichols, M. A., and Xiong, Y. (1994). Cell cycle expression and p53 regulation of the cyclin-dependent kinase inhibitor p21. *Oncogene* **9**, 2261–2268.

Lieber, A., and Kay, M. A. (1996). Adenovirus-mediated expression of ribozymes in mice. *J. Virol.* **70**, 3153–3158.

Lindgren, C. G., Thompson, J. A., Higuchi, C. M., and Fefer, A. (1993). Growth and autologous tumor lysis by tumor-infiltrating lymphocytes from metastatic melanoma expanded in interleukin-2 or interleukin-2 plus interleukin-4. *J. Immunother.* **14**, 322–328.

Lippe, R., and Graham, F. L. (1989). Adenoviruses with nonidentical terminal sequences are viable. *J. Virol.* **63**, 5133–5141.

Liu, T.-J., Zhang, W.-W., Taylor, D. L., Roth, J. A., Goepfert, H., and Clayman, G. L. (1994). Growth suppression of human head and neck cancer cells by introduction of a wild-type p53 gene via a recombinant adenovirus. *Cancer Res.* **54**, 3662–3667.

Livingstone, L. R., White, A., Sprouse, J., Livanos, E., Jacks, T., and Tlsty, T. D. (1992). Altered cell cycle arrest and gene amplification potential accompany loss of wild-type p53. *Cell (Cambridge, Mass.)* **70**, 923–935.

Lochmuller, H., Jani, A., Huard, J., Prescott, S., Simoneau, M., Massie, B., Karpati, G., and Ascadi, G. (1994). Emergence of early region 1-containing replication-competent adenovirus in stocks of replication-defective adenovirus recombinants (ΔE1 + ΔE3) during multiple passages in 293 cells. *Hum. Gene Ther.* **5**, 1485–1491.

Lochmuller, H., Petrof, B. J., Allen, C., Prescott, S., Massie, B., and Karpati, G. (1995). Immunosuppression by FK506 markedly prolongs expression of adenovirus-delivered transgene in skeletal muscles of adult dystrophic (mdx) mice. *Biochem. Biophys. Res. Commun.* **213**, 569–574.

Lotze, M. T., Grimm, E. A., Mazumder, A., Strausser, J. L., and Rosenberg, S. A. (1981). Lysis of fresh and cultured autologous tumor by human lymphocytes cultured in T-cell growth factor. *Cancer Res.* **41**, 4420–4425.

Lowenstein, P. R., Bain, D., Shering, A. F., Wilkinson, G. W., and Castro, M. G. (1995). Cell-type specific expression from viral promoters within replication-deficient adenovirus recombinants in primary neocortical cultures. *Restorative Neurol. Neurosci.* **8**, 37–39.

Lowenstein, P. R., Wilkinson, G. W. G., Castro, M. G., Shering, A. F., Fooks, A. R., and Bain, D. (1996). Non-neurotropic adenovirus: A vector for gene transfer to the brain and possible gene therapy of neurological disorders. *In* "Genetic Manipulation of the Nervous System," pp. 11–39. Academic Press, London.

Lowy, D. R., and Willumsen, B. M. (1993). Function and regulation of ras. *Annu. Rev. Biochem.* **62**, 851–891.

Maeda, H., Danel, C., and Crystal, R. G. (1994). Adenovirus-mediated transfer of human lipase complementary DNA to the gallbladder. *Gastroenterology* **106**, 1638–1644.

Magness, S. T., and Brenner, D. A. (1995). Ferrochelatase cDNA delivered by adenoviral vector corrects biochemical defect in protoporphyic cells. *Hum. Gene Ther.* **6**, 1285–1290.

Mansour, S. L., Grodzicker, T., and Tjian, R. (1985). An adenovirus vector system used to express polyoma virus tumor antigens. *Proc. Natl. Acad. Sci. U.S.A.* **82**, 1359–1363.

Mansour, S. L., Grodzicker, T., and Tjian, R. (1986). Downstream sequences affect transcription initiation from the adenovirus major late promoter. *Mol. Cell. Biol.* **6**, 2684–2694.

March, K. L., Madison, J. A., and Trapnell, B. C. (1995). Pharmokinetics of adenoviral vector-mediated gene delivery to vascular smooth muscle cells: Modulation by poloxamer 407 and implications for cardiovascular gene therapy. *Hum. Gene Ther.* **6**, 41–53.

Mason, B. B., Davis, R. R., Bhat, B. M., Chengalvala, M., Lubeck, M. D., Zandle, G., Kostek, B., Cholodofsky, S., Dheer, S., Molnar-Kimber, K., Mizutani, S., and Hung, P. P. (1990). Adenovirus vaccine vectors expressing hepatitis B surface antigen: Importance of regulatory elements in the adenovirus major late intron. *Virology* **177**, 452–461.

Massie, B., Gluzman, Y., and Hassell, J. A. (1986). Construction of a helper-free recombinant adenovirus that expresses polyomavirus large T antigen. *Mol. Cell. Biol.* **6**, 2872–2883.

Massie, B., Dionne, J., Lamarche, N., Fleurent, J., and Langelier, Y. (1995). Improved adenovirus vector provides herpes simplex virus ribonucleotide reductase R1 and R2 subunits very efficiently. *BioTechnology* **13**, 602–608.

Mastrangeli, A., Danel, C., Rosenfeld, M. A., Stratford-Perricaudet, L., Perricaudet, M., Pavirani, A., Lecocq, J. P., and Crystal, R. G. (1993). Diversity of airway epithelial cell targets for *in vivo* recombinant adenovirus-mediated gene transfer. *J. Clin. Invest.* **91**, 225–234.

Mastrangeli, A., O'Connell, B., Aladib, W., Fox, P. C., Baum, B. J., and Crystal, R. G. (1994). Direct *in vivo* adenovirus-mediated gene transfer to salivary glands. *Am. J. Physiol.* **266**, G1146–G1155.

Mazur, W., Ali, N. M., Grinstead, W. C., Schulz, D. G., Raizner, A. E., and French, B. A. (1994). Lipofectin-mediated versus adenovirus-mediated gene transfer *in vitro* and *in vivo:* Comparison of canine and porcine model systems. *Coronary Artery Dis.* **5**, 779–786.

McCoy, R. D., Davidson, B. L., Roessler, B. J., Huffnagle, G. B., Janich, S. L., Laing, T. J., and Simon, R. H. (1995). Pulmonary inflammation induced by incomplete or inactivated adenoviral particles. *Hum. Gene Ther.* **6**, 1553–1560.

McCray, P. B., Jr., Armstrong, K., Zabner, J., Miller, D. W., Koretzky, G. A., Couture, L., Robillard, J. E., Smith, A. E., and Welsh, M. J. (1995). Adenoviral-mediated gene transfer to fetal pulmonary epithelia *in vitro* and *in vivo*. *J. Clin. Invest.* **95**, 2620–2632.

McGrory, J., Bautista, D., and Graham, F. L. (1988). A simple technique for the rescue of early region I mutations into infectious human adenovirus type 5. *Virology* **163**, 614–617.

Mehra, M. R., Stapleton, D. D., Cook, J. L., Zhang, T., Ventura, H. O., Huang, C., Maldonado, B., Smart, F. W., Re, R. N., Murgo, J. P., and Barbee, R. W. (1996). Adenovirus-mediated *in vivo* gene transfer in a rabbit model of allograft vasculopathy. *J. Heart Lung Transplant* **15**, 51–57.

Mestril, R., Giordano, F. J., Conde, A. G., and Dillmann, W. H. (1996). Adenovirus mediated gene transfer of a heat shock protein 70 (hsp70i) protects against simulated ischemia. *J. Mol. Cell. Cardiology* **28**, 2351–2358.

Michael, S. I., and Curiel, D. T. (1994). Strategies to achieve targeted gene delivery via the receptor-mediated endocytosis pathway. *Gene Ther.* **1**, 223–232.

Michael, S. I., Hong, J. S., Curiel, D. T., and Engler, J. A. (1995). Addition of a short peptide ligand to the adenovirus fiber protein. *Gene Ther.* **2**, 660–668.

Mitani, K., Graham, F. L., and Caskey, C. T. (1994). Transduction of human bone marrow by adenoviral vector. *Hum. Gene Ther.* **5**, 941–948.

Mitani, K., Graham, F. L., Caskey, C. T., and Kochanek, S. (1995a). Rescue, propagation, and partial purification of a helper virus-dependent adenovirus vector. *Proc. Natl. Acad. Sci. U.S.A.* **92**, 3854–3858.

Mitani, K., Wakamiya, M., Hasty, P., Graham, F. L., Bradley, A., and Caskey, C. T. (1995b). Gene targeting in mouse embryoic stem cells with an adenoviral vector. *Somatic Cell Mol. Genet.* **21**, 221–231.

Mittal, S. K., McDermott, M. R., Johnson, D. C., Prevec, L., and Graham, F. L. (1993). Monitoring foreign gene expression by a human adenovirus-based vector using the firefly luciferase gene as a reporter. *Virus Res.* **28**, 67–90.

Mittereder, N., Yei, S., Bachurski, C., Cuppoletti, J., Whitsett, J. A., Tolstoshev, P., and Trapnell, B. C. (1994). Evaluation of the efficacy and safety of *in vitro,* adenovirus-mediated transfer of the human cystic fibrosis transmembrane conductance regulator cDNA. *Hum. Gene Ther.* **5**, 717–729.

Morris, B. D., Drazan, K. E., Csete, M. E., Werthman, P. E., Van Bree, M. P., Rosenthal, J. T., and Shaked, A. (1994). Adenoviral-mediated gene transfer to bladder *in vivo*. *J. Urol.* **152**, 506–509.

Morsy, M. A., Alford, E. L., Bett, A., Graham, F. L., and Caskey, C. T. (1993). Efficient adenoviral-mediated ornithine transcarbamylase expression in deficient mouse and human hepatocytes. *J. Clin. Invest.* **92**, 1580–1586.

Moullier, P., Friedlander, G., Calise, D., Ronco, P., Perricaudet, M., and Ferry, N. (1994). Adenoviral-mediated gene transfer to renal tubular cells *in vivo*. *Kidney Int.* **45**, 1220–1225.

Muhlhauser, J., Pili, R., Merrill, M. J., Maeda, H., Passaniti, A., Crystal, R. G., and Capogrossi, M. C. (1995a). *In vivo* angiogenesis induced by recombinant adenovirus vectors coding either for secreted or nonsecreted forms of acidic fibroblast growth factor. *Hum. Gene Ther.* **6**, 1457–1465.

Muhlhauser, J., Merrill, M. J., Pili, R., Maeda, H., Bacic, M., Bewig, B., Passaniti, A., Edwards, N. A., Crystal, R. G., and Capogrossi, M. C. (1995b). $VEGF_{165}$ expressed by a replication-deficient, recombinant adenovirus vector induces angiogenesis *in vivo*. *Circ. Res.* **77**, 1077–1086.

Muhlhauser, J., Jones, M., Yamada, I., Cirielli, C., Lemarchand, P., Gloe, T. R., Bewig, B., Signoretti, S., Crystal, R. G., and Capogrossi, M. C. (1996). Safety and efficacy of *in vivo* gene transfer into the porcine heart with replication-deficient, recombinant adenovirus vectors. *Gene Ther.* **3**, 145–153.

Muller, D. W., Gordon, D., San, H., Yang, Z., Pompili, V. J., Nabel, G. J., and Nabel, E. G. (1994). Catheter-mediated pulmonary vascular gene transfer and expression. *Circ. Res.* **75**, 1039–1049.

Muller, W. J., Sinn, E., Pattengale, P. K., Wallace, R., and Leder, P. (1988). Single-step induction of mammary adenocarcinoma in transgenic mice bearing the activated c-*neu* oncogene. *Cell (Cambridge, Mass.)* **54**, 105–115.

Munz, P. L., and Young, C. S. H. (1987). The creation of adenovirus genomes with viable, stable, internal redundancies centered about the E2b region. *Virology* **158**, 52–60.

Nevins, J. R. (1981). Mechanism of activation of early viral transcription by the adenovirus E1A gene product. *Cell (Cambridge, Mass.)* **26**, 213–220.

Nevins, J. R. (1987). Regulation of early adenovirus gene expression. *Microbiol. Rev.* **51**, 419–430.

Newman, K. D., Dunn, P. F., Owens, J. W., Schulick, A. H., Virmani, R., Sukhova, G., Libby, P., and Dichek, D. (1995). Adenovirus-mediated gene transfer into normal rabbit arteries results in a prolonged vascular cell activation, inflammation, and neointima hyperplasia. *J. Clin. Invest.* **96**, 2955–2965.

Nogee, L. M., DeMello, D. E., Dehner, L. P., and Colten, H. R. (1993). Brief report: Deficiency of pulmonary surfactant protein-B in congenital alveolar proteinosis. *N. Engl. J. Med.* **328**, 406–410.

Ohno, T., Gordon, D., San, H., Pompili, V. J., Imperiale, M. J., Nabel, G. J., and Nabel, E. G. (1994). Gene therapy for vascular smooth muscle cell proliferation after arterial injury. *Science* **265**, 781–784.

O'Malley, B. W., Jr., Chen, S.-H., Schwartz, M. R., and Woo, S. L. (1995). Adenovirus-mediated gene therapy for human head and neck squamous cell cancer in a nude model. *Cancer Res.* **55**, 1080–1085.

O'Neal, W. K., and Beaudet, A. L. (1994). Somatic gene therapy for cystic fibrosis. *Hum. Mol. Genet.* **3** (Spec. No.), 1497–1502.

Peault, B., Tirouvanziam, R., Sombardier, M. N., Chen, S., Perricaudet, M., and Gaillard, D. (1994). Gene transfer to human fetal pulmonary tissue developed in immunodeficient SCID mice. *Hum. Gene Ther.* **5**, 1131–1137.

Perez-Cruet, M. J., Trask, T. W., Chen, S.-H., Goodman, J. C., Woo, S. L. C., Grossman, R. G., and Shine, H. D. (1994). Adenovirus-mediated gene therapy of experimental gliomas. *J. Neurosci. Res.* **39**, 506–511.

Petrof, B. J., Acsadi, G., Jani, A., Massie, B., Bourdon, J., Matusiewicz, N., Yang, L., Lochmuller, H., and Karpati, G. (1995). Efficiency and functional consequences of adenovirus-mediated *in vivo* gene transfer to normal and dystrophic (mdx) mouse diaphragm. *Am. J. Respir. Cell Mol. Biol.* **13**, 508–517.

Philipson, L., Lonberg-Holm, K., and Pettersson, U. (1968). Virus–receptor interaction in an adenovirus system. *J. Virol.* **2**, 1064–1075.

Pilder, S., Moore, M., Logan, J., and Shenk, T. (1986). The adenovirus E1B-55k transforming polypeptide modulates transport or cytoplasmic stabilization of viral and host cell mRNAs. *Mol. Cell. Biol.* **6**, 470–476.

Pilewski, J. M., Engelhardt, J. F., Bavaria, J. E., Kaiser, L. R., Wilson, J. M., and Albelda, S. M. (1995a). Adenovirus-mediated gene transfer to human bronchial submucosal glands using xenografts. *Am. J. Physiol.* **268**, L657–L665.

Pilewski, J. M., Sott, D. J., Wilson, J. M., and Albelda, S. M. (1995b). ICAM-1 expression on bronchial epithelium after recombinant adenovirus infection. *Am. J. Respir. Cell Mol. Biol.* **12**, 142–148.

Qian, C., Idoate, M., Alzuguren, P., Vazquez, J., Bilbao, R., Sangro, B., Gil, A., and Prieto, J. (1995a). Gene therapy in an experimental model of hepatocellular carcinoma (HCC). *J. Hepatol.* **23**, 68.

Qian, C., Bilbao, R., Bruna, O., and Prieto, J. (1995b). Induction of sensitivity to ganciclovir in human hepatocellular carcinoma cells by adenovirus-mediated gene transfer of herpes simplex virus thymidine kinase. *Hepatology* **22**, 118–123.

Quantin, B., Perricaudet, L. D., Tajbakhsh, S., and Mandel, J. L. (1992). Adenovirus as an expression vector in muscle cells *in vivo*. *Proc. Natl. Acad. Sci. U.S.A.* **89**, 2581–2584.

Rade, J. J., Schulick, A. H., Virmani, R., and Dichek, D. A. (1996). Local adenoviral-mediated expression of recombinant hirudin reduces neointimal formation after arterial injury. *Nat. Med.* **2**, 293–298.

Ragot, T., Vincent, N., Chafey, P., Vigne, E., Gilgenkrantz, H., Couton, D., Cartaud, J., Briand, P., Kaplan, J.-C., Perricaudet, M., and Kahn, A. (1993). Efficient adenovirus-mediated transfer of a human minidystrophin gene to skeletal muscles of mdx mice. *Nature (London)* **361**, 647–650.

Rice, A. P., and Matthews, M. B. (1988a). Transcriptional but not translational regulation of HIV-1 by the *tat* gene product. *Nature (London)* **332**, 551–553.

Rice, A. P., and Matthews, M. B. (1988b). Trans-activation of the human immunodeficiency virus long terminal repeat sequences, expressed in an Ad vector, by the adenovirus E1A 13S protein. *Proc. Natl. Acad. Sci. U.S.A.* **85**, 4200–4204.

Rich, D. P., Couture, L. A., Cardoza, L. M., Guiggio, V. M., Armentano, D., Espino, P. C., Hehir, K., Welsh, M. J., Smith, A. E., and Gregory, R. J. (1993). Development and analysis of recombinant adenoviruses for gene therapy of cystic fibrosis. *Hum. Gene Ther.* **4**, 461–476.

Richards, C. D., Braciak, T., Xing, Z., Graham, F., and Gauldie, J. (1995). Adenovirus vectors for cytokine gene expression. *Ann. N. Y. Acad. Sci.* **762**, 282–292.

Ridoux, V., Robert, J. J., Zhang, X., Perricaudet, M., Mallet, J., and LeGal La Salle, G. (1994). The use of adenovirus vectors for intracerebral grafting of transfected nervous cells. *NeuroReport* **5**, 801–804.

Roberts, R. J., O'Neill, K. E., and Yen, C. I. (1984). DNA sequences from the adenovirus 2 genome. *J. Biol. Chem.* **259**, 13968–13985.

Roessler, B. J., Allen, E. D., Wilson, J. M., Hartman, J. W., and Davidson, B. L. (1993). Adenoviral-mediated gene transfer to rabbit synovium *in vivo*. *J. Clin. Invest.* **92**, 1085–1092.

Roessler, B. J., Hartman, J. W., Vallance, D. K., Latta, J. M., Janich, S. L., and Davidson, B. L. (1995). Inhibition of interleukin-1 induced effects in synoviocytes transduced with the human IL-1 receptor antagonist cDNA, using an adenoviral vector. *Hum. Gene Ther.* **6**, 307–316.

Rome, J. J., Shayani, V., Newman, K. D., Farrell, S., Lee, S. W., Virmani, R., and Dichek, D. A. (1994). Adenoviral vector-mediated gene transfer into sheep arteries using a double-balloon catheter. *Hum. Gene Ther.* **5**, 1249–1258.

Rosenfeld, M. A., Yoshimura, K., Trapnell, B. C., Yoneyama, K., Rosenthal, E. R., Dalemans, W., Fukayama, M., Bargon, J., Stier, L. E., Stratford-Perricaudet, L., Perricaudet, M., Guggino, W. B., Pavirani, A., Lecocq, J.-P., and Crystal, R. (1992). *In vivo* transfer of the human cystic fibrosis transmembrane conductance regulator gene to the airway epithelium. *Cell (Cambridge, Mass.)* **68**, 143–155.

Rosenfeld, M. A., Chu, C. S., Seth, P., Danel, C., Banks, T., Yoneyama, K., Yoshimura, K., and Crystal, R. G. (1994). Gene transfer to freshly isolated human respiratory epithelial cells *in vitro* using a replication-deficient adenovirus containing the human cystic fibrosis transmembrane conductance regulator cDNA. *Hum. Gene Ther.* **5**, 331–342.

Ruether, J. E., Maderious, A., Lavery, D., Logan, J., Fu, S. M., and Chen-Kiang, S. (1986). Cell-type-specific synthesis of murine immunoglobulin μ RNA from an adenovirus vector. *Mol. Cell. Biol.* **6**, 123–133.

Sabate, O., Horellou, P., Vigne, E., Colin, P., Perricaudet, M., Buc-Caron, M.-H., and Mallet, J. (1995). Transplantation to the rat brain of human neural progenitors that were genetically modified using adenoviruses. *Nat. Genet.* **9**, 256–260.

Saito, I., Oya, Y., Yamamoto, K., Yuasa, T., and Shimojo, H. (1985). Construction of nondefective adenovirus type 5 bearing a 2.8-kilobase hepatitis B virus DNA near the right end of its genome. *J. Virol.* **54**, 711–719.

Sarnow, P., Hearing, P., Anderson, C. W., Halbert, D. N., Shenk, T., and Levine, A. F. (1984). Adenovirus early region 1b 58,000-dalton tumor antigen is physically associated with an early region 4 25,000-dalton protein in productively infected cells. *J. Virol.* **49**, 692–700.

Scaria, A., Curiel, D. T., and Kay, M. A. (1995). Complementation of a human adenovirus early region 4 deletion mutant in 293 cells using adenovirus–polylysine–DNA complexes. *Gene Ther.* **2**, 295–298.

Schaack, J., Langer, S., and Guo, X. (1995a). Efficient selection of recombinant adenoviruses by vectors that express beta-galactosidase. *J. Virol.* **69**, 3920–3923.

Schaack, J., Guo, X., Ho, W. Y.-W., Karlok, M., Chen, C., and Ornelles, D. (1995b). Adenovirus type 5 precursor terminal protein-expressing 293 and HeLa cell lines. *J. Virol.* **69**, 4079–4085.

Schachtner, S. K., Rome, J. J., Hoyt, R. F., Newman, K. D., Virmani, R., and Dichek, D. A. (1995). *In vivo* adenovirus-mediated gene transfer via the pulmonary artery of rats. *Circ. Res.* **76**, 701–709.

Schneider, M., Graham, F. L., and Prevec, L. (1989). Expression of the glycoprotein of vesicular stomatitis virus by infectious adenovirus vectors. *J. Gen. Virol.* **70**, 417–427.

Schulick, A. H., Dong, G., Newman, K. D., Virmani, R., and Dichek, D. A. (1995a). Endothelium-specific *in vivo* gene transfer. *Circ. Res.* **77**, 475–485.

Schulick, A. H., Newman, K. D., Virmani, R., and Dichek, D. A. (1995b). *In vivo* gene transfer into injured carotid arteries. Optimization and evaluation of acute toxicity. *Circulation* **91**, 2407–2414.

Sene, C., Bout, A., Imler, J.-L., Schultz, H., Willemot, J.-M., Hennebel, V., Zurcher, C., Valerio, D., Lamy, D., and Pavirani, A. (1995). Aerosol-mediated delivery of recombinant adeonovirus to the airways of nonhuman primates. *Hum. Gene Ther.* **6**, 1587–1593.

Serrano, M., Gomez-Lahoz, E., Depinho, R. A., Beach, D., and Bar-Sagi, D. (1995). Inhibition of ras-induced proliferation and cellular transformation by p16[INK4]. *Science* **267**, 249–252.

Seth, P., Fitzgerald, D., Ginsberg, H., Willingham, M., and Pastan, I. (1984). Evidence that the penton base of adenovirus is involved in potentiation of toxicity of *Pseudomonas* exotoxin conjugated to epidermal growth factor. *Mol. Cell. Biol.* **4**, 1528–1533.

Seth, P., Rosenfeld, M., Higginbotham, J., and Crystal, R. G. (1994). Mechanism of enhancement of DNA expression consequent to cointernalization of a replication-deficient adenovirus and unmodified plasmid DNA. *J. Virol.* **68**, 933–940.

Setoguchi, Y., Jaffe, H. A., Danel, C., and Crystal, R. G. (1994a). *Ex vivo* and *in vivo* gene transfer to the skin using replication-deficient recombinant adenovirus vectors. *J. Invest. Dermatol.* **102**, 415–421.

Setoguchi, Y., Jaffe, H. A., Chu, C. S., and Crystal, R. G. (1994b). Intraperitoneal *in vivo* gene therapy to deliver α1-antitrypsin to the systemic circulation. *Am. J. Respir. Cell Mol. Biol.* **10**, 369–377.

Setoguchi, Y., Danel, C., and Crystal, R. G. (1994c). Stimulation of erythropoiesis by *in vivo* gene therapy: Physiologic consequences of transfer of the human erythropoietin gene to experimental animals using an adenovirus vector. *Blood* **84**, 2946–2453.

Shaked, A., Csete, M. E., Drazan, K. E., Bullington, D., Wu, L., Busuttil, R. W., and Berk, A. J. (1994). Adenovirus-mediated gene transfer in the transplant setting. II. Successful expression of transferred cDNA in syngeneic liver grafts. *Transplantation* **57**, 1508–1511.

Shaker, M., Hall, S. J., Mutchnik, S. E., Timme, T. L., Chen, S.-H., Woo, S. L. C., and Thompson, T. C. (1996). "Combination Therapy for Prostate Cancer: Androgen Ablation and Adenovirus-Mediated Herpes Simplex Virus Thymidine Kinase Gene Transduction and Ganciclovir Treatment, Ann. Meet. Abstr. American Urological Assoc., Orlando, Florida.

Sharp, P. A., Moore, C., and Haverty, J. L. (1976). The infectivity of adenovirus 5 DNA–protein complex. *Virology* **75**, 442–456.

Shering, A. F., Bain, D., Stewart, K., Epstein, A. L., Castro, M. G., Wilkinson, G. W. G., and Lowenstein, P. R. (1997). Cell-type specific expression in brain cell cultures from a short cytomegalovirus major immediate early promoter depends on whether it is inserted into herpesvirus or adenovirus vectors. *J. Gen. Virol.* **78**, 445–459.

Shy, M. E., Tani, M., Shi, Y.-J., Whyatt, S. A., Chbihi, T., Scherer, S. S., and Kamholz, J. (1995). An adenoviral vector can transfer *lacZ* expression into Schwann cells in culture and in sciatic nerve. *Ann. Neurol.* **38**, 429–436.

Siegfried, W., Rosenfeld, M., Stier, L., Stratford-Perricaudet, L., Perricaudet, M., Pavirani, A., Lecocq, J. P., and Crystal, R. G. (1995). Polarity of secretion of α1-antitrypsin by human respiratory epithelial cells after adenoviral transfer of a human α1-antitrypsin cDNA. *Am. J. Respir. Cell Mol. Biol.* **12**, 379–384.

Simon, R. H., Engelhardt, J. F., Yang, Y., Zepeda, M., Weber-Pendleton, S., Grossman, M., and Wilson, J. M. (1993). Adenovirus-mediated transfer of the CFTR gene to lung of nonhuman primates: Toxicity study. *Hum. Gene Ther.* **4**, 771–780.

Slamon, D. J., Clark, G. M., Wong, S. G., Levin, W. J., Ullrich, A., and McGuire, W. L. (1987). Human breast cancer: Correlation of relapse and survival with amplification of the HER-2/neu oncogene. *Science* **235**, 177–182.

Smith, C. L., Hager, G. L., Pike, J. W., and Marx, S. J. (1991). Overexpression of the human vitamin D3 receptor in mammalian cells using recombinant adenovirus vectors. *Mol. Endocrinol.* **5**, 867–878.

Smith, G. M., Hale, J., Pasnikowski, E. M., Lindsay, R., Wong, V., and Rudge, J. S. (1996). Astrocytes infected with replication-defective adenovirus containing a secreted form of CNTF or NT3 show enhanced support of neuronal population *in vitro*. *Exp. Neurol.* **139**, 156–166.

Smith, T. A. G., Mehaffey, M. G., Kayda, D. B., Saunders, J. M., Yei, S., Trapnell, B. C., McClelland, A., and Kaleko, M. (1993). Adenovirus mediated expression of therapeutic plasma levels of human factor IX in mice. *Nat. Genet.* **5**, 397–402.

Smith, T. A. G., White, B. D., Gardner, J. M., Kaleko, M., and McClelland, A. (1996). Transient immunosuppression permits successful repetitive intravenous administration of an adenovirus vector. *Gene Ther.* **3**, 496–502.

Smythe, W. R., Kaiser, L. R., Hwang, H. C., Amin, K. M., Pilewski, J. M., Eck, S. J., Wilson, J. M., and Albelda, S. M. (1994). Successful adenovirus-mediated gene transfer in an *in vivo* model of human malignant mesothelioma. *Ann. Thoracic Surg.* **57**, 1395–1401.

Smythe, W. R., Hwang, H. C., Elshami, A. A., Amin, K. M., Eck, S. L., Davidson, B. L., Wilson, J. M., Kaiser, L. R., and Albelda, S. M. (1995). Treatment of experimental human mesothelioma using adenovirus transfer of the herpes simplex thymidine kinase gene. *Ann. Surg.* **222**, 78–86.

Spady, D. K., Cuthbert, J. A., Willard, M. N., and Meidell, R. S. (1995). Adenovirus-mediated transfer of a gene encoding cholesterol 7α-hydroxylase into hamsters increases hepatic enzyme activity and reduces plasma total and low density lipoprotein cholesterol. *J. Clin. Invest.* **96**, 700–709.

Spector, D. J., and Samaniego, L. A. (1995). Construction and isolation of recombinant adenoviruses with gene replacements. *Methods Mol. Genet.* **7**, 31–44.

Spergel, J. M., Hsu, W., Akira, S., Thimmappaya, B., Kishimoto, T., and Chen-Kiang, S. (1992). NF-IL6, a member of the C/EBP family, regulates E1A-responsive promoters in the absence of E1A. *J. Virol.* **66**, 1021–1030.

Steg, P. G., Feldman, L. J., Scoazec, J. Y., Tahlil, O., Barry, J. J., Boulechfar, S., Ragot, T., Isner, J. M., and Perricaudet, M. (1994). Arterial gene transfer to rabbit endothelial and smooth muscle cells using percutaneous delivery of an adenoviral vector. *Circulation* **90**, 1648–1656.

Stevenson, S. C., Rollence, M., White, B., Weaver, L., and McClelland, A. (1995). Human adenovirus serotypes 3 and 5 bind to two different cellular receptors via the fiber head domain. *J. Virol.* **69**, 2850–2857.

Stewart, P. L., Burnett, R. M., Cyrklaff, M., and Fuller, S. D. (1991). Image reconstruction reveals the complex molecular organization of adenovirus. *Cell (Cambridge, Mass.)* **67**, 145–154.

Stewart, P. L., Fuller, S. D., and Burnett, R. M. (1993). Difference imaging of adenovirus: Bridging the resolution gap between X-ray crystallography and electron microscopy. *EMBO J.* **12**, 2589–2599.

St. George, J. A., Pennington, S. E., Kaplan, J. M., Peterson, P. A., Kleine, L. J., Smith, A. E., and Wadsworth, S. C. (1996). Biological response of nonhuman primates to long-term repeated lung exposure to Ad2/CFTR-2. *Gene Ther.* **3**, 103–116.

Stratford-Perricaudet, L. D., Levrero, M., Chasse, J.-F., Perricaudet, M., and Briand, P. (1990). Evaluation of the transfer and expression in mice of an enzyme-encoding gene using a human adenovirus vector. *Hum. Gene Ther.* **1**, 241–256.

Stratford-Perricaudet, L. D., Makeh, I., Perricaudet, M., and Briand, P. (1992). Widespread long-term gene transfer to mouse skeletal muscles and heart. *J. Clin. Invest.* **90**, 626–630.

Sullivan, D. M., Chung, D. C., Anglade, E., Nussenblatt, R. B., and Csaky, K. G. (1996). Adenovirus mediated gene transfer of ornithine δ-aminotransferase in cultured retinal pigment epithelial cells. *Invest. Ophthalmol. Visual Sci.* **37**, 766–774.

Tang, D. C., Johnston, S. A., and Carbone, D. P. (1994). Butyrate-inducible and tumor-restricted gene expression by adenoviral vectors. *Cancer Gene Ther.* **1**, 15–20.

Teng, B., Blumenthal, S., Forte, T., Navaratnam, N., Scott, J., Gotto, A. M., and Chan, L. (1994). Adenovirus-mediated gene transfer of rat apolipoprotein B mRNA-editing protein in mice virtually eliminates apolipoprotein B-100 and normal low density lipoprotein production. *J. Biol. Chem.* **269**, 29395–29404.

Teramoto, S., Johnson, L. G., Huang, W., Leigh, M. W., and Boucher, R. C. (1995). Effect of adenoviral vector infection on cell proliferation in cultured primary human airway epithelial cells. *Hum. Gene Ther.* **6**, 1045–1053.

Thimmappaya, B., Weinberger, C., Scheider, R. J., and Shenk, T. (1982). Adenovirus VAI RNA is required for efficient translation of viral mRNAs at late times after infection. *Cell (Cambridge, Mass.)* **31**, 543–551.

Thummel, C., Tjian, R., and Grodzicker, T. (1982). Construction of adenovirus expression vectors by site-directed *in vivo* recombination. *J. Mol. Appl. Genet.* **1**, 435–446.

Thummel, C., Tjian, R., Hu, S.-L., and Grodzicker, T. (1983). Translational control of SV40 T antigen expressed from the adenovirus late promoter. *Cell (Cambridge, Mass.)* **33**, 455–464.

Tracey, K. J., and Cerami, A. (1993). Tumor necrosis factor: An updated review of its biology. *Crit. Care Med.* **21**, S415–S422.

Trapnell, B. C. (1993). Adenoviral vectors for gene transfer. *Adv. Drug Delivery Rev.* **12**, 185–199.

Tripathy, S. K., Goldwasser, E., Lu, M.-M., Barr, E., and Leiden, J. M. (1994). Stable delivery of physiologic levels of recombinant erythropoeitin to the systemic circulation by intramuscular injection of replication-defective adenovirus. *Proc. Natl. Acad. Sci. U.S.A.* **91**, 11557–11561.

Tripathy, S. K., Black, H. B., Goldwasser, E., and Leiden, J. M. (1996). Immune responses to transgene-encoded proteins limit the stability of gene expression after injection of replication-defective adenovirus vectors. *Nat. Med.* **2**, 545–550.

Ueno, T., Miyamura, T., Saito, I., and Mizuno, K. (1993). Immortalization of differentiated human hepatocytes by a combination of a viral vector and collagen gel culture. *Hum. Cell* **6**, 126–136.

Van Doren, K., and Gluzman, Y. (1984). Efficient transformation of human fibroblasts by adenovirus-simian virus 40 recombinants. *Mol. Cell. Biol.* **4**, 1653–1656.

Van Doren, K., Hanahan, D., and Gluzman, Y. (1984). Infection of eucaryotic cells by helper-independent recombinant adenoviruses: Early region 1 is not obligatory for integration of viral DNA. *J. Virol.* **50**, 606–614.

Varley, A. W., Coulthard, M. G., Meidell, R. S., Gerard, R. D., and Munford, R. S. (1995). Inflammation-induced recombinant protein expression *in vivo* using promoters from acute-phase protein genes. *Proc. Natl. Acad. Sci. U.S.A.* **92**, 5346–5350.

Verhaagen, J., Hermens, W. T. J. M. C., Holtmaat, A. J. G. D., Oestreicher, A. B., Gispen, W. H., and Kaplitt, M. G. (1995). Viral vector-mediated gene transfer in the nervous system: Application to the reconstruction of neural circuits in the injured mammalian brain. *In* "Viral Vectors," pp. 119–132. Academic Press, San Diego, CA.

Vilquin, J.-T., Guerette, B., Kinoshita, I., Roy, B., Goulet, M., Gravel, C., Roy, R., and Tremblay, J. P. (1995). FK506 immunosuppression to control the immune reactions triggered by first-generation adenovirus-mediated gene transfer. *Hum. Gene Ther.* **6**, 1391–1401.

Vincent, A. J. P. E., Vogels, R., Someren, G., Esandi, M. C., Noteboom, J. L., Avezaat, C. J. J., Vecht, C., Bekkum, D. W., Valerio, D., Bout, A., and Hoogerbrugge, P. M. (1996). Herpes simplex virus thymidine kinase gene therapy for rat malignant brain tumors. *Hum. Gene Ther.* **7**, 197–206.

Vincent, M. C., Trapnell, B. C., Baughman, R. P., Wert, S. E., Whitsett, J. A., and Iwamoto, H. S. (1995). Adenovirus-mediated gene transfer to the respiratory tract of fetal sheep *in utero*. *Hum. Gene Ther.* **6**, 1019–1028.

Vincent, N., Ragot, T., Gilgenkrantz, H., Couton, D., Chafey, P., Grégoire, A., Briand, P., Kaplan, J.-C., Kahn, A., and Perricaudet, M. (1993). Long term correction of mouse dystrophic degeneration by adenovirus-mediated transfer of a minidystrophin gene. *Nat. Genet.* **5**, 130–134.

Wang, P., Anton, M., Graham, F. L., and Bacchetti, S. (1995). High frequency recombination between *loxP* sites in human chromosomes mediated by an adenovirus vector expressing Cre recombinase. *Somatic Cell Mol. Genet.* **21**, 429–441.

Wang, Q., and Taylor, M. W. (1993). Correction of a deletion mutant by gene targeting with an adenovirus vector. *Mol. Cell. Biol.* **13**, 918–927.

Welsh, M. J., Zabner, J., Graham, S. M., Smith, A. E., Moscicki, R., and Wadsworth, S. (1995). Adenovirus-mediated gene transfer for cystic fibrosis: Part A. Safety of dose and repeat administration in the nasal epithelium. Part B. Clinical efficacy in the maxillary sinus. *Hum. Gene Ther.* **6**, 205–218.

White, E., Cipriani, R., Sabbatini, P., and Denton, A. (1991). Adenovirus E1B 19-kilodalton protein overcomes the cytotoxicity of E1A proteins. *J. Virol.* **65**, 2968–2978.

Wickham, T. J., Mathias, P., Cheresh, D. A., and Nemerow, G. R. (1993). Integrins $\alpha_v\beta_3$ and $\alpha_v\beta_5$ promote adenovirus internalization but not virus attachment. *Cell (Cmabridge, Mass.)* **73**, 309–319.

Wilson, G. M., and Graham, F. L. (1989). The effect of E1 mutations on biochemical transformation by an adenovirus carrying the herpes simplex virus thymidine kinase gene in region E3. *Virus Res.* **13**, 29–44.

Wilson, J. M., principal investigator (1994). Clinical protocol: Gene therapy of cystic fibrosis lung disease using E1 deleted adenoviruses: A phase I trial. *Hum. Gene Ther.* **5**, 501–519.

Witmer, L. A., Rosenthal, K. L., Graham, F. L., Friedman, H. M., Yee, A., and Johnson, D. C. (1990). Cytotoxic T lymphocytes specific for herpes simplex virus (HSV) studied using adenovirus vectors expressing HSV glycoproteins. *J. Gen. Virol.* **71**, 387–396.

Wold, W. S., and Gooding, L. R. (1989). Adenovirus region E3 proteins that prevent cytolysis by cytotoxic T cells and tumor necrosis factor. *Mol. Biol. Med.* **6**, 433–452.

Wrighton, C. J., Hofer-Warbinek, R., Moll, T., Eytner, R., Bach, F. H., and deMartin, R. (1996). Inhibition of endothelial cell activation by adenovirus-mediated expression of $I\kappa B\alpha$, an inhibitor of the transcription factor NF-κB. *J. Exp. Med.* **183**, 1013–1022.

Xing, Z., Braciak, T., Jordana, M., Croitoru, K., Graham, F. L., and Gauldie, J. (1994). Adenovirus-mediated cytokine gene transfer at tissue sites. Overexpression of IL-6 induces lymphocytic hyperplasia in the lung. *J. Immunol.* **153**, 4059–4069.

Xing, Z., Braciak, T., Ohkawara, Y., Sallenave, J.-M., Foley, R., Sime, P. J., Jordana, M., Graham, F. L., and Gauldie, J. (1996a). Gene transfer for cytokine functional studies in the lung: The multifunctional role of GM-CSF in pulmonary inflammation. *J. Leukocyte Biol.* **59**, 481–488.

Xing, Z., Ohkawara, Y., Jordana, M., Graham, F. L., and Gauldie, J. (1996b). Transfer of granulocyte-macrophage colony-stimulating factor gene to rat lung induces eosinophilia, monocytosis, and fibrotic reactions. *J. Clin. Invest.* **97**, 1102–1110.

Xiong, Y., Hannon, G. J., Zhang, H., Casso, D., Kobayashi, R., and Beach, D. (1993). p21 is a universal inhibitor of cyclin kinase. *Nature (London)* **366**, 701–704.

Xu, Z. Z., Krougliak, V., Prevec, L., Graham, F. L., and Both, G. W. (1995). Investigation of promoter function in human and animal cells infected with human recombinant adenovirus expressing rotavirus antigen. VP7sc. *J. Gen. Virol.* **76**, 1971–1980.

Yamada, M., Lewis, J. A., and Grodzicker, T. (1985). Overproduction of the protein product of a non-selected foreign gene carried by an adenovirus vector. *Proc. Natl. Acad. Sci. U.S.A.* **82**, 3567–3571.

Yang, C., Cirielli, C., Capogrossi, M. C., and Passaniti, A. (1995). Adenovirus-mediated wild-type p53 expression induces apoptosis and suppresses tumorigenesis of prostatic tumor cells. *Cancer Res.* **55**, 4210–4213.

Yang, Y., and Wilson, J. M. (1995). Clearance of adenovirus-infected hepatocytes by MHC class I-restricted CD4+ CTLs *in vivo. J. Immunol.* **155**, 2564–2570.

Yang, Y., Raper, S. E., Cohn, J. A., Engelhardt, J. F., and Wilson, J. M. (1993). An approach for treating the hepatobiliary disease of cystic fibrosis by somatic gene transfer. *Proc. Natl. Acad. Sci. U.S.A.* **90**, 4601–4605.

Yang, Y., Nunes, F. A., Berencsi, K., Furth, E. F., Gonczol, E., and Wilson, J. M. (1994a). Cellular immunity to viral antigens limits E1-deleted adenoviruses for gene therapy. *Proc. Natl. Acad. Sci. U.S.A.* **91**, 4407–4411.

Yang, Y., Ertl, H. C., and Wilson, J. M. (1994b). MHC class I-restricted cytotoxic T lymphocytes to viral antigens destroy hepatocytes in mice infected with E1-deleted recombinant adenoviruses. *Immunity* **1**, 433–442.

Yang, Y., Nunes, F. A., Berencsi, K., Gonczol, E., Engelhardt, J. F., and Wilson, J. M. (1994c). Inactivation of E2a in recombinant adenoviruses improves the prospect for gene therapy in cystic fibrosis. *Nat. Genet.* **7**, 362–369.

Yang, Y., Li, Q., Ertl, H. C. J., and Wilson, J. M. (1995a). Cellular and humoral immune responses to viral antigens create barriers to lung-directed gene therapy with recombinant adenoviruses. *J. Virol.* **69**, 2004–2015.

Yang, Y., Trinchieri, G., and Wilson, J. M. (1995b). Recombinant IL-12 prevents formation of blocking IgA antibodies to recombinant adenovirus and allows repeated gene therapy to mouse lung. *Nat. Med.* **1**, 890–893.

Yang, Y., Jooss, K. U., Su, Q., Ertl, H. C. J., and Wilson, J. M. (1996). Immune responses to viral antigens vs. transgene product in the elimination of recombinant adenovirus infected hepatocytes *in vivo. Gene Ther.* **3**, 137–144.

Yang, Z.-Y., Perkins, N. D., Ohno, T., Nabel, E. G., and Nabel, G. J. (1995). The p21 cyclin-dependent kinase inhibitor suppresses tumorigenicity *in vivo. Nat. Med.* **1**, 1052–1056.

Yarosh, O. K., Wandeler, A. I., Graham, F. L., Campbell, J. B., and Prevec, L. (1996). Human adenovirus type 5 vectors expressing rabies glycoprotein. *Vaccine* **14**, 1257–1264.

Yeh, P., Dedieu, J. F., Orsini, C., Vigne, E., Denefle, P., and Perricaudet, M. (1996). Efficient dual transcomplementation of adenovirus E1 and E4 regions from a 293-derived cell line expressing a minimal E4 functional unit. *J. Virol.* **70**, 559–565.

Yei, S., Mittereder, N., Wert, S., Whitsett, J. A., Wilmott, R. W., and Trapnell, B. C. (1994a). *In vivo* evaluation of the safety of adenovirus-mediated transfer of the human cystic fibrosis transmembrane conductance regulator cDNA to the lung. *Hum. Gene Ther.* **5**, 731–744.

Yei, S., Mittereder, N., Tang, K., O'Sullivan, C., and Trapnell, B. C. (1994b). Adenovirus-mediated gene transfer for cystic fibrosis: Quantitative evaluation of repeated *in vivo* vector administration to the lung. *Gene Ther.* **1**, 192–200.

Yei, S., Bachurski, C. J., Weaver, T. E., Wert, S. E., Trapnell, B. C., and Whitsett, J. A. (1994c). Adenoviral-mediated gene transfer of human surfactant protein B to respiratory epithelial cells. *Am. J. Respir. Cell Mol. Biol.* **11**, 329–336.

Yoo, J.-H., Erzurum, S. C., Hay, J. G., Lemarchand, P., and Crystal, R. G. (1994). Vulnerability of the human airway epithelium to hyperoxia. *J. Clin. Invest.* **93**, 297–302.

York, I. A., Roop, C., Andrews, D. W., Riddell, S. R., Graham, F. L., and Johnson, D. C. (1994). A cytosolic herpes simplex virus protein inhibits antigen presentation to CD8+ T lymphocytes. *Cell (Cambridge, Mass.)* **77**, 525–535.

Yoshimura, K., Rosenfeld, M. A., Seth, P., and Crystal, R. G. (1993). Adenovirus-mediated augmentation of cell transfection with unmodified plasmid vectors. *J. Biol. Chem.* **268**, 2300–2303.

Yu, D., Matin, A., Xia, W., Sorgi, F., Huang, L., and Hung, M.-C. (1995). Liposome-mediated *in vivo* E1A gene transfer suppressed dissemination of ovarian cancer cells that overexpress HER-2/neu. *Oncogene* **11**, 1383–1388.

Zabner, J., Couture, L. A., Gregory, R. J., Graham, S. M., Smith, A. E., and Welsh, M. J. (1993). Adenovirus-mediated gene transfer transiently corrects the chloride transport defect in nasal epithelia of patients with cystic fibrosis. *Cell (Cambridge, Mass.)* **75**, 207–216.

Zabner, J., Petersen, D. M., Puga, A. P., Graham, S. M., Couture, L. A., Keyes, L. D., Lukason, M. J., St. George, J. A., Gregory, R. J., Smith, A. E., and Welsh, M. J. (1994a). Safety and efficacy of repetitive adenovirus-mediated transfer of CFTR cDNA to airway epithelia of primates and cotton rats. *Nat. Genet.* **6**, 75–83.

Zabner, J., Couture, L. A., Smith, A. E., and Welsh, M. J. (1994b). Correction of cAMP-stimulated fluid secretion in cystic fibrosis airway epithelia: Efficiency of adenovirus-mediated gene transfer *in vitro. Hum. Gene Ther.* **5**, 585–593.

Zhang, J.-F., Hu, C., Geng, Y., Selm, J., Klein, S. B., Orazi, A., and Taylor, M. W. (1996). Treatment of a human breast cancer xenograft with an adenovirus vector containing an interferon gene results in rapid regression due to viral oncolysis and gene therapy. *Proc. Natl. Acad. Sci. U.S.A.* **93**, 4513–4518.

Zhang, W.-W., Alemany, R., Wang, J., Koch, P. E., Ordonez, N. G., and Roth, J. A. (1995). Safety evaluation of Ad5CMV-p53 *in vitro* and *in vivo*. *Hum. Gene Ther.* **6**, 155–164.

Zhang, Y., Yu, D., Xia, W., and Hung, M.-C. (1995). HER-2/neu-targeting cancer therapy via adenovirus-mediated E1A delivery in an animal model. *Oncogene* **10**, 1947–1954.

Zhu, X., Young, C. S. H., and Silverstein, S. (1988). Adenovirus vector expressing functional herpes simplex virus ICP0. *J. Virol.* **62**, 4544–4553.

Zsengeller, Z. K., Wert, S. E., Hull, W. M., Hu, X., Yei, S., Trapnell, B. C., and Whitsett, J. A. (1995). Persistence of replication-deficient adenovirus-mediated gene transfer in lungs of immune-deficient (nu/nu) mice. *Hum. Gene Ther.* **6**, 457–467.

Akira Irie[*,†]
Hiroshi Kijima[*]
Tsukasa Ohkawa[*,†]
David Y. Bouffard[*,†]
Toshiya Suzuki[*,†]
Lisa D. Curcio[‡]
Per Sonne Holm[*]
Alex Sassani[*]
Kevin J. Scanlon[*,†]

[*]Section of Biochemical Pharmacology
Department of Medical Oncology
City of Hope National Medical Center
Duarte, California 91010

[†]Department of Cancer Research
Berlex Biosciences
Richmond, California 94804

[‡]Department of General and Oncologic Surgery
City of Hope National Medical Center
Duarte, California 91010

Anti-oncogene Ribozymes for Cancer Gene Therapy

I. Introduction

Ribozymes are small RNA molecules that possess specific catalytic activities and are being actively investigated for their therapeutic applications in the field of gene therapy. Ribozymes were initially discovered in the group I intervening sequence in the pre-rRNA of *Tetrahymena thermophilia*. This intervening sequence catalyzes its own excision, and has been called self-splicing (Cech *et al.*, 1981; Kruger *et al.*, 1982). The RNA portion of the RNase P enzyme purified from *Escherichia coli*, which cleaves molecules with the capacity for multiple turnovers, was reported as the first truly catalytic ribozyme (Guerrier-Takada *et al.*, 1983). In addition, self-cleaving reactions, which are characteristic of ribozymes, have been identified in the satellite RNA of the tobacco ring spot virus (sTobRV) (Buzayan *et al.*, 1986), the avocado sun blotch viroid (Hutchins *et al.*, 1986), and the virusoid lucerne transient streak virus (Forster and Symons, 1987). Additional studies

Advances in Pharmacology, Volume 40

have demonstrated that the plus strand of sTobRV possesses a hammerhead shape, and that the minus strand exhibits a hairpin shape. The human hepatitis δ virus (HDV) (Branch and Robertson, 1991) and a ribosomal RNA (Symons, 1992) have been reported to exhibit catalytic capabilities as ribozymes.

Ribozymes were originally found to act in *cis* for RNA cleavage, hammerhead ribozymes have been demonstrated to cleave their targets in *trans* in a truly catalytic manner (Uhlenbeck, 1987). Hairpin ribozymes were also shown to cleave their target RNAs in *trans* (Hampel *et al.*, 1990). Highly specific activities of hammerhead ribozymes were demonstrated and shown to inhibit the expression of specific genes by targeting their mRNAs (Haseloff and Gerlach, 1988). Ribozymes can be designed to target the mRNA for any disease process in which a specific protein has been linked to its etiology. Certain diseases may be caused by the undesirable expression of RNA. These diseases are candidates for ribozyme therapeutics and include neoplastic disorders and viral illnesses, especially those associated with human immunodeficiency virus type 1 (HIV-1) and chronic hepatitis B virus. There are several current reviews on ribozymes: for an overview of ribozymes (Bratty *et al.*, 1993; Castanotto *et al.*, 1994; Christoffersen and Marr, 1995), for applications of ribozymes (Cech, 1988; Kijima *et al.*, 1995), and for gene therapy (Sullivan, 1994; Kashani-Sabet and Scanlon, 1995).

Since the discovery of the protooncogene (Stehelin *et al.*, 1976), cancer has been defined as a genetic disease. Therefore, gene therapy could be a rational and promising strategy for the treatment of specific cancers. Extensive research has helped to clarify the mechanisms of tumorigenesis. Several oncogenes (Weinberg, 1989; Bishop, 1991) and suppressor genes (Weinberg, 1992) have been identified and linked with specific malignant processes. Cancer has been viewed as a multistep process that includes molecular alterations in the genome affected by tumor suppressor genes and/or oncogenes (Fearon and Vogelstein, 1990). Quiescent (latent) oncogenes may be activated by mutation, gene amplification, overexpression, or chromosomal translocation. This mutation subsequently alters the property of the corresponding protein and induces uncontrolled cell growth. The interference in the flow of information from the gene to the protein either during transcription or translation may be detrimental to cell growth. Various target sites and mechanisms of action have been investigated for their potential role as modulators of oncogene information. Several studies have demonstrated the reversion of a malignant phenotype by the elimination of a single oncogene or by restoration of a single suppressor gene (Kashani-Sabet *et al.*, 1992; Roth *et al.*, 1994). Examples of these modulators are ribozymes, DNA or RNA antisense, antigene oligonucleotides, aptamers (Table I; Scanlon *et al.*, 1995; Stull and Szoka, 1995), and tumor suppressor genes (Zhang *et al.*, 1995). The mRNA of activated oncogene can be viewed as an accessible and suitable target to interfere with the genetic basis of malignancy. Also,

TABLE I   Nucleic Acid Agents[a]

| | *Strategy* | | | |
	Ribozyme	*Antisense*	*Antigene*	*Aptamer*
Structure	RNA	DNA or RNA	DNA or RNA	DNA or RNA
Target	mRNA	mRNA	DNA	Protein
Mechanism	mRNA cleavage	Translation arrest; RNase H activation	Triplex formation; blockage of transcription	Binding; alteration of function

[a] Structure, target sites, and mechanisms of action for each nucleic acid agent are shown. Antigenes are nucleic acids targeting genomic DNA. Aptamers are nucleic acids that have the ability to bind proteins or small molecules.

activated oncogenes produce mRNAs that are distinguishable from those of their protooncogenes; thus mutated mRNAs may be inhibited selectively by ribozymes. The crucial role of ribozymes would be to inhibit the transfer of information from the gene to the protein by interfering with its intermediate mRNA function. Antisense oligonucleotides, like ribozymes, also interfere with the transfer of information from the mRNA to the protein. Several studies have demonstrated the superior efficacy of ribozymes over antisense oligonucleotides (Cameron and Jennings, 1989; Sioud *et al.*, 1992; Homann *et al.*, 1993). Also, ribozymes have been shown to be better inhibitors of gene expression than are noncatalytic RNAs, i.e., mutant ribozymes (Sioud and Drlica, 1991; Scanlon *et al.*, 1991a; Tone *et al.*, 1993; Kashani-Sabet *et al.*, 1994; Funato *et al.*, 1994). Presently, the use of ribozymes in the realm of cancer has been focused mainly on the inhibition of tumor-specific oncogene expression. Anti-oncogene ribozymes are currently being tested as potential agents for cancer gene therapy. This chapter focuses on a general overview of ribozymes and their applications for cancer gene therapy.

II. Biochemistry of Ribozymes

There are two general mechanisms for specifically inactivating RNAs. The first mechanism is termed *sterile blocking;* this describes the binding of an antisense compound to inhibit the translation or metabolism of RNA (Melton, 1985; Boiziau *et al.*, 1991). The second generally recognized mechanism is through the cleavage of the targeted RNA. A significant amount of interest has focused on ribozymes as more potent alternatives to antisense. Ribozymes are RNA molecules, typically small, that can catalyze a chemical cleavage reaction in the absence of a protein. Ribozymes were found to work in *cis* and subsequently have been engineered to work in *trans*. This modification has prompted their application to broader areas in gene regulation, especially in the field of gene therapy. Six types of RNA catalytic motifs are known: group I introns, RNase P, hammerhead ribozyme, hairpin ribozyme, and axehead ribozymes of the HDV and RNA transcripts of the mitochondrial DNA plasmid of *Neurospora* (Symons, 1994; Kijima *et al.*, 1995).

The catalytic activity of ribozymes has been demonstrated to occur through the rearrangement of phosphodiester bonds (van Tol *et al.*, 1990; Kumar and Ellington, 1995). The hammerhead, hairpin, and HDV motifs, which are typically found in virus or viroid RNAs, catalyze the cleavage of a particular phosphodiester bond by activating the adjacent 2′-hydroxyl to form a cyclic 2′,3′-phosphate with concomitant elimination of a new 5′ terminus. This helps viral replication by cleaving the newly synthesized multimers into monomers (Fig. 1). Ribozyme-mediated site-specific ligation as well as site-specific cleavage have been demonstrated by Sullenger and Cech (1994). They have shown the ability of a ribozyme to replace a defective portion of RNA. The functional utility of ribozyme has been increasing and may provide new approaches in the treatment of genetic disease. In the following sections the biochemistry of ribozymes is discussed in the context of their respective cleavage activity.

A. Hammerhead Ribozyme

I. Hammerhead Ribozyme

The hammerhead ribozyme (Fig. 2A) is found in the sTobRV RNA that cleaves itself for efficient replication (Gerlach *et al.*, 1987), and works in *trans*

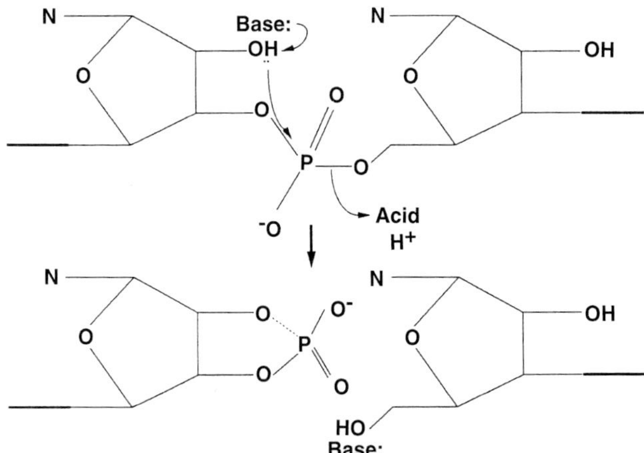

FIGURE 1   Mechanism of cleavage by hammerhead, hairpin, and HDV ribozymes (only the cleavage site of the substrate is shown). Ribozymes cleave a particular phosphodiester bond by activating the adjacent 2′-hydroxyl as a nucleophile. During the reaction a 5′-hydroxyl is displaced and a 2′,3′-cyclic phosphodiester is formed. [Modified from Kumar and Ellington, 1995.]

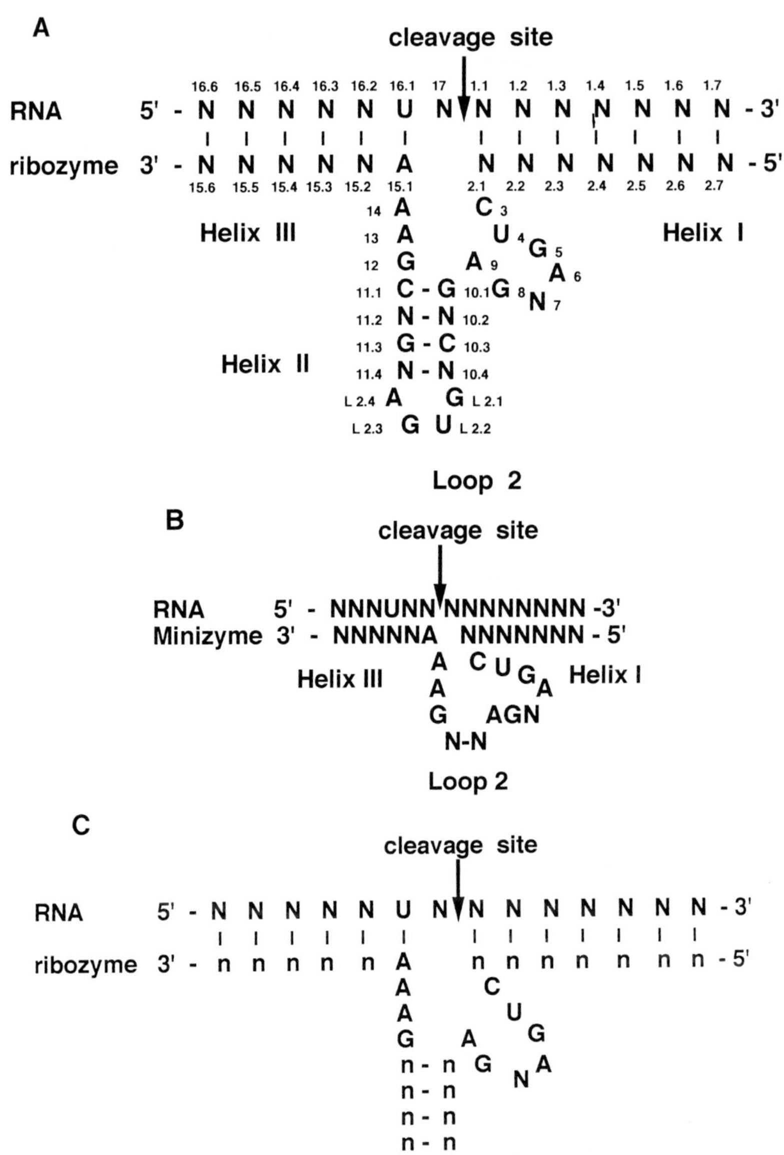

FIGURE 2   (A) Structure of hammerhead ribozyme. N, Any base. (B) Structure of minizyme. (C) Structure of chimeric ribozyme. N, RNA; n, DNA substituted for RNA.

(Uhlenbeck, 1987; Haseloff and Gerlach, 1988) through the nonhydrolytic transesterification of its substrates in the presence of divalent cations (Fig. 1; Buzayan *et al.*, 1988; van Tol *et al.*, 1990; Ohkawa *et al.*, 1995). A unified numbering system for the hammerhead ribozyme has been published in order to simplify data comparison (Hertel *et al.*, 1992). Mutational analysis, structural studies, and other biochemical experiments have determined consensus sequences, as well as structural and kinetic characteristics, for hammerhead ribozymes (Cech and Uhlenbeck, 1994). The three-dimensional conformation of ribozymes allows them to perform these catalytic reactions (Tuschl *et al.*, 1994). The consensus sequences of the hammerhead ribozyme and of its target RNA are also shown in Fig. 2A (Hertel *et al.*, 1992; Bratty *et al.*, 1993). The secondary structure of the ribozyme–substrate complex consists of the three helical regions, a catalytic core region (i.e., hammerhead domain), and a loop sequence. The ribozyme binds to its target RNA through two helices (helix I and helix III); the catalytic core region separates the two helices (helix I and II), as well as helix III formed by four complementary base pairs (i.e., eight nucleotides). The internal loop (loop 2) of four nucleotides is at the opposite end of helix II. Cleavage occurs in the 3′ site of the N^{17} residue. The hammerhead ribozyme can significantly discriminate substrate RNAs with a single base mutation (Koizumi *et al.*, 1989), as well as closely related RNAs (Bennett and Cullimore, 1992). Early studies of mutational analysis demonstrated that the cleavage reaction requires the $N^{16.2}U^{16.1}H^{17}$ triplet sequences 3′ of the cleavage site (N being any nucleotide and H being A, C, or U; Haseloff and Gerlach, 1988; Koizumi *et al.*, 1989; Ruffner *et al.*, 1990). Usually, the substrates containing the GUC, GUA, GUU, UUC, or CUC triplets are efficiently cleaved, and the GAC, GUG, AUC, CGC, GGC, AGC, or UGC triplet sequences are poorly cleaved (Perriman *et al.*, 1992). The AUA triplet is not cleaved; however, a hammerhead ribozyme with $G^{10.1}$–$C^{11.1}$ and a pyrimidine at position 7 has shown some efficiency for AUA cleavage (Nakamaye and Eckstein, 1994). X-Ray crystallography has shown that the catalytic core, consisting of a conserved $C^{3}U^{4}G^{5}A^{6}$, is involved in the sharp turn of the ribozyme strand at the base of helix I. This sharp turn is identical to the uridine turn of tRNA, suggesting the uniqueness of the hammerhead ribozyme (Pley *et al.*, 1994).

Several biochemical conditions are required for effective cleavage of the hammerhead ribozyme. Studies using DNA/RNA chimeric nucleotides have demonstrated that the cleavage reaction requires the 2′-hydroxyl (2′-OH) groups in the catalytic core region, especially at N^{17}, as well as $U^{16.1}$ (Perreault *et al.*, 1990; Yang *et al.*, 1990). The importance of the 2′-OH groups at $A^{15.1}$, G^{5}, G^{8}, and A^{9} has been shown for ribozyme catalysis (Perreault *et al.*, 1991; Yang *et al.*, 1992). Furthermore, other chemical modifications of ribozymes and their cleavage activities have been examined (Heidenreich *et al.*, 1993), including the modification of 2′-fluoro- and 2′-amino nucleotides (Pieken *et al.*, 1991), 2′-O-allyl and 2′-O-methyl nucleotides (Paolella *et al.*, 1992), 2′-O-methylation of flanking sequences (Goodchild, 1992), 2′-pyrimidine modifications such as 2-amino-2′-deoxyuridines (Heidenreich *et al.*, 1994), and isoguanosine substitution of conserved adenosines (A^{6}, A^{9}, A^{13}, $A^{15.1}$; Ng *et al.*, 1994). The successful chemical modification of a ribozyme has been reported. This is accomplished by substituting several RNA nucleotides simultaneously, including $2'-NH_2$ substitutions at U^4 and U^7, $2'-C$-allyl substitutions at U^4, and $2'-O$-methyl substitution at U^7 of the all-RNA ribozyme. The addition of a $3',3'$-linked thymidine to the ribozyme does not decrease the catalytic activity but increases nuclease resistance by 53,000- to 80,000-fold in serum in comparison with the all-RNA parent ribozyme (Beigelman *et al.*, 1995).

The *in vitro* kinetics of target RNA and cleavage reactions by hammerhead ribozymes have been studied (Heidenreich and Eckstein, 1992). With multiple turnover reactions, the rate of enzymatic kinetics using short substrates (25 bases) is about 11–400 nM. In contrast, low cleavage efficiencies have been reported using longer substrates (200 to 1000 bases) as targets of ribozymes. The length and base composition of flanking sequences (i.e., sequences in helices I and III) will affect ribozyme–substrate kinetics. The RNA sequence and secondary structure are also rate determining for maximizing the catalytic efficiency of ribozymes (Fedor and Uhlenbeck, 1990). The dissociation step of the ribozyme–substrate duplex is much slower when the flanking sequences of the ribozyme is increased (Herschlag, 1991). The maximum discrimination is expected to be greater with AU-rich sequences than with GC-rich sequences. Twelve bases have been reported to be the optimal length for the flanking sequences of a ribozyme (Bertrand *et al.*, 1994).

2. Minizyme

McCall and colleagues (1992) have characterized stem II, consisting of helix II and loop 2, of the hammerhead ribozyme in detail and have established the essential construct of the hammerhead ribozyme as a "minizyme" (Fig. 2B). They have made smaller ribozymes by decreasing stem II and replacing RNA with DNA. Cleavage activity of the minizyme is dependent on the number and sequence of nucleotides, and optimal activity is achieved with four or five deoxyribopyrimidines. This study suggests that helix II may not be essential for the active ribozyme structure. Another study reported that stem II with 2 base pairs (bp), rather than the conventional 4 bp, could maintain catalytic activity. However, when stem II was reduced to 1 bp or eliminated, the ribozyme activity was drastically reduced (Tuschl and Eckstein, 1993). In general, minizymes are thought to be less active than their comparable full-size ribozymes; however, minizymes targeted against the TAT transcript of HIV-1 have been reported to be more effective than full-size ribozymes (Hendry *et al.*, 1995). Minizymes are interesting molecules since they are capable of accessing their target sequence in complexly folded RNA and have less potential to bind with proteins that recognize RNA double helix (helix II).

3. DNA/RNA Chimeric Ribozyme

Studies substituting deoxynucleotides for ribonucleotides in the various stems of the hammerhead ribozyme have shown that chimeric ribozymes with DNA in helices I and III have a sixfold greater catalytic efficacy than the all-RNA ribozymes (Fig. 2C). Substitution for DNA in stem II yields a marked reduction in cleavage activity (Taylor *et al.*, 1992). These chimeric ribozymes, when transfected by cationic liposomes into human T lymphocytes, have a higher stability than their all-RNA counterparts. Another group has demonstrated that the cleavage ability of chimeric ribozymes with DNA in helices I and III was enhanced more than threefold as compared to the all-RNA ribozyme (Hendry *et al.*, 1992). Shimayama *et al.* (1993) introduced deoxyribonucleotides with phosphorothioate linkage in the regions of helices I and III and stem II of hammerhead ribozymes. They revealed that this thio-DNA/RNA chimeric ribozyme had a sevenfold higher cleavage activity than the all-RNA ribozyme. This study also suggested that thio-DNA/RNA chimeric ribozyme was more resistant to attack by nuclease than the all-RNA ribozyme *in vivo*. Several studies suggest that replacing the hybridizing arms of ribozymes with DNA could increase the cleavage rate and the overall turnover efficiency. However, one study has revealed that the presence of DNA in the hybridizing arms of hammerhead ribozymes may not enhance the cleavage rate but rather affects the turnover rate, resulting in the stimulation of the overall reaction (Hendry and McCall, 1995). A more recent study has also shown an enhanced rate of catalysis by the DNA hammerhead domain having only a single RNA base ($A^{15.1}$) compared to the all-RNA ribozyme (Chartrand *et al.*, 1995). These DNA/RNA chimeric ribozymes with modified DNA for RNA have the potential to be one of the most suitable structures for the transient applications of hammerhead ribozymes.

B. Hairpin Ribozyme

The hairpin ribozyme (Fig. 3) is derived from the minus strand of sTobRV (Hampel and Tritz, 1989; Feldstein *et al.*, 1989), and site-specifically cleaves its RNA substrates in *trans*. The cleavage reaction of the hairpin ribozyme is a multistep event involving the nonhydrolytic transesterification of the substrate in the presence of divalent cations (van Tol *et al.*, 1990; Kumar and Ellington, 1995). A biochemical difference between the hammerhead and hairpin ribozymes of the sTobRV is that a phosphorothioate modification strongly inhibits self-cleavage of the hammerhead ribozyme but not of the hairpin ribozyme in the presence of divalent cations (Buzayan *et al.*, 1988). Mutational analysis, computer modeling, and phylogenetic studies have proposed secondary structure models for the hairpin ribozyme–substrate complex (Haseloff and Gerlach, 1989; Hampel *et al.*, 1990). The ribozyme–substrate complex consists of four helical regions separated by two internal loop sequences. The substrate binds to the ribozyme through two helices (helix 1 and helix 2). Cleavage occurs at the 5′ side of a guanosine within the internal loop of the substrate (loop A) separating helices 1 and

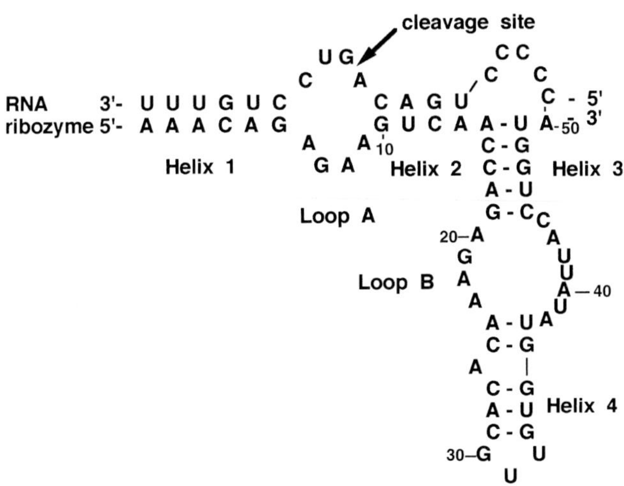

FIGURE 3 Structure of hairpin ribozyme.

2. A second internal loop (loop B) separates the two helices (helix 3 and helix 4) of the ribozyme.

Studies have described the essential nucleotide sequence for the catalytic activity of the hairpin ribozyme. Most of the nucleotides within loops A and B are essential (Chowrira *et al.*, 1991; Berzal-Herranz *et al.*, 1993). Within loop A, there are four essential bases: three in the ribozyme (G^8, A^9, and A^{10}; base number beginning at the RNA 5' end) and one in the substrate (G^6). Within loop B, 9 of 11 bases (all except A^{20} and U^{39}) are essential. In contrast, only one base (G^{11}) within the four helices is important for catalytic activity (Joseph *et al.*,1993). A structural bend of the ribozyme occurs at, or near, the A^{13}–A^{14} linkage (Feldstein and Bruening, 1993), and loop B is three-dimensionally present near the cleavage site within loop A. One study has reported that 2'-OH groups at A^{10}, G^{11}, A^{24}, and C^{25} are essential for catalysis because deletion of these hydroxyl groups results in severe inhibition of cleavage activity (Chowrira *et al.*, 1993). In addition, ultraviolet (UV)-sensitive tertiary structure has been identified within loop B (Butcher and Burke, 1994). The structure is thought to be a RNA-folding domain because it is similar to those found in several other RNA molecules, including loop E of eukaryotic 5S rRNA. From these studies, not only loop A, but also loop B, of the hairpin ribozyme may play an important role in its catalytic activity.

In vitro selection studies have described the substrate-targeting sequence $R^3Y^4N^5G^6H^7Y^8B^9$ (cleavage between $N^5 G^6$), as well as the ribozyme sequence $V^6A^7G^8A^9A^{10}G^{11}Y^{12}$ (R being A or G; Y being C or U; H being C, U, or A; B being C, U, or G; V being C, A, or G; N being C, U, A, or G) (Joseph *et al.*, 1993). On the other hand, *in vivo* mutagenesis studies have reported that the substrate requires only the $B^4N^5G^6U^7C^8$ sequence of the substrate, while the ribozyme–substrate complex preserves its original two-dimensional structure (Anderson *et al.*, 1994). These studies suggest the possibility of designing site-specific hairpin ribozymes against various substrate sequences to modulate target RNA expression.

C. Hepatitis δ Virus Axehead Ribozyme

Hepatitis δ virus is a satellite virus of the hepatitis B virus and contains a single minus-stranded RNA of about 1700 bp; the plus-strand (complementary) RNA has a coding region for 195 amino acids of the HDV antigen within the viral particle (Wang *et al.*, 1986; Taylor, 1990). The HDV RNA replicates by a rolling circle mechanism; the self-cleaving domains play a significant role in this mechanism and require divalent cations (Chen *et al.*, 1986; Wu *et al.*, 1989; Wu and Lai, 1990). Branch and Robertson (1991) have reported axehead structures for the HDV self-cleaving domains. The conserved sequences between the negative (genomic) strand and the positive (antigenomic) strand are demonstrated in Fig. 4. Recently, several studies have described detailed structural analysis and cleavage in trans of the ribozyme derived from the HDV sequence (Perrotta and Been, 1992; Thill *et al.*, 1993; Kumar *et al.*, 1994).

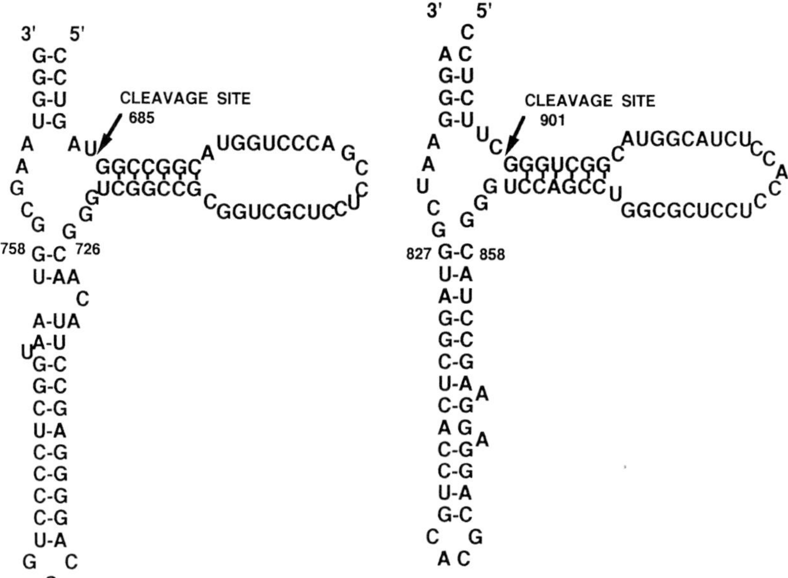

FIGURE 4   Axehead structure of hepatitis δ virus (HDV) ribozyme. (Left) genomic; (right) antigenomic (Modified from Branch and Robertson, 1991).

III. Strategies for Designing Ribozymes

Ribozymes can be designed to act as enzymes that cleave RNA site specifically, but several criteria must be defined for optimal cleavage *in vivo* (Fig. 5). The majority of intracellular ribozyme experiments have been designed with hammerhead ribozymes. Other types of ribozymes share many of the same basic requirements for intracellular efficacy (Bratty *et al.*, 1993; Christoffersen and Marr, 1995). Therefore, we focus on the criteria for optimal RNA cleavage by hammerhead ribozymes. The GUC triplet is frequently chosen for *trans*-acting hammerhead ribozyme as well as hairpin ribozyme because of its wide occurrence in natural ribozyme motifs (Haseloff and Gerlach, 1988). Other triplets are potential possibilities but the efficacy of cleavage seems to depend on the size and composition of the flanking sequence (Ruffner *et al.*, 1990).

Selecting a pivotal gene for specific cell growth or apoptosis in an appropriate target cell is one of the most important factors in the design of a ribozyme, i.e., understanding the biology of the system. Cancer gene therapy has focused on the control of oncogenes and tumor suppressor gene expression. Several studies have investigated the potential of antioncogene ribozymes for the treatment of cancer (Kashani-Sabet and Scanlon, 1995), as well as reversing drug resistance (Ishida *et al.*, 1995, Ohkawa *et al.*, 1996). *ras*, c-*fos*, and c-*jun* genes play a crucial role in the signal transduction pathways of neoplastic proliferation. Therefore, ribozymes could be used to inhibit this signal transduction pathway by suppressing the production of strategic molecules such as Ras and Fos (Kijima *et al.*, 1995). Other ribozymes have been targeted against the *bcr-abl* gene, which is thought to play a key role in the signal transduction pathway of chronic myelogenous

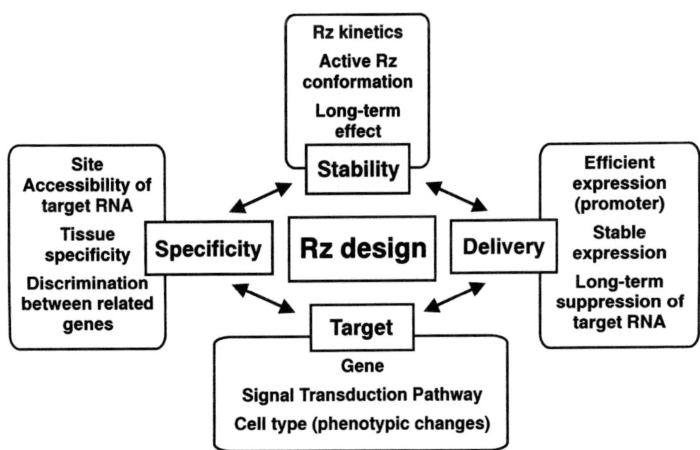

FIGURE 5   Strategies for designing ribozymes. Rz, Ribozyme.

leukemia (CML) (Snyder *et al.*, 1993; Wright *et al.*, 1993; Leopold *et al.*, 1995). Human carcinoma cells with aberrant tumor growth and drug resistance caused by point mutations or overexpression of genes are attractive targets for ribozymes. Ribozymes have been shown to act efficiently to overcome these malignant phenotypes (Kashani-Sabet and Scanlon, 1995).

The delivery of ribozymes to the desired tissue using nontoxic methods is critical for its therapeutic efficacy (see Section V, Delivery Systems for Gene Therapy). Advances in molecular engineering have made it possible to use plasmids, retroviruses, adenoviruses, adeno-associated viruses (AAVs), and cationic lipids to deliver ribozymes into cells (Jolly, 1994; Miller and Vile, 1995; Woolf and Budker, 1995). Expression of the ribozyme must be stable and expression efficient enough to reverse the phenotype. This can be achieved when the ribozyme accumulates within the cells, cleaves its target RNA, and suppresses protein expression. The stability of the target RNA, its degradation and synthesis, and its protein product stability are all important criteria in optimizing an efficient ribozyme. Also, ribozymes need to be resistant to cellular nucleases to maintain their activity within the intracellular compartment. A vector system that can optimally express high levels of ribozyme RNA with tissue specificity would be an ideal delivery system.

Efficient ribozyme expression requires an optimal promoter/enhancer system. Major considerations in the choice of expression systems include inducibility, tissue specificity, and whether it is constitutive. The preference for non-tissue-specific, constitutive promoters such as the β-actin promoter or the cytomegalovirus (CMV) promoter are important because of their high level of ribozyme RNA production (Kijima *et al.*, 1995). An example of this system would be either the pHβ Apr-1-neo plasmid driven by the human β-actin promoter (Gunning *et al.*, 1987) or the CMV promoter (Larsson *et al.*, 1994). A constitutive promoter does not optimally express only within the targeted cells, i.e., current delivery systems entail nonspecific delivery and expression in both normal and cancer tissue. Ribozymes could be designed with an inducible and/or a tissue-specific enhancer/promoter for the exclusive expression in diseased tissue only. The tyrosinase promoter is thought to be preferentially active in melanocytic cells. Melanoma cell-specific gene expression with tyrosinase promoters has been demonstrated (Vile and Hart, 1993). Another study has demonstrated the superior expression of anti-*ras* ribozyme within melanoma cells using a tyrosinase promoter rather than a non-tissue-specific promoter (Ohta *et al.*, 1996b). A heat-shock promoter has been used for the heat-inducible ribozyme against the *fushi tarazu* gene to study its role in *Drosophila* larval development (Zhao and Pick, 1993). A construct containing a mouse mammary tumor virus dexamethasone-inducible promoter has been used with an anti-*fos* ribozyme and has been found to express c-*fos* ribozyme transiently. In this case, transient expression of c-*fos* ribozyme was adequate to demonstrate reverse drug resistance (Scanlon *et al.*, 1991a, 1994).

The specific location of ribozyme activity in the cell is currently being investigated, i.e., cytoplasm, nucleus, or both. If a ribozyme acts in the cytoplasm, the binding rate must be fast enough to outpace the natural RNA half-life. If a ribozyme acts in the nucleus, the binding rate must be fast enough to cleave the target RNA substantially before it is processed and transported in the cytoplasm (Woolf, 1995). For the efficient activity of ribozymes, site-specific binding and efficient dissociation from the target RNAs are necessary. Once the target has been cleaved, the ribozyme must be able to dissociate from the cleaved products and repeat the cycle of binding, cleavage, and dissociation. The design of the flanking sequences is important for the kinetics of ribozymes. Although longer flanking sequences increase the specificity, dissociation from mismatched RNA becomes slower. Higher discrimination can be obtained by an AU-rich sequence rather than a GC-rich sequence (Herschlag, 1991). The ability of ribozymes to demonstrate multiple turnovers is required for optimal activity when the concentration of ribozyme is considerably lower than the concentration of the target. When targeting abundant RNAs (about 5000 copies per cell), ribozymes are effective at about 10-fold lower concentrations than antisenses (Woolf, 1995). However, when targeting less abundant or short half-life RNAs, hybrid formation becomes the rate-limiting step and catalytic turnover becomes less important.

All RNAs interact with cellular proteins from the moment they are synthesized until they are degraded. Protein/ribozyme interactions will influence the intracellular function of a particular ribozyme. When a heterogeneous nuclear ribonucleoprotein (hnRNP A1), associated with RNAs in the nucleus and cytoplasm, is incubated with a ribozyme, the hnRNP A1 readily associates with RNAs, facilitating the interaction between the ribozyme and its target and the release of the cleaved products from the ribozyme (Bertrand and Rossi, 1994; Herschlag *et al.*, 1994). hnRNP A1 facilitates ribozyme turnover by enhancing the release of the cleavage product. Other, similar RNA-binding proteins have been proposed to be good candidates for the enhancement of ribozyme activity (Woolf, 1995). Ribozymes have advantages over antisense in the cleavage of point mutation-activated genes. An antisense RNA short enough to discriminate between the mutant and the wild-type gene may not be able to inhibit translation, whereas an antisense RNA long enough to cover the point mutation and block translation may not be able to discriminate between wild-type and mutant gene (Monia *et al.*, 1992). The catalytic activity of a ribozyme requires a consensus sequence at the cleavage site and results in enhanced specific cleavage of the point mutation (Koizumi *et al.*, 1989; Kashani-Sabet *et al.*, 1992).

Sullenger and Cech (1993) have demonstrated that colocalization of a ribozyme with its target RNA greatly improves the efficacy of the ribozyme. The ribozyme and its target RNA were colocalized via the retroviral dimerization domain, forcing copackaging of the retroviral transcript encoding the ribozyme and the target RNA. The ribozyme was effective only when the RNA was copackaged. Colocalization of the ribozyme and the target RNA in the same intracellular compartment is critical for efficient ribozyme action. The localization of the ribozyme–mRNA complex is also critical for another reason: double-stranded RNases can cleave ribozyme–mRNA complexes, resulting in a loss of ribozyme activity (Scanlon *et al.*, 1991a). Some investigators have shown that a multiunit hammerhead ribozyme could cleave its target RNA more efficiently than a single-unit ribozyme (Chen *et al.*, 1992; Leopold *et al.*, 1995). The multiunit hammerhead ribozyme may help to minimize any loss of activity due to base-pairing mismatches or mutation at the cleavage site. The secondary structure of RNA is thought to be important for cleavage efficacy, and is difficult to determine. Some investigators have demonstrated the usefulness of computers for RNA structure modeling (Jaeger *et al.*, 1989; Holm *et al.*, 1996). These methods may become useful tools for designing ribozymes in the future.

The goal of cancer gene therapy is long-term inhibition of the translation of disease-related proteins and thus the blockage of aberrant proliferation. Stable long-term expression of a ribozyme will correct the phenotypic changes of cancer cells and will affect the disappearance of the targeted tumor cells. Currently such an optimal system for ribozyme expression is under active development. Appropriate controls are required to demonstrate the specific activity and selectivity of ribozymes (Stein and Krieg, 1995). Three types of control should be considered:

- Antisense control: The control consists of an antisense sequence that is common to the flanking sequence of the ribozyme
- Mutant ribozyme control: The ribozyme maintains its structural features but does not maintain its catalytic activity
- Mismatched target control: The cells do not have a specific site for the ribozyme

The latter control may demonstrate a lack of sequence specificity for the ribozyme. Since non-sequence-specific effects of oligodeoxynucleotides have been observed (Stein and Krieg, 1995), the possibility of nonspecific effects of the ribozyme should be eliminated. Also, the protein product of a target mRNA should be measured and found to be downregulated to demonstrate truly the activity of a ribozyme. Designing ribozymes is an unique process for each cell type and each gene. Therefore, optimal ribozyme design requires the understanding of the biology of each cell type, and the role of the target gene in the cell phenotype.

IV. Applications for Cancer Gene Therapy

Ribozyme strategy is applicable to any RNA/protein target responsible for a specific disease. At present, several mRNAs have been identified as obvious targets for ribozyme-mediated cancer gene therapy. These ribozymes include targets for tumor-specific oncogenes, growth factors, and drug-resistant genes (Table II and Fig. 6). The following section reviews progress in the applications of ribozymes to specific diseases.

A. Oncogenes

The understanding and ultimate manipulation of gene expression has become realized with the assistance of efficacious ribozyme systems demonstrated in cellular models and, more recently, in clinical protocols. Signal transduction pathways are crucial in understanding cell growth and proliferation (Cantley *et al.*, 1991; Hunter, 1991; Roberts, 1992; Egan and Weinberg, 1993). Alterations of normal pathways may quickly lead to the development of malignant phenotypes. Carcinogenesis is a process involving multiple alterations (Foulds, 1958) in suppressor genes and/or oncogenes (Fearon and Vogelstein, 1990). It is postulated that signal transduction is initiated by the binding of growth factors to their respective membrane-bound receptors. This binding causes phosphorylation of tyrosine residues in the receptor. These phosphorylations ultimately activate the Ras oncoprotein

TABLE II   Proto-oncogenes and Their Associated Neoplasms

Proto-oncogene/lesion	Neoplasm
abl/translocation	CML
erbB-1/amplification	Squamous cell carcinoma, astrocytoma
erbB-2 (*neu*)/amplification	Breast, prostate, orvarian, and gastric carcinoma
gip/point mutation	Ovarian and adrenal carcinoma
gsp/point mutation	Pituitary adenoma, thyroid carcinoma
myc/translocation, amplification	Burkitt's lymphoma; lung, breast, and cervical carcinoma
L-*myc*/amplification	Lung carcinoma
N-*myc*/amplification	Neuroblastoma, small cell carcinoma of the lung
H-*ras*/point mutation	Lung, pancreas, and colon carcinoma; melanoma
K-*ras*/point mutation	AML, ALL[a]; Thyroid carcinoma; Melanoma
N-*ras*/point mutation	Genitourinary tract and thyroid carcinoma melanoma
ret/rearrangement	Thyroid carcinoma
ros/?	Astrocytoma
K-*sam*/amplification	Gastric carcinoma
sis/?	Astrocytoma
src/?	Colon carcinoma

[a] AML, Acute myelogenous leukemia; ALL, acute lymphocytic leukemia.

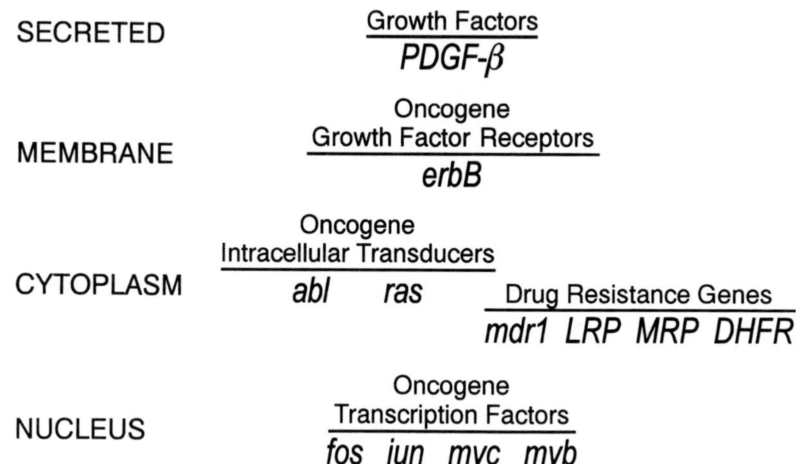

FIGURE 6   Targets of ribozymes for cancer gene therapy. Tumor-specific oncogenes, growth factors, and drug resistance genes have been targeted by ribozyme-mediated strategies for cancer gene therapy. PDGF-β, Platelet-derived growth factor β; LRP, lung resistance-related protein; MRP, multidrug resistance-related protein; DHFR, dihydrofolate reductase.

through the conversion of Ras–GDP to Ras–GTP (active form). The activation of these cytoplasmic phosphorylation cascades involves Raf proteins and mitogen-activated protein (MAP) kinase. With the assistance of multiple second-messenger systems (cAMP, Ca^{2+}, diacylglycerol) MAP kinase is translocated to the nucleus and may phosphorylate the *jun* and *fos* oncogenes.

Perturbations in these signal transduction cascades have been reported (Seemayer and Cavenee, 1989; Brunton and Workman, 1993). Suppressor genes have been identified in the oncogenic sequence and are generally related to deletions and mutations in the endogenous genome. Oncogenes may be the result of amplification, overexpression, and chromosomal translocations of endogenous proto-oncogenes. The understanding and delineation of these interrelated processes may lead to the identification of specific ribozyme systems that may correct the identified specific gene alterations and restore the normal phenotype. For any gene therapy to be efficacious and useful the following criteria must be met: (1) a proper clinical model must be identified; (2) the disease under consideration should have no successful therapy to date; (3) the disease should represent a significant clinical and epidemiologic problem; and (4) a successful delivery system must be constructed to allow for successful testing and treatment.

It is not surprising that much work in this field has been done with the *ras* oncogene family. Mutations in this group of genes may be identified in up to 20% of all human malignancies (Barbacid, 1987; Bos, 1989). The *ras* family mainly consists of functional genes such as K-*ras*, H-*ras,* and N-*ras*. The *ras* oncogene has been remarkably conserved throughout evolution and belongs to a larger family of oncogenes that encode guanine nucleotide-binding proteins, i.e., the G proteins. Studies have shown that the G proteins are molecular switches that serve to transduce a host of signal transduction pathways when it is in its active form, bound to a molecule of GTP. Analysis of human tumors with identified *ras* mutations has identified one of the three genes to harbor a mutation that produces an altered amino acid in the G protein in a critical position. Once a Ras protein is activated it loses its ability to return to its inactive form. Specific mutations have been identified in several disease processes. Pancreatic carcinoma has been found to have a *ras* mutation in 90% of cases (Almoguera *et al.*, 1988), colon cancer in 50% (Bos *et al.*, 1987; Forrester *et al.*, 1987), adenocarcinoma of the lung in 33% (Rodenhuis *et al.*, 1987), and advanced stages of melanoma in 45% (Ball *et al.*, 1994).

Anti-*ras* ribozymes have been tested in several tumor types. A specific anti-H-*ras* ribozyme has been designed to discriminate the mutated GUC sequence (from GGC) of the H-*ras* oncogene in codon 12 (Koizumi *et al.*, 1992). A human anti-activated *ras* ribozyme has been tested in an EJ human bladder carcinoma cell line (Kashani-Sabet *et al.*, 1992; Tone *et al.*, 1993) and has been found to specifically cleave the mutated gene while sparing the normal H-*ras* protooncogene (Funato *et al.*, 1994). Changes in the morphology of cells with mutated H-*ras* were clearly identified, and cell growth was markedly inhibited. This vector-mediated expression of the ribozyme was shown to downregulate the H-*ras* mRNA expression and its corresponding protein product (p21). *In vivo* studies demonstrated the suppressed tumorigenicity of EJ cells transfected with an anti-H-*ras* ribozyme in athymic mice (Tone *et al.*, 1993). The ribozyme was found to be more effective than a mutant ribozyme that did not possess catalytic activity. Other *in vivo* models were studied using the same cell line (Kashani-Sabet *et al.*, 1992). EJ cells were injected orthotopically, i.e., by transurethral inoculation, into the bladder of athymic mice. The mice inoculated with EJ cells transfected with anti-H-*ras* ribozyme showed less evidence of the malignant phenotype, with limited invasion and longer overall survival. Studies have shown that the anti-H-*ras* ribozyme was efficacious in tumor growth inhibition both *in vitro* and *in vivo*, using adenoviral delivery systems with no identifiable toxicity (Feng *et al.*, 1995). These studies have demonstrated the potential usefulness of anti-H-*ras* ribozyme in the treatment of human bladder cancer.

The FEM human melanoma cell line was tested with a ribozyme against the heterozygous mutation (GUC) of H-*ras* oncogene in codon 12. This mutation is said to cause the overexpression of H-*ras* mRNA and cause the aberrant tumor growth. Expression of H-*ras* mRNA and cell growth were found to be downregulated in the ribozyme-transfected cells (Ohta *et al.*, 1994, 1996a,b). This ribozyme caused the cells to become more differentiated with a more melanocytic phenotype. This was demonstrated by increased melanin synthesis, responsivity to 12-*O*-tetradecanoylphorbol-13-acetate (TPA), as well as a more dendritic morphology. *In vivo* study has also shown the tumor growth-inhibiting efficacy of anti-H-*ras* ribozyme (Kashani-Sabet *et al.*, 1994). In this study, NIH 3T3 cells transfected with heterozygous mutated H-*ras* codon 12 (GUC) of the human melanoma cell line FEMX-1 were injected into athymic mice. The overexpression of H-*ras* was shown by the transfection of the mutated H-*ras* gene and caused high tumorigenicity. Tumorigenicity was shown to be suppressed by the treatment with the anti-H-*ras* ribozyme. The 3T3 cells that did not contain the GUC cleavage site were not affected by the ribozyme (Koizumi *et al.*, 1992; Funato *et al.*, 1994).

Pancreatic cancer is one of the most difficult cancers to cure, or even to improve in terms of overall prognosis. K-*ras* point mutations are found in more than 90% of human pancreatic cancers, and 95% of these are located in codon 12 (Almoguera *et al.*, 1988; Grunewald *et al.*, 1989). Mutated K-*ras* oncogenes are thought to be adequate targets for ribozymes. A ribozyme was designed to cleave the mutated K-*ras* codon 12 (GUU) in the human pancreatic cancer cell line, Capan-1 (Kijima *et al.*, 1996). Anti-K-*ras* ribozymes were transfected by three different delivery systems, i.e., pHβ plasmid, retroviral plasmid, and adenovirus. The results demonstrated that the anti-K-*ras* ribozyme significantly inhibited the cell growth and reversed the malignant phenotype to varying degrees depending on the vector systems. In the three different delivery systems used, adenovirus-mediated delivery was shown to be the most efficacious in expressing the ribozyme in Capan-1 cells. Others have used plasmids expressing K-*ras* antisense, and transduced them into different pancreatic cell lines (ASPC-1 and MIA PaCa-2) with a K-*ras* point mutation, using liposome-mediated transfection (Aoki *et al.*, 1995). The antisense was found to suppress growth of the cell lines and cause a reduction in the p21 protein product. They have also demonstrated the *in vivo* efficacy of anti-K-*ras* antisense using a retroviral plasmid. These studies have demonstrated the usefulness of anti-K-*ras* oligonucleotide agents and the importance of selecting adequate delivery systems.

The *bcr–abl* gene has been actively studied as a potential target for ribozymes. This hybrid gene is formed when the protooncogene (*abl*) from chromosome 9 translocates to the breakpoint cluster region (*bcr*) of chromosome 22 (Rowley, 1973). This results in a new fusion gene, comprising portions of the *bcr* and *abl* genes, that encodes an 8.5-kb mRNA that translates into a 210-kDa protein (p210) with enhanced tyrosine kinase activity. This translocation, creating what is known as the Philadelphia chromosome, can be observed in up to 95% of cases of CML and 50% of cases of acute lymphocytic leukemia (Kurzock *et al.*, 1988). Thus, the *bcr–abl* transcript can become a potential target for ribozymes to treat this disease. An anti-*bcr-abl* ribozyme has been constructed and directed against the GUU triplet adjacent to the junction of the fused c-*bcr* and c-*abl* genes

(Snyder *et al.*, 1993; Shore *et al.*,1993). *In vitro* studies have shown the efficacy with diminished cell growth and transiently suppressed p210 protein activity. However, cleavage of the normal *bcr* gene by the ribozyme targeted to the *bcr–abl* fusion region was demonstrated (Wright *et al.*, 1993). Also, the *bcr–abl* fusion region is said to have an inauspicious secondary structure, and may be difficult to cleave with a ribozyme (Pachuk *et al.*, 1993). The efficacy of the multiunit ribozyme that targets *bcr–abl* mRNA has been demonstrated (Leopold *et al.*, 1995). This study showed increased cleavage efficacy of the multiunit ribozyme compared to single or double ribozymes. Delivery of the ribozyme has been problematic with liposome transfection, resulting in only transient expression of the ribozyme and intermittent down-regulation of the protein product. Lipofection of a modified ribozyme has been tested in a K562 human CML blast crisis cell line and found to be twice as efficacious as the noncatalytic antisense (Lange *et al.*, 1993, 1994). Other studies, however, have used antisense oligonucleotides against c-*myb* and the *bcr–abl* mRNA expression was found to be decreased or undetectable (Gewirtz, 1994). At present clinical studies are underway using antisense oligonucleotide in patients with CML.

The fusion gene *AML1/MTG8* is thought to play a role in the pathogenesis of some types of acute myelogenous leukemia (AML). This fusion gene is formed by a translocation between two genes, the *AML1* gene on 21Q22 and the *MTG8* (*ETO*) gene on 8q22. Two hammerhead ribozymes were designed against two separate cleavage sites: (1) CUC located 3 bases upstream from the fusion site, and (2) AUC located 3 bases downstream from the fusion site. These ribozymes were able to inhibit cell growth when transfected into Kasumi-1 cells (Matsushita *et al.*, 1995).

Other studies have looked at rearrangements and amplifications of other oncogenes in melanoma cell lines (Linnenbach *et al.*, 1988; Albino, 1992). One cell line demonstrated c-*myb* rearrangement, and another had a 1.5- to 3.0-fold amplification of the c-*myc* oncogene. The various oncogenes reported may represent the heterogeneity of malignant melanoma or more likely represent the multistep process necessary for expression of the malignant phenotype. The c-*myc* oncogene has shown alterations in such tumors as breast, colon, small cell lung carcinoma, ovarian carcinoma, squamous cell carcinoma, and various lymphomas. Clinically, amplification and/or elevated expression of the c-*myc* gene has been associated with a poorer prognosis and decreased survival. Animal studies with progressive rodent skin carcinomas have shown c-*myc* amplification to be directly related to the size and age of the tumor. Studies have implicated c-*myc* in both cell cycle progression and programmed cell death (Wurm *et al.*, 1986; Garte, 1993). Programmed cell death (apoptosis) was inhibited using a c-*myc* antisense sequence in a T cell hybridoma cell line (Shi *et al.*, 1992).

The human papillomaviruses (HPVs), particularly HPV-16 and HPV-18, have been associated with several carcinomas. Cervical carcinoma has been shown to possess a 90% incidence of HPV (Lowy *et al.*, 1994), oral cancers have been found to have up to 75%, and, less frequently, anogenital cancers have also been associated with HPV. The HPV transforming potential is indirect and mediated through other cellular proteins. E6 and E7 gene expression become unregulated after HPV infection owing to disruption of E2, a major regulatory gene. The E6 protein of HPV binds to the tumor suppressor protein, p53, and prevents its normal function (Werness *et al.*, 1990; Lechner *et al.*, 1992; Inouye, 1988). The E7 protein of HPV can cooperate with an activated *ras* oncogene or *trans*-activate the E2 adenovirus promoter, and prevent repression of c-*myc* by transforming growth factor β. E7 can also bind to the retinoblastoma (Rb) cellular suppressor gene and the retinoblastoma gene product and inhibit its function. One group has shown effective cleavage *in vitro* by hammerhead ribozymes of HPV-16 E6 and E7 open reading frames, which are associated with viral DNA regulation or cellular gene regulation (He *et al.*, 1993). An AAV vector has been constructed that encodes hammerhead ribozymes specifically designed to cleave the HPV-16 E6 and E7 mRNA, and these ribozymes were found to cleave their target RNAs *in vitro* (Lu *et al.*, 1994). These experiments need to be expanded into cell and animal studies before the gene-mediated therapy of HPV-associated malignancies can be attained.

Overexpression of platelet-derived growth factor β (PDGF-β) and PDGF-β receptor has been identified in malignant mesothelioma. The mesothelioma cells are thought to be stimulated by PDGF/PDGF receptor autocrine mechanisms (Versnel *et al.*, 1988, 1991). Transfection of a specific hammerhead ribozyme targeting PDGF-β (anti-c-*sis* ribozyme), with the help of a constitutive vector, leads to a decrease in the PDGF-β mRNA. Transfected clones expressing the anti-c-*sis* ribozyme displayed decreased cell growth (Dorai *et al.*, 1994). The ribozyme was found to be more active than an inactive ribozyme.

Pleiotrophin (PTN), a polypeptide growth factor, has been found to be overexpressed in many types of tumors. A PTN-targeted ribozyme was used against the *pleiotrophin* gene in a melanoma cell line that overexpresses PTN. Cotransfection of the ribozyme inhibited PTN-induced colony formation and was found to be more effective than an inactive ribozyme (Czubayko *et al.*, 1994). This demonstrates that growth factors may be downregulated when overexpressed and may prove to be helpful adjuncts in reversing the malignant phenotype.

Many investigators have demonstrated the efficacy of anti-oncogene ribozymes in efficiently inhibiting tumor growth *in vitro* (Table III). However, the efficacy and toxicity of ribozymes need to be defined further for *in vivo* applications. Many factors exist to alter the activity and specificity of ribozymes such as cell types, RNA targets, ribozyme design, and delivery systems. These factors must be investigated in detail to allow the *in vivo* use of ribozymes to be optimized. The clinical applications of anti-oncogene

TABLE III Therapeutic Applications of Anti-oncogene Ribozymes in Human Tumors

Targeted oncogene	Cancer cells	Vector	Promoter	Ref.
H-*ras*	EJ, bladder cancer	pHβ Apr-1-neo	β-Actin	Tone *et al.* (1993); Kashani-Sabet *et al.* (1994)
		Adenovirus	CMV	Feng *et al.* (1995)
	FEM, melanoma	pHβ Apr-1-neo	β-Actin	Ohta *et al.* (1994)
		pMAMneo	MMTV	Ohta *et al.* (1996a)
		pLNCX	CMV	Ohta *et al.* (1996a)
		pLNT	Tyrosinase	Ohta *et al.* (1996b)
K-*ras*	Capan-1, pancreatic cancer	pHβ Apr-1-neo	β-Actin	Kijima *et al.* (1996)
		pLNCX	CMV	Kijima *et al.* (1996)
		Adenovirus	CMV	Kijima *et al.* (1996)
bcr/abl	K562, CML	Lipofection		Lange *et al.* (1993)
		Retrovirus	β-Actin	Shore *et al.* (1993)
			Thymidine kinase	Shore *et al.* (1993)
	EM-2, CML	Lipofection		Snyder *et al.* (1993)
AML1/MTG8	Kasumi-1, AML	Lipofection		Matsushita *et al.* (1995)
c-*sis*	VAMT-1, mesothelioma	pHβ Apr-1-neo	β-Actin	Dorai *et al.* (1994)
c-*myc*	FEM, melanoma	pMAMneo	MMTV	Ohta *et al.* (1996a)
c-*fos*	FEM, melanoma	pMAMneo	MMTV	Ohta *et al.* (1996a)
	A2780, ovarian cancer	pMAMneo	MMTV	Scanlon *et al.* (1991b)
pleiotrophin	WM852, melanoma	pRc	CMW	Czubayko *et al.* (1994)

ribozymes for cancer gene therapy have only begun to be realized. The decoding of the human genome and discovery of new oncogenes will undoubtedly lead to a better understanding of the carcinogenic process and create new targets for ribozyme intervention.

B. Drug Resistance Genes

Ribozyme-mediated reversal of drug resistance has also been investigated extensively. The development of an intrinsic or acquired drug resistance mechanism is a major limitation for effective cancer chemotherapy. Multidrug resistance (MDR), defined as the resistance to a variety of different lipophilic compounds (Endicott and Ling, 1989; Gottesman and Pastan, 1993; Roninson, 1991), is difficult to overcome. P-Glycoprotein is overexpressed in many multidrug-resistant cells (Juliano and Ling, 1976; Kartner *et al.*, 1983, 1985). The gene encoding P-glycoprotein is the multidrug resistance gene (*mdr1*; Gros *et al.*, 1986; Roninson *et al.*, 1986; Fojo *et al.*, 1987). Amplification and overexpression of the *mdr1* gene through the cell signal transduction pathway confer overexpression of P-glycoprotein in cancer cells, thereby establishing the MDR phenotype (Ueda *et al.*, 1987b). Since the discovery of P-glycoprotein, a large number of pharmacologic agents have been shown to inhibit the function of this protein (Ford and Hait, 1993). However, lack of specificity and/or toxic side effects have resulted in a limited clinical application for even the most promising chemosensitizers. Therefore, more specific alternatives such as molecular approaches were investigated to reverse the MDR phenotype (Tidd, 1991; Marschall *et al.*, 1994; Kiehntopf *et al.*, 1994; Ohkawa *et al.*, 1996). Several groups have shown the efficacy of antisense oligonucleotides in the modulation of *mdr1* gene expression and reversal of the MDR phenotype (Vasanthakumar and Ahmed, 1989; Jaroszweski *et al.*, 1990; Corrias and Tonini, 1992; Efferth and Volm, 1993; Nakashima *et al.*, 1995). These studies have been controversial in reversing the MDR phenotype. Other studies have used ribozymes for the reversal of the MDR phenotype.

The development of hammerhead ribozymes against the MDR phenotype has been a goal of several investigations. Kobayashi *et al.* (1994) described the *in vitro* cleavage efficacy of two ribozymes, as well as the efficacy of one ribozyme in reducing resistance to vincristine in an acute leukemic cell line. Using a vector-mediated transfer system the expression of the ribozyme reduced resistance to vincristine by about 35-fold. Holm *et al.* (1994) have designed a ribozyme against the *mdr1* mRNA that cleaved the 3' end of the GUC triplet in exon 21. The target site was chosen between the two ATP-binding sites, which is suggested to be important for P-glycoprotein function (Teeter *et al.*, 1991). Detailed kinetic studies *in vitro* showed a high level of catalytic efficiency for the anti-*mdr1* ribozyme and suggested that the target mRNA is well exposed (Holm *et al.*, 1996). These data confirm that efficiently cleaved target sequences do exist and a search for them could be rewarding. The anti-*mdr1* ribozyme was then cloned into the pHβ Apr-1-neo plasmid (Gunning *et al.*, 1987) and transfected into the human pancreatic carcinoma cell line resistant to daunorubicin (Scanlon *et al.*, 1994). Clones expressing the ribozyme reduced the resistance to daunorubicin by 300-fold and P-glycoprotein and *mdr1* mRNA were not detectable. Scanlon and colleagues (1994) have demonstrated that an anti-*mdr1* ribozyme, designed to cleave the GUC sequence of triplet 880, as well as an anti-*fos* ribozyme (both cloned into the pHβ Apr-1-neo plasmid), reduced the resistance to actinomycin D in a human ovarian carcinoma cell line and effectively downregulated the expression of c-*fos*, *mdr1*, and topoisomerase II mRNAs. Because the promoter of the *mdr1* gene has an AP-1-binding site, the c-*fos* gene, in addition to *mdr1*, may play an important role in the MDR phenotype through its participation in the signal transduction pathway.

In contrast to the endogenous delivery of ribozymes, which requires a vector containing the ribozyme, Kiehntopf *et al.* (1994) described the reversal of the MDR phenotype in two different cell lines using a liposome-mediated transfer system. The incorporation of anti-*mdr1* ribozymes using liposomes nearly reversed the multidrug resistance. However, in this study the application of liposomes was not without problems, owing to their toxicity. Another similar approach to modulate the MDR phenotype was described by Bertram *et al.* (1995). To enhance the stability of their *mdr1* ribozyme against nuclease degradation the ribozyme was substituted with fluoro and allyl groups. An exogenous application of the fluoro-modified ribozyme reduced the chemoresistance to doxorubicin up to 50% in an *ex vivo* model of blast cells cultured from patients with AML (Palfner *et al.*, 1995).

Some multidrug-resistant cell lines do not overexpress the *mdr1* gene or display detectable levels of P-glycoprotein (Marsh *et al.*, 1986; Mirski *et al.*, 1987; McGrath and Center, 1987). Some non-P-glycoprotein-related cells have been found to overexpress the multidrug resistance-associated protein (MRP; McGrath *et al.*, 1989; Marquardt *et al.*, 1990; Cole *et al.*, 1992; Krishnamachary and Center, 1993; Müller *et al.*, 1994) or the lung resistance-related protein (LRP; Scheper *et al.*, 1993; Scheffer *et al.*, 1995). These MDR-related genes may also be adequate targets for ribozymes to reverse the MDR phenotype. Cancer chemotherapeutic agents have been shown to disrupt the signal transduction pathway that may contribute to the evolution of drug-resistant clones (Tritton and Hickman, 1990; Scanlon *et al.*, 1991a; Bruton and Workman, 1993; Ishida *et al.*, 1995; Ohkawa *et al.*, 1996). A heterodimeric complex, consisting of the Jun and Fos families, forms activator protein 1 (AP-1). AP-1 affects drug resistance through the transcriptional activation of genes containing AP-1 elements in their regulatory regions (Ransone and Verma, 1990). AP-1-responsive genes are important in DNA synthesis and repair as well as drug detoxification pathways involving proteins such as thymidylate synthase, DNA polymerase β, topoisomerase I, and metallothionein and glutathione S-transferase. Moreover, the promotor/enhancer element of the *mdr1* gene contains an AP-1-binding site and *mdr1* gene transcription is regulated by AP-1 (Ueda *et al.*, 1987a,c; Maddoen *et al.*, 1993). Therefore, downregulation of transcriptional factors may decrease the expression of genes associated with drug resistance in cancer cells. A ribozyme targeting the c-*fos* mRNA has been demonstrated to reverse resistance to chemotherapeutic agents (Scanlon *et al.*, 1991a,b). This anti-*fos* ribozyme downregulated not only c-*fos* expression, but also *mdr1*, c-*jun*, the topoisomerase I gene, and mutant *p53* through the signal transduction pathways (Scanlon *et al.*, 1991a,b, 1994; Funato *et al.*, 1992). Therefore, targeting pivotal genes in the signal transduction pathway may have an impact in reversing the MDR phenotype.

Methotrexate (MTX) is an important folate antagonist used in cancer chemotherapy; its mechanism of action is through the competitive inhibition of folate metabolism (Bertino, 1993). The clinical usefulness of MTX is limited by several mechanisms, including alteration in the affinity of dihydrofolate reductase (DHFR) for MTX caused by mutations of the *DHFR* gene (Ohnuma *et al.*, 1985). In an attempt to cleave the mutated *DHFR* mRNA from a human leukemia cell line resistant to both MTX and trimetrexate, one group designed a specific ribozyme (Kobayashi *et al.*, 1993). The specifically designed ribozyme did not discriminate between the mutated and the wild-type mRNA.

Ribozymes are new strategies for therapeutic intervention at the molecular level to reverse the MDR phenotype. The discovery of ribozymes and their ability to suppress gene expression by their catalytic potential seems to offer several advantages compared to antisense oligonucleotides. Although ribozymes offer an additional mechanism of action relative to antisense, the *in vitro* data may not be compatible with the *in vivo* data (Woolf, 1995). In spite of our knowledge of suitable delivery systems, methods of synthesizing stable ribozyme/antisense constructs and protein enhancement of ribozyme catalysis *in vivo* (Tsuchihashi *et al.*, 1993; Bertrand and Rossi, 1994), there still exist issues that must be resolved for ribozymes before they become more potent therapeutic agents (Bratty *et al.*, 1993).

V. Delivery Systems for Gene Therapy

The appearance of gene therapy as an alternative treatment for cancer and other diseases has led researchers toward the development of efficient delivery systems (Miller, 1990; Morgan and Anderson, 1993). Among the first methods developed for gene delivery, viral systems were and are potentially the most promising. Although viral transfer shows great potential for clinical settings, nonviral delivery systems are receiving increasing attention. Nonviral gene transfers include a broad range of techniques. Among these are particle bombardment, microinjection of DNA, liposomes, and receptor-mediated gene delivery. On the other hand, viral delivery systems have been derived from different viruses engineered to carry a foreign DNA fragment. A wide variety of viruses have been used including retroviruses, adenoviruses, AAVs, and others. Table IV shows the characteristics of viral vectors and liposomes.

A. Nonviral Delivery Systems

Particle bombardment can be seen as a physical method for the transfer of DNA into the host cell. To be incorporated into the cells, the DNA must first be coated onto the surface of mineral beads (gold or tungsten, 1–3 μm in diameter). These complexes are then accelerated by electric discharges and "shut" in the targeted tissue. The way by which the DNA is transferred into the cells is based on the principle that the physical force of impact overcomes the cell membrane barrier. Although this technique has shown the capacity to transfect cells efficiently both *in vivo* and *in vitro* (Yang *et al.*, 1991), the problem remains that invasive surgical settings are required to expose the targeted tissues. The skin remains the tissue most accessible for this technique, requiring the least invasion (Williams *et al.*, 1991). It is less probable that this method will achieve widespread use for systemic gene delivery since it is based on the local transfer of DNA and involves surgical procedures. To date, no publications have revealed this technique as being an efficient way to transfect antisense or ribozyme into cells. Direct injection of closed plasmid DNA or RNA is another means of physical gene transfer. Injection of DNA is safe, simple, and relatively nonimmunologic and appears to be a reasonable approach for local *in vivo* gene transfer. To date the only known susceptible tissue for this approach has been the striated muscle (Acsadi *et al.*, 1991b). Since this method is efficient in transfecting muscle cells, it has thus been used for the *in vivo* gene transfer of the Duchenne muscular dystrophy gene (Acsadi *et al.*, 1991a).

Among the nonviral techniques for gene transfer, cationic lipids have become important agents. One of the first described cationic lipids was DOTMA (*N*-[1-(2,3-dioleyloxy)propyl]-*N*,*N*,*N*-trimethylammonium chloride; Felgner *et al.*, 1987; Felgner and Ringold, 1989), and many other cationic lipids have been developed for gene transfer. Among these new reagents related to DOTMA are DMRIE (1,2-dimyristyloxypropyl-3-dimethyl-hydroxyethyl ammonium bromide; Felgner *et al.*, 1994) and DOTAP [1,2-dioleoyloxy-3-(trimethylammonio) propane; Stamatatos *et al.*, 1988]. The DNA or RNA bound to these cationic lipids does not form a true liposome structure. Rather, the cationic lipids form a particle in which

TABLE IV Characteristics of Different Delivery Systems

Characteristic	Retrovirus	Adenovirus	Adeno-associated virus	Liposomes
Titer	10^6–10^9 PFU/ml[a]	10^9–10^{10} PFU/ml	10^6–10^9 PFU/ml, possible	Efficient
Integration	Yes	No	Chromosome 19	No
Insertional mutagenesis	Possible (not yet observed)	No	Possible	No
Cell division requirement	Yes	No	No	No
Expression	Good, >1 year possible	Transient	Potentially long term	Transient
Production of replication-competent viruses	No	No	No	No
Delivery efficiency	Poor	Good	Good	Good
Insert capacity	8 kb	7–8 kb	4.5 kb	No limit

[a] PFU, Plaque-forming unit.

the DNA or RNA is trapped via ionic interactions between the negative charges of the DNA or RNA and the positive charges of the cationic lipids. These particles can achieve more than 90% transfection efficiency in some cell lines and primary cells *in vitro*. It was shown that a single injection of cationic lipid–DNA complexes in mice results in transfection of almost all tissues, with expression of the transgene for up to 9 weeks (Zhu *et al.,* 1993). Cationic lipid gene delivery appears to be a valid alternative to viral gene delivery since there is no limit for the size of the DNA or RNA, no immunogenicity, no carryover of viral proteins, easy preparation, and no integration of the DNA into the genome, eliminating the chance for insertional mutagenesis. Although it offers many advantages, cationic lipids also possess many disadvantages. Many injections will be required for treatment of a genetic disorder since the expression of the DNA is only transient. The DNA or RNA bound to the surface of cationic lipids is available for DNase or RNase degradation. Also, the DNA or RNA remains trapped in the endosome vesicles after penetration of the cell membrane, which for ribozyme delivery could be an important limitation in gene delivery to the nucleus. Many reports have demonstrated the use of cationic lipids for the delivery of oligonucleotides to cells (E. G. Nabel *et al.,* 1989, 1990; G. J. Nabel *et al.,* 1993; Plautz *et al.,* 1992, 1993; Stewart *et al.,* 1992; Zhu *et al.,* 1993). San *et al.* (1993) have shown that high concentrations of these cationic lipids can be toxic to mice following intravenous injection, but that modifications of the lipid carrier can reduce the toxicity. Although cationic lipids could be an effective delivery system for small pieces of DNA, this might not be the case for small pieces of RNA (fewer than 500 nucleotides), which do not seem to be taken up in the same fashion as large DNA (Christoffersen and Marr, 1995). The use of cationic lipids as an exogenous delivery system for oligonucleotides or ribozymes seems interesting but many aspects need to be resolved before this method can fill the "vector void," as shown by the fact that certain lipid preparations work in certain cell types but not in others. This method still remains empirical and investigations to ameliorate this system need to be performed.

Liposomes are spheres composed of bilamellar lipid membrane surrounding an aqueous milieu that can carry a large quantity of drugs or oligonucleotides (Lichtenberg, 1988; Litzinger and Huang, 1992). Liposomes offer the advantage of protecting these entrapped molecules, with no limit in terms of DNA size, from renal filtration, blood enzymatic degradation, and the effectors of the immunologic response. However, liposomes do not survive long in the systemic circulation and are less effective than cationic lipid complexes or other systems for transfection (Bertling *et al.,* 1991; Legendre and Szoka, 1992). Liposomes do not offer more advantages than cationic liposomes and are less probable as a delivery system for oligonucleotides and ribozymes, owing to these disadvantages.

Liposomes are more effective when viral protein, or proper antibody, is used on the liposomal surface to facilitate endocytosis into the cell and tissue-specific targeting (Leserman *et al.*, 1980; Berinstein *et al.*, 1987; Milhaud *et al.*, 1989; Litzinger and Huang, 1992). Receptor-mediated transfection of cells utilizes the association of DNA with molecules capable of binding to the surface of cells to facilitate endocytosis. Many types of molecules have been used in conjugation with DNA to facilitate transfection. Transferrin–DNA–polylysine complexes have demonstrated an acceptable level of transfection efficiency *in vitro* in hematopoietic cells and other cell types (Wagner *et al.*, 1990; Zenke *et al.*, 1990; Harris *et al.*, 1993). Liver-specific targeting has been obtained by complexing DNA with polylysine and asialoorosomucoid (Wu and Wu, 1987, 1988). The main concern in DNA delivery via a receptor-mediated process is the early release of the DNA from the endosomes quickly, before it is degraded. One approach to facilitate the release of the DNA molecules from endosomes is to use a replication-defective adenovirus. Adenoviral particles are able to break down the endosomes upon acidification, owing to the penton protein on the surface of this virus (Seth, 1994). The use of adenoviral particles with transferrin-, asialoglycoprotein-, or folate–polylysine–DNA complexes has increased the level of transfection by 100- to 1000-fold *in vitro* (Curiel *et al.*, 1991, 1992a,b; Cotten *et al.*, 1992; Wagner *et al.*, 1992; Cristiano *et al.*, 1993; Gottschalk *et al.*, 1993; Harris *et al.*, 1993; Wu *et al.*, 1994). The use of hemagglutinin, a protein from the pathogenic influenza virus that acts in a manner similar to the penton protein of adenovirus, also increases the transfection efficiency of transferrin–DNA–polylysine complexes, although the efficiency of this combination is lower than with adenovirus particles (Wagner *et al.*, 1992). Although this method of gene delivery offers the advantages of low immunogenicity and the ability to carry large fragments of DNA (up to 48 kb), more *in vivo* studies need to be performed in order to determine if receptor-mediated gene delivery offers better efficiency than viral vectors in clinical settings.

B. Viral Delivery Systems

Viral gene delivery, in contrast to exogenous gene delivery, is based on the concept of utilizing the cellular biochemistry to express a specific gene, ribozyme, or antisense molecule endogenously. To express a particular gene or ribozyme, the DNA is cloned into a plasmid or a viral vector and delivered into the cell by transfection or retroviral infection. To date, most of the studies have used retrovirus as the viral delivery system. The first retroviral vector was described more than 16 years ago (Wei *et al.*, 1981) and was the first vector to be used in gene therapy of patients with adenosine deaminase deficiency (Anderson *et al.*, 1990).

Wild-type retroviruses possess a diploid, positive-strand RNA genome of about 9.2 kb (Coffin, 1990). The retrovirus vector is constructed from the DNA form of its genome corresponding to the integrated form of the provirus (Cepko *et al.*, 1984; Temin, 1989; Miller *et al.*, 1993). This type of vector offers a capacity of 8 kb for the insertion of a specific piece of DNA, by removing the central portion of the retrovirus genome containing the *gag, pol,* and *env* genes. The major promoter for these retroviral-based vectors remains the long terminal report (LTR), although many other promoters have also been used (Hock *et al.*, 1989; Palmer *et al.*, 1993; Sullenger and Cech, 1993). Production of viral particles, containing the desired gene, is obtained by transfection of the vector into packaging cells expressing *gag, pol,* and *env,* which are essential for retroviral assembly and the lytic cycle. Replicant-defective retrovirus particles will be obtained from these packaging cells at a titer of about 10^4–10^5 colony-forming units (CFU)/ml and up to 10^9 CFU/ml in certain packaging cells. A wide variety of packaging cells now exist in which proper modifications have been made to eliminate the formation of replication-competent retroviruses owing to recombination between the vector and cellular genome (Danos and Mulligan, 1988; Markowitz *et al.*, 1988; Dougherty *et al.*, 1989).

Retroviral vectors infect replicating cells more efficiently than nonreplicating cells (Roe *et al.*, 1993). This could be an advantage for cancer gene therapy, where only the replicating cancer cells would be targeted by the retrovirus. Integration of the vector in the host cell genome could be of importance when ribozyme expression needs to be permanent in order to alter the genetic abnormality. To date, it has been difficult to obtain long-term expression of integrated retroviral vector in infected cells (Dai *et al.*, 1992; van Beusechem *et al.*, 1992; Blaese, 1993). Although no insertional mutagenesis has yet been observed with retroviral vectors, it remains a serious consideration as to the safety of its use in human. Although retroviral vectors are an efficient gene delivery system it remains that a high titer of the virus must be used in order to obtain acceptable transfection efficiency. This could be a major disadvantage for this type of vector in the clinical setting, where a high concentration of virus must be administered to obtain maximum efficacy. Most of the present clinical gene therapy protocols utilize retroviral vectors as the delivery system for various genes. To date, no significant adverse effects have been observed with the utilization of retroviral vectors in humans. At present more clinical trials are needed to determine if retroviral vectors will fill the "vector void."

Adenoviral vectors first appeared in the early 1980s (Solnick, 1981; Thummel *et al.*, 1981) and were utilized then to transfer genes that could induce cellular transformation (van Doren *et al.*, 1984; Berkner *et al.*, 1987; Sen *et al.*, 1988). In the adenovirus family, which contains 47 different serotypes, adenoviruses 2 and 5 have been well characterized and their genomic DNAs have been completely sequenced (Chroboczek *et al.*, 1992).

The linear adenovirus genome is around 36 kb in length and is divided into 100 map units (mu) of 360 bp each. At each end of the genome is found short inverted terminal repeat (ITRs), which are the origin for DNA replication. The adenovirus appears to rely completely on cellular machinery for transcription, maturation, and RNA processing. Following infection and migration of the adenoviral genome into the nucleus there is no notable integration into the cellular genome (van Doren *et al.*, 1984). Adenovirus vectors therefore exist extrachromosomally and gene expression is only transient. Most of the vectors used today represent modifications of the adenovirus 2 and 5 genomes. The method for constructing an adenoviral vector is based on the replacement of the E1a and E1b genes with the DNA of interest, creating a replication-defective adenovirus (Becker *et al.*, 1994). Although the adenovirus is not associated with human malignancies, the E1 genes appear to have oncogenic capacities; the E1a protein, in particular, acts like the simian virus 40 (SV40) and the HPV-16 E7 transforming proteins in binding the Rb growth suppressor gene (DeCaprio *et al.*, 1988; Whyte *et al.*, 1988). Thus the removal of the E1 regions in the adenoviral vectors contributes to their safety although no malignant transformation has been associated with adenoviruses to date. Deletion of both E1 and E3 genes in adenoviral vectors can permit up to 7.5 kb of foreign DNA to be inserted and up to 105% of the viral genome can be packaged in the adenoviral capsid, while larger viruses show instability (Bett *et al.*, 1993). In fact, only the ITRs and the encapsidation sequences would be necessary for replication and packaging, thus offering a capacity of 36 kb for insertion of foreign DNA into a vector. These defective vectors would then be dependent on a complex cell line expressing the proteins necessary for packaging. If such a system can be developed, this would eliminate the production of the viral proteins and decrease the immunity associated with adenoviruses. In fact, most of the population has been infected with wild-type adenoviruses and immunity directed against this type of virus could be a severe drawback for adenoviral-based gene therapy.

The large size of the adenovirus genome renders direct cloning of foreign DNA into the virus almost impossible. To achieve this goal, a bacterial plasmid is first constructed with adenoviral flanking regions. The high efficiency of adenoviral DNA recombination is then utilized to promote recombination with the bacterial plasmid into a transformed cell line containing the E1 adenoviral genes, such as 293 cells (Graham *et al.*, 1977). Only the product of homologous recombination will produce infectious virus in which the E1 regions are replaced by the foreign DNA. Adenoviruses are stable and can be obtained, in many systems, in higher titer than retrovirus (10^{10} compared to 10^8 CFU/ml, respectively), which gives adenovirus an advantage in clinical situations that require higher quantities of viral particles.

Because the adenovirus can infect both dividing and nondividing cells its efficacy of gene transfer is high in many types of human tumors (Kozarsky and Wilson, 1993; Brody *et al.*, 1994; Trapnell and Gorziglia, 1994; Chen *et al.*, 1995). Adenovirus-mediated transfection of the herpesvirus thymidine kinase in different tumors has resulted in the sensitization of tumors to gancyclovir *in vivo* (Smythe *et al.*, 1995; Chen *et al.*, 1994; Perez-Cruet *et al.*, 1994). Also, adenoviruses expressing interleukin 2 have been shown to induce regression and immunity of a murine breast cancer model *in vivo* (Addison *et al.*, 1995). Adenovirus has also been used to transfer the *p53* gene *in vitro* and resulted in higher apoptosis of the transfected cell line and increased sensitivity to cisplatin (Liu *et al.*, 1994; Clayman *et al.*, 1995). These examples demonstrate the potential of adenovirus as a vector for gene therapy. However, one of the most important concerns in the use of adenovirus remains its potential to elicit an immune response (Yang *et al.*, 1994b). Two important consequences can arise from this situation: First, this will limit repeated injection of the adenovirus; second, it will probably limit the expression of the foreign DNA and thus limit the effectiveness of the therapy (Yang *et al.*, 1994a). Cyclosporine has demonstrated the capacity to prolong adenovirus-mediated gene expression (Engelhardt *et al.*, 1994) and suggests that immunosuppressive conditions might assist in adenovirus-mediated gene therapy. The engineering of future nonimmunologic adenoviral vectors will certainly demonstrate their utility for gene therapy in future clinical trials.

Adeno-associated virus type 2 is an attractive vector system for gene therapy, owing primarily to its lack of association with any pathological human diseases (Berns *et al.*, 1982; Bartlett *et al.*, 1995) and to its ability to infect many types of cell lineages including hematopoietic cells (Laface *et al.*, 1988; Mendelson *et al.*, 1992). Adeno-associated virus is different from the adenovirus family and is in fact a nonautonomous parvovirus with a genome of 4.7 kb encapsidated as a single-stranded DNA molecule with ITRs of 145 bases (Berns, 1990). Adeno-associated virus requires the presence of a helper virus such as herpesvirus, adenovirus, or vaccinia virus in order to replicate in infected cells (Carter, 1990; Carter *et al.*, 1990; Berns, 1990) and also exists as an integrated double-stranded DNA form in the cellular genome when no helper virus is present (Samulski, 1993). Integration of the provirus occurs preferentially in chromosome 19 in about 70% of cases (Kotin *et al.*, 1990,1992; Samulski, 1993). Transduction of human cells by AAV vectors seems to occur mainly in dividing cells (Russell *et al.*, 1994). Current AAV vectors contain the two 145-base ITRs with a specific gene and promoter expression cassette in between (Tratschin *et al.*, 1985; Samulski *et al.*, 1989; Muzyczka, 1992; Zhou *et al.*, 1993). Virus particles are obtained by cotransfection with a helper plasmid containing the Rep and Cap proteins into adenovirus-infected cells, which are usually human KB or 293 cells (Nahreini *et al.*, 1993).

The major advantages of replication-defective (*rep$^-$*) AAV-based vectors are the stability of the viral particles (Muzyczka, 1992), high transduction frequency (McLaughlin *et al.*, 1988), integration into the cellular genome resulting in potential long-term expression of the transgene (Kotin *et al.*, 1990, 1992), and the high efficiency of infection of hematopoietic cells (Samulski *et al.*, 1989; Zhou *et al.*, 1993). Also, *rep*⁻ AAV vectors offer a cloning capacity of a foreign gene of about 4.7 kb (de la Maza and Carter, 1980 Muzyczka, 1992). This could limit the cloning of large DNA fragments into the AAV vectors but offers excellent potential for antisense or ribozyme strategies. Among the disadvantages of this type of vector: 40–80% of adults have existing immunity to AAV (Grossman *et al.*, 1992), some cells are altered after wild-type AAV infection (Walz and Schlehofer, 1992), and the *rep*⁻ viruses are sometimes inserted into other regions than chromosome 19 (Muzyczka, 1992), which could be a potential source of insertional mutagenesis. Adeno-associated virus-based vectors seem to offer good potential for a wide range of gene therapies but important considerations need to be defined further to determine the limits of this vector. Such considerations include the duration of gene expression, the potential risk for insertional mutagenesis, and the requirement for a helper virus for transduction.

In the herpesvirus family, herpes simplex 1 (HSV-1) has been most highly investigated as a potential vector system (for a review, see Efstathiou and Minson, 1995). Herpes simplex 1 is an enveloped virus with a double-stranded DNA genome, 152 kbp in size, comprising more than 70 genes (Roizman and Sears, 1990). Among the advantages of this type of vector, the genome offers a large cloning capacity for foreign gene (30 kb in theory) and does not normally integrate in the cellular genome but remains in a nonintegrated latent state (Kennedy and Steiner, 1993). This vector system also offers the capacity to infect a wide variety of both nonreplicating and replicating cells, especially neural cells (Geller and Federoff, 1991). Presently this type of vector does not offer all the characteristics essential for its use in humans, owing in part to the toxicity produced by defective HSV-1 vectors (Johnson *et al.*, 1992), and also to the uncontrolled latency of HSV infection and transient expression of the transgene (Glorioso *et al.*, 1992). Until these problems can be resolved it is uncertain what role this type of vector will play in gene therapy, especially with antisense sequences or ribozymes, although much progress has been achieved in generating systems free of replication-competent viruses (DeLuca *et al.*, 1985; Weir and Narayanan, 1988; Geller *et al.*, 1990).

Vaccinia virus systems are at an early stage of development. The vaccinia virus is part of the poxvirus family and possesses a 186-kb DNA genome that encodes more than 200 proteins (Goebel *et al.*, 1990) . This virus is one of the most complicated of all animal viruses and can infect both vertebrates and invertebrates (Moss, 1990). DNA fragments up to 25 kb have been inserted into the genome of vaccinia virus by homologous recombination (Smith and Moss, 1983), thus offering a large cloning capacity for gene therapy. Vaccination of human subjects has been performed with vaccinia vectors encoding HIV-1 envelope hybrids and no adverse effects have been observed, suggesting that these vectors could be safe for gene therapy (Cooney *et al.*, 1993; Graham *et al.*, 1993). Hence, the major use proposed for vaccinia vectors has been for vaccination. Further elucidation of the molecular biology and immunity of this virus will be required before it will have any potential as a vector for systemic gene therapy.

VI. Clinical Applications

It is important to select appropriate models for clinical cancer gene therapy. Neoplasms suitable for cancer gene therapy are those that have no successful therapy for their primary tumors or their metastatic lesions. The molecular bases of tumorigenesis have been partially clarified, and several cancer gene therapies have been developed clinically, *i.e.*, immunotherapy, replacement therapy (using suppressor genes), and the regression of onco-gene expression by the use of antisense oligonucleotides. Antisense strategies have several limitations related to their mechanisms of interaction with their target RNAs (Stein and Cheng, 1993). Although ribozymes may face some of these problems, ribozymes also have several advantages over antisense and could be more efficacious in clinical studies in the near future. It is anticipated that the use of ribozymes for cancer treatment will be focused mainly on the inhibition of tumor-specific mutated oncogene expression. For the clinical applications of ribozymes to become feasible, a number of issues should be addressed. First, the ribozyme-mediated therapeutic approach must effectively inhibit specific mRNA-associated with the neoplastic disease. Second, the specificity of ribozyme target recognition must be high, and nonspecific effects should be minimized. In addition, efficient delivery systems with minimal cytotoxicity should be designed (Blau and Springer, 1995).

Ribozymes have several advantages as a nucleic acid therapeutic agent, primarily their specificity and catalytic activity. Ribozymes need short flanking sequences for sufficient recognition of their targets compared with other nucleic acid agents (Hearst, 1988; Herschlag, 1991). The optimum length of recognition arms (*i.e.*, the flanking sequence of hammerhead ribozymes) is said to be six to seven nucleotides on each side of the catalytic core (Ruffner *et al.*, 1990). These sequences are sufficient to recognize target sequences uniquely with minimal nonspecific effects. Furthermore, ribozymes must have the ability to undergo multiple turnovers, increasing their potency to cleave their substrates without the need for a separate enzymatic component.

The method of ribozyme delivery is a key obstacle for further studies (Russell, 1994; Miller and Vile, 1995). Efficient cellular uptake, specific gene targeting, long-term ribozyme expression, and safety must be achieved.

Intracellular delivery of the ribozymes has been attempted by using both exogenous delivery and endogenous expression following transfection (Kiehntopf *et al.*, 1995a,b). In exogenous delivery, lipofection using cationic liposomes has been shown to achieve high cellular uptake. However, the transient expression and the high toxicity of lipofection formulations are major impediments for their clinical application. Viral vectors are considered to be promising technologies for gene delivery and endogenous expression. Several viral vectors have been exploited for ribozyme delivery, including retrovirus, adenovirus, AAV, and herpesvirus. For clinical use, the possibility of residual infectivity and toxicity of these vectors is of concern, and each viral-based vector has its own advantages and disadvantages as discussed above. For example, retrovirus, which has been used in clinical experiments, has the advantage of conferring stable long-term persistence due to integration into the host genome. However, their restricted usage to actively dividing cells, low vector titer, and lack of specific integration sites are potential concerns (San *et al.*, 1993). Adenoviral vectors are becoming more popular vectors for clinical trials. Advantages of the adenovirus vectors are their ability to produce large titers and their ability to infect both dividing and nondividing cells. At present, the adenovirus vector is a promising candidate for ribozyme-mediated gene therapy (Trapnell and Gorziglia, 1994).

Local injection and systemic administration, including intravenous injection, of ribozyme-containing vectors are currently being investigated. Local injection directly into the tumor mass may achieve a higher concentration of agents and lower toxicity by reducing the dilutional effects observed with systemic administration. Carcinomas of the respiratory system, digestive tract, skin, and urinary tract may be suitable systems for antioncogene ribozyme gene therapy using local injection.

Several clinical studies of cancer gene therapy that are targeting oncogenes have been approved. Roth *et al.* (1995) designed a protocol to inhibit expression of the mutant K-*ras* oncogene using intratumor injection of retroviruses encoding an antisense K-*ras*. Other studies targeting oncogenes (including c-*fos*, c-*myc*, *bcr–abl*, and c-*myb* antisense) have also been approved for clinical protocols. Although clinical protocols using ribozymes have not yet been approved, antisense studies could possibly reveal both failures and successes, which may help in the future application of ribozymes. The development of an effective delivery system with minimal toxicity, high transduction efficacy, and minimal production cost would be ideal. The safety and toxicity attendant on the administration of recombinant oligonucleotides, and the different efficacy rates related to the administration route, would also shed light on future ribozyme therapy.

VII. Conclusion

Ribozymes have the ability to modulate specific gene expression because of their site-specific cleavage activity. Ribozymes can be designed for any disease in which a specific protein has been linked to its etiology, and may offer some advantages over antisense oligonucleotide strategies. Cancer is considered a genetic disease, therefore oncogenes are obvious targets for the therapeutic application of anti-oncogene ribozymes. One of the problems for successful gene therapy is to define the role of specific oncogenes in specific tumors. Ribozyme technology can be used to help define and delineate the role of oncogenes in cancer and can be used as a therapeutic agent as well. Extensive studies have investigated the efficacy of antioncogene ribozymes, and have shown successful alteration of the human malignant phenotype *in vivo*. Effective delivery systems with minimal toxicity may advance ribozymes as important therapeutic modalities in the clinical field. Ribozymes could have an important impact on the field of gene therapy in the near future.

Acknowledgments

We thank Ms. Carol Polchow for preparing the manuscript. This research was supported by grants from Gene Shears Research Party, Ltd., Sydney, Australia and the State of California Tobacco-Related Disease Research Program (4RT-0297); Terry Fox Research Fellow Award (from the National Cancer Institute of Canada-Award No. 6597) supported with funds provided by the Terry Fox Run for Dr. David Y. Bouffard; Uehara Memorial Foundation for Research of Life Sciences, Japan for Dr. Toshiya Suzuki; Der Deutschen Forschungsgemeinschaft funds from Germany for Dr. Per Sonne Holm; and City of Hope Fellowship award in the Department of General and Oncologic Surgery for Dr. Lisa D. Curcio.

References

Acsadi, G., Dickson, G., Love, D. R., Jani, A., Walsh, F. S., Gurusignhe, A., Wolff, J. A., and Davies, K. E. (1991a). Human dystrophin expression in mdx mice after intracellular injection of DNA constructs. *Nature (London)* **352,** 815–818.

Acsadi, G., Jiao, S., Jani, A., Duke, D., Williams, P., Chong, W., and Wolf, J. A. (1991b). Direct gene transfer and expression into heart *in vivo*. *New Biol.* **3,** 71–81.

Addison, C., Braciak, T., Ralston, R., Muller, W. J., Gauldie, J., and Graham, F. L. (1995). Intratumoral injection of an adenovirus expressing interleukin 2 induces regression and immunity in a murine breast cancer model. *Proc. Natl. Acad. Sci. U.S.A.* **92,** 8522–8526.

Albino, A. P. (1992). The role of oncogenes and growth factors in progressive melanomagenesis. *Pigm. Cell Res.* **2,** 199–218.

Almoguera, C., Shibata, D., Forrester, K., Martin, J., Arnheim, N., and Perucho, M. (1988). Most human carcinomas of the exocrine pancreas contain mutant c-K-*ras* genes. *Cell (Cambridge, Mass.)* **53,** 549–554.

Anderson, P., Monforte, J., Tritz, R., Nesbitt, S., Hearst, J., and Hampel, A. (1994). Mutagenesis of the hairpin ribozyme. *Nucleic Acids Res.* **22,** 1096–1100.

Anderson, W. F., Blaese, R. M., and Culver, K. (1990). The ADA human gene therapy clinical protocol. *Hum. Gene Ther.* **1,** 331–362.

Aoki, K., Yoshida, T., Sugimura, T., and Terada, M. (1995). Liposome–mediated *in vivo* gene transfer of antisense K-*ras* construct inhibits pancreatic tumor dissemination in the murine peritoneal cavity. *Cancer Res.* **55,** 3810–3816.

Ball, N. J., Yohn, J. J., Morelli, J. G., Norris, D. A., Golitz, L. E., and Hoeffler, J. P. (1994). Ras mutations in human melanoma: A marker of malignant progression. *J. Invest. Dermatol.* **102**, 285–290.

Barbacid, M. (1987). *ras* genes. *Annu. Rev. Biochem.* **56**, 779–827.

Bartlett, J. S., Quattrocchi, K. B., and Samulski, R. J. (1995). The development of adeno-associated virus as a vector for cancer gene therapy. *In* "The Internet Book of Gene Therapy: Cancer Therapeutics" (R. E. Sobol and K. J. Scanlon, eds.), pp. 27–39. Appleton & Lange, Stamford, Connecticut.

Becker, T. C., Noel, R. J., Coats, W. S., Gomez-Foix, A. M., Alam, T., Gerard, R. D., and Newgard, C. B. (1994). Use of recombinant adenovirus for metabolic engineering of mammalian cells. *Methods Cell Biol.* **43**, 161–189.

Beigelman, L., McSwiggen, J. A., Draper, K. G., Gonzalez, C., Jensen, K., Karpeisky, A. M., Modak, A. S., Matulic-Adamic, J., DiRenzo, A. B., Haeberli, P., Sweedler, D., Tracz, D., Grimm, S., Wincott, F. E., Thackray, V. G., and Usman, N. (1995). Chemical modification of hammerhead ribozyme. *J. Biol. Chem.* **270**, 25702–25708.

Bennett, M. J., and Cullimore, J. V. (1992). Selective cleavage of closely-related mRNAs by synthetic ribozymes. *Nucleic Acids Res.* **20**, 831–837.

Berinstein, N., Matthay, K. K., Papaphadjopoulos, D., Levy, R., and Sikic, B. I. (1987). Antibody-directed targeting of liposomes to human cell lines: Role of binding and internalization of growth inhibition. *Cancer Res.* **47**, 5954–5959.

Berkner, K. L., Schaffhausen, B. S., Roberts, T. M., and Sharp, P. A. (1987). Abundant expression of polyomarvirus middle T antigen and dihydrofolate reductase in an adenovirus recombinant. *J. Virol.* **61**, 1213–1220.

Berns, K. I. (1990). Parvoviridae and their replication. *In* "Virology" (B. M. Fields and D. M. Knipe, eds.), pp. 1743–1759. Raven Press, New York.

Berns, K. I., Cheung, A., Ostrove, J., and Lewis, M. (1982). Adeno-associated virus latent infection. *In* "Virus Persistence" (B. W. J. Mahy, A. C. Minson, and G. K. Darby, eds.), p. 248. Cambridge University Press, Cambridge, UK.

Bertino, J. R. (1993). Karnofsky Memorial Lecture. Ode to methotrexate. *J. Clin. Oncol.* **11**, 5–14.

Bertling, W. M., Garis, M., Paspaleeva, V., Zimmer, A., Kreuter, J., Nurnberg, E., and Harrer, P. (1991). Use of liposomes, viral capsids, and nanoparticles as DNA carriers. *Biotechnol. Appl. Biochem.* **13**, 390–405.

Bertram, J., Palfner, K., Killian, M., Brysch, W., Schlingensiepen, K., Hiddemann, W., and Kneba, M. (1995). Reversal of multidrug resistance *in vitro* by phosphorothioate oligonucleotides and ribozymes. *Anti-Cancer Drugs* **6**, 124–134.

Bertrand, E., and Rossi, J. J. (1994). Facilitation of hammerhead ribozymes catalysis by the nucleocapsid protein of HIV-1 and the heterogeneous nuclear ribonucleoprotein A1. *EMBO J.* **13**, 2904–2912.

Bertrand, E., Pictet, R., and Grange, T. (1994). Can hammerhead ribozymes be efficient tools to inactivate gene function? *Nucleic Acids Res.* **22**, 293–300.

Berzal-Herranz, A., Joseph, S., Chowrira, B. M., Butcher, S. E., and Burke, J. M. (1993). Essential nucleotide sequences and secondary structure elements of the hairpin ribozyme. *EMBO J.* **12**, 2564–2567.

Bett, A. J., Prevec, L., and Graham, F. L. (1993). Packaging capacity and stability of human adenovirus type 5 vectors. *J. Virol.* **67**, 5911–5921.

Bishop, J. M. (1991). Molecular themes in oncogenesis. *Cell (Cambridge, Mass.)* **64**, 235–248.

Blaese, R. M. (1993). Development of gene therapy for immunodeficiency: Adenosine deaminase deficiency. *Pediatr. Res.* **33**, 549–555.

Blau, H. M., and Springer, M. L. (1995). Molecular medicine gene therapy—a novel form of drug delivery. *N. Engl. J. Med.* **333**, 1204–1207.

Boiziau, C., Kufurst, R., Cazenave, C., Roig, V., Thuong, N. T., and Toulmé, J.-J. (1991). Inhibition of translation by antisense oligonucleotides via an RNase-H independent mechanism. *Nucleic Acids Res.* **19**, 1113–1119.

Bos, J. L. (1989). *ras* oncogenes in human cancer: A review. *Cancer Res.* **49**, 4682–4689.

Bos, J. L., Fearon, E. R., Hamilton, S. R., Verlaan-de Vries, M., van Boom, J. H., van der Ed, A. J., and Vogelstein, B. (1987). Prevalence of *ras* gene mutations in human colorectal cancers. *Nature (London)* **27**, 293–297.

Branch, A. D., and Robertson, H. D. (1991). Efficient *trans* cleavage and a common structural motif for the ribozymes of the human hepatitis δ agent. *Proc. Natl. Acad. Sci. U.S.A.* **88**, 10163–10167.

Bratty, J., Chartrand, P., Ferbeyre, G., and Cedergren, R. (1993). The hammerhead RNA domain, a model ribozyme. *Biochim. Biophys. Acta* **1216**, 345–359.

Brody, S. L., Jaffe, H. A., Han, S. K., Wersto, R. P., and Crystal, R. G. (1994). Direct *in vivo* gene transfer and expression in malignant cells using adenovirus vectors. *Hum. Gene Ther.* **5**, 437–447.

Brunton, V. G., and Workman, P. (1993). Cell-signaling targets for antitumor drug development. *Cancer Chemother. Pharmacol.* **32**, 1–19.

Butcher, S. E., and Burke, J. M. (1994). A photo-cross-linkable tertiary structure motif found in functionally distinct RNA molecules is essential for catalytic function of the hairpin ribozyme. *Biochemistry* **33**, 992–999.

Buzayan, J. M., Hampel, A., and Bruening, G. (1986). Nucleotide sequence and newly formed phosphodiester bond of spontaneously limaged satellite tobacco ringspot virus RNA. *Nucleic Acids Res.* **14**, 9729–9743.

Buzayan, J. M., Feldstein, P. A., Bruening, G., and Eckstein, F. (1988). RNA mediated formation of a phosphorothioate diester bond. *Biochem. Biophys. Res. Commun.* **156**, 340–347.

Cameron, F. H., and Jennings, P. A. (1989). Specific gene suppression by engineered ribozymes in monkey cells. *Proc. Natl. Acad. Sci. U.S.A.* **86**, 9139–9143.

Cantley, L. C., Auger, K. R., Carpenter, C., Duckworth, B., Graziani, A., Kapeller, R., and Soltoff, S. (1991). Oncogenes and signal transduction. *Cell (Cambridge, Mass.)* **64**, 281–302.

Carter, B. J. (1990). The growth cycle of adeno-associated virus. *In* "Handbook of Parvoviruses" (P. Tjissen, ed.), Vol. 1, pp. 155–168. CRC Press, Boca Raton, Florida.

Carter, B. J., Mendelson, E., and Trempe, J. P. (1990). AAV DNA replication, integration, and genetics. *In* "Handbook of Parvoviruses" (P. Tjissen, ed.), Vol. 1, pp. 169–262. CRC Press, Boca Raton, Florida.

Castanotto, D., Rossi, J. J., and Sarver, N. (1994). Antisense catalytic RNAs as therapeutic agents. *Adv. Pharmacol.* **25**, 289–317.

Cech, T. R. (1988). Ribozymes and their medical implications. *JAMA, J. Am. Med. Assoc.* **260**, 3030–3034.

Cech, T. R., and Uhlenbeck, O. C. (1994). Hammerhead nailed down. *Nature (London)* **372**, 39–40.

Cech, T. R., Zaug, A. J., and Grabowski, P. J. (1981). *In vitro* splicing of the ribosomal RNA precursor of *Tetrahymena:* Involvement of a guanosine nucleotide in the excision of the intervening sequence. *Cell (Cambridge, Mass.)* **27**, 487–496.

Cepko, C. L., Roberts, B. E., and Mulligan, R. C. (1984). Construction and applications of a highly transmissible murine retrovirus shuttle vector. *Cell (Cambridge, Mass.)* **37**, 1053–1062.

Chartrand, P., Harvey, S. C., Febeyre, G., Usman, N., and Cedergren, R. (1995). An oligonucleotide that supports catalytic activity in the hammerhead ribozyme domain. *Nucleic Acids Res.* **23**, 4091–4096.

Chen, C.-J., Banerjea, A. C. A., Harmision, G. G., Haglund, K., and Schubert, M. (1992). Multitarget-ribozymes directed to cleave at up to nine highly conserved HIV-1 *env* RNA regions inhibits HIV-1 replication—potential effectiveness against most presently sequenced HIV-1 isolates. *Nucleic Acids Res.* **20**, 4581–4589.

Chen, P.-J., Kalpana, G., Goldberg, J., Mason, W., Werner, B., Gerin, J., and Taylor, J. (1986). Structure and replication of the genome of the hepatitis δ virus. *Proc. Natl. Acad. Sci. U.S.A.* **83**, 8774–8778.

Chen, S.-H., Shine, H. D., Goodman, J. C., Grossman, R. G., and Woo, S. L. (1994). Gene therapy for brain tumors: Regression of experimental gliomas by adenovirus-mediated gene transfer *in vivo. Proc. Natl. Acad. Sci. U.S.A.* **91,** 3054–3057.

Chen, S.-H., Chen, X. H., Wang, Y., Kosai, K., Fingegold, M. J., Rich, S. S., and Woo, S. L. (1995). Combination gene therapy for liver metastasis of colon carcinoma *in vivo. Proc. Natl. Acad. Sci. U.S.A.* **92,** 2577–2581.

Chowrira, B. M., Berzal-Herranz, A., and Burke, J. M. (1991). Novel guanosine requirement for catalysis by the hairpin ribozyme. *Nature (London)* **354,** 320–322.

Chowrira, B. M., Berzal-Herranz, A., Keller, C. F., and Burke, J. M. (1993). Four ribose 2′-hydroxyl groups essential for catalytic function of the hairpin ribozyme. *J. Biol. Chem.* **268,** 19458–19462.

Christoffersen, R. E., and Marr, J. J. (1995). Ribozymes as human therapeutic agents. *J. Med. Chem.* **38,** 2023–2037.

Chroboczek, J., Bieber, F., and Jacrot, B. (1992). The sequence of the genome of adenovirus type 5 and its comparison with the genome of adenovirus type 2. *Virology* **186,** 280–285.

Clayman, G. L., el-Naggar, A. K., Roth, J. A., Zhang, W. W., Goepfert, H., Taylor, D. L., and Liv, T. J. (1995). *In vivo* molecular therapy with p53 adenovirus for microscopic residual head and neck squamous carcinoma. *Cancer Res.* **55,** 1–6.

Coffin, J. M. (1990). Retroviridae and their replication. *In* "Virology" (B. N. Field, D. M. Knipe, R. M. Chanock, J .L. Melnick, M. S. Hirsch, T. P. Monath, and B. Roizman, eds.), pp. 1437–1489. Raven Press, New York.

Cole, S. P. C., Bhardwaj, G., Gerlach, J. H., Mackie, J. E., Grant, C. E., Almquist, K. C., Stewart, A. J., Kurz, E. U., Duncan, A. M. V., and Deeley, R. G. (1992). Overexpression of a transporter gene in multidrug-resistant human lung cancer cell line. *Science* **258,** 1650–1654.

Cooney, E. L., McElrath, M. J., Corey, L., Hu, S. L., Collier, A. C., Arolittia, D., Hoffman, M., Coombs, R. W., Smith, G. E., and Greenberg, P. D. (1993). Enhanced immunity to human immunodeficiency virus (HIV) envelope elicited by a combined vaccine regimen consisting of priming with a vaccinia recombinant expressing HIV envelope and boosting with gp160 protein. *Proc. Natl. Acad. Sci. U.S.A.* **90,** 1882–1886.

Corrias, M. V., and Tonini, G. P. (1992). An oligomer complementary to the 5′ end region of MDR1 gene decreases resistance to doxorubicin of human adenocarcinoma-resistant cells. *Anticancer Res.* **12,** 1431–1438.

Cotten, M., Wagner, E., Zatloukal, K., Phillips, S., Curiel, D. T., and Birnstiel, M. L. (1992). High-efficiency receptor-mediated delivery of small and large 48 kilobase gene constructs using the endosome-disruption activity of defective or chemically inactivated adenovirus particles. *Proc. Natl. Acad. Sci. U.S.A.* **89,** 6094–6098.

Cristiano, R. J., Smith, L. C., and Woo, S. L. (1993). Hepatic gene therapy: Adenovirus enhancement of receptor-mediated gene delivery and expression in primary hepatocytes. *Proc. Natl. Acad. Sci. U.S.A.* **90,** 2122–2126.

Curiel, D. T., Agarwal, S., Wagner, E., and Cotten, M. (1991). Adenovirus enhancement of transferrin–polylysine-mediated gene delivery. *Proc. Natl. Acad. Sci. U.S.A.* **88,** 8850–8854.

Curiel, D. T., Agarwal, S., Romer, M. U., Wagner, E., Cotten, M., Birnstiel, M. L., and Boucher, R. C. (1992a). Gene transfer to respirator epithelial cells via the receptor-mediated endocytosis pathway. *Am. J. Respir. Cell Mol. Biol.* **6,** 247–252.

Curiel, D. T., Wagner, E., Cotten, M., Birnstiel, M. L., Agarwal, S., Li, C. M., Loechel, S., and Hu, P. C. (1992b). High-efficiency gene transfer mediated by adenovirus coupled to DNA–polylysine complexes. *Hum. Gene Ther.* **3,** 147–154.

Czubayko, F., Riegel, A. T., and Wellstein, A. (1994). Ribozyme-targeting elucidates a direct role of pleiotrophin in tumor growth. *J. Biol. Chem.* **269,** 21358–21363.

Dai, Y., Roman, M., Naviaux, R. K., and Verma, I. M. (1992). Gene therapy via primary myoblasts: Long-term expression of factor IX protein following transplantation *in vivo. Proc. Natl. Acad. Sci. U.S.A.* **89,** 10892–10895.

Danos, O., and Mulligan, R. C. (1988). Safe and efficient generation of recombinant retroviruses with amphotropic and ecotropic host ranges. *Proc. Natl. Acad. Sci. U.S.A.* **85**, 6460–6464.

DeCaprio, J. A., Ludlow, J. W., Figge, J., Stein, J. Y., Huang, C. M., Lee, W. H., Marsilio, E., Paucha, E., and Livingstone, D. M. (1988). SV40 large tumor antigen forms a specific complex with the product of the retinoblastoma susceptibility gene. *Cell* (*Cambridge, Mass.*) **54**, 275–283.

de la Maza, L. M., and Carter, B. J. (1980). Molecular structure of adeno-associated virus variant DNA. *J. Biol. Chem.* **255**, 3194–3203.

DeLuca, N. A., McCarthy, A. M., and Schaffer, P. A. (1985). Isolation and characterization of deletion mutants of herpes simplex virus type 1 in the gene encoding immediate early regulatory protein ICP4. *J. Virol.* **56**, 558–570.

Dorai, T., Kobayashi, H., Holland, J. F., and Ohnuma, T. (1994). Modulation of platelet-derived growth factor-beta mRNA expression and cell growth in a human mesothelioma cell line by a hammerhead ribozyme. *Mol. Pharmacol.* **46**, 437–444.

Dougherty, J. P., Wisnewski, R., Yang, S., Yang, S. L., Rhoode, B. W., and Termin, H. M. (1989). New retrovirus helper cells with almost no nucleotide homology to retrovirus vectors. *J. Virol.* **63**, 3209–3212.

Efferth, T., and Volm, M. (1993). Modulation of P-glycoprotein-mediated multidrug resistance by monoclonal antibodies, immunotoxins or antisense oligodeoxynucleotides in kidney carcinoma and normal kidney cells. *Oncology* **50**, 303–308.

Efstathiou, S., and Minson, A. C. (1995). Herpes virus based vectors. *Br. Med. Bull.* **51**, 45–55.

Egan, S. E., and Weinberg, R. A. (1993). The pathway to signal achievement. *Nature* (*London*) **365**, 781–783.

Endicott, J. A., and Ling, V. (1989). The biochemistry of P-glycoprotein mediated multidrug resistance. *Annu. Rev. Biochem.* **58**, 131–171.

Engelhardt, J. F., Ye, X., Doranz, B., and Wilson, J. M. (1994). Ablation of E2A in recombinant adenovirus improves transgene persistence and decreases inflammatory response in mouse liver. *Proc. Natl. Acad. Sci. U.S.A.* **91**, 6196–6200.

Fearon, E. R., and Vogelstein, B. (1990). A genetic model for colorectal tumorigenesis. *Cell* (*Cambridge, Mass.*) **61**, 759–767.

Fedor, M. J., and Uhlenbeck, O. C. (1990). Substrate sequence effects on "hammerhead" RNA catalytic efficiency. *Proc. Natl. Acad. Sci. U.S.A.* **87**, 1668–1672.

Feldstein, P. A., and Bruening, G. (1993). Catalytically active geometry in the reversible circularization of "mini-monomer" RNAs derived from the complementary strand of tobacco ringspot virus satellite RNA. *Nucleic Acids Res.* **21**, 1991–1998.

Feldstein, P. A., Buzayan, J. M., and Bruening, G. (1989). Two sequences participating in the autolytic processing of satellite tobacco ringspot virus complementary RNA. *Gene* **82**, 53–61.

Felgner, J. H., Kumar, R., Sridhar, C. M., Wheeler, C. J., Tsai, Y. J., Border, R., Ramsey, P., Martin, M., and Felgner, P. L. (1994). Enhanced gene delivery and mechanism studies with a novel series of cationic lipid formulations. *J. Biol. Chem.* **269**, 2550–2561.

Felgner, P. L., and Ringold, G. M. (1989). Cationic liposome-mediated transfection. *Nature* (*London*) **337**, 387–388.

Felgner, P. L., Gadek, T. R., Holm, M., Roman, R., Chan, H. W., Wenz, M., Northrop, J. P., and Ringold, G. M. (1987). Lipofection: A highly efficient lipid-mediated DNA-transfection procedure. *Proc. Natl. Acad. Sci. U.S.A.* **84**, 7413–7417.

Feng, M., Cabrera, G., Deshane, J., Scanlon, K. J., and Curiel, D. T. (1995). Neoplastic reversion accomplished by high efficiency adenoviral mediated delivery of anti-*ras* ribozymes. *Cancer Res.* **55**, 2024–2028.

Fojo, A. T., Ueda, K., Slamon, D. J., Poplack, D. G., Gottesman, M. M., and Pastan, I. (1987). Expression of multidrug-resistance gene in human tumors and tissues. *Proc. Natl. Acad. Sci. U.S.A.* **84**, 265–269.

Ford, J. M., and Hait, W. N. (1993). Pharmacologic circumvention of multidrug resistance. *Cytotechnology* **12**, 171–212.

Forrester, K., Almoguera, C., Han, K., Grizzle, W. E., and Perucho, M. (1987). Detection of high incidence of K-*ras* oncogenes during human tumorigenesis. *Nature (London)* **327**, 298–303.

Forster, A. C., and Symons, R. H. (1987). Self-cleavage of plus and minus RNAs of a virusoid and structural model for the active sites. *Cell (Cambridge, Mass.)* **49**, 211–220.

Foulds, L. (1958). The natural history of cancer. *J. Chronic Dis.* **8**, 2–37.

Funato, T., Yoshida, E., Jiao, L., Tone, T., Kashani-Sabet, M., and Scanlon, K. J. (1992). The utility of the anti-*fos* ribozyme in reversing cisplatin resistance in human carcinomas. *Adv. Enzyme Regul.* **32**, 195–209.

Funato, T., Shitara, T., Tone, T., Jiao, L., Kashani-Sabet, M., and Scanlon, K. J. (1994). Suppression of H-*ras*-mediated transformation in NIH 3T3 cells by a *ras* ribozyme. *Biochem. Pharmacol.* **48**, 1471–1475.

Garte, S. J. (1993). The c-*myc* oncogene in tumor progression. *Crit. Rev. Oncog.* **4**, 435–449.

Geller, A. I. and Federoff, H. J. (1991). The use of HSV-1 vectors to introduce heterologous genes into neurons: Implications for gene therapy. *In* "Human Gene Transfer" (O. Cohen-Hauguenauer and M. Borion, eds.), pp. 63–73. John Libbey Eurotext, France.

Geller, A. I., Keyomarsi, K., Bryan, J., and Pardee, A. B. (1990). An efficient deletion mutant packaging system for defective herpes simplex virus vectors: Potential applications to human gene therapy and neuronal physicology. *Proc. Natl. Acad. Sci. U.S.A.* **87**, 8950–8954.

Gerlach, W. L., Llewellyn, D., and Haseloff, J. (1987). Construction of a plant disease resistance gene from the satellite RNA of tobacco ringspot virus. *Nature (London)* **328**, 802–805.

Gewirtz, A. M. (1994). Treatment of chronic myelogenous leukemia (CML) with c-*myb* antisense oligodeoxynucleotides. *Bone Marrow Transplant.* **14** (Suppl. 3), 57–61.

Glorioso, J. G., Goins, W. F., and Fink, D. J. (1992). Herpes simplex virus-based vectors. *Semin. Virol.* **3**, 265–276.

Goebel, S. J., Johnson, G. P., Perkus, M. E., Davis, S. W., Winslow, J. P., and Paoletti, E. (1990). The complete DNA sequence of vaccinia virus. *Virology* **179**, 247–266.

Goodchild, J. (1992). Enhancement of ribozyme catalytic activity by a contiguous oligodeoxynucleotide (facilitator) and by 2′-O-methylation. *Nucleic Acids Res.* **20**, 4607–4612.

Gottesman, M. M., and Pastan, I. (1993). Biochemistry of multidrug resistance mediated by the multidrug transporter. *Annu. Rev. Biochem.* **62**, 385–427.

Gottschalk, S., Cristiano, R. J., Smith, L. C., and Woo, S. L. C. (1993). Folate receptor mediated DNA delivery into tumor cells: Potosomal disruption results in enhanced gene expression. *Gene Ther.* **1**, 185–191.

Graham, B. S., Matthews, T. J., Belshe, R. B., Clements, M. L., Dolin, R., Wright, P. F., Gorse, G. J., Schwarts, D. H., Keefer, M. C., and Bologhesi, D. P. (1993). Augmentation of human immunodeficiency virus type 1 neutralizing antibody by priming with gp160 recombinant vaccinia and boosting with rgp160 in vaccinia-naive adults. *J. Infect. Dis.* **167**, 533–537.

Graham, F. L., Smiley, J., Russel, W. C., and Nairn, R. (1977). Characteristics of a human cell line transformed by DNA from human adenovirus 5. *J. Gen. Virol.* **36**, 59–74.

Gros, P., Croop, J., Roninson, I., Varshavsky, A., and Housman, D. E. (1986). Isolation and characterization of DNA sequences amplified in multidrug resistant hamster cells. *Proc. Natl. Acad. Sci. U.S.A.* **83**, 337–341.

Grossman, Z., Mendelson, E., Brok-Simoni, F., Milegvir, F., Leifner, Y., Rechavi, G., and Ramot, B. (1992). Detection of adeno-associated virus type 2 in human peripheral blood cells. *J. Gen. Virol.* **73**, 961–966.

Grunewald, K., Lyons, J., Frohlich, A., Feichtinger, H., Weger, R. A., Schwab, G., Janssen, J. W., and Bartram, C. R. (1989). High frequency of Ki-*ras* codon 12 mutations in pancreatic adenocarcinomas. *Int. J. Cancer* **43**, 1037–1041.

Guerrier-Takada, C., Gardiner, K., Marsh, T., Pace, N., and Altman, S. (1983). The RNA moiety of ribonuclease P is the catalytic subunit of the enzyme. *Cell (Cambridge, Mass.)* **35**, 849–857.

Gunning, P., Leavitt, J., Muscat, G., Ng, S., and Kedes, L. (1987). A human beta-actin expression vector system directs high-level accumulation of antisense transcripts. *Proc. Natl. Acad. Sci. U.S.A.* **84**, 4831–4835.

Hampel, A., and Tritz, R. (1989). RNA catalytic properties of the minimum (−)sTRSV sequence. *Biochemistry* **28**, 4929–4933.

Hampel, A., Tritz, R., Hicks, M., and Cruz, P. (1990). "Hairpin" catalytic RNA model: Evidence for helices and sequence requirement for substrate RNA. *Nucleic Acids Res.* **18**, 299–304.

Harris, C. E., Agarwal, S., Hu, P., Wagner, E., and Curiel, D. T. (1993). Receptor-mediated gene transfer to airway epithelial cells in primary culture. *Am. J. Respir. Cell Mol. Biol.* **9**, 441–447.

Haseloff, J., and Gerlach, W. L. (1988). Simple RNA enzymes with new and highly specific endoribonuclease activities. *Nature (London)* **334**, 585–591.

Haseloff, J., and Gerlach, W. L. (1989). Sequences required for self-catalyzed cleavage of the satellite RNA of tobacco ringspot virus. *Gene* **82**, 43–52.

He, Y.-K., Lu, C. D., and Qi, G. (1993). *In vitro* cleavage of HPV16 E6 and E7 RNA fragments by synthetic ribozymes and transcribed ribozymes from RNA-trimming plasmids. *FEBS Lett.* **322**, 21–24.

Hearst, J. E. (1988). A photochemical investigation of the dynamics of oligonucleotide hybridization. *Annu. Rev. Phys. Chem.* **39**, 291–315.

Heidenreich, O., and Eckstein, F. (1992). Hammerhead ribozyme-mediated cleavage of the long terminal repeat RNA of human immunodeficiency virus type 1. *J. Biol. Chem.* **267**, 1904–1909.

Heidenreich, O., Pieken, W., and Eckstein, F. (1993). Chemically modified RNA: Approaches and applications. *FASEB J.* **7**, 90–96.

Heidenreich, O., Benseler, F., Fahrenholz, A., and Eckstein, F. (1994). High activity and stability of hammerhead ribozymes containing 2′-modified pyrimidine nucleosides and phosphorothioates. *J. Biol. Chem.* **269**, 2131–2138.

Hendry, P., and McCall, M. J. (1995). A comparison of the *in vitro* activity of DNA-armed and all-RNA hammerhead ribozymes. *Nucleic Acids Res.* **23**, 3928–3936.

Hendry, P., McCall, M. J., Santiago, F. S., and Jennings, P. A. (1992). A ribozyme with DNA in the hybridizing arms displays enhanced cleavage ability. *Nucleic Acids Res.* **20**, 5737–5741.

Hendry, P., McCall, M. J., Santiago, F. S., and Jennings, P. A. (1995). *In vitro* activity of minimized hammerhead ribozyme. *Nucleic Acids Res.* **23**, 3922–3927.

Herschlag, D. (1991). Implications of ribozyme kinetics for targeting the cleavage of specific RNA molecules *in vivo:* More isn't always better. *Proc. Natl. Acad. Sci. U.S.A.* **88**, 6921–6925.

Herschlag, D., Khosla, M., Tsuchihashi, Z., and Karpel, R. L. (1994). An RNA chaperone activity of non-specific RNA binding proteins in hammerhead ribozyme catalysis. *EMBO J.* **13**, 2913–2914.

Hertel, K. J., Pardi, A., Uhlenbeck, O. C., Koizumi, M., Ohtsuka, E., Uesugi, S., Cedergren, R., Eckstein, F., Gerlach, W. L., Hodgson, R., and Symons, R. H. (1992). Numbering system for the hammerhead. *Nucleic Acids Res.* **20**, 3252.

Hock, R. A., Miller, A. D., and Osborne, W. R. A. (1989). Expression of human adenosine deaminase from various strong promoters after gene transfer into human hematopoietic cell lines. *Blood* **74**, 876–881.

Holm, P. S., Scanlon, K. J., and Dietel, M. (1994). Reversion of multidrug resistance in the P-glycoprotein-positive human pancreatic cell line (EPP85-181RDB) by introduction of a hammerhead ribozyme. *Br. J. Cancer* **70**, 239–243.

Holm, P. S., Dietel, M., and Krupp, G. (1995). Similar cleavage efficiencies of an oligoribonucleotide substrate and an *mdr1* mRNA segment by a hammerhead ribozyme. *Gene* **167**, 221–225.

Homann, M., Tzortzakaki, S., Rittner, K., Sczkiel, G., and Tabler, M. (1993). Incorporation of the catalytic domain of a hammerhead ribozyme into antisense RNA enhances its inhibitory effect on the replication of human immunodeficiency virus type 1. *Nucleic Acids Res.* **21**, 2809–2814.

Hunter, T. (1991). Cooperation between oncogenes. *Cell (Cambridge, Mass.)* **64**, 249–270.

Hutchins, C. J., Rathjen, P. D., Forster, A. C., and Symons, R. H. (1986). Self-cleavage of plus and minus RNA transcripts of avocado sunblotch viroid. *Nucleic Acids Res.* **14**, 3627–3640.

Inouye, M. (1988). Antisense RNA: Its functions and applications in gene regulation–a review *Gene* **72**, 25–34.

Ishida, H., Kijima, H., Ohta, Y., Kashani-Sabet, M., and Scanlon, K. J. (1995). Mechanisms of cisplatin resistance and its reversal in human tumors. *In* "Alternative Mechanisms of Multidrug Resistance in Cancer" (J. A. Kallen, ed.), pp. 225–264. Birkhaeuser, Boston.

Jaeger, J. A., Turner, D. H., and Zuker, M. (1989). Improved predictions of secondary structures for RNA. *Proc. Natl. Acad. Sci. U.S.A.* **86**, 7706–7710.

Jaroszweski, J. W., Kaplan, O., Syi, J., Sehested, M., Faustino, P. J., and Cohen, J. S. (1990). Concerning antisense inhibition of the multiple drug resistance gene. *Cancer Commun.* **2**, 287–294.

Johnson, P. A., Miyanohara, A., Levin, F., Cahill, T., and Friedman, T. (1992). Cytotoxicity of a replication-defective mutant of herpes simplex virus type 1. *J. Virol.* **66**, 2952–2965.

Jolly, D. (1994). Viral vector systems for gene therapy. *Cancer Gene Ther.* **1**, 51–64.

Joseph, S., Berzal-Herranz, A., Chowrira, B. M., Butcher, S. E., and Burke, J. M. (1993). Substrate selection rules for the hairpin ribozyme determined by *in vitro* selection, mutation, and analysis of mismatched substrates. *Genes Dev.* **7**, 130–138.

Juliano, R. L., and Ling, V. (1976). A surface glycoprotein modulating drug permeability in Chinese hamster ovary cell mutants. *Biochim. Biophys. Acta* **455**, 152–162.

Kartner, N., Riordan, J. R., and Ling, V. (1983). Cell surface P-glycoprotein associated with multidrug resistance in mammalian cell lines. *Science* **221**, 1285–1288.

Kartner, N., Evernden-Porelle, D., Bradley, G., and Ling, V. (1985). Detection of P-glycoprotein in multidrug resistant cell lines by monoclonal antibodies. *Nature (London)* **316**, 820–823.

Kashani-Sabet, M., and Scanlon, K. J. (1995). Application of ribozymes to cancer gene therapy. *Cancer Gene Ther.* **2**, 213–223.

Kashani-Sabet, M., Funato, T., Tone, T., Jiao, L., Wang, W., Yoshida, E., Kashfian, B. I., Shitara, T., Wu, A. M., Moreno, J. G., Traweek, S. T., Ahlering, T. E., and Scanlon, K. J. (1992). Reversal of the malignant phenotype by an anti-*ras* ribozyme. *Antisense Res. Dev.* **2**, 3–15.

Kashani-Sabet, M., Funato, T., Florenes, V. A., Fodstad, O., and Scanlon, K. J. (1994). Suppression of the neoplastic phenotype *in vivo* by an anti-*ras* ribozyme. *Cancer Res.* **54**, 900–902.

Kennedy, P. G. E., and Steiner, I. (1993). The use of herpes simplex virus vectors for gene therapy in neurological diseases. *Q. J. Med.* **86**, 697–702.

Kiehntopf, M., Brach, M. A., Licht, T., Petschauer, S., Karawajew, L., Kirschning, C., and Hermann, F. (1994). Ribozyme-mediated cleavage of the MDR-1 transcript restores chemosensitivity in previously resistant cancer cells. *EMBO J.* **13**, 4645–4652.

Kiehntopf, M., Esquivel, E. L., Brach, M. A., and Herrmann, F. (1995a). Clinical applications of ribozymes. *Lancet* **345**, 1027–1031.

Kiehntopf, M., Esquivel, E. L., Brach, M. A., and Herrmann, F. (1995b). Ribozymes: Biology, biochemistry, and implications for clinical medicine. *J. Mol. Med.* **73**, 65–71.

Kijima, H., Ishida, H., Ohkawa, T., Kashani-Sabet, M., and Scanlon, K. J. (1995). Therapeutic application of ribozymes. *Pharmacol. Ther.* **68**, 247–267.

Kijima, H., Bouffard, D. Y., and Scanlon, K. J. (1996). Ribozyme-mediated reversal of human pancreatic carcinoma phenotype. *In* "Proceedings of International Symposium on Bone Marrow Transplantation" (S. Ikehara, ed.). pp. 153–163. Springer, Berlin.

Kobayashi, H., Kim, N., Halatsch, M., and Ohnuma, T. (1993). Specificity of ribozyme designed for mutated *DHFR* mRNA. *Biochem. Pharmacol.* **47,** 1607–1613.

Kobayashi, H., Dorai, T., Holland, J. F., and Ohnuma, T. (1994). Reversal of drug sensitivity in multidrug-resistant tumor cells by MDR-1(PGY-1) ribozyme. *Cancer Res.* **54,** 1271–1275.

Koizumi, M., Hayase, Y., Iwai, S., Kamiya, H., Inoue, H., and Ohtsuka, E. (1989). Design of RNA enzymes distinguishing a single base mutation in RNA. *Nucleic Acids Res.* **17,** 7059–7071.

Koizumi, M., Kamiya, H., and Ohtsuka, E. (1992). Ribozymes designed to inhibit transformation of NIH 3T3 cells by the activated c-Ha-*ras* gene. *Gene* **117,** 179–184.

Kotin, R. M., Siniscalco, M., Samulski, R. J., Zhu, X. D., Hunter, L., Laughlin, C. A., McLaughlin, S., Muzyczka, N., Rocchi, M., and Berns, K. I. (1990). Site-specific integration by adeno-associated virus. *Proc. Natl. Acad. Sci. U.S.A.* **87,** 2211–2215.

Kotin, R. M., Linden, R. M., and Berns, K. I. (1992). Characterization of a preferred site on human chromosome 19q for integration of adeno-associated virus DNA by non-homologous recombination. *EMBO J.* **11,** 5071–5078.

Kozarsky, K. F., and Wilson, J. M. (1993). Gene therapy adenvirus vectors. *Curr. Opin. Genet. Dev.* **3,** 499–503.

Krishnamachary, N., and Center, M. S. (1993). The MRP gene associated with a non-P-glycoprotein multidrug resistance encodes a 190-kDa membrane bound glycoprotein. *Cancer Res.* **53,** 3658–3661.

Kruger, K., Grabowski, P. J., Zaug, A. J., Sands, J., Gottschling, D. E., and Cech, T. R. (1982). Self-splicing RNA: Autoexcision and autocyclization of the ribosomal RNA intervening sequence of *Tetrahymena. Cell (Cambridge, Mass.)* **31,** 147–157.

Kumar, P. K. R., and Ellington, A. D. (1995). Artifical evolution and natural ribozymes. *FASEB J.* **9,** 1183–1195.

Kumar, P. K. R., Taira, K., and Nishikawa, S. (1994). Chemical probing studies of variants of the genomic hepatitis delta virus ribozyme by primer extension analysis. *Biochemistry* **33,** 583–592.

Kurzock, R., Gutterman, J. U., and Talpaz, M. (1988). The molecular genetics of Philadelphia chromosome-positive leukemias. *N. Engl. J. Med.* **319,** 990–998.

Laface, D., Hermonat, P., Wakeland, E., and Peck, A. (1988). Gene transfer into hematopoietic progenitor cells mediated by an adeno-associated virus vector. *Virology* **162,** 483–486.

Lange, W., Canten, E. M., Finke, J., and Dolken, G. (1993). *In vitro* and *in vivo* effects of synthetic ribozymes targeted against BCR/ABL mRNA. *Leukemia* **7,** 1786–1794.

Lange, W., Daskalakis, M., Finke, J., and Dolken, G. (1994). Comparision of different ribozymes for efficient and specific cleavage of BCR/ABL related mRNAs. *FEBS Lett.* **338,** 175–178.

Larsson, S., Hotchkiss, G., Andang, M., Nyholm, T., Inzunza, J., Jansoon, I., and Ahrlund-Richter, L. (1994). Reduced β-microglobulin mRNA levels in transgenic mice expressing a designed hammerhead ribozyme. *Nucleic Acids Res.* **22,** 2242–2248.

Lechner, M. S., Mack, D. H., Finicle, A. B., Crook, T., Vousden, K. H., and Laimins, L. A. (1992). Human papillomavirus E6 proteins bind *p53 in vivo* and abrogate *p53*-mediated repression of transcription. *EMBO J.* **11,** 3045–3052.

Legendre, J. Y., and Szoka, F. C. (1992). Delivery of plasmid DNA into mammalian cell lines using pH-sensitive liposomes: Comparison with cationic lipsomes. *Pharmacol. Res.* **9,** 1235–1242.

Leopold, L. H., Shore, S. K., Newkirk, T. A., Reddy, R. M. V., and Reddy, E. P. (1995). Multi-unit ribozyme mediated cleavage of *bcr-abl* mRNA in myeloid leukemias. *Blood* **85,** 2162–2170.

Leserman, L. D., Weinstein, J. N., Blumenthal, R., and Terry, W. D. (1980). Receptor-mediated endocytosis of antibody-opsonized liposomes by tumor cells. *Proc. Natl. Acad. Sci. U.S.A.* **77**, 4089–4093.

Lichtenberg, D. (1988). Liposomes: Preparation, characterization, and preservation. *Methods Biochem. Anal.* **33**, 337–468.

Linnenbach, A. J., Huebner, K., Reddy, P., Herlyn, M., Parmiter, A. H., Nowell, P. C., and Koprowski, H. (1988). Structural alteration in the *MYB* proto-oncogene and deletion within the gene encoding α-type protein kinase C in human melanoma cell lines. *Proc. Natl. Acad. Sci. U.S.A.* **85**, 74–78.

Litzinger, D. C., and Huang, L. (1992). Phosphatidylethanolamine liposomes: Drug delivery, gene transfer and immunodiagnostic applications. *Biochim. Biophys. Acta* **1113**, 201–227.

Liu, T. J., Zhang, W. W., Taylor, D. L., Goephert, H., and Clayman, G. L. (1994). Growth suppression of human head and neck cancer cells by the introduction of a wild-type *p53* gene via a recombinant adenovirus. *Cancer Res.* **54**, 2287–2291.

Lowy, D. R., Kirnbauer, R., and Schiller, J. T. (1994). Genital human papillomavirus infection. *Proc. Natl. Acad. Sci. U.S.A.* **91**, 2436–2440.

Lu, C. D., Chatterjee, S., Brar, D., and Wong, K. K., Jr. (1994). Ribozyme-mediated *in vitro* cleavage of transcripts arising from the major transforming genes of human papillomavirus type 16. *Cancer Gene Ther.* **1**, 267–277.

Maddoen, C. S., Morrow, C. S., Nakagawa, M., Goldsmith, M. E., Fairchild, C. R., and Cowan, K. H. (1993). Identification of the 5′ and 3′ sequences involved in the regulation of transcription of the human *mdr1* gene *in vivo*. *J. Biol. Chem.* **268**, 8290–8297.

Markowitz, D., Goff, S., and Bank, A. (1988). Construction and used of a safe and efficient amphotropic packaging cell line. *Virology* **167**, 400–406.

Marquardt, D., McCrone, S., and Center, M. S. (1990). Mechanisms of multidrug resistance in HL-60 cells: Detection of resistance-associated proteins with antibodies against synthetic peptides that correspond to the deduced sequence of P-glycoprotein. *Cancer Res.* **50**, 1426–1430.

Marschall, P., Thomson, J. B., and Eckstein, F. (1994). Inhibition of gene expression with ribozymes. *Cell. Mol. Neurobiol.* **14**, 523–538.

Marsh, W., Sicheri, R., and Center, M. S. (1986). Isolation and characterization of Adriamycin-resistant HL-60 cells which are not defective in the initial intracellular accumulation of drug. *Cancer Res.* **46**, 4053–4057.

Matsushita, H., Kobayashi, H., Mori, S., Kizaki, M., and Ikeda, Y. (1995). Ribozymes cleave the AML1/MTG8 fusion transcript and inhibit proliferation of leukemic cells with t(8;21). *Biochem. Biophys. Res. Commun.* **215**, 431–437.

McCall, M. J., Hendry, P., and Jennings, P. A. (1992). Minimal sequence requirements for ribozyme activity. *Proc. Natl. Acad. Sci. U.S.A.* **89**, 5710–5714.

McGrath, T., and Center, M. S. (1987). Adriamycin resistance in HL-60 cells in the absence of detectable P-glycoprotein. *Biochem. Biophys. Res. Commun.* **145**, 1171–1176.

McGrath, T., Latoud, C., Arnold, S. T., Safa, A. R., Felsted, R. L., and Center, M. S. (1989). Mechanisms of multidrug resistance in HL60 cells: Analysis of resistance associated membrane proteins and levels of *mdr* gene expression. *Biochem. Pharmacol.* **38**, 3611–3619.

McLaughlin, S. K., Collis, P., Hermonat, P. L., and Muzyczk, N. (1988). Adeno-associated virus general trasnduction vectors: Analysis of proviral structures. *J. Virol.* **62**, 1963–1973.

Melton, D. A. (1985). Injected antisense RNAs specifically block messenger RNA translation *in vivo*. *Proc. Natl. Acad. Sci. U.S.A.* **82**, 144–148.

Mendelson, E., Grossman, Z., Mileguir, F., Rechavi, G., and Carter, B. J. (1992). Replication of adeno-associated virus type 2 in human lymphocytic cells and interaction with HIV-1. *Virology* **187**, 453–463.

Milhuad, P. G., Machy, P., Lebleu, B., and Leserman, L. (1989). Antibody-targeted liposomes containing poly(rI)–poly(rC) exert a specific antiviral and toxic effect on cells primed with interferon alpha/beta or gamma. *Biochim. Biophys. Acta* **1135**, 269–274.

Miller, A. D., Miller, D. G., Garcia, J. V., and Lynch, C. M. (1993). Use of retroviral vectors for gene transfer and expression. *In* "Methods in Enzymology" (R. Wu, ed.), vol. 217, pp. 581–599. Academic Press, San Diego, California.

Miller, D. (1990). Progress toward human gene therapy. *Blood* **76**, 271–599.

Miller, N., and Vile, R. (1995). Targeted vectors for gene therapy. *FASEB J.* **9**, 190–199.

Mirski, S. E. L., Gerlach, J. H., and Cole, S. P. C. (1987). Multidrug resistance in a human small cell lung cancer cell line selected in Adriamycin. *Cancer Res.* **47**, 2594–2598.

Monia, B. P., Johnston, J. F., Ecker, D. J., Zounes, M. A., Lima, W. F., and Freier, S. M. (1992). Selective inhibition of mutant Ha-*ras* messenger RNA expression by antisense oligonucleotides. *J. Biol. Chem.* **267**, 19954–19962.

Morgan, R. A., and Anderson, W. F. (1993). Human gene therapy. *Annu. Rev. Biochem.* **62**, 191–217.

Moss, B. (1990). Herpes simplex viruses and their replication. *In* "Virology" (B. N . Field, D. M. Knipe, R. M. Chanock, J. L. Melnick, M. S. Hirsch, T. P. Monath, and B. Roziman, eds.), pp. 2079–2099. Raven Press, New York.

Müller, M., Meijer, C., Zaman, G. J. R., Borst P., Scheper, R. J., Mulder, N. H., de Vries, E. G., and Jansen, P. L. M. (1994). Overexpression of the gene encoding the multidrug resistance-associated protein results in increased ATP-dependent glutathione *S*-conjugate transport. *Proc. Natl. Acad. Sci. U.S.A.* **91**, 13033–13037.

Muzyczka, N. (1992). Use of adeno-associated virus as a general transduction vector for mammalian cells. *Curr. Top. Microbiol. Immunol.* **158**, 97–129.

Nabel, E. G., Plautz, G., Boyce, F. M., Stanley, J. C., and Nabel, G. J. (1989). Recombinant gene expression *in vivo* within endothelial cells of the arterial wall. *Science* **244**, 1342–1344.

Nabel, E. G., Plautz, G., and Nabel, G. J. (1990). Site-specific gene expression *in vivo* by direct gene transfer into the arterial wall. *Science* **249**, 1285-1288.

Nabel, G. J., Nabel, E. G., Yang, Z.-Y., Fox, B. A., Plautz, G. E., Gao, X., Huang, L., Shu, S., Gordon, D., and Chang, A. E. (1993). Direct gene transfer with DNA–liposome complexes in melanoma: Expression, biologic activity and a lack of toxicity in humans. *Proc. Natl. Acad. Sci. U.S.A.* **90**, 11307–11311.

Nahreini, P., Woody, M. J., Zhou, S. Z., and Srivastava, A. (1993). Versatile adeno-associated virus 2-based vectors for constructing recombinant virions. *Gene* **124**, 257–262.

Nakamaye, K. L., and Eckstein, F. (1994). AUA-cleaving hammerhead ribozymes: Attempted selection for improved cleavage. *Biochemistry* **33**, 1271–1277.

Nakashima, E., Matsushita, R., Negishi, H., Nomura, M., Harada, S., Yamamoto, H., Miyamoto, K., and Ichimura, F. (1995). Reversal of drug sensitivity in MDR subline of P388 leukemia by gene-targeted antisense oligonucleotide. *J. Pharm. Sci.* **84**(10), 1205.

Ng, M. M. P., Benseler, F., Tuschl, T., and Eckstein, F. (1994). Isoguanosine substitution of conserved adenosines in the hammerhead ribozyme. *Biochemistry* **33**, 12119–12126.

Ohkawa, J., Koguma, T., Okhda, T., and Taira, K. (1995). Ribozyme: From mechanistic studies to applications *in vivo*. *J. Biochem.* (*Tokyo*) **118**, 251–258.

Ohkawa, T., Kijima, H., Irie, A., Horng, G., Kaminski, A., Tsai, J., Kashfian, B. I., and Scanlon, K. J. (1996). Oligonucleotide modulation of multidrug resistance gene expression. *In* "Multidrug Resistance in Cancer Cells: Cellular, Biochemical, Molecular and Biological Aspects" (S. Gupta and T. Tsuruo, eds.) pp. 413–433. Wiley, London.

Ohnuma, T., Lo, R. J., Scanlon, K. J., Kamen, B. A., Ohnoshi, T., Wolman, S. R., and Holland, J. F. (1985). Evolution of methotrexate resistance of human acute lymphoblastic leukemia cells *in vitro*. *Cancer Res.* **45**, 1815–1822.

Ohta, Y., Tone, T., Shitara, T., Funato, T., Jiao, L., Kashfian, B. I., Yoshida, E., Horng, M., Tsai, P., Lauterbach, K., Kashani-Sabet, M., Florenes, V. A., Fodstad, O. Y., and Scanlon, K. J. (1994). H-*ras* ribozyme mediated alteration of the human melanoma phenotype. *Ann. N. Y. Acad. Sci.* **716**, 242–253.

Ohta, Y., Kijima, H., Kashani-Sabet, M., and Scanlon, K. J. (1996a). Suppression of the malignant phenotype of melanoma cells by anti-oncogene ribozymes. *J. Invest. Dermatol.* **106**, 1–6.

Ohta, Y., Kijima, H., Kashani-Sabet, M., and Scanlon, K. J. (1996b). Tissue-specific expression of an anti-*ras* ribozyme inhibits proliferation of human malignant melanoma cells. *Nucleic Acids Res.,* **24**, 938–942.

Pachuk, C. J., Yoon, K., Moelling, K., and Coney, L. R. (1993). Selective cleavage of bcr-abl chimeric RNAs by a ribozyme targeted to noncontiguous sequences. *Nucleic Acids Res.* **22**, 301–307.

Palfner, K., Kneba, M., Hiddenmann, W., and Bertram, J. (1995). Improvement of hammerhead ribozymes cleaving *mdr-1* mRNA. *Biol. Chem. Hoppe-Seyler* **376**, 289–295.

Palmer, T. D., Miller, A. D., Reeder, R. H., and Mestay, B. (1993). Efficient expression of a protein coding gene under the control of an RNA polymerase I promoter. *Nucleic Acids Res.* 21, 3451–3457.

Paolella, G., Sproat, B. S., and Lamond, A. L. (1992). Nuclease resistant ribozymes with high catalytic activity. *EMBO J.* **11**, 1913–1919.

Perez-Cruet, M. J., Trask, T. W., Chen, S. H., Goodman, J. C., Woo, S. L., Grossman, R. G., and Shine, H. D. (1994). Adenovirus-mediated gene therapy of experimental gliomas. *J. Neurosci. Res.* **39**, 506–511.

Perreault, J.-P., Wu, T., Cousineau, B., Ogilvie, K. K., and Cedergren, R. (1990). Mixed deoxyribo- and ribo-oligonucleotides with catalytic activity. *Nature (London)* **344**, 565–567.

Perreault, J.-P., Labuda, D., Usman, N., Yang, J.-H., and Cedergren, R. (1991). Relationship between 2′-hydroxyls and magnesium binding in the hammerhead RNA domain: A model for ribozyme catalysis. *Biochemistry* **30**, 4020–4025.

Perriman, R., Delves, A., and Gerlach, W. L. (1992). Extended target-site specificity for a hammerhead ribozyme. *Gene* **113**, 157–163.

Perrotta, A. T., and Been, M. D. (1992). Cleavage of oligoribonucleotides by a ribozyme derived from the hepatitis δ virus RNA sequence. *Biochemistry* **31**, 16–21.

Pieken, W. A., Olsen, D. B., Benseler, F., Aurup, H., and Eckstein, F. (1991). Kinetic characterization of ribonuclease-resistant 2′-modified hammerhead ribozymes. *Science* **253**, 314–317.

Plautz, G. E., Wu, B.-Y., Gao, X., Huang, L., and Nabel, G. J. (1992). Gene transfer *in vivo* with DNA–liposome complexes: Lack of autoimmunity and gonadal localization. *Hum. Gene Ther.* 3, 649–656.

Plautz, G. E., Yang, Z., Wu, B., Gao, X., Huang, L., and Nabel, G. J. (1993). Immunotherapy of malignancy by *in vivo* gene transfer into tumors. *Proc. Natl. Acad. Sci. U.S.A.* **90**, 4645–4649.

Pley, H. W., Flaherty, K. M., and McKay, D. B. (1994). Three-dimensional structure of a hammerhead ribozyme. *Nature (London)* **372**, 68–74.

Ransone, L. J., and Verma, I. M. (1990). Nuclear proto-oncogenes Fos and Jun. *Annu. Rev. Biol.* **6**, 539–557.

Roberts, T. M. (1992). A signal chain of events. *Nature (London)* **360**, 534–535.

Rodenhuis, S., van de Wetering, M. L., Mooi, W. J., Evers, S. G., Van Zandwij, K. N., and Bos, J. L. (1987). Mutational activation of the K-*ras* oncogene; a possible pathogenic factor in adenocarcinoma of the lung. *N. Engl. J. Med.* **317**, 929–935.

Roe, T. Y., Reynolds, T. C., Yu, G., and Brown, P. O. (1993). Integration of murine leukemia virus DNA depends on mitosis. *EMBO J.* **12**, 2099–2108.

Roizman, B. and Sears, A.E. (1990). Herpes simplex viruses and their replication. *In* "Virology" (B. N. Field, D. M. Knipe, R. M. Chanock, J. L. Melnick, M. S. Hirsch, T. P. Monath, and B. Roizman, eds.), pp. 1795–1841. Raven Press, New York.

Roninson, I. B. (1991). "Molecular and Cellular Biology of Multidrug Resistance in Tumor Cells." Plenum, New York.

Roninson, I. B., Chin, J. E., Choi, K., Gros, P., Housman, D. E., Fojo, A., Gottesman, M. M., and Pastan, I. (1986). Isolation of human mdr DNA sequences amplified in multidrug resistant KB carcinoma cells. *Proc. Natl. Acad. Sci. U.S.A.* **83**, 4538–4542.

Roth, J. A., Mukhopadhyay, T., Zhang, W.-W., Fujiwara, T., and Georges, R. (1994). Gene replacement strategies for the prevention and therapy of cancer. *Eur. J. Cancer* **30A,** 20032–20037.

Roth, J. A., Mukhopadhyay, T., Zhang, W., Fujiwara, T., and Georges, R. (1995). Gene replacement strategies for the prevention and therapy of cancer. *In* "The Internet Book of Gene Therapy: Cancer Therapeutics" (R. E. Sobol, and K. J. Scanlon, eds.), pp. 229–233. Appleton & Lange, Stamford, Connecticut.

Rowley, J. D. (1973). A new consistent chromosomal abnormality in chronic myelogenous leukaemia identified by quinacrine fluorescence and Giemsa staining. *Nature (London)* **243,** 290–293.

Ruffner, D. E., Stormo, G. D., and Uhlenbeck, O. C. (1990). Sequence requirements of the hammerhead RNA self-cleavage reaction. *Biochemistry* **29,** 10695–10702.

Russell, D. W. (1994). Replicating vectors for gene therapy of cancer: Risks, limitations and prospects. *Eur. J. Cancer* **30A,** 1165–1171.

Russell, D. W., Miller, A. D., and Alexander, I. E. (1994). Adeno-associated virus vectors preferentially transduce cells in S phase. *Proc. Natl. Acad. Sci. U.S.A.* **91,** 8915–8919.

Samulski, R. J. (1993). Adeno-associated virus: integration at a specific chromosomal locus. *Curr. Opin. Biotech.* 3, 74–80.

Samulski, R. J., Chang, L.-S., and Shenk, T. (1989). Helper-free stocks of recombinant adeno-associated viruses: Normal integration does not require viral gene expression. *J. Virol.* **63,** 3822–3828.

San, H., Yang, Z.-H., Pompili, V. J., Jaffe, M. L., Plautz, G. E., Xu, L., Felgner, J. H., Wheeler, C. J., Felgner, P. L., Gao, X., Huang, L., Gordon, D., Nabel, G. J., and Nabel, E. G. (1993). Safety and short-term toxicity of novel cationic lipid formulation for human gene therapy. *Hum. Gene Ther.* **4,** 781–788.

Scanlon, K. J., Jiao, L., Funato, T., Wang, W., Tone, T., Rossi, J. J., and Kashani-Sabet, M. (1991a). Ribozyme-mediated cleavage of c-*fos* mRNA reduces gene expression of DNA synthesis enzymes and metallothionein. *Proc. Natl. Acad. Sci. U.S.A.* **88,** 10591–10595.

Scanlon, K. J., Kashani-Sabet, M., Tone, T., and Funato, T. (1991b). Cisplatin resistance in human cancers. *Pharmacol. Ther.* **52,** 385–406.

Scanlon, K. J., Ishida, H., and Kashani-Sabet, M. (1994). Ribozyme-mediated reversal of the multidrug-resistant phenotype. *Proc. Natl. Acad. Sci. U.S.A.* **91,** 11123–11127.

Scanlon, K. J., Ohta, Y., Ishida, H., Kijima, H., Ohkawa, T., Kaminski, A., Tsai, J., Horng, G., and Kashani-Sabet, M. (1995). Oligonucleotide-mediated modulation of mammalian gene expression. *FASEB J.* **9,** 1288–1296.

Scheffer, G. L., Wijngaard, P. L. J., Flens, M. J., Izquierdo, M. A., Slovak, M. L., Pinedo, H. M., Meijer, C. J. L. M., Clevers, H. C., and Scheper, R. J. (1995). The drug resistance-related protein LRP is the human major vault protein. *Nat. Med.* **1,** 578–582.

Scheper, R. J., Broxterman, H. J., Scheffer, G. L., Kaaijk, P., Dalton, W. S., van Heijningen, T. H. M., van Kalken, C. K., Slovak, M. L., de Vries, E. G. E., vander Valk, P., Meijer, C. J. L. M., and Pinedo, H. M. (1993). Overexpression of a Mr 110,000 vesicular protein in non-P-glycoprotein-mediated multidrug resistance. *Cancer Res.* **53,** 1475–1479.

Seemayer, T. A., and Cavenee, W. K. (1989). Biology of diseases: Molecular mechanisms of oncogenesis. *Lab. Invest.* **60,** 585–599.

Sen, A., Dunnmon, S. A., Gerard, R. D., and Chien, K. R. (1988). Terminally differentiated neonatal rat myocardial cells proliferate and maintain specific differentiated functions following expression of SV40 large T antigen. *J. Biol. Chem.* **263,** 19132–19136.

Seth, P. (1994). Adenovirus-dependent release of choline from plasma membrane vesicles at an acidic pH is mediated by penton base protein. *J. Virol.* **68,** 1204–1206.

Shi, Y., Glynn, J. M., Guilbert, L. J., Cotter, T. G., Bissonnette, R. P., and Green, D. R. (1992). Role of c-myc in activation-induced apoptotic cell death in T cell hybridomas. *Science* **257,** 212–214.

Shimayama, T., Wishikawa, F., Nishidawa, S., and Taira, K. (1993). Nuclease-resistant chimeric ribozymes containing dexoyribonucleotides and phosphorothioate linkages. *Nucleic Acids Res.* **21**, 2605–2611.

Shore, S. K., Nabriss, P. M., and Reddy, E. P. (1993). Ribozyme-mediated cleavage of the *BCR/ABL* oncogene transcript; *in vitro* cleavage of RNA and *in vivo* loss of P210 protein kinase activity. *Oncogene* **8**, 3183–3188.

Sioud, M., and Drlica, K. (1991). Prevention of human immunodeficiency virus type 1 integrase expression in *Escherichia coli* by a ribozyme. *Proc. Natl. Acad. Sci. U.S.A.* **88**, 7303–7307.

Sioud, M., Natvig, J. B., and Forre, O. (1992). Preformed ribozyme destroys tumor necrosis factor mRNA in human cells. *J. Mol. Biol.* **223**, 831–835.

Smith, G. L., and Moss, B. (1983). Infectious poxvirus vectors have a capacity for at least 25,000 base pairs of foreign DNA. *Gene* **25**, 21–28.

Smythe, W. R., Hwang, H. C., Amin, K. M., Elshami, A. A., Eck, S. L., Davidson, B. L., Wilson, J. M., Kaiser, L. R., and Albelda, S. M. (1995). Treatment of experimental human mesothelioma using adenovirus transfer of the herpes simplex–thymidine kinase gene. *Ann. Surg.* **222**, 78–86.

Snyder, D. S., Wu, Y., Wang, J. L., Rossi, J. J., Swiderski, P., Kaplan, B. E., and Forman, S. J. (1993). Ribozyme-mediated inhibition of *bcr-abl* gene expression in a Philadelphia chromosome-positive cell line. *Blood* **82**, 600–605.

Solnick, D. (1981). Construction of an adenovirus-SV40 recombinant producing SV40 T-antigen from adenovirus late promoter. *Cell (Cambridge, Mass.)* **24**, 135–143.

Stamatatos, L., Leventis, R., Zuckermann, M. J., and Silivius, J. R. (1988). Interactions of cationic lipid vesicles with negatively charged phospholipid vesicles and biological membranes. *Biochemistry* **27**, 3917–3925.

Stehelin, D., Varmus, H. E., Bishop, J. M., and Vogt, P. K. (1976). DNA related to the transforming gene(s) of avian sarcoma virus is present in normal avian DNA. *Nature (London)* **260**, 170–173.

Stein, C. A., and Cheng, Y.-C. (1993). Antisense oligonucleotides as therapeutic agents—Is the bullet really magical? *Science* **261**, 1165–1171.

Stein, C. A., and Krieg, A. M. (1995). Problems in interpretation of data derived from *in vitro* and *in vivo* use of antisense oligodeoxynucleotides. *Antisense Res. Dev.* **4**, 67–69.

Stewart, M. J., Plautz, G. E., Del Buono, L., Yang, Z. -Y., Xu, L., Gao, X., Huang, L., Nabel, E. G., and Nabel, G. J. (1992). Gene transfer *in vivo* with DNA–liposomes complexes: Safety and acute toxicity in mice. *Hum. Gene Ther.* **3**, 267–275.

Stull, R. A., and Szoka, F. C. J. (1995). Antigene, ribozyme and aptamer nucleic acid drugs: Progress and prospects. *Pharm. Res.* **12**, 465–483.

Sullenger, B. A., and Cech, T. R. (1993). Tethering ribozymes to a retroviral packaging signal for destruction of viral RNA. *Science* **262**, 1566–1569.

Sullenger, B. A., and Cech, T. R. (1994). Ribozyme-mediated repair of defective mRNA by targeted trans-splicing. *Nature (London)* **371**, 619–622.

Sullivan, S. M. (1994). Development of ribozymes for gene therapy. *J. Invest. Dermatol.* **103**, 85S–89S.

Symons, R. H. (1992). Small catalytic RNAs. *Annu. Rev. Biochem.* **61**, 641–671.

Symons, R. H. (1994). Ribozymes. *Curr. Opin. Struct. Biol.* **4**, 322–330.

Taylor, J. M. (1990). Hepatitis delta virus: *Cis* and *trans* functions required for replication. *Cell (Cambridge, Mass.)* **61**, 371–373.

Taylor, N. R., Kaplan, B. E., Swiderski, T., Li, H., and Rossi, J. J. (1992). Chimeric DNA–RNA hammerhead ribozymes have enhanced *in vitro* catalytic efficiency and increased stability *in vivo*. *Nucleic Acids Res.* **20**, 4559–4565.

Teeter, L. D., Eckersberg, T., Tsai, Y., and Kuo, M. T. (1991). Analysis of the Chinese hamster P-glycoprotein/multidrug resistance gene *pgp1* reveals that the AP-1 site is essential for full promoter activity. *Cell Growth Differ.* **2**, 429–437.

Temin, H. M. (1989). Retrovirus vectors: Promise and reality. *Science* **246**, 983.

Thill, G., Vasseur, M., and Tanner, N. K. (1993). Structural and sequence elements required for the self-cleaving activity of the hepatitis delta virus ribozyme. *Biochemistry* **32**, 4254–4262.

Thummel, C. R., Tjian, R., and Grodzicker, T. (1981). Expression of SV40 T antigen under control of adenovirus promoter. *Cell (Cambridge, Mass.)* **23**, 825–836.

Tidd, D. M. (1991). Synthetic oligonucleotides as therapeutic agents. *Br. J. Cancer* **63**, 6–8.

Tone, T., Kashani-Sabet, M., Funato, T., Shitara, T., Yoshida, E., Kashfian, B. I., Horng, M., Fodstad, O., and Scanlon, K. J. (1993). Suppression of EJ cells tumorigenicity. *In Vivo* **7**, 471–476.

Trapnell, B. C., and Gorziglia, M. (1994). Gene therapy using adenoviral vectors. *Curr. Opin. Biotechnol.* **5**, 617–625.

Tratschin, J.-D., Miller, I. L., and Smith, M. G. (1985). Adeno-associated virus vector for high-frequency intergration, expression, and rescue of genes in mammalian cells. *Mol. Cell. Biol.* **5**, 3251–3260.

Tritton, T. R., and Hickman, J. A. (1990). How to kill cancer cells: Membranes and cell signaling as targets in cancer chemotherapy. *Cancer Cells* **2**, 95–105.

Tsuchihashi, Z., Khosla, M., and Herschlag, D. (1993). Protein enhancement of hammerhead ribozymes catalysis. *Science* **262**, 99–102.

Tuschl, T., and Eckstein, F. (1993). Hammerhead ribozymes: Importance of stem–loop II for activity. *Proc. Natl. Acad. Sci. U.S.A.* **90**, 6991–6994.

Tuschl, T., Gohlke, C., Jovin, T. M., Westhof, E., and Eckstein, F. (1994). A three-dimensional model for the hammerhead ribozyme based on fluorescence measurements. *Science* **266**, 785–788.

Ueda, K., Cardarelli, C., Gottesman, M. M., and Pastan, I. (1987a). Expression of a full length cDNA for the human *MDR1* gene confers resistance to colchicine, doxorubicin and vinblastine. *Proc. Natl. Acad. Sci. U.S.A.* **84**, 3004–3008.

Ueda, K., Clark, D. P., Chen, C.-J., Roninson, I. B., Gottesman, M. M., and Pastan, I. (1987b). The human multidrug resistance (*mdr1*) gene. cDNA cloning and transcription initiation. *J. Biol. Chem.* **262**, 505–508.

Ueda, K., Pastan, I., and Gottesman, M. M. (1987c). Isolation and sequence of the promoter region of the human multidrug-resistance (P-glycoprotein) gene. *J. Biol. Chem.* **262**, 17432–17436.

Uhlenbeck, O. C. (1987). A small catalytic oligoribonucleotide. *Nature (London)* **328**, 596–600.

van Beusechem, V. M., Kukler, A., Heidt, P. J., and Valerio, D. (1992). Long-term expression of human adenosine deaminase in rhesus monkeys transplanted with retrovirus-infected bone marrow cells. *Proc. Natl. Acad. Sci. U.S.A.* **89**, 7640–7644.

van Doren, K., Hanahan, D., and Gluzman, Y. (1984). Infection of eukaryotic cells by helper-independent recombinant adenoviruses: Early region I is not obligatory for integration of viral DNA. *J. Virol.* **50**, 606–614.

van Tol, H., Buzayan, J. M., Feldstein, P. A., Eckstein, F., and Bruening, G. (1990). Two autolytic processing reactions of a satellite RNA proceed with inversion of configuration. *Nucleic Acids Res.* **18**, 1971–1975.

Vasanthakumar, G., and Ahmed, N. K. (1989). Modulation of drug resistance in a daunorubicin resistant subline with oligonucleoside methyl phosphonates. *Cancer Commun.* **1**, 225–232.

Versnel, M. A., Hagemeijer, A., Bouts, M. J., van der Kwast, T. H., and Hoogesteden, H. C. (1988). Expression of c-sis (PDGF-B chain) and PDGF-A chain genes in ten human malignant mesothelioma cell lines derived from primary and metastatic tumors. *Oncogene* **2**, 601–605.

Versnel, M. A., Claesson-Welsh, L., Hammacher, A., Bouts, M. J., van der Kwast, T. H., Eriksson, A., Willemsen, R., Weima, S. M., Hoogesteden, H. C., Hagemeijer, A., and

Heldin, C.-H. (1991). Human malignant mesothelioma cell lines express PDGF-beta receptors whereas cultured normal mesothelioma cells express predominately PDGF-alpha receptors. *Oncogene* **6**, 2005–2011.

Vile, R. G., and Hart, I. G. (1993). *In vitro* and *in vivo* targeting of gene expression to melanoma cells. *Cancer Res.* **53**, 962–967.

Wagner, E., Zenke, M., Cotten, M., Beug, H., and Birnstiel, M. L. (1990). Transferrin–polycation conjugates as carriers for DNA uptake into cells. *Proc. Natl. Acad. Sci. U.S.A.* **87**, 3410–3414.

Wagner, E., Plank, C., Zatloukal, K., Cotten, M., and Birnstiel, M. L. (1992). Influenza virus hemagglutinin HA-2N-terminal fusogenic peptides augment gene transfer by transferrin–polylysine–DNA complexes: Toward a synthetic virus-like gene transfer vehicle. *Proc. Natl. Acad. Sci. U.S.A.* **89**, 7934–7938.

Walz, C., and Schlehofer, J. R. (1992). Modification of some biological properties of HeLa cells containing adeno-associated virus DNA integrated into chromosome 17. *J. Virol.* **66**, 2990–3002.

Wang, K.-S., Choo, Q.-L., Weiner, A. J., Ou, J.-H., Najarian, R. C., Thayer, R. M., Mullenbach, G. T., Denniston, K. J., Gerin, J. L., and Houghton, M. (1986). Structure, sequence and expression of the hepatitis delta (δ) viral genome. *Nature (London)* **323**, 508–514, corrigendum, **328**, 456.

Wei, C.-M., Gibson, M., Spears, P. G., and Scolnik, E. M. (1981). Construction and isolation of a transmissible retrovirus containing the *src* gene of Harvey murine sarcoma virus and the thymidine kinase gene of herpes simplex virus type 1. *J. Virol.* **39**, 935–944.

Weinberg, R. A. (1989). Oncogenes, anti-oncogenes, and the molecular basis of multistep carcinogenesis. *Cancer Res.* **49**, 3713–3721.

Weinberg, R. A. (1992). Tumor suppressor genes. *Science* **254**, 1138–1145.

Weir, J. P., and Narayanan, P. R. (1988). The use of beta galactosidase as a marker gene to define the regulatory sequences of the herpes simplex virus type 1 glycoprotein C in recombinant herpes virus. *Nucleic Acids Res.* **16**, 10267–10282.

Werness, B. A., Levine, A. J., and Howley, P. M. (1990). Association of human papillomavirus type 16 and 18 E6 protein with *p53*. *Science* **248**, 76–79.

Whyte, D., Buchkovich, K. J., Horowitz, J. M., Friend, S. H., Raybuck, M., Weingberg, R. A., and Harlow, E. (1988). Association between an oncogene and an anti-oncogene: The adenovirus E1A proteins bind to the retinoblastoma gene product. *Nature (London)* **334**, 124–129.

Williams, R. S., Johnston, S. A., Riedy, M., DeVit, J. M., McElligott, S. G., and Sanford, J. C. (1991). Introduction of foreign genes into tissues of living mice by DNA-coated microprojectiles. *Proc. Natl. Acad. Sci. U.S.A.* **88**, 2726–2730.

Woolf, T. M. (1995). To cleave or not to cleave: Ribozymes and antisense. *Antisense Res. Dev.* **5**, 227–232.

Woolf, T. M., and Budker, V. (1995). Cationic lipid-mediated gene transfer. *In* "The Internet Book of Gene Therapy: Cancer Therapeutics" (R. E. Sobol and K. J. Scanlon, eds.), pp. 65–73. Appleton & Lange, Stamford, Connecticut.

Wright, L., Wilson, S. B., Milliken, S., Biggs, J., and Kearney, P. (1993). Ribozyme-mediated cleavage of the *bcr/abl* transcript expressed in chronic myeloid leukemia. *Exp. Hemortol. (N.Y.)* **21**, 1714–1718.

Wu, G. Y., and Wu, C. H. (1987). Receptor-mediated *in vitro* gene transformation by a soluble DNA carrier system. *J. Biol. Chem.* **262**, 4429–4432.

Wu, G. Y., and Wu, C. H. (1988). Evidence for targeted gene delivery to hepG2 hepatoma cells *in vitro*. *Biochemistry* **27**, 887–892.

Wu, G. Y., Zhan, P., Sze, L. L., Rosenberg, A. R., and Wu, C. H. (1994). Incorporation of adenovirus into a ligand-based DNA carrier system results in retention of original receptor specificity and enhances targeted gene expression. *J. Biol. Chem.* **269**, 11542–11546.

Wu, H.-N., and Lai, M. M. C. (1990). RNA conformational requirements of self-cleavage of hepatitis delta virus RNA. *Mol. Cell. Biol.* **10,** 5575–5579.

Wu, H.-N., Lin, Y.-J., Lin, F.-P., Makino, S., Chang, M.-F., and Lai, M. M. C. (1989). Human hepatitis δ virus RNA subfragments contain an autocleavage activity. *Proc. Natl. Acad. Sci. U.S.A.* **86,** 1831–1835.

Wurm, F. M., Gwinn, K. A., and Kingston, R. E. (1986). Inducible overproduction of the mouse c-*myc* protein in mammalian cells. *Proc. Natl. Acad. Sci. U.S.A.* **83,** 5414.

Yang, J. H., Perreault, J.-P., Labuda, D., Usman, N., and Cedergren, R. (1990). Mixed DNA/RNA polymers are cleaved by the hammerhead ribozyme. *Biochemistry* **29,** 11156–11160.

Yang, J. H., Usman, N., Chartrand, P., and Cedergren, R. (1992). Minimum ribonucleotide requirement for catalysis by the RNA hammerhead domain. *Biochemistry* **31,** 5005–5009.

Yang, N.-S., Burkholder, J., Roberts, B., Martinell, B., and McCabe, D. (1991). *In vivo* and *in vitro* gene transfer to mammalian somatic cells by particle bombardment. *Proc. Natl. Acad. Sci. U.S.A.* **88,** 2726–2730.

Yang, Y., Ertle, H. C., and Wilson, J. M. (1994a). MHC class I-restricted cytotoxic T lymphocytes to viral antigens destroy hepatocytes in mice infected with E1-deleted recombinant adenoviruses. *Immunity* **1,** 433–442.

Yang, Y., Nunes, F. A., Berencsi, K., Furth, E. E., Gohezd, E., and Wilson, J. M. (1994b). Cellular immunity to viral antigens limits E1-deleted adenoviruses for gene expression. *Proc. Natl. Acad. Sci. U.S.A.* **91,** 4407–4411.

Zenke, M., Steinlein, P., Wagner, E., Cotten, M., Beug, H., and Birnstiel, M. L. (1990). Receptor-mediated endocytosis of transferrin–polycation conjugates: An efficient way to introduce DNA into hematopoietic cells. *Proc. Natl. Acad. Sci. U.S.A.* **87,** 3655–3659.

Zhang, W.-W., Fujiwara, T., Grimm, E. A., and Roth, J. A. (1995). Advances in cancer gene therapy. *Adv. Pharmacol.* **32,** 289–341.

Zhao, J. J., and Pick, L. (1993). Generating loss-of-function phenotypes of the *fushi tarazu* gene with targeted ribozyme in *Drosophila*. *Nature* (*London*) **365,** 448–451.

Zhou, S. Z., Broxmeyer, H. E., Cooper, S., Harrington, M. A., and Srivastara, A. (1993). Adeno-associated virus 2-mediated gene transfer in murine hematopoietic progenitor cells. *Exp. Hematol.* (*N.Y.*) **21,** 928–933.

Zhu, N., Liggit, D., Liu, Y., and Debs, R. (1993). Systemic gene expression after intravenous DNA delivery into adult mice. *Science* **261,** 209–211.

Giorgio Parmiani
Mario P. Colombo
Cecilia Melani
Flavio Arienti
Gene Therapy Program
Division of Experimental Oncology D
Istituto Nazionale Tumori
Milan, Italy

Cytokine Gene Transduction in the Immunotherapy of Cancer

I. Introduction and Background

Considerable progress has been made in understanding the mechanism(s) of the immune response in general and of the immune response against tumors in particular. Information has been obained in three crucial areas: (1) the role of cytokines in the regulation of the immune response (Paul and Seder, 1994), (2) the molecular characterization of tumor antigens in both mouse and human neoplasms (Boon *et al.*, 1994), and (3) the molecular mechanism of T cell activation and antigen presentation (Robey and Allison, 1995). Such information has provided new impetus to research in the field of tumor immunology and immunotherapy, the history of which is characterized by a series of successes and disappointments. Furthermore, recombinant DNA technology now allows modification of the genome of mammalian cells for therapeutic purposes in several diseases. This improved technology is being particularly exploited in cancer immunotherapy in a

Advances in Pharmacology, Volume 40

combined approach that has been termed *immunogene therapy* of tumors. Crucial to this approach has been the ability to transfer into normal or neoplastic cells genes of cytokines known to increase the immunogenicity of the recipient cells, which subsequently can be used as a vaccine in tumor-bearing mice or in cancer patients. We discuss how these new findings are being translated into a clinical setting and elaborate on the major developments listed above, which have allowed the planning of preclinical studies of immunogene therapy on which the designs for new clinical protocols are based.

A. Cytokines

Cytokines include heterogeneous groups of glycoproteins that act locally in minute concentrations to mediate intercellular relationships by either activating or inhibiting a large array of cell functions. Cytokines encompass lymphokines (usually produced by lymphocytes), monokines (produced by monocytes), and other factors synthesized and released by fibroblasts, endothelial cells, or even epithelial cells. Cytokines are key agents not only in modulating the immune system but also in regulating the growth and differentiation of several tissues (e.g., the hematopoietic system); in addition, cytokines are crucial in processes such as tissue repair, inflammation, and carcinogenesis. The essential role played by several cytokines in the tumor–host relationship is becoming more and more evident. It should be noted that cytokines not only can modulate the immune response to tumor antigens, as they do for many other antigens, but also can influence tumor growth independently from any immune recognition. In fact, neoplastic cells may constitutively produce cytokines, express a series of cytokine receptors, or do both, leading to (1) paracrine or even autocrine growth circuits (Herlyn *et al.*, 1990; Colombo *et al.*, 1992a; Kerbel, 1992; Mattei *et al.*, 1994; Schadendorf *et al.*, 1994), (2) inhibition or stimulation of inflammatory reactions (Yamashiro *et al.*, 1994; Melani *et al.*, 1995a), (3) induction of migratory and invasive activity (Wang *et al.*, 1990; Singh *et al.* 1994), and (4) neoangiogenesis (Folkman, 1995).

As for the immune system, cytokines play an essential role in practically every step of the process leading to recognition of tumor antigens by T lymphocytes and the subsequent events that may result in biologically opposite effects, namely, tumor cell destruction or T cell tolerance and even tumor growth promotion. The essential steps of this complex pathway and the cytokines that may be involved in its regulation are summarized in Table I, which shows that T cell responses can be activated but also downregulated by cytokines. For example:

1. Transforming growth factor β_1 (TGF-β_1) is known to inhibit early steps of lymphocyte activation (Kehrl *et al.*, 1986).

TABLE I   Involvement of Cytokines in the Generation of a T Cell Response against Tumor

	Cytokine	
Step	*Upregulation*	*Downregulation*
MHC, ICAM-1[a] expression by tumor cells	IFN-α, IFN-γ, TNF-α	IL-10
APC differentiation and antigenic peptide processing and presentation	GM-CSF, IL-4, IL-6, IL-12	IL-10, TNF-α
Antigen recognition by T_H1 or T_H2 cells	TM-CSF, IFN-γ, IL-2, IL-4, IL-10, IL-12	
CTL differentiation and triggering	IFN-γ, IL-2, IL-6, IL-7, IL-12	IL-10, TGF-β
Tumor cell lysis by CTLs	GM-CSF, IFN-γ, TNF-α	

[a] ICAM-1, Intercellular adhesion molecule 1.

2. Antigen-presenting cell (APC) activity is upregulated by granulocyte-macrophage colony-stimulating factor (GM-CSF) but downregulated by interleukin 10 (IL-10) (Grabbe *et al.*, 1995).

3. Antigen recognition by helper T lymphocytes is diverted to a cellular immune response by IL-2, interferon γ (IFN-γ), and IL-12 and to a humoral immune response by IL-4 and IL-10 (Fitch *et al.*, 1993).

4. Activation and differentiation of cytotoxic T lymphocytes (CTLs) require an interplay of IL-2, IFN-γ, IL-6, IL-7, and IL-12 (Gajewski *et al.*, 1995).

5. Lysis of tumor cells by CTLs may be mediated by the release of tumor necrosis factor α (TNF-α), IFN-γ, or GM-CSF (Mazzocchi *et al.*, 1991).

It is clear that, in addition to the regulation of antigen presentation by APCs and of recognition by T lymphocytes shared with other antigenic systems, and inflammatory type of reaction resulting from release of cytokines at the tumor site can significantly increase the major histocompatibility complex (MHC) and costimulatory expression on tumor cells, thereby favoring tumor antigen recognition, tumor cell killing, or both by *in situ* recruited T cells.

B. Tumor Antigens

It is now well established that human tumors may express antigens that can be recognized by autologous or HLA-restricted CTLs, the immune cells that are considered to play a major role in the rejection of tumors *in vivo* as suggested by work in animal models (Greenberg, 1991). Several classes of human tumor antigens recognized by T cells are known and can be distinguished according to their origin and tissue distribution (Table II). The first group includes differentiation antigens (i.e., normal proteins that are synthesized by both neoplastic and normal cells of the same lineage but not by normal or neoplastic cells of different histotypes). These antigens have been previously described as normal proteins recognized by antibodies in the cell cytoplasm, but now it appears that peptides derived from these same or similar proteins are intracellularly processed and presented on the cell surface within the groove of the MHC molecule, as occurs for other antigens (Germain, 1994). The paradigm of such a situation is melanoma, since the same peptides can be recognized by CTLs on melanoma and on normal melanocytes in the context of the same MHC class I-restricting molecule (Anichini *et al.*, 1993a; Bakker *et al.*, 1994). Several genes cloned from melanoma cells encode proteins (e.g., tyrosinase, Melan-A/Mart-1, gp75, gp100) involved in the biosynthesis of melanin, a lineage-specific protein (Table II).

The second group of tumor antigens includes those expressed on tumor cells but usually not on normal tissues. This is exemplified by the MAGE, BAGE, and GAGE families of proteins, again first discovered in melanoma and then found to be expressed by several other human tumors (although with different frequency), but not by normal tissues except in the testis, where only germinal cells appear to have these antigens (Van Pel *et al.*, 1995).

The third group of antigens results from point mutations of a variety of proteins that usually occur in only one or very few tumors; these mutations give rise to a new peptide epitope that can be recognized by T cells. Interestingly enough, one of these mutations was described in melanoma to affect the cycline-dependent, cell cycle–regulating protein CDK4, whose modification may thus be involved in neoplastic transformation (Wölfel *et al.*, 1995).

The important point here is that several peptides (typically 8–10 amino acids in length) have been identified as being derived from different protein antigens (Table II) that can be recognized by T cells. It should be noted that a given protein can provide more than one peptide recognized by the same or different HLA alleles. Some of these peptides have already been shown, when tested *in vitro*, to stimulate lymphocytes of melanoma patients, leading to the generation of specific (i.e., peptide-restricted) CTLs able to kill melanoma cells bearing the given peptide, such as Melan-A/MART-1, tyrosinase, or gp100 and the appropriate MHC molecule (Salgaller *et al.*, 1994, 1995; Rivoltini *et al.*, 1995; Spagnoli *et al.*, 1995).

Another group of human tumor antigens is represented by the products of oncogenes or tumor suppressor genes whose alterations (mutations, deletions, etc.) are thought to be involved in the genesis of several important human epithelial cancers (Table II). The relevant genes are members of the *RAS* family, *Her-2/neu*, and *P53*, in which point mutations can result in new cancer-specific proteins and, therefore, new peptides whose epitopes, when in a proper molecular configuration to bind the MHC molecule, can be recognized by T cells (Houbiers *et al.*, 1993; Cheever *et al.*, 1995). Likewise, potentially new antigens are generated by chromosomal translocation and the resulting fusion of different genes leading to the synthesis of a fusion protein whose joining region includes a unique sequence different from that of the two normal proteins. Because this process occurs during neoplastic transformation and because fusion proteins are often necessary to maintain the neoplastic state, the new antigens are expressed by the majority of cells of that particular tumor. Examples of this are (1) the translocation of the c-*abl* protooncogene on chromosome 9 to the breakpoint cluster region on chromosome 22, which occurs in 95% of chronic myelogenous leukemias, resulting in a 210-kDa fusion protein, and (2) a chromosome 15:17 translocation, in which the gene encoding the α receptor of retinoic acid (RARα) on chromosome 17 is joined to the *pml* gene on chromosome 15, resulting in a fusion protein in 90% of patients with acute promyelocytic leukemia. These new tumor antigens can generate peptides that can be recognized by either CD4 or CD8 T cells of normal donors in MHC class II or class I restriction (Gambacorti-Passerini *et al.*, 1993; ten Bosch *et al.*, 1995) but less frequently by the T cells of autologous patients (Dermime *et al.*, 1995, 1996).

The final group of human tumor antigens that is worth mentioning is that represented by mucins. These glycosylated proteins are expressed on the cell surface of epithelial tumors. Mucins are high molecular weight glycoproteins with a large number of O-glycosylated tandem repeat domains, which may vary in number, length, and glycosylation. These proteins are also presented on normal cells, but they become aberrantly glycosylated in tumor cells, thus making new epitopes available for the recognition of T cells, which, however, occurs through the T cell receptor but without MHC restriction (Finn, 1993).

C. Immunotherapy: Limitations to Be Overcome

Many cancer immunotherapy studies have been carried out in the last 10 years with essentially two approaches: adoptive immunotherapy, which consists of the infusion into the individuals of autologous lymphocytes after their *in vitro* sensitization against the tumor; and active immunotherapy or vaccination, which implies the administration of tumor antigens under various formulations in an attempt to generate or increase an effective antitumor immune response. Although these studies have been considered altogether disappointing, in a few cases patients have benefitted from treatment with both (1) adoptive immunotherapy with autologous nonspecific lymphokine (IL-2)–activated lymphocytes and IL-2 (Rosenberg *et al.*, 1989) or with more specific tumor-infiltrating T lymphocytes and IL-2 (Arienti *et al.*, 1993; Rosenberg *et al.*, 1994) and (2) vaccination with irradiated autologous or allogeneic tumor cells or subcellular fractions or lysates (Hersey, 1992;

TABLE II   Human Tumor Antigens Recognized by T Cells

Gene encoding:	Expression	Peptide	HLA-A, -B, -C, -DR restriction
Tyrosinase	Melanoma and melanocytes	MLLAVLYCLY	A2.1
		YMNGTMSOV	A2.1
		YMDGTMSQV	
		AFLPWHRLF	A24
		SEIWRDIDF	B44
		?	DR4
gp100	Melanoma and melanocytes	YLEPGPVTA	A2.1
		LLDGTATLRL	A2.1
		KTWGQYWQV	A2.1
		ITDQVPFSV	A2.1
		VLYRYGSFSV	A2.1
		ALLAVGATK	A3.1
gp75 (TRP-1)	Melanoma and melanocytes	MSLQRQFLR	A31
TRP-2		LLPGGRPYR	A31
Melan-A/ MART-1	Melanoma and melanocytes	AAGIGILTV	A2.1
		ILTVILGVL	A2.1
		AEEAAGIGIL	B45
MAGE-1	Different tumors and testis	EADPTGHSY	A1
		SAYGEPRKL	Cw1601
MAGE-2	Different tumors and testis	EVVPISHLY	A1
MAGE-3	Different tumors and testis	EVDPIGHLY	A1, B44
		FLWGPRALV	A2.1
BAGE	Different tumors and testis	AARAVFLAL	Cw1601
GAGE	Different tumors and testis	YRPRPRRY	Cw6
NA-17	Melanoma	VLPDVFIRCV	A2.1
RAGE 1	Renal cancer	SPSSNRIRNT	B7
HLA-A2.1/ R1701	Renal cancer	ND[a]	A2.1
P15	Many normal tissues	AYGLDFYIL	A24
Mutated CDK4	Melanoma (2/28)[b]	ACDPHSGHFV[c]	A2.1
MUM-1	Melanoma	EEKLIVVLF	B44
β-Catenin	Melanoma	SYLDSGIHF	A24
RAS-D12	Pancreatic and colon carcinoma	YKLVVVGADGVGKSALTI	ND
RAS-V12	ND[a]	KLVVVGAVGVGK	DR1
K-RAS-D13	Colon carcinoma	MTEYKLVVVGAGDVGK	DQ7

(continues)

TABLE II (continued)

Gene encoding:	Expression	Peptide	HLA-A, -B, -C, -DR restriction
P53	Different carcinomas, colon carcinoma	LLGRNSFEV HMTEVVR<u>H</u>C	A2.1 A2.1
HER-2/Neu	Breast and ovarian carcinomas	HLYQGCQVV CLTSTVQLV	A2.1
pml/RARα	Acute promyelocyte leukemia	NSNHVASGAGCAAIETQSSSSEEIV	DR11
bcr/abl	CML	ATGFKQSS<u>K</u>ALQRPVAS	DR2
HPV16 E7	Cervical carcinoma	YMLDLQPETT LLMGTLGIV TLGIVCPI	A2.1
MUC-1	Breast, colon, pancreatic carcinomas	PDTRPAPGSTAPPAHGVTSA	NA[d]

[a] ND, Not determined.

[b] In parentheses: No. of affected melanomas/No. of melanomas examined.

[c] Underlined letter indicates mutation.

[d] NA, Not applicable because of MHC-unrestricted recognition.

Morton *et al.*, 1992; Berd *et al.*, 1994; Mitchell, 1995). The major limitations of these studies were the lack of information on the antigen(s) contained in the vaccine (if any) and involved in the antitumor immune response, and on the mechanism of such a response *in vivo*. Furthermore, results of these trials and other *in vitro* studies with human tumor cells led to the conclusion that immunogenicity of tumor cells was rather weak and probably unable to trigger helper T lymphocytes.

Therefore, in 1988, mouse experiments were initiated in an attempt to increase the immunogenicity of tumor cells by inserting into them genes of cytokines, such as IL-2 (Bubenik *et al.*, 1988; Fearon *et al.*, 1990; Gansbacher *et al.*, 1990a) or IL-4 (Tepper *et al.*, 1989). It was hypothesized that such manipulation would have allowed a better immunization through activation of helper T cells that were usually not triggered by tumor cells themselves because of the lack of appropriate factors. Simultaneously, the first gene encoding a human tumor antigen recognized by T cells was cloned by Boon's group (van der Bruggen *et al.*, 1991), but it was three more years before other human tumor antigens could be molecularly defined (Kawakami and Rosenberg, 1995; Van Pel *et al.*, 1995). Therefore the two main research approaches that would have allowed at least some important limitations of the previous studies of cancer immunotherapy to be overcome (i.e., use of well-defined antigens and better immunogens) developed quite independently during the last few years. The ability to clone genes encoding tumor antigens, to define their protein sequence, and to identify the motifs that are known to bind the different MHC alleles, together with the possibility of assessing the expression of the known antigens in tumor tissues by monoclonal antibodies, by a polymerase chain reaction analysis, or by both, now allows us to establish whether neoplastic cells of a certain lesion express a given antigen and, therefore, whether that individual can potentially benefit from an immunotherapy based on active immunization (vaccination) with well-defined peptides. Meanwhile, molecular techniques were also used to modify tumor or normal cells with the aim of constructing new cytokine-modified (i.e., more immunogenic) cellular vaccines to be employed for the immunogene therapy of animals and humans.

In this system, the use of tumor cells that express a whole array of antigens that are known to potentially immunize the host can be advantageous. In fact, the loss of one antigen, which may occur during *in vivo* tumor growth (Lehman *et al.*, 1995; Mäurer *et al.*, 1996), may not be sufficient to prevent killing of tumor cells when the immune response is generated against the whole spectrum of tumor antigens expressed by the cellular component of vaccine. Here the limitation may lie in the weak immunogenicity of tumor cells in general because of the lack or low expression of immunologically relevant molecules such as MHC classes I and II (Garrido *et al.*, 1993), adhesion molecules such as intercellular adhesion molecules 1, 2, and 3 (ICAM-1, -2, and -3; Anichini *et al.*, 1993b), and B7-1 or B7-2, necessary for a direct stimulation of T lymphocytes (Salvadori *et al.*, 1995; Sulé-Suso *et al.*, 1995). To provide tumor cells with all these immunogenic factors, one can select tumor lines expressing high levels of MHC class I and II molecules or upregulate these and other (e.g., ICAM) molecules by IFN-γ and transfect these cells with genes encoding B7. To improve further the immunogenicity of tumor cell lines, additional genes should be transduced, in particular those encoding cytokines.

II. Strategies and Techniques of Cytokine Gene Transfer

Strategies for gene transfer are usually classified according to the site where the genetic manipulation occurs, *in vivo* or *ex vivo*, and according to the duration of the transgene expression, which can be stable or transient. Similar strategies can make use of different vectors and techniques (Table III). Different strategies respond to different aims. *"Ex vivo–in vitro"* gene transduction, which requires the removal of target cells (neoplastic or normal cells) from the host and their *in vitro* manipulation, allows the phenotypic characterization of the engineered cells (immunogenicity, growth rate, transgene expression, tumorigenicity and metastatic potential, or safety) and, more important, the evaluation of the transgene product expression. The efficiency and duration of transgene expression will vary according to the gene transfer technique used. For instance, the use of retroviral vectors that infect replicating cells and integrate into their genome will result in a stable phenotype of the transduced cell, while the use of nonviral vectors, such as various formulations of liposomes or liposomes complexed with viral subunits, will result in the transduction of both proliferating and nonproliferating cells, with a transient but possibly higher level of transgene expression. A precise characterization of the transduced cells is important in designing protocols of vaccination with engineered cells; therefore *ex vivo* gene manipulation has been widely preferred for the transduction of cytokine genes to prepare antitumor vaccines. Moreover, comparable and reproducible results are expected with the use of engineered vaccines releasing similar amounts of a given cytokine.

In vivo gene transfer would appear more promising in view of a simpler administration of therapy and limitation of genetic manipulation of target cells. However, this approach faces several problems, such as specificity and efficiency of *in vivo* gene targeting and the degree of transgene expression; therefore, the reproducibility of the results obtained by employing such transfer techniques is still unpredictable. Most studies using *in vivo* gene transfer deal with the modification of tumor phenotype (restoration of defective gene function, downmodulation of activated oncogene, expression of suicide gene products, etc.), and studies have demonstrated the feasibility of *in vivo* gene transduction to elicit an antitumor immune response (Nabel *et al.*, 1993).

With few exceptions, both *ex vivo* and *in vivo* gene-transfer techniques use the same vectors, which can be defined as viral and nonviral. In the following sections we review the results obtained with the different vectors in the transduction of immunotherapeutic genes, while other reviews better summarize studies of different approaches to the gene therapy of cancer (A. D. Miller *et al.*, 1993; Mulligan, 1993; Jolly, 1995).

A. Retroviral Vectors

Replication of defective retroviral vectors is currently the system of choice when prolonged expression of the gene of interest is required. Most of the vectors used are derived from the Moloney murine leukemia virus (MoMLV), a diploid RNA retrovirus that, once having infected the host cell, is reverse transcribed into double-stranded DNA and integrated into the host genome. The inserted provirus encodes the viral proteins Gag, Pol, and Env and is transcribed into the viral RNA genome; the proteins are assembled around two copies of viral RNA to generate a complete retroviral

TABLE III Properties of Viral Vectors for Gene Transfer[a]

Characteristic	Retrovirus	Adenovirus	Adeno-associated virus	Poxvirus	Herpesvirus
Wild-type virus	Diploid positive-strand RNA	Double-strand DNA	Mostly single-strand DNA	Double-strand DNA	Double-strand DNA
Cell location	Nuclear	Nuclear	Nuclear	Cytoplasmic	Nuclear
Cell range	Replicating only	Replicating and nonreplicating	Replicating and probably nonreplicating	Replicating and nonreplicating	Replicating and nonreplicating
Administration	*Ex vivo*, direct injection *in vivo*	*Ex vivo*, direct injection *in vivo*	*Ex vivo, in vivo* associated with liposomes	Direct injection *in vivo*	*Ex vivo*, direct injection *in vivo*
Duration of expression	Stable	Transient	Transient, potentially stable	Transient	Transient
Level of expression	Moderate	High	Moderate	High	Moderate
Safety issues	Insertional mutagenesis	Inflammatory responses, possible insertional mutagenesis?	Toxicity of viral Rep protein, insertional mutagenesis	Dangerous in immunosuppressed hosts	Neurovirulence, insertional mutagenesis?

Advantages	Efficient entry into cells, host genome integration, no viral genes in the vector, no preexisting immunity, no wild-type viruses in patients, no possible rescue or recombination with host or patient viruses	Efficient entry into cells, high titer, very high expression, infection of quiescent cells, no genomic integration	Possible genomic insertion at specific sites	Large insert size, high titer, no genomic integration, unlikely rescue by recombination with host viruses	Very large insert size, high titer
Disadvantages	Limited insert size, low titer, infection of replicating targets, expensive validation of safety	Limited insert size, vectors containing many viral genes, highly immunogenic, generation of replication competent viruses, preexisting host immunity, possible recombination with host viruses	Permanent producer cell lines not available, limited insert size, preexisting host immunity	Immunity in smallpox-vaccinated people, high immunogenic	Complex construction, permanent producer lines not available, preexisting host immunity, possible recombination with host viruses

[a] Modified from D. Jolly: Viral Vector Systems for Gene Therapy. *The Internet Book of Gene Therapy* 1:3–16, 1995. Reprinted by permission of Appleton & Lange, Inc.

particle that is then released from the infected cell. Retroviral vectors are appropriately modified to remove the *gag, pol,* and *env* genes, responsible for the replication and infectivity of the virus, and to introduce useful cloning sites and selectable marker genes.

The resulting viral backbone maintains the retroviral long terminal repeats (LTRs) that promote RNA transcription and the Ψ region for the encapsidation, and may contain splice donor and acceptor sites 5' of the cloning site for maturation of mRNA (Cepko *et al.*, 1984; A. D. Miller *et al.*, 1993). The easy manipulation of the retroviral genome resulted in a variety of vector constructs designed to improve the expression of the cloned gene, to select the transduced cells easily, or to improve the safety of the viral vector. Strong internal transcriptional promoters have been inserted in the direct or in the opposite orientation (to obviate the phenomenon of the so-called interference of promoters) in order to obtain higher expression of the gene of interest (Dzierzak *et al.*, 1988; Overell *et al.*, 1988). Viral LTRs have been modified to allow the transgene expression in particular cell types (Grez *et al.*, 1990; Hawley *et al.*, 1992), and double-copy vectors that contain the expression cassette inserted into the LTR have been designed so that, on proviral integration, two copies of the transgene will be inserted into the genome of the host cell (Hantzopoulos *et al.*, 1989). Various selectable marker genes have been inserted in the vectors for easy rescue of the infected cells, even though vectors designed without selection genes appear to give higher titers and improved efficiency of transduction (Jaffee *et al.*, 1993). Other modifications, such as self-inactivating LTRs, have been designed to increase the overall safety of the retroviral vectors (Yu *et al.*, 1986).

Packaging cell lines are used to obtain infective retroviral particles from such defective vectors. These cells have been engineered to produce the viral Gag, Pol, and Env proteins but no viral RNA molecules, so that on infection with a defective retroviral vector, they are able to assemble its genome into an infective viral particle. However, the recombinant virus is defective in the genes encoding the viral proteins and, therefore, is unable to replicate itself outside the packaging cell. The development of "safe" packaging cell lines, which lack the possibility of recombination among the viral genes, allowed the production of stable, high-titer retroviral preparations suitable for infection of target cells and lacking replication-competent retroviral contaminants (Danos and Mulligan, 1988; Markowitz *et al.*, 1988; Miller, 1990).

The insert maximal capacity of retroviral vectors is roughly 8 kb, sufficient for cloning small genes such as those encoding cytokines or immunostimulatory cofactors. More genes can be cloned, however, into the so-called polycistronic vectors by the insertion of internal ribosome entry site (IRES) sequences in between. These particular sequences allow ribosome attachment and translation of the gene located at their 3' end (Morgan *et al.*, 1992).

Therefore, polycistronic retroviral vectors are useful for the expression of multiple genes with the same efficiency as for the two genes encoding the IL-12 heterodimer (Martinotti *et al.*, 1995; Tahara *et al.*, 1995).

Since retroviral vectors infect and integrate only into dividing cells, some limitations to their use come from their inability to infect quiescent cells (Roe *et al.*, 1993) and from the risk of insertional mutagenesis resulting from random integration (Moolten and Cupples, 1992). Because of their characteristics, retroviral vectors have been preferred over other systems for *ex vivo* gene transduction. The level and duration of the transgene expression vary according to the nature of the promoter used, the structure of the vector, and the target cell used, but generally the result is the expression of a moderate amount of the transgene over time (more than a burst), which is not significantly inhibited for at least 10–15 days after irradiation of the target cell (Gansbacher *et al.*, 1992a; Belldegrun *et al.*, 1993; Arienti *et al.*, 1994). For these reasons, gene transfer with retroviral vectors has been successfully employed for the generation of stocks of vaccine cells used in ongoing clinical trials (see Section III,B).

Direct *in vivo* administration of retroviral constructs is under study. This approach suffers from the limitations that are also common to other *in vivo* delivery systems, mainly involving the specificity of targeting and the efficiency of gene transduction; moreover, there are concerns about the safety of the *in vivo* delivery of retroviruses, mostly because of the possibility of insertional mutagenesis. The inability to infect nonreplicating cells can represent an advantage *in vivo*, since it can improve the targeting of actively proliferating tumor cells. Different strategies have been tried to achieve specific tissue targeting, including chemical modification of the virus, modification of the envelope gene to confer specificity for the desired cell surface molecules, production of pseudotype viruses that express new ligands or portions of antibodies on their surface, and expression of vector-encoded genes under the control of tissue-specific enhancers and promoters (Russel *et al.*, 1993; Salmons and Gunzburg, 1993; Han *et al.*, 1995; Somia *et al.*, 1995). The efficiency of targeting *in vivo* is also dependent on the route of administration, since retroviruses are easily destroyed in the circulation by complement. Injection of packaging cells into brain tumors or introduction of retroviral particles into intestinal lumen has been shown to result in the infection of target cells (Culver *et al.*, 1992; Lau *et al.*, 1995). To date, attempts at targeting *in vivo* with retroviral vectors have been performed mainly with reporter genes as preliminary experiments for somatic gene therapy. However, intrasplenic injection of packaging cells releasing retroviral vectors that carry IL-2- and IL-4-encoding genes has been shown to target hepatic metastasis efficiently and stimulate a cytokine-mediated inflammatory reaction that inhibited tumor formation (Hurford *et al.*, 1995).

B. Adenoviral or Adeno-Associated Viral Vectors

I. Adenoviral Vectors

The attractive features of adenoviral vectors developed for gene transfer are represented by the high titer of viral preparation, high but transient gene expression, and low toxicity (Table III). Adenoviruses are a family of viruses that cause benign infections of the respiratory tract in humans; more than 40 serotypes are known, but strains 5 and 2 are the best studied and are preferentially used for making vectors. The viral genome is represented by 36 kb of double-stranded DNA, with short inverted terminal repeats at both ends acting as the origin of viral replication, repeated specific sequences for viral encapsidation and a series of genes, divided into early and late regions, that encode multiple proteins. Usually adenoviral vectors are made by replacing the early genes, particularly the E1a, E1b, and the E3 genes, with the gene of interest by recombination between pieces of modified and wild-type viral DNA, or by molecular biological techniques (Stratford-Perricaudet *et al.*, 1990; Bett *et al.*, 1994; Weitzman *et al.*, 1995). These vectors are propagated on a complementing cell line that contains an integrated copy of the E1 gene and allows the full replicative cycle of the defective viral vector; stocks of adenoviral vectors can then be prepared whose titer range is 10^{11}–10^{12} plaque-forming units (PFU)/ml.

Adenoviruses infect dividing as well as quiescent cells of many histotypes. On infection, the virus replicates in the cell nucleus without integration into the host genome; viral DNA and proteins are then assembled in the cytoplasm, and the virions are released by cell lysis. Because of these characteristics and because of the high titer of viral preparations, adenoviral vectors have been proposed mainly for *in vivo* gene transfer and used mostly to correct genetic disorders, to replace antioncogenic functions in malignant cells, or to express suicide genes in tumor cell masses (Brody *et al.*, 1994; Chen *et al.*, 1995; Kaneko *et al.*, 1995). In fact, the efficiency of gene transduction and the level of gene expression obtained with adenoviral vectors are high although transient. However, there are some concerns about their use that must be taken into account. Regarding safety, it should be noted that illegitimate recombination and integration of adenoviral DNA sequences have been reported, therefore suggesting that insertional mutagenesis, as well as recombination with wild-type adenoviruses, can potentially occur within the patient (Doerfler, 1991). Although defective in replication, adenoviral vectors still express a number of viral proteins that can be recognized by the host as antigens, resulting in an inflammatory reaction and immune response. Since adenoviruses also infect nondividing cells, the antiviral immune response may be useful in controlling the infection of nonmalignant cells consequent to *in vivo* injection. However, both inflammation and immune reactions impair the duration of gene expression and limit the efficacy of repeated *in vivo* administrations of adenoviral vectors. Again, one should consider that immune suppression due to malignancy or therapy may balance these effects in patients, and transient expression of a high amount of an immunomodulatory agent can be less toxic although sufficient to obtain the desired effect. However, the antiviral immune response may complicate the interpretation of results obtained in cancer immunotherapy with adenoviral vectors encoding cytokines, immune costimulatory factors, or tumor antigens.

Although there are limitations to the use of adenoviral vectors in clinical protocols of cancer immunotherapy, modifications to improve their *in vivo* targeting to avoid the host immune response or to induce a tissue-specific gene expression by using appropriate promoters are under study (Kaneko *et al.*, 1995). Moreover, unique properties of the adenoviruses, such as the ability to dissociate from the endosome after entry into the cell, have been usefully employed to improve gene delivery by nonviral systems (see Section II,D).

2. Adeno-Associated Viral Vectors

Adeno-associated viruses (AAVs) are nonautonomous parvoviruses that need a helper, usually adenovirus or herpesvirus, to replicate. Their use as vectors for gene transfer is supported by the lack of any associated pathological condition and the ability of AAV to integrate in the host genome (Table III). Their genome consists of a single strand of DNA with internal terminal repeats (ITRs) working as promoters and necessary for integration, and two major coding regions, *cap* and *rep*, associated with structural proteins and viral replication function, respectively. Except for the ITRs, all the internal segments can be replaced with the gene of interest to obtain an AAV vector, although the maximal insert length is limited to 4–5 kb. Since the ITR is a weak promoter, strong internal promoters or tissue-specific promoters can be inserted to regulate gene expression, thereby allowing for a cell-specific expression of the gene of interest. Viral particles are made by cotransfection of adenovirus-carrying cells with both recombinant AAV vector and a helper plasmid encoding Cap and Rep proteins; the AAV resulting from cell lysis must be purified from contaminant adenoviruses (Bartlett *et al.*, 1995). This system does not yet allow preparation of a stable packaging cell line; therefore, viral titer may change from one stock to another, but usually ranges between 10^6 and 10^7 PFU/ml. The two major advantages of AAV vectors lie in their ability to infect a wide range of cell types, including hematopoietic cells and lymphocytes (Muro-Cacho *et al.*, 1992), regardless of their proliferative status, and in their possible integration in the host genome, resulting in stable and long-term expression (see Bartlett *et al.*, 1995).

Although promising, AAV vectors are still under study and their use in the gene therapy of cancer is limited. Moreover, their possible use for *in vivo* gene transfer is made difficult by the preexisting immune response against wild-type AAV in 80% of individuals, and by the risk of recombination with possibly integrated AAV provirus with rescue of replication-competent AAV.

Adeno-assisted virus vector DNAs encoding IL-2 have been successfully combined with liposomes for the *in vitro* transduction of lymphocytes and primary tumor cells (Philip *et al.*, 1994; Vieweg *et al.*, 1995). The efficiency of gene transduction obtained with this system overlaps that obtained by infection with AAV vectors; however, the IL-2 release declined with time and the level of expression was low in both studies, thus raising questions about the real integration of the AAV DNA.

C. Other Viral Vectors

I. Vaccinia Viral Vectors

The use of smallpox viruses as vectors for gene transfer is at an early stage. The advantages of using vaccinia viruses (VVs) as vectors lie in their large insert size, which allows the cloning of multiple genes; in their ability to infect replicating and nonreplicating cells, with a moderate to high expression level; and in their cytoplasmic location, which avoids the risk of insertional mutagenesis (Table III). For their use *in vivo*, it should be considered that VVs are highly immunogenic and that a large part of the population has been vaccinated against smallpox; therefore their efficacy could be impaired by a strong host immune response. Vaccinia virus vectors are also replication competent and cytopathic, a characteristic that limits their use for *ex vivo* gene transduction of normal or tumor cells and their use in immune-compromised hosts, in whom VV vectors could induce a disseminated smallpox. Toxicity can be reduced by using fowlpox or canarypox strains that infect human cells abortively, and improved tumor targeting can be obtained by using attenuated strains that preferentially infect actively dividing cells.

Vaccinia viruses are double-stranded DNA viruses with ITR sequences that direct the viral replication. Almost all of the other viral genes can be replaced by the gene(s) of interest, and a marker gene can be included for the easy identification of recombinants; the gene expression is then regulated by internal late or early viral promoters. Recombinant vectors are obtained by homologous recombination following transfection of the plasmid construct into cells infected with vaccinia virus (Moss and Flexner, 1987).

Taking advantage of the features of such viruses, including the immunogenicity of their proteins [which may help to elicit an immune response against the inserted tumor-associated antigens (TAAs)], recombinant VV vectors expressing TAAs or their derivative MHC-presented peptides have been used as vaccines to activate an antitumor response *in vivo* (Estin *et al.*, 1988; Kantor *et al.*, 1992; McCabe *et al.*, 1995). Such vectors were also usefully employed to express high levels of cytokines by direct injection at the tumor site, taking advantage of the wide range of target infection and of the cytopathic effect on the infected tumor cells, which helps in inhibition of tumor growth (Meko *et al.*, 1995).

2. Herpes Viral Vectors

The advantages of the herpes simplex vector system include the wide host range of these viruses, their ability to infect quiescent as well as replicating cells, their neurotropism, the availability of high-titer viral stocks, and a large insert size capacity (Table III). Herpes viral vectors are derived from herpes simplex type 1 (HSV-1), and enveloped DNA virus whose genome has been completely sequenced and whose life cycle is well known. On target cell infection, HSV-1 can undergo a replicative cycle that leads to cell lysis and new infection, or can become latent and survive within the cell nucleus, normally without integration in the host genome. Recombinant HSV-1 vectors can contain the transgene and various mutated viral genes, or can consist only of the HSV-1 origin of DNA replication, the packaging signal, and the transgene (the so-called amplicon vectors). Both constructs are converted into infective vectors by homologous recombination using the complementing functions of cells infected with HSV-1 (Geller *et al.*, 1990). However, the resulting viral stocks contain replication-competent herpesviruses, contributing to the toxicity of this vector system. Altogether, this system needs a more complete characterization to improve safety, and attempts have been made with attenuated or temperature-sensitive viral mutants (Fink *et al.*, 1992). The overall interest in HSV-1 vectors resides in their ability to transduce CNS cells. By direct injection into the brain, and by taking advantage of vector mutants that preferentially replicate in proliferating cells, tumor cells can be efficiently targeted (Martuza *et al.*, 1991). More than in protocols of immunotherapy, HSV-1 vector–mediated gene therapy has been employed in direct *in vivo* tumor transduction of the gene encoding viral thymidine kinase (HSV TK), which works as a suicide gene on treatment with the cytotoxic drug ganciclovir (normally ineffective in nontransduced cells) (Boviatsis *et al.*, 1994). The potential therapeutic applications of this strategy in cancer have been discussed (Blaese *et al.*, 1994).

As with the vaccinia and adenoviral vectors, the intrinsic antigenicity of the herpes viral proteins should be considered when protocols for immunotherapy of cancer are designed using this system for gene transduction. A preexisting antiviral immune response may be helpful in limiting the spread of replication-competent HSV-1 vectors and in controlling their toxicity, while its deficiency can be harmful in immune-compromised hosts. Moreover, when evaluating an antitumor response resulting from the employment of this vector system, one should take into account the role of the antiviral response.

D. Liposomes and Plasmid DNA

Nonviral systems have been studied to obviate the risks of virus-mediated gene delivery, that is, immunogenicity, toxicity, cytopathic effects, and risk of genetic recombination. The aims of the various formulations of nonviral delivery systems are (1) targeting of a gene (i.e., a negatively charged DNA or RNA) to a cell surface that is also negatively charged, thus overcoming the physical rejection; (2) entry of the gene into the cell, avoiding its destruction by lysosomal and cytoplasmic enzymes; and (3) entry into the cell nucleus with the subsequent expression of the transgene for as long as possible. Although *in vitro* gene transfer can be easily obtained with nonviral delivery systems (DNA transfection mediated by $CaPO_4$, by liposomes, by polylysine-bound specific cell ligand, etc.), the goal of these systems is the *in vivo* gene therapy (Ledley, 1995). The more promising vehicles appear to be cationic liposomes, prepared in various formulations and complexed to DNA or RNA. More complex formulations include the presence of adeno-viral proteins to enhance the endosome disruption after endocytosis, the binding of specific ligands or antibody for cell targeting, and the presence of adeno-associated viral vectors to increase the possibility of stable integration of the transgene (Philip *et al.*, 1994; Vieweg *et al.*, 1995).

A wide variety of genes have been targeted *in vivo* by liposome conjugation, particularly in cancer gene therapy where liposomes have been used to express cytokines, allogeneic HLA molecules, immune costimulatory factors, and suicide genes. The liposome transfer system has been used mostly for noncytokine genes, with the exception of human IFN-β, which has been effectively targeted to orthotopically xenotransplanted human gliomas growing in nude mice (Yagi *et al.*, 1994). Although all these systems have been shown to be safe and without toxicity (Stewart *et al.*, 1992; Nabel *et al.*, 1994), the rate and duration of gene expression achieved remain lower than that obtained with viral vectors.

More recently, naked plasmid DNA has been shown to be effective in inducing an immune response when given by direct injection or "gene gun" (Pardoll and Beckerleg, 1995). In fact, DNA is easily taken up, particularly by muscle cells, albeit by an unknown mechanism, and expressed by them in a form that stimulates an efficient immune response. This approach, however, may be of limited use for transferring cytokine genes except for the targeting of subcutaneously growing tumors. In fact, naked DNA has been administered by direct injection or by "gene gun" technique into liver, skin, or muscle; transient gene expression has been reported, as well as an immune response against the gene product, usually a viral determinant (Raz *et al.*, 1994). Reduced tumor growth without complete regression was obtained with subcutaneous tumor by "gene gun" delivery of cytokine genes (IL-2, IL-6, TNF-α, or IFN-γ) (Sun *et al.*, 1995).

III. Active Immunogene Therapy

A. Animal Models

Modification of tumor–host interactions by cytokines injected at the tumor site or around draining lymph nodes has been widely studied, first with IL-2 (Forni *et al.*, 1986) and then with IFN-γ (Giovarelli *et al.*, 1986), IL-1β (Forni *et al.*, 1989), and IL-4 (Bosco *et al.*, 1990). As a whole, these results concordantly indicate that local injection of cytokines stimulates nonspecific inflammation-like reactions that impair tumor growth or even lead to rejection of incipient tumors, and stimulates helper T functions, often resulting in the induction of systemic immunity. A technological improvement of those earlier studies was pioneered by Bubenik and coworkers (1988) and by Tepper and coworkers (1989), who transfected the genes encoding IL-2 and IL-4, respectively, into cancer cells, which acquired the ability to stimulate a strong antitumor immune response resulting in local growth inhibition. Since then, many cytokine genes have been transduced into a variety of rodent tumors. Altogether, these studies established that all the transduced cytokines induce local recruitment of leukocytes, whose type depends on the given cytokine. In most cases the infiltrating leukocytes produce secondary cytokines, thus creating an entirely new environment that builds up the condition for tumor rejection and development of CD4 and CD8 T cell-dependent systemic antitumor immunity (Colombo *et al.*, 1992b). In a few cases, cytokine gene transfer is without effect; this may depend on the tumor histotype, on the dose of cytokine released, or on unknown factors. These few cases include the macrophage colony-stimulating factor (M-CSF) whose gene, transduced into the J558L plasmacytoma, was shown to recruit MAC-1$^+$ leukocytes without determining tumor growth inhibition (Dorsch *et al.*, 1993). In contrast, when the gene encoding M-CSF was transduced into B16F10 murine melanoma intravenous injection of transduced cells led to establishment of lung metastases that, however, were eliminated after 2–3 weeks. Monocytes and lymphocytes were shown to mediate such immune response (Walsh *et al.*, 1995).

Tumors constitutively produce many cytokines, including chemotactic factors, and macrophages may either help or impair tumor growth, depending on the production of paracrine growth factor acting on the tumor or on the tumor vessels, or on the production of inhibitory factors, including prostaglandins, in addition to some antitumor effect (Mantovani *et al.*, 1992). Thus IL-10, by inhibiting macrophage functions, may alternatively promote or impair tumor growth (Richter *et al.*, 1993). Indeed, when released at high dose, IL-10-triggered tumor rejection involves infiltration of neutrophils, CD8$^+$ T lymphocytes, and natural killer (NK) cells, each of them shown to be instrumental in tumor growth inhibition. The cytokine that has never been found to induce tumor growth inhibition is IL-5, despite its ability to recruit eosinophils (Krüger-Krasagakes *et al.*, 1993), which, in contrast, were indicated as the major factor responsible for rejection of IL-4-transduced tumor cells (Tepper *et al.*, 1992). This apparent contradiction is probably due to the fact that the conclusion by Tepper and coworkers was based on depletion of granulocytes, resulting in tumor growth; they did not conclusively distinguish the role of eosinophils from that of other granulocytes. All other cytokine genes transduced into different neoplasms usually led to tumor growth inhibition.

Hereafter, for each cytokine gene transduced, we give a brief description of the effect on local tumor growth, host cells infiltrating the tumor, and induction of systemic immunity. For a complete list of tumors transduced with cytokine genes, see Forni and associates (1995).

IL-1α: Gene transfer has been performed in fibrosarcoma cells; tumor growth inhibition occurred through the activation of both helper T cells and CTLs. Surviving mice develop an immune memory, which confers protection against a challenge of parental untransduced cells (Douvdevani *et al.*, 1992).

IL-2: Interleukin 2 is the cytokine most used for gene-transfer experiments. Because IL-2 is an active cross-species, both human and murine genes have been introduced into a variety of murine tumors, including colon, bladder, lung, and mammary carcinomas, melanoma, fibrosarcoma, mastocytoma, lymphoma, plasmacytoma, thymoma, myeloma, and neuroblastoma. A rat fibrosarcoma was also studied (Russel *et al.*, 1991). In all cases, the released IL-2 induced local tumor growth inhibition mostly by activating CD8$^+$ T and NK cells and macrophages (Fearon *et al.*, 1990; Gansbacher *et al.*, 1990b; Hock *et al.*, 1993a; Bannerji *et al.*, 1994; Maass *et al.*, 1995); a role for neutrophils has also been reported (Cavallo *et al.*, 1992). CD4 and CD8 T cell-dependent immune memory generally followed tumor growth inhibition. Of importance, studies employing IL-2 gene transfer established the concept that the amount of cytokine released is crucial since it can differentially affect tumor growth inhibition and induction of immune memory (Cavallo *et al.*, 1992; Zatloukal *et al.*, 1995). In the presence of high levels of IL-2 (>6000 U) a strong and fast tumor rejection occurs so as to prevent the induction of memory T cells, thus explaining the lack of therapeutic effects of such high doses (Schmidt *et al.*, 1995). A low level (<50 U) is only partially active on tumor inhibition, and an intermediate level (3000 U) is effective in both tumor inhibition and induction of memory T cells (Cavallo *et al.*, 1992). The gene encoding IL-2 was also used to transduce allogeneic C3Hf fibroblasts along with cDNA of the B16 melanoma. These transfectants were able to immunize C57BL/6 mice against a challenge of syngeneic B16 tumor and, when given as therapeutic vaccine, to delay significantly the survival of tumor-bearing mice (Kim *et al.*, 1993; Kim and Cohen, 1994). This is one of the few examples of vaccination with allogeneic, cytokine gene-transduced tumor lines in animal models. Conversely, the allogeneic lines are those most frequently used in clinical trials, compared with autologous lines (see Section III,B,2).

IL-3: Gene-transfer experiments are limited to the alveolar lung carcinoma line 1. This transduced tumor can induce CTLs to the same extent as IL-2 transduced counterparts, but IL-3 has less inhibitory activity than IL-2 on local tumor growth (McAdam *et al.*, 1995).

IL-4: Like IL-2, IL-4 has been transduced in many different tumors, including mammary, lung, and renal cell carcinomas, fibrosarcoma, plasmacytoma, and melanoma. According to Tepper and associates (1992), local tumor growth inhibition depends on infiltration of eosinophils, but macrophages and other granulocytes may also be involved (Golumbek *et al.*, 1991), even though the precise role of each subpopulation of leukocytes has not been clearly defined. CD8$^+$ T lymphocytes are also needed if tumor continues to grow and when a systemic immunity is generated (Golumbek *et al.*, 1991; Hock *et al.*, 1993a; Pericle *et al.*, 1994). It is noteworthy that antibodies of the IgA, IgE, and IgG$_1$ classes were found in sera of mice immunized with IL-4-transduced TSA mammary carcinoma. Such antibodies can bind tumor cells but, in order to transfer the immunological memory, both immune sera and CD4$^+$ T lymphocytes were needed (Pericle *et al.*, 1994).

IL-6: The effect of IL-6 on tumor growth inhibition is controversial and appears to depend on the tumor model. Apart from plasmacytoma, for which IL-6 is a growth factor (Vink *et al.*, 1990), growth of the mammary carcinoma TSA was not inhibited after gene transfer (Allione *et al.*, 1994), whereas growth of sarcomas, melanoma, and 3LL lung carcinoma was inhibited (Mullen *et al.*, 1992; Porgador *et al.*, 1992). Induction of T cell-mediated systemic immunity followed tumor inhibition, but this was tested only in sarcomas (Mullen *et al.*, 1991).

IL-7: Growth of glioblastoma, fibrosarcoma, and mammary carcinoma was inhibited after IL-7 gene transfer, mainly by a CD8$^+$ T cell response, and the surviving mice developed a strong immune memory (McBride *et al.*, 1992; Forni *et al.*, 1996). Growth of plasmacytoma J558L was also inhibited by IL-7 transduction, but in this case CD4$^+$ and macrophages were responsible for tumor rejection (Hock *et al.*, 1993a); induction of immune memory was not tested (Hock *et al.*, 1991).

IL-12: Unlike other cytokines, IL-12 showed a clear antitumor activity when given systemically as a recombinant protein (Brunda *et al.*, 1993). Gene-transfer experiments with this cytokine are more complex because it is the product of two distinct genes encoding the p35 and the p40 chains, respectively. After gene transfer, both tumor inhibition and induction of systemic immunity occurred (Tahara *et al.*, 1995). An exception is the C-26 colon carcinoma, which is resistant to recombinant IL-12 given systemically; transduced with IL-1 genes, it also escaped rejection when the level of IL-12 released was low (30 pg/ml/10^6 cells/24 hr). It is of interest that host depletion of $CD4^+$ T cells allowed tumor rejection to occur in mice injected with low-cytokine-producing C-26 cells (Martinotti *et al.*, 1995). In terms of tumor rjeection, this indicates that repeated systemic injections of 1 μg of recombinant IL-12 are likely to cause, at the tumor site, the same effect as 30 pg released locally.

G-CSF: Experiments are limited to C-26, which, when transduced with the human G-CSF (granulocyte colony-stimulating factor) cDNA, induced a massive infiltration and activation of neutrophils that destroyed the transduced cells in less than 7 days (Colombo *et al.*, 1991). Infiltrating neutrophils expressed IL-1α and -β as well as TNF-α. Injection of transduced cells into sublethally irradiated mice allowed tumor growth to a large size; nevertheless, tumor rejection took place after the leukocyte functions were self-reconstituted 10–15 days after irradiation. In this case, tumor rejection was still dependent on neutrophils, but $CD8^+$ T cells producing IFN-γ were also required (Stoppacciaro *et al.*, 1993).

GM-CSF: GM-CSF has been credited with being one of the best inducers of antitumor immunity when transduced into tumors. This conclusion was based on a comparison with other cytokines in experiments originally performed by Dranoff and associates (1993) but not confirmed by others with a different tumor (Allione *et al.*, 1994). The activity of GM-CSF seems mainly due to upregulation of dendritic cell survival and functions (Grabbe *et al.*, 1995). The role of GM-CSF in the function of dendritic cells in the priming of the CTL response has been confirmed (Paglia *et al.*, 1996). When compared with IL-2 or IL-3, GM-CSF was a poor stimulator for generation of primary CTLs against the alveolar lung carcinoma line 1 (McAdam *et al.*, 1995). The level of cytokine production and the inherent immunogenicity of different tumors may explain such differences in activity.

IFN-α: The growth of a Friend virus–induced leukemia and a mammary adenocarcinoma was inhibited by IFN-α gene transduction via a mechanism involving T lymphocytes and neutrophils (Ferrantini *et al.*, 1993). A weak tumor inhibition without induction of immune memory in surviving mice was found in B16 melanoma-transduced cells (Kaido *et al.*, 1995).

IFN-γ: Interferon γ was extensively studied after gene transfer in many tumor systems (Forni *et al.*, 1996). Tumor growth inhibition varied from strong to weak, and the antitumor activity was associated with macrophages and $CD8^+$ T cell activation (Gansbacher *et al.*, 1990a; Porgador *et al.*, 1993a). An important feature of IFN-γ activity is the upregulation of MHC class I expression, which favors recognition by T cells but may also augment the metastatic properties as shown for the TSA carcinoma (Lollini *et al.*, 1993).

TNF-α: Human or mouse TNF-α genes were transduced into sarcomas, a mammary carcinoma, a plasmacytoma, and ultraviolet (UV)-induced skin tumors (Asher *et al.*, 1991; Blankenstein *et al.*, 1991; Allione *et al.*, 1994). Tumor growth inhibition occurred in all transduced tumors but the nonimmunogenic MCA-102 sarcoma, which, however, was inhibited by IL-2 gene transduction in the same group of experiments (Asher *et al.*, 1991). Induction of systemic immunity was obtained in only a few cases and was dependent on an inherent expression of transplantation antigens by tumor cells. This suggests that local tumor destruction induced by nonspecific accumulation of immune cells may elicit release of antigen, which can then stimulate a systemic immunity (Colombo *et al.*, 1992b). In addition, TNF-α released by tumor cells may determine opposite effects, as in the EB/ESB lymphoma, the first subline growing as a local, solid tumor and the second one growing as a metastatizing tumor. In this system, TNF-α release was found to inhibit local growth of EB while promoting metastasis formation of ESB (Qin *et al.*, 1993).

Multiple cytokine gene transduction was carried out in an attempt to increase the immunogenicity of tumor cells, but without significant advantage (Hock *et al.*, 1993b). The lack of advantage probably stems from the fact that, independent from the cytokine released by the tumor, other cytokines will be locally produced by the incoming host cells recruited by the first cytokine (Colombo and Forni, 1994), thus making the release of the second cytokine by tumor cells irrelevant.

In addition to tumor cells, even normal fibroblasts have been transduced with cytokines such as IL-2 (Fakhrai *et al.*, 1995) and IL-12 (Tahara *et al.*, 1994). In these studies, transduced fibroblasts admixed with syngeneic, irradiated tumor cells given subcutaneously were shown to impair local tumor growth, to generate systemic immunity, and, in the case of IL-2, to cure mice bearing established tumors. A similar effect can also be obtained by using biodegradable polymers, which can slowly release cytokines at the site of tumor growth (Golumbek *et al.*, 1993). If the inhibitory effect on local tumor growth is achieved, biopolymers may represent an advantage over the more cumbersome preparation of engineered fibroblasts or tumor cells. These studies provide the rationale for a similar approach in humans and overcome the need to generate autologous tumor cell lines. However, even the culture of autologous fibroblasts, their gene transduction, and safety assays may require too long a time for metastatic tumors that can progress so that the patients become no longer eligible for the study.

I. Mechanism(s) of Tumor Growth Inhibition versus Regression of Established Tumors: The Lesson from Animal Studies

The consensus emerging from these studies is that the cytokine released by transduced tumor cells triggers a complex inflammatory response with induction of secondary cytokines and leukocyte infiltration whose nature depends on the type of cytokine gene transduced (Colombo and Forni, 1994). The strength of the reaction appears to depend on the amount of released cytokine; the local concentration of cytokine may thus increase as the tumor expands and decrease as long as it activates the debulking immune reaction. Therefore, the local reaction is in some way regulated in a feedback fashion by its own efficacy. This has implications for the induction of a T cell-mediated systemic immunity, which has been shown to be stronger if it follows the rejection of tumors that underwent an initial growth. Thus, injection of viable cytokine-transduced tumor cells induces a stronger immune memory than injection of the same cells made nonreplicating by irradiation or mitomycin (Hock *et al.*, 1993b; Allione *et al.*, 1994). With a view to designing new clinical protocols, however, it is important that even irradiated cells be effective as therapeutic vaccines (Dranoff *et al.*, 1993; Zatloukal *et al.*, 1995). It is likely that initially growing tumors cause a continuous and quantitatively more relevant antigen release. Accordingly, local tumor rejection after initial growth occurring on transduction with a suicide gene (e.g., a gene encoding an enzyme converting a prodrug to a cytotoxic agent) results in systemic memory against a challenge with parental tumor cells as effective as that induced by cytokine-transduced tumor cell injection (Vile *et al.*, 1994). Although tumor rejection is in itself a strong immunogenic stimulus, cytokine-induced tumor regression has the advantage of deflecting the type of immune memory to either type 1 helper T cell (T_H1) and T_H2, depending on the cytokine involved (Table I). Animal studies also indicate that, despite the immune memory that follows the regression of an incipient tumor and that can inhibit the growth of a challenge of parental tumor cells injected as a cell suspension, only in a few cases in which selected cytokine genes (particularly IL-2, IL-4, GM-CSF) were transduced was such immunity able to prevent the progression of established metastatic tumors (Golumbek *et al.*, 1991; Cavallo *et al.*, 1993; Dranoff *et al.*, 1993; Porgador *et al.*, 1993a, b; Zatloukal *et al.*, 1995). Of particular interest is the curative effect of vaccination with IL-2 gene-transduced tumor cells in mice with orthotopically implanted bladder carcinomas (Connor *et al.*, 1993). Cure of established tumors by vaccination with cytokine-transduced tumor cells, therefore, appears to be difficult to achieve, even with strongly antigenic mouse tumors. This is likely to depend on many factors, including size, growth rate, invasiveness, location of the tumor, the amount and type of cytokine released, and, of paramount importance, the immune status of the host, which is known to be often compromised in cancer patients.

Although any vaccination approach is, therefore, more likely to be successful in the presence of a minimal tumor burden, combination therapies may widen the stage of tumors that can be effectively treated. For example, vaccination with cytokine gene-transduced tumor cells can be combined with adoptive immunotherapy. In fact, lymphocytes from individuals treated with vaccines are expected to contain a larger number of antitumor CTL precursors. These lymphocytes could be specifically expanded with autologous tumor cells transduced with IL-2, IL-7, or IL-12 genes and then reinfused to boost the effector phase. Another approach already successfully tested in a murine model is vaccination with IL-2-transduced tumor plus recombinant IL-12 given systemically as adjuvant (Vagliani *et al.*, 1996).

Further improvements in immunogene therapy may also depend on the possibility of directly transducing cytokine genes into neoplastic lesions *in vivo*. This can cause at least a partial regression of tumors expressing well-defined antigens with a subsequent release of antigen, resulting in the local recruitment of inflammatory cells, T cells, or both. For such an approach, vectors other than retroviral ones have been used to avoid stable integration in proliferating cells, a process that carries the risk of continuous cytokine release and immune stimulation that could result in autoimmune diseases. Adenoviral and vaccinia viral vectors were therefore employed to transduce subcutaneously growing tumors of mice with IL-2 (Addison *et al.*, 1995; Cordier *et al.*, 1995) and IL-12 genes, respectively (Meko *et al.*, 1995), resulting in efficient cytokine gene transduction, local tumor growth inhibition, and systemic immunity.

Molecularly defined tumor antigens now allow one to test whether coexpression of antigens and cytokines may result in a better curative effect. To overcome the relative lack of cloned murine tumor antigens, the β-galactosidase (β-Gal) foreign gene of *Escherichia coli* has been used as an operational tumor antigen. The *in vivo* administration of VV vector expressing IL-2 and β-Gal resulted in a reduction in the number of experimental metastases and increased the survival of mice bearing a β-Gal-transduced tumor; substitution of IL-2 with other cytokines, such as GM-CSF or TNF, weakened the immune response (Bronte *et al.*, 1995).

Although murine studies have been considered sufficient to test the potential efficacy of tumor cells engineered with cytokine genes in the clinical setting, further experiments in the mouse model are still necessary to provide much of the key information needed for optimization of clinical trials. In fact, despite the fact that animal models suggest IL-2, IL-4, and GM-CSF as the best therapeutic cytokines, there is no consensus yet on which cytokine gene-engineered vaccine will provide the strongest curative potential, mainly because the tumors used in the experimental system and the cytokines produced have often been different. Finally, information gained from the use of the same tumor transduced with different cytokines (Dranoff *et al.*, 1993; Hock *et al.*, 1993b; Allione *et al.*, 1994) is still to be considered valid for that tumor and must be substantiated in other neoplasms with similar vaccines.

2. Toxicity and Pharmacokinetics Issues

Studies performed on murine models showed that injection with cytokine gene-transduced, regressing, engineered tumor cells or with DNA–liposome complexes carries no acute toxicity, as evaluated by the serum enzymatic activities, renal and cardiac functions, and hematologic parameters (Stewart *et al.*, 1992; Dranoff *et al.*, 1993; Jaffee *et al.*, 1995). A concern about treatment with transduced tumor cells is that cytokine release could continuously stimulate the immune response and activate an autoimmune reaction because of the break in tolerance to common tissue antigens. This has been shown to occur with IL-2 given systemically (Parmiani, 1990). Histological and functional studies performed in mice vaccinated with GM-CSF-transduced renal cancer cells, however, showed no activation of autoimmune reactivity, even when a large excess of normal antigen (i.e., normal renal cells) was administered (Jaffee *et al.*, 1995).

Few studies, however, have analyzed the pharmacokinetics of cytokine release in mice vaccinated with *ex vivo* transduced tumor cells. Although the amount of released cytokine varied widely according to the gene-transfer technique used, different studies showed that the local level of cytokine is high whereas in the serum it is frequently undetectable or is detectable over a short period of time and at a level unable to induce the side effects associated with the systemic administration of the cytokine (Jaffee *et al.*, 1995; Zatluokal *et al.*, 1995).

B. Clinical Studies

I. Cytokine Gene-Transduced Human Cell Lines

Mouse studies have shown that insertion of cytokine genes into different tumors may significantly increase their immunogenicity *in vitro* and *in vivo*. Using mostly retroviral vectors, several groups have demonstrated that human neoplastic lines can be easily transduced with genes encoding IL-2 or other cytokines. Table IV lists most of these studies.

Characterization of the lines was the first step in the preparation of these new vaccines. In the case of melanoma, by and large the most frequently studied neoplasm, it is now possible to satisfy all of the major requirements for the construction of a potentially immunogenic vaccine containing molecularly well-defined antigens recognizable by T cells. Thus, cytokine gene-transduced melanoma lines have been prepared and shown to maintain, after gene transduction, the necessary antigenic phenotype (Parmiani and Colombo, 1995). However, only a few examples have been reported (and with IL-2 or IFN-γ gene-transduced melanoma lines only) of the capacity of these transduced cells to increase, at least *in vitro*, their tumor-specific, MHC-restricted T cell-stimulatory activity in comparison with the parental, untransduced tumor (Ogasawara and Rosenberg, 1993; Uchiyama *et al.*, 1993; Arienti *et al.*, 1994). Genes encoding other cytokines, in addition to IL-2 (e.g., IL-4, IL-7, IL-10, IL-12, IFN-α, IFN-γ, and TNF-α), were transduced into human tumor cells or fibroblasts to prepare more immunogenic vaccines (Table IV). However, even with nonmelanoma lines, the biological activity of the cytokine released on transduction was demonstrated in the majority of cases *in vitro*, whereas evidence of a better *in vitro* stimulation

TABLE IV   Cytokine Gene-Transduced Human Tumor Cells

Gene encoding:	Tumor histotype	Amount of cytokine released (ng/10⁵ cells/24 hr)	Ref.
IL-2	Melanoma	0.1–3.6	Yannelli *et al.* (1993)
IL-2	Melanoma	2.6–6.6	Uchiyama *et al.* (1993)
IL-2	Melanoma	0.1–4.0[a]	Gansbacher *et al.* (1992a)
IL-2	Melanoma	0.95–3.83	A. R. Miller *et al.* (1993)
IL-2	Melanoma	20–600[a]	Osanto *et al.* (1993)
IL-2	Melanoma	2.28–2.33	Arienti *et al.* (1994)
IL-2	Melanoma	15–250[a,b]	Patel *et al.* (1994)
IL-4	Melanoma	200–600	Melani *et al.* (1995b)
IL-7	Melanoma	1.42–9.26	A. R. Miller *et al.* (1993)
IFN-α	Melanoma	2060–2100	Ogasawara and Rosenberg (1993)
IFN-γ	Melanoma	0.1–0.8	Gansbacher *et al.* (1992a)
IFN-γ	Melanoma	~470	Ogasawara and Rosenberg (1993)
TNF-α	Melanoma	0.2–5.8[b]	Yannelli *et al.* (1993)
IL-2	Renal cancer	22[a]	Belldegrun *et al.* (1993)
IL-2	Renal cancer	0.2–1.5[a]	Gastl *et al.* (1992)
IFN-γ	Renal cancer	≤10[a]	Gastl *et al.* (1992)
IFN-γ	Renal cancer	ND[c]	Ogasawara and Rosenberg (1993)
GM-CSF	Renal cancer	5.2–7.9	Jaffee *et al.* (1993)
GM-CSF	Colon carcinoma	1.6–8.1	Jaffee *et al.* (1993)
IL-2	Acute leukemia	0.03–0.34[a]	Cignetti *et al.* (1994)
IL-2	Neuroblastoma	>0.015	Brenner *et al.* (1992)
IFN-α	Renal cancer	102.4[a]	Belldegrun *et al.* (1993)
IL-4	Fibroblats	0.01–10[a]	Lotze and Rubin (1994)
IL-12	Melanoma, renal cancer	2–6	Zitvogel *et al.* (1994)

[a] Value refers to units/10⁵ cells/24 hr.
[b] Depending on different lines, number of subsequent infections, or both.
[c] ND, Not detectable.

of MHC-restricted antitumor T lymphocytes by cytokine gene-transduced lines was scanty.

2. Autologous versus Allogeneic Tumor Lines

Animal studies showing the therapeutic effectiveness of cytokine gene-transduced tumor cells were carried out by and large with syngeneic tumor cells, the equivalent of autologous tumor cells in humans. The use of autologous lines offers the advantage that only tumor antigens can be recognized by T cells of the patient, whereas allogeneic lines, even when partially HLA compatible, can generate antiallogeneic HLA or even antiminor histocompatibility antigen responses that could impair or compete for the recognition of tumor antigens. The choice of autologous transduced lines, however, has many distinct disadvantages, which tend to discourage their use. These can be summarized as follows: (1) the need to have patients bearing a resectable tumor mass of a certain size in order to generate the line with a frequency that, apart from melanoma, is usually low (10–30%), (2) the variability in gene transfer from culture to culture in nonstabilized cell lines, (3) the labor intensity and economic cost of preparing the necessary amount of vaccine and of performing the required safety assays before injection into patients, and (4) the need to assess the expression of tumor antigens, MHC, and other molecules on the cell line of each patient and the likelihood of discarding several lines because of a reduction in or even lack of expression of MHC molecules on neoplastic cells caused by a variety of molecular alterations (Angelini *et al.*, 1986; Garrido *et al.*, 1993; Ferrone and Marincola, 1995). All these problems can be avoided by the use of allogeneic lines that, in the case of melanoma, can also be selected or genetically manipulated in such a way that they can express high levels of antigens known to be shared by the majority of tumors and be presented by a known HLA allele (e.g., HLA-A2.1 for Melan-A/MART-1, gp100, or tyrosinase and HLA-A1 for MAGE-1, -2, and -3). These lines can be expanded, gene transduced, evaluated for the stability of the cytokine released, and used for safety assays and melanoma cells can be irradiated in enough quantity that vials containing the desired number of cells to be administered to patients when necessary can be prepared and frozen in advance. The ultimate choice of autologous vs allogeneic cells, however, may also depend on the mechanism of vaccination (see Section III,B,4).

On the basis of results in animal tumor models, several clinical protocols have been proposed, approved, and initiated to test the hypothesis that cytokine gene-transduced tumor cells may represent a vaccine better than that described in the previous studies in which unmodified tumor cells and adjuvants were used. A list of these protocols is shown in Table V. The choice of the cytokine gene to be transduced was based on previous animal studies and the potential mechanism of its therapeutic activity that indicated IL-2 and GM-CSF as the two most effective cytokines in several different tumors (Dranoff *et al.*, 1993; Allione *et al.*, 1994; Schmidt *et al.*, 1995). As for the tumor histotype, the most frequent choice was melanoma, both for its resistance to conventional anticancer drugs and for the advanced knowledge of the antigens expressed by these neoplastic cells (Table II). Table V, however, indicates that attempts to vaccinate with tumor or fibroblast gene-transduced cells are also ongoing for renal, colon, and prostate cancer and for neuroblastoma, although no clear evidence has been provided that T cell-defined antigens are frequently expressed in such tumor histotypes. In fact, T cell recognition of tumor-restricted antigens has occasionally been reported for renal cancers (Belldegrun *et al.*, 1988; Alexander *et al.*, 1990; Kim *et al.*, 1990; Schendel *et al.*, 1993). Recently, however, two genes have been cloned encoding antigens recognized by autologous CTL on renal cancers. One of them (designed RAGE) appears to be expressed in less than 5% of renal cancers (Gaugler *et al.*, 1996), and the other, resulting from a point mutation of the HLA-A2.1 gene itself, is unique to that given tumor (Brändle *et al.*, 1996). In our opinion, it is instead mandatory to document that the neoplastic cells to be used as vaccine express a known antigen(s) that can be measured in tumor cells of the prospective recipients. The phenotypic features of transduced cell lines to be used as vaccine are summarized in Table VI and discussed in Section III,B,4.

3. Features of Immunogene Therapy Protocols

The first protocols for gene therapy of cancer patients can be traced back to 1990, when the groups of both Rosenberg and Lotze prepared TNF-α and IL-2/IL-4 gene-transduced tumor-infiltrating lymphocytes to infuse autologous patients (Rosenberg *et al.*, 1990). The rationale was to increase the cytotoxic activity of tumor-infiltrating lymphocytes (TILs), which would then reach the distant lesions and destroy neoplastic cells. These studies, however, were made difficult by the low efficiency of transduction of TILs, which prevented the generation of high enough numbers of these cells to target tumor lesions effectively *in vivo*.

Clinically protocols (usually pilot or phase I–II studies) were then initiated, first in the United States (Gansbacher *et al.*, 1992b; Rosenberg, 1992) and then in Europe (Osanto *et al.*, 1993), mainly with gene-transduced allogeneic cells and in melanoma patients. The common objectives of these protocols were the evaluation of toxicity and clinical response and, more important, of the induction or increase of CTL-specific antitumor rseponse. The last represents a crucial parameter that needs to be assessed if we are to be able to interpret any clinical result and to further optimize the protocols.

However, while in at least some mouse systems the availability of unlimited numbers of tumor and T cells (which can be easily expanded in culture, transplanted *in vitro*, or obtained from syngeneic animals) allows us to carry out different immunological assays to evaluate the induction or augmentation of the antitumor immunity caused by vaccination, in a clinical setting the autologous line is available in a negligible number of patients (if any) and the number of T lymphocytes to be used in the assays is limited. At least with melanoma, however, it is now possible to use, as *in vitro* stimulators and APCs, autologous peripheral blood monocytes or B lymphocytes pulsed with the different known antigenic peptides (e.g., those derived from Melan-A/MART-1, gp100, tyrosinase, gp75, MAGE, BAGE, and GAGE) and then test the activated CTLs on targets pulsed with the same peptide used for *in vitro* stimulation and expressed by the vaccinating line. These targets include either Epstein–Barr virus (EBV)-transformed autologous lines pulsed with peptides or special tumor lines in which mutations of genes associated with the cytoplasmic transport of peptides for antigen presentation (TAP) prevent presentation of endogenous peptides, thus resulting in empty MHC mole-

TABLE V   Approved and Initiated Clinical Protocols for Cytokine Immunogene Therapy of Cancer[a]

Tumor population	Gene transduced	Ex-vivo target cells	Responsible investigator(s) and institution
Melanoma	IL-2	Autologous melanoma	S. A. Rosenberg (NCI, NIH, Bethesda, MD)
Melanoma	IL-2	Allogeneic melanoma	B. Gansbacher (Memorial Sloan Kettering Cancer Center, New York, NY)
Melanoma	IL-2	Allogeneic melanoma	S. Osanto, P. Schrier (University Hospital, Leiden, The Netherlands)
Melanoma	IL-2	Autologous melanoma	J. Economou (UCLA, Los Angeles, CA)
Melanoma	IL-2	Allogeneic melanoma	T. K. Das Gupta (University of Illinois, Chicago, IL)
Melanoma	IL-2 or IL-4	Allogeneic melanoma	N. Cascinelli, G. Parmiani (National Tumor Institute, Milan, Italy)
Melanoma	IFN-γ	Autologous or allogeneic melanoma	H. F. Seigler (Duke University, Durham, NC)
Melanoma	TNF-α	Autologous melanoma	S. A. Rosenberg (NCI, NIH, Bethesda, MD)
Melanoma	GM-CSF	Autologous melanoma	E. Rinkin (The Netherlands Cancer Center, Amsterdam, The Netherlands)
Melanoma	GM-CSF	Autologous melanoma	G. Dranoff (Harvard Medical School, Boston, MA)
Melanoma	IL-7	Allogeneic melanoma	G. Schmidt-Wolf (Free University, Berlin, Germany)
Melanoma	IL-4	Autologous fibroblasts[b]	M. T. Lotze (Pittsburgh Cancer Institute, Pittsburgh, PA)
Melanoma	IL-2	Autologous fibroblasts[b]	R. Mertelsman (University Medical Center, Freiburg, Germany)
Neuroblastoma	IL-2	Autologous neuroblastoma	M. K. Brenner (St. Jude Children Hospital, Memphis, TN)
Renal carcinoma	IL-2	Allogeneic renal carcinoma	B. Gansbacher (Memorial Sloan Kettering Cancer Center, New York, NY)
Renal carcinoma	GM-CSF	Autologous renal carcinoma	J. W. Simons (Johns Hopkins University, Baltimore, MD)

(*continues*)

TABLE V (*continued*)

Tumor population	Gene transduced	Ex-vivo target cells	Responsible investigator(s) and institution
Prostate carcinoma	GM-CSF	Autologous prostate carcinoma	J. W. Simons (Johns Hopkins University, Baltimore, MD)
Colon cancer	IL-2	Autologous fibroblasts	R. E. Sobol, J. Royston (Regional Cancer Center, San Diego, CA)

[a] Approved and initiated since 1994. Sources: The RAC Report—*Human Gene Therapy* issues 1994, 1995.
[b] Fibroblasts are admixed with irradiated autologous tumor cells and used as vaccine.

cules on the cell surface that can be easily occupied by the exogenously added peptides (Salter and Cresswell, 1986). Although this *in vitro* assay has been widely used to measure lymphocyte stimulation by known peptides (Carbone *et al.*, 1988; Celis *et al.*, 1994; Rivoltini *et al.*, 1995; Salgaller *et al.*, 1995), its use in evaluating the immune response in patients vaccinated with cytokine gene-transduced tumor cells remains to be assessed. That APCs pulsed with melanoma peptides can also generate an *in vivo* antigen-specific CTL response has been demonstrated (Mukherji *et al.*, 1995).

At our institute we have initiated clinical protocols for vaccination with IL-2 or IL-4 gene-transduced allogeneic HLA-A2 melanoma cells using Melan-A/MART-1-, tyrosinase-, gp100-, and MAGE-3-positive cells in advanced-stage, HLA-A2$^+$ melanoma patients (Cascinelli *et al.*, 1994). The scheme of these protocols is summarized in Fig. 1. In the first protocol, HLA-A2–positive patients were injected subcutaneously on days 1, 13, and 26 with IL-2 gene-transduced and irradiated melanoma cells at a dose of 5×10^7 cells (three patients) and 15×10^7 cells (four patients). Since vaccinating cells and patients shared only HLA-A2 but no other HLA class I alleles, the alloreactive response was also evaluated after vaccination. HLA-A, HLA-B, and HLA-C antibodies against alloantigens expressed on injected melanoma cells were undetectable in serum of samples taken before

TABLE VI Requirements for Use of Transduced Cell Lines as Vaccine

Expression of molecularly defined, common antigens recognized by CTLs with the knowledge of the MHC-restricting allele
Expression of MHC classes I and II and T cell costimulatory molecules (ICAM-1, B7-1 and -2)
Stable release over time (even after irradiation) of a sufficient amount of the cytokine encoded by the transduced gene

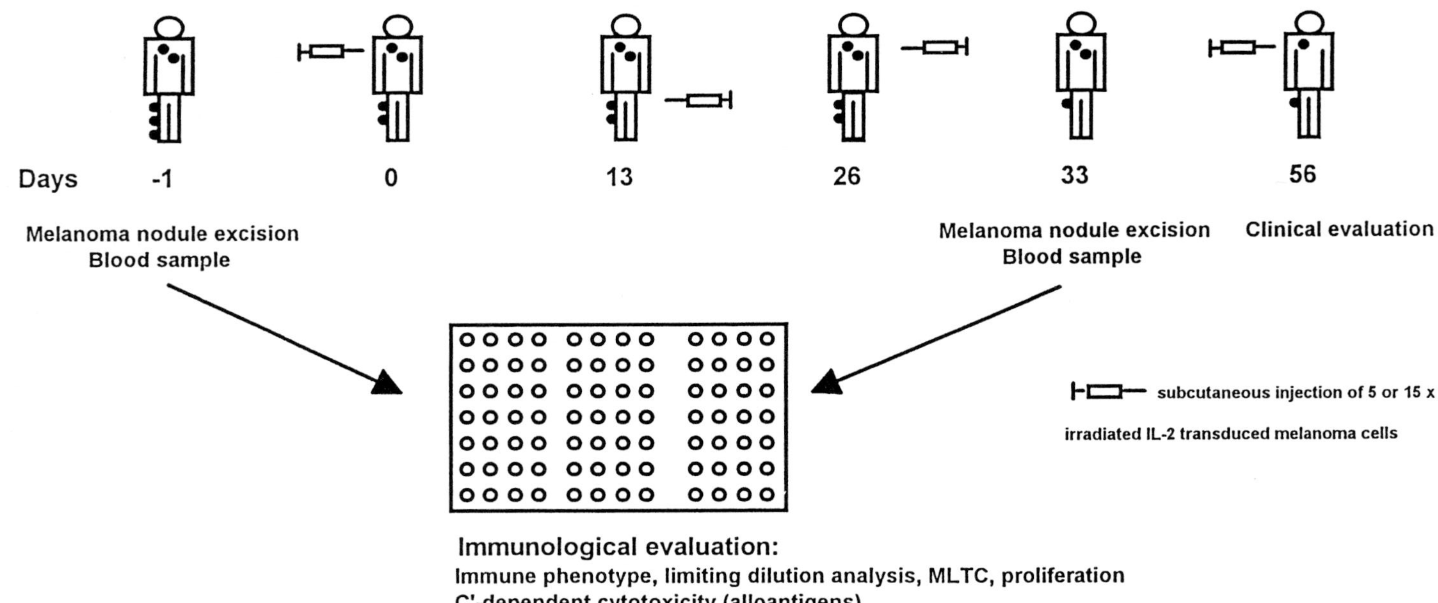

FIGURE I Scheme of the clinical protocol of vaccination with allogeneic, HLA-A2[+], Melan-A/MART-1[+], MAGE-3[+], gp100[+] melanoma cells. (Top) The treatment schedule of Stage IV melanoma patients with IL-2 transduced melanoma cells. The immunological evaluation involving different assays is also shown (see text for details).

and after vaccination. The lack of allo-HLA antibodies was probably due to the limited number of injections (three or four), the low immunological responsiveness of these patients bearing a significant tumor burden, or to both. To evaluate the specific CTL response, mixed lymphocyte–tumor cultures (MLTCs) and limiting dilution analysis were performed, comparing peripheral blood lymphocytes (PBLs) obtained before and after immunization. Although MLTCs revealed an increased but MHC-unrestricted cytotoxicity, in two cases the frequencies of melanoma-specific CTL precursors were clearly augmented as a result of vaccination (Arienti *et al.*, 1996). The evidence was particularly clear with respect to a patient from whom a tumor cell line could be obtained to carry out the assays. This indicates that vaccination with cells bearing the appropriate antigens and releasing IL-2 locally can expand a T cell response against antigens of autologous, untransduced tumor but only in a minority of patients. Similar evidence has been found in two other clinical protocols involving IL-2 and IL-4 genetransduced cells (Table VII). However, these data are still preliminary and need to be confirmed in further studies.

4. Critical Issues in Clinical Trials

There are several critical issues that are being examined in the ongoing clinical trials of cytokine gene-transduced vaccination and that need to be considered. They have also been discussed by Pardoll (1995), and we list them again with some of our additional comments.

1. Taking into account the difficulties of obtaining enough fresh tumor cells from each patient and of generating a line (even a short-lived one) that can release a constant amount of cytokine and express the appropriate antigenic profile, it is preferable to choose allogeneic, standardized, and well-characterized lines as vaccine (see also Section III,B,2). Retroviral vec-

TABLE VII Evidence for Induction or Augmentation of a Tumor-Specific MHC-Restricted T Cell Response after Patient Vaccination with Cytokine Gene-Transduced Tumors

Cytokine gene	Target cells	MHC restricted	MHC unrestricted
IL-2[a]	Neuroblastoma	3/7[b]	+
IL-4[a]	Fibroblasts	+[c]	+
IL-2[d]	Melanoma	2/6[b]	+

[a] Results presented at the RAC Report (Brenner *et al.*, 1992; Lotze and Rubin, 1995).
[b] Number of positive cases/number of patients tested.
[c] Frequency of responding patients not mentioned.
[d] F. Arienti, M. P. Colombo, C. Melani, and G. Parmiani, unpublished.

tors remain the vectors of choice for this system, but a variety of nonviral vectors that are under study may change the situation in the near future.

2. The amount of cytokine released on a per-cell basis appears to be an important parameter in conferring immunogenicity to the transduced cells, at least in the mouse system (see Section III,A), where too-high amounts of IL-2 were shown to inhibit the systemic antitumor immune response (Cavallo *et al.*, 1992; Schmidt *et al.*, 1995). For IL-4 and GM-CSF, no such bell-shaped curve was found (Pardoll, 1995). Clinical protocols therefore need to be designed to compare groups of patients vaccinated with the same tumor line releasing different amounts of the same cytokine.

3. The number of cells injected, the number and time of injections, and the route of immunizations are crucial parameters that need to be investigated in humans. Although only subcutaneous spaces can receive a number of cells on the order of 10^8–10^9, intradermal injections should also be attempted because of the presence of Langerhans cells, which are known to have a potent antigen-presenting activity. Also, the kinetics of the T cell response to vaccine should be investigated by testing patient PBLs at different times after vaccination.

4. To avoid proliferation of transduced neoplastic cells, the vaccine needs to be irradiated. It is already known that for each neoplasm there exists a dose of irradiation that allows the release of a biologically significant amount of cytokines for at least 10–15 days (Gastl *et al.*, 1992; Belldegrun *et al.*, 1993; Arienti *et al.*, 1994), such a possibility should be checked *in vivo* with appropriate biopsies at the site of vaccination.

5. The major issue in vaccination with cytokine gene-transduced cells is the knowledge of the mechanism(s) by which such modified cells can increase their immunogenicity. *In vitro* studies with human melanomas or renal carcinomas indicate that the released cytokine can improve the stimulatory activity of autologous tumor-specific T cells after MLTC (Schendel *et al.*, 1993; Uchiyama *et al.*, 1993; Arienti *et al.*, 1994). However, whether the *in vitro* system is representative of what occurs *in vivo* at the site of vaccine injection is questionable. In fact, the *in vitro* data obtained from MLTCs can be explained by either a direct or an indirect mechanism. The direct mechanism implies the presentation of tumor antigens by tumor cells themselves, and this can be especially true with melanomas. These are known to express class II MHC (in our study the IL-2 gene-transduced line was purposely selected for such a feature) and can present antigens to autologous or allogeneic MHC class II-compatible lymphocytes (Radrizzani *et al.*, 1991), although this function can be compromised in metastatic cells (Alexander *et al.*, 1989). Alternatively, an indirect mechanism of cross-priming (Huang *et al.*, 1994) can be operative by which monocytes, B cells, or dendritic cells serve as APCs that process tumor-derived antigenic proteins for their own T cells. Such a mechanism will not require that tumor cells express MHC class I or II or any costimulatory molecules that are provided

by APCs (Huang *et al.*, 1994). Can either of these stimulation pathways occur *in vivo* when the cellular vaccine is given subcutaneously?

An important difference between the mouse and the human systems is the requirement that humans must be injected with nonreplicating cell vaccines. Thus, it is not clear whether the nonreplicating cell vaccine, once injected subcutaneously or intradermally into patients, will recruit first inflammatory cells and then T lymphocytes, as occurs in the mouse system with cytokine gene-transduced replicating tumor cells (see Section III,B,1). Most likely, nonreplicating cells will secrete a limited amount of cytokine before dying, as a result of previous irradiation and host reaction against alloantigens; the released antigens can now be taken up by local, bone-marrow-derived APCs which then migrate into draining lymph nodes where the interaction with T cell precursors may occur. Such a process has been elegantly demonstrated in the mouse, first by Huang and associates (1994) with GM-CSF gene-transduced tumors and then by Maass and associates (1995) with the IL-2 gene-transduced, irradiated mouse tumors. This scenario implies that the patient is not already primed against tumor antigens or cannot become primed during the first administration of the vaccine. If, on the contrary, the patient has circulating T cell precursors in large enough numbers, as may occur in melanoma (Mazzocchi *et al.*, 1994), T lymphocytes can directly reach the cellular vaccine given subcutaneously with the help of the cytokine released locally and of other cytokines produced by the influx of inflammatory cells. T cells recruited at the tumor site can then be directly restimulated with the help of costimulatory signals provided by tumor cells themselves (e.g., when the B7-1 gene is transduced), by cytokines, or by both. Such T cells can then recirculate and target other metastases growing at distant sites. In the mouse model, cytokine gene-transduced tumors can be allowed to grow and to pump the cytokine as long as the tumor grows. Accumulation of inflammatory cells occurs and, by the release of different inflammatory-type cytokines (e.g., TNF-α, IL-1, IL-6), these cells then induce the recruitment of T cells, at least with tumors that express a reasonable amount of antigen (Stoppacciaro *et al.*, 1993).

Altogether, the different situations can be summarized as indicated in Table VIII. It is clear that systemic immunity, a crucial condition for a clinical response, is likely to take place only when local tumor destruction causes a release of tumor antigens that APCs can process and present to T cells either locally (when tumor is allowed to grow as in mouse models) or in the lymph nodes. In the clinical setting where only nonreplicating transduced tumor cells can be used as vaccines, T cells will encounter the APCs in the lymph nodes, the possible exception being represented by those patients who, for different reasons, are already primed against the tumor antigens contained in the vaccine. Thus, as previously discussed, the mechanism can be different according to the immune status of the vaccinated host, naive

TABLE VIII Effect of Transient Local Growth and Nature of Infiltrate on Induction of Systemic Immunity and Therapeutic Response of IL-2 Gene-Transduced, Antigenic Tumor Cell Vaccines

Local growth	IL-2 production	Induction of: Nonspecific infiltrate	T cell infiltrate	Systemic immunity	Therapeutic efficacy[a]
−	+	+	−	+	+
+[b]	+	+	+	+	+
−	−	±	−	−	−
+	−	±	−	±[c]	±

[a] Depends also on the amount of cytokine released and number of cells injected.
[b] Transient local growth can be terminated by host reactions due to cytokine release, by surgical resection, or destruction by different cytotoxic drugs.
[c] Depends on the constitutive antigenic strength of the tumor.

vs primed. In the latter case the expression of MHC classes I and II and of costimulatory molecules can confer potent immunogenicity as also shown in mouse systems (Salvadori *et al.*, 1995). Because of this uncertainty about the mechanism of vaccination by subcutaneous injection of cytokine-engineered tumor cells, it may be wise to use, at present, cells that express MHC and costimulatory molecules as well.

IV. Concluding Remarks

After 5–6 years of studies, the somatic gene therapy of cancer can still be considered in its early phase. The research effort in this field has been considerable, more in the United States than in Europe, probably because of the more aggressive behavior of American than European companies and because of regulatory or ethical problems.

The success has been considerable as indicated by (1) the tremendous amount of new, basic information gathered through the preclinical studies evaluable by the number and quality of published papers, (2) the boost in the interactions between investigators from different disciplines, and (3) the initial translation of principles defined in these studies into informative clinical protocols. However, as pointed out by the ad hoc National Institutes of Health panel, some biomedical researchers and their sponsors may have "oversold" the results of their studies, leading to an excessively optimistic reaction by the public. In particular, it is our belief that, of the more than 60 trials of gene therapy of cancers approved worldwide, some appear to have a weak rationale and will probably fail. In conclusion, we deem it necessary, on the one hand, to continue our effort in preclinical studies particularly aimed at solving the problems of delivery and *in vivo* targeting and, on the other hand, to design clinical protocols aimed at answering the many questions that animal models cannot answer by themselves. If a high scientific standard will characterize our work, then in the years ahead we will ultimately understand and define the potential therapeutic effect of cancer gene therapy.

Acknowledgments ____

We thank Ms. Grazia Barp for typing and editing. Our work was supported by grants from the Italian Association for Cancer Research (Milan) and from the Italian Ministry of Health (Finalized Projects).

References ____

Addison, C. L., Braciak, T., Ralston, R., Muller, W. J., Gauldie, J., and Graham, F. L. (1995). Intratumoral injection of an adenovirus expressing interleukin 2 induced regression and immunity in a murine breast cancer model. *Proc. Natl. Acad. Sci. U.S.A.* **92**, 8522–8526.

Alexander, J., Rayman, P., Edinger, M., Connelly, R., Tubbs, R., Bukowski, R., Pontes, E., and Finke, J. (1990). TIL from renal-cell carcinoma: Restimulation with tumor influences proliferation and cytolytic activity. *Int. J. Cancer* **45**, 119–124.

Alexander, M. A., Bennicelli, J., and Guerry, D. IV (1989). Defective antigen presentation by human melanoma cell lines cultured from advanced, but not biologically early, disease. *J. Immunol.* **142**, 4070–4078.

Allione, A., Consalvo, M., Nanni, P., Lollini, P. L., Cavallo, F., Giovarelli, M., Forni, M., Gulino, A., Colombo, M. P., Dellabona, P., Hock, H., Blankenstein, T., Rosenthal, F. M., Gansbacher, B., Bosco, M. C., Musso, T., Gusella, L., and Forni, G. (1994). Immunizing and curative potential of replicating and nonreplicating murine mammary adenocarcinoma cells engineered with interleukin (IL)-2, IL-4, IL-6, IL-7, IL-10, tumor necrosis factor α, granulocyte-macrophage colony-stimulating factor, and γ-interferon gene or admixed with conventional adjuvants. *Cancer Res.* **54**, 6022–6026.

Angelini, G., Fossati, G., Radice, P., Longo, A., Piccioli, P., Pierotti, M. A., Ferrara, G. B., and Parmiani, G. (1986). Loss of polymorphic restriction fragments of class I and class II MHC genes in a malignant melanoma. *J. Immunogenet.* **13**, 241–246.

Anichini, A., Maccalli, C., Mortarini, R., Salvi, S., Mazzocchi, A., Squarcina, P., Herlyn, M., and Parmiani, G. (1993a). Melanoma cells and normal melanocytes share antigens recognized by HLA-A2 restricted cytotoxic T-cell clones from melanoma patients. *J. Exp. Med.* **177**, 989–998.

Anichini, A., Mortarini, R., Alberti, S., Mantovani, A., and Parmiani, G. (1993b). T-cell receptor engagement and tumor ICAM-1 up-regulation are required to by-pass low susceptibility of melanoma cells to autologous CTL-mediated lysis. *Int. J. Cancer* **53**, 994–1001.

Arienti, F., Belli, F., Gambacorti-Passerini, C., Furlan, L., Mascheroni, L., Prada, A., Rizzi, M., Marchesi, E., Vaglini, M., Parmiani, G., and Cascinelli, N. (1993). Adoptive immunotherapy of advanced melanoma patients with interleukin-2 (IL2) and tumor-infiltrating lymphocytes selected *in vitro* with low doses of IL2. *Cancer Immunol. Immunother.* **36**, 315–322.

Arienti, F., Sulé-Suso, J., Melani, C., Maccalli, C., Belli, F., Illeni, M. T., Anichini, A., Cascinelli, N., Colombo, M. P., and Parmiani, G. (1994). Interleukin-2 gene-transduced human melanoma cells efficiently stimulate MHC-unrestricted and MHC-restricted autologous lymphocytes. *Hum. Gene Ther.* **5**, 1139–1150.

Arienti, F., Sulé-Suso, J., Belli, F., Mascheroni, L., Rivoltini, L., Melani, C., Maio, M., Caseinelli, N., Colombo, M. P., and Parmiani, G. (1996). Limited antitumor T cell response in melanoma patients vaccinated with interleukin-2 gene-transduced allogeneic melanoma cells. *Hum. Gene Ther.* **7**, 1955–1963.

Asher, A. L., Mulé, J. J., Kasid, A., Restifo, N. P., Salo, J. C., Reichert, C. M., Jaffe, G., Fendly, B., Kriegel, M., and Rosenberg, S. A. (1991). Murine tumor cells transduced with Th gene for tumor necrosis factor-α. *J. Immunol.* **146**, 3227–3234.

Bakker, A. B. H., Schreurs, M. W. J., de Boer, A. J., Kawakami, Y., Rosenberg, S. A., Adema, G., and Figdor, C. (1994). Melanocyte lineage-specific antigen gp100 is recognized by melanoma-derived tumor-infiltrating lymphocytes. *J. Exp. Med.* **179**, 1005–1009.

Bannerji, R., Arroyo, C. D., Cordon-Cardo, C., and Gilboa, E. (1994). The role of IL-2 secreted from genetically modified tumor cells in the establishment of antitumor immunity. *J. Immunol.* **152**, 2324–2333.

Bartlett, J. S., Quattrocchi, K. B., and Samulski, R. J. (1995). The development of adeno-associated virus as a vector for cancer gene therapy. *In* "The Internet Book of Gene Therapy: Cancer Therapeutics" (R. E. Sobol and K. J. Scanlon, eds.), pp. 27–39. Appleton & Lange, Stamford, Connecticut.

Belldegrun, A., Muul, L. M., and Rosenberg, S. A. (1988). Interleukin 2 expanded tumor-infiltrating lymphocytes in human renal cell cancer: Isolation, characterization, and antitumor activity. *Cancer Res.* **48**, 206–214.

Belldegrun, A., Tso, C.-L., Sakata, T., Duckett, T., Brunda, M. J., Barsky, S. H., Chai, J., Kaboo, R., Lavey, R. S., McBride, W. H., and deKernion, J. B. (1993). Human renal carcinoma line transfected with interleukin-2 and/or interferon α gene(s): Implications for liver cancer vaccines. *J. Natl. Cancer Inst.* **85**, 207–216.

Berd, D., Sato, T., Lattime, E. D., Maguire, H. C., Jr., and Mastrangelo, M. J. (1994). Immunization with hapten-modified tumor cells: A strategy for the treatment of human melanoma. *Proc. Am. Assoc. Cancer Res.* **35**, 667–668.

Bett, A. J., Haddara, W., Prevec, L., and Graham, F. L. (1994). An efficient and flexible system for construction of adenovirus vectors with insertion or deletions in early regions 1 and 3. *Proc. Natl. Acad. Sci. U.S.A.* **91**, 8802–8806.

Blaese, R. M., Ishii-Morita, H., Mullen, C., Ramsey, J., Ram, Z., Oldfield, E. O., and Culver, K. (1994). *In situ* delivery of suicide genes for cancer treatment. *Eur. J. Cancer* **30A**, 1190–1193.

Blankenstein, T., Qin, Z., Uberla, K., Muller, W., Rosen, H., Volk, H.-D., and Diamantstein, T. (1991). Tumor suppression after tumor cell-targeted tumor necrosis factor α gene transfer. *J. Exp. Med.* **173**, 1047–1052.

Boon, T., Cerottini, J.-C., Van den Eynde, B., van der Bruggen, P., and Van Pel, A. (1994). Tumor antigens recognized by T lymphocytes. *Annu. Rev. Immunol.* **12**, 337–365.

Bosco, M., Giovarelli, M., Forni, M., Modesti, A., Scarpa, S., Masuelli, L., and Forni, G. (1990). Low doses of IL-4 injected perilymphatically in tumor-bearing mice inhibit the growth of poorly and apparently nonimmunogenic tumors and induce a tumor-specific immune memory. *J. Immunol.* **145**, 3136–3143.

Boviatsis, E. F., Park, J. S., Sena-Esteves, M., Kramm, C. M., Chase, M., Efird, J. T., Wei, M. X., Breakefield, X. O., and Chiocca, A. E. (1994). Long-term survival of rats harboring brain neoplasms treated with ganciclovir and a herpes simplex virus vector that retains an intact thymidine kinase gene. *Cancer Res.* **54**, 5745–5751.

Brändle, D., Brasseur, F., Weynants, P., Boon, T., and Van den Eynde, B. (1996). A mutated HLA-A2 molecule recognized by autologous cytotoxic T lymphocytes on a human renal cell carcinoma. *J. Exp. Med.*, **183**, 2501–2508.

Brenner, M. K., Furman, W. L., Santana, V. M., Bowman, L., and Meyer, W. (1992). Phase I study of cytokine-gene modified autologous neuroblastoma cells for treatment of relapsed/refractory neuroblastoma. *Hum. Gene Ther.* **3**, 665–676.

Brody, S. L., Jaffe, A., Han, S. K., Werstor, P., and Crystal, R. G. (1994). Direct *in vivo* gene transfer and expression in malignant cells using adenovirus vectors. *Hum. Gene Ther.* **15**, 437–447.

Bronte, V., Tsung, K., Rao, J. B., Chen, P. W., Wang, M., Rosenberg, S. A., and Restifo, N. P. (1995). IL-2 enhances the function of recombinant poxvirus-based vaccines in the treatment of established pulmonary metastasis. *J. Immunol.* **154**, 5282–5292.

Brunda, M. J., Luistro, L., Warrier, R. R., Wright, R. B., Hubbard, B. R., Murphy, M., Wolf, S. F., and Gately, M. K. (1993). Antitumor and antimetastatic activity of interleukin-12 against murine tumors. *J. Exp. Med.* **178**, 1223–1230.

Bubenik, J., Voitenok, N. N., and Kieler, J. (1988). Local administration of cells containing an inserted IL-2 gene and producing IL-2 inhibits growth of human tumors in nu/nu mice. *Immunol. Lett.* **19**, 279–282.

Carbone, F. R., Moore, M. W., Sheil, J. M., and Bevan, M. J. (1988). Induction of cytotoxic T lymphocytes by primary *in vitro* stimulation with peptides. *J. Exp. Med.* **167**, 1767–1779.

Cascinelli, N., Foà, R., and Parmiani, G. (1994). Active immunization of metastatic melanoma patients with interleukin-4 transduced, allogeneic melanoma cells: A phase I–II study. *Hum. Gene Ther.* **5**, 1059–1064.

Cavallo, F., Giovarelli, M., Gulino, A., Vacca, A., Stoppacciaro, A., Modesti, A., and Forni, G. (1992). Role of neutrophils and CD4+ T lymphocytes in the primary and memory response to nonimmunogenic murine mammary adenocarcinoma made immunogenic by IL-2 gene transfection. *J. Immunol.* **149**, 3627–3635.

Cavallo, F., Di Pierro, F., Giovarelli, M., Gulino, A., Vacca, A., Stoppacciaro, A., Forni, M., Modesti, A., and Forni, G. (1993). Protective and curative potential of vaccination with interleukin-2-gene-transfected cells from a spontaneous mouse mammary adenocarcinoma. *Cancer Res.* **53**, 5067–5070.

Celis, E., Tsai, V., Crimi, C., DeMars, R., Wentworth, P. A., Chesnut, R. W., Grey, H. M., Sette, A., and Serra, H. M. (1994). Induction of anti-tumor cytotoxic T lymphocytes in normal humans using primary cultures and synthetic peptide epitopes. *Proc. Natl. Acad. Sci. U.S.A.* **91**, 2105–2109.

Cepko, C. L., Roberts, B. E., and Mulligan, R. C. (1984). Construction and application of a highly transmissible murine retrovirus shuttle vector. *Cell (Cambridge, Mass.)* **37**, 1053–1062.

Cheever, M. A., Disis, M. L., Bernhard, H., Gralow, J. R., Hand, S. L., Huseby, E. S., Qin, H. L., Takahashi, M., and Chen, W. (1995). Immunity to oncogenic proteins. *Immunol. Rev.* **145**, 33–59.

Chen, S. H., Chen, H. X., Wang, Y., Kosai, K., Finegold, M. J., Rich, S., and Woo, S. L. C. (1995). Combination gene therapy for liver metastasis of colon carcinoma *in vivo*. *Proc. Natl. Acad. Sci. U.S.A.* **92**, 2577–2581.

Cignetti, A., Guarini, A., Carbone, A., Forni, M., Cronin, K., Forni, G., Gansbacher, B., and Foà, R. (1994). Transduction of the IL2 gene into human acute leukemia cells: Induction of tumor rejection without modifying cell proliferation and IL2 receptor expression. *J. Natl. Cancer Inst.* **86**, 785–791.

Colombo, M. P., and Forni, G. (1994). Cytokine gene transfer in tumor inhibition and tumor therapy: Where are we now? *Immunol. Today* **15**, 47–51.

Colombo, M. P., Ferrari, G., Stoppacciaro, A., Parenza, M., Rodolfo, M., Mavilio, F., and Parmiani, G. (1991). Granulocyte colony-stimulating factor gene suppresses tumorigenicity of a murine adenocarcinoma *in vivo*. *J. Exp. Med.* **173**, 1093–1102.

Colombo, M. P., Maccalli, C., Mattei, C., Melani, C., Radrizzani, M., and Parmiani, G. (1992a). Expression of cytokines genes, including IL-6, in human malignant melanoma cell lines. *Melanoma Res.* **2**, 181–189.

Colombo, M. P., Modesti, A., Parmiani, G., and Forni, G. (1992b). Local cytokine availability elicits tumor rejection and systemic immunity through granulocyte-T-lymphocyte cross-talk. *Cancer Res.* **52**, 4853–4857.

Connor, J., Bannerji, R., Saito, S., Heston, W., Fair, W., and Gilboa, E. (1993). Regression of bladder tumors in mice treated with interleukin 2 gene-modified tumor cells. *J. Exp. Med.* **177**, 1127–1134.

Cordier, L., Duffour, M. T., Sabourin, J. C., Lee, M. G., Cabannes, J., Ragot, T., Perricaudet, M., and Haddada, H. (1995). Complete recovery of mice from a pre-established tumor by direct intratumoral delivery of an adenovirus vector harboring the murine IL-2 gene. *Gene Ther.* **2**, 16–21.

Culver, K. W., Ram, Z., Wallbridge, S., Ishii, H., Oldfield, E. M., and Blaese, M. (1992). *In vivo* gene transfer with retroviral-vector producer cells for treatment of experimental brain tumors. *Science* **256**, 1550–1552.

Danos, O., and Mulligan, R. C. (1988). Safe and efficient generation of recombinant retroviruses with amphotropic and ecotropic host range. *Proc. Natl. Acad. Sci. U.S.A.* **85**, 6460–6464.

Dermime, S., Molldrem, J., Parker, K. C., Jiang, Y. Z., Mavroudis, D., Hensel, N., Couriel, D., Mahonev, M., Coligan, J. E., and Barrett, A. J. (1995). Human CD8+ T lymphocytes recognize the fusion region of BCR/ABL hybrid protein present in chronic myelogenous leukemia. *Blood* **86s1**, 158a.

Dermime, S., Bertazzoli, C., Marchesi, E., Ravagnani, F., Blaser, K., Corneo, G. M., Parmiani, G., and Gambacorti-Passerini, C. (1996). Lack of T-cell mediated recognition of the fusion region of the PML/RARα hybrid protein by lymphocytes of acute promyelocytic leukemia patients. *Clin. Cancer Res.* **2**, 593–600.

Doerfler, W. (1991). The abortive infection and malignant transformation by adenoviruses: Integration of viral DNA and control of viral gene expression by specific patterns of DNA methylation. *Adv. Virus Res.* **39**, 89–128.

Dorsch, M., Hock, H., Kuzendorf, U., Diamantstein, T., and Blankenstein, T. (1993). Macrophage colony-stimulating factor gene transfer into tumor cells induces macrophage infiltration but not tumor suppression. *Eur. J. Immunol.* **23**, 186–190.

Douvdevani, A., Huleihel, M., Zoller, M., Segal, S., and Apte, R. N. (1992). Reduced tumorigenicity of fibrosarcomas which constitutively generate IL-1α either spontaneously or following IL-1α gene transfer. *Int. J. Cancer* **51**, 822–830.

Dranoff, G., Jaffee, E. M., Lazenby, A., Golumbek, P., Levitsky, H., Brose, K., Jackson, V., Hamada, H., Pardoll, D., and Mulligan, R. C. (1993). Vaccination with irradiated tumor cells engineered to secrete murine granulocyte-macrophage colony-stimulating factor stimulates potent, specific, and long lasting anti-tumor immunity. *Proc. Natl. Acad. Sci. U.S.A.* **90**, 3539–3543.

Dzierzak, E. A., Papayannopoulou, T., and Mulligan, R. C. (1988). Lineage-specific expression of a human B-globin gene in murine bone marrow transplant reconstituted with retrovirus-transduced stem cells. *Nature (London)* **331**, 35–41.

Estin, C. D., Stevenson, U. S., Plowman, G. D., Hu, S. L., Sriddar, P., Hellström, I., Brown, J. P., and Hellström, K. E. (1988). Recombinant vaccinia virus vaccine against human melanoma antigen p97 for use in immunotherapy. *Proc. Natl. Acad. Sci. U.S.A.* **85**, 1052–1056.

Fakhrai, H., Shawler, D. L., Gjerset, R., Naviaux, R. K., Koziol, J., Royston, I., and Sobol, R. E. (1995). Cytokine gene therapy with interleukin-2-transduced fibroblasts: Effects of IL-2 dose on anti-tumor immunity. *Hum. Gene Ther.* **6**, 591–601.

Fearon, E. R., Pardoll, D. M., Itaya, T., Golumbeck, P., Levitsky, H. I., Simons, J. W., Karasuyama, H., Vogelstein, B., and Frost, P. (1990). Interleukin-2 production by tumor cells bypasses T helper function in the generation of an antitumor response. *Cell (Cambridge, Mass.)* **60**, 397–403.

Ferrantini, M., Proietti, E., Santodonato, L., Gabriele, L., Peretti, M., Plavec, I., Meyer, F., Kaido, T., Gresser, I., and Belardelli, F. (1993). α1-interferon gene transfer into metastatic Friend leukemia cells abrogated tumorigenicity in immunocompetent mice: Antitumor therapy by means of interferon-producing cells. *Cancer Res.* **53**, 1107–1112.

Ferrone, S., and Marincola, F. (1995). Loss of HLA class I antigens by melanoma cells: Molecular mechanisms, functional significance and clinical relevance. *Immunol. Today* **16**, 487–494.

Fink, D. J., Sternberg, L. R., Weber, P. C., Mata, M., Goins, W. F., and Glorioso, J. C. (1992). *In vivo* expression of β-galactosidase in hippocampal neurons by HSV-mediated gene transfer. *Hum. Gene Ther.* **3**, 11–19.

Finn, O. J. (1993). Tumor rejection antigens recognized by T lymphocytes. *Curr. Opin. Immunol.* **5**, 701–708.

Fitch, F. W., McKisic, M. D., Laucki, D. W., and Gajewski, T. F. (1993). Differential regulation of murine T lymphocyte subsets. *Annu. Rev. Immunol.* **11**, 29–48.

Folkman, J. (1995). Angiogenesis in cancer, vascular, rheumatoid and other diseases. *Nat. Med.* **1**, 27–31.

Forni, G., Giovarelli, M., Santoni, A., Modesti, A., and Forni, M. (1986). Tumour inhibition by interleukin-2 at the tumour/host interface. *Biochim. Biophys. Acta* **865**, 307–327.

Forni, G., Musso, T., Jemma, C., Boraschi, D., Tagliabue, A., and Giovarelli, M. (1989). Lymphokine-activated tumor inhibition (LATI) in mice: Ability of a nonapeptide of the human interleukin-1 to recruit antitumor reactivity in recipient mice. *J. Immunol.* **142**, 712–718.

Forni, G., Cavallo, F., Consalvo, M., Allione, A., Dellabona, P., Casorati, G., and Giovarelli, M. (1995). Molecular approaches to cancer immunotherapy. *Cytokine Mol. Ther.* **1**, 225–248.

Gajewski, T., Renauld, J.-C., Van Pel, A., and Boon, T. (1995). Costimulation with B7-1, IL-6, and IL-12 is sufficient for primary generation of murine antitumor cytolytic T lymphocytes *in vitro. J. Immunol.* **154**, 5637–5648.

Gambacorti-Passerini, C., Grignani, F., Arienti, F., Pandolfi, P. P., Pelicci, P. G., and Parmiani, G. (1993). Human CD4 lymphocytes specifically recognize a peptide representing the fusion region of the hybrid protein pml/RARα present in acute promyelocytic leukemia cells. *Blood* **81**, 1369–1375.

Gansbacher, B., Bannerji, R., Daniels, B., Zier, K., Cronin, K., and Gilboa, E. (1990a). Retroviral vector-mediated γ interferon gene transfer to tumor cells generates potent and long lasting antitumor immunity. *Cancer Res.* **50**, 7820–7825.

Gansbacher, B., Zier, K., Daniels, B., Cronin, K., Bannerji, R., and Gilboa, E. (1990b). Interleukin-2 gene transfer into tumor cells abrogates tumorigenicity and induces protective immunity. *J. Exp. Med.* **172**, 1217–1224.

Gansbacher, B., Zier, K., Cronin, K., Hantzopoulos, P. A., Bouchard, B., Houghton, A., Gilboa, E., and Golde, D. (1992a). Retroviral transfer induced constitutive expression of interleukin-2 or interferon-γ in irradiated human melanoma cells. *Blood* **80**, 2817–2825.

Gansbacher, B., Houghton, A., and Livingston, P. (1992b). A pilot study of immunization with HLA-A2 matched allogeneic melanoma cells that secrete interleukin-2 in patients with metastatic melanoma. *Hum. Gene Ther.* **3**, 677–690.

Garrido, F., Cabrera, T., Concha, A., Glew, S., Ruiz-Cabello, F., and Stern, P. (1993). Natural history of HLA expression during tumour development. *Immunol. Today* **14**, 491–499.

Gastl, G., Finstad, C. L., Guarini, A., Bosl, G., Gilboa, E., Bander, N. H., and Gansbacher, B. (1992). Retroviral vector-mediated gene transfer into human renal cancer cells. *Cancer Res.* **52**, 6229–6236.

Gaugler, B., Brouwenstijn, N., Ventomme, V., Szikora, J.-P., Van der Speck, C. W., Patard, J.-J., Boon, T., Schrier, P., Van den Eynde, B. J. (1996). A new gene coding for an antigen recognized by autologous cytolytic T lymphocytes on a human renal carcinoma. *Immunogenetics,* **44**, 323–330.

Geller, A. I., Keyomarsi, K., Bryan, J., and Pardee, A. B. (1990). An efficient deletion mutant packaging system for defective herpes simplex viral vectors: Potential application to human gene therapy and neuronal physiology. *Proc. Natl. Acad. Sci. U.S.A.* **87**, 8950–8954.

Germain, R. N. (1994). MHC-dependent antigen processing and peptide presentation: Providing ligands for T lymphocyte activation. *Cell (Cambridge, Mass.)* **76**, 287–299.

Giovarelli, M., Cofano, F., Vecchi, A., Forni, M., Landolfo, S., and Forni, G. (1986). Interferon-activated tumor inhibition *in vivo*. *Int. J. Cancer* **37**, 141–147.

Golumbek, P. T., Lazenby, A. L., Levitsky, H., Jaffee, L. M., Karasuyama, H., Baker, M., and Pardoll, D. M. (1991). Treatment of established renal cancer by tumor cells engineered to secrete interleukin-6. *Science* **254**, 713–716.

Golumbek, P. T., Azahari, R., Jaffee, E. M., Levitsky, H., Lazenby, A., Leong, K., and Pardoll, D. M. (1993). Controlled release, biodegradable cytokine depots: A new approach in cancer vaccine design. *Cancer Res.* **53**, 5841–5844.

Grabbe, S., Beissert, S., Schwarz, T., and Granstein, R. D. (1995). Dendritic cells as initiators of tumor immune response: A possible strategy for tumor immunotherapy? *Immunol. Today* **16**, 117–121.

Greenberg, P. D. (1991). Adoptive T-cell therapy of tumors: Mechanisms operative in the recognition and elimination of tumor cells. *Adv. Immunol.* **49**, 281–355.

Grez, M., Akgun, E., Hilberg, F., and Ostertag, W. (1990). Embryonic stem cell virus, a recombinant murine retrovirus with expression in embryonic stem cells. *Proc. Natl. Acad. Sci. U.S.A.* **87**, 9202–9206.

Han, X., Kasahara, N., and Kan, Y. W. (1995). Ligand directed retroviral targeting of human breast cancer cells. *Proc. Natl. Acad. Sci. U.S.A.* **92**, 9747–9751.

Hantzopoulos, P. A., Sullenger, B. A., Ungers, G., and Gilboa, E. (1989). Improved gene expression upon transfer of the adenosine deaminase minigene outside the transcriptional unit of a retroviral vector. *Proc. Natl. Acad. Sci. U.S.A.* **86**, 3519–3523.

Hawley, R. G., Fong, A. Z. C., Burns, B. F., and Hawley, T. S. (1992). Transplantable myeloproliferative disease induced in mice by an interleukin 6 retrovirus. *J. Exp. Med.* **176**, 1149–1163.

Herlyn, M., Kath, R., Williams, N., Valyi-Nagy, I., and Rodeck, U. (1990). Growth regulatory factors for normal, premalignant, and malignant human cells *in vitro*. *Adv. Cancer Res.* **54**, 213–254.

Hersey, P. (1992). Active immunotherapy with viral lysates of micrometastases following surgical removal of high risk melanoma. *World J. Surg.* **16**, 251–260.

Hock, H., Dorsch, M., Diamantstein, T., and Blankenstein, T. (1991). Interleukin-7 induces CD4+ T cell-dependent tumor rejection. *J. Exp. Med.* **174**, 1291–1298.

Hock, H., Dorsch, M., Kunzendorf, U., Qin, Z., and Diamantstein, T. (1993a). Mechanism of rejection induced by tumor cells targeted gene transfer of interleukin-2, interleukin-4, interleukin-7, tumor necrosis factor or interferon-gamma. *Proc. Natl. Acad. Sci. U.S.A.* **90**, 2774–2778.

Hock, H., Dorsch, M., Kunzendorf, U., Überla, K., Qin, Z., Diamantstein, T., and Blankenstein, T. (1993b). Vaccinations with tumor cells genetically engineered to produce different cytokines: Effectively not superior to a classical adjuvant. *Cancer Res.* **53**, 714–716.

Houbiers, J. G. A., Nijman, N. W., Van der Burg, S. H., Drijfhout, J. W., Kenemans, P., Van de Velde, C. J. N., Braud, A., Momburg, F., Kasy, M. W., and Melief, C. J. M. (1993). *In vitro* induction of human cytotoxic T lymphocyte responses against peptides of mutant and wild type p53. *Eur. J. Immunol.* **23**, 2072–2077.

Huang, A. Y. C., Golumbek, P., Ahmadzadeh, M., Jaffee, E., Pardoll, D. M., and Levitsky, H. (1994). Role of bone marrow-derived cells in presenting MHC class I-restricted tumor antigens. *Science* **264**, 961–965.

Hurford, R. K., Jr., Dranoff, G., Mulligan, R. C., and Tepper, R. I. (1995). Gene therapy of metastatic cancer by *in vivo* retroviral gene targeting. *Nat. Genet.* **10**, 430–435.

Jaffee, E. M., Dranoff, G., Cohen, L. K., Hauda, K. M., Clift, S., Marshall, F. F., Mulligan, R. C., and Pardoll, D. M. (1993). High efficiency gene transfer into primary human tumor explants without cell selection. *Cancer Res.* **53**, 2221–2226.

Jaffee, E. M., Lazenby, A., Meurer, J., Marshall, F., Hauda, K. M., Counts, C., Hurwitz, H, Simons, J. W., Levitsky, H. I., and Pardoll, D. M. (1995). Use of murine models of cytokine-secreting tumor vaccines to study feasibility and toxicity issues critical to designing clinical trials. *J. Immunother.* **18,** 1–9.

Jolly, D. (1995). Viral vector system for gene therapy. *In* "The Internet Book of Gene Therapy: Cancer Therapeutics" (R. E. Sobol and K. J. Scanlon, eds), pp. 3–16. Appleton & Lange, Stamford, Connecticut.

Kaido, T., Bandu, M.-T., Maury, C., Ferrantini, M., Belardelli, F., and Gresser, I. (1995). IFN-α1 gene transfection completely abolishes the tumorigenicity of murine B16 melanoma cells in allogeneic DBA/2 mice and decreases their tumorigenicity in syngeneic C57BL/6 mice. *Int. J. Cancer* **60,** 221–229.

Kaneko, C., Hallenbeck, P., Nakabayashi, H., McGarrity, G., Tamaoki, T., Anderson, F. W., and Chiang, Y. L. (1995). Adenovirus-mediated gene therapy of hepatocellular carcinoma using cancer-specific gene expression. *Cancer Res.* **55,** 5283–5287.

Kantor, J., Irvine, K., Abrams, S., Kaufman, H., DiPietro, J., and Schlom, J. (1992). Antitumor activity and immune response induced by a recombinant carcinoembryonic antigen–vaccinia virus vaccine. *J. Natl. Cancer Inst.* **84,** 1084–1091.

Kawakami, Y., and Rosenberg, S. A. (1996). T-cell recognition of self peptides as tumor rejection antigens. *Immunol. Res.* **15,** 179–190.

Kehrl, J. H., Wakefield, L. M., Roberts, A. B., Jakowlew, S. B., Alvarez-Mon, M., Derynck, R., Sporn, M. B., and Fauci, A. S. (1986). Production of transducing growth factor β by human lymphocytes and its potential role in regulation of T-cell growth. *J. Exp. Med.* **163,** 1037–1050.

Kerbel, R. (1992). Expression of multi-cytokine resistance and multi-growth factor independence in advanced stage metastatic cancers. *Am. J. Pathol.* **141,** 510–524.

Kim, T. S., and Cohen, E. P. (1994). Interleukin-2-secreting mouse fibroblasts transfected with genomic DNA from murine melanoma cells prolong the survival of mice with melanoma. *Cancer Res.* **54,** 2531–2535.

Kim, T. S., Russel, S. J., Collins, M. K. L., and Cohen, E. P. (1993). Immunization with interleukin-2-secreting allogeneic mouse fibroblasts expressing melanoma-associated antigens prolongs the survival of mice with melanoma. *Int. J. Cancer* **55,** 865–872.

Kim, T.-Y., von Eschenbach, A. C., Filaccio, M. D., Hayakawa, K., Parkinson, D. R., Balch, C. M., and Itoh, K. (1990). Clonal analysis of lymphocytes from tumor, peripheral blood, and nontumorous kidney in primary renal cell carcinoma. *Cancer Res.* **50,** 5263–5268.

Krüger-Krasagakes, S., Li, W., Richter, G., Diamantstein, T., and Blankenstein, T. (1993). Eosinophils infiltrating interleukin-5 gene-transfected tumors do not suppress tumor growth. *Eur. J. Immunol.* **23,** 992–995.

Lau, C., Soriano, H. E., Ledley, F. D., Finegold, M. J., Wolfe, J. H., Birkenmeier, E. H., and Henning, S. J. (1995). Retroviral gene transfer into the intestinal epithelium. *Hum. Gene Ther.* **6,** 1145–1151.

Ledley, F. D. (1995). Nonviral gene therapy: The promise of genes as pharmaceutical products. *Hum. Gene Ther.* **6,** 1129–1144.

Lehman, F., Marchand, M., Hainaut, P., Pouillart, P., Sastre, X., Ikeda, H., Boon, T., and Coulie, P. G. (1995). Differences in the antigens recognized by cytolytic T-cells on two successive metastases of a melanoma patient are consistent with immune selection. *Eur. J. Immunol.* **25,** 340–347.

Lollini, P.-L., Bosco, M. C., Cavallo, F., De Giovanni, C., Giovarelli, M., Landuzzi, L., Musiani, P., Modesti, A., Nicoletti, G., Palmieri, G., Santoni, A., Young, H. A., Forni, G., and Nanni, P. (1993). Inhibition of tumor growth and enhancement of metastasis after transfection of the γ-interferon gene. *Int. J. Cancer* **55,** 320–329.

Lotze, M. T., and Rubin, J. T. (1994). Gene therapy of cancer: A pilot study of IL-4-gene-modified fibroblasts admixed with autologous tumor to elicit an immune response. *Hum. Gene Ther.* **5,** 41–55.

Lotze, M. T., and Rubin, J. T. (1995). Protocol #9209-033 (Gene therapy of cancer: A pilot study of IL-4-gene-modified fibroblasts admixed with autologous tumor to elicit an immune response). *RAC Rep.* **June,** 16.

Maass, G., Schmidt, W., Berger, M., Schilcher, F., Koszik, F., Schneeberger, A., Stingl, G., Birnstiel, M. L., and Schweighoffer, T. (1995). Priming of tumor-specific T cells in the draining lymph nodes after immunization with interleukin 2-secreting tumor cells: Three consecutive stages may be required for successful tumor vaccination. *Proc. Natl. Acad. Sci. U.S.A.* **92,** 5540–5544.

Mantovani, A., Bottazzi, B., Colotta, F., Sozzani, S., and Ruco, L. (1992). The origin and function of tumor-associated macrophages. *Immunol. Today* **13,** 265–270.

Markowitz, D., Goff, S., and Bank, A. (1988). Construction and use of a safe and efficient amphotropic packaging cell line. *Virology* **167,** 400–406.

Martinotti, A., Stoppacciaro, A., Vagliani, M., Melani, C., Spreafico, F., Wysocka, M., Parmiani, G., Trinchieri, G., and Colombo, M. P. (1995). CD4 T-cells inhibit in vivo the CD8-mediated immune response against murine colon carcinoma cells transduced with interleukin-12 genes. *Eur. J. Immunol.* **25,** 137–146.

Martuza, R. L., Malik, A., Markert, J. M., Ruffner, K. J., and Coen, D. M. (1991). Experimental therapy of human glioma by means of a genetically engineered virus mutant. *Science* **252,** 854–856.

Mattei, S., Colombo, M. P., Melani, C., Silvani, A., Parmiani, G., and Herlyn, M. (1994). Expression of cytokine/growth factors and their receptors in human melanoma and melanocytes. *Int. J. Cancer* **56,** 853–857.

Mäurer, M. J., Gollin, S., Martin, D., Swaney, W., Bryant, J., Castelli, C., Robbins, P., Parmiani, G., Storkus, W., and Lotze, M. T. (1996). Escape from immune recognition: Lethal recurrent melanoma in a patient associated with downregulation of the peptide transporter protein TAP-1 and loss of expression of the immunodominant MART-1/Melan-A antigen. *J. Clin. Invest.* **98,** 1633–1641.

Mazzocchi, A., Rodolfo, M., Parmiani, G., and Anichini, A. (1991). An autologous T-cell clone overcomes intra-melanoma heterogeneity for susceptibility to cell-mediated lysis by using multiple lytic mechanisms: *In vitro* and *in vivo* analysis. *Melanoma Res.* **1,** 169–176.

Mazzocchi, A., Belli, F., Mascheroni, L., Vegetti, C., Parmiani, G., and Anichini, A. (1994). Frequency of cytotoxic T lymphocyte precursors (CTLp) interacting with autologous tumor via the T-cell receptor: Limiting dilution analysis of specific CTLp in peripheral blood and tumor-invaded lymph nodes of melanoma patients. *Int. J. Cancer* **58,** 330–339.

McAdam, A. J., Pulaski, B. A., Storozynsky, E., Yeh, K.-Y., Sickel, J. Z., Frelinger, J. G., and Lord, E. M. (1995). Analysis of the effect of cytokines (interleukins 2, 3, 4, and 6, granulocyte-monocyte colony stimulating factor, and interferon-γ) on generation of primary cytotoxic T lymphocytes against a weakly immunogenic tumor. *Cell. Immunol.* **165,** 183–192.

McBride, W. H., Thacker, J. D., Comora, S., Economou, J. S., Kelley, D., Hogge, D., Dubinett, S. M., and Dougherty, G. J. (1992). Genetic modification of a murine fibrosarcoma to produce interleukin 7 stimulates host cell infiltration and tumor immunity. *Cancer Res.* **52,** 3931–3937.

McCabe, B. J., Irvine, K. R., Nishimura, M. I., Yang, J. C., Spiess, P. J., Shulman, E. P., Rosenberg, S. A., and Restifo, N. P. (1995). Minimal determinant expressed by a recombinant vaccinia virus elicits therapeutic antitumor cytolytic T lymphocyte response. *Cancer Res.* **55,** 1741–1747.

Meko, J. B., Yim, J. H., Tsung, K., and Norton, J. A. (1995). High cytokine production and effective antitumor activity of a recombinant vaccinia virus encoding murine IL-12. *Cancer Res.* **55,** 4765–4770.

Melani, C., Pupa, S. M., Stoppacciaro, A., Ménard, S., Colnaghi, M. I., Parmiani, G., and Colombo, M. P. (1995a). An *in vivo* model to compare human leukocyte infiltration in carcinoma xenografts producing different chemokines. *Int. J. Cancer* **62,** 572–578.

Melani, C., Sulé-Suso, J., Arienti, F., Maccalli, C., Passerini, F., Colombo, M. P., and Parmiani, G. (1995b). A human melanoma cell line transduced with an interleukin-4 gene by a retroviral vector releases biologically active IL-4 and maintains the original tumor antigenic phenotype. *Hum. Gene Ther.* **6,** 1427–1436.

Miller, A. D. (1990). Retrovirus packaging cells. *Hum. Gene Ther.* **1,** 5–14.

Miller, A. D., Miller, D. G., Garcia, J. V., and Lynch, C. M. (1993). Use of retroviral vectors for gene transfer and expression. *In* "Methods in Enzymology" (R. Wu, ed.), Vol. 217, pp. 581–599. Academic Press, San Diego, California.

Miller, A. R., McBride, W. H., Dubinett, S. M., Dougherty,, G. J., Thacker, J. D., Shau, H., Kohn, D. B., Moen, R. C., Walker, M. J., Chiu, R., Schuk, B. L., Rosenblatt, J. A., Huang, M., Dhanani, S., Rhoades, K., and Economou, J. S. (1993). Transduction of human melanoma cell lines with the human interleukin-7 gene using retroviral-mediated gene transfer: Comparison of immunologic properties with interleukin-2. *Blood* **82,** 3686–3694.

Mitchell, M. S. (1995). Active specific immunotherapy of melanoma. *Br. Med. Bull.* **51,** 631–646.

Moolten, F. L., and Cupples, A. L. (1992). A model for predicting the risk of cancer consequent to retroviral gene therapy. *Hum. Gene Ther.* **3,** 479–486.

Morgan, R. A., Couture, L., Eloroy-Stein, O., Ragheb, J., Moss, B., and Anderson, W. F. (1992). Retroviral vectors containing putative internal ribosome entry sites: Development of a polycistronic gene transfer system and applications to human gene therapy. *Nucleic Acids Res.* **20,** 1293–1299.

Morton, D. L., Foshag, L. J., Hoon, D. S. B., Nizze, A., Wanek, L. A., Chang, C., Davtyan, D. G., Gupta, R. K., Elashoff, R., and Irie, R. F. (1992). Prolongation of survival in metastatic melanoma after active specific immunotherapy with a new polyvalent melanoma vaccine. *Ann. Surg.* **216,** 463–482.

Moss, B., and Flexner, C. (1987). Vaccinia virus expression vectors. *Annu. Rev. Immunol.* **5,** 305–324.

Mukherji, B., Chakraborty, N. G., Yamasaki, S., Okino, T., Yamase, H., Sporn, J. R., Kurtzman, S. K., Ergin, M. T., Ozols, J., Meehan, J., and Mauri, F. (1995). Induction of antigen-specific cytolytic T-cells *in situ* in human melanoma by immunization with synthetic peptide-pulsed autologous antigen presenting cells. *Proc. Natl. Acad. Sci. U.S.A.* **92,** 8078–8082.

Mullen, C. A., Coale, M., Levy, A. T., Stetler-Stevenson, W. G., Liotta, L. A., Brandt, S., and Blaese, R. M. (1992). Fibrosarcoma cells transduced with the IL-6 gene exhibit reduced tumorigenicity, increased immunogenicity and decreased metastatic potential. *Cancer Res.* **52,** 6020–6024.

Mulligan, R. C. (1993). The basic science of gene therapy. *Science* **260,** 926–932.

Muro-Cacho, C. A., Sumulsky, R. L., and Kaplan, D. (1992). Gene transfer in human lymphocytes using a vector based on adeno-associated virus. *J. Immunother.* **11,** 231–237.

Nabel, E. G., Yang, Z., Muller, D., Chang, A., Gao, X., Huang, L., Cho, K. J., and Nabel, G. J. (1994). Safety and toxicity of catheter gene delivery to the pulmonary vasculature in a patient with metastatic melanoma. *Hum. Gene Ther.* **5,** 1089–1094.

Nabel, G. J., Nabel, E. G., Yang, Z.-Y., Fox, B. A., Plautz, G. E., Gao, X., Huang, L., Shu, S., Gordon, D., and Chang, A. E. (1993). Direct gene transfer with DNA–liposome complexes in melanoma: Expression, biologic activity, and lack of toxicity in humans. *Proc. Natl. Acad. Sci. U.S.A.* **90,** 11307–11311.

Ogasawara, M., and Rosenberg, S. A. (1993). Enhanced expression of HLA molecules and stimulation of autologous human tumor infiltrating lymphocytes following transduction of melanoma cells with γ-interferon genes. *Cancer Res.* **53,** 3561–3568.

Osanto, S., Brouwenstyn, N., Figdor, C. G., Melief, C. J. M., and Schrier, P. (1993). Immunization with interleukin-2 transfected melanoma cells: A phase I–II study in patients with metastatic melanoma. *Hum. Gene Ther.* **4,** 323–330.

Overell, R. W., Weisser, K. E., and Cosman, D. (1988). Stably transmitted triple promoter retroviral vectors and their use in transformation of primary mammalian cells. *Mol. Cell. Biol.* **8**, 1803–1808.

Paglia, P., Chiodoni, C., Rodolfo, M., and Colombo, M. P. (1996). Murine dendritic cells loaded *in vitro* with soluble protein prime cytotoxic T lymphocytes against tumor antigen *in vivo. J. Exp. Med.* **183**, 317–322.

Pardoll, D. M. (1995). Paracrine cytokine adjuvants in cancer immunotherapy. *Annu. Rev. Immunol.* **13**, 399–415.

Pardoll, D. M., and Beckerleg, A. M. (1995). Exposing the immunology of naked DNA vaccines. *Immunity* **3**, 165–169.

Parmiani, G. (1990). An explanation of the variable clinical response to interleukin 2 and LAK cells. *Immunol. Today* **11**, 113–115.

Parmiani, G., and Colombo, M. P. (1995). Somatic gene therapy of human melanoma: Preclinical studies and early clinical trials. *Melanoma Res.* **5**, 295–301.

Patel, P. M., Flemming, C. L., Fisher, C., Porter, C. D., Thomas, J. M., Gore, M. E., and Collins, M. K. L. (1994). Generation of interleukin-2 secreting melanoma cell populations from resected metastatic tumors. *Hum. Gene Ther.* **5**, 577–584.

Paul, W. E., and Seder, R. A. (1994). Lymphocyte response and cytokines. *Cell (Cambridge, Mass.)* **76**, 241–251.

Pericle, F., Giovarelli, M., Colombo, M. P., Ferrari, G., Musiani, P., Modesti, A., Cavallo, F., Di Piero, F., Novelli, F., and Forni, G. (1994). An efficient Th2-type memory follows CD8+ lymphocyte-driven and eosinophil-mediated rejection of a spontaneous mouse mammary adenocarcinoma engineered to release IL-4. *J. Immunol.* **153**, 5659–5673.

Philip, R., Brunette, E., Kilinski, L., Murugesh, D., McNally, M. A., Ucar K., Rosenblatt, J., Okarma, T. B., and Lebkowski, J. S. (1994). Efficient and sustained gene expression in primary T. lymphocytes and primary and cultured tumor cells mediated by adeno-associated virus plasmid DNA complexed to cationic liposomes. *Mol. Cell. Biol.* **14**, 2411–2418.

Porgador, A., Tzehoval, E., Katz, A., Vadai, E., Revel, M., Feldman, M., and Eisenbach, L. (1992). Interleukin-6 gene transfection into Lewis lung carcinoma tumor cells suppresses the malignant phenotype and confers immunotherapeutic competence against parental metastatic cells. *Cancer Res.* **52**, 3679–3686.

Porgador, A., Bannerji, R., Watanabe, Y., Feldman, M., Gilboa, E., and Eisenbach, L. (1993a). Antimetastatic vaccination of tumor-bearing mice with two types of IFN-γ gene-inserted tumor cells. *J. Immunol.* **150**, 1458–1470.

Porgador, A., Gansbacher, B., Bannerji, R., Tzehoval, E., Gilboa, E., Feldman, M., and Eisenbach, L. (1993b). Anti-metastatic vaccination of tumor-bearing mice with IL-2-gene-inserted tumor cells. *Int. J. Cancer* **53**, 471–477.

Qin, Z., Krüger-Krasagakes, S., Kunzendorf, U., Hock, H., Diamantstein, T., and Blankenstein, T. (1993). Expression of tumor necrosis factor by different tumor cell lines results either in tumor suppression or augmented metastasis. *J. Exp. Med.* **178**, 355–360.

Radrizzani, M., Benedetti, B., Castelli, C., Longo, A., Ferrara, G. B., Herlyn, M., Parmiani, G., and Fossati, G. (1991). Human allogeneic melanoma-reactive T-helper lymphocyte clones: Functional analysis of lymphocyte-melanoma interactions. *Int. J. Cancer* **49**, 1–8.

Raz, E., Carson, D. A., Parker, S. E., Parr, T. B., Abal, A. M., Aichinger, G., Gromkowski, S. H., Sing, M., Lew, D., Yankaukas, M. A., Baird, A. M., and Rhodes, G. H. (1994). Intradermal gene immunization: The possible role of DNA uptake in the induction of cellular immunity to viruses. *Proc. Natl. Acad. Sci. U.S.A.* **91**, 9519–9523.

Richter, G., Krüger-Krasagakes, S., Hein, G., Huls, C., Schmitt, E., Diamantstein, T., and Blankenstein, T. (1993). Interleukin 10 transfected into Chinese hamster ovary cells prevents tumor growth and macrophage infiltration. *Cancer Res.* **53**, 4134–4137.

Rivoltini, L., Kawakami, Y., Sakaguchi, K., Southwood, S., Sette, A., Robbins, P. F., Marincola, F. M., Salgaller, M. L., Yannelli, J. R., Appella, E., and Rosenberg, S. A. (1995). Induction of tumor-reactive CTL from peripheral blood and tumor-infiltrating lymphocytes of melanoma patients by *in vitro* stimulation with an immunodominant peptide of the human melanoma antigen MART-1. *J. Immunol.* **154,** 2257–2265.

Robey, E., and Allison, J. P. (1995). T-cell activation: Integration of signals from the antigen receptor and costimulatory molecules. *Immunol. Today* **16,** 306–310.

Roe, T. Y., Reynolds, T. C., Yu, G., and Brown, P. O. (1993). Integration of murine leukemia virus DNA depends on mitosis. *EMBO J.* **12,** 2099–2108.

Rosenberg, S. A. (1992). Immunization of cancer patients using autologous cancer cells modified by insertion of the gene for tumor necrosis factor. *Hum. Gene Ther.* **3,** 57–73.

Rosenberg, S. A., Lotze, M. T., Yang, J. C., Aebersold, P. M., Linehan, W. M., Seipp, C., and White, D. E. (1989). Experience with the use of high-dose interleukin-2 in the treatment of 652 cancer patients. *Ann. Surg.* **210,** 474–485.

Rosenberg, S. A., Aebersold, P. M., Cornetta, K., Kasid, A., Morgan, R. A., Moen, R., Karson, E. M., Lotze, M. T., Yang, J. C., Topalian, S. L., Merino, M. J., Culver, K., Miller, A. D. E., Blaese, R. M., and Anderson, W. F. (1990). Gene transfer into humans—immunotherapy of patients with advanced melanoma, using tumor-infiltrating lymphocytes modified by retroviral gene transduction. *N. Engl. J. Med.* **323,** 570–578.

Rosenberg, S. A., Yannelli, J. R., Yang, J. C., Topalian, S. L., Schwartzentruber, D. J., Weber, J. S., Parkinson, D. R., Seipp, C. A., Einhorn, J. H., and White, D. E. (1994). Treatment of patients with metastatic melanoma with autologous tumor-infiltrating lymphocytes and interleukin 2. *J. Natl. Cancer Inst.* **6,** 1159–1166.

Russell, S. J., Eccles, S. A., Flemming, C. L., Johnson, C. A., and Collins, M. K. L. (1991). Decreased tumorigenicity of a transplantable rat sarcoma following transfer and expression of an IL-2 cDNA. *Int. J. Cancer* **47,** 24–251.

Russell, S. J., Hawkins, R. E., and Winter, G. (1993). Retroviral vectors displaying functional antibody fragments. *Nucleic Acids Rres.* **21,** 1081–1085.

Salgaller, M. L., Weber, J. S., Koening, S., Yannelli, J. R., and Rosenberg, S. A. (1994). Generation of specific anti-melanoma activity by stimulation of human tumor-infiltrating lymphocytes with MAGE-1 synthetic peptides. *Cancer Immunol. Immunother.* **39,** 105–116.

Salgaller, M. L., Afshar, A., Marincola, F. M., Rivoltini, L., Kawakami, Y., and Rosenberg, S. A. (1995). Recognition of multiple epitopes in the human antigen gp100 by peripheral blood lymphocytes stimulated *in vitro* with synthetic peptides. *Cancer Res.* **55,** 4972–4979.

Salmons, B., and Gunzburg, W. H. (1993). Targeting of retroviral vectors for gene therapy. *Hum. Gene Ther.* **4,** 129–141.

Salter, R. D., and Cresswell, P. (1986). Impaired assembly and transport of HLA-A and -B antigens in a mutant T × B cell hybrid. *EMBO J.* **5,** 943–949.

Salvadori, S., Gansbacher, B., Wernick, I., Tirelli, S., and Zier, K. (1995). B7-1 amplifies the response to interleukin-2-secreting tumor vaccines *in vivo*, but fails to induce a response by naive cells *in vitro*. *Hum. Gene Ther.* **6,** 1299–1306.

Schadendorf, D., Moller, A., Algermissen, B., Worm, M., Sticherling, M., and Czarnetzki, B. M. (1994). IL-8 produced by human malignant melanoma cells *in vitro* is an essential autocrine growth factor. *J. Immunol.* **151,** 2667–2675.

Schendel, D. J., Gansbacher, B., Oberneder, R., Kriegmair, M., Hofstetter, A., Reithmuller, G., and Segurado, O. G. (1993). Tumor-specific lysis of human renal cell carcinomas by tumor-infiltrating lymphocytes. *J. Immunol.* **151,** 4209–4220.

Schmidt, W., Schweighoffer, T., Herbst, E., Maass, G., Berger, M., Schilcher, F., Schaffner, G., and Bristiel, M. L. (1995). Cancer vaccines: The interleukin 2 dosage effect. *Proc. Natl. Acad. Sci. U.S.A.* **92,** 4711–4714.

Singh, R. K., Gutman, M., Radinsky, R., Buacana, C. D., and Fidler, I. J. (1994). Expression of interleukin-8 correlates with the metastatic potential of human melanoma cells in nude mice. *Cancer Res.* **54,** 3242–3247.

Somia, N. V., Zoppè, M., and Verma, I. (1995). Generation of targeted retroviral vectors by using single-chain variable fragment: An approach to *in vivo* gene delivery. *Proc. Natl. Acad. Sci. U.S.A.* **92,** 7570–7574.

Spagnoli, G. C., Schaefer, C., Willimann, T. E., Kocher, T., Amoroso, A., Juretic, A., Zuber, M., Luscher, U., Harder, F., and Heberer, M. (1995). Peptide-specific, CTL in tumor-infiltrating lymphocytes from metastatic melanomas expressing MART-1/Melan-A, gp100 and tyrosinase genes: A study in an unselected group of HLA-A2.1-positive patients. *Int. J. Cancer (Pred. Oncol.)* **64,** 309–315.

Stewart, M. J., Plautz, G. E., Del Buono, L., Yang, Z. Y., Xu, L., Gao, X., Huang, L., Nabel, E. G., and Nabel, G. J. (1992). Gene transfer *in vivo* with DNA–liposome complexes: Safety and acute toxicity in mice. *Hum. Gene Ther.* **3,** 267–275.

Stoppacciaro, A., Melani, C., Parenza, M., Mastracchi, A., Bassi, C., Baroni, C., Parmiani, and Colombo, M. P. (1993). Regression of an established tumor genetically modified to release a granulocyte-colony stimulating factor requires granulocyte T-cell cooperation and T-cell produced interferon γ. *J. Exp. Med.* **178,** 151–161.

Stratford-Perricaudet, L. D., Levrero, M., Chasse, J. F., Perricaudet, M., and Briand, P. (1990). Evaluation of the transfer and expression in mice of an enzyme-encoding gene using a human adenovirus vector. *Hum. Gene Ther.* **1,** 241–256.

Sulé-Suso, J., Arienti, F., Melani, C., Colombo, M. P., and Parmiani, G. (1995). A B7-1-transfected human melanoma line stimulates proliferation and cytotoxicity of autologous and allogeneic lymphocytes. *Eur. J. Immunol.* **25,** 2737–2742.

Sun, W. H., Burkholder, J. K., Sun, J., Culp, J., Turner, J., Lu, X., Pugh, T., and Yang, N. S. (1995). *In vivo* cytokine gene transfer by gene gun reduces tumor growth in mice. *Proc. Natl. Acad. Sci. U.S.A.* **92,** 2889–2893.

Tahara, H., Zeh, H. J., Storkus, W. J., Pappo, I., Watkins, S. C., Gubler, U., Wolf, S. F., Robbins, P. D., and Lotze, M. T. (1994). Fibroblasts genetically engineered to secrete interleukin 12 can suppress tumor growth and induce antitumor immunity to a murine melanoma *in vivo*. *Cancer Res.* **54,** 182–189.

Tahara, H., Zitvogel, L., Storkus, W. J., Zeh, H. J., III, McKinney, T. G., Schreiber, R. D., Gubler, U., Robbins, P. D., and Lotze, M. T. (1995). Effective eradication of established murine tumors with IL-12 gene therapy using a polycistronic retroviral vector. *J. Immunol.* **154,** 6466–6474.

ten Bosch, G. J. A., Toornvliet, A. C., Friede, T., Melief, C. J. M., and Leeksma, O. C. (1995). Recognition of peptides corresponding to the joining region of p210 bcr-abl protein by human T cells. *Leukemia* **9,** 1344–1348.

Tepper, R. I., Pattengale, P. K., and Leder, P. (1989). Murine interleukin-4 displays potent antitumor activity *in vivo*. *Cell (Cambridge, Mass.)* **57,** 503–512.

Tepper, R. I., Coffman, R. L., and Leder, P. (1992). An eosinophil-dependent mechanism for the antitumor effect of interleukin-4. *Science* **257,** 548–551.

Uchiyama, A., Hoon, D. S. B., Morisaki, T., Kaneda, Y., Yuzuki, D., and Morton, D. L. (1993). Transfection of interleukin-2 gene into human melanoma cells augments cellular immune response. *Cancer Res.* **53,** 949–952.

Vagliani, M., Rodolfo, M., Cavalo, F., Parenza, M., Melani, C., Parmiani, G., Forni, G., and Colombo, M. P. (1996). Interleukin 12 potentiates the curative effect of a vaccine based on interleukin 2-transduced tumor cells. *Cancer Res.* **56,** 467–470.

van der Bruggen, P., Traversari, C., Chomez, P., Lurquin, C., De Plaen, E., Van den Eynde, B., Knuth, A., and Boon, T. (1991). A gene encoding an antigen recognized by cytolytic T lymphocytes on a human melanoma. *Science* **254,** 1643–1647.

Van Pel, A., van der Fruggen, P., Coulie, P. G., Brichard, V. G., Lethé, B., Van den Eynde, B., Uyttenhove, C., Renauld, J.-C., and Boon, T. (1995). Genes coding for tumor antigens recognized by cytolytic T lymphocytes. *Immunol. Rev.* **145,** 229–250.

Vieweg, J., Boczkowski, D., Roberson, K. M., Edwards, D. W., Philip, M., Philip, R., Rudoll, T., Smith, C., Robertson, C., and Gilboa, E. (1995). Efficient gene transfer with adeno-associated virus-based plasmid complexed to cationic liposomes for gene therapy of human prostate cancer. *Cancer Res.* **55**, 2366–2372.

Vile, R. G., Nelson, J. A., Castleden, S., Chong, H., and Hart, I. R. (1994). Systemic gene therapy of murine melanoma using tissue specific expression of the HSVtk gene involves an immune component. *Cancer Res.* **54**, 6228–6234.

Vink, A., Coulie, P., Warnier, G., Renault, J.-C., Stevens, M., Donckers, D., and Van Snick, J. (1990). Mouse plasmacytoma growth *in vivo:* Enhancement by interleukin-6 (IL-6) and inhibition by antibodies directed against IL-6 or its receptor. *J. Exp. Med.* **172**, 997–1000.

Walsh, P., Dorner, A., Duke, R. C., Su, L.-J., and Glode, L. M. (1995). Macrophage colony-stimulating factor complementary DNA: A candidate for gene therapy in metastatic melanoma. *J. Natl. Cancer Inst.* **87**, 809–816.

Wang, J. M., Taraboletti, G., Matsushima, K., Van Damme, J., and Mantovani, A. (1990). Induction of hepatotactic migration of melanoma cells by neutrophil activating protein/IL-8. *Biochem. Biophys. Res. Commun.* **169**, 165–170.

Weitzman, M. D., Wilson, J. M., and Eck, S. L. (1995). Adenovirus vectors in cancer gene therapy. *In* "The Internet Book of Gene Therapy: Cancer Therapeutics" (R. E. Sobol and K. J. Scanlon, eds.), pp. 17–25. Appleton & Lange, Stamford, Connecticut.

Wölfel, T., Hauer, M., Schneider, J., Serrano, M., Wölfel, C., Klehmann-Hieb, E., DePlaeu, E., Hankein, T., Meyer zum Buschenfelde, K.-H., and Beach, D. (1995). A p16INK4a-insensitive CDK4 mutant targeted by cytolytic T lymphocytes in a human melanoma. *Science* **269**, 1281–1284.

Yagi, K., Hayashi, Y., Ishida, N., Ohbayashi, M., Ohishi, N., Mizuno, M., and Yoshida, J. (1994). Interferon-beta endogenously produced by intratumoral injection of cationic liposome-encapsulated gene: Cytocidal effect on glioma transplanted into nude mouse brain. *Biochem. Mol. Biol. Int.* **32**, 167–171.

Yamashiro, S., Takeya, M., Nishi, T., Kuratsu, J., Yoshimura, T., Ushio, Y., and Takahashi, K. (1994). Tumor-derived monocyte chemoattractant protein-1 induces intratumoral infiltration of monocyte-derived macrophages subpopulation in transplanted rat tumors. *Am. J. Pathol.* **145**, 856–867.

Yannelli, J. R., Hyatt, C., Johnson, S., Hwu, P., and Rosenberg, S. A. (1993). Characterization of human tumor cell lines transduced with the cDNA encoding either tumor necrosis factor α (TNF-α) or interleukin-2 (IL2). *J. Immunol. Methods* **161**, 77–90.

Yu, S. F., von Ruden, T., Kantoff, P. W., Garber, C., Seinberg, M., Ruther, U., Anderson, W. F., Wagner, E., and Gilboa, E. (1986). Self inactivating retroviral vectors designed for transfer of whole genes into mammalian cells. *Proc. Natl. Acad. Sci. U.S.A.* **83**, 3194–3198.

Zatloukal, K., Schneeberger, A., Berger, M., Schmidt, W., Koszik, F., Kutil, R., Cotten, M., Wagner, E., Buschle, M., Maass, G., Payer, R., Stingl, G., and Birnstiel, M. L. (1995). Elicitation of a systemic and protective anti-melanoma immune response by an IL-2-based vaccine. *J. Immunol.* **154**, 3406–3419.

Zitvogel, L., Tahara, H., Cai, Q., Storkus, W. J., Muller, G., Wolf, S. F., Gately, M., Robbins, P. D., and Lotze, M. T. (1994). Construction and characterization of retroviral vectors expressing biologically active human interleukin-12. *Hum. Gene Ther.* **5**, 1493–1506.

Daniel L. Shawler*
Habib Fakhrai*
Charles Van Beveren*
Dan Mercola*
Daniel P. Gold*
Richard M. Bartholomew†
Ivor Royston*
Robert E. Sobol*

*Sidney Kimmel Cancer Center
San Diego, California 92121

†The Immune Response Corporation
Carlsbad, California 92008

Gene Therapy Approaches to Enhance Antitumor Immunity

I. Introduction

Advances have contributed to the rapid development of immunogene therapy of cancer. Significant strides in this young field have been driven by an improved understanding of the biology of antitumor immunity, particularly those mechanisms mediated by T lymphocytes, combined with strides in molecular biology techniques that allow safe and effective gene transfer for both *in vivo* and *ex vivo* applications. This chapter summarizes these advances and concentrates on several types of genetic manipulations that have been explored to enhance the efficacy of cancer immunotherapies: (1) gene transfer of immunostimulatory cytokines, (2) inhibition of immunosuppressive and differentiation factors by antisense vectors, (3) expression of costimulatory molecules, (4) genetic modification of tumor-infiltrating lymphocytes (TILs), and (5) DNA tumor antigen vaccines. Evaluation of these approaches in clinical trials has been supported by investigations in animal

Advances in Pharmacology, Volume 40

tumor models indicating the efficacy of these different types of therapies utilizing genetically engineered products. These gene transfer approaches are summarized and their respective advantages and disadvantages are considered in this chapter.

II. Gene Transfer of Immunostimulatory Cytokines

As the biology of the immune system has unfolded, numerous cytokines that modulate immune responses have been identified (Borden and Sondel, 1990; Gabrilove and Jakubowski, 1990; Kelso, 1989). Many of the immune responses important for antitumor immunity are mediated by these proteins. A number of cytokines, produced in purified form by recombinant DNA methodology, have been evaluated for their antitumor effects. In several clinical trials, cytokines and related immunomodulators have produced objective tumor responses in some patients afflicted with a variety of neoplasms (Borden and Sondel, 1990; Lotze *et al.*, 1986; Rosenberg *et al.*, 1988).

Interleukin 2 (IL-2) plays a central role in the generation of antitumor immunity (Rosenberg *et al.*, 1988). Responding to tumor antigens, the T_H1 helper T lymphocytes secrete small amounts of IL-2 to activate cytotoxic T cells and natural killer cells, which in turn mediate systemic tumor cell destruction at the site of tumor antigen presentation. Interleukin 2 delivered by an intravenous or intralymphatic method has generated clinically significant responses in several types of cancer (Gandolfi *et al.*, 1989; Rosenberg *et al.*, 1988), but severe toxicities such as edema and hypotension have necessarily limited the dose and efficacy of dispensing IL-2 by these routes (Lotze *et al.*, 1986; Sarna *et al.*, 1990). Since cytokines are normally secreted in small quantities that mediate cellular interactions over short distances, it is not surprising that systemically delivered cytokines have proved to be toxic. To circumvent this toxicity, several investigators have examined the efficacy of direct, intralesional injection of IL-2 (Bubenik *et al.*, 1988; Gandolfi *et al.*, 1989). While this approach eliminates the toxicity associated with systemic IL-2 administration, multiple intralesional injections are required for optimal therapeutic efficacy (Bubenik *et al.*, 1988; Gandolfi *et al.*, 1989). Furthermore, such injections are impractical for patients when tumor sites are not accessible for direct injection without potentially significant morbidity.

Rather than treating patients with these powerful cytokines themselves, a number of investigators have investigated the effect of transferring a gene expressing the cytokine into tumor cells or into cells such as fibroblasts that are then injected together with tumor cells. The development of antitumor immune responses as a consequence of cytokine gene transfer in the treatment of cancer has been demonstrated in several animal tumor models (Fearon *et al.*, 1990; Pattengale and Leder, 1989). Expression of cytokine genes decreased or abolished the tumorigenicity of cells implanted into syngeneic host animals. The transfer of cDNA encoding IL-2 (Fearon *et al.*, 1990; Gansbacher *et al.*, 1990), IL-3 (McBride *et al.*, 1994), IL-4 (Pattengale and Leder, 1989), IL-6 (Mackiewicz *et al.*, 1995), IL-7 (Allione *et al.*, 1994), IL-12 (Tahara *et al.*, 1994), granulocyte-macrophage colony-stimulating factor (GM-CSF; Dranoff *et al.*, 1993), G-CSF (Stoppacciaro *et al.*, 1993), and gamma interferon (IFN-γ; Watanabe *et al.*, 1989) significantly reduced or eliminated the ability of the cells to form various histological types of murine tumors. Furthermore, with transfer of several of these cytokine genes, animals also developed systemic antitumor immunity and were protected against subsequent challenge with the unmodified parental tumor (Fearon *et al.*, 1990; Gansbacher *et al.*, 1990). Protection could also be demonstrated when animals were immunized with a mixture of unmodified tumor cells and tumor cells engineered to express the cytokine gene.

A. Genetically Modified Fibroblasts for Cytokine Gene Therapy

For tumors that are difficult to maintain in culture, it is impractical to modify genetically such autologous tumor cells to express cytokines. It is feasible, however, to modify primary autologous fibroblasts obtained from skin biopsies and cultured *in vitro,* or established human allogeneic fibroblast cell lines, to express and secrete cytokines (Fakhrai *et al.*, 1995; Tahara *et al.*, 1994). The genetically modified fibroblasts may then be combined with autologous or allogeneic tumor cells, and the irradiated mixture of cells may be used to induce systemic antitumor immunity. Use of genetically modified fibroblasts in therapeutic vaccines facilitates titration of single or multiple cytokine doses independent of tumor cell doses and permits other forms of genetic manipulation to be performed on the tumor cell component of the vaccines to enhance further their immunogenicity. The efficacy of active tumor immunotherapy with cytokine-transduced syngeneic or allogeneic fibroblasts has been demonstrated by several groups of investigators, including our laboratory (Fakhrai *et al.*, 1995; Tahara *et al.*, 1994). In our studies on a murine colon tumor model, immunizations with an irradiated mixture of tumor cells and IL-2-transduced fibroblasts produced systemic immunity that eradicated established tumors and rejected a subsequent tumor challenge (Fakhrai *et al.*, 1995).

We have demonstrated the efficacy and equivalence of IL-2 cytokine gene therapy with IL-2-transduced tumor cells or syngeneic and allogeneic transduced fibroblasts in the CT-26 BALB/c colorectal carcinoma tumor model (Fakhrai *et al.*, 1995; Shawler *et al.*, 1995). The results of these studies can be summarized here. Our initial series of experiments documented the efficacy of coadministration of irradiated tumor cells and IL-2-transduced syngeneic fibroblasts to induce systemic immunity and we were the first to indicate important inhibitory effects of high levels of IL-2 transgene expression (Fakhrai *et al.*, 1995). In our studies, immunization with a mixture of irradiated unmodified tumor cells and IL-2-transduced fibroblasts induced significantly greater protection against a live tumor challenge compared to immunization with irradiated tumor cells alone. Protective effects were observed with doses of IL-2-transduced fibroblasts secreting from 5 to 100 units of IL-2/24 hr. Parallel experiments in nude mice produced no protection, indicating that the effects of immunization were mediated by a T cell-dependent mechanism. In animals with established tumors, complete tumor remissions were observed following immunization with a mixture of irradiated tumor cells and IL-2-transduced fibroblasts secreting 100 units of IL-2/24 hr but not after immunization with irradiated tumor cells alone. Importantly, fibroblasts secreting higher doses of IL-2 were ineffective in generating systemic immunity. These findings indicated important relationships between IL-2 dose, immune system effector mechanisms, and antitumor efficacy that we have considered in the design of our IL-2 gene therapy trials by initiating our escalation of IL-2 doses at low levels.

Standard chromium release assays were utilized to evaluate cell-mediated cytotoxicity in these studies. To measure antitumor lytic activity, mice were immunized with irradiated CT-26 tumor cells alone or irradiated tumor cells mixed with different IL-2-secreting doses of transduced fibroblasts. Splenocytes from immunized and nonimmunized mice were used in these assays. Antitumor lytic activity in mice immunized with a mixture of irradiated tumor cells and transduced fibroblasts secreting 100 units of IL-2/24 hr was approximately 10- to 20-fold greater than the activity in mice immunized with irradiated tumor cells alone, or with a mixture of irradiated tumor cells and transduced fibroblasts secreting 1700 units of IL-2/24 hr. Antitumor lytic activity in all groups was inhibited by the incubation of effector cells with anti-CD8 monoclonal antibody, while anti-CD4 antibody had no effect on lytic activity. The results of these studies suggest that the systemic antitumor effects of this form of IL-2 gene therapy were mediated by CD8$^+$ cytotoxic T cells (Fakhrai *et al.*, 1995).

We have more recently completed additional animal studies in the BALB/c CT-26 colon tumor model indicating equivalent protective immunity against a subsequent tumor challenge following immunization with IL-2-transduced tumor cells or irradiated tumor cells mixed with either syngeneic (BALB/c) or allogeneic (C3H) IL-2-transduced fibroblasts. These findings support the use of an "off-the-shelf" IL-2-transduced allogeneic fibroblast cell line as a more practical alternative to the "customized" transduction of autologous fibroblasts in IL-2 immunogene therapy. This alternative will be utilized in the clinical trials to be performed on the completion of the initial phase I study of immunizations with irradiated autologous tumor and autologous IL-2-transduced fibroblasts that are currently being performed at our center (see Section VII, Clinical Cytokine Gene Therapy Experience).

Lotze and coworkers obtained similar results in a murine melanoma model using tumor cells with allogeneic fibroblasts genetically modified to express IL-12 (Tahara *et al.*, 1994). In animal studies using a slightly different approach, Kim and Cohen (1994; Russell *et al.*, 1992) induced systemic antitumor immunity by immunization with IL-2-modified fibroblasts that had also been transfected with DNA prepared from tumor tissue. In these latter studies, the results of immunizations with allogeneic and syngeneic fibroblasts transfected with tumor DNA were compared with and without concomitant IL-2 gene transfer. They found that different types of effector cells were induced by immunizations with IL-2-transduced autologous versus allogeneic fibroblasts and that combined IL-2 gene transfer and allogeneic stimulation had synergistic effects with enhanced survival compared to immunization with either approach alone (Kim and Cohen, 1994; Russell *et al.*, 1992). The data derived from these studies bolster the development of vaccines including cytokine-secreting fibroblasts as a means to enhance antitumor immune responses. The results of preclinical animal studies suggest both practical and potential therapeutic advantages to the application of allogeneic fibroblasts for cytokine gene transfer.

B. Genetically Modified, Partially HLA-Matched, Allogeneic Tumor Cells

The applicability of immunogene therapy in the treatment of a broad range of tumors would be more practical if immunizations could be performed using allogeneic cells, thus obviating the need to establish primary fibroblast and colon tumor cultures for each patient. Allogeneic tumor cells can be effective as antitumor vaccines if they express tumor-associated antigens (TAAs) that are shared by the patients' tumors. The HLA-A1, HLA-A2, and HLA-A3 haplotypes, expressed by approximately 25, 50, and 20% of the North American population, respectively, play a major role in presentation of shared TAAs to mediate MHC-restricted tumor destruction by cytolytic T lymphocytes (CTLs) (Chen *et al.*, 1994; Crowley *et al.*, 1990). Several TAAs defined by CTLs have been identified in colon carcinomas (De Plaen *et al.*, 1994; Finn, 1993). The protein components of tumor mucin (MUC-1) and the MAGE gene family are TAAs expressed by many colon carcinomas and other adenocarcinomas (De Plaen *et al.*, 1994; Finn, 1993). Additional TAAs expressed by the majority of carcinomas include the carcinoembryonic antigen (CEA) and the glycoprotein recognized by the monoclonal antibodies CO-17-1A and GA733 (Herlyn *et al.*, 1991, 1994).

An alternative approach is to inject cytokine or related immunostimulatory genes directly into tumors. Direct injection of allogeneic MHC or cytokine gene vectors into tumors has been successful in animal tumor models and this approach is being evaluated in phase I clinical trials (Nabel *et al.*, 1993). Antitumor cellular immune responses were found following direct tumor injection of allogeneic HLA-B7 cDNA in melanoma patients (Nabel *et al.*, 1993). However, application of this approach could be problematic in the surgical adjuvant setting where all clinically detectable tumor is removed and only microscopic residual disease remains.

III. Antisense Inhibition of Immunosuppressive and Differentiation Factors

In general, cancer patients have a compromised immune system. Several factors secreted by tumors have been identified that suppress antitumor immune responses (Sulitzeanu, 1993). One of these immunosuppressive factors that has been extensively characterized is transforming growth factor β (TGF-β), a protein produced by many normal cells and overexpressed by most common cancer cells (Anzano *et al.*, 1989; Coffey *et al.*, 1987). Transforming growth factor β is a potent agent that principally inhibits *afferent* immune system functions required for the generation of effective antitumor immunity. In particular, TGF-β inhibits the induction of cytotoxic T cells and other immune cells known to mediate antitumor immune responses (Sulitzeanu, 1993). Expression of TGF-β may inhibit the efficacy of whole-cell vaccine preparations and account for the equivocal results obtained in previous active immunotherapy studies employing autologous or allogeneic tumor cell vaccines. In particular, the suppressive effect of TGF-β suggests that genetically modified tumor vaccines may be most effective when used in patients with a small tumor burden. Our studies in animal tumor models indicate that antisense inhibition of TGF-β expression significantly increases the efficacy of tumor cell vaccines and that immunizations with tumor cells genetically modified to suppress TGF-β are efficacious against established tumors that express TGF-β (Fakhrai *et al.*, 1996).

Tumor cells have derived other strategies to evade immune surveillance. The expression of insulin-like growth factor I (IGF-I) can affect tumor cell differentiation, with a resultant decrease in immunogenicity (Resnicoff *et al.*, 1994; Trojan *et al.*, 1993). In Wistar rats, C6 glial tumor cells carrying antisense vectors that inhibited the expression of either IGF-I or its receptor were able to induce antitumor immune responses that rejected otherwise lethal doses of unmodified parental tumor (Resnicoff *et al.*, 1994; Trojan *et al.*, 1993). Unmodified tumor cells were unable to confer this protective immune response. Results such as these, and those from our own laboratory, suggest that mechanisms that allow tumor cells to escape immune destruction may be circumvented by inhibiting the expression of immunosuppressive or differentiation factors, or their receptors. It is possible that other tumor-derived agents, such as IL-10 or prostaglandin E_2 (Alleva *et al.*, 1994; Sawamura *et al.*, 1990), may exert similar negative effects on active tumor immunotherapy. The development of strategies to inhibit the expression of immunosuppressive and differentiation factors may play an important role in the development of effective immunogene therapies for cancer.

IV. Costimulatory Molecules and Antitumor Immunity

It is becoming increasingly clear that antigen recognition alone is not sufficient for T cell activation to effector functions. "Second signals" such as coligation of auxiliary molecules are also critical for generating T cell-mediated immunity (June *et al.*, 1990; Mondino and Jenkins, 1994). Antigen recognition in the absence of these second signals can lead to tolerance or "anergy" (June *et al.*, 1990; Mondino and Jenkins, 1994). Two costimulatory molecules in particular, B7.1 (CD80) and B7.2 (CD86), the ligands for CD28 and CTLA-4, respectively, have received a great deal of attention as potent costimulators for T cell function. In humans, B7 is expressed on dendritic cells and is induced on activated B cells, T cells, natural kill (NK) cells, and macrophages (Azuma *et al.*, 1993; Freeman *et al.*, 1989). Northern analysis for mRNA expression of B7 revealed that most carcinomas, leukemias of B cell origin [including non-T cell acute lymphoblastic leukemia (ALL)], prolymphocytic leukemia, hairy cell leukemia, and chronic lymphocytic leukemia were B7 negative while some non-Hodgkin's lymphomas were positive (Freeman *et al.*, 1989). These results suggest that lack of B7 expression by many tumors may contribute to their poor immunogenicity.

In previous studies, transfection of the B7.1 gene into murine melanoma and sarcoma models caused the transfected tumors to be rejected *in vivo* (Baskar *et al.*, 1993; Townsend and Allison, 1993). In both cases, once immunity was induced, the animals were protected from challenge with the unmodified tumor. Since this immunity was dependent on the presence of cytolytic T cells, an important conclusion that can be drawn is that the presence of B7 on the tumor is critical for T cell induction but not for effector cell function. These studies also suggest that the absence of appropriate costimulatory molecules on tumors could be a critical factor allowing escape from immune attack despite the expression of potentially strong tumor-associated antigens. In more recent studies, combined IL-2 cytokine and B7.1 gene transfer has demonstrated synergistic effects in generating efficacious antitumor immunity in animal tumor models (Hollingsworth *et al.*, 1995).

V. Genetic Modification of Tumor-Infiltrating Lymphocytes

The first human gene transfer studies on cancer patients examined the expression of a transferred gene, neomycin phosphotransferase (*neo*[R]), in tumor-infiltrating lymphocytes (TILs) expanded *ex vivo* and genetically modified by retroviral transduction (Rosenberg *et al.*, 1990). In patients with metastatic cancer, polymerase chain reaction analyses consistently demonstrated the presence of genetically modified cells that persisted in the circulation for as long as 2 months after administration. Antitumor effects were detected in a subset of patients, and significantly, no infectious retroviruses or side effects induced by gene transfer were noted in any patient (Rosenberg *et al.*, 1990). The data generated in this study provided important feasibility and safety experience to support the progression to other gene transfer clinical studies. In animal tumor models (Marincola *et al.*, 1994), it was demonstrated that TILs transduced with the gene for tumor necrosis factor α (TNF-α) were able to confer impressive antitumor immunity. In this cytokine gene therapy approach, the tumor-targeting ability of TILs was exploited to deliver cytokines to metastatic tumor sites. In related approaches, some investigators have generated TILs following intratumoral injection of immunostimulatory genes in an effort to enhance their antitumor activity (Wahl *et al.*, 1995).

To broaden the utility of adoptive cellular immunotherapy, investigators have employed chimeric antibody/T cell receptor (TCR) genes to enhance the targeting ability of transferred immune effector cells (Eshhar and Gross, 1990; Gross and Eshhar, 1992; Hwu *et al.*, 1995). These chimeric receptors are composed of the variable domains of tumor antigen-specific monoclonal antibodies joined to T cell receptor-signaling chains. T cells transduced with these chimeric genes recognize antibody-defined antigens, resulting in T cell activation, tumor-specific cell lysis, and cytokine release (Eshhar and Gross, 1990; Gross and Eshhar, 1992; Hwu *et al.*, 1995). Specific lysis of transformed cells overexpressing HER2/Neu has been reported utilizing cytotoxic T cell hybridomas genetically modified with chimeric genes composed of a single-chain Fv domain (scFv) of an anti-HER2/Neu antibody linked with the ζ signal-transducing subunit of the TCR/CD3 complex or the γ signal-transducing subunit of the immunoglobulin (Ig) Fc receptor complex (Stancovski *et al.*, 1993). Similar results were observed *in vivo* with murine T cells transduced with a chimeric receptor gene (*MOv-γ*) derived from the monoclonal antibody (MAb) MOv18, which binds to a folate-binding protein overexpressed on most human ovarian adenocarcinomas. Nude mice bearing human ovarian cancer cells treated with *MOv-γ*-transduced TILs had significantly increased survival compared to mice treated with saline only, nontransduced TILs, or TILs transduced with a control antitrinitrophenyl chimeric receptor gene (Hwu *et al.*, 1995). Similar chimeric receptor genes have been expressed as functional surface receptors in a mast cell line where the chimeric receptors exhibited binding properties of an antibody molecule and triggered degranulation of transfected mast cells on stimulation with antigen (Bach *et al.*, 1994). These studies indicate that immune effector cells such as T lymphocytes, mast cells, or natural killer cells can be genetically modified to react *in vivo* against tumor antigens defined by MAbs.

However, clinical application of these approaches may be hindered by difficulties associated with efficient genetic modification of immune effector cells and by the time and expense required for *ex vivo* expansion of these preparations. Improved vector systems for the transfer and expression of genes in hematopoietic cells are being developed and should prove useful in further advancing this form of immunogene therapy.

VI. DNA Vaccines

A number of approaches have been developed to circumvent the need to establish cell lines or cultures *ex vivo* for immunogene therapy. Direct tumor injection of cytokine or allogeneic MHC gene vectors has been successful in animal models and this strategy is being evaluated in phase I clinical trials (Nabel *et al.*, 1993). One problem with using this approach in the surgical adjuvant setting is that all clinically detectable tumor has been removed and tumor sites are no longer available for injection. The successful identification and cloning of tumor-associated antigens (TAAs) provides an alternative approach for *in vivo* immunogene therapy. Expression of these TAAs in vaccinia, adenovirus, liposomal, and "naked" DNA vectors has been proposed for clinical applications based on the generation of antigen-directed immune responses following immunizations in animal models (Nabel *et al.*, 1993; Ulmer *et al.*, 1993; Vieweg *et al.*, 1995). Candidate TAAs suitable for application in these approaches are listed in Table I and may be classified into four general categories: (1) TAAs recognized by cellular immune responses, (2) TAAs identified by antibodies, (3) mutated tumor suppressor and oncogenes, and (4) tumor-associated viral antigens.

The ability to identify TAAs recognized by cellular immune responses has been advanced by the availability of CTL clones, derived from TILs or peripheral blood T cells of cancer patients, that are specific for different tumor types and that lyse tumor cells expressing the target TAA in an MHC-restricted manner (De Plaen *et al.*, 1994; Pandolfi *et al.*, 1991). The TAAs recognized by these CTL clones have been identified by the transfection of cDNA from lysis-sensitive tumor cells into MHC-matched lysis-insensitive recipient cells. The isolation of tumor cell lines converted to lysis sensitivity following cDNA transfection permitted the cloning and sequencing of the cDNA responsible for lysis sensitivity. The MAGE family of TAAs was identified in this manner as well as other melanoma-associated TAAs including MART and tyrosinase (Brichard *et al.*, 1993; Gaugler *et al.*, 1994; Nabel *et al.*, 1993; Sensi *et al.*, 1995; Topalian *et al.*, 1994; van der Bruggen *et al.*, 1991). Subsequent studies have identified MAGE TAA expression by a wide variety of tumor types, including other neuroectodermal tumors such as glioblastoma and subsets of more common malignancies such as breast, colon, and lung carcinomas (Rimoldi *et al.*, 1993). MAGE family members

TABLE I Human Tumor-Associated Antigens

Antigen	Tumor	MHC restriction	Comments	Ref.
Tumor-associated antigens identified by T cell reactivity				
BAGE	Melanoma	HLA-Cw1601	Antigen also found in bladder, mammary, and squamous cell carcinomas	Boel *et al.* (1995)
CEA	Colon	Not described	Vaccination with admixtures of vaccinia virus/CEA and vaccinia virus/B7-generated tumor-protective CTL in mice	Hodge *et al.* (1995)
Decapeptide 810	Melanoma	HLA-A2 and HLA-A11	Same antigen presented by two different class I molecules	Morioka *et al.* (1994)
gp100	Melanoma	HLA-A2.1	Antigen is a 9-mer with poor affinity for MHC	Cox *et al.* (1994)
MAGE 1	Melanoma and others	HLA-A2	Antigen silent on normal tissue except testes. Antigen found in multiple tumor types	Rimoldi *et al.* (1993); van der Bruggen *et al.* (1991)
MAGE 3	Melanoma	HLA-A2	Antigen silent on normal tissue except testes	Gaugler *et al.* (1994)
MART 1	Melanoma	HLA-A2	Antigen also found on normal melanocytes CTLs secrete GM-CSF, TNF, and IFN in response to antigen Conserved TCR usage found in antigen-dependent CTLs	Kawakami *et al.* (1994); Sensi *et al.* (1995)
MUC-1	Pan-carcinoma	Non-MHC	Antigen revealed through aberrant glycosylation of normal protein on tumor cells. Antigen found in multiple tumor types	Finn *et al.* (1995); Jerome *et al.* (1991)
TAG-72	Colon	Non-MHC	CD4$^+$ T cells secrete IL-2, IL-4, TNF, and IFN in response to antigen	Kim *et al.* (1995)
Tyrosinase	Melanoma	HLA-A2	Antigen also found on normal melanocytes	Brichard *et al.* (1993)

Antigen	Tumor	HLA restriction	Comments	Reference
Tyrosinase	Melanoma	HLA-DR	$CD4^+$ T cells secrete IL-4, TNF, IFN, and GM-CSF in response to antigen	Topalian *et al.* (1994)
Unnamed	Melanoma	HLA-A2.1	Six distinct antigenic epitopes described	Slingluff *et al.* (1993)
Unnamed	Squamous cell	HLA-A*2601	Antigen found on both esophageal and lung squamous cell carcinomas	Nakao *et al.* (1995)
Unnamed	Squamous cell	HLA-Aw68	Two antigens identified from one peptide	Slingluff *et al.* (1994)
Tumor-associated antigens identified by antibodies				
CEA	Colon	Not applicable	Vaccination with CEA plasmid DNA induces cellular and humoral responses in mice	Conry *et al.* (1995)
TAG-72	Pan-carcinoma	Not applicable	TAA with wide tumor distribution	Myers *et al.* (1995)
CA 125	Ovary	Not applicable	Antiidiotypic anti-CA 125 antibody can induce non-MHC-restricted CTLs	Schlebusch *et al.* (1995)
17-1A	Colon	Not applicable	Antiidiotypic antibody can induce immune response	Herlyn *et al.* (1994)
Oncogenes and mutated tumor suppressor genes				
HER2/neu	Breast, ovary, lung	HLA-A2.1, HLA-DR, HLA-A2	Antigen is an oncogene 9-mer with high affinity for MHC	Disis *et al.* (1995)
			$CD4^+$ T cells exhibit antigen-dependent proliferation	Disis *et al.* (1994)
			Antigen found in multiple tumor types	Yoshino *et al.* (1994)
p53	Pan-carcinoma	HLA-A2.1	Transgenic mice used to identify antigens	Theobald *et al.* (1995)
Tumor-associated viral antigens				
EBV	B cell lymphoma		Epstein–Barr virus	Khanna *et al.* (1994, 1995)
HBV	Hepatoma		Hepatitis B virus	Jubelirer *et al.* (1991)
HPV16 E7	Cervical cancer	HLA A*0201	Human papillomavirus oncogene product	Ressing *et al.* (1995)

contain amino acid motifs suitable for presentation by frequently expressed HLA haplotypes (HLA-A1, HLA-A2, and HLA-A3), supporting their development as tumor vaccines (Robbins *et al.*, 1994). The MAGE-related gene product tyrosinase is capable of inducing antigen-specific $CD4^+$ T cells (Topalian *et al.*, 1994). Another family of TAAs identified by these methods includes tumor mucins (Finn *et al.*, 1995; Jerome *et al.*, 1991; Kim *et al.*, 1995; Myers *et al.*, 1995). One of the better characterized members of this family, MUC-1, is expressed by a wide variety of adenocarcinomas (Finn *et al.*, 1995). cDNA transfection studies have indicated that this TAA can mediate CTL lysis in an MHC-independent manner, presumably owing to the presence of tandemly repeated amino acid sequences capable of cross-linking T cell receptors (Finn *et al.*, 1995; Jerome *et al.*, 1991). MUC-1 is also capable of inducing a non-MHC-restricted $CD4^+$ T cell response (Kim *et al.*, 1995). These characteristics of MHC nonrestriction make MUC-1 an attractive candidate for vaccine development.

Tumor-associated antigens identified by antibodies include the TAG-72 and 17-1A/GA733 antigens, which are expressed by many types of carcinomas (Herlyn *et al.*, 1994; Myers *et al.*, 1995). The carcinoembryonic antigen (CEA) and CA125 antigens are preferentially expressed by adenocarcinomas derived from the gastrointestinal tract and ovary, respectively (Conry *et al.*, 1995; Schlebusch *et al.*, 1995). In melanomas, monoclonal antibodies identified the p97, gp240/480, and gp100 TAAs (Cox *et al.*, 1994; Slingluff *et al.*, 1993). The genes encoding the protein components of these antigens have been cloned, permitting their evaluation in tumor vaccine preparations. CEA, the protein backbone of tumor mucin (MUC-1) and the melanoma antigens MAGE and MART, have been expressed in a variety of vectors for use in immunization (Freeman *et al.*, 1989; Townsend and Allison, 1993). Antiidiotypic antibodies that mimic the antigenic epitopes of CEA and 17-1A have been used to induce cellular immune responses in some studies (Conry *et al.*, 1995; Herlyn *et al.*, 1994). Studies in animals have indicated that these vaccinations are able to induce immune responses against the TAA immunogene (Conry *et al.*, 1995).

Additional targets for immunogene therapy approaches include mutated oncogenes, tumor suppressor genes, and viral antigens. These potential TAAs have been suggested as appealing targets for immunotherapy as they are not expressed by normal tissues. Several investigators have evaluated whether mutated oncogenes and tumor suppressor proteins are recognized by the immune system. Preliminary studies have suggested that mutated *ras* and *p53* sequences may be immunogenic and capable of mediating antitumor immune responses in model systems (Fossum *et al.*, 1994; Peace *et al.*, 1991; Theobald *et al.*, 1995). Several tumors have presumed causal relationships with viruses including certain nasopharyngeal and lymphoma neoplasms with Epstein–Barr viral infections, cervical cancer with papillomaviruses, and hepatomas with the hepatitis B virus (Jubelirer *et al.*, 1991; Khanna *et *al.*, 1994, 1995; Ressing *et al.*, 1995). Hence, there exists a broad array of candidate TAAs for evaluation in immunogene therapy and related approaches. However, further studies will be required to demonstrate that vaccination of patients with preparations derived from any of the candidate types of TAAs listed in Table I can induce clinically meaningful antitumor immune responses.

VII. Clinical Cytokine Gene Therapy Experience

A. Glioblastoma

Ours was one of the first groups to evaluate the effects of IL-2 gene transfer in human subjects (Sobol *et al.*, 1995). The results of our treatment of a patient with glioblastoma multiforme (GBM) are summarized below. The patient was a 52-year-old female with GBM of the right temporal lobe. She was initially treated with surgical resection, conventional radiotherapy, and PCV chemotherapy [procarbazine, N-(2-chloroethyl)-N'-cyclohexyl-N-nitrosourea (CCNU), and vincristine]. Nine months later, a second resection was performed for tumor recurrence. Tumor pathology revealed a GBM at reresection. The tumor in this patient progressed after experimental treatment with accutane and with an ^{131}I radioisotope-labeled anti-tenacin monoclonal antibody. Subsequently, the patient was treated with experimental stereotactic radiation therapy designed to encompass the site of tumor involvement. Interleukin 2 gene therapy was initiated approximately 1 year after the first tumor resection. The patient received 10 subcutaneous immunizations at approximately 2- to 4-week intervals with either autologous, irradiated IL-2-transduced tumor cells or a mixture of irradiated tumor cells and irradiated IL-2-transduced fibroblasts. The total administered IL-2 dose ranged from 3 to 440 units/24 hr. The total tumor cell dose for each immunization was 10^7 cells.

Patient peripheral blood mononuclear cells and serum were analyzed to assess the development of cellular and humoral antitumor immune responses against autologous cultured tumor cells. Peripheral blood mononuclear cells obtained after the third and subsequent immunizations frequently demonstrated three- to fourfold greater tumor lytic activity compared to the pretreatment control. This tumor lytic activity could be inhibited by incubation of the effector cells with autologous tumor or K562 cells. Partial inhibition of tumor lytic activity was observed with anti-CD8 but not with anti-CD4 antibody. These findings are consistent with the generation of a cellular antitumor immune response. The data suggest that this cellular immune response was composed primarily of cells with natural killer activity and contains a component of $CD8^+$ cytotoxic T cells.

There were no significant adverse reactions at the immunization sites and no treatment-related abnormalities have been observed on monitoring

TABLE II Clinical Protocols Worldwide: Cytokine/Immunotherapy

No.	Cancer patient population	Transferred nucleic acids	Method of transfer	Title/cell target	Principal Investigator	Institution/country
1	Advanced cancer	Carcinoembryonic antigen cDNA	Canarypox	A Phase I Study of Recombinant ALVAC Virus that Expresses Carcinoembryonic Antigen in Patients with Advanced Cancers	M. J. Hawkins J. L. Marshall	Georgetown University Medical Center (Washington, D.C.)
2	Advanced cancer	Carcinoembryonic antigen cDNA	Vaccinia	A Phase I Study of Recombinant CEA Vaccinia Virus Vaccine with Postvaccination CEA Peptide Challenge	D. J. Cole	Medical University of South Carolina (Charleston, SC)
3	Advanced cancer	HLA-B7 cDNA	Lipofection	Immunotherapy for Cancer by Direct Gene Transfer into Tumors	G. J. Nabel	University of Michigan (Ann Arbor, MI)
4	Advanced cancer	HLA-B7 cDNA	Lipofection	Immunotherapy of Malignancy by *in vivo* Gene Transfer into Tumors	G. J. Nabel	University of Michigan (Ann Arbor, MI)
5	Advanced cancer	HLA-B7 cDNA	Lipofection	Phase I Study of Tumor-Infiltrating Lymphocytes Derived from *in vivo* HLA-B7 Gene-Modified Tumors in the Adoptive Immunotherapy of Melanoma	A. E. Chang G. J. Nabel	University of Michigan Medical Center (Ann Arbor, MI)
6	Advanced cancer	HLA-B7 and β_2-microglobulin cDNA	Lipofection	Adoptive Cellular Therapy of Cancer Combining Direct HLA-B7/β_2-Microglobulin Gene Transfer with Autologous Tumor Vaccination for the Generation of Vaccine-Primed Anti-CD3 Activated Lymphocytes	B. A. Fox W. J. Urba	Earle A. Chiles Research Institute, Providence Portland Medical Center (Portland, OR)
7	Advanced cancer	HLA-B7 and β_2-microglobulin cDNA	Lipofection	Phase II Study of Immunotherapy of Metastatic Cancer by Direct Gene Transfer	A. E. Chang E. Hersh N. Vogelzang R. Levy B. Redman R. A. Figlin J. Rubin	Multicenter trial

8	Advanced cancer	IL-2 cDNA	Retrovirus	Immunization of Cancer Patients Using Autologous Cancer Cells Modified by Insertion of the Gene for Interleukin 2 (IL-2)	J. J. Rinehart J. H. Doroshow H. Silver S. A. Rosenberg	National Institutes of Health (Bethesda, MD)
9	Advanced cancer	IL-2 cDNA	Lipofection	A Phase I Trial of IL-2 Plasmid DNA/DMRIE/DOPE Lipid Complex as an Immunotherapeutic Agent in Solid Malignant Tumors or Lymphomas by Direct Gene Transfer	E. M. Hersh	Arizona Cancer Center, University of Arizona (Tuscon, AZ)
10	Advanced cancer	IL-2 cDNA	Lipofection	Fibroblasts and Autologous Tumor	R. Mertelsmann A. Lindemann	University Medical Center (Freiburg, Germany)
11	Advanced cancer	IL-4 cDNA	Retrovirus	Gene Therapy of Cancer: A Pilot Study of IL-4 Gene-Modified Antitumor Vaccines	M. T. Lotze	University of Pittsburgh (Pittsburgh, PA)
12	Advanced cancer	IL-12 cDNA	Retrovirus	IL-12 Gene Therapy Using Direct Injection of Tumor with Genetically Engineered Autologous Fibroblasts	M. T. Lotze H. Tahara	University of Pittsburgh (Pittsburgh, PA)
13	Advanced cancer	T cell receptor antibody	Retrovirus	Cytotoxic T lymphocytes	Z. Eshhar	Weizman Institute of Science (Rehovot, Israel)
					S. Slavin	Hadassah University Hospital (Jerusalem, Israel)
					R. L. Bolhuis	Daniel den Hoed Cancer Center (Rotterdam, The Netherlands)
					S. A. Rosenberg	National Institutes of Health (Bethesda, MD)
14	Advanced cancer	TNF cDNA	Retrovirus	Immunization of Cancer Patients Using Autologous Cancer Cells Modified by Insertion of the Gene for Tumor Necrosis Factor (TNF)	S. A. Rosenberg	National Institutes of Health (Bethesda, MD)
15	Brain tumor	IGF-I antisense	Lipofection	Gene Therapy for Human Brain Tumors Using Episome-Based Antisense cDNA Transcription of Insulin-Like Growth Factor I	J. Ilan	Case Western Reserve (Cleveland, OH)

(*continues*)

TABLE II (continued)

No.	Cancer patient population	Transferred nucleic acids	Method of transfer	Title/cell target	Principal Investigator	Institution/country
16	Brain tumor	IL-2 cDNA	Retrovirus	Injection of a Glioblastoma Patient with Tumor Cells and Fibroblasts Genetically Modified to Secrete Interleukin 2 (IL-2)	R. E. Sobol I. Royston	Sidney Kimmel Cancer Center (San Diego, CA)
17	Brain tumor	TGF–β_2 antisense	Retrovirus	Injection of Glioblastoma Patients with TGF-b2 Antisense Gene-Modified Autologous Tumor Cells—A Phase I Study	K. L. Black H. Kakhrai	UCLA School of Medicine (Los Angeles, CA)
18	Brain tumor	IL-4 cDNA	Retrovirus	A Phase I Study of IL-4 Gene-Modified Autologous Tumor to Elicit an Immune Response	M. Bozik M. Gilbert M. T. Lotze	University of Pittsburgh (Pittsburgh, PA)
19	Breast	IL-2 cDNA	Lipofection	A Pilot Study of Autologous Human Interleukin 2 Gene-Modified Tumor Cells in Patients with Refractory or Recurrent Metastatic Breast Cancer	H. Kim Lyerly	Duke University Medical Center (Durham, NC)
20	Colon	CEA cDNA	Naked DNA	A Phase I Trial of a Polynucleotide Augmented Antitumor Immunization to Human Carcinoembryonic Antigen in Patients with Metastatic Colorectal Cancer	D. T. Curiel	University of Alabama at Birmingham (Birmingham, AL)
21	Colon	HLA-B7 cDNA	Lipofection	A Phase I Study of Immunotherapy of Advanced Colorectal Carcinoma by Direct Gene Transfer into Hepatic Metastases	J. Rubin	Mayo Clinic (Rochester, MN)
22	Colon	IL-2 cDNA	Retrovirus	Injection of Colon Carcinoma Patients with Autologous Irradiated Tumor Cells and Fibroblasts Genetically Modified to Secrete Interleukin 2 (IL-2): A Phase I Study	R. E. Sobol I. Royston	Sidney Kimmel Cancer Center (San Diego, CA)

23	Colon	IL-2 cDNA	Adenovirus	Intratumoral *in Vivo*	B. Gilly	Lyon, France
24	Colon, lymphoma, melanoma, renal	IL-7 cDNA	Electroporation	Phase I Study: Interleukin 7 Gene Therapy for Patients with Metastatic Colon Cancer, Renal Cell Cancer, Malignant Melanoma, or Lymphoma	I. Schmidt-Wolf	Free University of Berlin (Berlin)
25	Head and neck	HLA-B7 and β_2-microglobulin cDNA	Lipofection	A Phase II Study of Allovectin 7 in the Treatment of Squamous Cell Carcinoma of the Head and Neck	J. L. Gluckman	University of Cincinnati Medical Center (Cincinnati, OH)
26	Lung	IL-2 cDNA	Retrovirus	Interleukin 2 Gene Transfer in Lung Carcinoma Patients with Pleural Effusions	N. Mao	Institute of Basic Medical Sciences (Beijing)
27	Lung	IL-2 cDNA	Adenovirus	Intratumoral *in Vivo*	T. Tursz M. Perricauder	Institut Gustave-Roussy (Villejuif, France)
28	Lung, small cell	IL-2 cDNA	Lipofection	Phase I Study of Transfected Cancer Cells Expressing the Interleukin 2 Gene Product in Limited-Stage Small-Cell Lung Cancer	P. Cassileth	University of Miami and University Veterans Administration Hospital (Miami, FL)
29	Lymphoma	Ig cDNA	Naked DNA	A Pilot Study of Idiotypic Vaccination for Follicular B Cell Lymphoma Using a Genetic Approach	R. Hawkins	MRC Cambridge and Centre for Protein Engineering (Cambridge, UK)
30	Melanoma	HLA-B7 cDNA	Lipofection	A Phase I Trial of B7-Transfected Lethally Irradiated Allogeneic Melanoma Cell Lines to Induce Cell-Mediated Immunity against Tumor-Associated Antigens Presented by HLA-A2 or HLA-A1 in Patients with Stage IV Melanoma	M. Sznol R. Fenton	Clinical Research Branch, Biological Response Modifiers Program, National Institutes of Health (Bethesda, MD)
31	Melanoma	GM-CSF cDNA	Retrovirus	A Phase I Study of Vaccination with Autologous, Irradiated Melanoma Cells Engineered to Secrete Human Granulocyte-Macrophage Colony-Stimulating Factor	G. Dranoff	Dana Farber Cancer Institute (Boston, MA)

(*continues*)

TABLE II (continued)

No.	Cancer patient population	Transferred nucleic acids	Method of transfer	Title/cell target	Principal Investigator	Institution/country
32	Melanoma	GM-CSF cDNA	Retrovirus	Autologous Tumor	E. M. Rankin	The Netherlands Cancer Institute (Amsterdam, The Netherlands)
33	Melanoma	HLA-B7 cDNA	Lipofection	Phase I Study of Immunotherapy of Malignant Melanoma by Direct Gene Transfer	E. Hersh	Arizona Cancer Center and University of Arizona (Tucson, AZ)
34	Melanoma	IFN-γ cDNA	Retrovirus	A Phase I Trial of Human Gamma Interferon-Transduced Autologous Tumor Cells in Patients with Disseminated Malignant Melanoma	H. F. Seigler	Duke University (Durham, NC)
35	Melanoma	IL-2 cDNA	Retrovirus	A Pilot Study of Immunization with HLA-A2-Matched Allogeneic Melanoma Cells that Secrete Interleukin 2 in Patients with Metastatic Melanoma	B. Gansbacher	Memorial Sloan-Kettering Cancer Center (New York, NY)
36	Melanoma	IL-2 cDNA	Transfection	Immunization with Interleukin 2-Transfected Melanoma Cells. A Phase I–II Study in Patients with Metastatic Melanoma	S. Osanto	University Hospital (Leiden, The Netherlands)
37	Melanoma	IL-2 cDNA	Retrovirus	Pilot Study of Toxicity of Immunization of Patients with Unresectable Melanoma with IL-2-Secreting Allogeneic Human Melanoma Cells	T. K. Das Gupta	University of Illinois at Chicago (Chicago, IL)
38	Melanoma	IL-2 cDNA	Retrovirus	Active Immunization of Metastatic Melanoma Patients with IL-2-Transduced, Allogeneic Melanoma Cells; a Phase I/II Study	N. Cascinelli R. Foa G. Parmiani	National Cancer Institute (Milan, Italy) Univeristy of Torino (Torino, Italy)
39	Melanoma	IL-2 cDNA and β-galactosidase cDNA	Naked DNA	Gene Therapy for Metastatic Melanoma: Assessment of Expression of DNA Constructs Directly Injected into Metastases	A. L. Harris I. Hart	ICRF Molecular Oncology Laboratory (Oxford, UK)

	Melanoma	IL-2 cDNA	Retrovirus	The Treatment of Metastatic Malignant Melanoma with Autologous Melanoma Cells Genetically Engineered to Secrete Interleukin 2; a Phase 1B Trial	M. Gore M. Collins	Royal Marsden Hospital, Institute of Cancer Research (London, UK)
41	Melanoma	IL-2 cDNA	Retrovirus	Genetically Engineered Autologous Tumor Vaccines Producing Interleukin 2 for the Treatment of Metastatic Melanoma	J. S. Economou J. Glaspy W. H. McBride	University of California Medical Center (Los Angeles, CA)
42	Melanoma	IL-2 cDNA	Transfection	Autologous Tumor	G. Stingl E. B. Brocker R. Mertelsmann M. L. Birnstiel	University of Vienna (Vienna, Austria) University of Wurzburg (Wurzburg, Germany) University Medical Center (Freiburg, Germany) Research Institute of Molecular Pathology (Vienna, Austria)
43	Melanoma	IL-4 cDNA	Retrovirus	Adoptive Immunotherapy of Melanoma with Activated Lymph Node Cells Primed *in Vivo* with Autologous Tumor Cells Transduced with the IL-4 Gene	A. E. Chang	University of Michigan (Ann Arbor, MI)
44	Melanoma	IL-4 cDNA	Retrovirus	Active Immunization of Metastatic Melanoma Patients with IL-4-Transduced, Allogeneic Melanoma Cells; a Phase I/II Study	N. Cascinelli R. Foa G. Parmiani	National Cancer Institute (Milan, Italy) University of Torino (Torino, Italy)
45	Melanoma	IL-7 cDNA	Retrovirus	A Phase I Testing of Genetically Engineered Interleukin 7 Melanoma Vaccines	J. S. Economou	University of California Medical Center (Los Angeles, CA)
46	Melanoma	IL-7, IL-12, GM-CSF cDNA	Ballistic	Autologous Tumor	D. Schadendorf E. M. Czarnetzki	Humbolt University of Berlin (Berlin, Germany)
47	Melanoma	IL-7, IL-12, GM-CSF cDNA	Ballistic	Lymphokine-Activated Killer Cells	D. Schadendorf E. M. Czarnetzki	Humbolt University of Berlin (Berlin, Germany)
48	Melanoma	MART-1 cDNA	Adenovirus	A Phase I Trial in Patients with Metastatic Melanoma of Immunization with a Recombinant Adenovirus Encoding the MART-1 Melanoma Antigen	S. A. Rosenberg	National Institutes of Health (Bethesda, MD)

(*continues*)

TABLE II (continued)

No.	Cancer patient population	Transferred nucleic acids	Method of transfer	Title/cell target	Principal Investigator	Institution/country
49	Neuroblastoma	IFN-γ cDNA	Retrovirus	A Phase I study of Immunization with Gamma Interferon-Transduced Neuroblastoma Cells	J. Rosenblatt R. Seeger	University of California, and Children's Hospital (Los Angeles, CA)
50	Neuroblastoma	IL-2 cDNA	Adenovirus	A Phase I Study of Cytokine Gene-Modified Autologous Neuroblastoma Cells for Treatment of Relapsed-Refractory Neuroblastoma Using an Adenoviral Vector	M. K. Brenner D. Dilloo L. Bowman	St. Jude Children's Research Hospital (Memphis, TN)
51	Neuroblastoma	IL-2 cDNA	Retrovirus	Phase I Study of Cytokine Gene-Modified Autologous Neuroblastoma Cells for Treatment of Relapsed/Refractory Neuroblastoma	M. K. Brenner	St. Jude Children's Research Hospital (Memphis, TN)
52	Ovary	Chimeric antibody/TCR cDNA	Retrovirus	Treatment of Patients with Advanced Epithelial Ovarian Cancer Using Anti-CD3-Stimulated Peripheral Blood Lymphocytes Transduced with a Gene Encoding a Chimeric T Cell Receptor Reactive with Folate-Binding Protein	P. Hwu	National Institutes of Health (Bethesda, MD)
53	Ovary	IL-2 cDNA	AAV plasmid	A Phase I Study of Autologous Human Interleukin 2 Gene-Modified Tumor Cells in Patients with Refractory Metastatic Ovarian Cancer	A. Berchuck H. K. Lyerly	Duke University Medical Center (Durham, NC)

54	Ovary	IL-2 cDNA	Lipofection	A Phase I Study of Autologous Human Interleukin 2 Gene-Modified Tumor Cells in Patients with Refractory Metastatic Ovarian Cancer	A. Berchuck H. K. Lyerly	Duke University Medical Center (Durham, NC)
55	Prostate	GM-CSF cDNA	Retrovirus	A Phase I/II Study of Autologous Human GM-CSF Gene-Transduced Prostate Cancer Vaccines in Patients with Metastatic Prostate Carcinoma	J. Simons	Johns Hopkins Oncology Center (Baltimore, MD)
56	Prostate	IL-2 and IFN-γ cDNA	Retrovirus	A Phase I/II Study of Immunization with MHC Class I-Matched Allogeneic Human Prostatic Carcinoma Cells Engineered to Secrete Interleukin 2 and Interferon γ	B. Gansbacher	Memorial Sloan-Kettering Cancer Center (New York, NY)
57	Prostate	IL-2 cDNA	AAV plasmid	A Phase I Study of Autologous Human Interleukin 2 Gene-Modified Tumor Cells in Patients with Locally Advanced or Metastatic Prostate Cancer	D. F. Paulson H. K. Lyerly	Duke University Medical Center (Durham, NC)
58	Prostate	IL-2 cDNA	Lipofection	A Phase I Study of Autologous Human Interleukin 2 Gene-Modified Tumor Cells in Patients with Locally Advanced or Metastatic Prostate Cancer	D. F. Paulson H. K. Lyerly	Duke University Medical Center (Durham, NC)
59	Prostate	Prostate-specific antigen cDNA	Vaccinia	A Phase I Study of Recombinant Vaccinia that Expresses Prostate Specific Antigen in Adult Patients with Adenocarcinoma of the Prostate	A. P. Chen	National Naval Medical Center (Bethesda MD)

(continues)

TABLE II (continued)

No.	Cancer patient population	Transferred nucleic acids	Method of transfer	Title/cell target	Principal Investigator	Institution/country
60	Renal cell	GM-CSF cDNA	Retrovirus	Phase I Study of Nonreplicating Autologous Tumor Cell Injections Using Cells Prepared with or without Granulocyte-Macrophage Colony-Stimulating Factor Gene Transduction in Patients with Metastatic Renal Cell Carcinoma	J. Simons	Johns Hopkins Oncology Center (Baltimore, MD)
61	Renal cell	HLA-B7 cDNA	Lipofection	Phase I Study of Immunotherapy for Metastatic Renal Cell Carcinoma by Direct Gene Transfer into Metastatic Lesions	N. Vogelzang	University of Chicago (Chicago, IL)
62	Renal cell	HLA-B7 cDNA	Lipofection	Phase I Study of HLA-B7 Plasmid DNA/DMRIE/DOPE Lipid Complex as an Immunotherapeutic Agent in Renal Cell Carcinoma by Direct Transfer with Concurrent Low-Dose Bolus IL-2 Protein Therapy	R. A. Figlin	University of California, Los Angeles Medical Center (Los Angeles, CA)
63	Renal cell	HLA-B7, β_2-microglobulin cDNA	Lipofection	Phase I: Adoptive Cellular Therapy of Cancer Combining Direct HLA-B7/β_2-Microglobulin Gene Transfer with Autologous Tumor Vaccination for the Generation of Vaccine-Primed Anti-CD3-Activated Lymphocytes	B. A. Fox W. J. Walter	Earle A. Chiles Research Institute, Providence Medical Center (Portland, OR)
64	Renal cell	IL-2 cDNA	Retrovirus	A Pilot Study of Immunization with Interleukin 2-Secreting Allogeneic HLA-A2-Matched Renal Cell Carcinoma in Patients with Advanced Renal Cell Carcinoma	B. Gansbacher	Memorial Sloan-Kettering Cancer Center (New York, NY)

of complete blood counts, serum chemistries, and urinalyses. Transient, mild erythema (<24 hr) was observed at the injection site with immunizations at IL-2 doses >100 units/24 hr. Magnetic resonance imaging (MRI) scans performed at approximately 4-week intervals during the first 5 months of treatment revealed modest changes in overall tumor size, with waxing and waning of peritumoral edema associated with alterations in Decadron (dexamethasone) doses. The MRI scan performed 6 months after the initiation of treatment (4 weeks after the ninth and highest dose IL-2 immunization) revealed marked tumor necrosis with significant peritumoral edema (Sobol *et al.*, 1995). Clinically, these MRI findings were associated with an exacerbation of the baseline left-sided weakness of the patient. This partially resolved following the administration of increased Decadron doses, which were gradually tapered. The patient received no further treatment until 2 months later when the MRI scan revealed renewed tumor growth (8 months after the initiation of treatment). At that time, the patient received the tenth immunization with a combination of transduced tumor cells and fibroblasts and nontransduced tumor cells. The patient developed progressive weakness, became withdrawn, and did not wish further therapy. Decadron administration was withdrawn. Her clinical condition continued to deteriorate. She became increasingly somnolent and expired 1 month later, approximately 10 months from the initiation of IL-2 gene therapy.

In summary, IL-2 gene therapy resulted in no significant toxicity at the sites of immunization and was associated with the generation of a cellular antitumor immune response. Similar cellular antitumor immune responses were reported by Nabel *et al.* (1993) in melanoma patients following intratumoral injection of DNA encoding a foreign MHC protein. Marked tumor necrosis was observed in our patient following the highest IL-2 immunization dose. It is not possible to draw meaningful conclusions regarding safety, efficacy, or immune response induction from the results of a single patient. However, we were encouraged by these preliminary clinical findings and, in conjunction with the supporting data from our animal colon carcinoma studies, we commenced evaluation of IL-2 gene therapy in patients with colon cancer as outlined below.

B. Colorectal Carcinoma

We have initiated a phase I clinical trial in patients with colorectal carcinoma, comprising subcutaneous immunizations with a mixture of autologous irradiated tumor cells and IL-2-transduced autologous fibroblasts. In the phase I study, the dose of transduced cells will be escalated when three patients at each IL-2 dosage level (100, 200, and 400 units/24 hr) have been treated and followed for 1 month without $\geq$ grade 3 toxicity. The patients will be monitored for toxicity, antitumor responses, and the induction of antitumor immunity.

VIII. Clinical Trials of Immunogene Therapy Worldwide

Table II provides a summary of immunogene therapy clinical protocols submitted to regulatory agencies worldwide. The results of these trials should provide insights regarding which of the immunogene therapy approaches are most appropriate for further clinical evaluation.

References

Alleva, D. G., Burger, C. J., and Elgert, K. D. (1994). Tumor-induced regulation of suppressor macrophage nitric oxide and TNF-alpha production. Role of tumor-derived IL-10, TGF-beta, and prostaglandin E2. *J. Immunol.* **153**, 1674–1686.

Allione, A., Consalvo, M., Nanni, P., Lollini, P. L., Cavallo, F., Giovarelli, M., Forni, M., Gulino, A., Colombo, M. P., Dellabona, P., Hock, H., Blankenstein, T., Rosenthal, F. M., Gansbacher, B., Bosco, M. C., Musso, T., Gusella, L., and Forni, G. (1994). Immunizing and curative potential of replicating and nonreplicating murine mammary adenocarcinoma cells engineered with interleukin (IL)-2, IL-4, IL-6, IL-7, IL-10, tumor necrosis factor alpha, granulocyte-macrophage colony-stimulating factor, and gamma-interferon gene or admixed with conventional adjuvants. *Cancer Res.* **54**, 6022–6026.

Anzano, M. A., Rieman, D., Prichett, W., Bowen-Pope, D. F., and Greig, R. (1989). Growth factor production by human colon carcinoma cell lines. *Cancer Res.* **49**, 2898–2904.

Azuma, M., Yssel, H., Phillips, J. H., Spits, H., and Lanier, L. L. (1993). Functional expression of B7/BB1 on activated T lymphocytes. *J. Exp. Med.* **177**, 845–850.

Bach, N. L., Waks, T., Schindler, D. G., and Eshhar, Z. (1994). Functional expression in mast cells of chimeric receptors with antibody specificity. *Cell Biophys.* **24–25**, 229–236.

Baskar, S., Ostrand-Rosenberg, S., Nabavi, N., Nadler, L. M., Freeman, G. J., and Glimcher, L. H. (1993). Constitutive expression of B7 restores immunogenicity of tumor cells expressing truncated major histocompatibility complex class II molecules. *Proc. Natl. Acad. Sci. U.S.A.* **90**, 5687–5690.

Boel, P., Wildmann, C., Sensi, M. L., Brasseur, R., Renauld, J. C., Coulie, P., Boon, T., and van der Bruggen, P. (1995). BAGE: A new gene encoding an antigen recognized on human melanomas by cytolytic T lymphocytes. *Immunity* **2**, 167–175.

Borden, E. C., and Sondel, P. M. (1990). Lymphokines and cytokines as cancer treatment. Immunotherapy realized. *Cancer (Philadelphia)* **65** (Suppl. 3), 800–814.

Brichard, V., van Pel, A., Wolfel, T., Wolfel, C., De Plaen, E., Lethe, B., Coulie, P., and Boon, T. (1993). The tyrosinase gene codes for an antigen recognized by autologous cytolytic T lymphocytes on HLA-A2 melanomas. *J. Exp. Med.* **178**, 489–495.

Bubenik, J., Viotenok, N. N., Kieler, J., Prassolov, V. S., Chumakov, P. M., Bubenikova, D., Simova, J., and Jandlova, T. (1988). Local administration of cells containing an inserted IL-2 gene and producing IL-2 inhibits growth of human tumors in nu/nu mice. *Immunol. Lett.* **19**, 279–282.

Chen, Q., Smith, M., Nguyen, T., Maher, D. W., and Hersey, P. (1994). T cell recognition of melanoma antigens in association with HLA-A1 on allogeneic melanoma cells. *Cancer Immunol. Immunother.* **38**, 385–393.

Coffey, R. J.,Jr., Goustin, A. S., Soderquist, A. M., Shipley, G. D., Wolfshohl, J., Carpenter, G., and Moses, H. L. (1987). Transforming growth factor alpha and beta expression in human colon cancer lines: Implications for an autocrine model. *Cancer Res.* **47**, 4590–4594.

Conry, R. M., LoBuglio, A. F., Loechel, F., Moore, S. E., Sumerel, L. A., Barlow, D. L., Pike, J., and Curiel, D. T. (1995). A carcinoembryonic antigen polynucleotide vaccine for human clinical use. *Cancer Gene Ther.* **2,** 33–38.

Cox, A. L., Skipper, J., Chen, Y., Henderson, R. A., Darrow, T. L., Shabanowitz, J., Engelhard, V. H., Hunt, D. F., and Slingluff, C. L., Jr. (1994). Identification of a peptide recognized by five melanoma specific human cytotoxic T cell lines. *Science* **264,** 716–719.

Crowley, N. J., Slingluff, C. L., Jr., Darrow, T. L., and Seigler, H. F. (1990). Generation of human autologous melanoma-specific cytotoxic T-cells using HLA-A2-matched allogeneic melanomas. *Cancer Res.* **50,** 492–498.

De Plaen, E., Arden, K., Traversari, C., Gaforio, J. T., Szikora, J. P., De Smet, C., Brasseur, F., van der Bruggen, P., Lethé, B., Lurquin, C., Brasseur, R., Chomez, P., De Backer, O., Cavnee, W., and Boon, T. (1994). Structure, chromosomal localization, and expression of 12 genes of the MAGE family. *Immunogenetics* **40,** 360–369.

Disis, M. L., Calenoff, E., McLaughlin, G., Murphy, A. E., Chen, W., Groner, B., Jeschke, M., Lydon, N., McGlynn, E., and Livingston, R. B. (1994). Existent T cell and antibody immunity to HER2/neu protein in patients with breast cancer. *Cancer Res.* **54,** 16–20.

Disis, M. L., Smith, J. W., Murphy, A. E., Chen, W., and Cheever, M. A. (1995). *In vitro* generation of human cytolytic T cells specific for peptides derived from the HER2/neu protooncogene protein. *Cancer Res.* **54,** 1071–1076.

Dranoff, G., Jaffee, E., Lazenby, A., Golumbek, P., Levitsky, H., Brose, K., Jackson, V., Hamada, H., Pardoll, D., and Mulligan, R. C. (1993). Vaccination with irradiated tumor cells engineered to secrete murine granulocyte-macrophage colony-stimulating factor stimulates potent, specific, and long-lasting anti-tumor immunity. *Proc. Natl. Acad. Sci. U.S.A.* **90,** 3539–3543.

Eshhar, Z., and Gross, G. (1990). Chimeric T cell receptor which incorporates the anti-tumour specificity of a monoclonal antibody with the cytolytic activity of T cells: A model system for immunotherapeutical approach. *Br. J. Cancer* **10** (Suppl.), 27–29.

Fakhrai, H., Shawler, D. L., Gjerset, R., Naviaux, R. K., Koziol, J., Royston, I., and Sobol, R. E. (1995). Cytokine gene therapy with interleukin-2-transduced fibroblasts: Effects of IL-2 dose on anti-tumor immunity. *Hum. Gene Ther.* **6,** 591–601.

Fakhrai, H., Dorigo, O., Shawler, D. L., Lin, H., Mercola, D., Black, K. L., Royston, I., and Sobol, R. E. (1996). Eradication of established intracranial rat gliomas by transforming growth factor β antisense gene therapy. *Proc. Natl. Acad. Sci. U.S.A.* **93,** 2909–2914.

Fearon, E. R., Pardoll, D. M., Itaya, T., Golumbek, P., Levitsky, H. I., Simons, J. W., Karasu-yama, H., Vogelstein, B., and Frost, P. (1990). Interleukin-2 production by tumor cells bypasses T helper function in the generation of an anti-tumor response. *Cell (Cambridge, Mass.)* **60,** 387–403.

Finn, O. J. (1993). Tumor-rejection antigens recognized by T lymphocytes. *Curr. Opin. Immunol.* **5,** 701–708.

Finn, O. J., Jerome, K. R., Henderson, R. A., Pecher, G., Domenech, N., Magarian-Blander, J., and Barratt-Boyes, S. M. (1995). MUC-1 epithelial tumor mucin-based immunity and cancer vaccines. *Immunol. Rev.* **145,** 61–89.

Fossum, B., Gedde-Dahl, T., 3rd, Breivik, J., Eriksen, J. A., Spurkland, A., Thorsby, E., and Gaudernack, G. (1994). p21-ras-peptide-specific T-cell responses in a patient with colorectal cancer. CD4+ and CD8+ T cells recognize a peptide corresponding to a common mutation (13Gly → Asp). *Int. J. Cancer* **56,** 40–45.

Freeman, G. J., Freedman, A. S., Segil, J. M., Lee, G., Whitman, J. F., and Nadler, L. M. (1989). B7, a new member of the Ig superfamily with unique expression on activated and neoplastic B cells. *J. Immunol.* **143,** 2714–2722.

Gabrilove, J. L., and Jakubowski, A. (1990). Hematopoietic growth factors: Biology and clinical application. *Monogr. J. Natl. Cancer Inst.* **10,** 73–77.

Gandolfi, L., Solmi, L., Pizza, G. C., Bertoni, F., Muratori, R., DeVinci, C., Bacchini, P., Morelli, M. C., and Corrado, G. (1989). Intratumoral echo-guided injection of interleukin-

2 and cytokine-activated killer cells in hepatocellular carcinoma. *Hepato-Gastroenterology* **36**, 352–356.

Gansbacher, B., Zier, K., Daniels, B., Cronin, K., Bannerji, R., and Gilboa, E. (1990). Interleukin-2 gene transfer into tumor cells abrogates tumorigenicity and induces protective immunity. *J. Exp. Med.* **172**, 1217–1223.

Gaugler, B., Van den Eynde, B., van der Bruggen, P., Romero, P., Gaforio, J. J., De Plaen, E., Lethe, B., Brasseur, F., and Boon, T. (1994). Human gene MAGE-3 codes for an antigen recognized on a melanoma by autologous cytolytic T lymphocytes. *J. Exp. Med.* **179**, 921–930.

Gross, G., and Eshhar, Z. (1992). Endowing T cells with antibody specificity using chimeric T cell receptors. *FASEB J.* **6**, 3370–3378.

Herlyn, D., Linnenbach, A., Koprowski, H., and Herlyn, M. (1991). Epitope- and antigen-specific cancer vaccines. *Int. Rev. Immunol.* **7**, 245–257.

Herlyn, D., Harris, D., Zaloudik, J., Sperlagh, M., Maruyama, H., Jacob, L., Kieny, M.-P., Scheck, S., Somasundaram, R., Hart, E., Ertl, H., and Mastrangelo, M. (1994). Immunomodulatory activity of monoclonal anti-idiotypic antibody to anti-colorectal carcinoma antibody CO17-1A in animals and patients. *J. Immunother.* **15**, 303–311.

Hodge, J. W., McLaughlin, J. P., Abrams, S. I., Shupert, W. L., Schlom, J., and Kantor, J. A. (1995). Admixture of a recombinant vaccinia virus containing the gene for the costimulatory molecule B7 and a recombinant vaccinia virus containing a tumor-associated antigen gene results in enhanced specific T-cell responses and antitumor immunity. *Cancer Res.* **55**, 3598–3603.

Hollingsworth, S., Gäken, J., Darling, D., Hirst, W., Kuiper, M., Buggins, A., Barnard, A., Peakman, M., Humphries, S., Mufti, G. J., and Farzaneh, F. (1995). Induction of tumour rejection by combination B7.1/IL-2 expressing tumor cells. *Cancer Gene Ther.* **2**, 240.

Hwu, P., Yang, J. C., Cowherd, R., Treisman, J., Shafer, G. E., Eshhar, Z., and Rosenberg, S. A. (1995). *In vivo* antitumor activity of T cells redirected with chimeric antibody/T-cell receptor genes. *Cancer Res.* **55**, 3369–3373.

Jerome, K. R., Barnd, D. L., Bendt, K. M., Boyer, C. M., Taylor-Papadimitriou, J., McKenzie, I. F., Bast, R. C., Jr., and Finn, O. J. (1991). Cytotoxic T-lymphocytes derived from patients with breast adenocarcinoma recognize an epitope present on the protein core of a mucin molecule preferentially expressed by malignant cells. *Cancer Res.* **51**, 2908–2916.

Jubelirer, S. J., Talkington, A., and Bailey, D. (1991). Hepatocellular carcinoma: A review of 30 years of experience. *W. Va. Med. J.* **87**, 400–402.

June, C. H., Ledbetter, J. A., Linsley, P. S., and Thompson, C. B. (1990). Role of the CD28 receptor in T-cell activation. *Immunol. Today* **11**, 211–216.

Kawakami, Y., Eliyahu, S., Delgado, C. H., Robbins, P. F., Rivoltini, L., Topalian, S. L., Miki, T., and Rosenberg, S. A. (1994). Cloning of the gene coding for a shared human melanoma antigen recognized by autologous T cells infiltrating into tumor. *Proc. Natl. Acad. Sci. U.S.A.* **91**, 3515–3519.

Kelso, A. (1989). Cytokines: Structure function and synthesis. *Curr. Opin. Immunol.* **2**, 215–225.

Khanna, R., Burrows, S. R., Argaet, V., and Moss, D. J. (1994). Endoplasmic reticulum signal sequence facilitated transport of peptide epitopes restores immunogenicity of an antigen processing defective tumour cell line. *Int. Immunol.* **6**, 639–645.

Khanna, R., Burrows, S. R., and Moss, D. J. (1995). Immune regulation in Epstein–Barr virus-associated diseases. *Microbiol. Rev.* **59**, 387–405.

Kim, J. A., Martin, E. W., Jr., E. W., Morgan, C. J., Aldrich, W., and Triozzi, P. L. (1995). Expansion of mucin-reactive T-helper lymphocytes from patients with colorectal cancer. *Cancer Biother.* **10**, 115–123.

Kim, T. S., and Cohen, E. P. (1994). Interleukin-2-secreting mouse fibroblasts transfected with genomic DNA from murine melanoma cells prolong the survival of mice with melanoma. *Cancer Res.* **54**, 2531–2535.

Lotze, M. T., Chang, A. E., Seipp, C. A., Simpson, C., Vetto, J. J., and Rosenberg, S. A. (1986). High-dose recombinant interleukin 2 in the treatment of patients with disseminated cancer: Responses, treatment-related morbidity and histologic findings. *JAMA, J. Am. Med. Assoc.* **256**, 3117–3124.

Mackiewicz, A., Wiznerowicz, M., Roeb, E., Nowak, J., Pawlowski, T., Baumann, H., Heinrich, P. C., and Rose-John, S. (1995). Interleukin-6-type cytokines and their receptors for gene therapy of melanoma. *Ann. N.Y. Acad. Sci.* **762**, 361–373.

Marincola, F. M., Ettinghausen, S., Cohen, P. A., Cheshire, L. B., Restifo, N. P., Mulé, J. J., and Rosenberg, S. A. (1994). Treatment of established lung metastases with tumor-infiltrating lymphocytes derived from a poorly immunogenic tumor engineered to secrete human TNF-alpha. *J. Immunol.* **152**, 3501–3513.

McBride, W. H., Dougherty, G. D., Wallis, A. E., Economou, J. S., and Chiang, C. S. (1994). Interleukin-3 in gene therapy of cancer. *Folia Biol. Praha.* **40**, 62–73.

Mondino, A., and Jenkins, M. K. (1994). Surface proteins involved in T cell costimulation. *J. Leukocyte Biol.* **55**, 805–815.

Morioka, N., Kikumoto, Y., Hoon, D. S., Morton, D. L., and Irie, R. F. (1994). A decapeptide (Gln-Asp-Leu-Thr-Met-Lys-Tyr-Gln-Ile-Phe) from human melanoma is recognized by CTL in melanoma patients. *J. Immunol.* **153**, 5650–5658.

Myers, R. B., Schlom, J., Srivastava, S., and Grizzle, W. E. (1995). Expression of tumor-associated glycoprotein 72 in prostatic intraepithelial neoplasia and prostatic adenocarcinoma. *Mod. Pathol.* **8**, 260–265.

Nabel, G. J., Nabel, E. G., Yang, Z.-Y., Fox, B. A., Plautz, G. E., Gao, X., Huang, L., Shu, S., Gordon, D., and Chang, A. E. (1993). Direct gene transfer with DNA–liposome complexes in melanoma: Expression, biologic activity, and lack of toxicity in humans. *Proc. Natl. Acad. Sci. U.S.A.* **90**, 11307–11311.

Nakao, M., Yamana, H., Imai, Y., Toh, Y., Toh, U., Kimura, A., Yanoma, S., Kakegawa, T., and Itoh, K. (1995). HLA-A2601 restricted CTLs recognize a peptide antigen expressed on squamous cell carcinoma. *Cancer Res.* **55**, 4248–4252.

Pandolfi, F., Boyle, L. A., Trentin, L., Kurnick, J. T., Isselbacher, K. J., and Gattoni-Celli, S. (1991). Expression of HLA-A2 antigen in human melanoma cell lines and its role in T-cell recognition. *Cancer Res.* **51**, 3164–3170.

Pattengale, P. K., and Leder, P. (1989). Murine interleukin-4 displays potent anti-tumor activity *in vivo*. *Cell (Cambridge, Mass.)* **57**, 503–512.

Peace, D. J., Chen, W., Nelson, H., and Cheever, M. A. (1991). T cell recognition of transforming proteins encoded by mutated *ras* proto-oncogenes. *J. Immunol.* **146**, 2059–2065.

Resnicoff, M., Sell, C., Rubini, M., Coppola, D., Ambrose, D., Baserga, R., and Rubin, R. (1994). Rat glioblastoma cells expressing an antisense RNA to the insulin-like growth factor-1 (IGF-1) receptor are nontumorigenic and induce regression of wild-type tumors. *Cancer Res.* **54**, 2218–2222.

Ressing, M. E., Sette, A., Brandt, R. M. P., Ruppert, J., Wentworth, P. A., Hartman, M., Oseroff, C., Grey, H. M., Melief, C. J. M., and Kast, W. M. (1995). Human CTL epitopes encoded by human papillomavirus type 16 E6 and E7 identified through *in vivo* and *in vitro* immunogenicity studies of HLA-A*0201-binding peptides. *J. Immunol.* **154**, 5934–5943.

Rimoldi, D., Romero, P., and Carrel, S. (1993). The human melanoma antigen encoding gene, MAGE-1, is expressed by other tumour cells of neuroectodermal origin such as glioblastomas and neuroblastomas. *Int. J. Cancer* **54**, 527–528.

Robbins, P. F., el Gamil, M., Kawakami, Y., Stevens, E., Yannelli, J. R., and Rosenberg, S. A. (1994). Recognition of tyrosinase by tumor-infiltrating lymphocytes from a patient responding to immunotherapy. *Cancer Res.* **54**, 3124–3126.

Rosenberg, S. A., Lotze, M. T., and Mulé, J. J. (1988). New approaches to the immunotherapy of cancer. *Ann. Intern. Med.* **108**, 853–864.

Rosenberg, S. A., Aebersold, P., Cornetta, K., Kasid, A., Morgan, R. A., Moen, R., Karson, E. M., Lotze, M. T., Yang, J. C., Topalian, S. L., Merino, M. J., Culver, K., Miller,

A. D., Blaese, R. M., and Anderson, W. F. (1990). Gene transfer into humans—immunotherapy of patients with advanced melanoma, using tumor-infiltrating lymphocytes modified by retroviral gene transduction. *N. Engl. J. Med.* **323**, 570–578.

Russell, S. J., Collins, M. K., and Cohen, E. P. (1992). Immunity to B16 melanoma in mice immunized with IL-2-secreting allogeneic mouse fibroblasts expressing melanoma-associated antigens. *Int. J. Cancer* **51**, 283–289.

Sarna, G., Collins, J., Figlin, R., Robertson, P., Altrock, B., and Abels, R. (1990). A pilot study of intralymphatic interleukin-2. II. Clinical and biological effects. *J. Biol. Response Modifiers* **9**, 81–86.

Sawamura, Y., Diserens, A. C., and de Tribolet, N. (1990). *In vitro* prostaglandin E2 production by glioblastoma cells and its effect on interleukin-2 activation of oncolytic lymphocytes. *J. Neuro-Oncol.* **9**, 125–130.

Schlebusch, H., Wagner, U., Grunn, U., and Schultes, B. (1995). A monoclonal antiidiotypic antibody ACA 125 mimicking the tumor-associated antigen CA 125 for immunotherapy of ovarian cancer. *Hybridoma* **14**, 167–174.

Sensi, M., Traversare, C., Radrizzani, M., Salvi, S., MacCalli, C., Mortarini, R., Rivoltini, L., Farina, C., Nicolini, G., Wolfel, T., Brichard, V., Boon, T., Bordignon, C., Anichini, A., and Parmiani, G. (1995). Cytotoxic T-lymphocyte clones from different patients display limited T-cell-receptor variable-region gene usage in HLA-A2-restricted recognition of the melanoma antigen Melan-A/MART-1. *Proc. Natl. Acad. Sci. U.S.A.* **92**, 5674–5678.

Shawler, D. L., Dorigo, O., Gjerset, R., Royston, I., Sobol, R. E., and Fakhrai, H. (1995). Comparison of gene therapy with interleukin-2 (IL-2) gene modified fibroblasts and tumor cells in the murine CT-26 model of colorectal carcinoma. *J. Immunother.* **17**, 201–208.

Slingluff, C. L., Jr., Cox, A. L., Henderson, R. A., Hunt, D. F., and Engelhard, V. H. (1993). Recognition of human melanoma cells by HLA-A2.1 restricted cytotoxic T lymphocytes is mediated by at least six shared peptide epitopes. *J. Immunol.* **150**, 2955–2963.

Slingluff, C. L., Jr., Cox, A. L., Stover, J. M., Jr., Moore, M. M., Hunt, D. F., and Engelhard, V. H. (1994). Cytotoxic T lymphocyte response to autologous human squamous cell cancer of the lung: Epitope reconstitution with peptides extracted from HLA-Aw68. *Cancer Res.* **54**, 2731–2737.

Sobol, R. E., Fakhrai, H., Shawler, D. L., Gjerset, R., Dorigo, O., Carson, C., Khaleghi, T., Koziol, J., Shiftan, T. A., and Royston, I. (1995). Interleukin-2 gene therapy in a patient with glioblastoma. *Gene Ther.* **2**, 164–167.

Stancovski, I., Schindler, D. G., Waks, T., Yarden, Y., Sela, M., and Eshhar, Z. (1993). Targeting of T lymphocytes to Neu/HER2-expressing cells using chimeric single chain Fv receptors. *J. Immunol.* **151**, 6577–6582.

Stoppacciaro, A., Melani, C., Parenza, M., Mastracchio, A., Bassi, C., Baroni, C., Parmiani, G., and Colombo, M. P. (1993). Regression of an established tumor genetically modified to release granulocyte colony-stimulating factor requires granulocyte-T cell cooperation and T cell-produced interferon gamma. *J. Exp. Med.* **178**, 151–161.

Sulitzeanu, D. (1993). Immunosuppressive factors in human cancer. *Adv. Cancer Res.* **60**, 247–267.

Tahara, H., Zeh, H. J., 3rd, Storkus, W. J., Pappo, I., Watkins, S. C., Gubler, U., Wolf, S. F., Robbins, P. D., and Lotze, M. T. (1994). Fibroblasts genetically engineered to secrete interleukin 12 can suppress tumor growth and induce antitumor immunity to a murine melanoma *in vivo*. *Cancer Res.* **54**, 182–189.

Theobald, M., Biggs, J., Dittmer, D., Levine, A. J., and Sherman, L. A. (1995). Targeting p53 as a general tumor antigen. *Proc. Natl. Acad. Sci. U.S.A.* **92**, 11993–11997.

Topalian, S. L., Rivoltini, L., Mancini, M., Markus, N. R., Robbins, P. F., Kawakami, Y., and Rosenberg, S. A. (1994). Human CD4+ T cells specifically recognize a shared melanoma-associated antigen encoded by the tyrosinase gene. *Proc. Natl. Acad. Sci. U.S.A.* **91**, 9461–9465.

Townsend, S. E., and Allison, J. P. (1993). Tumor rejection after direct costimulation of CD8+ T cells by B7-transfected melanoma cells. *Science* **259**, 368–370.

Trojan, J., Johnson, T. R., Rudin, S. D., Ilan, J., and Tykocinski, M. L. (1993). Treatment and prevention of rat glioblastoma by immunogenic C6 cells expressing antisense insulin-like growth factor I RNA. *Science* **259**, 94–97.

Ulmer, J. B., Donnelly, J. J., Parker, S. E., Rhodes, G. H., Felgner, P. L., Dwarki, V. J., Gromkowski, S. H., Deck, R. R., DeWitt, C. M., Friedman, A., Hawe, L. A., Leander, K. R., Martinez, D., Perry, H. C., Shiver, J. W., Montgomery, D. L., and Liu, M. A. (1993). Heterologous protection against influenza by injection of DNA encoding a viral protein. *Science* **259**, 1745–1749.

van der Bruggen, P., Traversare, C., Chmez, P., Lurquin, C., De Plaen, E., van den Eyde, B., Knuth, A., and Boon, T. (1991). A gene encoding an antigen recognized by cytotoxic T lymphocytes on a human melanoma. *Science* **254**, 1643–1645.

Vieweg, J., Boczkowski, D., Roberson, K. M., Edwards, D. W., Philip, M., Philip, R., Rudoll, T., Smith, C., Robertson, C., and Gilboa, E. (1995). Efficient gene transfer with adeno-associated virus-based plasmids complexed to cationic liposomes for gene therapy of human prostate cancer. *Cancer Res.* **55**, 2366–2372.

Wahl, W. L., Strome, S. E., Nabel, G. J., Plautz, G. E., Cameron, M. J., San, H., Fox, B. A., Shu, S., and Chang, A. E. (1995). Generation of therapeutic T-lymphocytes after *in vivo* tumor transfection with an allogeneic class I major histocompatibility complex gene. *J. Immunother.* **17**, 1–11.

Watanabe, Y., Kuribayashi, K., Miyatake, S., Nishihara, K., Nakayama, E. L., Taniyama, T., and Sakata, T. A. (1989). Exogenous expression of mouse interferon gamma cDNA in mouse neuroblastoma C1300 cells results in reduced tumorigenicity by augmented antitumor immunity. *Proc. Natl. Acad. Sci. U.S.A.* **86**, 9456–9460.

Yoshino, I., Goedegebuure, P. S., Peoples, G. E., Parikh, A. S., DiMaio, J. M., Lyerly, H. K., Gazdar, A. F., and Eberlein, T. J. (1994). HER-2/neu derived peptides are shared antigens among human non-small cell lung cancer and ovarian cancer. *Cancer Res.* **54**, 3387–3390.

John H. White

Department of Physiology
McGill University
Montreal, Quebec, Canada H3G 1Y6

Modified Steroid Receptors and Steroid-Inducible Promoters as Genetic Switches for Gene Therapy

I. Overview

Gene expression programs of cells in both the developing organism and the adult are modified in response to external signals. These signals stimulate intracellular signal transduction pathways that modulate the expression of specific genes. The superfamily of nuclear receptors represents the primary response to a wide range of extracellular signals, and, unlike membrane-bound receptors, is directly implicated in the transcriptional control of genes whose products are active in virtually all aspects of physiology and metabolism. The mechanisms of action of nuclear receptors make them attractive targets for use in engineered gene expression systems with applications to gene therapy.

The purpose of this chapter is two-fold. A background to the mechanisms of signal transduction by nuclear receptors is provided, with the

Advances in Pharmacology, Volume 40

emphasis placed on the action of steroid receptors. This is then used as a basis for discussing the potential utility of modified steroid receptors in controlling expression of transgenes in gene therapy protocols. Before detailing the mechanisms of action of nuclear receptors, a general introduction to regulation of gene transcription is provided.

II. Structure of Eukaryotic Promoters and Regulation of Transcription

A. The TATA Box and Transcription Initiation

The DNA sequences that control the transcription of a given protein-coding gene are known collectively as the promoter, or promoter sequences. While promoters are variable in sequence and organization, a number of common elements have been characterized (Fig. 1). Most promoters contain a sequence element with the consensus TATAAAA, known for obvious reasons as the TATA box. In mammalian cells, TATA boxes are located 25–30 bp upstream of the transcription start site (also known as the cap site). The TATA box represents the focus of regulatory events controlling initiation of transcription, as it is the site where the components of the multicomponent transcription preinitiation complex are assembled. The preinitiation complex, which is composed of upward of 50 polypeptides, is assembled in an ordered fashion, commencing with the binding of the factor TFIID to the TATA box. TFIID is composed of several subunits of which one, TATA-binding protein (TBP), binds to the TATA box. The binding of TFIID is followed by the assembly of the other components of the preinitiation complex, including RNA polymerase (itself a multisubunit protein). For a detailed review of preinitiation complex components and assembly the reader is referred to Zawel and Reinberg (1995).

B. Regulation of Transcription

I. TATA-Binding Protein-Associated Factors

The other components of TFIID aside from TBP are known as TAFs (TBP-associated factors). TAFs have been studied intensively, and with good

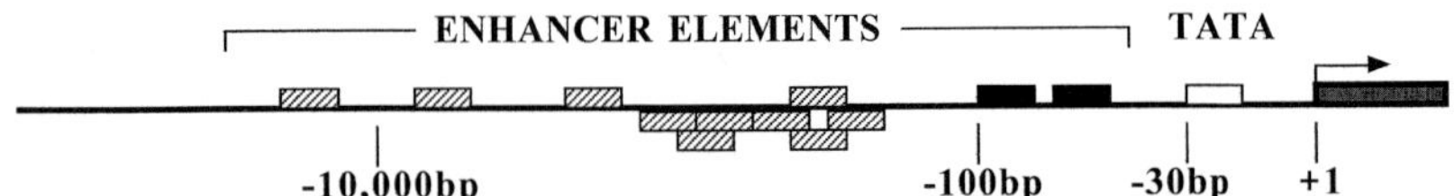

FIGURE I   Schematic representation of the structure of an RNA polymerase II promoter. The transcription start site (cap site) is represented by the bent arrow. The TATA box is represented by the white box, and upstream elements are represented by black boxes. Enhancer sequences, represented by striped boxes, can be spread out over long distances (above the line), or tightly clustered (below the line).

reason. TAFs are essential for the functional liaison between the preinitiation complex and the several families of regulatory proteins that bind to specific DNA sequence motifs, usually located upstream of the TATA box. *In vitro* studies with reconstituted components of the preinitiation complex have shown that TAFs are not required for initiation per se but are required for stimulation of initiation by regulatory proteins (Dynlacht *et al.*, 1991; Goodrich and Tjian, 1994).

2. Enhancers and Enhancer Factors

The sequence motifs recognized by proteins that regulate transcriptional initiation are known as enhancers (Fig. 1), and the proteins that bind to them are variously termed enhancer factors, transacting transcription factors, or transactivators. Biochemical studies have suggested that enhancer factors act by increasing the number of preinitiation complexes formed rather than by increasing the rate of their formation (White *et al.*, 1992; Choy and Green, 1993). These factors act, at least in part, by recruiting components of the preinitiation complex to the site of initiation or by stabilizing interactions of components with the growing preinitation complex.

There are no strict rules governing the location and arrangement of enhancers in a given promoter. Enhancers can be tightly clustered, as in the simian virus 40 (SV40) early promoter or the *fos* gene, or they can be spread over large distances as in globin promoters, or the steroid-regulated tyrosine aminotransferase gene (Jantzen *et al.*, 1987). Generally speaking, however, promoters will contain one or more proximal promoter elements recognized by regulatory factors, which are located within 50–100 bp upstream of the TATA box, in addition to more distal enhancer elements (Fig. 1).

A number of different classes of enhancer factors have been identified to date, many of which have been found to share similar types of functional organization. Generally, they are composed of separate domains responsible for site-specific DNA binding, and for activation of transcription (transactivating domains) (Keegan *et al.*, 1986). Several different types of DNA-binding domains have been characterized to date. For example, all nuclear receptors share a common DNA-binding domain composed of tandem motifs organized around coordinated zinc ions, known as zinc fingers (see below). Several different types of transactivating domains have also been described. The herpes simplex virus activator VP16, as well as a number of yeast transcription factors, have domains that are particularly rich in acidic amino acids. Other domains, for example that of the mammalian transcription factor Sp1, have domains that are rich in glutamine and proline residues (Pascal and Tjian, 1991).

3. Regulating the Regulators

Transcription of a given gene may be controlled by the activity of one or more different types of enhancer factors. The action of enhancer factors is under tight control. Many enhancer factors are regulated by phosphorylation by specific kinases, which can control any one of a number of aspects of enhancer factor function (Hunter and Karin, 1992). Phosphorylation controls the subcellular localization, and hence activity, of NF-KB, and members of the signal transducer and activator of transcription (STAT) family of enhancer factors (Montminy, 1993), the DNA-binding affinity of c-Jun (Lin *et al.*, 1992), and the activity of the transactivating domain of the cyclic AMP-response element binding protein CREB (Gonzalez and Montminy, 1989). The kinases that control these events are themselves responsive to specific intracellular phosphorylation cascades, which in turn are controlled by extracellular signals regulating the function of a given cell. Thus, the gene expression pattern of a given cell is responsive to the regulatory signals impinging on it.

III. Nuclear Receptors and Signal Transduction

Extracellular signals propagated by peptide hormones, growth factors, and neurotransmitters are transmitted to target cells through binding to specific membrane receptors. Ligand binding to a given receptor stimulates intracellular signal transduction cascades, modulating the activity of myriad protein kinases or protein phosphatases. These phosphorylation (or dephosphorylation) events can affect the function not only of transcription factors, but also of membrane proteins, including receptors, ion channels and transporters, and components of the cytoplasm such as metabolic enzymes and translation factors.

In contrast to the complexity of signal transduction stimulated by membrane receptors, signaling by nuclear receptors is focused in the nucleus. The nuclear receptors are a family of enhancer factors that regulate the expression of genes controlling a wide range of processes such as reproduction, differentiation, development, homeostasis, and oncogenesis. They are activated by binding small lipophilic molecules such as steroid and thyroid hormones, retinoids, vitamin D_3, and specific prostaglandins (Green and Chambon, 1988; Beato, 1989; Wahli and Martinez, 1991; Gronemeyer, 1991; Mangelsdorf and Evans, 1995; Beato *et al.*, 1995; Mangelsdorf *et al.*, 1995; Kliewer *et al.*, 1995; Forman *et al.*, 1995). The nuclear receptor family also includes a large number of so-called orphan receptors with unidentified ligands. While some of these may have specific ligands, it is also possible that some orphan receptors may be modulated exclusively by phosphorylation events. Nuclear receptor cDNAs have been isolated from a wide variety of organisms from humans to *Drosophila* and *Caenorhabditis elegans* (Beato, 1989; Green and Chambon, 1988; Wahli and Martinez, 1991; Gronemeyer, 1991; Mangelsdorf *et al.*, 1995).

IV. Mechanisms of Action of Nuclear Receptors ___________

A. The Steroid Receptor Family

Steroid receptors can be grouped into two subfamilies on the basis of their ligand- and DNA-binding specificities. The glucocorticoid receptor (GR) subfamily is composed of the GR and receptors for progesterone, androgens, and mineralocorticoids. These receptors are well conserved, and recognize ligand with a similar core structure (Fig. 2). They also recognize similar palindromic hormone response elements composed of half-sites with the consensus AGAACA (Beato, 1989; Green and Chambon, 1988; Wahli and Martinez, 1991; Gronemeyer, 1991). The estrogen receptor (ER) is distinct from the GR subfamily. Its cognate ligand, 17β-estradiol, differs from those of the GR subfamily by virtue of its aromatic A ring (Fig. 2). Moreover, the ER recognizes distinct response elements, composed of PuGGTCA half-sites (Fig. 2; and see below). In many respects, however, the molecular properties of the ER and the GR subfamilies are similar. The following sections describe in detail the mechanisms of steroid receptors, concentrating on the ER as model.

B. The Estrogen Receptor

I. Physiological and Pathophysiological Action of the Estrogen Receptor

The estrogen receptor is the primary intracellular target of estrogen, which plays a central role in the development, regulation, and function of female vertebrate reproductive tissue (George and Wilson, 1988). Normal

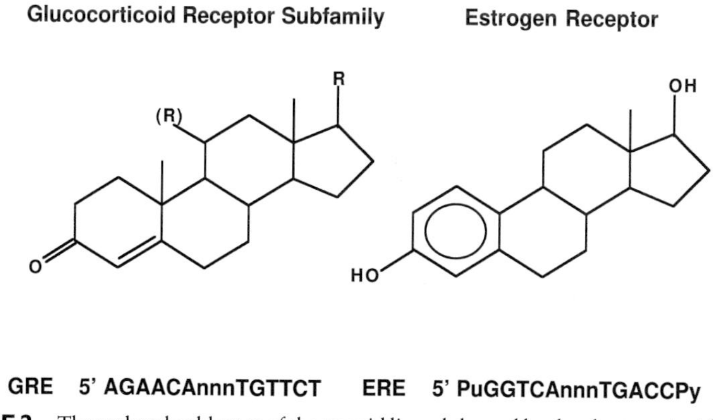

FIGURE 2   The carbon backbones of the steroid ligands bound by the glucocorticoid receptor subfamily (left) and the estrogen receptor (17β-estradiol; right) are shown. The top strand of a perfectly palindromic glucocorticoid response element (GRE), and estrogen response element (ERE), are shown below.

postnatal development in females has been linked to the action of estrogen at several sites including the breast, reproductive tract, and neuroendocrine tissues. The importance of the ER in normal physiology of vertebrate reproductive tissue was eloquently demonstrated with the production of transgenic mice lacking a functional ER gene (Lubahn *et al.*, 1993; Korach, 1994). Females are sterile and display a number of phenotypic changes associated with breast and reproductive tract function. Strikingly, males of this strain are also infertile, and are defective in normal testicular function, including sperm production. An estrogen insensitivity syndrome has been described in humans with the molecular genetic analysis of a 28-year-old, fully masculinized male lacking a functional ER (Lubahn *et al.*, 1993; Smith *et al.*, 1994). It is not clear whether the patient is fertile, and therefore the relevance to human physiology of the observations of sterility among male ER null mice is unclear at this time.

Significantly, however, the major phenotype of the ER-negative patient is a lack of closure of the epiphyses of long bones and abnormally low bone density, similar to the condition in ER null mice (Lubahn *et al.*, 1993; Smith *et al.*, 1994; Korach, 1994). These observations correlate well with loss of bone density, or osteoporosis, in postmenopausal women. Osteoporosis is associated with a decline in estrogen production after menopause, and several studies have indicated that estrogen replacement is an effective therapy (Hutchinson *et al.*, 1979; Weiss *et al.*, 1980; Kreiger *et al.*, 1982; Munk-Jenson *et al.*, 1988; Christiansen, 1991; Stevenson *et al.*, 1990). However, hormonal replacement therapy is not recommended for women at risk for endometrial or breast cancer. Estrogen and the ER can play central roles in controlling proliferation of breast and endometrial carcinomas. Indeed, the nonsteroidal antiestrogen tamoxifen is the most widely used endocrine treatment for all stages of breast cancer in pre- and postmenopausal women. Moreover, its effectiveness as a preventative agent is currently being tested in clinical trials (Jordan, 1993; Powles *et al.*, 1994). However, tamoxifen is a partial estrogen agonist in endometrial tissue, and has been shown to mimic the action of estrogen in bone (Freidl and Jordan, 1994; Vanleeuwen *et al.*, 1994).

Given the central role in estrogen signaling in these diseases of major clinical importance, the task for the future will be to design ER ligands that will act in a tissue-specific manner, for example, compounds blocking the action of estrogen in breast without the proliferative effects of estrogen on endometrial tissue. To this end, it is essential to define fully the molecular basis of ligand to the ER, in order to understand the mechanisms of action of receptor ligands, and to design more effective antiestrogenic compounds.

2. Structure of the Estrogen Receptor

The primary structure of the estrogen receptor, like those of all nuclear receptors, is organized into a series of conserved domains (A–F; Fig. 3)

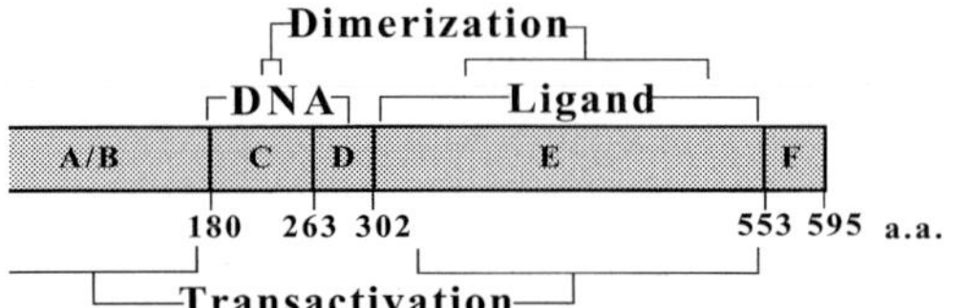

FIGURE 3   Schematic representation of the primary structure of the human estrogen receptor is shown. Domains of the receptor required for DNA and ligand binding, activation of transcription (transactivation), and dimerization are indicated.

(Krust *et al.*, 1986). The DNA-binding domain is composed of the highly conserved 66- to 68-amino acid region C, containing two zinc fingers that fold to create a single structural domain (Evans and Hollenberg, 1988; Hard *et al.*, 1990; Schwabe *et al.*, 1990), followed by a series of basic amino acids extending into region D. The C regions of the chicken and human ERs are 100% conserved (Krust *et al.*, 1986). The ligand-binding domains (region E) are less highly conserved among the receptors. In addition to a ligand-binding site, region E also contains a ligand-dependent transcriptional activation domain (Beato, 1989; Green and Chambon, 1988; Webster *et al.*, 1988; Fawell *et al.*, 1990). The least well-conserved domain, the N-terminal A/B region, varies widely among the receptors. The A/B region of numerous nuclear receptors, including the ER, contains transactivating domains that cooperate with transcriptional activating domains located in the hormone-binding domain (Kumar *et al.*, 1987; Godowski *et al.*, 1988; Hollenberg and Evans, 1988; Tora *et al.*, 1988, 1989a; Bocquel *et al.*, 1989). The A/B region of the human ER is 180 amino acids long, and acts in a cell-specific manner. Whereas truncated ER mutants containing an A/B region but lacking a functional ligand-binding domain are virtually inactive in HeLa cells, the A/B region acts independently as a transactivator in chicken embryo fibroblasts (CEFs) and, interestingly, in yeast (Webster *et al.*, 1988; Tora *et al.*, 1989a). This provides an explanation for the cell-specific action of tamoxifen, which acts as an estrogen antagonist in HeLa cells, but is an efficient agonist in CEFs and yeast (Berry *et al.*, 1990).

Generally, nuclear receptors stimulate transcription by binding directly to specific cognate DNA sequences (see below). One notable exception is the activation of transcription of the chicken ovalbumin gene, where the ER apparently can act by associating with DNA-bound transcription factor AP-1 (Gaub *et al.*, 1990). In addition, several examples have been reported of receptors acting as transcriptional repressors by blocking the action of c-Fos and c-Jun (Jonat *et al.*, 1990; Nicholson *et al.*, 1990; Schüle *et al.*, 1990; Yang-Yen *et al.*, 1990; Love *et al.*, 1992).

3. DNA Binding and Transactivation

Nuclear receptors activate transcription through binding to specific DNA sequences known collectively as hormone response elements (HREs), which have been generally defined as being composed of two 5- or 6-bp sequence motifs arranged in a palindrome or a direct repeat. Three amino acids adjacent to the N-terminal zinc finger of the DNA-binding domain, known as the P box, are critical for DNA sequence recognition (Mader *et al.*, 1989; Danielson *et al.*, 1989; Umesomo and Evans, 1989). The estrogen receptor (ER), which recognizes palindromic response elements composed of PuGGTCA half-sites, has a P box containing glutamate, glycine, and alanine (Fig. 4; and see Mader *et al.*, 1989, and references therein), whereas members of the glucocorticoid receptor family all contain P boxes composed of glycine, serine, and valine, and recognize response elements with AGAACA half-sites.

While examples of consensus response elements for steroid receptors have been described (Walker *et al.*, 1984; Klein-Hitpass *et al.*, 1986, 1988; Martinez *et al.*, 1987; Burch *et al.*, 1988), most HREs contain one or more nonconsensus nucleotides. Nonconsensus HREs are often found in multiple arrays, usually upstream of the site of transcriptional initiation, and the individual elements act synergistically to augment the response to hormone (Payvar *et al.*, 1983; Van het Schip *et al.*, 1986; Jantzen *et al.*, 1987; Burch *et al.*, 1988; Kaplan *et al.*, 1988; Martinez and Wahli, 1989; Berry *et al.*, 1989; Weisz and Rosales, 1990; Richard and Zingg, 1990; Slater *et al.*, 1990; Liu and Teng, 1991, 1992; Chang *et al.*, 1992; Kato *et al.*, 1992). The degree of synergism between paired estrogen response elements (EREs) is dependent on their sequence, the spacing between them, and their distance from the TATA box of the promoter (Ponglikitmongkol *et al.*, 1990).

Although the response elements described above are generally defined as being palindromic, work has suggested that the repertoire of sequences acting as steroid response elements is not limited to palindromes. Studies of the far-upstream ERE of the chicken ovalbumin gene (Kato *et al.*, 1992) and molecular genetic studies in yeast (Nawaz *et al.*, 1992; Dana *et al.*, 1994) have suggested that direct repeats of cognate half-sites are capable of acting as response elements. Moreover, glucocorticoid response elements (GREs) identified in the promoter of the widely studied mouse mammary tumor virus (MMTV) long terminal repeat, and in Epstein–Barr virus, are more direct repeats than palindromes (Payvar *et al.*, 1983; Tur-Kaspa *et al.*, 1988). We have completed DNA-binding and transactivation studies defining the function of direct repeats as response elements for steroid hormone receptors (Aumais *et al.*, 1996; and see below). These experiments suggested that steroid receptors can bind with low affinity to direct repeats. Unlike binding to palindromic elements, recognition of direct repeats is not strictly dependent on the spacing between each half-site. Our studies have suggested that direct repeats function as response elements only in multiple arrays, and thus behave similarly to weak nonconsensus palindromic elements (Aumais *et al.*, 1996).

4. Dimerization and DNA Binding

Several studies have shown that nuclear receptors generally bind to their response elements as dimers (Forman and Samuels, 1990; however, see Wilson *et al.*, 1992). Gel retardation experiments performed with full-length and truncated derivatives of the human ER have demonstrated cooperative DNA binding by receptor dimers (Kumar and Chambon, 1998). The hormone-binding domains of the human and mouse ERs contain strong dimerization domains (Kumar and Chambon, 1998; Fawell *et al.*, 1990), composed of a series of heptad repeats conserved among nuclear receptors (Figs. 3 and 4). Different truncated mutants of the human ER lacking all or most of the A/B domain and the hormone-binding domain (E/F region) bind cooperatively to palindromic elements as homodimers and as heterodimers (Kumar and Chambon, 1988; Mader *et al.*, 1993a), indicating the presence of a second dimerization domain in the ER DNA-binding domain (Fig. 4). However, kinetic experiments have shown that the ligand-binding domain greatly stabilizes receptor binding (our unpublished results), emphasizing its importance in stable dimerization. Initially, it was thought that ER dimerization was ligand dependent (Metzger *et al.*, 1988; Kumar and Chambon, 1988). However, later work showed that the original cDNA isolate contained a point mutation, G400V, which destabilizes the structure of the ligand-binding domain. Estrogen receptors containing Gly-400 form stable homodimers *in vitro* in the absence of ligand (Tora *et al.*, 1989b; Salomonsson *et al.*, 1994a), indicating the hormone is not necessary to stabilize dimerization.

Dimerization by the isolated ER DNA-binding domain on palindromes is controlled by a 5-amino acid region that is adjacent to the C-terminal

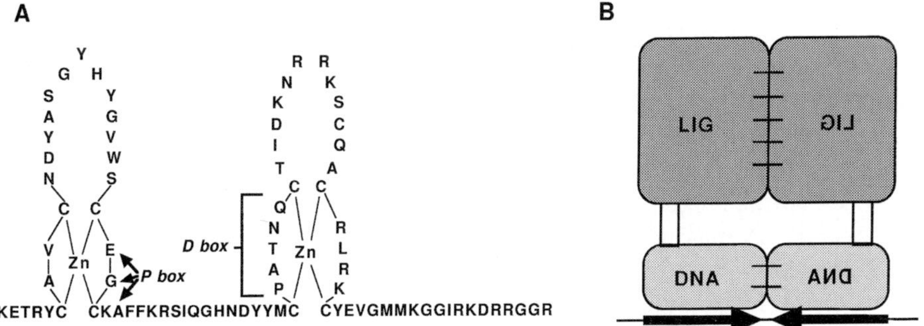

FIGURE 4   DNA binding by the estrogen receptor. (A) Schematic representation of the two zinc fingers of the estrogen receptor. The amino acids of the P box and D box, required for site-specific DNA binding and DNA-binding domain dimerization, respectively, are indicated. (B) Schematic representation of an estrogen receptor homodimer bound to an ERE palindrome is indicated. For clarity the N-terminal A/B domain of the receptor is not shown.

zinc finger (Mader *et al.*, 1993a), the so-called D box (Fig. 4). Structural studies have suggested that dimerization of the ER closely juxtaposes D-box amino acids of adjacent DNA-binding domains (Hard *et al.*, 1990; Schwabe *et al.*, 1990; Luisi *et al.*, 1991). Although the D box stabilizes DNA binding to palindromic response elements, and likely controls the specificity of receptors for palindromes with 3-bp inter-half-site spacing (Umesomo and Evans, 1989; Dahlman-Wright *et al.*, 1991; Aumais *et al.*, 1996), our studies indicate that the D box does not function in recognition of direct repeats by steroid receptors (Aumais *et al.*, 1996). We have shown that the dimerization interface in the ligand-binding domain is essential for this binding. In contrast to palindromes, no evidence for cooperative binding of isolated DNA-binding domains to direct repeats was observed.

5. Ligand Binding and Receptor Activation

a. The Estrogen Receptor in the Absence of Ligand   While dimerization, DNA binding, and transcriptional activation by nuclear receptors have received much attention, relatively little is known about ligand binding. In the absence of ligand, steroid receptors are associated with a number of accessory proteins, including three members of the family of heat shock proteins (HSPs) (Smith and Toft, 1993; Church-Landel *et al.*, 1994). *In vivo* studies in yeast have suggested that HSP90 acts to stabilize the GR to maintain it in a state in which it can be activated by hormone (Yamamoto *et al.*, 1988; Picard *et al.*, 1990; Smith and Toft, 1993), and other work indicates that hormone binding induces dissociation of the GR from HSP90, leading to dimerization and DNA binding (Smith and Toft, 1993). However, the role of the interactions of the ER *in vivo* with HSP members, particularly HSP90, is less clear (Yamamoto *et al.*, 1988; Picard *et al.*, 1990; Church-Landel *et al.*, 1994).

Numerous studies have indicated that the ER interacts with HSP90 *in vitro*, suggesting that similar interactions may occur *in vivo*. However, *in vivo* studies including immunocytochemistry and gene transfer experiments have provided evidence that, in the absence of ligand, homodimers of the ER are present in the nucleus (Press *et al.*, 1985; Greene and Press, 1986; Fuxe *et al.*, 1987; Ishibashi *et al.*, 1989; Tzukerman *et al.*, 1990; Zhuang *et al.*, 1995), which would be inconsistent with stable interaction with HSP90. Gene transfer experiments using mutated ERs have provided evidence that the receptor can bind DNA in the absence of hormone (Tzukerman *et al.*, 1990; Zhuang *et al.*, 1995). These findings are inconsistent with a stable interaction with HSP90. Taken together, the above results suggest that if the ER interacts with HSP90 *in vivo*, this interaction is transient, and that a significant concentration of hormone-free receptor is present in the nucleus. Accessory proteins other than HSP90 may remain bound to the ER in the absence of hormone. Experiments with chimeric ERs have suggested that factors other than HSP90 retain the hormone-free ER in a transcriptionally repressed state (Lee *et al.*, 1996). Whether these putative repressors represent other members of the heat shock or immunophilin family found to be associated with the receptor, or as yet unidentified factors, is unclear at this stage.

b. Structure of the Ligand-Binding Domain   Relatively little information has been obtained regarding the key amino acids necessary for high-affinity ligand binding by steroid receptors. Classically, chemical affinity labeling has been used to identify residues of receptors that interact with small ligands. However, only a handful of residues of steroid receptor ligand-binding domains have been identified by electrophilic affinity labeling (Hotjo and Dall, 1993). For example, dexamethasone mesylate was used to label the rat GR at Cys-656, and the ER was labeled with tamoxifen aziridine at Cys-530 (Katzenellenbogen *et al.*, 1993). However, attempts to generate a family of more versatile labeling agents have proved unsuccessful (Katzenellenbogen, 1993). Such an approach is further complicated by the relatively large size [~250 amino acids (aa)] of receptor ligand-binding domains.

Model studies of the structure of the ER ligand-binding domain have been given a tremendous boost with the publication of the crystal structures of the ligand-binding domains of the hormone-free retinoid X receptor α (RXRα) and hormone-bound retinoic acid receptor γ (RARγ) (Bourget *et al.*, 1995; Renaud *et al.*, 1995; Wurtz *et al.*, 1996). Although these receptors share only 27% homology, the structures of their ligand-binding domains are remarkably similar, being composed of a series of conserved α helices. These results also support the notion that ligand binding by receptors leads to a conformational change in the ligand-binding domain, which would be an integral part of the process of activation of the receptor. Further modeling studies have suggested that ligand-binding domains of several nuclear receptors, including steroid receptors, share a canonical structure (Wurtz *et al.*, 1996). These results will greatly facilitate the modeling of the ER ligand-binding domain and the determination of key points of receptor–ligand interactions.

C. Nonsteroid Receptors

Receptors for retinoids, thyroid hormone, and vitamin D_3, the PPAR receptors, and a number of orphan receptors can be grouped together on the basis of their molecular characteristics (see below). Similar to the steroid hormone receptors, this family of nonsteroid receptors controls a wide range of physiological responses.

1. *Physiological Actions*

Retinoids are characterized initially for their role in vision, but also act as morphogenic agents, and are essential for reproduction and regulation and differentiation of a number of cell types (Giguère, 1994; Kastner *et al.*,

1995). Thyroid hormones L-thyroxine and L-triiodothyronine act in a number of tissues to control growth and development. Thyroid hormone affects basal metabolic rate and metabolism of carbohydrates and proteins (Pangaro, 1990). The secosteroid 1α,25-dihydroxyvitamin D_3 (D_3) is one of the principal hormones controlling calcium homeostasis, influencing calcium uptake in the intestine, calcium reabsorption in the kidney, and bone formation (Clemens and Riordan, 1990). J-type prostaglandins have been shown to be specific ligands for PPARγ (peroxisome proliferator-activated receptor γ), a member of a class of receptors activated by a number of industrial lipophilic compounds that stimulate proliferation of peroxisomes (Isseman and Green, 1990). Ligand-activated PPARγ has been shown to stimulate adipogenesis *in vitro* (Kliewer *et al.*, 1995; Forman *et al.*, 1995). Interestingly, PPARγ is also activated specifically by a number of adipogenic antidiabetic drugs known as thiazolodine diones (Lehmann *et al.*, 1995).

2. Structure–Function Studies

The nonsteroid receptors can be distinguished from the nuclear receptors described previously, not only by their ligand specificity but also by their mechanisms of action. Unlike the steroid receptors, nonsteroid receptors do not preferentially recognize palindromic response elements, but rather bind to half-sites arrayed in direct repeats (Fig. 5). Moreover, numerous studies have shown that, unlike steroid receptors, these receptors do not bind to their cognate response elements as homodimers. Receptors for retinoic acid (RARα, -β, and -γ), thyroid hormone (TR), and vitamin D_3 (VDR) and PPARs (PPARα, -γ, and -δ) bind as heterodimers coupled to the retinoid X receptors (RXRα, -β, and -γ) (Yu *et al.*, 1991; Leid *et al.*, 1992; Kliewer *et al.*, 1992; Zhang *et al.*, 1992a). Heterodimerization occurs through ligand-dependent interactions between ligand-binding domains, and specific dimerization interfaces in DNA-binding domains (Yu *et al.*, 1991; Leid *et al.*, 1992).

The P boxes of RARs, TRs, the VDR, RXRs, and PPARs are identical Glu-Gly-Gly motifs. It is not surprising, then, that these receptors recognize response elements with similar half-sites composed of the sequence PuG(G/T)TCA (Fig. 5). Generally speaking, response elements for different heterodimeric combinations are distinguished not on the basis of their sequence, but on the relative spacing between half-sites. VDR/RXR and TR/RXR heterodimers recognize direct separated by 3, or 4 bp, respectively (Umesomo *et al.*, 1991). RAR/RXR heterodimers bind to direct repeats with 1-, 2-, or 5-bp spacing (Fig. 5; Umesomo *et al.*, 1991; Smith *et al.*, 1991; Durand *et al.*, 1992), and PPAR/RXRs bind repeats separated by 1 bp. Structure–function studies have demonstrated that RXR binds to the 5′ half-site of the direct repeat, and that interaction between DNA-binding domains controls discrimination between response elements with different spacings (Perlmann *et al.*, 1993; Kurokawa *et al.*, 1993; Mader *et al.*, 1993b;

<pre>
DR1   5'AGGTCAAAGGTCA   CRPB II GENE
          TCCAGTTTCCAGT 3'   RARE

DR2   5'AGGTCAAAAGGTCA   CRBP GENE
          TCCAGTTTTCCAGT 3'   RARE

DR3   5'GGTTCACGAGGTTCA   OSTEOPONTIN
          CCAAGTGCTCCAAGT 3'   GENE VDRE

DR4   5'AGGTGACAGGAGGACA   MHC α GENE
          TCCACAGTCCTCCTCT 3'   TRE

DR5   5'GGTTCACCGAAAGTTCA   MOUSE RARβ
          CCTTCTGGCTTTCAAGT 3'   GENE RARE

            PuG(G/T)TCA   CONSENSUS
                                    HALF-SITE
</pre>

FIGURE 5   Hormone response elements recognized by receptors for vitamin D_3, thyroid hormone, prostaglandins, and retinoids. The response elements are organized as direct repeats (DRs) of the consensus sequence (shown at bottom). Typical elements are shown, arranged in order of increasing inter-half-site spacing. The DR1 element is from the retinoid-regulated cellular retinol-binding protein II (CRBP II) gene. DR1 elements are recognized by RXR/RAR heterodimers, RXR/RXR homodimers, and RXR/PPAR heterodimers. The DR2 element is from the cellular retinol-binding protein I (CRBP I) gene, and is recognized by RXR/RAR heterodimers. The DR3 element is from the mouse osteopontin gene and is recognized by RXR/VDR heterodimers. The DR4 element is from the major histocompatibility complex α gene and is recognized by RXR/TR receptor heterodimers. The DR5 element is from the mouse retinoic acid receptor β gene and is recognized by RXR/RAR heterodimers.

Predki *et al.*, 1994; Towers *et al.*, 1993). Work has indicated that RXRs and their heterodimeric partners recognize direct repeats with different spacings by virtue of the presence of multiple possible dimerization interfaces in their DNA-binding domains (Zechel *et al.*, 1994a,b).

RXRs are cognate receptors for 9-*cis*-retinoic acid (9-*cis* RA), and ligand binding has been shown to stimulate homodimerization, and binding to direct repeats separated by 1 bp (Zhang *et al.*, 1992b). Numerous studies have shown that 9-*cis* RA can modulate the expression of reporter genes regulated by all-*trans* RA, thyroid hormone, and vitamin D_3 (Rosen *et al.*, 1992; MacDonald *et al.*, 1993; Ferrara *et al.*, 1994; Roy *et al.*, 1995). Modulatory effects of 9-*cis* RA thus increase the complexity of control of gene expression by these receptors.

V. Function of Natural and Synthetic Steroid-Responsive Promoters

A. Natural Steroid-Responsive Promoters

One of the key observations to have emerged from functional analyses of steroid receptor-regulated promoters is that enhancer factors bound to different sites on a given promoter can combine to activate transcriptional initiation synergistically. For example, the level of glucocorticoid-induced transcription from the mouse mammary tumor virus (MMTV) promoter is strongly dependent not only on the integrity of its glucocorticoid response elements (GREs), but also on the function of sites adjacent to the TATA box that bind the activators nuclear factor 1 (NF-1) and octomer factor (Bruggemeier *et al.*, 1990, 1991). Similarly, the maximal response to estrogen of the human pS2 promoter is dependent on the ER and other classes of enhancer factors (Berry *et al.*, 1989). Studies with model promoters have shown that the glucocorticoid receptor can combine synergistically with a number of different types of enhancer factors (Strahle *et al.*, 1988).

Synergistic activation of transcription can also be achieved by a given type of enhancer factor bound to multiple arrays of binding sites. Two GREs located far upstream of the tyrosine aminotransferase gene mediate a synergistic response to the glucocorticoid receptor (Jantzen *et al.*, 1987). One of the sites is inactive in isolation, and the other site is only 25% as active as the two sites combined. Similarly, the chicken vitellogenin promoter contains two EREs that combine to give a synergistic response to estrogen (Burch *et al.*, 1988).

B. Synthetic Steroid-Responsive Promoters

1. Glucocorticoid-Responsive Promoters

Natural steroid-responsive promoters generally contain binding sites for a number of different classes of transcription factors in addition to steroid receptors. In the late 1980s a number of studies were performed with synthetic promoter systems to determine what constituted the minimal number of elements necessary to render a promoter responsive to steroid hormone. One of these showed that a GRE and a TATA box were sufficient to generate a glucocorticoid-responsive promoter. Moreover, two GREs placed upstream of a TATA box functioned far better than a single element, consistent with the synergistic action of multiple arrays elements in natural promoters. DNA-binding studies suggested that this synergism was due, at least in part, to cooperative binding of glucocorticoid receptor homodimers to adjacent response elements (Schmid *et al.*, 1989).

2. Estrogen-Responsive Promoters

Studies similar to those described previously were performed with synthetic promoters containing EREs, and, not surprisingly, an ERE and a TATA box were sufficient to reconstitute a transcriptional response to estrogen. In addition, two EREs combined to mediate a synergistic response to estrogen (Ponglikitmongkol *et al.*, 1990). However, in contrast to studies with the glucocorticoid receptor, no evidence was found for cooperative binding of the ER to adjacent EREs. This suggested that the synergism arose through interactions between the transactivating domains of DNA-bound receptors, rather than through cooperative DNA binding.

The importance of the experiments with these so-called minimal promoters lies in the observation that a physiological response to steroid hormone does not require the input from additional classes of transcription factors whose activity would be controlled by other signal transduction pathways. This suggested that the expression of synthetic steroid-responsive promoters can be strictly regulated exclusively by controlling the level of steroid. Moreover, the studies raised the possibility that synthetic promoters composed of a TATA box and extended arrays of response elements could be powerfully induced by ligand.

3. Expression Vectors Containing Synthetic Steroid-Inducible Promoters

The capacity of steroid receptors to function independently of other classes of enhancers and synergistic action of multiple arrays of hormone response elements raised the possibility that synthetic steroid-inducible promoters would be useful in synthetic vectors where regulated gene expression was required. We constructed a series of minimal promoters (Mader and White, 1993) containing one or more GREs from the rat tyrosine aminotransferase gene placed upstream of the TATA region of the adenovirus 2 major late promoter (Ad2MLP). A glucocorticoid-inducible promoter was chosen for these studies because the glucocorticoid receptor is widely expressed, and because receptors for androgens, mineralocorticoids, and progesterone also recognize GREs. This maximizes the range of cell types in which the vector could be used.

a. Transient Expression Studies In initial studies, the GREs were placed upstream of a rabbit β-globin gene, and gene expression in the transiently transfected human cervical carcinoma cell line HeLa was monitored by S1 nuclease of total RNA with a globin-specific probe. No globin mRNA was detected under any conditions in cells transfected with a control plasmid containing an Ad2MLP TATA region and no GREs or 1 GRE (Fig. 6A). Moreover, no expression was detected from any of the GRE-containing promoters in cells cultured in the absence of dexamethasone. However, dexamethasone-dependent expression was detected from a promoter containing two GREs, and at least eight-fold more expression was observed from a promoter containing five response elements, which was dubbed GRE5 (Fig. 7). Thus, these types of minimal promoters provide a combination of low background expression, and efficient hormone inducibility.

To place the levels of hormone-inducible expression observed with minimal promoters in perspective, they were compared to that observed with a vector in which the GREs were replaced by an SV40 enhancer, which contains binding sites for several different types of factors, and the steroid-

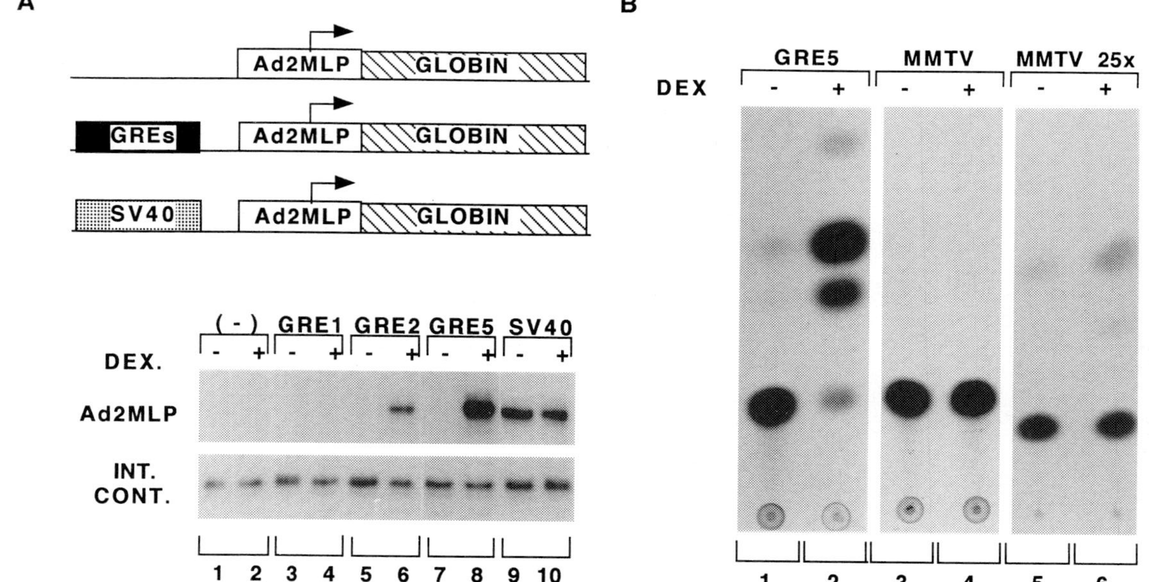

FIGURE 6 Dexamethasone-inducible transcription from synthetic glucocorticoid-responsive promoters. (A) Ligand-inducible transcription from promoters composed of one or more GREs, and the TATA box from the adenovirus 2 major late promoter (Ad2MLP). A schematic representation of synthetic promoter–globin gene reporter recombinants used for S1 nuclease analysis is given at the top. The results of S1 nuclease analysis of RNA extracted from HeLa cells transiently transfected with reporter plasmids is presented. The number of GREs in the test promoter is indicated above, where appropriate. Transcripts driven from the Ad2MLP are indicated, along with transcripts from an internal control plasmid (int. cont.) containing a constitutively active promoter. (B) Comparison of dexamethasone-inducible CAT activity driven by the GRE5 promoter (see Fig. 7), and a mouse mammary tumor virus (MMTV)–CAT recombinant.

inducible MMTV promoter. In the presence of dexamethasone, approximately threefold higher levels of globin mRNA were expressed from the minimal promoter containing five GREs than from the constitutively active promoter containing the SV40 enhancer (Fig. 6A). In other experiments (Fig. 6B), at least 100-fold more dexamethasone-inducible expression of the chloramphenicol acetyltransferase reporter (CAT) gene was observed from the GRE5 promoter of pGRE5-CAT (Fig. 4) than from the MMTV promoter. It is noteworthy that these experiments were performed in the absence of a transfected glucocorticoid receptor expression vector. Therefore, the GRE5 promoter does not require overexpression of steroid receptors to function efficiently.

b. Activity of Minimal Promoters in Stably Transfected Cells

While experiments in transiently transfected cells are easy to perform, and indicated the potential utility of steroid-inducible minimal promoters, it was important to verify that hormone-inducible expression could be achieved under conditions in which the promoter was propagated in chromatin, either through integration into the chromosome, or through stable replication on an episomal plasmid such as an Epstein–Barr virus (EBV)-based vector (Fig. 7). Several Hela cell lines were isolated containing the GRE5-CAT cassette integrated into the chromosome (Fig. 8A). These lines all displayed dexamethasone-inducible expression of CAT activity, indicating function of the promoter. However, this expression was clearly dependent on the site of integration as the levels or hormone-induced expression varied considerably; CAT activity was detected in the absence of dexamethasone in four of the eight lines tested (Fig. 8A).

More compelling results were obtained in either HeLa cells or the human breast carcinoma cell line T47D when the GRE5-CAT cassette was propagated on an EBV episomal vector (White *et al.*, 1994). Unlike HeLa cells, which express the glucocorticoid receptor, T47D cells express receptors for androgens and progesterone, which also recognize GREs. No or very low levels of CAT activity were observed in the absence of inducer in either cell line, while high levels of CAT activity were expressed in the presence of dexamethasone in HeLa, or progesterone in T47D, cells (Fig. 8B). Strong induction was also observed with dihydrotestosterone in T47D cells. Taken together, these results indicate that minimal promoters can function when stably propagated in cells. Moreover, if placed in a transcriptionally silent environment minimal promoters display low levels of basal activity. This suggests that these types of promoters may be useful for tightly controlling the expression of downstream genes under a variety of conditions. The GRE5 vectors have been used by a number of other researchers for controlled expression of foreign genes in a number of tissue culture systems (Philpott *et al.*, 1994; Samaniego *eg al.*, 1994; Choi and Ballerman, 1995; Tong *et al.*, 1995; Derusso *et al.*, 1995).

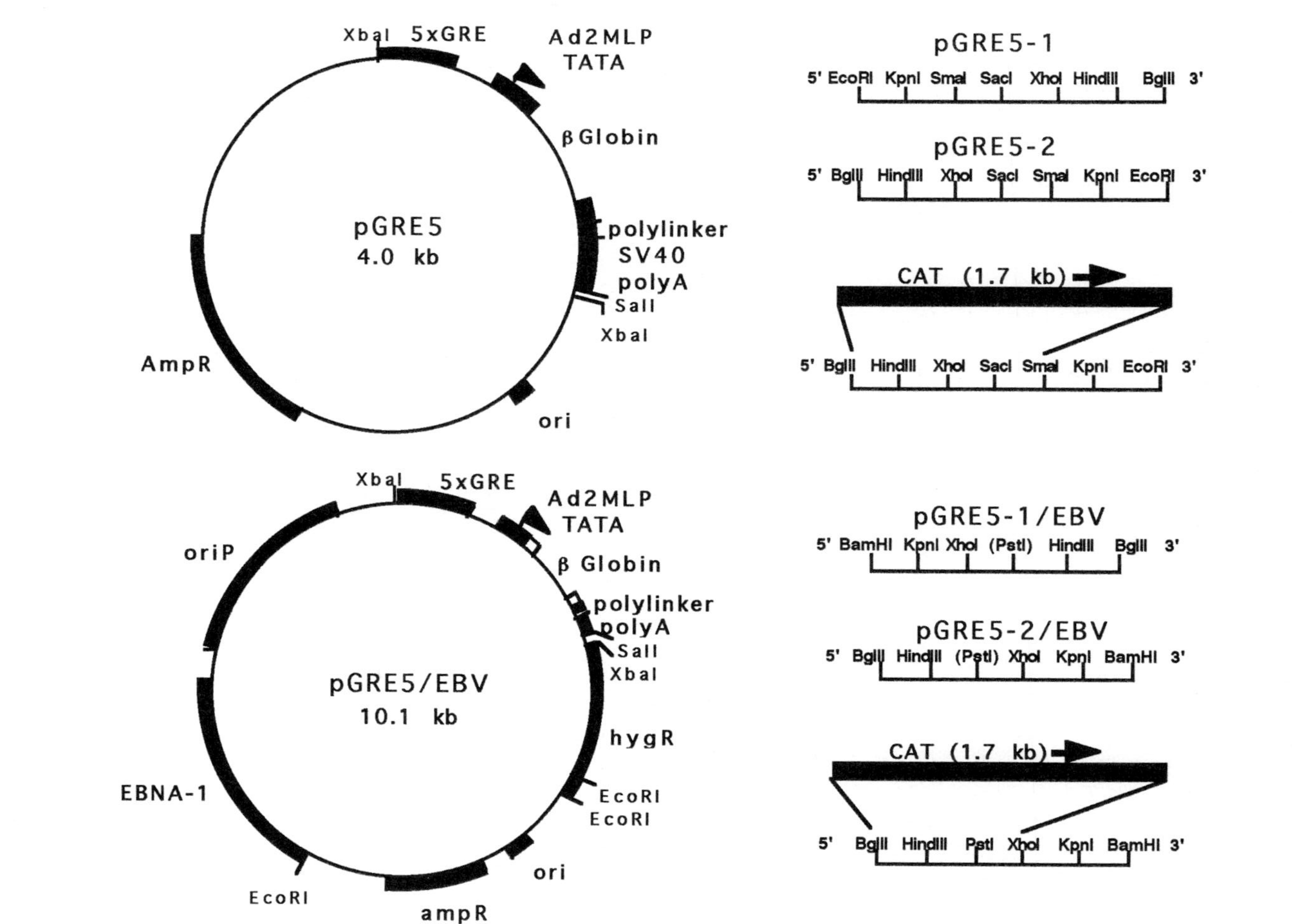

Xbal 5xGRE Ad2MLP TATA β Globin polylinker SV40 polyA Sall Xbal
pGRE5 4.0 kb
AmpR ori
pGRE5-1
5' EcoRI KpnI SmaI SacI XhoI HindIII BgIII 3'
pGRE5-2
5' BgIII HindIII XhoI SacI SmaI KpnI EcoRI 3'
CAT (1.7 kb)
5' BgIII HindIII XhoI SacI SmaI KpnI EcoRI 3'
Xbal 5xGRE Ad2MLP TATA β Globin polylinker polyA Sall Xbal
oriP pGRE5/EBV 10.1 kb hygR
EcoRI EcoRI ori EBNA-1 EcoRI ampR
pGRE5-1/EBV
5' BamHI KpnI XhoI (PstI) HindIII BgIII 3'
pGRE5-2/EBV
5' BgIII HindIII (PstI) XhoI KpnI BamHI 3'
CAT (1.7 kb)
5' BgIII HindIII PstI XhoI KpnI BamHI 3'

VI. Uses of Modified Steroid-Inducible Expression Systems in Gene Therapy

A. Regulated Expression of Genes in Gene Therapy

As discussed in Section II, the expression of a given gene is under the control of regulatory signals, which can control both its tissue-specific and temporal expression. These parameters will be dependent on the distribution and levels of expression of the enhancer factors that bind to the promoter of the gene. Ideally, for appropriate expression of an exogenous gene in a gene therapy protocol where replacement of a defective gene is involved, the gene would be best placed under control of the same regulatory signals that control the endogenous gene. However, in a number of cases this may not be practical. Enhancer sequences may not be fully defined, or may be spread out over large regions of DNA, rendering recapitulation of expression patterns of the endogenous gene problematic. In addition, with some viral delivery systems the onus is on minimizing the DNA sequences required to regulate gene expression. Under these conditions, a short synthetic promoter whose expression can be regulated may provide a useful alternative. A similar regulatory system may also be applicable for the expression of a gene that is normally foreign to a cell.

B. Potential Use of Steroid-Inducible Promoters for Controlled Expression of Genes

Engineered steroid-inducible expression systems may be ideally suited to the applications described previously. Steroid receptors are both primary targets of specific intercellular signals, and transcription factors, and can function independently of other signal transduction pathways, rendering expression strictly dependent on the presence of steroid hormone. In this regard, steroid receptors are better suited for expression systems that their nonsteroid counterparts since they recognize response elements as homodimers (Kumar and Chambon, 1988), whereas receptors for thyroid hormone, vitamin D_3, or retinoic acid bind DNA as heterodimers with endogenous retinoid X receptors (Yu *et al.*, 1991; Leid *et al.*, 1992; Kliewer *et al.*, 1992; Zhang *et al.*, 1992a). Steroid-inducible promoters can be constructed that are less than 300 bp in length, thus maximizing the space in a vector for other genetic elements (Mader and White, 1993; White *et al.*, 1994). The

---

FIGURE 7 Structure of expression vectors containing the GRE5 promoter, composed of five glucocorticoid response elements (GREs) placed upstream of the adenovirus 2 major late promoter (Ad2MLP). The plasmid GRE5/EBV contains sequences encoding the Epstein–Barr virus origin of replication, and the Epstein–Barr nuclear antigen 1 (EBNA-1) coding sequence, and replicates as an episome in human and primate cells (see White *et al.*, 1994).

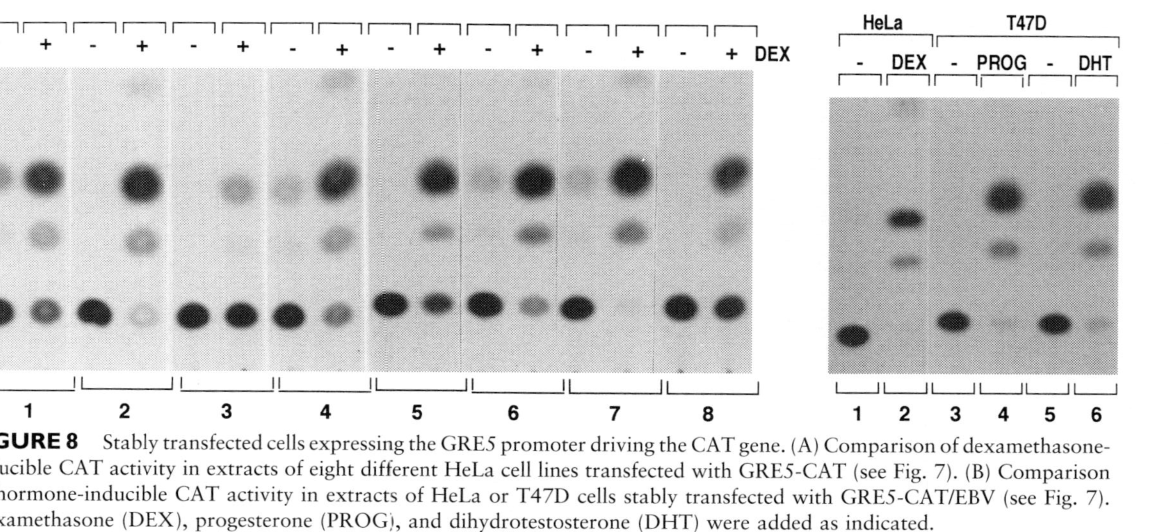

FIGURE 8   Stably transfected cells expressing the GRE5 promoter driving the CAT gene. (A) Comparison of dexamethasone-inducible CAT activity in extracts of eight different HeLa cell lines transfected with GRE5-CAT (see Fig. 7). (B) Comparison of hormone-inducible CAT activity in extracts of HeLa or T47D cells stably transfected with GRE5-CAT/EBV (see Fig. 7). Dexamethasone (DEX), progesterone (PROG), and dihydrotestosterone (DHT) were added as indicated.

level of steroid-inducible gene expression can be regulated either by controlling the level of inducer, or by varying the number of response elements in the promoter. In addition, steroids or steroid-like molecules are already widely used in clinical protocols, and are ideally suited as potential reagents for gene therapy.

A prototypical steroid-inducible expression system based on the GRE5 promoter has already been set up in tissue culture cells infected with a herpes simplex virus (HSV) vector (Lu and Federoff, 1995). In this study, the GRE5 promoter was used to drive expression of a *lacZ* reporter gene. Lu and Federoff (1995) found that the GRE5 promoter could function in HSV-infected cells, and that reporter gene expression was inducible 50-fold by dexamethasone. Inducible *lacZ* activity was similar to that obtained under similar conditions with a promoter controlled by the constitutively active cytomegalovirus enhancer, indicating that the GRE5 promoter can direct high levels of expression. It is noteworthy that high levels of dexamethasone-independent expression were obtained with a preliminary version of the vector. This background activity was found to be due to transcripts encoding lacZ that arose from a constitutively active promoter located upstream of GRE5 sequences (Lu and Federoff, 1995). This result is reminiscent of the basal activity observed in some HeLa cell lines containing a GRE5 reporter cassette integrated into the chromosome, and emphasizes the importance of assuring that inducible promoter sequences be placed in a transcriptionally silent environment.

C. Engineering Steroid-Inducible Expression Systems for Use in Gene Therapy

Ideally, an inducible expression system for use in gene therapy should be exquisitely specific, i.e., a ligand should activate a receptor that controls the expression of a defined gene, or genes, and does not affect the expression of other genes in the organism (Fig. 9). By these criteria, the GRE5 system, although it serves as a useful model, is unsuitable. Dexamethasone will affect the expression of all glucocorticoid-inducible genes in all cells expressing the receptor. An appropriate ligand would not activate any steroid receptors other than that controlling the expression of the transgene. This necessitates modification of the ligand-binding specificity of the receptor, and its expression along with the transgene to be regulated. In addition, the inducible promoter regulating transgene expression cannot contain response element sequences, such as GREs, recognized by endogenous nuclear receptors. This requires the use of novel response element sequences recognized by a receptor with altered DNA-binding specificity. Thus, a tightly controlled expression system would require engineering both the ligand- and DNA-binding specificities of the receptor. Several published experiments suggest that both of these modifications are feasible.

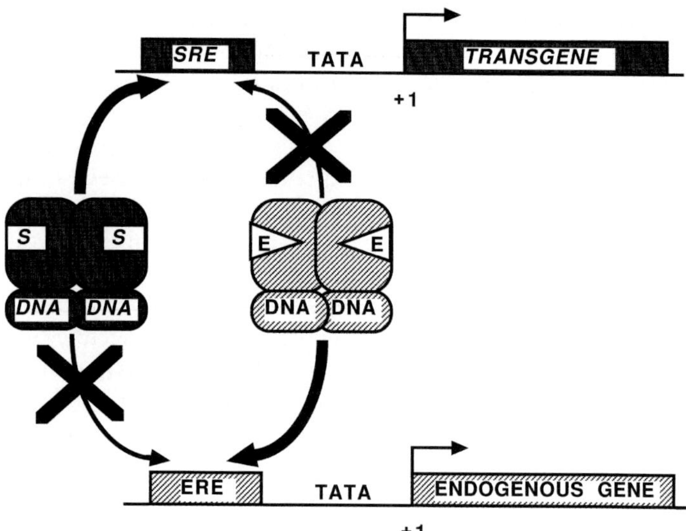

FIGURE 9   Model for the action of a modified estrogen receptor that functions as an exquisitely specific signal transducer. The modified receptor would bind a modified estrogen-like steroid molecule (S), and recognize exclusively a steroid response element (SRE) not bound by the ER or any other endogenous receptor.

I. Engineering Receptors with Modified DNA-Binding Specificities

There are several potential options available for modification, or indeed replacement, of the DNA-binding domains of steroid receptors. It has been known for at least 10 years that functional chimeric enhancer factors can be created by exchanging DNA-binding and transactivating domains of different proteins (Keegan *et al.*, 1986). Many structure–function studies have been performed with chimeric nuclear receptors in which the receptor DNA-binding domain has been replaced by that of the yeast transactivator GAL4 (Webster *et al.*, 1988; Brou *et al.*, 1993; Lehman *et al.*, 1995). GAL4 is a member of a large family of yeast transcriptional regulators that possess DNA-binding domains containing so-called zinc cluster motifs. These are structurally distinct from the zinc finger domains of nuclear receptors. GAL4–nuclear receptor chimeras are potentially appealing for use in human cells because there are no known homologs of GAL4 in human cells, and the DNA-binding specificity of GAL4 is distinct from those of all mammalian transcriptional regulators described to date. Therefore, chimeric receptors would activate specifically promoters containing cognate GAL4-binding sites, known as 17-mers.

While the use of GAL4 chimeras represents a simple solution to the problem of controlling the specificity of action of transactivators, it may raise the possibility that cells expressing the chimeras are recognized by the immune system as foreign owing to the presentation of GAL4-specific peptides on the cell surface. While the use of modified receptors to control gene expression precludes elimination of this problem entirely, it may be more appropriate to introduce more subtle modifications into receptor DNA-binding domains. The most likely target for modification would be the 3-amino acid P box, which is important for site-specific DNA recognition by nuclear receptors.

Two approaches to modifying receptor P boxes are immediately obvious. In the first, a modified response element sequence is chosen that is not recognized by endogenous receptors. For example, a palindrome composed of AGcTCA half-sites may be appropriate since a C residue at the third position of the half-site is generally not found in response elements recognized by the ER, other steroid receptors, or other nonsteroid receptors. A library of receptors mutagenized in the P box could then be expressed in yeast, along with a reporter gene under control of a promoter containing the modified response elements, and candidate receptors isolated from clones displaying hormone-dependent reporter gene expression. In an alternative approach, the P box of a steroid receptor could be replaced by any one of a number of novel P boxes found in nuclear receptors expressed in more primitive organisms such as *C. elegans* (Mangelsdorf *et al.*, 1995). The DNA-binding specificities controlled by these P boxes could be determined using polymerase chain reaction (PCR)-based techniques for identifying cognate DNA sequences for DNA-binding proteins (Blackwell and Weintraub, 1990).

2. Engineering the Ligand-Binding Specificity of Steroid Receptors

The observation that the ER functions as a ligand-inducible transactivator in yeast (Metzger *et al.*, 1988) opened the way to using efficient phenotypic screens in yeast to probe nuclear receptor function. Yeast has since been widely used to study the function of the ER and its derivatives (Nawaz *et al.*, 1992; Dana *et al.*, 1994; Pham *et al.*, 1992; McDonnell *et al.*, 1992; Ince *et al.*, 1993; Louvion *et al.*, 1993; Wrenn and Katzenellenbogen, 1993; Castles *et al.*, 1993; Salomonsson *et al.*,1994b; Pierrat *et al.*, 1994). Residues in the ligand-binding domain have been identified, through screening for mutants of the ER, that display impaired hormone-dependent transactivation (Ince *et al.*, 1993; Wrenn and Katzenellenbogen, 1993). However, this type of study suffers from using a loss of function approach to identify residues implicated in ligand binding.

It would be useful to develop a gain of function approach to probe for receptors with enhanced affinity for ligands modified at specific positions on the steroid backbone. One study that screened for progesterone receptors with altered responses to ligands isolated a truncated receptor that was activated by the antiprogestin RU486 (Vegato *et al.*, 1992). A similar screen with the ER isolated a number of truncated receptors that were constitutively active (Pierrat *et al.*, 1994). While truncated receptors are of potential interest, screens that tend to isolate exclusively deletion mutants severely limit the number of potentially useful modifications.

The isolation of truncated receptors could be eliminated by designing a screen on the basis of a rearranged receptor in which the ligand-binding domain is placed at the N terminus of the molecule (Fig. 10). In this way, introduction of stop codons into the ligand-binding domain would produce proteins that lacked DNA-binding domains and would, therefore, be nonfunctional. A screen of this kind has been developed (J. White, unpublished results). Preliminary results indicate that it is possible to isolate receptors bearing single amino acid substitutions that have enhanced affinity for estradiol derivatives modified in the aromatic A ring of the molecule. None of these modified receptors displayed reduced sensitivity to 17β-estradiol. Therefore, it is likely that, in a number of instances, the engineering of a mutant receptor with an enhanced affinity for a modified estrogen, and a reduced affinity for physiological hormone, would require at least two rounds of mutagenesis.

While screens of this type will likely be useful for generation of receptors specific for a number of modified steroids, the range of potential compounds that could serve as ligand may be limited. In preliminary model experiments we have failed to isolate modified receptors with enhanced affinity for the androgen dihydrotestosterone (DHT), which activates the ER at micromolar concentrations. Dihydrotestosterone differs from 17β-estradiol principally in its A ring, which is not aromatic. Whether this result is specific to DHT or generally applicable to compounds lacking the steroid skeleton of estrogen remains to be seen.

VII. Conclusions

Since nuclear receptors are both primary targets of specific intercellular signals, and transcription factors, they are ideally suited for engineering as specific transducers of signals that regulate the expression of a single gene. Steroids or steroid-like molecules are already in use in clinical protocols, and are therefore potential reagents for gene therapy. To be practical, steroid receptors must be converted into exquisitely specific signal transducers; i.e.,

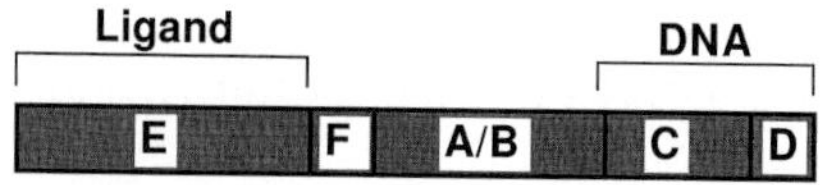

FIGURE 10   Schematic representation of the structure of a "reverse" ER. Note that regions E and F have been placed at the N terminus of the molecule.

a transcription factors with unique DNA-binding specificities (and thus capable of regulation of a single gene), which are activated by an otherwise physiologically inactive signal. Indeed, the need for such transducers in gene therapy has been spelled out (Hodson, 1995).

Many of the required experimental precedents for the production of an appropriately modified steroid-inducible expression system are already in place. Steroid receptors can act as powerful and specific signal transducers by activation of synthetic promoters composed solely of hormone response elements and a TATA region in stably transfected cells. These "minimal" promoters are <300 bp in length, they are responsive to only one signal (the steroid hormone), and their degree of induction can be modulated by changing the number of hormone response elements in the promoter and/ or the concentration of hormone in the growth medium. It is well established that chimeric steroid receptors with altered DNA-binding specificities can activate genes in a ligand-dependent manner. The combination of all these results, along with the use of a receptor bearing a modified ligand-binding domain, will result in the engineering of a protein that acts as an exquisitely specific signal transducer.

References

Aumais, J., Lee, H. S., DeGannes, C., Horsford, J., and White, J. H. (1996). *J. Biol. Chem.* **271**, 12568–12577.

Beato, M. (1989). *Cell (Cambridge, Mass.)* **56**, 335–344.

Beato, M., Herrlich, P., and Schutz, G. (1995). *Cell (Cambridge, Mass.)* **83**, 851–857.

Berry, M., Nunez, A.-M., and Chambon, P. (1989). *Proc. Natl. Acad. Sci. U.S.A.* **86**, 1218–1222.

Berry, M., Metzger, D., and Chambon, P. (1990). *EMBO J* **9**, 2811–2818.

Blackwell, T. K., and Weintraub, H. (1990). *Science* **250**, 1104–1110.

Bocquel, M.-T., Kumar, V., Stricker, C., Chambon, P., and Gronemeyer, H. (1989). *Nucleic Acids Res.* **17**, 2581–2595.

Bourget, W., Ruff, M., Chambon, P., Gronemeyer, H., and Moras, D. (1995). *Nature (London)* **375**, 377–382.

Brou, C., Wu, J., Ali, S., Scheer, E., Lang, C., Davidson, I., Chambon, P., and Tora, L. (1993) *Nucleic Acids Res.* **21**, 5–12.

Bruggemeier, U., Rogge, L., Winnacker, E. L., and Beato, M. (1990). *EMBO J* **9**, 2233–2239.

Bruggemeier, U., Kalff, M., Franke, S., Scheidereit, C., and Beato, M. (1991). *Cell (Cambridge, Mass.)* **64**, 565–572.

Burch, J. B. E., Evans, M. I., Friedman, T. M., and O'Malley, P. J. (1988). *Mol. Cell. Biol.* **8**, 1123–1131.

Castles, C. G., Fuqua, S. A., Klotz, D. M., and Hill, S. M. (1993). *Cancer Res* **53**, 5934–5939.

Chang, T.-C., Nardulli, A. M., Lew, D., and Shapiro, D. J. (1992). *Mol. Endocrinol.* **6**, 346–354.

Choi, M. E., and Ballerman, B. J. (1995). *J. Biol. Chem.* **270**, 21144–21150.

Choy, B., and Green, M. R. (1993). *Nature (London)* **366**, 531–536.

Christiansen, C. (1991). *Am. J. Med.* **90**, 107–115.

Church-Landel, C., Kushner, P. J., and Greene, G. (1994). *Mol. Endocrinol.* **8**, 1407–1419.

Clemens, T. L., and Riordan, J. L. (1990). *In* "Principles and Practice of Endocrinology and Metabolism" (K. L. Becker, ed.), pp. 417–423. Lippincott, Philadelphia.

Dahlman-Wright, K., Wright, A., Gustafsson, J.-A., and Carlstedt-Duke, J. (1991). *J. Biol. Chem.* **266**, 3107–3112.

Dana, S. L., Hoener, P. A., Wheeler, D. A., Lawrence, C. B., and McDonnell, D. P. (1994). *Mol. Endocrinol.* **8**, 1193–1207.

Danielson, M., Hinck, L., and Ringold, G. M. (1989). *Cell (Cambridge, Mass.)* **57**, 1131–1138.

Derusso, P. A., Philpott, C. C., Iwai, K., Mostowski, H. S., Klausner, R., and Rouault, T. A. (1995). *J. Biol. Chem.* **270**, 15451–15454.

Durand, B., Saunders, M., Leroy, P., Leid, M., and Chambon, P. (1992). *Cell (Cambridge, Mass.)* **71**, 73–85.

Dynlacht, B. D., Hoey, T., and Tjian, R. (1991). *Cell (Cambridge, Mass.)* **66**, 563–576.

Evans, R. M., and Hollenberg, S. M. (1988). *Cell (Cambridge, Mass.)* **52**, 1–3.

Fawell, S. E., Lees, J. A., White, R., and Parker, M. G. (1990). *Cell (Cambridge, Mass.)* **60**, 953–962.

Ferrara, J., McCuaig, K. A., Hendy, G. N., Uskokovic, M., and White, J. H. (1994). *J. Biol. Chem.* **269**, 2971–2981.

Forman, B. M., and Samuels, H. H. (1990). *New Biol.* **2**, 587–594.

Forman, B. M., Tontonoz, P., Chen, J., Brun, R., Spiegelman, B. M., and Evans, R. M. (1995). *Cell (Cambridge, Mass.)* **83**, 803–812.

Freidl, A., and Jordan, V. C. (1994). *Breast Cancer Res. Treat.* **31**, 27–35.

Fuxe, K., Cintra, A., Agnati, L. F., Harfstrand, A., Wikström, A. C., Okret, S., Zoli, M., Miller, L. S., Greene, J. L., and Gustaffson, J.-A. (1987). *J. Steroid Biochem.* **27**, 159–170.

Gaub, M.-P., Bellard, M., Scheuer, I., Chambon, P., and Sassone-Corsi, P. (1990). *Cell (Cambridge, Mass.)* **63**, 1267–1276.

George, F. W., and Wilson, J. D. (1988). *In* "Physiology of Reproduction" (E. Knobil *et al.*, eds.), pp. 2–27. Raven Press, New York.

Giguère, V. (1994). *Endocr. Rev.* **15**, 61–79.

Godowski, P. J., Picard, D., and Yamamoto, K. R. (1988). *Science* **241**, 812–816.

Gonzalez, G. A., and Montminy, M. R. (1989). *Cell (Cambridge, Mass.)* **59**, 675–680.

Goodrich, J. A., and Tjian, R. (1994). *Curr. Opin. Cell Biol.* **6**, 403–409.

Green, S., and Chambon, P. (1988). *Trends Genet.* **4**, 309–314.

Greene, G. L., and Press, M. F. (1986). *Steroid Biochem.* **26**, 1–7.

Gronemeyer, H. (1991). *Annu. Rev. Genet.* **25**, 89–123.

Hard, T., Kellenbach, E., Boelens, R., Maler, B. A., Dahlman, K., Freedman, L. P., Carlstedt-Duke, J., Yamamoto, K. R., Gustafsson, J.-A., and Kaptein, R. (1990). *Science* **249**, 157–160.

Hodson, C. P., (1995). *Bio/Technology* **13**, 222–226.

Hollenberg, S. M., and Evans, R. M. (1988). *Cell (Cambridge, Mass.)* **55**, 899–906.

Hotjo, H. D., and Dall, N. (1993). *Pharmazie* **48**, 243–249.

Hunter, T., and Karin, M. (1992). *Cell (Cambridge, Mass.)* **70**, 375–387.

Hutchinson, T. A., Polansky, S. M., and Feinstein, A. R. (1979). *Lancet* **2**, 705–709.

Ince, B. A., Zuang, Y., Wrenn, C. K., Shapiro, D. J., and Katzenellenbogen, B. S. (1993). *J. Biol. Chem.* **268**, 14026–14032.

Ishibashi, T., Nakabeppu, Y., and Sekigushi, M. (1989). *Biol. Reprod.* **40**, 1275–1285.

Isseman, I., and Green, S. (1990). *Nature (London)* **347**, 645–650.

Jantzen, H. M., Strahle, U., Gloss, B., Stewart, F., Schmid, W., Boshart, M., Miksicek, R., and Schutz, G. (1987). *Cell (Cambridge, Mass.)* **49**, 29–38.

Jonat, C., Rahmsdorf, H. J., Park, K. K., Cato, A. C. B., Gebel, S., Ponta, H., and Herrlich, P. (1990). *Cell (Cambridge, Mass.)* **62**, 1189–1204.

Jordan, V. C. (1993). *Br. J. Pharmacol* **110**, 507–520.

Kaplan, F. S., Fallon, M. D., Boden, S. D., Schmidt, R., Senior, M., and Haddad, J. G. (1988). *N. Engl. J. Med.* **319**, 421–426.

Kastner, P., Mark, M., and Chambon, P. (1995). *Cell (Cambridge, Mass.)* **83**, 859–869.

Kato, S., Tora, L., Yamauchi, J., Masushige, S., Bellard, M., and Chambon, P. (1992). *Cell (Cambridge, Mass.)* **68**, 731–742.

Katzenellenbogen, B. B., Bhardwaj, B., Fang, H., Ince, B. A., Pakdel, F., Reese, J. C., Schodin, D., and Wrenn, C. K. (1993). *J. Steroid Biochem. Mol. Biol.* **47**, 39–48.

Katzenellenbogen, J. A. (1993). *Pharmacochem. Libr.* **20**, 243–257.

Keegan, L., Gill, G., and Ptashne, M. (1986). *Science* **231**, 699–704.

Klein-Hitpass, L., Schorpp, M., Wagner, U., and Ryffel, G. U. (1986). *Cell (Cambridge, Mass.)* **46**, 1053–1061.

Klein-Hitpass, L., Kailing, M., and Ryffel, G. U. (1988). *J. Mol. Biol.* **201**, 537–544.

Kliewer, S. A., Umesomo, K., Mangelsdorf, D. J., and Evans, R. M. (1992). *Nature (London)* **350**, 446–449.

Kliewer, S. A., Lenhard, J. M., Willson, T. M., Patel, I., Morris, D. C., and Lehmann, J. (1995). *Cell (Cambridge, Mass.)* **83**, 813–819.

Korach, K. S. (1994). *Science* **266**, 1524–1527.

Kreiger, N., Kelsey, J. L., Holford, T. R., and O'Connor, S. (1982). *Am. J. Epidemiol.* **116**, 141–148.

Krust, A., Green, S., Argos, P., Kumar, V., Walter, P., Bornert, J. M., and Chambon, P. (1986). *EMBO J* **5**, 891–897.

Kumar, V., and Chambon, P. (1988). *Cell (Cambridge, Mass.)* **55**, 145–156.

Kumar, V., Green, S., Stack, G., Berry, M., Jin, J. R., and Chambon, P. (1987). *Cell (Cambridge, Mass.)* **51**, 941–951.

Kurokawa, R., Yu., V. C., Naar, A., Kyakumoto, S., Han, Z., Silverman, S., Rosenfeld, M. G., and Glass, C. K. (1993). *Genes Dev.* **7**, 1423–1435.

Lee, H. S., Aumais, J., and White, J. H. (1996). *J. Biol. Chem.* **271**, 25727–25730.

Lehman, J., Moore, L. B., Smith-Oliver, T. A., Wilkison, W. O., Willson, T. M., and Kliewer, S. A. (1995). *J. Biol. Chem.* **270**, 12953–12956.

Leid, M., Kastner, P., Lyons, R., Nakshatri, H., Saunders, M., Zacharewski, T., Chen, J.-Y., Staub, A., Garnier, J.-M., Mader, S., and Chambon, P. *Cell (Cambridge, Mass.)* **68**, 377–395.

Lin, A., Frost, J., Deng, T., Smeal, T., Al-Alawi, N., Kikkawa, U., Hunter, T., Brenner, D., and Karin, M. (1992). *Cell (Cambridge, Mass.)* **70**, 777–789.

Liu, Y., and Teng, C. (1991). *J. Biol. Chem.* **266**, 2180–2185.

Liu, Y., and Teng, C. (1992). *Mol. Endocrinol.* **6**, 355–364.

Louvion, J. F., Havaux-Copf, B., and Picard, D. (1993). *Gene* **131**, 129–134.

Love, R. R., Mazess, R. B., Barden, H. S., Epstein, S., Newcomb, P. A., Jordan, V. C., Carbone, P. P., and DeMets, D. L. (1992). *N. Engl. J. Med.* **320**, 852–856.

Lu, Y., and Federoff, H. J. (1995). *Hum. Gene Ther.* **6**, 419–428.

Lubahn, D. B., Moyer, J. S., Golding, T. S., Couse, J. F., Korach, K. S., and Smithies, O. (1993). *Proc. Natl. Acad. Sci. U.S.A.* **90**, 11162–11167.

Luisi, B. F., Xu, W. X., Otwinowski, Z., Freedman, L. P., Yamamoto, K. R., and Sigler, P. B. (1991). *Nature (London)* **352**, 497–505.

MacDonald, P. N., Dowd, D. R., Nakajima, S., Galligan, M. A., Reeder, R. C., Haussler, C. A., Ozato, K., and Haussler, M. R. (1993). *Mol. Cell. Biol. (London)* **13**, 5907–5917.

Mader, S., and White, J. H. (1993). *Proc. Natl. Acad. Sci. U.S.A.* **90**, 5603–5607.

Mader, S., Kumar, V., de Verneuil, H., and Chambon, P. (1989). *Nature (London)* **338**, 271–274.

Mader, S., Chambon, P., and White, J. H. (1993a). *Nucleic Acids Res.* **23**, 1125–1132.

Mader, S., Chen, J. Y., Chen, Z., White, J. H., Chambon, P., and Gronemeyer, H. (1993b). *EMBO J.* **12**, 5029–5041.

Mangelsdorf, D. J., and Evans, R. M. (1995). *Cell (Cambridge, Mass.)* **83**, 841–850.

Mangelsdorf, D. J., Thummel, C., Beato, M., Herrlich, P., Shutz, G., Umesono, K., Blumberg, B., Kastner, P., Mark, M., Chambon, P., and Evans, R. M. (1995). *Cell (Cambridge, Mass.)* **83**, 835–839.

Martinez, E., and Wahli, W. (1989). *EMBO J.* **8**, 3781–3791.

Martinez, E., Givel, F., and Wahli, W. (1987). *EMBO J.* **6**, 3719–3727.

McDonnell, D. P., Vegato, E., and O'Malley, B. W. (1992). *Proc. Natl. Acad. Sci. U.S.A.* **89**, 10563–10567.

Metzger, D., White, J. H., and Chambon, P. (1988). *Nature (London)* **334**, 31–36.

Montminy, M. (1993). *Science* **261**, 1694–1695.

Munk-Jenson, N., Nielson, S. P., Obel, E. B., and Erikson, P. B. (1988). *Br. Med. J.* **296**, 1150–1152.

Nawaz, Z., Tsai, M.-J., McDonnell, D. P., and O'Malley, B. W. (1992). *Gene Express.* **2**, 39–47.

Nicholson, R. C., Mader, S., Nagpal, S., Leid, M., Rochette-Egly, C., and Chambon, P. (1990). *EMBO J.* **9**, 4442–4454.

Pangaro, L. N. (1990). *In* "Principles and Practice of Endocrinology and Metabolism" (K. L. Becker, ed.), pp. 271–278. Lippincott, Philadelphia.

Pascal, E., and Tjian, R. (1991). *Genes Dev.* **5**, 1646–1656.

Payvar, F., DeFranco, D., Firestone, G. L., Edgar, Wrange, O., Okret, S., Gustafsson, J.-A., and Yamamoto, K. R. (1983). *Cell (Cambridge, Mass.)* **35**, 381–392.

Perlmann, T., Rangarajan, P. N., Umesono, K., and Evans, R. M. (1993). *Genes Dev.* **7**, 1411–1422.

Pham, T. A., Hwung, Y. P., Santiso-Mere, D., McDonnell, D. P., and O'Malley, B. W. (1992). *Mol. Endocrinol.* **6**, 1043–1050.

Philpott, C. C., Klausner, R. D., and Rouault, T. A. (1994). *Proc. Natl. Acad. Sci. U.S.A.* **91**, 7321–7325.

Picard, D., Khursheed, B., Garabedian, M., Fortin, M. G., Lindquist, S., and Yamamoto, K. R. (1990). *Nature (London)* **348**, 166–168.

Pierrat, B., Heery, D. M., Chambon, P., and Losson, R. (1994). *Gene* **143**, 193–200.

Ponglikitmongkol, M., White, J. H., and Chambon, P. (1990). *EMBO J.* **9**, 2221–2231.

Powles, T. J., Jones, A. L., Ashley, S. E., O'Brien, M. E., Tidy, V. A., Treleaven, J., Cosgrove, D., Nash, A. G., Sacks, N., Baum, M., McKinna, J. A., and Davey, J. B. (1994). *Breast Cancer Res. Treat.* **31**, 73–82.

Predki, P. F., Zamble, D., Sarkar, B., and Giguère, V. (1994). *Mol. Endocrinol.* **8**, 31–39.

Press, M. F., Nousek-Goebl, N. A., and Greene, G. L. (1985). *J. Histochem. Cytochem.* **33**, 915–924.

Renaud, J. P., Rochel, N., Chambon, P., Gronemeyer, H., and Moras, D. (1995). *Nature (London)* **378**, 681–689.

Richard, S., and Zingg, H. (1990). *J. Biol. Chem.* **265**, 6098–6103.

Rosen, E. D., O'Donnell, A. L., and Koenig, R. J. (1992). *J. Biol. Chem.* **267**, 22010–22013.

Roy, B., Taneja, R., and Chambon, P. (1995). *Mol. Cell Biol.* **15**, 6481–6487.

Salomonsson, M., Haggblad, J., O'Malley, B. W., and Sitbon, G. M. (1994a). *J. Steroid Biochem. Mol. Biol.* **48**, 447–452.

Salomonsson, M., Carlsson, B., and Haggblad, J. (1994b). *J. Steroid Biochem. Mol. Biol.* **50**, 313–318.

Samaniego, F., Chin, J., Iwai, K., Rouault, T. A., and Klausner, R. D. (1994). *J. Biol. Chem.* **269**, 30904–30910.

Schmid, W., Strahle, U., and Schutz, G. (1989). *EMBO J.* **8**, 2257–2263.

Schüle, R., Rangarajan, P., Kliewer, S., Ransone, L. J., Bolado, J., Yang, N., Verma, I., and Evans, R. M. (1990). *Cell (Cambridge, Mass.)* **62**, 1217–1226.

Schwabe, J. W. R., Neuhaus, D., and Rhodes, D. (1990). *Nature (London)* **348**, 458–461.

Slater, E. P., Redeuhl, G., Theis, K., Suske, G., and Beato, M. (1990). *Mol. Endocrinol.* **4**, 604–610.

Smith, D., and Toft, D. (1993). *Mol. Endocrinol.* **7**, 4–11.

Smith, E. P., Boyd, J., Frank, G. R., Takahashi, H., Cohen, R. M., Specker, B. M., Williams, T. C., Lubahn, D. B., and Korach, K. S. (1994). *N. Engl. J. Med.* **331**, 1056–1060.

Smith, W. C., Nakshrati, H., Leroy, P., Rees, J., and Chambon, P. (1991). *EMBO J.* **10**, 2223–2230.

Stevenson, J. C., Cust, M. P., Ganger, K. F., Hillard, T. C., Lees, B., and Whitehead, M. I. (1990). *Lancet* **335**, 265–269.

Strahle, U., Schmid, W., Schutz, G., Schmitt, J., and Stunnenberg, H. (1988). *EMBO J.* **7**, 3389–3395.

Tong, X., Drapkin, R., Yalamanchili, R., Mosialos, G., and Kieff, E. (1995). *Mol. Cell. Biol.* **15**, 4735–4744.

Tora, L., Gronemeyer, H., Turcotte, B., Gaub, M.-P., and Chambon, P. (1988). *Nature (London)* **333**, 185–188.

Tora, L., White, J. H., Brou, C., Tasset, D., Webster, N., Scheer, E., and Chambon, P. (1989a). *Cell (Cambridge, Mass.)* **59**, 477–487.

Tora, L., Mullick, A., Metzger, D., Ponglikitmongkol, M., Park, I., and Chambon, P. (1989b). *EMBO J.* **8**, 1981–1986.

Towers, T. L., Luisi, B. F., Asianov, A., and Freedman, L. P. (1993). *Proc. Natl. Acad. Sci. U.S.A.* **90**, 6310–6314.

Tur-Kaspa, R., Shaul, Y., Moore, D,. Burk, R., Okret, S., Poellinger, L., and Shafritz, D. A. (1988). *Virology* **167**, 630–633.

Tzukerman, M., Zhang, X.-K., Herman, T., Wills, K. N., Graupner, G., and Pfahl, M. (1990). *New Biol.* **2**, 613–620.

Umesomo, K., and Evans, R. M. (1989). *Cell (Cambridge, Mass.)* **57**, 1139–1146.

Umesomo, K., Murkakami, K. K., Thompson, C. C., and Evans, R. M. (1991). *Cell (Cambridge, Mass.)* **65**, 1255–1266.

Van het Schip, F., Strijker, R., Samallo, J., Gruber, M., and Ab, G. (1986). *Nucleic Acids Res.* **14**, 8669–8680.

Vanleeuwen, F. E., Benraadt, J., Coebergh, J. W. W., Kiemeney, L. A. L. M., Gimbrere, C. H. F., Otter, R., Shouten, L. T., Damhuis, R. A. M., Bontenbal, M., Diepenhorst, F. W., van den Belt-Dusebout, A. W., and van Tinteren, H. (1994). *Lancet* **343**, 448–452.

Vegato, E., Allan, G. F., Schrader, W. T., Tasi, M.-J., McDonnell, D. P., and O'Malley, B. W. (1992). *Cell (Cambridge, Mass.)* **69**, 703–712.

Wahli, W., and Martinez, M. (1991). *FASEB J.* **5**, 2243–2249.

Walker, P., Germond, J.-E., Brown-Leudi, M., Givel, F., and Wahli, W. (1984). *Nucleic Acids Res.* **12**, 8611–8626.

Webster, N. J. G., Green, S., Jin, J. R., and Chambon, P. (1988). *Cell (Cambridge, Mass.)* **54**, 199–207.

Weiss, N. S., Ure, C. L., and Ballard, J. H. (1980). *N. Engl. J. Med.* **303**, 1195–1198.

Weisz, A., and Rosales, R. (1990). *Nucleic Acids Res.* **18**, 5097–5106.

White, J. H., Brou, C., Lutz, Y., Moncollin, V., and Chambon, P. (1992). *EMBO J.* **11**, 2229–2240.

White, J. H., McCuaig, K. A., and Mader, S. (1994). *Bio/Technology* **12**, 1003–1007.

Wilson, T. E., Paulson, R. E., Padjett, K. A., and Milbrandt, J. (1992). *Science* **256**, 107–110.

Wrenn, C. K., and Katzenellenbogen, B. S. (1993). *J. Biol. Chem.* **268**, 24089–24098.

Wurtz, J. M., Bourget, W., Renaud, J.-P., Vivat, V., Chambon, P., Moras, D., and Gronemeyer, H. (1996). *Nat. Struct. Biol.* **3**, 87–94.

Yamamoto, K. R., Godowski, P. J., and Picard, D. (1988). *Cold Spring Harbor Symp. Quant. Biol.* **53**, 803–811.

Yang-Yen, H.-F., Chambard, J.-C., Sun, Y.-L., Smeal, T., Schmidt, T. J., Drouin, J., and Karin, M. (1990). *Cell (Cambridge, Mass.)* **62**, 1205–1215.

Yu, V. C., Delsert, C., Andersen, B., Holloway, J., Devary, O. V., Naar, A., Kim, S. Y., Boutin, J.-M., Glass, C. K., and Rosenfeld, M. G. (1991). *Cell (Cambridge, Mass.)* **67**, 1251–1266.

Zawel, L., and Reinberg, D. (1995). *Annu. Rev. Biochem.* **64**, 533–561.

Zechel, C., Shen, X.-Q., Chambon, P., and Gronemeyer, H. (1994a). *EMBO J.* **13**, 1414–1424.

Zechel, C., Shen, X.-Q., Chen, J.-Y., Chen, Z.-P., Chambon, P., and Gronemeyer, H. (1994b). *EMBO J.* **13**, 1425–1433.

Zhang, X.-K., Hoffmann, B., Tran, P. B.-V., Graupner, G., and Pfahl, M. (1992a). *Nature (London)* **355**, 441–444.

Zhang, X.-K., Lehman, J., Hoffman, B., Dawson, M. I., Cameron, J., Graupner, G., Hermann, T., Tran, P. B.-V., and Pfahl, M. (1992b). *Nature (London)* **358**, 587–591.

Zhuang, Y., Katzenellenbogen, B. S., and Shapiro, D. S. (1995). *Mol. Endocrinol.* **9**, 457–466.

Hong-Ji Xu

Department of Molecular Oncology
The University of Texas M.D. Anderson Cancer Center
Houston, Texas 77030

Strategies for Approaching Retinoblastoma Tumor Suppressor Gene Therapy

I. Introduction

Among the many genetic changes in neoplastic cells, inactivation of tumor suppressor genes plays an important role in the development of human cancer. Examples of tumor suppressor genes and candidate tumor suppressor genes include, but are not limited to, the retinoblastoma gene (*RB*) (Friend *et al.*, 1986; Fung *et al.*, 1987; Lee *et al.*, 1987a), the wild-type *p53* gene (Finlay *et al.*, 1989; Baker *et al.*, 1990), the deleted in colon carcinoma gene (*DCC*) (Fearon *et al.*, 1990), the neurofibromatosis type 1 gene (*NF-1*) (Wallace *et al.*, 1990; Viskochil *et al.*, 1990; Cawthon *et al.*, 1990), the Wilms' tumor gene (*WT-1*) (Call *et al.*, 1990; Gessler *et al.*, 1990; Pritchard-Jones *et al.*, 1990), the von Hippel-Lindau disease tumor suppressor gene (*VHL*) (Duan *et al.*, 1995), the Maspin (Zou *et al.*, 1994), *Brush-1* (Schott *et al.*, 1994), and *BRCA-1* genes (Miki *et al.*, 1994; Futreal *et al.*, 1994) for breast cancer, and the multiple tumor suppressor (*MTS*) or *p16* gene

Advances in Pharmacology, Volume 40

(Serrano *et al.*, 1993; Kamb *et al.*, 1994). The total number of tumor suppressor genes is expected to be well beyond 50 (Knudson, 1993). These suppressor genes apparently act to keep cell growth in check and their inactivation has been linked to the development of a wide variety of human cancers (Weinberg, 1991). The key observation, i.e., that correction of one tumor suppressor gene defect alone in tumors carrying multiple genetic alterations was sufficient to revert their malignant phenotypes, sparked off the hopes for cancer gene therapy (Huang *et al.*, 1988; Baker *et al.*, 1990; Klein, 1990).

The first tumor suppressor gene identified was the retinoblastoma gene (*RB*), which causes hereditary retinoblastoma (Knudson, 1971; 1985; Murphree and Benedict, 1984). The retinoblastoma gene (*RB*) was cloned in the mid-1980s. After years of intense scrutiny, the biological functions of the *RB* gene are beginning to be understood. The *RB* gene encodes a nuclear phosphoprotein of ~110 kDa (pRB), which is differentially phosphorylated during the cell cycle. The established components of the pRB pathway include the E2F transcription factors, which are involved in the transcriptional control of numerous cellular genes responsible for advancing cells through the cell cycle (Nevins, 1992; La Thangue, 1994). The pRB also interacts with certain G_1-phase cyclins (Koff *et al.*, 1992; Resnitzky and Reed, 1995; Geng *et al.*, 1996). Therefore, the *RB* gene apparently plays a key role in cell growth regulation, being involved in those major decisions during the G_1 phase of the cell cycle that govern cell proliferation, quiescence, and differentiation (Weinberg, 1995).

Mutations in *RB* are seen in virtually all cases of retinoblastoma; in addition, the *RB* gene products could potentially be inactivated by hyperphosphorylation, and by viral oncoprotein-like cellular protein binding. Consequently, not only are mutations of the *RB* gene causally related to the retinoblastoma and tumors often occur as the second malignancies in patients with hereditary retinoblastoma (such as osteosarcoma and soft-tissue sarcomas), but loss of *RB* gene function has now been implicated in the progression of many common human cancers, including carcinomas of the bladder, lung, breast, and prostate. Although the literature remains controversial, there is growing evidence suggesting that the RB protein status is potentially a prognostic marker in urothelial carcinoma, non-small-cell lung carcinoma, and perhaps also in some other types of human neoplasms (Xu, 1995).

Moreover, a number of studies have indicated that replacement of the normal *RB* gene in *RB*-defective tumor cells from disparate types of human cancers could suppress their tumorigenic activity in nude mice (Huang *et al.*, 1988; Goodrich and Lee, 1993; Zhou *et al.*, 1994). Although the molecular mechanism of the *RB*-mediated tumor suppression has remained unclear, studies suggest that *RB* may also play a role in elicitation of immunogenicity of tumor cells (Lu *et al.*, 1994a,b, 1996a), antiangiogenesis (Dawson *et al.*, 1995), and suppression of tumor invasiveness (Li *et al.*, 1996), which make the emerging *RB* gene therapy even more attractive. In this regard, preclinical studies have demonstrated that treatment of established human xenograft tumors in nude mice by recombinant adenovirus vectors expressing either wild-type or an N terminal-truncated retinoblastoma protein resulted in regression of the treated tumors (Xu *et al.*, 1996). In addition, a constitutively active form of the pRB protein has been tested in a rat artery model of restenosis to inhibit vascular proliferative disorders following balloon angioplasty (Chang *et al.*, 1995).

II. Basis for Considering the *RB* Tumor Suppressor Gene as a Therapeutic Target

A. Special Role Played by the Retinoblastoma Protein in Regulation of Cell Proliferation

The functional aspects of the *RB* gene and the RB proteins (pRB) have been reviewed frequently (for example, Cooper and Whyte, 1989; Hamel *et al.*, 1993; Horowitz, 1993; Riley *et al.*, 1994; Wang *et al.*, 1994; Weinberg, 1995). The *RB* gene is one of the best-studied tumor suppressor genes or antioncogenes. This gene encodes a nuclear phosphoprotein of 928 amino acids (Lee *et al.*, 1987b). The normal RB protein pattern seen on a Western blot demonstrates a major underphosphorylated RB protein band with apparent relative molecular mass (M_r) of 110 kDa and a more variable region above this band with an apparent relative molecular mass ranging from 110 to 116 kDa, representing the phosphorylated forms of the RB protein (Xu *et al.*, 1989). We initially reported that there was a striking difference in the ratio of underphosphorylated to phosphorylated pRB forms between normal fibroblasts growing exponentially and those arrested in G_1 phase. More underphosphorylated pRB was observed in G_1-arrested cells, suggesting the change in ratio of phosphorylated to underphosphorylated RB proteins was related to the fluctuation in cell cycle (Xu *et al.*, 1989). Four subsequent papers have described the cell cycle-dependent phosphorylation of RB protein in detail (DeCaprio *et al.*, 1989; Buchkovich *et al.*, 1989; Chen *et al.*, 1989; Mihara *et al.*, 1989). It is now widely accepted that the product of the *RB* gene has a key role in cell cycle control.

Cell proliferation depends on transcriptional activation of genes that are responsible for the onset of DNA synthesis as well as other critical events in the G_1 phase of the cell cycle. As demonstrated by Pardee, transition of cells from a serum mitogen-dependent to a serum mitogen-independent state is separated by a distinct time point occurring several hours before the onset of S phase, namely the R (restriction) point (Pardee, 1989). By passing through the R point, the cell commits itself to complete the remainder of the cell cycle through the M phase. Therefore, the R point between the middle G_1 and late G_1 phases of the cell cycle represents a transition in the life of the cell that is as important as the G_1/S boundary.

The phosphorylation status of pRB undergoes a readily distinguishable alteration at a time close to and perhaps contemporaneous with the R point transition of the cell cycle (Weinberg, 1995). During the mid-G_1 phase, the only pRB species detected is an underphosphorylated form. When cells progress through the cell cycle, the pRB content increases gradually. However, the majority of pRB synthesized after the mid-G_1 phase is hyperphosphorylated. In other words, pRB hyperphosporylation occurs in late G_1, preceding the G_1/S boundary (Xu *et al.*, 1991a; Mittnacht *et al.*, 1994). The RB protein maintains this hyperphosphorylated status throughout the remainder of the cell cycle, becoming dephosphorylated only on evolution from the M to early G_1 phase (Ludlow *et al.*, 1990; Xu *et al.*, 1991a; Mittnacht *et al.*, 1994).

The underphosphorylated form of pRB is able to form complexes with the transcription factor E2Fs or directly interact with the E2F site, and switches the E2F site from a positive to negative element in transcriptional control. The E2F site (variants of the consensus nucleotide sequence TTTCG-CGC) is present in the promoters of diverse cellular genes that are responsible for advancing cells through the cell cycle, including, for instance, c-*myc*, B-*myb*, *cdc2*, dihydrofolate reductase, thymidine kinase, and *RB* as well as the *E2F-1* gene itself (Chellappan *et al.*, 1991; Nevins, 1992; Weintraub *et al.*, 1992; La Thangue, 1994; Shan *et al.*, 1994, 1996; Sardet *et al.*, 1995). Since hyperphosphorylated pRB appears to have lost the ability to interact with E2Fs, the inhibitory effect of pRB on cell growth can be abrogated by hyperphosphorylation.

The timing of pRB phosphorylation led to an attractive functional model, although it is still largely unproven (Weinberg, 1995). This model suggests that pRB is an R point guardian. The RB protein exerts most of its growth inhibitory effects in the first two-thirds of the G_1 phase. A cell that has progressed through the early and mid-G_1 phase encounters the R point gate. Should conditions be ready for advancement into the remainder of the cell cycle, pRB will undergo phosphorylation and functional inactivation, causing it to open the gate and permit the cell to proceed into late G_1. Cells that lack normal pRB function for various reasons will proceed freely into late G_1. Without pRB, the upstream components of the cell cycle clock that regulate pRB phosphorylation, such as cyclin D, cyclin E, and their corresponding cyclin-dependent kinases (CDKs) (Kato *et al.*, 1993; Ewen *et al.*, 1993) lose much of their influence in the decision of the cell to pass through the R point gate. Taken together, pRB allows the cell cycle clock to control the expression of numerous genes that mediate advancement of the cell through a critical phase of its growth cycle being involved in the major decisions concurrent with the R point transition. Functional loss of pRB deprives the clock and thus the cell of an important mechanism for braking cell proliferation.

B. Association of the Loss of Retinoblastoma Protein Function with Many Common Human Malignancies

Although the *RB* gene was initially named because deletions or mutations within the gene caused the rare childhood ocular tumor, retinoblastoma, loss of pRB function is not only causally related to the retinoblastoma, but is also linked to the progression of many common human cancers as summarized in Table I.

In addition, with the revolutionary antigen retrieval technique and the available specific anti-pRB antibodies, immunohistochemistry has become a highly sensitive and reliable method for detection of pRB inactivation in routinely processed pathological specimens (Xu, 1995). Altered pRB expression as determined by immunohistochemical analysis appears to signal a poor prognosis in a subset of human malignancies. It was initially reported by Cance *et al.*, (1990) that loss of functional pRB was a statistically significant negative prognostic factor in high-grade adult soft-tissue sarcomas.

TABLE I The Role of Altered Retinoblastoma Protein Expression/Function in Human Malignancies

Role	Tumor type	Ref.
Tumor initiation[a]	Retinoblastoma	Friend *et al.* (1986); Lee *et al.* (1987a); Fung *et al.* (1987)
	Osteosarcoma	Fung *et al.* (1987); Toguchida *et al.* (1988)
	Soft tissue sarcomas	Reissmann *et al.* (1989)
	Small-cell lung carcinoma	Harbour *et al.* (1988); Yokota *et al.* (1988)
	Parathyroid carcinoma	Cryns *et al.* (1994)
Tumor progression[b]	Bladder carcinoma	Presti *et al.* (1991); Xu *et al.* (1993)
	Non-small-cell lung carcinoma	Yokata *et al.* (1988); Xu *et al.* (1991b)
	Breast carcinoma	T'Ang *et al.* (1988); Lee *et al.* (1988)
	Prostate carcinoma	Bookstein *et al.* (1990a)
	Hepatocellular carcinoma	Zhang *et al.* (1994)
	Astrocytoma	Henson *et al.* (1994)
	Acute myelogenous leukemia	Kornblau *et al.* (1992)

[a] Altered pRB expression/function occurs as an early event directly related to tumor formation.
[b] Altered pRB expression/function occurs as a later event in tumor development and is seen more frequently in advanced disease.

Subsequently, two independent studies done concurrently concluded that altered pRB expression was a prognostic factor among patients with transitional cell carcinoma of the bladder (Cordon-Cardo *et al.*, 1992; Logothetis *et al.*, 1992). For lung cancer patients, the initial pilot studies have also been promising, implying that altered RB and p53 protein status could be a synergistic prognostic factor in early-stage non-small-cell lung carcinomas (Xu *et al.*, 1994a). A much worse survival pattern has been reported as well for acute myelogenous leukemia patients who have low or absent levels of pRB protein in their peripheral blood leukemic cells (Kornblau *et al.*, 1994). Since all studies done so far to investigate association between pRB status in human cancer and the clinical outcome of patients have been retrospective, and the number of cases in each cohort was fairly small, definitive retrospective and prospective studies with an adequate sample size for statistical calculations are ow underway to determine whether or not loss of pRB function can be considered as a prognostic factor in clinical practice.

C. RB Gene-Mediated Tumor Suppression

I. Suppression of Tumorigenicity of RB-Defective Tumor Cells by Wild-Type RB Gene Replacement

The most direct proof that the cloned *RB* gene is indeed a tumor suppressor gene must come from introduction of a cloned intact copy of the gene into cancer cells with observed tumor suppression function. The authenticity of the *RB* tumor suppressor gene and the suggestion that pRB inactivation may play a broad role in human malignancies have been reinforced by abundant studies indicating replacement of the normal *RB* gene into *RB*-defective tumor cells could suppress their tumorigenic activity in immunodeficient mice. The tumor cell lines studied were derived from widely disparate types of human cancers such as the retinoblastoma, osteosarcoma, and carcinomas of the bladder, prostate, breast, and lung (Table II).

Of note, there has been a tendency in the literature to separate the inhibition of cell growth by *RB* replacement in *RB*-defective tumor cells from its tumor suppression function (Takahashi *et al.*, 1991; Chen *et al.*, 1992; Goodrich *et al.*, 1992; Zhou *et al.*, 1994). After transient transduction with a wild-type pRB-expressing retrovirus or plasmid, as documented in several early studies, the *RB*-deficient retinoblastoma and osteosarcoma tumor cells in culture displayed striking changes, including cell enlargement, senescent phenotype, and lower growth rate (Huang *et al.*, 1988; Templeton *et al.*, 1991). Subsequently, it was found that long-term stable clones of the *RB*-reconstituted tumor cells can be isolated that grew just as rapidly as the parental or matched RB⁻ revertant clones. The majority of RB⁺ clones obtained, however, were nontumorigenic or with significantly reduced tumorigenicity in nude mice. The mechanisms for the dissociation of suppression of tumorigenicity in nude mice from inhibition of tumor cell growth

TABLE II Suppression of Tumorigenicity by Replacing Wild-Type *RB* Gene into *RB*-Defective Human Tumor Cell Lines

Tumor type	Cell line	Type of RB reconstitution	Suppression of tumorigenicity	Ref.
Retinoblastoma	WERI-Rb-27	Transient	Complete	Huang *et al.* (1988)
		Transient	Complete	Sumegi *et al.* (1990)
		Stable	Partial	Xu *et al.* (1991c)
		Stable	Complete	Madreperla *et al.* (1991)
		Stable	Complete	Chen *et al.* (1992)
		Stable	Partial	Zhou *et al.* (1994)
	WERI-Rb-1	Stable	No effect	Muncaster *et al.* (1992)
	Y79	Stable	No effect	Muncaster *et al.* (1992)
Osteosarcoma	Saos-2	Transient	Complete	Huang *et al.* (1988)
		Stable	Partial	Zhou *et al.* (1994)
Bladder carcinoma	5637	Stable	Partial	Takahashi *et al.* (1991)
		Stable	Partial	Goodrich *et al.* (1992)
		Stable[a]	Partial	Banerjee *et al.* (1992)
		Stable	Partial	Zhou *et al.* (1994)
	HT1376	Stable	Partial	Goodrich *et al.* (1992)
Prostate carcinoma	DU145	Stable	Partial	Bookstein *et al.* (1990b)
		Stable[a]	Partial	Banerjee *et al.* (1992)
Breast carcinoma	MDA-MB-468	Stable	Partial	Wang *et al.* (1993)
		Stable	Partial	Li *et al.* (1996)
	MDA-468-S4	Stable	No effect	Muncaster *et al.* (1992)
Lung carcinoma	H2009[b]	Stable	Partial	Kratzke *et al.* (1993)
		Stable	Partial	Li *et al.* (1996)
	Lu-135[c]	Stable	Partial	Ookawa *et al.* (1993)
	N417[c]	Stable	Partial	Ookawa *et al.* (1993)

[a] By transfer of a normal chromosome 13 via microcell fusion.
[b] Non-small-cell lung carcinoma.
[c] Small-cell lung carcinoma.

in culture by *RB* replacement are unclear. It is certainly possible that *RB* replacement restores the sensitivity of tumor cells to a variety of physiologic growth inhibitory signals that may be present *in vivo* when the tumorigenicity assay is done in nude mice. Such external growth-inhibitory agents would be absent under regular cell culture conditions, leading to rapid cell growth (Chen *et al.*, 1992).

Nevertheless, suppression of tumorigenicity of RB⁻ tumor cells *in vivo* by reexpressing the wild-type pRB implies that the *RB* gene could be a potential therapeutic target for human cancer.

2. Broad Biological Basis of RB-Mediated Tumor Suppression

Although the molecular mechanisms of *RB*-mediated tumor suppression have remained unclear, publications have suggested that, in addition to its well-known antiproliferative effects, pRB may also play a role in elicitation of antiangiogenesis and in enhancing the immunogenicity of tumor cells. It was reported that conditioned medium (CM) collected from RB⁻ retinoblastoma, osteosarcoma, and non-small-cell lung carcinoma cell lines were able to induce angiogenesis as shown by their ability to stimulate endothelial cell migration and proliferation *in vitro* and by neovascularization in the rat cornea *in vivo*. In contrast, after these cell lines were reverted to nontumorigenic status by wild-type *RB* gene replacement, their CM became antiangiogenic (Dawson *et al.*, 1995). It has also been documented in the literature that HLA class II induction by interferon γ (IFN-γ) in the *RB*-defective breast carcinoma cell line MDA-468-S4 and the non-small-cell lung carcinoma cell line H2009 requires reconstitution of the wild-type *RB* gene expression (Lu *et al.*, 1994a, 1996a). The class II proteins present peptides derived from proteolytically processed antigens to CD4[+] T lymphocytes as part of the immune response. Therefore, pRB likely has a role in enhancing tumor immunogenicity.

To determine if replacement of the retinoblastoma (*RB*) tumor suppressor gene could inhibit invasion of *RB*-defective tumor cells, we have completed studies using the Boyden chamber assay (Li *et al.*, 1996). The studies were done in a diverse group of stable *RB*-reconstituted human tumor cell lines, including those derived from osteosarcoma and carcinomas of the lung, breast, and bladder. The expression of the exogenous wild-type RB protein in these tumor cell lines was driven by either a constitutively active promoter or an inducible promoter. These studies provided a novel insight into the biological basis of *RB*-mediated tumor suppression in *RB*-defective tumor cells.

The Boyden chamber assay is a rapid chemoinvasion assay for quantitating the invasive potential of tumor cells *in vitro*, and many tumor cells characterized as invasive and metastatic *in vivo* are able to invade Matrigel (Becton, Bedford, MA) *in vitro* (Albini *et al.*, 1987; Karikó *et al.*, 1993; Sato *et al.*, 1994). The assay measures the capability of cultured cells to (1) attach to the matrix, (2) degrade the matrix, and (3) migrate toward a chemoattractant. These events are considered to be important steps in tumor metastasis through basement membrane *in vivo*. Since nonmetastatic and highly metastatic cell lines often give comparable chemotactic responses in the chemotaxis assay (without Matrigel coatings), whereas their responses differ in the chemoinvasion assay (with Matrigel coatings; only the metastatic cells invade), it appears that the ability to degrade the basement membrane barrier is essential for penetration of tumor cells through Matrigel (Albini *et al.*, 1987). In our present studies, we used a relatively thin coating of Matrigel (0.5 μg/mm^2) and overnight incubation (~16 hr) for the invasion assay. Under these conditions, the great majority of tumor cells studied remained attached to the filter after washing with phosphate-buffered saline (PBS) at the end of each incubation. There were no substantial changes in chemotactic migration of *RB*-defective tumor cells after wild-type *RB* replacement. A differential Matrigel invasion, however, was clearly observed, with significantly more tumor cells from the parental *RB*-defective cell lines and the RB⁻ revertants than from the *RB*-reconstituted RB⁺ cell lines penetrating through the Matrigel ($p < 0.001$, two-tailed t test). The results suggest that pRB may directly or indirectly regulate expression of certain cellular genes that are capable of degrading the basement membrane, and thus are crucial for tumor cell invasion and metastasis. A list of such candidate cellular genes includes, for example, those encoding gelatinase A (type IV collagenase) and/or membrane-type matrix metalloproteinase (MT-MMP) (Liotta *et al.*, 1991; Azzam *et al.*, 1993; Brown *et al.*, 1993; Sato *et al.*, 1994). Expression of these gene products enhances tumor cell invasion and metastasis by degrading extracellular matrix macromolecules. Such issues are now under further investigation in our laboratory.

Furthermore, the inhibition of invasiveness of *RB*-defective tumor cells by pRB was apparently well correlated with suppression of tumorigenicity *in vivo*. In contrast, although replacing either wild-type *p53* or *RB* gene into RB⁻/p53ⁿᵘˡˡ Saos-2 osteosarcoma cells significantly suppressed their tumorigenicity in nude mice, introduction of wild-type *p53* into Saos-2 had much less impact on invasiveness of the tumor cells as compared to *RB* replacement (Li *et al.*, 1996). Therefore, the mechanisms for tumor suppression by individual tumor suppressor genes may be quite different from each other.

These findings, taken together, may intimate that *RB*-mediated tumor suppression has a broader biological basis, which certainly makes the emerging *RB* tumor suppressor gene therapy for human cancer even more attractive.

3. Tumor Suppressor Resistance

In general, the results of previous studies, except for one report (Muncaster *et al.*, 1992), indicated that replacement of the *RB* gene reduced tumorigenicity of *RB*-defective tumor cells. However, there were inconsistent statements on the tumorigenic potential of *RB*-reconstituted tumor cells (Table II). This has raised questions as to whether correction of the *RB* gene defect alone in tumors usually carrying multiple genetic alterations is sufficient to completely revert their malignant phenotypes. A solution to this question may need to be considered carefully when evaluating the feasibility of *RB* gene therapy for human cancers.

We have found that, although the *RB*-mediated tumor suppression was substantial, it was often incomplete anda portion of the *RB*-reconstituted retinoblastoma, osteosarcoma, and bladder carcinoma tumor cells were able to survive and form RB⁺ xenograft tumors in nude mice after a prolonged latency period (Xu *et al.*, 1991c; Zhou *et al.*, 1994). Similar observations have been reported by others investigating carcinomas of the breast, prostate, and lung (Table II). The *RB*-mediated partial (incomplete) tumor suppression may reflect the biological heterogeneity of the tumor cell population, i.e., some tumor cells in a given *RB*-defective tumor cell population are resistant to the tumor suppression function rendered by *RB* gene replacement. A fundamental aspect of tumor cell biology behind this phenomenon is the fact that the *RB* gene may play a role in tumor progression of only a portion of the tumor cells, for example, in carcinomas of the bladder, lung, prostate, and breast (Table I). Furthermore, even though the *RB* gene may play a critical role in the initiation of all or the majority of the tumor cells in retinoblastomas and osteosarcomas (Table I), these tumor cells are likely to acquire additional random and/or specific genetic defects late in the tumor progression. Therefore, some tumor cells in a given tumor cell population may have inherited or acquired the ability to surmount/deregulate the downstream effector pathway(s) in which the *RB* gene normally functions, or the ability simply to inactivate the *RB* gene product by hyperphosphorylation. In this connection, a shorter survival, comparable to survival in those lacking pRB expression, has been reported for acute myelogenous leukemia patients with evidence of hyperphosphorylated pRB in their leukemic cells (Kornblau *et al.*, 1994).

Consequently, as documented in the literature, a significant percentage of *RB*-reconstituted tumor cells still formed tumors in the nude mouse assay. Such tumors, although retaining normal *RB* expression, were histologically malignant and invasive. This phenomenon is referred to by us as *tumor suppressor resistance* (TSR) (Zhou *et al.*, 1994), which is equivalent to multiple drug resistance (MDR) in chemotherapeutics. Since inactivation of the *RB* gene is restricted to a subset of human tumors, in a broad sense, besides a portion of the *RB*-defective tumor cells, other tumors with a normal endogenous *RB* gene probably have also inherited or acquired TSR to pRB. In fact, TSR could be a general phenomenon extending to include other tumor suppressor genes. For instance, *p53* gene-mediated tumor suppression may be abrogated by changing the protein conformation in a certain cellular environment (Zhang *et al.*, 1992). The TSR phenotype of a tumor cell may result from amplified/activated oncogene products (such as cyclins D and E and CDKs, in the case of pRB resistance), tumor suppressor-associated proteins, or other physiological growth signals.

From a practical point of view, since tumor suppressor resistance may reduce the efficacy of, and pose obstacles to, the *RB* gene or any other tumor suppressor gene therapy for human cancers, new strategies would be required to improve the potential treatment.

4. Enhanced Tumor Suppression by an Amino-Terminal Truncated Retinoblastoma Protein

The effort in pRB research has been focused on the C-terminal moiety; little is known so far about its N terminal. Notably, a bacterium-expressed 56-kDa RB protein segment lacking the N-terminal 393 amino acid residues was initially considered as a functional equivalent of the full-length pRB (Goodrich *et al.*, 1991). Nevertheless, low molecular mass proteins immunoreactive to several anti-pRB antibodies, including RB-WL-1, RB-PMG3-245, C36, and RB1-AB20, were often observed in human fibroblasts and hematopoietic cells (Xu *et al.*, 1989; Mihara *et al.*, 1989; Furukawa *et al.*, 1990; Stein *et al.*, 1990; Thomas *et al.*, 1991; Rogalsky *et al.*, 1993). It initially was proposed by us that these pRB-like proteins could represent translation from the second in-frame AUG codon on the *RB* messenger RNA (Xu *et al.*, 1989). We have subsequently compared the biological properties of the full-length pRB (pRB^{110}) with an artificial N terminal-truncated RB protein of about 94 kDa (pRB^{94}). pRB^{94} was initiated from the second in-frame AUG codon of the *RB* transcript and lacked the N-terminal 112 amino acid residues of the full-length RB protein (Xu *et al.*, 1994b).

To compare the effects of pRB^{94} and pRB^{110} on tumor cells, three types of RB plasmid vectors were constructed. The vectors contained pRB^{94}- or pRB^{110}-coding sequences whose expression was driven, respectively, by the β-actin gene promoter, the cytomegalovirus (CMV) promoter, and the long terminal repeats (LTRs) of Moloney murine leukemia virus (MoMLV). A combined procedure was employed involving immunocytochemical staining and [^{3}H]thymidine *in situ* labeling of the tumor cells following transfection with the pRB^{94}- or pRB^{110}-encoding plasmid vectors. A series of RB$^-$ tumor cell lines with diverse tissue/organ origins were examined. The data are summarized in Table III. The studies demonstrated that the RB-defective tumor cells expressing exogenous pRB^{94} did not progress through the cell cycle, as evidenced by their failure to incorporate [^{3}H]thymidine into DNA. However, the percentage of tumor cells undergoing DNA replication were only slightly lower in cells producing the exogenous pRB^{110} than in cells that were RB$^-$ (Table III).

Perhaps even more striking was the observation that pRB^{94} expression also significantly reduced colony formation in two RB$^+$ (with normal *RB* alleles) tumor cell lines examined, namely the fibrosarcoma cell line HT1080 and the cervical carcinoma cell line HeLa (Xu *et al.*, 1994b), while no such effects were observed when an additional pRB^{110}-coding gene(s) was introduced by transfection using plasmid vectors (Fung *et al.*, 1993) or by microcell fusion (Anderson *et al.*, 1994).

In an attempt to elucidate the basis for the increased tumor cell growth inhibition by pRB^{94} as compared to pRB^{110}, the pRB^{94} protein expressed in transient-transfected 5637 tumor cells was pulse–chase labeled with [^{35}S]methionine. It was determined that the half-life of pRB^{94} protein was approximately 12 hr, which was two to three times longer than the half-life of pRB^{110} in the same tumor cell line. Of particular interest was the fact that the *RB*-defective bladder carcinoma cell line 5637 failed to phosphory-late the pRB^{94} protein as determined by the banding pattern on Western immunoblots, although in a parallel study pRB^{110} expressed in transfected 5637 cells was hyperphosphorylated. In addition, using an *in vitro* kinase

TABLE III   DNA Synthesis in Various *RB*-Defective Tumor Cells Expressing Exogenous pRB or pRB[94] Proteins[a,b]

Recipient cell type	Tumor origin	Promoter in vector[c]	Percentage of cells incorporating [³H]thymidine			
			pRB^{94}		pRB^{110}	
			RB^+	RB^-	RB^+	RB^-
5637	Bladder carcinoma	I	2	43	34	45
		II	2	41	21	39
		III	3	39	25	32
MDA-MB-468	Breast carcinoma	II	0[d]	39	14	40
		III	1	31	16	28
H2009	Lung carcinoma	I	0[d]	19	19	26
DU145	Prostate carcinoma	II	1	33	23	33
Hs913T	Fibrosarcoma	II	1	36	18	34
Saos-2	Osteosarcoma	II	1	35	19	32

[a] Reproduced from Xu *et al.* (1994b), by copyright permission of the *National Academy of Sciences of the United States of America.*

[b] Tumor cells ~24 hr after transfection with pRB[94] or the full-length pRB[110] expression vectors containing a β-actin, CMV, or long terminal repeat promoter were labeled with [³H]thymidine for 2 hr and then fixed and immunostained with MAb-1 anti-RB antibody. Stained slides were subsequently coated with autoradiographic emulsion and exposed for 5–7 days. Detailed methods have been described (Xu *et al.*, 1991a). About 400 to 1600 pRB[94]- or pRB[110]-expressing tumor cells and 600 RB⁻ tumor cells in the same slides were assessed for each entry of [³H]thymidine uptake. Lack of cellular DNA synthesis as determined by failure of the vast majority of the tumor cells to incorporate thymidine implies the indicated tumor cell population tended not to progress through the cell cycle (Baker *et al.*, 1990). Differences in percent tumor cells incorporating [³H]thymidine were statistically significant between pRB[94]- or pRB[110]-expressing tumor cells (two-tailed *t* test, $p < 0.0001$), whereas no such differences were observed between the RB⁻ tumor cells ($p = 0.57$ by *t* test).

[c] Promoters in vector: I, β-actin gene promoter; II, CMV promoter/enhancer; III, long terminal repeat of Moloney murine leukemia virus.

[d] Less than 0.5%.

reaction, we have also found that the baculovirus-produced un- and hypo-phosphorylated pRB[94] obtained from insect cells was a less effective substrate for the human cdc2 kinase when compared to pRB[110] (Xu *et al.*, 1994b).

The mechanism for enhanced tumor suppression by an N terminal-truncated RB protein is not clear yet. To better understand the functional difference between the N terminal-truncated pRB[94] and the full-length pRB[110], a modified tetracycline-responsive gene expression system has been used in our laboratory to establish several stable tumor cell lines, in which expression of the full-length or the N terminal-truncated RB protein can be reversibly turned on and off (Gossen and Bujard, 1992). In this *tet* repressor/operator-based regulatory system, tetracycline is an "inhibitor" rather than an "inducer." By using this system, we found that, for example, at about 8 hr after removal of tetracycline from the culture medium, the tumor cells clearly started accumulation of pRB[94], followed by morphological changes and failure of the vast majority of the tumor cells to incorporate [^{3}H]thymidine; while at the same time point most of the pRB[110]-reconstituted tumor cells remained RB$^-$ and had normal DNA synthesis. At 24 hr after removal of tetracycline, however, both cell lines became RB$^+$ (Zhou *et al.*, 1996). Since it is known that the N terminal-truncated RB protein has a longer half-life and tends to remain in an active un- or hypophosphorylated form (Xu *et al.*, 1994b), we suggest that fast accumulation of mostly the active form of the RB protein in the tumor cells may account for the enhanced tumor cell growth suppression by the N terminal-truncated RB protein. In this regard, other investigators have reported that a mutant murine RB protein, altered at eight potential phosphorylation sites and not able to be phosphorylated, represses the E2-containing promoters with 50- to 80-fold greater efficiency than its wild-type counterparts (Hamel *et al.*, 1992).

III. *RB* Tumor Suppressor Gene Therapy in Animal Models: *In Vivo* Efficacy Tests

A. *RB* Gene Therapy of Established RB$^-$ and RB$^+$ Human Xenograft Tumors in Nude Mice via Replication-Deficient Adenovirus Vectors

We have constructed several replication-deficient recombinant adenovirus (Ad) vectors expressing either N terminal-truncated pRB[94], or the full-length pRB[110], and compared their tumor suppression effects in laboratory animal models (Xu *et al.*, 1996). The desired human *RB* cDNA fragments encoding pRB[94] (or pRB[110]) were originally inserted into the plasmid vector, pRc/CMV (Invitrogen, San Diego, CA). The pRB[94] or the pRB[110] minigene cassette from the corresponding pRc/CMV plasmid vector, which contains the human cytomegalovirus promoter/enhancer (CMVp), the *RB* cDNA fragment, and the polyadenylation site of the rabbit β-globin gene, was then recovered and transferred into the shuttle plasmid pXCJL.1 between the sequences representing 0–1.3 and 9.2–16 map units (mu) of the Ad5 genome. The resultant recombinant shuttle plasmids, pXCJL.RB94 and pXC-JL.RB110, respectively, were further cotransfected with the pJM17 plasmid into 293 cells using the lipofectin reagent (GIBCO/BRL life Technologies, Gaithersburg, MD). Infectious *RB* adenoviruses were generated after *in vivo* recombination (McGrory *et al.*, 1988). The resultant recombinant viruses, in which a large portion of the E1a and E1b region (1.3–9.2 mu) of the Ad5dl309 genome was replaced by the pRB[94] or pRB[110] expression cassette, were named AdCMVpRB94 and AdCMVpRB110, respectively (Fig. 1).

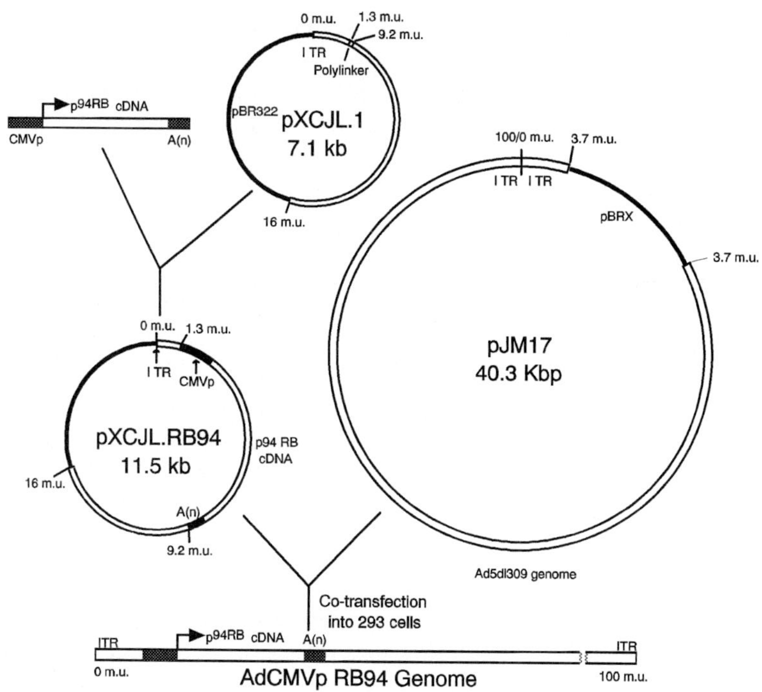

FIGURE 1   Construction of the replication-deficient recombinant adenovirus AdCMV-pRB94 and schematic representation of its genome. [The E1 shuttle plasmid pXCJL.1 and the master adenovirus type 5 (Ad5) plasmid pJM17, which contains the backbone of the circular adenovirus Ad5dl309 genome, were generously provided by F. L. Graham, McMaster University, Ontario, Canada (McGrory *et al.*, 1988).]

AdCMVpRB94 and AdCMVpRB110-infected *RB*-defective (RB⁻) non-small-cell lung carcinoma H2009, bladder carcinoma 5637, HT1376, breast carcinoma MDA-MB-468, and osteosarcoma Saos-2 cells expressed high levels of exogenous RB proteins as determined by immunocytochemical staining. Western blot analysis of cell lysates prepared from virus-transduced tumor cells demonstrated that the exogenous RB proteins accumulated in the tumor cells were mostly hypophosphorylated or unphosphorylated with corresponding molecular masses of 94 and 10 kDa, respectively (Fig. 2).

To assess the effect of pretreating the tumor cells with the recombinant RB viruses on tumorigenicity *in vivo*, H2009 and 5637 tumor cells were infected with AdCMVpRB94, AdCMVpRB110, and a control adenovirus vector, AdCMVpβ-gal, respectively, at a multiplicity of infection (MOI) of 10 or 100, and then implanted into the dorsal flanks of athymic nude mice. The human lung and bladder carcinoma cells following treatment with PBS and the control AdCMVpβ-gal virus consistently formed large, progressively growing tumors in nude mice. In contrast, even at a lower MOI of 10, no

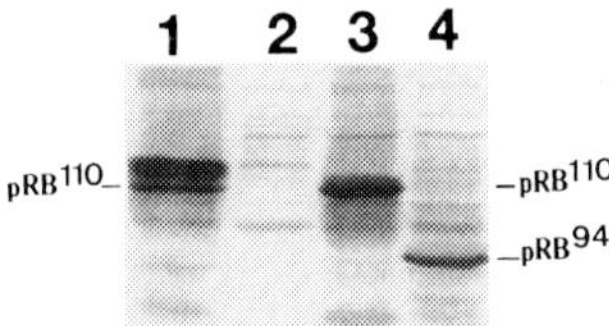

FIGURE 2   Detection of pRB^{94} and pRB^{110} protein expression in the recombinant adenovirus-infected tumor cells by Western immunoblotting. *RB*-defective human lung carcinoma H2009 tumor cells were infected with AdCMVpβ-gal, AdCMVpRB94, or AdCMVpRB110 recombinant adenoviruses at an MOI of 10 PFU/cell. Cell lysates were prepared from tumor cells at 24 hr after virus infection and subjected to Western blot analysis (Xu *et al.*, 1991a). The RB-defective (RB$^-$) tumor cells after infection with AdCMVpRB94 or AdCMVpRB110 viruses produced, respectively, pRB^{94} (lane 4) or pRB^{110} (lane 3) proteins with the anticipated molecular mass and banding pattern, while no RB protein was detectable in the AdCMVpβ-gal virus-infected tumor cells (lane 2). A cell lysate made from the human diploid fibroblast cell line WI-38, expressing the normal endogenous RB protein (lane 1), was used in the Western blot analysis to provide a comparison between RB protein patterns. [Reproduced from Xu *et al.* (1996), by copyright permission of the American Association for Cancer Research, Inc.].

tumor growth was observed in the flanks of nude mide 5 to 10 weeks after injection of tumor cells pretreated with the AdCMVpRB94 recombinant adenovirus (Fig. 3). In addition, although pre-treatment with either the AdCMVpRB94 or the AdCMVpRB110 viruses suppressed tumorigenicity of *RB*-defective tumor cells in nude mice, there were differences in effectiveness between the two RB adenoviruses. When the H2009 and 5637 tumor cells were infected with the AdCMVpRB110 virus at MOI of 100, no tumors were formed following subcutaneous injection of the tumor cells into nude mice. However, about half of the mice injected with the H2009 or the 5637 tumor cells pre-treated with the AdCMVpRB110 virus at the lower MOI of 10 still formed small tumors (Fig. 3).

In simulated cancer gene therapy trials, mice bearing established subcutaneous tumors derived from RB$^+$ and RB$^-$ human 5637 bladder carcinoma cells received intra- and peritumoral injection of either PBS alone, or AdCMVpβ-gal, AdCMVpRB110, and AdCMVpRB94 virus supernatant, respectively. In the RB$^-$ tumor-bearing mice treated with buffer alone, tumors continued to grow aggressively (Fig. 4, top). While the majority of tumors treated with the AdCMVpRB110 virus grew at a reduced rate, all tumors that received six doses of AdCMVpRB94 treatment stopped growing and three of five tumors partially shrank (Fig. 4, top). Notably, all RB$^+$ tumors derived from the *RB*-reconstituted bladder carcinoma 5637R cells, which expressed the full-length pRB^{110} protein (Zhou *et al.*, 1994), also partially or completely regressed in response to the initial AdCMVpRB94 adenovirus treatment, whereas no such effect was observed when the tumors were treated with the AdCMVpβ-gal virus (Fig. 4, bottom). Therefore, pRB^{94} is evidently a more potent tumor-suppressing reagent than is the full-length, wild-type pRB^{110} (Xu *et al.*, 1996).

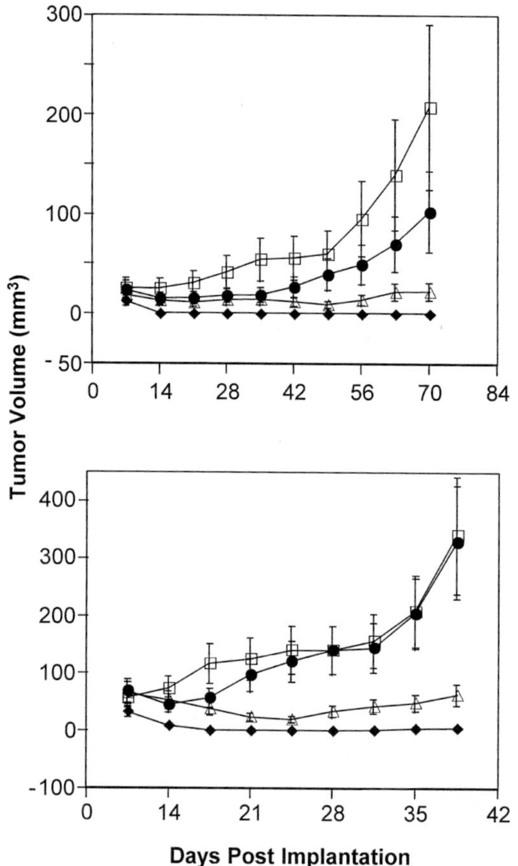

FIGURE 3   Suppression of tumor formation of human lung and bladder carcinoma cells in nude mice by pretreatment with the replication-deficient adenovirus vectors AdCMVpRB94 and AdCMVpRB110. Human non-small-cell lung carcinoma H2009 cells (*top*) and bladder carcinoma 5637 cells (*bottom*) were incubated with PBS alone (□), or infected with AdCMVpβ-gal (●), AdCMVpRB94 (◆), and AdCMVpRB110 (△), respectively, at an MOI of 10. The pretreated tumor cells were injected subcutaneously into the flanks of nude mice at 24 hr after treatment, and the tumor volume was measured weekly. Bars, standard deviations. [Reproduced from Xu *et al.* (1996), by copyright permission of the American Association for Cancer Research, Inc.].

Another truncated version of pRB, pRB[56], beginning at amino acid 379, has also been reported in the literature as a more potent inhibitor of cell cycle progression compared to the full-length pRB (Wills *et al.*, 1995). When recombinant adenoviruses expressing either pRB[56] or the full-length pRB wee tested for growth suppression in several human tumor cell lines, both viral vectors were able to inhibit cell growth of tumor cells as measured by [³H]thymidine incorporation assay. The inhibition of cell growth by pRB[56], however, occurred at lower pRB[56] doses as compared to the effective doses

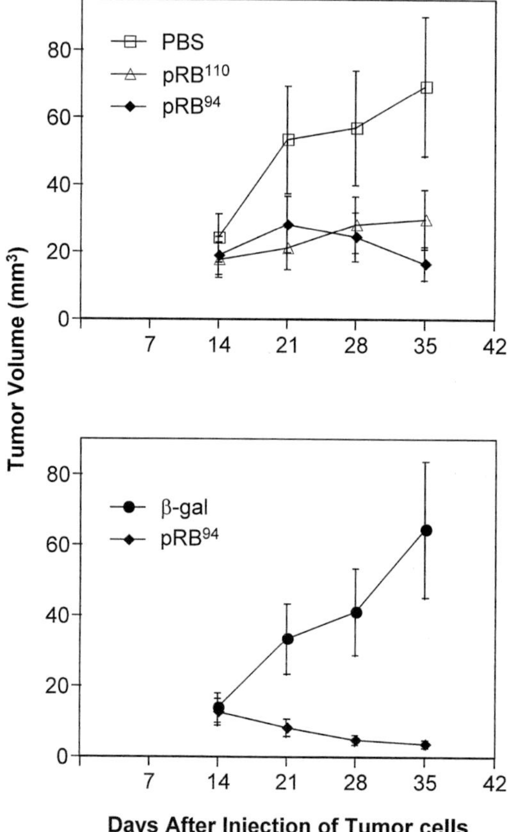

FIGURE 4   *RB* gene therapy of human cancers *in vivo* in nude mice xenograft tumor models via AdCMVpRB94 and AdCMVpRB110 adenoviral vectors. The human RB⁻ 5637 (*top*) and RB⁺ 5637R (*bottom*) (Zhou *et al.*, 1994) bladder carcinoma cells were injected subcutaneously into the dorsal flanks of nude mice. Prior to therapy, the established tumors were randomized and regrouped by volume. Intra- and peritumoral injections of either PBS alone, the control AdCMVpβ-gal adenovirus, the AdCMVpRB110 adenovirus, or the AdCMVpRB94 adenovirus were administered twice a week for a total of six doses per animal (5×10^8 PFU/dose). Tumor volume was measured once a week for 4 weeks. Bars, standard deviations. [Reproduced from Xu *et al.*(1996), by copyright permission of the American Association for Cancer Research, Inc.]

for the full-length pRB. Notably, by a chloramphenicol acetyltransferase (CAT) reporter gene assay, it was also demonstrated *in vitro* that transcriptional represssion of the E2 promoter by the N terminal-truncated pRB[56] was greater than by the full-length pRB (Wills *et al.*, 1995). The E2 promoter contains two consensus sequence sites (the E2F sites) that are the common targets of the E2F transcriptional factors. Variants of the consensus sequence are present in the promoters of many cellular genes that are known to be important for cell growth control (Nevins, 1992; Weinberg, 1995). Therefore, the finding further suggests that pRB^{56} controls the cell cycle clock with greater efficiency than its wild-type counterparts.

B. Cytostatic Gene Therapy of Vascular Proliferative Disorders with a Modified Retinoblastoma Protein in Rat and Pig Artery Models of Restenosis after Balloon Angioplasty

As discussed in Section II,A, pRB functions as an R point guardian and allows the cell cycle clock to control the expression of numerous genes that are responsible for cell cycle progression. Hence, pRB-mediated cell growth inhibition is not tumor specific by nature. In this regard, it has been reported that different types of normal cells in culture were also growth arrested by overexpression of an additional wild-type *RB* gene (Fung *et al.*, 1993).

Vascular smooth muscle cells (SMCs) located in the arterial tunica media are normally maintained in a nonproliferative state *in vivo*. Arterial injury results in the migration of SMCs into the intimal layer of the arterial wall, where they begin to proliferate and synthesize redundant exracellular matrix components. Arterial injury after percutaneous balloon angioplasty of the coronary arteries results in neointimal SMC proliferation and restenosis in 30 to 50% of patients (Schwartz *et al.*, 1992). Neointimal SMC proliferation has also been implicated in the pathogenesis of atherosclerosis (Forrester *et al.*, 1991).

Chang *et al.* (1995) constructed a replication-deficient adenovirus vector, AdΔRb, encoding a hemagglutinin (HA) N-terminal epitope-tagged, nonphosphorylatable mutant form of the human *RB* gene product (HAΔRb). The HaΔRb fusion potein with point mutations at potential phosphorylation sites of pRB is nonphosphorylatable and is considered to be a constitutively active form of pRB (Hamel *et al.*, 1992). In the AdΔRb vector, expression of HAΔRb is under control of the human elongation factor 1α gene promoter. Two well-characterized animal models, the rat carotid artery model (Simons *et al.*, 1992; Morishita *et al.*, 1993) and the porcine femoral artery model (Prescott *et al.*, 1991; Ohno *et al.*, 1994) of restenosis, were used to determine the effects of AdΔRb treatment on restenosis. The arteries were exposed to approximately 2×10^9 (rat) or 10^{10} (pig) plaque-forming units (PFU) of AdΔRb viral vectors for 20 min immediately after balloon angioplasty. Restenosis as judged by the neointima-to-tunica media ratio was scored 3 weeks after surgery. The AdΔRb-infected arteries showed a 42 to 47% decrease in the neointima-to-tunica media ratio as compared to the Adβ-gal virus-infected controls ($p < 0.01$). Therefore, overexpression of a constitutively active form of pRB via adenovirus-mediated gene transfer *in vivo* significantly reduced restenosis in both rat and pig animal models of balloon angioplasty (Chang *et al.*, 1995).

IV. Prospects for Emerging *RB* Tumor Suppressor Gene Therapy

There seems more than one paradox in the emerging *RB* tumor suppressor gene therapy. The cytostatic effects of overexpression of pRB on normal and tumor cells are well known. The transient nature of the cytostatic effect, however, often implies that tumor cells may recommence malignant growth once pRB treatment is completed. On the other hand, if we assume the effect of pRB to be cytotoxic, then why is *RB* gene therapy any better or more potent than the conventional chemotherapy, radiation therapy, or suicidal gene (such as the herpes simplex virus thymidine kinase gene) therapy? Moreover, pRB expression or replacement often render tumor cells more resistant to apoptosis. What does this mean to cancer therapy (Morgenbesser *et al.*, 1994; Almasan *et al.*, 1995; Lu *et al.*, 1996b; Fan *et al.*, 1996)?

By examining several long-term tumor cell clones with tetracycline-regulatable wild-type pRB expression, we have demonstrated that tumor cells constitutively reexpressing functional pRB were irreversibly growth arrested (senescent). In contrast, human young diploid fibroblasts that were arrested in G_1 phase by pRB overexpression did not senesce, and they subsequently reentered the cell proliferative cycle after depletion of additional pRB (H.-J. Xu, unpublished data). The finding, if it can be confirmed in a variety of human malignancies, implies that the cytostatic effect of *RB* gene therapy will result in differential elimination of tumor cells through cellular senescence, and the replicative life space of normal cells *in vivo* may not be affected. We now believe that *RB* gene therapy may have the advantage over other conventional anticancer therapies since the *RB* gene product itself could differentially kill tumor cells, and it is likely to be nontoxic to normal somatic cells. The pRB-mediated senescence is also consistent with its apparent inhibitory effect on apoptosis. Since the pRB-mediated reduction in apoptosis is probably required for pRB-dependent IFN-γ induction of MHC class II gene expression and regulation of surface CD74 (invariant chain) expression in some tumor cells (Lu *et al.*, 1994a, 1996a), the inhibitory effect of pRB on apoptosis may coordinate with the host immune system. In certain physiological situations, however, there is also a pathway responsible for p53-independent, pRB-mediated apoptosis (Dou *et al.*, 1995; Bing and Dou, 1996). In brief, we believe that we are now close to having the answers for most of the puzzle, and the rationale behind *RB* gene therapy for human cancer appears impeccable.

There have been remarkable advancements in our understanding of the molecular basis of tumor suppressor genes, but many obstacles remain to be overcome before tumor suppressor gene therapy can be more effective in cancer treatment. Currently, one of the main problems is the difficulty in delivering the target tumor suppressor gene efficiently and specifically into the majority of tumor cells in a patient. Our preclinical studies demonstrated that adenovirus vectors expressing the functional RB protein were capable of transducing human tumor cells and resulted in partial or complete regression of newly established, small xenograft tumors in nude mice. The same treatment, however, has been less effective with well-established (>1 month old) and larger tumors (>50 mm³), which may directly imply that developing more efficacious vectors, preferably with tumor cell-specific or tissue-specific promoters, is of vital importance to any successful tumor suppressor gene therapy (Xu *et al.*, 1996). Despite this problem, several clinical protocols in which retrovirus or adenovirus vectors expressing the wild-type *p53* gene are being administered bronchoscopically or intratumorally to patients with lung cancer and head and neck tumors have been conditionally approved by the Recombinant DNA Advisory Committee of the National Institutes of Health (NIH, Bethesda, MD) (Fujiware *et al.*, 1994; Zhang *et al.*, 1995; Liu *et al.*, 1995; Clayman *et al.*, 1996). It is expected that, since many of the preclinical efficacy and safety data have been generated *in vitro* and in animal models, *RB* gene therapy will soon be applied to patients with bladder and prostate cancer for phase I clinical trials as well. It was our experience that following infusion into the mouse urinary bladder, replication-deficient adenovirus vectors were capable of penetrating the transitional epithelia *in vivo*, and expression of the reporter gene encoding β-galactosidase (β-Gal) or the human *RB* tumor suppressor gene was readily detectable in the majority of transitional epithelium cells as well as in the inner longitudinal smooth muscle cells of the muscularis (Y. Zhou and H.-J. Xu, unpublished data). Other authors have previously demonstrated that, by intratracheal instillation and aerosol delivery of viral vectors or plasmid constructs complexed with cationic lipids, high levels of reporter gene expression can be achieved in all cell types forming the airway epithelium, including alveolar cells of the lung (Rosenfeld *et al.*, 1992; Stribling *et al.*, 1992; Mastrangeli *et al.*, 1993; Engelhardt *et al.*, 1993; Bout *et al.*, 1994). Therefore, using the currently available vector systems, *RB* gene therapy may be beneficial in treating postsurgery residue tumors, superficial cancers, and premalignancies in hollow fluid (or air)-filled organs. In addition, *RB* gene-mediated cytostatic gene therapy may have broad utilities for cancer prevention as well as for treatment of nonmalignant hyperproliferative diseases such as restenosis after balloon angioplasty (Chang *et al.*, 1995). It is especially true since the *RB* gene, whose naturally occurring mutant forms are often nonfunctional, is a relatively safe therapeutic target as compared to another common tumor suppressor gene, *p53*. The frequent mutant forms of the *p53* gene are well-known dominant oncogenes (Lane and Benchimol, 1990).

Another obstacle to *RB* gene therapy, and potentially also to other kinds of tumor suppressor gene therapy, is tumor suppressor resistance (TSR). As we have discussed, reexpression of the wild-type RB protein in a variety of

RB-defective tumor cells significantly suppress their tumorigenicity in nude mice. However, the suppression is often incomplete because a portion of the tumor cells may have inherited or acquired the ability to inactivate the *RB* gene product or deregulate its effector pathway. Accordingly, to date the majority of preclinical studies reported on *RB* gene therapy used modified pRB constructs with enhanced cell growth-suppressing function rather than the wild-type *RB* gene itself. The modified pRB constructs documented in the literature include pRB[94] (Xu *et al.*, 1996), pRB[56] (Wills *et al.*, 1995), and a nonphosphorylatable, constitutively active form of the wild-type *RB* gene product, HAΔRb (Chang *et al.*, 1995). Of importance, these modified pRB proteins are able to inhibit effectively the proliferation not only of *RB*-deficient tumor cells but also of those with normal *RB* allele(s). The findings have significantly increased the potential of the *RB* gene as a therapeutic target. Furthermore, we expect the same strategy could be worked out for *p53* gene therapy. We have identified a conservative N-terminal segment between *p53* and *RB* with 48% identity. Studies on several modified *p53* gene constructs with intragenic deletions following this conservative region have demonstrated strong tumor cell growth suppression (Y. Zhou and H.-J. Xu, unpublished data). The studies are now in progress in our laboratory.

Alternative strategies might be used in the near future to surmount possible tumor suppressor resistance and to make *RB* anticancer gene therapy more effective. For example, it was known that the titer and quality of pRB[94]-expressing adenovirus vector decreased rapidly after large-scale propagation in 293 producer cells. There was also a trend toward reduction in the yields of recombinant adenovirus vectors expressing the full-length pRB[110] or the wild-type p53 if the viruses were subjected to sequential propagation in 293 cells (H.-J. Xu, unpublished data). We reason that this was because expression of pRB[94] as well as other tumor suppressor proteins in 293 cells severely restricted viral vector DNA synthesis. Therefore, the ability to develop a recombinant pRB[94] adenoviral genome that can be efficiently propagated in 293 cells is vital to the success of pRB[94] for adenovirus-mediated gene therapy. In this connection, generation of higher titer and quality of the recombinant adenovirus vector through regulatable pRB[94] gene expression in 293 producer cells would ensure delivery of high-dose-pRB[94] gene therapy. Second, it is known that transduction frequencies through replication-deficient adenovirus vectors are inadequate in large tumors to modify a sufficient percentage of tumor cells. A method to circumvent this issue would be to amplify tumor transduction *in vivo* in the tumor sites (Miller and Curiel, 1996). In this context, prior to treatment with the therapeutic *RB* adenovirus vector, the actively dividing tumor cells as well as the proliferating endothelial cells of the angiogenic blood vessels in the tumor sites might be premodified by E1a-expressing retrovirus vectors and converted into recombinant adenovirus-producing cells. This would provide a means of amplifying *RB* gene transduction and its therapeutic effects *in vivo*. Finally, combination *RB* and other tumor suppressor gene therapy may have synergistic effects of pRB-mediated, tumor-specific senescence, and therefore will provide augmentation of the antitumor effect of *RB* gene therapy. In conclusion, we are currently in the earliest stages of applying *RB* and other tumor suppressor genes to anticancer therapy, but great potential exists for the future development of improved tumor suppressor gene therapies for human cancer.

Acknowledgments

I thank Dr. Clifford J. Steer for helpful comments on this manuscript. The work done in this laboratory was supported in part by grants from the National Institutes of Health (CA67274) and the Texas Higher Education Coordinating Board (ATP 4949018) to H.-J.X.

References

Albini, A., Iwamoto, Y., Kleinman, H. K., Martin, G. R., Aaronson, S. A., Kozlowski, J. M., and McEwan, R. N. (1987). A rapid *in vitro* assay for quantitating the invasive potential of tumor cells. *Cancer Res.* **47,** 3239–3245.

Almasan, A., Yin, Y., Kelly, R. E., Lee, E. Y., Bradley, A., Li, W., Bertino, J. R., and Wahl, G. M. (1995). Deficiency of retinoblastoma protein leads to inappropriate S-phase entry, activation of E2F-responsive genes, and apoptosis. *Proc. Natl. Acad. Sci. U.S.A.* **92,** 5436–5440.

Anderson, M. J., Fasching, C. L., Xu, H. J., Benedict, W. F., and Stanbridge, E. J. (1994). Chromosome 13 transfer provides evidence for regulation of RB1 protein expression. *Genes, Chromosomes, Cancer* **9,** 251–260.

Azzam, H. S., Arand, G., Lippman, M. E., and Thompson, E. W. (1993). Association of MMP-2 activation potential with metastatic progression in human breast cancer cell lines independent of MMP-2 production. *J. Natl. Cancer Inst.* **85,** 1758–1764.

Baker, S. J., Markowitz, S., Fearon, E. R., Wilson, J. K., and Vogelstein, B. (1990). Suppression of human colorectal carcinoma cell growth by wild-type p53. *Science* **249,** 912–915.

Banerjee, A., Xu, H. J., Hu, S. X., Araujo, D., Takahashi, R., Stanbridge, E. J., and Benedict, W. F. (1992). Changes in growth and tumorigencity following reconstitution of retinoblastoma gene function in various human cancer cell types by microcell transfer of chromosome 13. *Cancer Res.* **52,** 6297–6304.

Bing, A., and Dou, Q. P. (1996). Cleavage of retinoblastoma protein during apoptosis: An interleukin 1 beta-converting enzyme-like protease as candidate. *Cancer Res.* **56,** 438–442.

Bookstein, R., Rio, P., Madreperla, S. A., Hong, F., Allred, C., Grizzle, W. E., and Lee, W. H. (1990a). Promoter deletion and loss of retinoblastoma gene expression in human prostate carcinoma. *Proc. Natl. Acad. Sci. U.S.A.* **87,** 7762–7766.

Bookstein, R., Shew, J., Chen, P., Scully, P., and Lee, W. H. (1990b). Suppression of tumorigenicity of human prostate carcinoma cells by replacing a mutated RB gene. *Science* **247,** 712–715.

Bout, A., Perricaudet, M., Baskin, G., Imler, J. L., Scholte, B. J., Pavirani, A., and Valerio, D. (1994). Lung gene therapy: *In vivo* adenovirus-mediated gene transfer to rhesus monkey airway epithelium. *Hum. Gene Ther.* **5,** 3–10.

Brown, P. D., Bloxidge, R. E., Stuart, N. S., Gatter, K. C., and Carmichael, J. (1993). Association between expression of activated 72-kilodalton gelantinase and tumor spread in non-small-cell lung carcinoma. *J. Natl. Cancer Inst.* **85,** 574–578.

Buchkovich, K., Duffy, L. A., and Harlow, E. (1989). The retinoblastoma protein is phosphory-lated during specific phases of the cell cycle. *Cell (Cambridge, Mass.)* **58**, 1097–1105.

Call, K. M., Glaser, T., Ito, C. Y., Buckler, A. J., Pelletier, J., Haber, D. A., Rose, E. A., Kral, A., Yeger, H., Lewis, W. H., Jones, C., and Housman, D. E. (1990). Isolation and characterization of a zinc finger polypeptide gene at the human chromosome 11 Wilms' tumor locus. *Cell (Cambridge, Mass)* **60**, 509–520.

Cance, W. G., Brennan, M. F., Dudas, M. E., Huang, C.-M., and Cordon-Cardo, C. (1990). Altered expression of the retinoblastoma gene product in human sarcomas. *N. Engl. J. Med.* **323**, 1457–1462.

Cawthon, R. M., Weiss, R., Xu, G. F., Viskochil, D., Culver, M., Stevens, J., Robertson, M., Dunn, D., Gesteland, R., O'Connell, P., and White, R. (1990). A major segment of the neurofibromatosis type 1 gene: cDNA sequence, genomic structure, and point mutations. *Cell (Cambridge, Mass.)* **62**, 193–201.

Chang, M. W., Barr, E., Seltzer, J., Jiang, Y. Q., Nabel, G. J., Nabel, E. G., Parmaceck, M. S., and Leiden, J. M. (1995). Cytostatic gene therapy for vascular proliferative disorders with a constitutively active form of the retinoblastoma gene product. *Science* **267**, 518–522.

Chellappan, S. P., Hiebert, S., Mudryj, M., Horowitz, J. M., and Nevins, J. R. (1991). The E2F transcription factor is a cellular target for the RB protein. *Cell (Cambridge, Mass.)* **65**, 1053–1061.

Chen, P. L., Scully, P., Shew, J. Y., Wang, J. Y., and Lee, W. H. (1989). Phosphorylation of the retinoblastoma gene product is modulated during the cell cycle and cellular differentia-tion. *Cell (Cambridge, Mass.)* **58**, 1193–1198.

Chen, P. L., Chen, Y., Shan, B., Bookstein, R., and Lee, W. H. (1992). Stability of retinoblastoma gene expression determines the tumorigenicity of reconstituted retinoblastoma cells. *Cell Growth Differ.* **3**, 119–125.

Clayman, G. L., Liu, T. J., Overholt, S. M., Mobley, S. R., Wang, M., Janot, F., and Goepfert, H. (1996). Gene therapy for head and neck cancer—comparing the tumor suppressor gene p53 and a cell cycle regulator WAF1/CIP1 (p21). *Arch. Otolaryngol. Head Neck Surg.* **122**, 489–493.

Cooper, J. A., and Whyte, P. (1989). RB and the cell cycle: Entrance or exit? *Cell (Cambridge, Mass.)* **58**, 1009–1011.

Cordon-Cardo, C., Wartinger, D., Petrylak, D., Dalbagni, G., Fair, W. R., Fuks, Z., and Reuter, V. E. (1992). Altered expression of the retinoblastoma gene product: Prognostic indicator in bladder cancer. *J. Natl. Cancer Inst.* **84**, 1251–1256.

Cryns, V. L., Thor, A., Xu, H. J., Xu, S. X., Wierman, M. E., Vickery, A. L., Jr., Benedict, W. F., and Arnold, A. (1994). Loss of the retinoblastoma tumor-suppressor gene in parathyroid carcinoma. *N. Engl. J. Med.* **330**, 757–761.

Dawson, D. W., Tolsma, S. S., Volpert, O. V., Polverini, P. J., and Bouck, N. P. (1995). Retinoblastoma gene expression alters angiogenic phenotype. *Proc. Am. Assoc. Cancer Res.* **36**, 88 (abstr.).

DeCaprio, J. A., Ludlow, J. W., Lynch, D., Furukawa, Y., Griffin, J., Piwnica Worms, H., Huang, C. M., and Livingston, D. M. (1989). The product of the retinoblastoma suscepti-bility gene has properties of a cell cycle regulatory element. *Cell (Cambridge, Mass.)* **58**, 1085–1095.

Dou, Q. P., An, B., and Will, P. L. (1995). Induction of a retinoblastoma phosphatase activity by anticancer drugs accompanies p53-independent G1 arrest and apoptosis. *Proc. Natl. Acad. Sci. U.S.A.* **92**, 9019–9023.

Duan, D. R., Pause, A., Burgess, W. H., Aso, T., Chen, D. Y., Garrett, K. P., Conaway, R. C., Conaway, J. W., Linehan, W. M., and Klausner, R. D. (1995). Inhibition of transcription elongation by the VHL tumor suppressor protein. *Science* **269**, 1402–1406.

Engelhardt, J. F., Simon, R. H., Yang, Y., Zepeda, M., Weber Pendleton, S., Doranz, B., Grossman, M., and Wilson, J. M. (1993). Adenovirus-mediated transfer of the CFTR gene to lung of nonhuman primates: Biological efficacy study. *Hum. Gene Ther.* **4**, 759–769.

Ewen, M. E., Sluss, H. K., Sherr, C. J., Matsushime, H., Kato, J., and Livingston, D. M. (1993). Functional interactions of the retinoblastoma protein with mammalian D-type cyclins. *Cell (Cambridge, Mass.)* **73**, 487–497.

Fan, G., Ma, X., Kren, B. T., and Steer, C. J. (1996). The retinoblastoma gene product inhibits TGF-beta1 induced apoptosis in primary rat hepatocytes and human HuH-7 hepatoma cells. *Oncogene* **12**, 1909–1919.

Fearon, E. R., Cho, K. R., Nigro, J. M., Kern, S. E., Simons, J. W., Ruppert, J. M., Hamilton, S. R., Preisinger, A. C., Thomas, G., Kinzler, K. W., and Vogelstein, B. (1990). Identification of a chromosome 18q gene that is altered in colorectal cancers. *Science* **247**, 49–56.

Finlay, C. A., Hinds, P. W., and Levine, A. J. (1989). The p53 proto-oncogene can act as a suppressor of transformation. *Cell (Cambridge, Mass.)* **57**, 1083–1093.

Forrester, J. S., Fishbein, M., Helfant, R., and Fagin, J. (1991). A paradigm for restenosis based on cell biology: Clues for the development of new preventive therapies (review). *J. Am. Coll. Cardiol.* **17**, 758–769.

Friend, S. H., Bernards, R., Rogeji, S., Weinberg, R. A., Rapaport, J. M., Albert, D. M., and Dryja, T. P. (1986). A human DNA segment with properties of the gene that predisposes to retinoblastoma and osteosarcoma. *Nature (London)* **323**, 643–646.

Fujiwara, T., Grimm, E. A., and Roth, J. A. (1994). Gene therapeutics and gene therapy for cancer (review). *Curr. Opin. Oncol.* **6**, 96–105.

Fung, Y. K., Murphree, A. L., T'Ang, A., Qian, J., Hinrichs, S. H., and Benedict, W. F. (1987). Structural evidence for the authenticity of the human retinoblastoma gene. *Science* **236**, 1657–1660.

Fung, Y. K., T'Ang, A., Murphree, A. L., Zhang, F. H., Qui, W. R., Wang, S. W., Shi, X. H., Lee, L., Driscoll, B., and Wu, K. J. (1993). The Rb gene suppresses the growth of normal cells. *Oncogene* **8**, 2659–2672.

Furukawa, Y., DeCaprio, J. A., Freedman, A., Kanakura, Y., Nakamura, M., Ernst, T. J., Livingston, D. M., and Griffin, J. D. (1990). Expression and state of phosphorylation of the retinoblastoma susceptibility gene product in cycling and noncycling human hematopoietic cells. *Proc. Natl. Acad. Sci. U.S.A.* **87**, 2770–2774.

Futreal, P. A., Liu, Q., Shattuck-Eidens, D., Cochran, C., Harshman, K., Tavtigian, S., Bennett, L. M., Haugen-Strano, A., Swensen, J., Miki, Y., Eddington, K., McClure, M., Frye, C., Weaver-Feldhaus, J., Ding, W., Gholami, Z., Söderkvist, P., Terry, L., Jhanwar, S., Berchuck, A., Iglehart, J. D., Marks, J., Ballinger, D. G., Barrett, J. C., Skolnick, M. H., Kamb, A., and Wiseman, R. (1994) BRCA1 mutations in primary breast and ovarian carcinomas. *Science* **266**, 120–122.

Geng, Y., Eaton, E. N., Picon, M., Roberts, J. M., Lundberg, A. S., Gifford, A., Sardet, C., and Weinberg, R. A. (1996). Regulation of cyclin E transcription by E2Fs and retinoblastoma protein. *Oncogene* **12**, 1173–1180.

Gessler, M., Poustka, A., Cavenee, W., Neve, R. L., Orkin, S. H., and Bruns, G. A. (1990). Homozygous deletion in Wilms tumours of a zinc-finger gene identified by chromosome jumping. *Nature (London)* **343**, 774–778.

Goodrich, D. W., and Lee, W. H. (1993). Molecular characterization of the retinoblastoma susceptibility gene (review). *Biochim. Biophys. Acta* **1155**, 43–61.

Goodrich, D. W., Wang, N. P., Qian, Y. W., Lee, E. Y. H. P., and Lee, W. H. (1991). The retinoblastoma gene product regulates progression through the G1 phase of the cell cycle. *Cell (Cambridge, Mass.)* **67**, 293–302.

Goodrich, D. W., Chen, Y., Scully, P., and Lee, W. H. (1992). Expression of the retinoblastoma gene product in bladder carcinoma cells associates with a low frequency of tumor formation. *Cancer Res.* **52**, 1968–1973.

Gossen, M., and Bujard, H. (1992). Tight control of gene expression in mammalian cells by tetracycline-responsive promoters. *Proc. Natl. Acad. Sci. U.S.A.* **89**, 5547–5551.

Hamel, P. A., Gill, R. M., Phillips, R. A., and Gallie, B. L. (1992). Transcriptional repression of the E2-containing promoters EIIaE, c-myc, and RB1 by the product of the RB1 gene. *Mol. Cell. Biol.* **12**, 3431–3438.

Hamel, P. A., Phillips, R. A., Muncaster, M., and Gallie, B. L. (1993). Speculations on the roles of RB1 in tissue-specific differentiation, tumor initiation, and tumor progression. *FASEB J.* **7**, 846–854.

Harbour, J. W., Lai, S. L., Whang Peng, J., Gazdar, A. F., Minna, J. D., and Kaye, F. J. (1988). Abnormalities in structure and expression of the human retinoblastoma gene in SCLC. *Science* **241**, 353–357.

Henson, J. W., Schnitker, B. L., Correa, K. M., von Deimling, A., Fassbender, F., Xu, H. J., Benedict, W. F., Yandell, D. W., and Louis, D. N. (1994). The retinoblastoma gene is involved in malignant progression of astrocytomas. *Ann. Neurol.* **36**, 714–721.

Horowitz, J. M. (1993). Regulation of transcription by the retinoblastoma protein. *Genes, Chromosomes, Cancer* **6**, 124–131.

Huang, H. J., Yee, J. K., Shew, J. Y., Chen, P. L., Bookstein, R., Friedmann, T., Lee, E. Y., and Lee, W. H. (1988). Suppression of the neoplastic phenotype by replacement of the RB gene in human cancer cells. *Science* **242**, 1563–1566.

Kamb, A., Gruis, N. A., Weaver-Feldhaus, J., Liu, Q., Harshman, K., Tavtigian, S. V., Stockert, E., Day, R. S., 3rd, Johnson, B. E., and Skolnick, M. H. (1994). A cell cycle regulator potentially involved in genesis of many tumor types. *Science* **264**, 436–440.

Karikó, K., Kuo, A., Boyd, D., Okada, S. S., Cines, D. B., and Barnathan, E. S. (1993). Overexpression of urokinase receptor increases matrix invasion without altering cell migration in a human osteosarcoma cell line. *Cancer Res.* **53**, 3109–3117.

Kato, J., Matsushime, H., Hiebert, S. W., Ewen, M. E., and Sherr, C. J. (1993). Direct binding of cyclin D to the retinoblastoma gene product (pRb) and pRb phosphorylation by the cyclin D-dependent kinase CDK4. *Genes Dev.* **7**, 331–342.

Klein, G. (1990). Multistep emancipation of tumors from growth control: Can it be curbed in a single step? *BioEssays* **12**, 347–350.

Knudson, A. G., Jr. (1971). Mutation and cancer: Statistical study of retinoblastoma. *Proc. Natl. Acad. Sci. U.S.A.* **68**, 820–823.

Knudson, A. G., Jr. (1985). Hereditary cancer, oncogenes, and antioncogenes. *Cancer Res.* **45**, 1437–1443.

Knudson, A. G., Jr. (1993). Antioncogenes and human cancer. *Proc. Natl. Acad. Sci. U.S.A.* **90**, 10914–10921.

Koff, A., Giordano, A., Desai, D., Yamashita, K., Harper, J. W., Elledge, S., Nishimoto, T., Morgan, D. O., Franza, B. R., and Roberts, J. M. (1992). Formation and activation of a cyclin E-cdk2 complex during the G1 phase of the human cell cycle. *Science* **257**, 1689–1694.

Kornblau, S. M., Xu, H. J., del Giglio, A., Hu, S. X., Zhang, W., Calvert, L., Beran, M., Estey, E., Andreeff, M., Trujillo, J., Cork, A., Smith, T. L., Benedict, W. F., and Deisseroth, A. B. (1992). Clinical implications of decreased retinoblastoma protein expression in acute myelogenous leukemia. *Cancer Res.* **52**, 4587–4590.

Kornblau, S. M., Xu, H. J., Zhang, W., Hu, S. X., Beran, M., Smith, T. L., Hester, J., Estey, E., Benedict, W. F., and Deisseroth, A. B. (1994). Levels of retinoblastoma protein expression in newly diagnosed acute myelogenous leukemia. *Blood* **84**, 256–261.

Kratzke, R. A., Shimizu, E., Geradts, J., Gerster, J. L., Segal, S., Otterson, G. A., and Kaye, F. J. (1993). RB-mediated tumor suppression of a lung cancer cell line is abrogated by an extract enriched in extracellular matrix. *Cell Growth Differ.* **4**, 629–635.

Lane, D. P., and Benchimol, S. (1990). p53: Oncogene or anti-oncogene? (Review). *Genes Dev.* **4**, 1–8.

La Thangue, N. B. (1994). DRTF1/E2F: An expanding family of heterodimeric transcription factors implicated in cell-cycle control (review). *Trends Biochem. Sci.* **19**, 108–114.

Lee, E. Y., To, H., Shew, J. Y., Bookstein, R., Scully, P., and Lee, W. H. (1988). Inactivation on the retinoblastoma susceptibility gene in human breast cancers. *Science* **241**, 218–221.

Lee, W. H., Bookstein, R., Hong, F., Young, L., Shew, J., and Lee, E. (1987a). Human retinoblastoma susceptibility gene: Cloning, identification and sequence. *Science* **235**, 1394–1399.

Lee, W. H., Shew, J. Y., Hong, F. D., Sery, T. W., Donoso, L. A., Young, L. J., Bookstein, R., and Lee, E. Y. (1987b). The retinoblastoma susceptibility gene encodes a nuclear phosphoprotein associated with DNA binding activity. *Nature (London)* **329**, 642–645.

Li, J., Hu, S.-X., Xu, K., Perng, G.-S., Zhou, Y., Xu, K., Zhang, C., Benedict, W. F., and Xu, H.-J. (1996). Expression of the retinoblastoma (*RB*) tumor suppressor gene inhibits tumor cell invasion *in vitro*. *Oncogene* **13**, 2379–2386.

Liotta, L. A., Steeg, P. S., and Stetler-Stevenson, W. G. (1991). Cancer metastasis and angiogenesis: An imbalance of positive and negative regulation (review). *Cell (Cambridge, Mass.)* **64**, 327–336.

Liu, T. J., El-Naggar, A. K., McDonnell, T. J., Steck, K. D., Wang, M., Taylor, D. L., and Clayman, G. L. (1995). Apoptosis induction mediated by wild-type p53 adenoviral gene transfer in squamous cell carcinoma of the head and neck. *Cancer Res.* **55**, 3117–3122.

Logothetis, C., Xu, H., Ro, J. Y., Sahin, A., Ordonez, N., and Benedict, W. F. (1992). Altered retinoblastoma protein expression and known prognostic variables in locally advanced bladder cancer. *J. Natl. Cancer Inst.* **84**, 1256–1261.

Lu, Y. M., Ussery, G. D., Muncaster, M. M., Gallie, B. L., and Blanck, G. (1994a). Evidence for retinoblastoma protein (RB) dependent and independent IFN-gamma responses: RB coordinately rescues IFN-gamma induction of MHC class II gene transcription in noninducible breast carcinoma cells. *Oncogene* **9**, 1015–1019.

Lu, Y. M., Ussery, G. D., Jacim, M., Tschickardt, M., Boss, J. M., and Blanck, G. (1994b). Retino-blastoma protein regulation of surface CD74 (invariant chain) expression in breast carcinoma cells. *Mol. Immunol.* **31**, 1365–1368.

Lu, Y. M., Boss, J. M., Hu, S. X., Xu, H.-J., and Blanck, G. (1996a). Apoptosis-independent retinoblastoma protein rescue of HLA class II messenger RNA IFN-gamma inducibility in non-small cell lung carcinoma cells—lack of surface class II expression associated with a specific defect in HLA-DRA induction. *J. Immunol.* **156**, 2495–2502.

Lu, Y. M., Marler, D. F., Hu, S.-H., Xu, H.-J., and Blanck, G. (1996b). Retinoblastoma protein inhibits IFN-γ induced apoptosis. *Oncogene* **12**, 1809–1819.

Ludlow, J. W., Shon, J., Pipas, J. M., Livingston, D. M., and DeCaprio, J. A. (1990). The retinoblastoma susceptibility gene product undergoes cell cycle-dependent dephosphorylation and binding to and release from SV40 large T. *Cell (Cambridge, Mass.)* **60**, 387–396.

Madreperla, S. A., Whittum Hudson, J. A., Prendergast, R. A., Chen, P. L., and Lee, W. H. (1991). Intraocular tumor suppression of retinoblastoma gene-reconstituted retinoblastoma cells. *Cancer Res.* **51**, 6381–6384.

Mastrangeli, A., Danel, C., Rosenfeld, M. A., Stratford Perricaudet, L., Perricaudet, M., Pavirani, A., Lecocq, J. P., and Crystal, R. G. (1993). Diversity of airway epithelial cell targets for *in vivo* recombinant adenovirus-mediated gene transfer. *J. Clin. Invest.* **91**, 225–234.

McGrory, W. J., Bautista, D. S., and Graham, F. L. (1988). A simple technique for the rescue of early region I mutations into infectious human adenovirus type 5. *Virology* **163**, 614–617.

Mihara, K., Cao, X. R., Yen, A., Chandler, S., Driscoll, B., Murphree, A. L., T'Ang, A., and Fung, Y. K. (1989). Cell cycle-dependent regulation of phosphorylation of the human retinoblastoma gene product. *Science* **246**, 1300–1303.

Miki, Y., Swensen, J., Shattuck-Eidens, D., Futreal, P. A., Harshman, K., Tavtigian, S., Liu, Q., Cochran, C., Bennett, L. M., Ding, W., Bell, R., Rosenthal, J., Hussey, C., Tran, T., McClure, M., Frye, C., Tom, H., Phelps, R., Haugen-Strano, A., Katcher, H., Yakumo, K., Gholami, Z., Shaffer, D., Stone, S., Bayer, S., Wray, C., Bogden, R., Dayananth, P., Ward, J., Tonin, P., Narod, S., Bristow, P. K., Norris, F. H., Helvering, L., Morrison, P., Rosteck, P., Lai, M., Barrett, J. C., Lewis, C., Neuhausen, S., Cannon-Albright, L., Goldgar, D., Wiseman, R., Kamb, A., and Skolnick, M. H. (1994). A strong candidate for the breast and ovarian cancer susceptibility gene BRCA1. *Science* **266**, 66–71.

Miller, R., and Curiel, D. T. (1996). Towards the use of replicative adenoviral vectors for cancer gene therapy (editorial). *Gene Ther.* **3**, 557–559.

Mittnacht, S., Lees, J. A., Desai, D., Harlow, E., Morgan, D. O., and Weinberg, R. A. (1994). Distinct sub-populations of the retinoblastoma protein show a distinct pattern of phosphorylation. *EMBO J.* **13**, 118–127.

Morgenbesser, S. D., Williams, B. O., Jacks, T., and DePinho, R. A. (1994). p53-dependent apoptosis produced by Rb-deficiency in the developing mouse lens. *Nature (London)* **371**, 72–74.

Morishita, R., Gibbons, G. H., Ellison, K. E., Nakajima, M., Zhang, L., Kaneda, Y., Ogihara, T., and Dzau, V. J. (1993). Single intraluminal delivery of antisense cdc2 kinase and proliferating-cell nuclear antigen oligonucleotides results in chronic inhibition of neointimal hyperplasia. *Proc. Natl. Acad. Sci. U.S.A.* **90**, 8474–8478.

Muncaster, M. M., Cohen, B. L., Phillips, R. A., and Gallie, B. L. (1992). Failure of RB1 to reverse the malignant phenotype of human tumor cell lines. *Cancer Res.* **52**, 654–661.

Murphree, A. L., and Benedict, W. F. (1984). Retinoblastoma: Clues for human oncogenesis. *Science* **223**, 1028–1033.

Nevins, J. R. (1992). E2F: A link between the Rb tumor suppressor protein and viral oncoproteins, *Science* **258**, 424–429.

Ohno, T., Gordon, D., San, H., Pompili, V. J., Imperiale, M. J., Nabel, G. J., and Nabel, E. G. (1994). Gene therapy for vascular smooth muscle cell proliferation after arterial injury. *Science* **265**, 781–784.

Ookawa, K., Shiseki, M., Takahashi, R., Yoshida, Y., Terada, M., and Yokota, J. (1993). Reconstitution of the RB gene suppresses the growth of small-cell lung carcinoma cells carrying multiple genetic alterations. *Oncogene* **8**, 2175–2181.

Pardee, A. B. (1989). G1 events and regulation of cell proliferation (review). *Science* **246**, 603–608.

Prescott, M. F., McBride, C. H., Hasler-Rapacz, J., von Linden, J., and Rapacz, J. (1991). Development of complex atherosclerotic lesions in pigs with inherited hyper-LDL cholesterolemia bearing mutant alleles for apolipoprotein B. *Am. J. Pathol.* **139**, 139–147.

Presti, J. C., Jr., Reuter, V. E., Galan, T., Fair, W. R., and Cordon-Cardo, C. (1991). Molecular genetic alterations in superficial and locally advanced human bladder cancer. *Cancer Res.* **51**, 5405–5409.

Pritchard-Jones, K., Fleming, S., Davidson, D., Bickmore, W., Porteous, D., Gosden, C., Bard, J., Buckler, A., Pelletier, J., Housman, D., van Heyningen, V., and Hastie, N. (1990). The candidate Wilms' tumour gene is involved in genitourinary development. *Nature (London)* **346**, 194–197.

Reissmann, P. T., Simon, M. A., Lee, W. H., and Slamon, D. J. (1989). Studies of the retinoblastoma gene in human sarcomas. *Oncogene* **4**, 839–843.

Resnitzky, D., and Reed, S. I. (1995). Different roles for cyclins D1 and E in regulation of the G1-to-S transition. *Mol. Cell Biol.* **15**, 3463–3469.

Riley, D. J., Lee, E. Y., and Lee, W. H. (1994). The retinoblastoma protein: More than a tumor suppressor (review). *Annu. Rev. Cell Biol.* **10**, 1–29.

Rogalsky, V., Todorov, G., and Moran, D. (1993). Translocation of retinoblastoma protein associated with tumor cell growth inhibition. *Biochem. Biophys. Res. Commun.* **192**, 1139–1146.

Rosenfeld, M. A., Yoshimura, K., Trapnell, B. C., Yoneyama, K., Rosenthal, E. R., Dalemans, W., Fukayama, M., Bargon, J., Stier, L. E., Stratford Perricaudet, L., and Crystal, R. G. (1992). *In vivo* transfer of the human cystic fibrosis transmembrane conductance regulator gene to the airway epithelium. *Cell (Cambridge, Mass.)* **68**, 143–155.

Sardet, C., Vidal, M., Cobrinik, D., Geng, Y., Onufryk, C., Chen, A., and R. A. Weinberg, (1995). E2F-4 and E2F-5, two members of the E2F family, are expressed in the early phases of the cell cycle. *Proc. Natl. Acad. Sci. U.S.A.* **92**, 2403–2407.

Sato, H., Takino, T., Okada, Y., Cao, J., Shinagawa, A., Yamamoto, E., and Seiki, M. (1994). A matrix metalloproteinase expressed on the surface of invasive tumour cells (see comments). *Nature (London)* **370**, 61–65.

Schott, D. R., Chang, J. N., Deng, G., Kurisu, W., Kuo, W. L., Gray, J., and Smith, H. S. (1994). A candidate tumor suppressor gene in human breast cancers. *Cancer Res.* **54,** 1393–1396.

Schwartz, R. S., Holmes, D. R., Jr., and Topol, E. J. (1992). The restenosis paradigm revisited: An alternative proposal for cellular mechanisms (editorial) (review). *J. Am. Coll. Cardiol.* **20,** 1284–1293.

Serrano, M., Hannon, G. J., and Beach, D. (1993). A new regulatory motif in cell-cycle control causing specific inhibition of cyclin D/CDK4 (see comments). *Nature (London)* **366,** 704–707.

Shan, B., Chang, C. Y., Jones, D., and Lee, W. H. (1994). The transcription factor E2F-1 mediates the autoregulation of RB gene expression. *Mol. Cell Biol.* **14,** 299–309.

Shan, B., Durfee, T., and Lee, W. H. (1996). Disruption of RB/E2F-1 interaction by single point mutations in E2F-1 enhances S-phase entry and apoptosis. *Proc. Natl. Acad. Sci. U.S.A.* **93,** 679–684.

Simons, M., Edelman, E. R., DeKeyser, J. L., Langer, R., and Rosenberg, R. D. (1992). Antisense c-myb oligonucleotides inhibit intimal arterial smooth muscle cell accumulation *in vivo. Nature (London)* **359,** 67–70.

Stein, G. H., Beeson, M., and Gordon, L. (1990). Failure to phosphorylate the retinoblastoma gene product in senescent human fibroblasts. *Science* **249,** 666–669.

Stribling, R., Brunette, E., Liggitt, D., Gaensler, K., and Debs, R. (1992). Aerosol gene delivery *in vivo. Proc. Natl. Acad. Sci. U.S.A.* **89,** 11277–11281.

Sumegi, J., Uzvolgyi, E., and Klein, G. (1990). Expression of the RB gene under the control of MuLV-LTR suppresses tumorigenicity of WERI-Rb-27 retinoblastoma cells in immuno-defective mice. *Cell Growth Differ.* **1,** 247–250.

Takahashi, R., Hashimoto, T., Xu, H., Hu, S., Matsui, T., Miki, T., Migo-Marshall, H., Aaronson, S. A., and Benedict, W. F. (1991). The retinoblastoma gene functions as a growth and tumor suppressor in human bladder carcinoma cells. *Proc. Natl. Acad. Sci. U.S.A.* **88,** 5257–5261.

T'Ang, A., Varley, J. M., Chakraborty, S., Murphree, A. L., and Fung, Y. K. (1988). Structural rearrangement of the retinoblastoma gene in human breast carcinoma. *Science* **242,** 263–266.

Templeton, D. J., Park, S. H., Lanier, L., and Weinberg, R. A. (1991). Nonfunctional mutants of the retinoblastoma protein are characterized by defects in phosphorylation, viral onco-protein association, and nuclear tethering. *Proc. Natl. Acad. Sci. U.S.A.* **88,** 3033–3037.

Thomas, N. S., Burke, L. C., Bybee, A., and Linch, D. C. (1991). The phosphorylation state of the retinoblastoma (RB) protein in G0/G1 is dependent on growth status. *Oncogene* **6,** 317–322.

Toguchida, J., Ishizaki, K., Sasaki, M. S., Ikenaga, M., Sugimoto, M., Kotoura, Y., and Yamamuro, T. (1988). Chromosomal reorganization for the expression of recessive muta-tion of retinoblastoma susceptibility gene in the development of osteosarcoma. *Cancer Res.* **48,** 3939–3943.

Viskochil, D., Buchberg, A. M., Xu, G., Cawthon, R. M., Stevens, J., Wolff, R. K., Culver, M., Carey, J. C., Copeland, N. G., Jenkins, N. A., White, R., and O'Connell, P. (1990). Deletions and a translocation interrupt a cloned gene at the neurofibromatosis type 1 locus. *Cell (Cambridge, Mass.)* **62,** 187–192.

Wallace, M. R., Marchuk, D. A., Andersen, L. B., Letcher, R., Odeh, H. M., Saulino, A. M., Fountain, J. W., Brereton, A., Nicholson, J., Mitchell, A. L., Brownstein, B. H., and Collins, F. S. (1990). Type 1 neurofibromatosis gene: Identification of a large transcript disrupted in three NF1 patients. *Science* **249,** 181–186.

Wang, J. Y., Knudsen, E. S., and Welch, P. J. (1994). The retinoblastoma tumor suppressor protein (review). *Adv. Cancer Res.* **64,** 25–85.

Wang, N. P., To, H., Lee, W. H., and Lee, E. Y. (1993). Tumor suppressor activity of RB and p53 genes in human breast carcinoma cells. *Oncogene* **8,** 279–288.

Weinberg, R. A. (1991). Tumor suppressor genes. *Science* **254**, 1138–1146.

Weinberg, R. A. (1995). The retinoblastoma protein and cell cycle control (review). *Cell* (*Cambridge, Mass.*) **81**, 323–330.

Weintraub, S. J., Prater, C. A., and Dean, D. C. (1992). Retinoblastoma protein switches the E2F site from positive to negative element. *Nature* (*London*) **358**, 259–261.

Wills, K. N., Barff, J., Souza, D., Antelman, D., Smith, R., Walsh, K., and Gregory, R. J. (1995). P56, a truncated form of RB, functions as a more potent regulator of cell cycle growth than full-length RB. *Cancer Gene Ther.* **2**, 339 (abstr.).

Xu, H. J. (1995). Altered retinoblastoma (RB) protein expression in human malignancies. *Adv. Anat. Pathol.* **2**, 213–226.

Xu, H. J., Hu, S. X., Hashimoto, T., Takahashi, R., and Benedict, W. F. (1989). The retinoblastoma susceptibility gene product: A characteristic pattern in normal cells and abnormal expression in malignant cells. *Oncogene* **4**, 807–812.

Xu, H. J., Hu, S. X., and Benedict, W. F. (1991a). Lack of nuclear RB protein staining in G0/ middle G1 cells: Correlation to changes in total RB protein level. *Oncogene* **6**, 1139–1146.

Xu, H. J., Hu, S. X., Cagle, P. T., Moore, G. E., and Benedict, W. F. (1991b). Absence of retinoblastoma protein expression in primary non-small cell lung carcinomas. *Cancer Res.* **51**, 2735–2739.

Xu, H. J., Sumegi, J., Hu, S. X., Benerjee, A., Uzvolgyi, E., Klein, G., and Benedict, W. F. (1991c). Intraocular tumor formation of RB reconstituted retinoblastoma cells. *Cancer Res.* **51**, 4481–4485.

Xu, H. J., Cairns, P., Hu, S. X., Knowles, M. A., and Benedict, W. F. (1993). Loss of RB protein expression in primary bladder cancer correlates with loss of heterozygosity at the RB locus and tumor progression. *Int. J. Cancer* **53**, 781–784.

Xu, H. J., Quinlan, D. C., Davidson, A. G., Hu, S. X., Summers, C. L., Li, J., and Benedict, W. F. (1994a). Altered retinoblastoma protein expression and prognosis in early-stage non-small-cell lung carcinoma. *J. Natl. Cancer Inst.* **86**, 695–699.

Xu, H. J., Xu, K., Zhou, Y., Li, J., Benedict, W. F., and Hu, S. X. (1994b). Enhanced tumor cell growth suppression by an N-terminal truncated retinoblastoma protein. *Proc. Natl. Acad. Sci. U.S.A.* **91**, 9837–9841.

Xu, H. J., Zhou, Y. L., Seigne, J., Perng, G. S., Mixon, M., Zhang, C. Y., Li, J., Benedict, W. F., and Hu, S. X. (1996). Enhanced tumor suppressor gene therapy via replication-deficient adenovirus vectors expressing an N-terminal truncated retinoblastoma protein. *Cancer Res.* **56**, 2245–2249.

Yokota, J., Akiyama, T., Fung, Y. K., Benedict, W. F., Namba, Y., Hanaoka, M., Wada, M., Terasaki, T., Shimosato, Y., Sugimura, T., and Terada, M. (1988). Altered expression of the retinoblastoma (RB) gene in small-cell carcinoma of the lung. *Oncogene* **3**, 471–475.

Zhang, W., Hu, G., Estey, E., Hester, J., and Deisseroth, A. (1992). Altered conformation of the p53 protein in myeloid leukemia cells and mitogen-stimulated normal blood cells. *Oncogene* **7**, 1645–1647.

Zhang, W. W., Fujiwara, T., Grimm, E. A., and Roth, J. A. (1995). Advances in cancer gene therapy (review). *Adv. Pharmacol.* **32**, 289–341.

Zhang, X., Xu, H. J., Murakami, Y., Sachse, R., Yashima, K., Hirohashi, S., Hu, S. X., Benedict, W. F., and Sekiya, T. (1994). Deletions of chromosome 13q, mutations in Retinoblastoma 1, and retinoblastoma protein state in human hepatocellular carcinoma. *Cancer Res.* **54**, 4177–4182.

Zhou, Y., Li, J., Xu, K., Hu, S.-X., Benedict, W. F., and Xu, H.-J. (1994). Further characterization of retinoblastoma gene-mediated cell growth and tumor suppression in human cancer cells. *Proc. Natl. Acad. Sci. U.S.A.* **91**, 4165–4169.

Zhou, Y., Hu, S.-X., Seigne, J., Kong, C.-T., Perng, G.-S., Benedict, W. F., and Xu, H.-J. (1996). Mechanisms for the enhanced tumor cell growth suppression by an N-terminal truncated RB protein. *Proc. Am. Assoc. Cancer Res.* **37**, 594–595 (abstr.).

Zou, Z., Anisowicz, A., Hendrix, M. J., Thor, A., Neveu, M., Sheng, S., Rafidi, K., Seftor, E., and Sager, R. (1994). Maspin, a serpin with tumor-suppressing activity in human mammary epithelial cells. *Science* **263**, 526–529.

John W. Park*
Keelung Hong†
Dmitri B. Kirpotin†
Demetrios Papahadjopoulos†
Christopher C. Benz*

*Department of Medicine
Division of Hematology-Oncology
University of California, San Francisco
San Francisco, California 94143

†Department of Cellular and Molecular Pharmacology
University of California, San Francisco
San Francisco, California 94143

Immunoliposomes for Cancer Treatment

I. Introduction

The notion of targeted drug delivery has long tantalized investigators in many fields, perhaps most of all in the treatment of cancer. Immunoliposomes represent a rational strategy to achieve targeted drug delivery for cancer treatment, and work suggests that optimization of immunoliposome design may finally lead to realization of this goal.

Despite more than a decade of preclinical work, immunoliposomes have yet to appear sufficiently promising to be evaluated in clinical trials. Early attempts to develop immunoliposomes as a targeted drug delivery system were plagued by multiple obstacles involving all of the components of immunoliposome construction: target antigen, antibody, antibody–liposome linkage, liposome composition and structure, and drug. In particular, progress in immunoliposome development ultimately had to await critical advances in the design of liposomes as a drug delivery system for cancer treatment.

Advances in Pharmacology, Volume 40

These advances included liposomes with much improved pharmacologic properties, including long circulation and selective extravasation in tumors. Fortunately, advances in the liposome field were also paralleled by steady progress in monoclonal antibody-based therapy and in antibody engineering. As a result, immunoliposome development can now benefit from improved targeting technologies as well as important lessons learned from the extensive preclinical and clinical experience with antibody-based therapeutics.

This chapter reviews developments in the use of immunoliposomes for cancer treatment. The use of immunoliposomes for other therapeutic applications, such as the treatment of infectious or inflammatory disorders, is not discussed. Similarly, the use of immunoliposomes for diagnostic applications, such as imaging of solid tumors, also is not discussed. So-called ligandoliposomes, in which targeting ligands other than antibodies are linked to liposomes, also have been omitted from this chapter.

II. Lipsomes as a Drug Delivery System for Cancer Treatment

So-called "conventional" liposomes have been used in cancer treatment for more than two decades to deliver a number of anticancer agents, sometimes resulting in an improved therapeutic index owing to reduced toxicity to normal tissues. However, their clinical utility has been severely limited by the lack of specific tumor targeting, and by rapid clearance by phagocytic cells of the reticuloendothelial system (RES) (Gregoriadis, 1976). To circumvent this rapid clearance, we and other investigators have developed modified liposomes with altered lipid components, inclusion of substituents such as polyethylene glycol (PEG), small diameter, and reduction of surface charge (Gabizon and Papahadjopoulos, 1988). These so-called "stealth" or "sterically stabilized" liposomes possess much reduced reactivity with serum proteins and are less susceptible to RES uptake, resulting in significant prolongation of circulation time. In addition, sterically stabilized liposomes have been shown to accumulate preferentially within tumors in animal models and in humans (Papahadjopoulos *et al.*, 1991; Huang *et al.*, 1992b; Lasic and Martin, 1995). It appears that the mechanism for tumor localization of liposomes involves enhanced liposome extravasation from tumor-associated vasculature, which occurs because of endothelial fenestrations and other structural abnormalities associated with tumor angiogenesis. Because of their improved pharmacologic properties, sterically stabilized liposomes have generated renewed interest in liposomes as drug carriers (Lasic and Papahadjopoulos, 1995). Sterically stabilized liposomes containing doxorubicin have shown encouraging clinical activity (Lasic and Martin, 1995); and two versions of doxorubicin-loaded liposomes have been approved by the Food and Drug Administration (FDA) for the treatment of acquired immunodeficiency syndrome (AIDS)-associated Kaposi's sarcoma. Clinical trials of sterically stabilized liposomes containing doxorubicin for the treatment of other solid tumors, including breast, lung, and liver cancer, are in progress.

III. Antibodies Useful for Targeting

A. General Considerations

Immunoliposomes (ILs) represent a further strategy to enhance liposomal drug delivery, by linking liposomes to monoclonal antibodies (MAbs) directed against tumor-associated antigens. Choice of target antigen and antibody is a critical aspect of immunoliposome design.

In the past, target antigens have often been chosen on the basis of availability of antibodies and model systems rather that rational therapeutic design. However, it is obvious that target antigen selection for monoclonal antibody-based therapies must involve careful consideration (Table I). Most tumor-associated antigens are quantitatively overexpressed on tumor cells as opposed to normal cells and tissues, but in cancer patients overexpression often does not occur homogeneously in all cells of a given tumor or in

TABLE I   Components of Immunoliposome Design

Component	Considerations for optimal design
Target antigen	*Expression:* Homogeneously overexpressed
	Function: Vital to tumor progression, so that downmodulation does not occur or is associated with therapeutic benefit
	Shed antigen: Limited, to avoid binding to soluble antigen and accelerated clearance
Antibody	*Immunogenicity:* Humanized MAb, to remove murine sequences. MAb fragment (e.g., Fab, Fab′, Fv), to remove Fc region
	Internalization: Efficiently endocytosed
	Biological activity: Intrinsic antitumor activity, to enhance antitumor effect
Linkage	*Stability:* Antibody covalently attached to hydrophobic anchor
	Attachment site: Specific sites on antibody and liposome. Avoids steric hindrance (e.g., due to PEG) of MAb binding and internalization
Liposome	*Stability:* Stable as intact construct *in vivo*
	Pharmacokinetics: Long circulating
	Tumor penetration: Selective extravasation in tumors (e.g., sterically stabilized). Small diameter to improve tumor penetration
Drug	*Encapsulation:* Efficient and high capacity
	Bystander toxicity: Small molecule drug to diffuse throughout tumor once released, or other mechanism to affect tumor cells not directly targeted
	Interaction with target antigen, MAb: Anticancer effect particularly suited to target cell population. Cytotoxicity enhanced by binding of MAb

cells from different metastatic sites. In addition, clinical studies involving monoclonal antibody therapies have shown that the expression of some target antigens can become significantly downmodulated during treatment or disease progression, including loss of the overexpression phenotype in some or all tumor cells (for review, see Waldmann, 1991). Finally, some antigens are shed or secreted by tumor cells in some patients, leading to potentially high circulating levels of soluble antigen that can bind antibody.

Monoclonal antibodies or antibody fragments to be used in immuno-liposomes, even those directed against the same target antigen, vary in ways that are crucial to immunoliposome development. The immunogenicity of MAb-based therapeutic agents has constituted a major barrier to successful therapy, particularly owing to the use of murine MAbs and/or immunoconju-gates containing immunogenic components. Similarly, immunoliposomes containing heterologous IgG have elicited high levels of anti-IgG antibodies in mice, resulting in marked reduction in circulation time with repeated administration (Phillips and Emili, 1991; Phillips *et al.*, 1994). In addition to the use of heterologous IgG, exposure of the Fc region and the use of an avidin–biotin linkage were also associated with increased immunogenicity in these studies.

Antibodies that are efficiently internalized after antigen binding, and that are capable of mediating immunoliposome internalization as well, offer a unique advantage for immunoliposomal drug delivery. Although immuno-liposomes that do not efficiently internalize have been reported to produce enhanced cytotoxicity *in vitro* and *in vivo*, these immunoliposomes appar-ently act by slowly releasing encapsulated drug while bound at the cell membrane (Ahmad and Allen, 1992; Ahmad *et al.*, 1993). In contrast, internalizing immunoliposomes can directly deliver drug intracellularly as well as release drug extracellularly, leading in principle to more efficient delivery of drug and enhanced cytotoxicity. Furthermore, internalizing im-munoliposomes allow strategies in which therapeutic agents otherwise un-able to reach intracellular sites of action can be delivered intracellularly.

Another exploitable aspect of MAb-mediated targeting of immunolipo-somes is the potential contribution of the biological activity of the antibody itself. Monoclonal antibodies that have anticancer properties, such as growth factor receptor antagonism or induction of apoptosis, may enhance the cytotoxic effect of encapsulated drug.

B. Antibodies Directed against the HER2/*neu* (c-*erb*B-2) Oncogene Product

1. The HER2 Oncogene Product as a Target for Cancer Treatment

The *HER2* (c-*erb*B-2, *neu*) proto-oncogene appears to play an important role in the development and progression of many breast and other cancers. *HER2* encodes p185^{HER2} (ErbB-2), a 1255-amino acid, 185-kDa receptor tyrosine kinase (RTK) that is a member of the class I RTK family, along with the proteins encoded by the epidermal growth factor receptor (*EGFR*), *HER3* (*erb*B-3), and *HER4* (*erb*B-4) genes (for review, see Carraway and Cantley, 1994). HER2 overexpression occurs in 20–30% of breast cancers, most commonly via gene amplification, and is associated with poor prognosis for these patients (Slamon *et al.*, 1987, 1989). HER2 overexpression also occurs in ≥40% of cases of ductal carcinoma *in situ* (DCIS) of the breast, and up to 80% of comedo-type DCIS lesions, which are associated with higher risk of recurrence (VandeVijver *et al.*, 1988; Liu *et al.*, 1992). In addition to breast cancer, HER2 overexpression also occurs frequently in other malignant diseases, including ovarian cancer (Slamon *et al.*, 1989; Berchuck *et al.*, 1990), endometrial cancer (Berchuck *et al.*, 1991), non-small-cell lung cancer (Kern *et al.*, 1990), gastric cancer (Yokota *et al.*, 1988; Park *et al.*, 1989; Yonemura *et al.*, 1991), bladder cancer (Zhau *et al.*, 1990), and prostate cancer (Zhau *et al.*, 1992). Much experimental evidence indicates that HER2 overexpression directly contributes to transformation and tumor progression, and suggests that its prognostic significance arises from the particularly aggressive phenotype it confers (for review, see Hynes and Stern, 1994).

The HER2 receptor is a logical focus for the development of targeted cancer therapies. First, such therapies may be able to antagonize or interfere with this important mediator of tumor growth. Second, HER2 provides an ideal target antigen. It is a readily accessible cell surface receptor, and, when overexpressed, provides a means for immunotherapies to target the tumor population. In contrast, in normal adult tissues, HER2 occurs only at low levels in certain epithelial cell types (Press *et al.*, 1990). In addition, HER2 overexpression occurs relatively homogeneously within primary breast tumors, and is maintained at synchronous or metachronous metastatic sites, suggesting a continuous requirement for HER2 overexpression throughout the malignant process (Niehans *et al.*, 1993). This is in contrast to most other tumor-associated antigens, which are often heterogeneously expressed and/or can undergo downmodulation of expression without significantly affecting tumor growth.

2. Anti-HER2 Monoclonal Antibodies

A number of monoclonal antibodies directed against HER2 have been developed for use in various forms of cancer immunotherapy (for review, see Park *et al.*, 1992). The use of "naked" anti-HER2 monoclonal antibodies for immunotherapy potentially offers two distinct mechanisms of action against HER2-overexpressing tumors: first, the antiproliferative (cytostatic) effect that some anti-HER2 monoclonal antibodies exert against HER2-overexpressing cells, which appears to be mediated by partial activation of HER2-dependent signal transduction and/or HER2 receptor downregulation; and second, the activation or facilitation of endogenous host immune effector functions against antibody-bound tumor cells, e.g., via antibody-dependent cellular cytotoxicity. One such antibody, muMab4D5, reacts specifically with an epitope on the extracellular domain of p185^{HER2} (Fendly *et al.*, 1990), and inhibits the growth of HER2-overexpressing breast cancer cells *in vitro* (Lewis *et al.*, 1993) and in animal models (Park *et al.*, 1992). Treatment with muMAb4D5 also renders HER2-overexpressing breast cancer cells more sensitive to the cytotoxic action of cisplatin (Shepard *et al.*, 1991; Pietras *et al.*, 1994) and doxorubicin (Baselga *et al.*, 1994). A fully humanized version of this antibody, rhuMAbHER2, has been developed to retain these properties while reducing the potential for immunogenicity (Carter *et al.*, 1992). In a phase II clinical trial, treatment with rhuMAbHER2 alone was well tolerated and associated with objective antitumor responses in 12% of patients with metastatic breast cancer (Baselga *et al.*, 1996). In another phase II study, treatment of metastatic breast cancer patients with rhuMAbHER2 and cisplatin resulted in an objective response rate of 25% (Pegram *et al.*, 1995). Use of rhuMAbHER2 has entered phase III clinical trials for the treatment of advanced breast cancer in conjunction with chemotherapy.

In addition to the use of "naked" anti-HER2 monoclonal antibodies by themselves or in combination with chemotherapy, therapeutic strategies in which anti-HER2 monoclonal antibodies or antibody fragments are used to target potent anticancer effectors of HER2-overexpressing cells have been developed. These strategies include anti-HER2 immunotoxins (Batra *et al.*, 1992; Wels *et al.*, 1992), bispecific antibodies directed against the HER2 receptor as well as against lymphocytes or other immune effector cells (Shalaby *et al.*, 1992; Weiner *et al.*, 1993; Valone *et al.*, 1995), constructs linking an anti-HER2 antibody domain with a prodrug-activating enzyme for antibody-dependent enzyme-prodrug therapy (ADEPT) (Rodrigues *et al.*, 1995), intracellular anti-HER2 single-chain antibody (scFv) expressed via adenoviral gene transfer (Deshane *et al.*, 1995), and constructs linking an anti-HER2 antibody domain with a DNA-binding domain for gene delivery (Fominaya and Wels, 1996).

IV. Immunoliposome Design and Construction

A. Conjugation Strategies

To form an immunoliposome, antibody or antibody fragments are linked to the liposome surface. The linkage must be stable enough to resist dissociation *in vivo*, should not impair the antigen-binding properties of the antibody, and should not disrupt the liposome and encapsulated drug (Table I). Importantly, the linkage procedure must be simple, efficient, and economical.

Considerable work has been performed to optimize antibody conjugation to liposomes (Heath, 1987; Leserman and Machy, 1987; Ranade, 1989; Torchilin, 1994). The solution offered by nature itself is to incorporate hydrophobic domains into antibodies to allow tight association with biomembranes. Early attempts to use the natural hydrophobicity of immunoglobulins to associate with liposomes helped to establish the concept of immunoliposome targeting, but were impractical owing to the weakness of the antibody association with the lipid bilayer (Gregoriadis and Neerunjun, 1975). Currently immunoliposomes are produced by linkage of antibody to a hydrophobic "anchor" stably rooted in the lipid bilayer of the liposome, sometimes with a spacer "arm" between the antibody and the anchor. Hence, antibody conjugation involves multiple elements, each of which can be addressed by a variety of strategies.

I. Antibody

Development of monoclonal antibodies, recombinant MAbs and their fragments, and single-chain antibody fragments (scFvs) generated by phage-display libraries has greatly expanded the scope of antigen-seeking molecules for immunoliposomes. Intact MAbs provide bivalent interaction with the target; however, major disadvantages include larger size and the presence of the Fc region, which is associated with higher immunogenicity and recognition by macrophage Fc receptors. Although IgG has been the most common type of antibody used for immunoliposomes, reduced IgM has two thiol groups conveniently located in the Fc region, and has also been used in immunoliposomes (Hashimoto *et al.*, 1986b). Monoclonal antibody fragments such as Fab′ offer a number of important advantages over intact MAb. Although Fab′ can be generated enzymatically from intact antibody. Fab′ can now be obtained far more efficiently using cloned antibody genes and high-level expression of Fab′ fragments in *Escherichia coli* (up to 2 g/liter)(Carter *et al.*, 1996). Fab′ fragments lack the Fc region, and their smaller size and greater flexibility relative to intact MAbs entail a reduction in the effective diameter of the immunoliposome. The thiol group in the Fab′ hinge region provides a single specific site for covalent coupling to a thiol-reactive anchor, and ensures that the antigen-binding site is oriented away from the liposome (Martin and Papahadjopoulos, 1982; Park *et al.*, 1995). Single-chain Fv fragments are becoming increasingly available through phage-display library screening, can also be expressed efficiently, and have also been linked to immunoliposomes (Laukkanen *et al.*, 1994). Attachment of several different antibodies (antibody "cocktail") to the same liposome can be used to create immunoliposomes with multiple specificities.

2. Hydrophobic Anchor

Early studies used long-chain fatty acids, such as palmitic acid, coupled to antibody (Harsch *et al.*, 1981; Huang *et al.*, 1982; Weissig *et al.*, 1986).

More recent and much more frequently used anchors include phospholipids [phosphatidylethanolamine (PE), phosphatidylinositol (PI)] with modified head groups to accommodate covalent binding, which are then linked either directly to the antibody or to an additional linker such as streptavidin (Leserman and Machy, 1987; Torchilin, 1994). Interest in fatty acid anchors has revived with the introduction of recombinant antibodies containing a lipophilic domain, thus avoiding the necessity for chemical coupling to a separate hydrophobic anchor (Keinanen and Laukkanen, 1994; Laukkanen *et al.,* 1994).

3. Linkage

As discussed, antibody can be covalently or noncovalently linked to the hydrophobic anchor. Noncovalent linkages have included biotinylated antibody bound to avidin-containing liposomes (Phillips and Tsoukas, 1990), or to streptavidin, which is in turn linked to biotinylated PE in the liposome bilayer (Loughrey *et al.,* 1987). The extremely high binding constant of the avidin (streptavidin)–biotin complex, as well as the commercial availability of many biotinylated antibodies, make this a feasible alternative to covalent binding; however, the introduction of additional bulky and potentially immunogenic proteins on the liposome surface is clearly undesirable. Another noncovalent linkage between liposome and antibody utilizes polyhistidine sequences, which are frequently added to recombinant proteins to facilitate their purification: a terminal polyhistidine sequence added to recombinant antibody was attached to the lipid bilayer via coordination complex with bivalent nickel ions immobilized on a membrane-anchored amphiphilic chelator (Dietrich *et al.,* 1996).

To achieve covalent linkage between antibody and hydrophobic anchor, a variety of biochemical strategies have been used. The coupling chemistry is similar to that of protein immobilization and modification. The functional group on the antibody molecule reacts with an active group on a liposome-associated hydrophobic anchor, or vice versa. Amino groups of lysine side chains on the antibody can be coupled to liposome-associated N-glutaryl-phosphatidylethanolamine activated with water-soluble carbodiimide, to palmitic acid via N-hydroxysuccinimide ester, or to periodate-oxidized PI through a Schiff base, further reduced by borohydride to stable amine (Torchilin, 1994). Thiol groups of reduced antibody or antibody fragment can be coupled to liposomes via thiol-reactive anchors such as pyridylthiopropionyl-phosphatidylethanolamine (PDP-PE), which forms a disulfide bond (Martin *et al.,* 1981; Wolff and Gregoriadis, 1984), maleimido-functionalized PE (Martin and Papahadjopoulos, 1982; Park *et al.,* 1995), or N-iodoacetyl derivatives of PE (Wolff and Gregoriadis, 1984; Hashimoto *et al.,* 1986a), which form a stable thioether bond. Thiol-reactive chemistry gives better prediction of the linking position and less possibility of lysine modification in the antigen-binding site. A convenient single thiol group is present in the hinge region of Fab' fragments, or may be introduced into the antibody molecule by treatment with heterobifunctional reagents, homocysteine thiolactone, or iminothiolane (Trout's reagent) (Thomas *et al.*, 1994). Alternatively, the antibody may be activated for linkage to a functional group exposed on the liposome surface. For example, oxidation of carbohydrate moieties in the antibody produces dialdehydes that can react with liposome-incorporated PE-hydrazide to form stable hydrazone linkages (Chua *et al.*, 1984). An important consideration with all of these approaches is the presence of excess active groups that remain after the coupling reaction; quenching of these groups may be required to avoid nonspecific reactivity of the remaining active groups with cell surface (Kirpotin *et al.*, 1997).

4. Linkage and Liposome Formation

Another vital consideration concerns when to perform antibody conjugation in relation to liposome formation: i.e., linkage of antibody to preformed liposomes containing activated anchor vs liposome formation in the presence of an anchor-modified antibody. To ensure stable "rooting" of the hydrophobic anchor in the lipid bilayer of the liposome, it should be present during bilayer formation. Since antibodies are often sensitive to conditions used in liposome preparation (e.g., organic solvents, shear forces, and elevated temperature), the most common strategy has been first to prepare liposomes containing a hydrophobic anchor with an active head group (see Section IV,A,2). Antibody is then allowed to react with the anchor head group, and uncoupled protein is separated from the resulting immunoliposomes by dialysis or size-exclusion chromatography (Chua *et al.*, 1984; Leserman and Machy, 1987; Torchilin, 1994). An exception to this rule is afforded by preparation of immunoliposomes using the detergent dialysis method, in which antibody is first coupled to a hydrophobic anchor molecule, and then solubilized in the presence of bilayer lipids under mild conditions using dialyzable detergents, such as octylglucoside; the hydrophobic anchor attached to the antibody becomes integrated into the liposome bilayer during detergent removal. This method can be useful when the accessibility of attachment sites on the hydrophobic anchor may be compromised by other components (e.g, surface-grafted polymers such as PEG) of the preformed liposome. Long-circulating immunoliposomes modified with polyethylene glycol chains and targeted to infarcted myocardium were prepared by this method, using anti-myosin antibody linked to the hydrophobic anchor *N*-glutaryl-phosphatidylethanolamine via lysine residues (Torchilin, 1994).

5. Site of Attachment on Sterically Stabilized (Polyethylene Glycol-Modified) Liposomes

An important case is represented by long-circulating sterically stabilized liposomes, which contain up to 6–7 mol% (of total lipid) of surface-grafted PEG. As discussed, the pharmacologic advantages of sterically stabilized liposomes make them particularly suitable for use as immunoliposomes; however, surface-grafted PEG chains can inhibit the interaction between antibody and target antigen (Klibanov *et al.*, 1991). This inhibition depends on the chain length of PEG (Mori *et al.*, 1991), and is more pronounced when Fab′ fragments are used (Kirpotin *et al.*, 1997). To avoid this problem, several methods have been developed to allow linkage of antibodies to the distal ends of the liposome-grafted PEG chains themselves (Hansen *et al.*, 1995; Zalipsky *et al.*, 1996). These methods were facilitated by the development of functionalized polyethylene glycol-1,2-distearoyl-3-*sn*-glycerophosphoethanolamine (PEG-DSPE) derivatives containing hydrazido, 2-pyridyldithio-propionamide bromacetamido, or carboxy functions on the free terminus of PEG. These derivatives have been coupled to antibody via periodate-oxidized carbohydrate moiety, thiol groups, or active esters derived from amino groups (Hansen *et al.*, 1995; Zalipsky *et al.*, 1996). The resulting immunoliposomes retained the properties of sterically stabilized liposomes *in vivo* (Allen *et al.*, 1995). Two novel anchors, N-maleimidomethylcyclohexylcarboxamido-polyethylene glycol-1,2-distearoyl-3-*sn*-glycerophosphoethanolamine (MMC-PEG-DSPEC) and N-maleimidopropionylcarboxamido-polyethylene glycol-1,2-distearoyl-3-*sn*-glycerophosphoethanolamine (MP-PEG-DSPE), were developed for conjugation of humanized recombinant anti-HER2 Fab′ fragments, resulting in sterically stabilized immunoliposomes with nearly quantitative conjugation yield, extremely low nonspecific binding *in vitro*, and long circulation time *in vivo* (Kirpotin *et al.*, 1997; and below). The use of DSPE in these PEG-containing anchors was important, since PEG-PE with shorter acyl chains tended to leave the bilayer under *in vivo* conditions (Parr *et al.*, 1994).

B. Anti-HER2 Immunoliposomes

I. Construction of Anti-HER2 Immunoliposomes

We have described the development of sterically stabilized anti-HER2 immunoliposomes (ILs), in which the anti-HER2 specificity and inhibitory activity of the humanized monoclonal antibody rhuMAbHER2 are combined with the pharmacokinetic and drug delivery advantages of sterically stabilized ("stealth") liposomes (Park *et al.*, 1995). This strategy attempts to increase the likelihood of successful immunoliposome-mediated drug delivery to HER2-overexpressing cells by (1) use of humanized Fab′ to lessen the potential for immunogenicity, by removal of murine antibody sequences and of the antibody Fc region, (2) development of internalizing immunoliposomes for efficient intracellular delivery of encapsulated agents, (3) use of Fab′ derived from rhuMAbHER2, which has intrinsic antiproliferative and chemotherapy-sensitizing activities, and developing immunoliposomes that retain these properties, (4) use of PEG-containing sterically stabilized liposomes for prolongation of circulation and selective tumor extravasation, and (5) encapsulation of a cytotoxic agent (doxorubicin) that provides an additive or synergistic antitumor effect when combined with antibody treatment, and that can readily diffuse through tumors when released to provide "bystander" cytotoxicity.

We have produced several versions of anti-HER2 immunoliposomes (ILs) containing covalently linked rhuMAbHER2-Fab' fragments and PEG$_{1900}$ (polyethylene glycol)-containing phosphatidylethanolamine (PEG-PE) included in varying amounts (0–12 mol% of total phospholipid). Immunoliposomes were prepared by conjugation of "conventional" (phosphatidylcholine plus cholesterol; PC/Chol) or "sterically stabilized" (PC/Chol/PEG-PE) small unilamellar liposomes with rhuMAbHER2-Fab' (Carter *et al.*, 1996). Unlike previous so-called conventional liposomes, the PC used was hydrogenated soy phosphatidylcholine (HSPC), allowing for more rigid drug encapsulation as well as longer circulation *in vivo*. Although the inclusion of PEG-PE in these small HSPC liposomes and immunoliposomes was not required for long circulation, the addition of PEG-PE did provide further resistance to clearance and correspondingly improved pharmacokinetic properties (see Section IV,B,4). Recombinant Fab' fragments were used because of their efficient expression in *E. coli*, single site attachment, smaller size, and lack of Fc sequences. In addition, rhuMAbHER2-Fab' has much less antiproliferative activity than intact rhuMAbHER2, and thus it was of interest to see whether attachment of rhuMAbHER2-Fab' to liposomes could reconstitute this activity (see Section IV,B,3). Initially, Fab' was conjugated directly to maleimido-phosphatidylethanolamine (M-PE), resulting in Fab' directly linked to the liposome surface (Ls-Fab'; Fig. 1), and hence alongside or in parallel with PEG (if present) (Park *et al.*, 1995). We have also prepared immunoliposomes in which Fab' was conjugated to maleimide-terminated PEG-PE (MMC-PEG-DSPE or MP-PEG-DSPE) (Kirpotin *et al.*, 1997), resulting in Fab' linked to the distal end of PEG chains (PEG-Fab'; Fig. 1). Both procedures were highly efficient, typically yielding 50–100 Fab' fragments per liposome particle.

2. Binding and Internalization of Anti-HER2 Immunoliposomes

Specific binding of anti-HER2 immunoliposomes to HER2-overexpressing cancer cells in culture has been determined by flow cytometric assay, competitive binding assay, and spectrofluorometric assay. A flow cytometric assay showed binding of anti-HER2 immunoliposomes to cells with HER2 overexpression (BT-474 and SK-BR-3 breast cancer cells, 10^5–10^6 receptors/cell), but no detectable binding to cells lacking HER2 overexpression (MCF-7 breast cancer cells, 10^4 receptors/cell) (Park *et al.*, 1995). A competitive binding assay, in which breast cancer cells were simultaneously incubated with ^{125}I-labeled rhuMAbHER2 and increasing concentrations of anti-HER2 immunoliposomes, showed single-component binding of immunoliposomes to SK-BK-3 and BT-474 cells. An approximation of binding affinity was

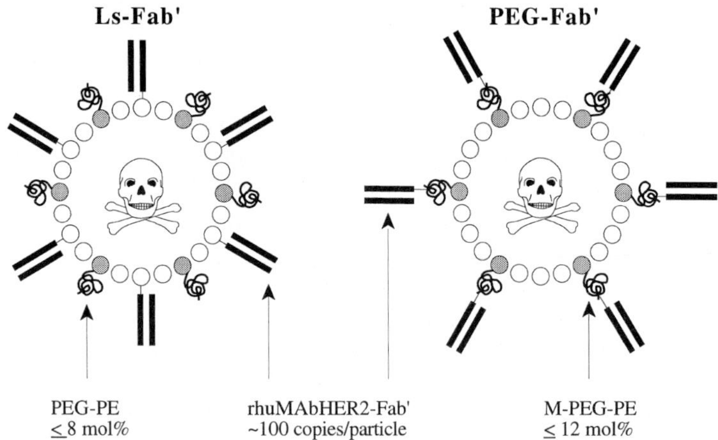

FIGURE I   Schematic representation of anti-HER2 immunoliposomes (ILs), including immunoliposomes with Fab' linked directly to the liposome surface (Ls-Fab') or immunoliposomes with Fab' linked to PEG (PEG-Fab'). M-PEG-PE, Maleimide-terminated PEG-PE [either *N*-maleimidomethylcyclohexylcarboxamido-PEG-DSPE (MMC-PEG-DSPE) or *N*-maleimido-propionylcarboxamido-PEG-DSPE (MP-PEG-DSPE)].

obtained via Scatchard analysis by assuming that Fab' on immunoliposomes behaved as free ligand. With this model, immunoliposomes showed high-affinity binding comparable to that of free Fab' or intact antibody. Binding affinity of immunoliposomes with the Ls-Fab' linkage was reduced with increasing PEG content, indicating that high concentrations of PEG could interfere with Fab' binding. Control liposomes lacking Fab' showed negligible binding (Park *et al.*, 1995).

Monoclonal antibody rhuMAbHER2 is rapidly internalized by HER2-overexpressing tumor cells via receptor-mediated endocytosis (Sarup *et al.*, 1991). To assess whether anti-HER2 immunoliposomes internalize within target cells *in vitro,* a series of studies using conventional and confocal fluorescence microscopy of rhodamine-labeled immunoliposomes was performed. SK-BR-3 cells treated with rhodamine-labeled anti-HER2 immunoliposomes (Ls-Fab' linkage, 0 mol% PEG) demonstrated intense foci of fluorescence both at the cell surface and intracellularly by 5 min (Park *et al.*, 1995), indicating rapid internalization. Treatment with anti-HER2 immunoliposomes containing high PEG concentrations (≥4 mol%) resulted in decreased intracellular fluorescence, indicating that the rate of internalization was inhibited by the PEG component of immunoliposomes containing the Ls-Fab' linkage. Significantly, the inhibitory effect of PEG on internalization was not observed with immunoliposomes containing the PEG-Fab' linkage rather than the Ls-Fab' linkage: rapid internalization was observed regardless of PEG concentration (2–12 mol%). These results indicated that while PEG can inhibit binding and internalization of Ls-Fab'-linked immunoliposomes, shifting the Fab′ attachment site to the terminus of PEG (PEG-Fab′ linkage) prevents inhibition by PEG.

The specificity of immunoliposome uptake in SK-BR-3 cells was confirmed by preincubation of SK-BR-3 cells with rhuMAbHER2 at 10-fold excess over immunoliposomes, which totally blocked uptake of immunoliposomes but not of coadministered fluorescein isothiocyanate (FITC)-labeled transferrin (Kirpotin *et al.*, 1997). In addition, MCF-7 cells coincubated with immunoliposomes and transferrin showed efficient uptake of transferrin, but no detectable uptake of immunoliposomes.

For quantitative studies of immunoliposomes uptake, internalization, and intracellular drug delivery, immunoliposomes were loaded with pyranine (1-hydroxypyrene-3,6,8-trisulfonic acid; HPTS), a pH-sensitive fluorophore that can be readily encapsulated in liposomes (Daleke *et al.*, 1990). Intracellular disposition was determined by measurement of the pH-dependent fluorescence of HPTS, allowing quantitation of immunoliposome/HPTS in neutral compartments (surface bound) vs acidic compartments (endocytosis associated). Immunoliposomes were taken up rapidly into a neutral environment, with subsequent accumulation in acidic compartments, consistent with surface binding followed by receptor-mediated endocytosis (Kirpotin *et al.*, 1997). Total uptake of immunoliposomes in SK-BR-3 cells when present at saturating concentrations was 7.21 nmol of phospholipid per milligram of cell protein (SEM, 0.45), which corresponds to 23,000 immunoliposomes per cell. Selectivity of uptake was also extremely high: total uptake in non-HER2-overexpressing MCF-7 cells was <0.01 nmol of phospholipid per milligram of cell protein, a reduction of more than 700-fold. In addition, uptake of anti-HER2 immunoliposomes was 600-fold higher than uptake of nontargeted control liposomes in SK-BR-3 cells, but identical in MCF-7 cells. In immunoliposomes containing Fab′ linked at the liposome surface (Ls-Fab′ linkage), increasing concentrations of PEG-PE (2–10 mol%) reduced cell binding by up to 100-fold and endocytosis by 2-fold; however, when Fab′ fragments were instead conjugated to the termini of PEG-PE (PEG-Fab′ linkage), binding and endocytosis were unaffected by total PEG-PE content. Binding and endocytosis depended on quantity of conjugated Fab′, reaching a plateau at ~40 Fab′ per immunoliposome for binding and ~10 Fab′ per immunoliposome for internalization.

The intracellular disposition of immunoliposomes and their contents was also studied by electron microscopy, using immunoliposomes containing colliodal gold particles (Park *et al.*, 1995). SK-BR-3 cells treated with anti-HER2 immunoliposomes (0 mol% PEG) showed gold-laden immunoliposomes at the cell surface and intracellularly in coated pits, coated vesicles, endosomes, multivesicular bodies, and lysosomes. This intracellular distribution is consistent with internalization occurring via the coated pit pathway. However, observations were also made of immunoliposomes that appeared to fuse with the cell membrane, without coated pit formation. In addition, gold particles were noted free within the cytoplasm, not associated with a membrane-bound organelle. Gold particles free within the cytoplasm might have resulted from fusion events between immunoliposomes and the cell membrane. Alternatively, they may have arisen following endocytosis, with escape of the encapsulated gold particles from the endosomal pathway. The ability of anti-HER2 immunoliposomes to internalize and efficiently deliver their contents intracellularly, including delivery outside of endolysosomal compartments, may be exploitable for the delivery of therapeutic genes. Anti-HER2 immunoliposomes may be particularly advantageous because of cytosolic delivery, thus reducing the potential for endosomal sequestration or lysosomal degradation that can accompany receptor-mediated endocytosis.

3. Antiproliferative Effects of Anti-HER2 Immunoliposomes

While the intact and bivalent antibody rhuMAbHER2 inhibits the growth of HER2-overexpressing breast cancer cells in monolayer culture, monovalent Fab and Fab′ fragments of rhuMAbHER2 are much less effective at inhibiting growth (O'Connell *et al.*, 1993). This observation suggested that cross-linking of HER2 receptors by bivalent antibody is important for the antiproliferative effect, and further suggested that liposomal anchoring of multiple rhuMAbHER2-Fab′ fragments might reconstitute antiproliferative activity comparable to that of intact rhuMAbHER2. The effect of empty anti-HER2 immunoliposomes on SK-BR-3 cells in monolayer culture was tested and compared to that of rhuMAbHER2 and rhuMAbHER2-Fab′ (Park *et al.*, 1995). Anti-HER2 immunoliposomes inhibited growth in a dose-dependent manner, indicating that empty anti-HER2 immunoliposomes retain the antiproliferative activity of rhuMAbHER2. Treatment with control liposomes lacking Fab′ did not significantly affect cell growth. Thus, anti-HER2 immunoliposomes possess intrinsic antiproliferative activity against HER2-overexpressing cells, and this unique property may act additively or synergistically with the cytotoxic activity of encapsulated drug.

4. Pharmacokinetics of Anti-HER2 Immunoliposomes

Pharmacokinetic (PK) studies of doxorubicin (dox)-loaded anti-HER2 immunoliposomes were performed in healthy, non-tumor-bearing adult rats. Dox-loaded immunoliposomes were prepared with 0–12 mol% PEG and with Fab′ linked to the liposome surface (Ls-Fab′) or to the distal end of PEG chains (PEG-Fab′). Doxorubicin levels obtained by sampling of venous blood following intravenous (i.v.) injection of immunoliposomes demonstrated greatly prolonged circulation of immunoliposomal dox. In comparison, blood and plasma dox levels following administration of free dox under these conditions were undetectable at 5 min, indicating a marked PK advantage for delivery by immunoliposomes. Pharmacokinetic analyses of the different versions of dox-loaded immunoliposomes revealed biexponential pharmacokinetics, with terminal plasma half-lives of approximately 10 hr (Table II) and mean residence times of 16–24 hr. These results are consistent with the PK of sterically stabilized liposomes, and are markedly superior to those of so-called "conventional" liposomes (Gabizon *et al.*, 1989). The addition of PEG to anti-HER2 immunoliposomes was associated with additional prolongation of mean residence time, but importantly was not required for long circulation. That is, immunoliposomes containing 0 mol% PEG still displayed a mean residence time of 16 hr. Anti-HER2 immunoliposomes and similarly constituted sterically stabilized liposomes lacking Fab' showed similar plasma PK, indicating that the presence of Fab' did not significantly increase clearance in normal, non-tumor-bearing rats.

To evaluate the integrity of dox-loaded anti-HER2 immunoliposomes in circulation, two-component PK studies were performed. For these studies, plasma PK of dox and of rhuMAbHER2-Fab' were codetermined on identical plasma samples following a single i.v. injection of dox-loaded immunoliposomes . Levels of rhuMAbHER2-Fab' were measured by enzyme-linked immunosorbent assay (ELISA), using HER2 extracellular domain (ECD)-coated wells for capture and horseradish peroxidate (HRP)-linked goat anti-human IgG for detection. Levels of rhuMAbHER2-Fab' in plasma following immunoliposome administration indicated significantly prolonged circulation, in contrast to the relatively rapid clearance of free rhuMAbHER2-Fab'. As indicated in Table II, the terminal plasma half-life for rhuMAbHER2-Fab' following immunoliposome administration was approximately 10 hr, as compared to 1–2 hr for free rhuMAbHER2-Fab'. Similarly, the terminal plasma half-life for dox in the same plasma samples following immunoliposome administration was also approximately 10 hr. These results indicate that dox-loaded immunoliposomes remain relatively stable in circulation, yielding equally long circulation times for both Fab' and dox components, and suggesting negligible dissociation or drug leakage.

TABLE II Plasma Pharmacokinetics of Anti-HER2 Immunoliposomes in Rats[a]

Agent	Terminal $t_{1/2}$
Anti-HER2 immunoliposomes	
Doxorubicin component (Spectrophotometric assay)	10 hr
Anti-HER2 Fab' component (ELISA)	10 hr
Free doxorubicin	<5 min
Free rhuMAbHER2-Fab'	1–2 hr

[a] Anti-HER2 immunoliposomes loaded with doxorubicin were administered i.v. via jugular vein catheter at a single dose of 5.0 μmol of total phospholipid (500 μg of doxorubicin), and plasma was collected serially via catheter for up to 48 hr postinjection. Pharmacokinetic analysis was performed using the ADAPT2 program.

5. Localization of Anti-HER2 Immunoliposomes to Tumors in Vivo

The biodistribution and tumor localization of dox-loaded anti-HER2 immunoliposomes were evaluated in an HER2-overexpressing tumor xenograft model. In this model, nude mice carrying established subcutaneous (s.c.) BT-474 tumor xenografts were treated with immunoliposomes by single intraperitoneal (i.p.) or i.v. injection, and tumor localization of dox determined by spectrofluorimetric assay of tissue extracts (Park et al., 1995). Twenty-four hours after i.p. injection, dox delivered by anti-HER2 immunoliposomes had accumulated within tumor xenografts, with significantly lower levels of doxorubicin found in surrounding muscle and in blood (Park *et al.*, 1995). However, we have subsequently observed that immunoliposomes administered i.p. in nude mice are erratically absorbed, resulting in highly variable systemic levels. When dox-loaded anti-HER2 immunoliposomes were administered by single i.v. (retroorbital) injection, dox attained consistently high levels in the blood, with a concentration of 6% of injected dose per gram of blood still present at 24 hr. This long circulation was consistent with formal pharmacokinetic studies performed in rats (see Section IV,B,4). High levels of dox were also observed in liver at 24 and 48 hr, consistent with its role in liposome clearance, with much reduced levels by 67 hr. Tumor localization was evident at 24 hr, with tumor-to-muscle and tumor-to-lung ratios of greater than 1. By 67 hr, the concentration of dox in tumor tissue was significantly higher than in all other tissues except liver. The tumor-to-blood ratio was greater than 22-fold.

Since anti-HER2 immunoliposomes produce efficient intracellular delivery of encapsulated agents to target cells *in vitro*, studies were performed to evaluate whether systemic treatment with anti-HER2 immunoliposomes results in intracellular delivery *in vivo*, using the NCF7/HER2 tumor xenograft-nude mouse model. The MCF7/HER2 cell line (MCF7 cells stably transfected with HER2) expresses high levels of HER2 *in vitro* and as xenografts in nude mice (10^6 receptors/cell). Anti-HER2 immunoliposomes loaded with gold particles were administered i.v. in nude mice containing s.c. MCF7/HER2 xenografts, and were visualized by light microscopy using the silver enhancement technique (Huang *et al.*, 1992a). These studies showed abundant accumulation of silver grains (indicating the presence of immunoliposomes) throughout tumor tissue, both in perivascular areas and within cellular regions of the tumor. Significantly, silver grains were frequently observed within the cytoplasm of individual tumor cells (Fig. 2). In contrast, treatment with control sterically stabilized liposomes, prepared identically to immunoliposomes but without Fab', showed silver grains accumulating predominantly in the extracellular space of perivascular areas, consistent with previous reports of the tumor interstitial localization of sterically stabilized liposomes (Huang *et al.*, 1992a). Control liposomes,

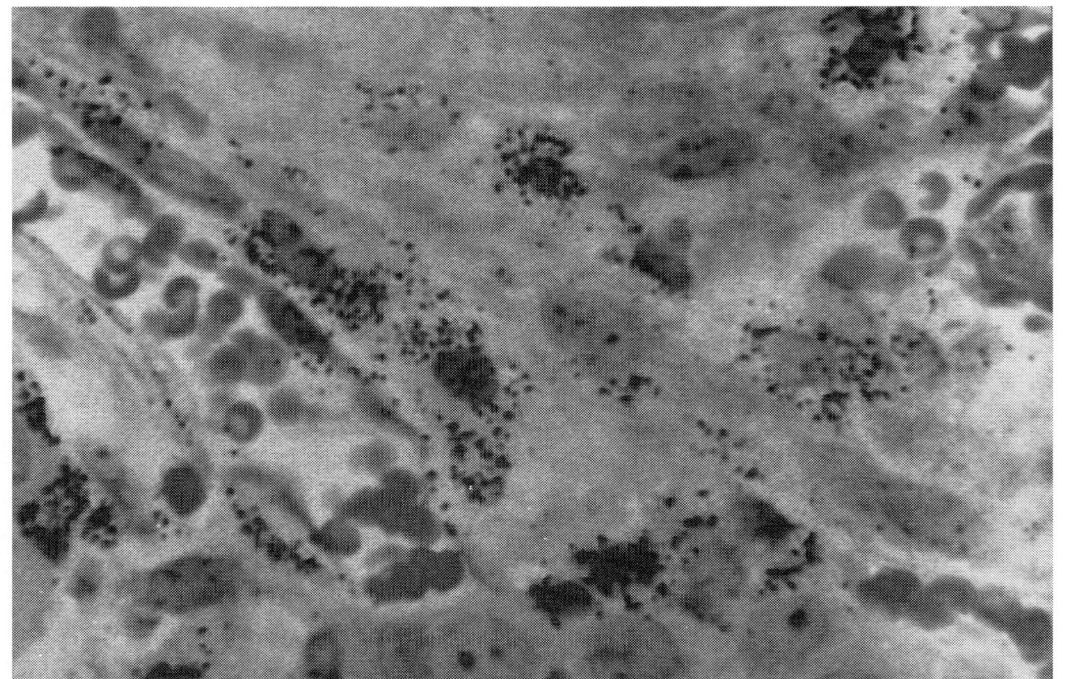

FIGURE 2 Localization of colloidal gold-loaded anti-HER2 immunoliposomes in tumor xenografts 24 hr following i.v. injection. Sterically stabilized immunoliposomes were prepared as described (Kirpotin *et al.*, 1997), labeled with entrapped colloidal gold as described (Huang *et al.*, 1992a), and injected i.v. in nude mice bearing established s.c. MCF7/HER2 tumor xenografts. Twenty-four hours after injection, mice were sacrificed and tumors excised. Tissues were fixed, embedded, and sections cut for silver enhancement (Huang *et al.*, 1992a). Silver grains, indicative of immunoliposome localization, can be seen throughout this representative section, including within the cytoplasm of individual tumor cells.

unlike immunoliposomes, did not produce any silver grains visible within individual tumor cells.

To confirm the intracellular localization of immunoliposomes *in vivo*, anti-HER2 immunoliposomes and control sterically stabilized liposomes (prepared exactly like immunoliposomes but without Fab′) were radiolabeled with ^{67}Ga-diethylenetriamine pentaacetic acid (DTPA) chelates and administered by single i.v. injection in nude mice containing s.c. MCF7/HER2 xenografts. γ camera imaging of these mice 48 and 72 hr after treatment showed accumulation of immunoliposomes primarily in tumor xenografts and in the liver. Control sterically stabilized liposomes, which also preferentially localize to sites of tumor (see Section II), showed similar biodistribution but with less tumor localization. Similarly, γ counting of excised tumors after sacrifice at 72 hr confirmed high levels of ^{67}Ga accumulation in tumors following treatment with immunoliposomes (2.0% injected dose per gram of tissue) as well as with control liposomes (1.0% injected dose per gram of tissue). To evaluate intracellular delivery of ^{67}Ga to tumor xenografts *in vivo*, excised tumors were subjected to electron microscopic autoradiography. In tumors from mice treated with ^{67}Ga-loaded anti-HER2 immunoliposomes, autoradiographic grains signifying ^{67}Ga emission were frequently detected intracellularly within tumor cells (Fig. 3); autoradiographic grains were detected at or near the cell surface, within the cytoplasm, and within the nucleus. The intracellular distribution of ^{67}Ga in this *in vivo* study was consistent with the intracellular distribution observed by electron microscopy after colloidal gold delivery in SK-BR-3 cells *in vitro* (Park *et al.*, 1995); and in preceding discussion). In contrast, tumors from mice treated with ^{67}Ga-loaded control liposomes showed no intracellular ^{67}Ga localization.

C. Other Immunoliposomes

Immunoliposomes incorporating advances in liposome design have been developed for a variety of cancer treatment strategies. Sterically stabilized immunoliposomes directed against tumor-associated antigens have been used to target murine squamous cell lung cancer cells *in vitro* and *in vivo* (Ahmad and Allen, 1992; Ahmad *et al.*, 1993) (see subsequent discussion), and murine fibrosarcoma cells *in vitro* (Emanuel *et al.*, 1996). Immunoliposomes designed for intraperitoneal therapy have been used to target human ovarian cancer cells *in vitro* and in ascites fluid *in vivo* (Straubinger *et al.*, 1988; Nassander *et al.*, 1992, 1995).

In addition to targeting tumor-associated antigens, immunoliposomes have been developed to target endothelial cells. Huang and co-workers have developed sterically stabilized immunoliposomes directed against thrombomodulin, which is overexpressed by murine lung endothelium, for targeted

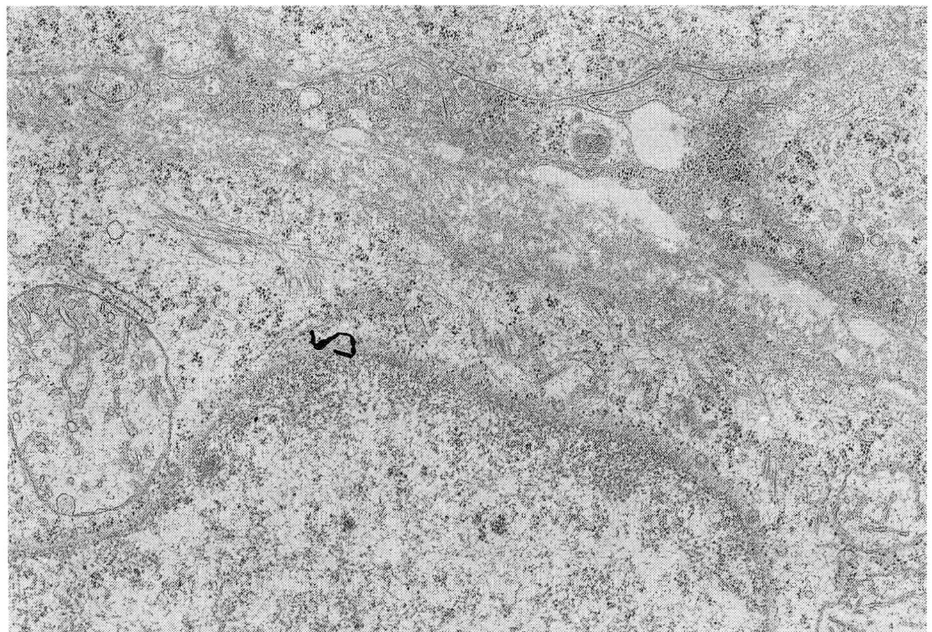

FIGURE 3   Intracellular localization in tumor xenograft cells *in vivo* of ^{67}Ga-loaded anti-HER2 immunoliposomes 72 hr following single i.v. injection. ^{67}Ga-loaded anti-HER2 immunoliposomes were injected i.v. in nude mice bearing established s.c. MCF7/HER2 tumor xenografts. Seventy-two hours after injection, mice were sacrificed and tumors excised. Tissues were fixed and ultrathin sections cut and processed for electron microscopic autoradiography. An autoradiographic grain can be seen in the juxtanuclear region of the cytoplasm. Original magnification: ×18,900.

delivery of a lipophilic prodrug of floxuridine (FUDR) to the lung (Mori *et al.*, 1993, 1995). These immunoliposomes were long circulating, and accumulated to high levels in mouse lung endothelium. Immunoliposome targeting of endothelial cells led to indirect delivery of prodrug to murine breast cancer cells within the lung in a mouse model of lung metastasis. In this strategy, immunoliposomes bound to thrombomodulin on endothelial cells subsequently release prodrug extracellularly, which then diffuses across the endothelium to nearby tumor cells. Strategies such as this, which target tumor-associated endothelium rather than the tumor cells themselves, although indirect, have the advantage of a highly accessible target cell population. Since tumor angiogenesis appears to be a critical determinant of tumor growth, immunoliposomes directed against tumor-associated endothelium could potentially be used to deliver cytotoxic agents to the endothelial cells.

V. Drug or Small-Molecule Delivery via Immunoliposomes ___

A. Doxorubicin

1. Doxorubicin-Loaded Immunoliposomes

Allen and co-workers have developed doxorubicin-loaded, sterically stabilized immunoliposomes that contain noncovalently linked murine MAb 174H.64, which is directed against an epitope associated with some squamous cell cancers (Ahmad and Allen, 1992; Ahmad *et al.*, 1993). These immunoliposomes specifically bound to murine squamous cell lung cancer cells expressing the epitope (although immunoliposome binding yielded only threefold higher uptake than control liposomes lacking antibody), and when loaded with doxorubicin were highly cytotoxic against these cells *in vitro* (Ahmad and Allen, 1992). *In vivo* therapy studies were performed in a murine model of nascent lung metastases: the same murine squamous cell lung cancer cells were administered i.v. in mice, where they begin to form lung metastases within 3 days, at which time dox-loaded immunoliposomes were administered i.v. Treatment with dox-loaded immunoliposomes produced a reduction in subsequent tumor burden, and prolongation of survival that was significantly superior to that associated with free doxorubicin or dox-loaded sterically stabilized liposomes without antibody.

2. Doxorubicin-Loaded Anti-HER2 Immunoliposomes

Anti-HER2 immunoliposome-mediated delivery of doxorubicin may represent a particularly advantageous strategy for the treatment of breast and other cancers with frequent HER2 overexpression. Increased tumor exposure to doxorubicin via immunoliposome delivery may be particularly beneficial in the case of HER2-overexpressing tumors, since HER2-overexpressing breast cancers appear to possess an especially steep dose–response relationship to doxorubicin-based therapy (Muss *et al.*, 1994). In addition, anti-HER2 immunoliposome delivery of doxorubicin provides a means of limiting the toxicity of doxorubicin in normal tissues. HER2 receptor expression is negligible in myocardium and among hematopoietic cells (Press *et al.*, 1990), the critical sites of doxorubicin toxicity in humans. Finally, the antiproliferative effect of anti-HER2 immunoliposomes themselves (derived from rhuMAbHER2) may act synergistically with doxorubicin, since rhuMAbHER2 augments the efficacy of doxorubicin (Baselga *et al.*, 1994), as well as cisplatin (Pietras *et al.*, 1994) and taxol (Baselga *et al.*, 1994).

We have evaluated the cytotoxicity of dox-loaded anti-HER2 immunoliposomes against HER2-overexpressing breast cancer cells and nonoverexpressing nonmalignant cells *in vitro* (Park *et al.*, 1995). Because of the efficient internalization of anti-HER2 immunoliposomes in HER2-overexpressing breast cancer cells, we reasoned that doxorubicin (dox) delivered by immunoliposomes might be just as effective at killing target cells *in vitro* as free dox, a small (M_r 544), amphipathic molecule that readily passes through the cell membrane. Treatment of SK-BR-3 cells for 1 hr with dox-loaded anti-HER2 immunoliposomes yielded dose-dependent cytotoxicity equivalent to that of free dox, indicating that immunoliposomes delivered dox as efficiently as the rapid diffusion of free dox into cells *in vitro.* Doxorubicin-loaded anti-HER2 immunoliposomes were 10- to 30-fold more cytotoxic than dox-loaded immunoliposomes bearing irrelevant Fab'. The specificity of targeting was further confirmed by dox-loaded anti-HER2 immunoliposomes treatment of WI-38 cells, a nonmalignant lung fibroblast cell line that expresses minimal levels of HER2. While free dox produced significant dose-dependent cytotoxicity against WI-38 cells, dox-loaded anti-HER2 immunoliposomes demonstrated 20-fold less cytotoxicity, equivalent to dox-loaded irrelevant immunoliposomes.

Similar results were reported by Suzuki and co-workers, who developed their own doxorubicin-loaded anti-HER2 immunoliposome construct and have studied its cytotoxicity *in vitro* (Suzuki *et al.,* 1995). These small unilamellar immunoliposomes did not contain PEG, and were covalently coupled to intact murine MAb SER4. These dox-loaded anti-HER2 immunoliposomes showed slightly less cytotoxicity than free dox against SK-BR-3 cells *in vitro,* but much reduced cytotoxicity against non-HER2-overexpressing bladder cancer cells or peripheral blood mononuclear cells. When compared to dox-loaded immunoliposomes directed against gp125, a tumor-associated antigen abundantly present on SK-BR-3 cells, dox-loaded anti-HER2 immunoliposomes demonstrated significantly higher cytotoxicity than anti-gp125 immunoliposomes in SK-BR-3 cells despite lower total binding. Instead, cytotoxicity was correlated with internalization: anti-HER2 immunoliposomes were more efficiently internalized (90% at 1-hr incubation) than anti-gp125 immunoliposomes (60% at 1 hr). Interestingly, these investigators had previously shown that endocytosis of dox-loaded anti-gp125 immunoliposomes did not correlate with cytotoxicity (Suzuki *et al.,* 1994). Thus, these investigators concluded that dox-loaded anti-HER2 immunoliposomes can exploit efficient internalization to achieve effective intracellular drug delivery with enhanced cytotoxicity, while dox-loaded immunoliposomes directed against other antigens such as gp125 may not provide effective intracellular delivery, either because of less efficient endocytosis or different intracellular routing and processing.

We have performed a series of *in vivo* therapy studies using dox-loaded anti-HER2 sterically stabilized immunoliposomes in multiple tumor xenograft-nude mice models bearing established HER2-overexpressing tumors. Two independent versions of the BT-474 tumor xenograft–nude mouse model were used, one developed by our collaborators (J. Baselga and J. Mendelsohn) at Memorial Sloan-Kettering Cancer Center (New York, NY) and the other established by our collaborators (G. Colbern and H. Smith) at the Geraldine Brush Cancer Research Institute (San Francisco, CA). In addition, therapy studies were performed using the MCF7/HER2 xenograft model described in Section IV,B,5 as well as an MDA-MB-453 xenograft model that expresses lower HER2 receptor levels ($\sim 10^5$ receptors per cell). In each study, experimental treatment was initiated 1–2 weeks after tumor implantation, at which time tumors were approximately 200–1000 mm^3 in volume. Immunoliposomes were administered at a total dox dose of 15 mg/kg, which was divided into weekly i.v. injection for a total of three doses. Therapeutic effects were compared to those of dox-loaded control liposomes, prepared exactly like immunoliposomes but without inclusion of Fab' and given at the same dose and schedule. Additional control arms included treatment with free dox, which was given at its maximum tolerated dose (MTD) in these animals (7.5 mg/kg; see following discussion), and treatment with saline. Treatment with dox-loaded anti-HER2 immunoliposomes containing the 10% PEG/PEG-Fab' linkage produced superior mean tumor growth suppression when compared with immunoliposomes containing the 2% PEG/Ls-Fab' linkage (data not shown). With either immunoliposome construction, immunoliposomes produced significantly superior antitumor

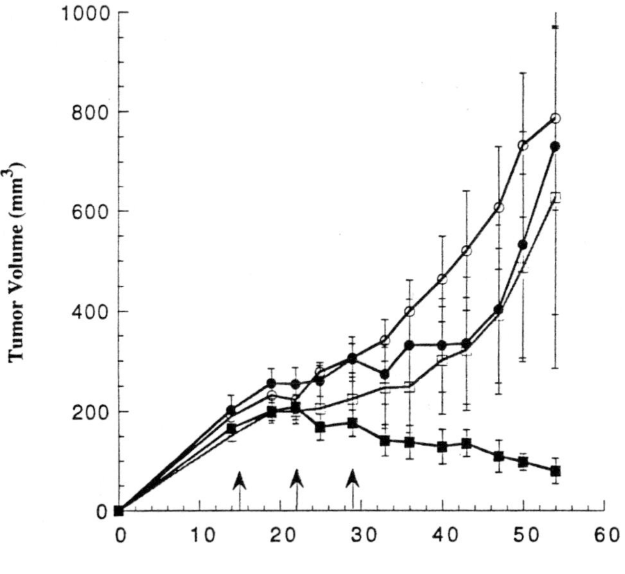

Time Post-Tumor Injection (Days)

FIGURE 4 Antitumor efficacy of dox-loaded anti-HER2 immunoliposomes following i.v. injections in nude mice bearing established BT-474 tumor xenografts. Dox-loaded anti-HER2 immunoliposomes (■, anti-HER2 IL-dox) containing 2 mol% PEG were administered by i.v. (retroorbital) injection at a total doxorubicin dose of 15 mg/kg on days 15, 22, and 28 posttumor injection (arrows). Other groups included saline treated (○, control), free dox at its maximum tolerated dose (MTD) of 7.5 mg/kg (□), or dox-loaded liposomes (L-dox) containing 2 mol% PEG but no Fab' at a total doxorubicin dose of 15 mg/kg (●). Data represent the mean (8 mice per group); bars, SEM.

efficacy than matched dox-loaded control liposomes, free dox, or saline (Fig. 4). Immunoliposome treatment yielded the greatest therapeutic efficacy in each of the animal models. In addition to regression of established tumors, treatment with immunoliposomes was occasionally associated with apparent cures (defined as complete regression of established tumor during the study without any regrowth up to time of sacrifice), which were not observed in any of the other treatment conditions (free dox, dox-loaded control sterically stabilized liposomes).

Treatment with anti-HER2 immunoliposomes resulted in no apparent acute toxicities or significant weight loss in the nude mice. In studies designed to determine MTD in nude mice, varying doses of free dox, dox-loaded liposomes, and dox-loaded anti-HER2 immunoliposomes were compared. The MTD was defined as the highest tested dose level that gave less than 20% weight loss and no treatment-related deaths. Free dox was associated with an MTD of 50 μg/dose over three doses (7.5 mg/kg). In contrast, dox-loaded anti-HER2 immunoliposomes and dox-loaded liposomes were both associated with a 2.5-fold increase in MTD, with an MTD of 100 μg dox/dose over four doses (20 mg/kg) or 125 μg of dox per dose over three doses (19 mg/kg). These results indicated that drug delivery via anti-HER2 immunoliposomes significantly increased the therapeutic index of dox, both by increasing antitumor efficacy and by reducing systemic toxicity. On the basis of these promising results, phase I clinical trials involving doxorubicin-loaded anti-HER2 immunoliposomes for the treatment of breast cancer are under consideration.

B. Other Cytotoxic Drugs

Although doxorubicin and other anthracyclines have been the most widely used drugs in liposomes and immunoliposomes, a number of other cytotoxic drugs have been encapsulated into immunoliposomes for targeted drug delivery. Immunoliposomes directed against placental alkaline phosphatase were used to deliver methotrexate to A431 cells *in vitro* (Jones and Hudson, 1993). As discussed above, sterically stabilized antithrombomodulin immunoliposomes have been used to deliver lipophilic prodrugs, including 3′,5′-O-dipalmitoyl-5-fluoro-2′-deoxyridine (dpFUDR), a prodrug of floxuridine (FUDR) (Mori *et al.*, 1993, 1995). In a mouse model of lung metastasis, dpFUDR-containing immunoliposomes were administered shortly after murine tumor cell injection, resulting in prlongation of median survival in the group of treated mice (Mori *et al.*, 1995). In another strategy, anti-carcinoembryonic antigen (CEA) immunoliposomes containing a [10]B-labeled compound were prepared and evaluated in an *in vitro* model of boron neutron capture therapy (Yanagie *et al.*, 1991).

VI. Macromolecule Delivery via Immunoliposomes

Immunoliposomes may prove useful as a tumor-targeted delivery system for a variety of anticancer agents, such as doxorubicin, by increasing tumor exposure and reducing toxicity to normal cells and tissues. For certain applications, internalizing immunoliposomes may be particularly advantageous, such as for the delivery of agents that otherwise cannot readily gain intracellular access or that are susceptible to lysosomal degradation. Thus, potentially therapeutic macromolecules with intracellular sites of action (e.g., toxins) can in principle be encapsulated and delivered via internalizing immunoliposomes.

Several critical obstacles remain to be overcome for successful implementation of this strategy. First, macromolecular encapsulation must be efficient and reproducible. Methods of liposomal encapsulation, such as passive encapsulation, have in fact been developed for a number of therapeutic proteins. Use of these methods in conjunction with internalizing immunoliposome delivery may be feasible. Second, internalization must lead to release of the therapeutic agent in a biologically active state. Delivery to the endolysosomal compartment without escape into the cytoplasm or other intracellular site of action will obviously compromise any therapeutic effect, and for most proteins will result in degradation. For example, gelonin-loaded anti-HER2 immunoliposomes were able to efficiently deliver gelonin intracellularly in SK-BR-3 cells, but were not cytotoxic probably owing to gelonin remaining sequestered within the endocytic compartment (data not shown). However, certain proteins appear able to reach their sites of action following immunoliposome endocytosis; tumor necrosis factor (TNF) encapsulated in anti-CD4 immunoliposomes was cytotoxic *in vitro* against human T cell leukemia cells that had been TNF resistant (Morishige *et al.*, 1993).

VII. Nucleic Acid Delivery via Immunoliposomes ⸺⸺⸺

A. Delivery Systems for Cancer Gene Therapy

A number of strategies for the gene therapy of cancer have been proposed, and some of these have entered phase I clinical trials (for review, see Roth and Cristiano, 1997). Presently, *in vivo* gene therapy has been limited to locoregional routes of administration, typically by injection of vector directly into or around an identified tumor site. However, the clinical utility of such locoregional approaches, even if efficient gene transfer can be achieved, is extremely limited. Direct vector injection is unlikely to become an important modality for locoregional cancer treatment, in place of existing methods such as surgery or radiation therapy. Furthermore, in most cases of advanced cancer, direct injection of all actual or potential tumor sites in a patient is not usually feasible. Thus local gene therapy, while of scientific interest, is unlikely to become an important modality for cancer treatment or prevention. Ideally, *in vivo* gene therapy of cancer would involve systemic treatment to control or eradicate potentially all tumor cells, including cells at known and unknown metastatic sites.

To develop a feasible strategy for systemic *in vivo* gene therapy, what is needed is a vector or gene delivery system that is safe, nontoxic, stable in circulation, specifically targeted to tumor cells, and capable of mediating efficient gene transfer (Anderson, 1992; Vile and Russell, 1994). Current gene therapy vector technologies, which include modified viral vectors and cationic lipid complexes or liposomes, have not yet yielded vectors that satisfy all (or even most) of these requirements, and remain unsuitable for systemic gene therapy. Limitations of viral vectors include important safety concerns, the potential emergence of replication-competent virus, possible contaminating pathogens, induction of inflammation, and immunogenicity (Miller, 1992). Immunogenicity, in particular, is likely to be a critical obstacle to successful systemic gene therapy, as induction of a host immune response may severely compromise transduction efficiency as well as lead to additional toxicities. Since the efficiency of gene transfer is currently much less than 100% with any vector technology, it is likely that repeat administrations will be required to achieve gene transfer in a reasonable proportion of the tumor cell population.

Recombinant retroviral vectors have additional disadvantages, which include limited DNA insert size (<8 kb), potential for insertional mutagenesis, rapid complement-mediated inactivation following systemic administration, and difficulty in producing in high titers. This last problem has necessitated the local implantation of murine "producer cells" to release a greater amount of recombinant retroviral vector, rather than injection of the retroviral vector itself (Culver *et al.*, 1992). Retroviral vectors expressing ligand molecules on the viral envelope have been developed for tumor targeting (Han *et al.*, 1995). However, retroviral transduction is restricted to actively dividing cells (Miller *et al.*, 1990). This represents an important limitation in the treatment of most solid tumors, which typically have a low growth fraction and significant subpopulation of quiescent cells. Disadvantages of recombinant adenoviral vectors include its particularly high immunogenicity, as most patients are already seropositive for antibodies against the commonly used subgroup C adenovirus, and its lack of selectivity for tumor cells.

Lipid-based or liposomal gene delivery offers a number of potential advantages over viral vectors, particularly with regard to safety, immunogenicity, and ease of preparation. Debs and co-workers used cationic liposome–plasmid DNA complexes to achieve sustained reporter gene expression in multiple tissue sites following intravenous administration in mice (Zhu *et al.*, 1993). However, cationic lipid-based gene delivery technology must overcome problems of nonspecific reactivity and lack of tumor targeting. Molecular conjugate vectors have been developed to provide targeted delivery via attached ligand; these constructs contain a DNA-binding component, plasmid DNA, a ligand for receptor-mediated endocytosis, and additional components to provide subsequent escape from the endocytic pathway. Although interesting, this approach is not yet suitable for systemic administration owing to problems with stability, immunogenicity, rapid clearance, and use of viral components (for review, see Michael and Curiel, 1994).

Immunoliposomes represent a potentially ideal gene delivery vehicle for the systemic gene therapy of cancer, provided they can be constructed so as to package plasmid DNA: retain favorable drug delivery properties, such as stability, long circulation, minimal nonspecific reactivity, and reduced immunogenicity; selectively deliver genes to tumor cells; and provide intracellular delivery that achieves reasonable gene expression.

B. Candidate Therapeutic Genes for Systemic Cancer Gene Therapy

An efficient and tumor-targeted immunoliposome vector suitable for systemic use would greatly improve the prospects for the gene therapy of cancer, and could be used to deliver many types of therapeutic genes, some examples, of which are discussed in the following sections.

I. Prodrug-Converting Enzyme Genes

Sometimes referred to as *suicide genes*, these genes have been used to subsequently activate cytotoxic drugs selectively in cells that have been transduced. For example, the herpes simplex virus thymidine kinase (HSVtk) gene, the most frequently used in gene therapy studies, converts the relatively nontoxic prodrug ganciclovir (GCV) to a phosphorylated metabolite that is highly cytocidal. This strategy has been used in the local treatment of gliomas, via direct implantation of retroviral-releasing producer cells into tumors, followed by subsequent i.v. GCV (Culver *et al.*, 1992). Since retroviral vectors only transduce dividing cells, tumor cells but not non-dividing normal cells will express HSVtk and so become susceptible to GCV. This treatment produced extensive tumor cell death in an animal model, even though only a small fraction of tumor cells actually expressed HSVtk. The mechanism for this phenomenon, sometimes referred to as the "bystander effect," has been attributed to toxic GCV metabolites moving from transduced cells via gap junctions to surrounding nontransduced cells (Ram *et al.*, 1993). Bystander toxicity may enable this strategy to produce significant tumor reduction even with suboptimal gene transfer efficiency. Stereotactic injection of retroviral vector-HSVtk producer cells into human gliomas is currently in phase I clinical trials (Oldfield *et al.*, 1993). A number of clinical studies of HSVtk gene therapy administered by other locoregional routes have now been approved by the Recombinant DNA Advisory Committee of the National Institutes of Health (NIH-RAC).

2. Tumor Suppressor Genes

Loss of function of the *p53* tumor suppressor gene by mutation and/or allelic loss occurs frequently in many types of human cancer, and appears to play a critical role in the pathogenesis of these diseases (for review, see Frebourg and Friend, 1993). Delivery of the wild-type *p53* gene to tumors lacking p53 function may be a useful therapeutic strategy, resulting in restoration of normal growth control and/or induction of apoptosis. For example, introduction of the wild-type *p53* gene in a colon cancer xenograft model has been shown to induce tumor regression due to apoptosis (Shaw *et al.*, 1992). In addition, introduction of the wild-type *p53* gene in an animal model of lung cancer produced moderately efficient gene transfer, which was associated with a disproportionately large antitumor effect consistent with a bystander effect (Fujiwara *et al.*, 1994). A recent report of a phase I clinical trial of *p53* gene therapy, which involved direct injection of endobronchial tumors of non-small-cell lung cancer with a retroviral vector containing the *p53* gene, described apoptosis and evidence of bystander effect following treatment (Roth *et al.*, 1996). Another locoregional *p53* gene strategy uses a modified adenoviral vector containing the wild-type *p53* gene, which is administered via hepatic artery infusion for the treatment of malignant liver tumors, including hepatocellular cancer and metastatic colorectal cancer (Bookstein *et al.*, 1996). In a rat model of orthotopic hepatocellular cancer, *p53* gene therapy via hepatic artery infusion markedly inhibited tumor progression. Formal toxicology studies demonstrated no significant p53-associated toxicities in animals treated either intravenously or by hepatic artery infusion; observed toxicities were tolerable and adenovirus related (data not shown). A clinical gene therapy protocol using this approach has been initiated at the University of California at San Francisco (UCSF).

3. Immunoregulatory Genes

A number of immunoregulatory genes have been proposed or are being included in gene therapy studies. Many of these strategies have the theoretical advantage that efficacy does not presuppose highly efficient gene transfer, as the resulting immune response may be effective against nontransduced tumor cells as well.

4. Antisense Oligonucleotides

The therapeutic use of antisense aptameric or triplex-forming oligonucleotides directed against oncogenes has been limited by their lack of stability *in vivo*, inefficient delivery to and uptake within tumor cells, and routing to lysosomes with resultant degradation (for review, see Calabretta, 1991). While their stability can be signfciantly improved by the use of derivatized nucleic acids, the problems of efficient cytoplasmic delivery remain. Immunoliposomes with the capacity for intracellular and cytoplasmic drug delivery represent a potentially advantageous delivery system for therapeutic oligonucleotides.

C. Strategies for the Development of Immunoliposomes for Gene Therapy

I. Immunoliposomes as a Targeted Gene Delivery System

It is possible to hypothesize that immunoliposome delivery of genes may produce specific and efficient expression of therapeutic genes within tumor cells, while avoiding many of the problems associated with current vector technology. A targeted gene delivery system of this kind may provide the means to achieve systemic gene therapy. Two type of immunoliposome–DNA construct designs can be envisioned: neutral immunoliposomes, with demonstrated long-circulating and tumor-localizing properties; and cationic immunoliposomes, in which neutral and cationic lipids are complexed with plasmid DNA, with further modification to block nonspecific reactivity and to allow inclusion of targeting antibody.

Neutral immunoliposomes, such as the sterically stabilized anti-HER2 immunoliposomes used to deliver doxorubicin (see Section V,A,2), can in principle be used to delivery DNA for systemic gene therapy. Indeed, anti-HER2 immunoliposomes possess highly desirable pharmacologic properties as gene carriers, including reproducibility of preparation, stability, long circulation, minimal reactivity with serum proteins, resistance to RES uptake, and ability to internalize and deliver contents cytoplasmically. However, efficient procedures for the loading of plasmid DNA into neutral liposomes or immunoliposomes have yet to be developed. Liposomes can be formed in the presence of concentrated solutions of supercoiled DNA to achieve encapsulation (Hoffman *et al.*, 1978; Fraley *et al.*, 1980; Wang and Huang, 1987). Although this method has been reported to be efficient, it is likely that highly efficient loading of plasmid DNA will require further technical developments beyond passive encapsulation. One approach may be to use polyamine polymers to condense DNA molecules for more efficient encapsulation (Arscott *et al.*, 1990; Plum *et al.*, 1990; Marquet and Houssier, 1991).

An alternative strategy to produce immunoliposome–DNA constructs is to package DNA within cationic liposomes or lipid complexes rather than neutral liposomes. These cationic liposomes can efficiently coat, via electrostatic interactions, DNA of virtually unlimited size. In addition, cationic liposomes appear to undergo fusion with the cell membrane, facilitating intracellular delivery of DNA (Felgner *et al.*, 1987), although the predominant entry pathway seems to involve endocytosis (Friend *et al.*, 1996). However, such cationic liposomes are limited by poor stability, high nonspecific reactivity, and lack of targeting. We were therefore interested in whether we could overcome the pharmacologic limitations of cationic liposomes via structural modifications and addition of targeting antibody, while retaining their favorable properties. For example, we have developed formulations of cationic liposome–plasmid DNA complexes that are significantly more stable by manipulation of lipid composition, inclusion of PEG-PE, and condensation of plasmid DNA with polyamines (Hong *et al.*, 1997). These stable constructs showed high gene transfer efficiency *in vitro*, and in multiple tissues sites *in vivo* following systemic administration, with up to 1000-fold greater reporter gene expression than in previously published reports.

2. In Vitro Gene Transfer Using Anti-HER2 Cationic Immunoliposomes

To produce a targeted gene delivery vector, rhuMAbHER2-Fab' fragments were conjugated covalently to PEG-PE on cationic liposomes (Fab'-PEG linkage), thus generating anti-HER2 cationic immunoliposome–DNA complexes (Park *et al.*, 1997). PEG-PE was included to reduce significantly the nonspecific reactivity of the cationic liposome component as well as to improve construct stability. Although cationic immunoliposome construction has yet to be optimized, anti-HER2 cationic immunoliposomes were prepared and tested for specific gene transfer *in vitro*. Cationic liposomes were prepared as described (Hong *et al.*, 1997), and loaded with an expression plasmid DNA containing the firefly luciferase (*lux*) gene as a reporter. Cationic liposomes yielded variable transfection in different cell types, including highly efficient transfection of SK-BR-3 cells. However, when cationic liposomes were prepared with the addition of PEG-PE to reduce nonspecific reactivity, the resulting PEG-containing cationic liposomes did yield significantly reduced Lux expression (17-fold) in SK-BR-3 cells. Next, rhuMAbHER2-Fab' was conjugated to these PEG-containing cationic liposomes via M-PEG-PE (PEG-Fab' linkage), thereby generating anti-HER2 cationic immunoliposomes for targeted gene delivery to HER2-overexpressing cells (Park *et al.*, 1997). Anti-HER2 cationic immunoliposomes mediated highly efficient transfection of SK-BR-3 cells: the addition of Fab' was associated with an 18-fold increase in Lux expression as compared to PEG-containing cationic liposomes lacking Fab'. The specificity of immunoliposome-mediated transfection was confirmed by treatment of non-HER2-overexpressing MCF-7 cells with the same constructs. Unmodified cationic liposomes transfected MCF-7 cells with relatively low efficiency. Incorporation of PEG-PE inhibited transfection by four-fold. As expected, addition of rhuMAbHER2-Fab' to PEG-containing cationic liposomes to form anti-HER2 cationic immunoliposomes failed to improve transfection in MCF-7 cells. These results suggest that the reactivity of cationic liposomes can be modulated by PEG-PE, and that the use of cationic immunoliposomes containing PEG-PE and specific Fab' may then provide tumor-targeted gene transfer.

VIII. Conclusion

Successful development of immunoliposomes for targeted drug delivery will clearly require optimized immunoliposome design and construction. Critical parameters include (1) appropriate choice of target antigen, including expression pattern *in vivo*, cellular function, and presence of soluble antigen; (2) antibody, including reduced potential for immunogenicity, ability to promote internalization, and intrinsic biological activity; (3) antibody–liposome linkage, including stable attachment and specific attachment sites; (4) liposome composition and structure, including *in vivo* stability, long circulation, and ability to extravasate and penetrate at tumor sites; and (5) drug, including efficient encapsulation, ability to diffuse beyond the site of delivery, and usefulness in conjunction with target antigen and/or antibody.

Experience in manipulating these parameters to produce doxorubicin-loaded anti-HER2 immunoliposomes has resulted in an apparently optimized preparation that appears to have superior *in vitro* and *in vivo* therapeutic properties over free doxorubicin and nontargeted liposomes. Anti-HER2 immunoliposomes can deliver doxorubicin efficiently and specifically to HER2-overexpressing breast cancer cells while sparing nonoverexpressing normal cells, and, like sterically stabilized liposomes, are stable and long circulating *in vivo*. In HER2-overexpressing tumor xenograft–nude mouse models, anti-HER2 immunoliposomes localize in tumor tissue, deliver encapsulated agents intracellularly, and are associated with significantly increased antitumor cytotoxicity compared with free doxorubicin or doxorubicin-loaded sterically stabilized liposomes. In these animal models, delivery via doxorubicin-loaded anti-HER2 immunoliposomes greatly extends the therapeutic index of doxorubicin, both by increasing antitumor efficacy and by reducing systemic toxicity.

Although promising in animal models thus far, this new generation of immunoliposomes must still overcome questions regarding the therapeutic utility of immunoliposome-mediated drug delivery in cancer patients. Critical questions relating to immunoliposome biodistribution, tumor penetration, and immunogenicity are not fully evaluable in currently available animal models. On the basis of the preclinical experience with doxorubicin-loaded anti-HER2 immunoliposomes, targeted drug delivery by immunoliposomes may perhaps finally be ready for clinical evaluation in cancer patients.

In addition to targeted delivery of small molecule drugs, improvements in immunoliposome design and construction may lead to new therapeutic applications, such as gene therapy. Because of their advantageous properties of targeted intracellular delivery, along with new lipid composition enabling efficient DNA complex formation, immunoliposomes represent a potentially powerful strategy to achieve targeted gene delivery for the *in vivo* gene therapy of cancer.

Acknowledgments

The authors sincerely thank collaborating scientists Gail T. Colbern and Helene S. Smith (Geraldine Brush Cancer Research Institute, San Francisco, CA); Jose Baselga and John Mendelsohn (Memorial Sloan-Kettering Cancer Center, New York, NY); Gilbert-Andre Keller, William I. Wood, and Paul Carter (Genentech, Inc., South San Francisco, CA); and Weiwen Zhang and Yvonne S. Shao (UCSF, San Francisco, CA). This work was partially supported by grants from the SPORE Program of the National Cancer Institute and National Institutes of Health (P50-CA 58207-01); the U.S. Army Medical Research and Materiel Command (DAMD17-94-J-4195); and the American Society of Clinical Oncology Young Investigator Award (J.W.P) sponsored by the Don Shula Foundation.

References

Ahmad, I., and Allen, T. M. (1992). Antibody-mediated specific binding and cytotoxicity of liposome-entrapped doxorubicin to lung cancer cells *in vitro*. *Cancer Res.* **52**, 4817–4820.

Ahmad, I., Longenecker, M., Samuel, J., and Allen, T. M. (1993). Antibody-targeted delivery of doxorubicin entrapped in sterically stabilized liposomes can eradicate lung cancer in mice. *Cancer Res.* **55**, 1484–1488.

Allen, T. M., Brandeis, E., Hansen, C. B., Kao, G. Y., and Zalipsky, S. (1995). A new strategy for attachment of antibodies to sterically stabilized liposomes resulting in efficient targeting to cancer cells. *Biochim. Biophys. Acta* **1237**, 99–108.

Anderson, W. F. (1992). Human gene therapy. *Science* **256**, 808–813.

Arscott, P. G., Li, A.-Z., and Bloomfield, A. (1990). Condensation of DNA by trivalent cations. 1. Effects of DNA length and topology on the size and shape of condensed particles. *Biopolymers* **30**, 619–630.

Baselga, J., Norton, L., Shalaby, R., and Mendelsohn, J. (1994). Anti-HER2 humanized monoclonal antibody (MAb) alone and in combination with chemotherapy against breast carcinoma xenografts. *Proc. Am. Soc. Clin. Oncol.*, p. 63.

Baselga, J., Tripathy, D., Mendelsohn, J., Baughman, S., Benz, C. C., Dantis, L., Sklarin, N. T., Seidman, A. D., Hudis, C. A., Moore, J., Rosen, P. P., Twaddell, T., Henderson, I. C., and Norton, L. (1996). Phase II study of weekly intravenous recombinant humanized anti-p185[HER2] monoclonal antibody in patients with HER/*neu*-overexpressing metastatic breast cancer. *J. Clin. Oncol.* **14**, 737–744.

Batra, J. K., Kasprzyk, P. G., Bird, R. E., Pastan, I., and King, C. R. (1992). Recombinant anti-*erb*B2 immunotoxins containing *Pseudomonas* exotoxin. *Proc. Natl. Acad. Sci. U.S.A.* **89**, 5867–5871.

Berchuck, A., Kamel, A., Whitaker, R., Kerns, B., Olt, G., Kinney, R., Soper, J. T., Dodge, R., Clarke-Pearson, D. L., Marks, P., McKenzie, S., Yin, S., and Bast, R. C. (1990). Overexpression of HER2/*neu* is associated with poor survival in advanced epithelial ovarian cancer. *Cancer Res.* **50**, 4087–4091.

Berchuck, A., Rodriguez, G., Kinney, R. B., Soper, J. T., Dodge, R. K., Clarke-Pearson, D. L., and Bast, R. C. (1991). Overexpression of HER-2/*neu* in endometrial cancer is associated with advanced stage disease. *Am J. Obstet. Gynecol.* **164**, 15–21.

Bookstein, R., Demers, W., Gregory, R., Maneval, D., Park, J. W., and Wills, K. (1996). p53 gene therapy *in vivo* of hepatocellular and liver-metastatic colorectal cancer. *Semin. Oncol.* **23**, 66–77.

Calabretta, B. (1991). Inhibition of protooncogene expression by antisense oligodeoxynucleotides: Biological and therapeutic implications. *Cancer Res.* **51**, 4505–4510.

Carraway, K. L., and Cantley, L. C. (1994). A neu acquaintance for ErbB3 and ErbB4: A role for receptor heterodimerization in growth signaling. *Cell* **78**, 5–8.

Carter, P., Presta, L., Gorman, C. M., Ridgway, J. B. B., Henner, D., Wong, W. L. T., Rowland, A. M., Kotts, C., Carver, M. E., and Shepard, H. M. (1992). Humanization of an anti-p185HER2 antibody for human cancer therapy. *Proc. Natl. Acad. Sci. U.S.A.* **89**, 4285–4289.

Carter, P., Rodrigues, M. L., Park, J. W., and Zapata, G. (1996). Preparation and uses of Fab' fragments from *E. coli*. *In* "Antibody Engineering: A Practical Approach" (H. H. R. McCafferty and D. J. Chiswell, eds.), pp. 291–308. IRL Press, Oxford, UK.

Chua, M. M., Fan, S. T., and Karush, F. (1984). Attachment of immunoglobulin to liposomal membrane via protein carbohydrate. *Biochim. Biophys. Acta* **800**, 291–300.

Culver, K. W., Ram, Z., Wallbridge, S., Ishii, H., Oldfield, E. H., and Blaese, R. M. (1992). *In vivo* gene transfer with retroviral vector-producer cells for treatment of experimental brain tumors. *Science* **256**, 1550–1552.

Daleke, D. L., Hong, K., and Papahadjopoulos, D. (1990). Endocytosis of liposomes by macrophages: Binding, acidification and leakage of liposomes monitored by a new fluorescence assay. *Biochim. Biophys. Acta* **1024**, 352–366.

Deshane, J., Siegal, G. P., Alvarez, R. D., Wang, M. H., Feng, M., Cabrera, G., Liu, T., Kay, M., and Curiel, D. T. (1995). Targeted tumor killing via an intracellular antibody against erbB-2. *J. Clin. Invest.* **96**, 2980–2989.

Dietrich, C., Boscheinen, O., Scharf, K.-D., Schmitt, L., and Tampe, R. (1996). Functional immobilization of a DNA-binding protein at a membrane interface via histidine tag and synthetic chelator lipids. *Biochemistry* **35**, 1100–1105.

Emanuel, N., Kedar, E., Bolotin, E. M., Smorodinsky, N. I., and Barenholz, Y. (1996). Preparation and characterization of doxorubicin-loaded sterically stabilized immunoliposomes. *Pharm. Res.* **13**, 352–359.

Felgner, P. L., Gadek, T. R., Holm, M., Roman, R., Chan, H. W., Wenz, M., Northrop, J. P., Ringold, G. M., and Danielsen, M. (1987). Lipofection: A highly efficient, lipid mediated DNA-transfection procedure. *Proc. Natl. Acad. Sci. U.S.A.* **84**, 7413–7417.

Fendly, B. M., Winget, M., Hudziak, R. M., Lipari, M. T., Napier, M. A., and Ullrich, A. (1990). Characterization of murine monoclonal antibodies reactive to either the human epidermal growth factor receptor or HER2/*neu* gene product. *Cancer Res.* **50**, 1550–1558.

Fominaya, J., and Wels, W. (1996). Target cell-specific DNA transfer mediated by a chimeric multidomain protein: Novel non-viral gene delivery system. *J. Biol. Chem.* **371**, 1–9.

Fraley, R., Subramani, S., Berg, P., and Papahadjopoulos, D. (1980). Introduction of liposome-encapsulated SV40 DNA into cells. *J. Biol. Chem.* **255**, 10431–10435.

Frebourg, T., and Friend, S. H. (1993). The importance of p53 gene alterations in human cancer: Is there more than circumstantial evidence? *J. Natl. Cancer Inst.* **85**, 1554–1557.

Friend, D. S., Papahadjopoulos, D., and Debs, R. J. (1996). Endocytosis and intracellular processing accompanying transfection mediated by cationic liposomes. *Biochim. Biophys. Acta* **1278**, 41–50.

Fujiwara, T., Cai, D. W., Georges, R. N., Mukhopadhyay, T., Grimm, E. A., and Roth, J. A. (1994). Therapeutic effect of a retroviral wild-type p53 expression vector in an orthotopic lung cancer model. *J. Natl. Cancer Inst.* **86**, 1458–1462.

Gabizon, A., and Papahadjopoulos, D. (1988). Liposome formulations with prolonged circulation time in blood and enhanced uptake by tumors. *Proc. Natl. Acad. Sci. U.S.A.* **85**, 6949–6953.

Gabizon, A., Shiota, R., and Papahadjopoulos, D. (1989). Pharmacokinetics and tissue distribution of doxorubicin encapsulated in stable liposomes with long circulation times. *J. Natl. Cancer Inst.* **81**, 1484–1488.

Gregoriadis, G. (1976). The carrier potential of liposomes in biology and medicine. *N. Engl. J. Med.* **295**, 704–710.

Gregoriadis, G., and Neerunjun, E. D. (1975). Homing of liposomes to target cells. *Biochem. Biophys. Res. Commun.* **65**, 537–544.

Han, X., Kasahara, N., and Kan, Y. W. (1995). Ligand-directed retroviral targeting of human breast cancer cells. *Proc. Natl. Acad. Sci. U.S.A.* **92**, 9747–9751.

Hansen, C. B., Kao, G. Y., Moase, E. H., Zalipsky, S., and Allen, T. M. (1995). Attachment of antibodies to sterically stabilized liposomes: Evaluation, comparison and optimization of coupling procedures. *Biochim. Biophys. Acta* **1239**, 133–144.

Harsch, M., Walther, P., Weder, H. G., and Hengartner, H. (1981). Targeting of monoclonal antibody-coated liposomes to sheep red blood cells. *Biochem. Biophys. Res. Commun.* **103**, 1069–1076.

Hashimoto, K., Loader, J. E., and Kinsky, S. C. (1986a). Iodoacetylated and biotinylated liposomes: Effect of spacer length on sulfhydryl ligand binding and avidin precipitibility. *Biochim. Biophys. Acta* **856**, 556–565.

Hashimoto, Y., Sugawara, M., Kamiya, T., and Suzuki, S. (1986b). Coating of liposomes with subunits of monoclonal IgM antibody and targeting of the liposomes. *In* "Methods in Enzymology" (J. J. Langone and H. Van Vunakis, eds.), Vol. 121, pp. 817–828. Academic Press, New York.

Heath, T. D. (1987). Covalent attachment of proteins to lipopsomes. *In* "Methods in Enzymology" (K. J. Widder and R. Green, eds.), Vol. 149, pp. 111–119. Academic Press, Orlando, Florida.

Hoffman, R. M., Margolis, L. B., and Bergelson, L. D. (1978). Binding and entrapment of high molecular weight DNA by lecithin liposomes. *FEBS Lett.* **93**, 365–368.

Hong, K., Zheng, W., Baker, A., and Papahadjopoulos, D. (1997). Stabilization of cationic liposome–plasmid DNA complexes by polyamines and poly(ethylene glycol)-phospholipid conjugates for efficient *in vivo* gene delivery. *FEBS Lett.* **400**, 233–237.

Huang, A., Tsao, Y. S., Kennel, S., and Huang, L. (1982). Characterization of antibody covalently coupled to liposomes. *Biochim. Biophys. Acta* **716**, 140–150.

Huang, S. K., Lee, K. D., Hong, K., Friend, D. S., and Papahadjopoulos, D. (1992a). Microscopic localization of sterically stabilized liposomes in colon carcinoma-bearing mice. *Cancer Res.* **52**, 5135–5143.

Huang, S. K., Mayhew, E. M., Gilani, S., Lasic, D. D., Martin, F. J., and Papahadjopoulos, D. (1992b). Pharmacokinetics and therapeutics of sterically stabilized liposomes in mice bearing C-26 colon carcinoma. *Cancer Res.* **52**, 6774–6781.

Hynes, N. E., and Stern, D. F. (1994). The biology of *erb*B-2/*neu*/HER-2 and its role in cancer. *Biochim. Biophys. Acta* **1198**, 165–184.

Jones, M. N., and Hudson, M. J. H. (1993). The targeting of immunoliposomes to tumour cells (A431) and the effects of encapsulated methotrexate. *Biochim. Biophys. Acta* **1152**, 231–242.

Keinanen, K., and Laukkanen, M.-L. (1994). Biosynthetic lipid-tagging of antibodies. *FEBS Lett.* **346**, 123–126.

Kern, J. A., Schwartz, D. A., Nordberg, J. E., Weiner, D. B., Greene, M. I., Torney, L., and Robinson, R. A. (1990). p185*neu* expression in human lung adenocarcinomas predicts shortened survival. *Cancer Res.* **50**, 5184–5191.

Kirpotin, D., Park, J. W., Hong, K., Zalipsky, S., LI, W. L., Carter, P., Benz, C. C., and Papahadjopoulos, D. (1997). Sterically stabilized anti-HER2 immunoliposomes: Design and targeting to human breast cancer cell *in vitro*. *Biochemistry* **36**, 66–75.

Klibanov, A. L., Maruyama, K., Kecklerberg, A. M., Torchilin, V. P., and Huang, L. (1991). Activity of amphipathic poly(ethylene glycol) 5000 to prolong the circulation time of liposomes depends on the liposome size and is unfavorable for immunoliposome binding to target. *Biochim. Biophys. Acta* **1062**, 142–148.

Lasic, D. D., and Martin, F. J. (1995). "Stealth Liposomes." CRC Press, Boca Raton, Florida.

Lasic, D. D., and Papahadjopoulos, D. (1995). Liposomes revisited. *Science* **267**, 1275–1276.

Laukkanen, M.-L., Alfthan, K., and Keinanen, K. (1994). Functional immunoliposomes harboring a biosynthetically lipid-tagged single-chain antibody. *Biochemistry* **33**, 11664–11670.

Leserman, L., and Machy, P. (1987). Ligand targeting of liposomes. *In* "Liposomes: From Biophysics to Therapeutics" (M. J. Ostro, ed.), pp. 157–194. Dekker, New York.

Lewis, G. D., Figari, I., Fendly, B., Wong, W. L., Carter, P., Gorman, C., and Shepard, H. M. (1993). Differential responses of human tumor cell lines to anti-p185HER2 monoclonal antibodies. *Cancer Immunol. Immunother.* **37**, 255–263.

Liu, E., Thor, A., He, M., Barcos, M., Ljung, B.-M., and Benz, C. (1992). The *HER2* (c-*erbB*-2) oncogene is frequently amplified in *in situ* carcinomas of the breast. *Oncogene* **7**, 1027–1032.

Loughrey, H., Bally, M. B., and Cullis, P. R. (1987). A non-covalent method of attaching antibodies to liposomes. *Biochim. Biophys. Acta* **901**, 157–160.

Marquet, R., and Houssier, C. (1991). Thermodynamics of cation-induced DNA condensation. *J. Biomol. Struct. Dyn.* **9**, 159–167.

Martin, F. J., and Papahadjopoulos, D. (1982). Irreversible coupling of immunoglobulin fragments to preformed vesicles: An improved method for liposome targeting. *J. Biol. Chem.* **257**, 286–288.

Martin, F. J., Hubbell, L. W., and Papahadjopoulos, D. (1981). Immunospecific targeting of liposomes to cells: A novel and efficient method for covalent attachment of Fab′ fragments via disulfide bonds. *Biochemistry* **20**, 4429–4238.

Michael, S. I., and Curiel, D. T. (1994). Strategies to achieve targeted gene delivery via the receptor-mediated endocytosis pathway. *Gene Ther.* **1**, 223–232.

Miller, A. D. (1992). Human gene therapy comes of age. *Nature (London)* **357**, 455–460.

Miller, D. G., Adam, M. A., and Miller, A. D. (1990). Gene transfer by retrovirus vectors occurs only in cells that are actively replicating at the time of infection. *Mol. Cell. Biol.* **10**, 4239–4242.

Mori, A., Klibanov, A. L., Torchilin, V. P., and Huang, L. (1991). Influence of the steric barrier of amphipathic poly(ethylene glycol) and ganglioside GM1 on the circulation time of liposomes and on the target binding of immunoliposomes *in vivo*. *FEBS Lett.* **284**, 263–266.

Mori, A., Kennel, S. J., and Huang, L. (1993). Immunotargeting of liposomes containing lipophilic antitumor prodrugs. *Pharm. Res.* **10**, 507–514.

Mori, A., Kennel, S. J., Waalkes, M. van B., Scherphof, G. L., and Huang, L. (1995). Characterization of organ-specific immunoliposomes for delivery of 3′,5′-O-dipalmitoyl-5-fluoro-2′-deoxyridine in a mouse lung-metastasis model. *Cancer Chemother. Pharmacol.* **35**, 447–456.

Morishige, H., Ohkuma, T., and Kaji, A. (1993). *In vitro* cytostatic effect of TNF (tumor necrosis factor) entrapped in immunoliposomes on cells normally insensitive to TNF. *Biochim. Biophys. Acta* **1151**, 59–68.

Muss, H. B., Thor, A. D., Berry D. A., Kute, T., Liu, E. T., Koerner, F., Cirrincione, C. T., Budman, D. R., Wood, W. C., Barcos, M., and Henderson, I. C. (1994). c-*erb*B-2 expression and response to adjuvant chemotherapy in women with node-positive early breast cancer. *N. Engl. J. Med.* **330**, 1260–1266.

Nassander, U. K., Steerenberg, P. A., Poppe, H., Storm, G., Poels, L. G., Jong, W. H. D., and Crommelin, D. J. A. (1992). *In vivo* targeting of OV-TL 3 immunoliposomes to ascitic ovarian carcinoma cells (OVCAR-3) in athymic nude mice. *Cancer Res.* **52**, 646–653.

Nassander, U. K., Steerenberg, P. A., Jong, W. H. D., Overveld, W. O. W. M. V., Boekhorst, C. M. E. T., Poels, L. G., Jap, P. H. K., and Storm, G. (1995). Design of immunoliposomes directed against human ovarian carcinoma. *Biochim. Biophys. Acta* **1235**, 126–139.

Niehans, G. A., Singleton, T. P., Dykoski, D., and Kiang, D. T. (1993). Stability of HER-2/*neu* expression over time and at multiple metastatic sites. *J. Natl. Cancer Inst.* **85**, 1230–1235.

O'Connell, R. P., Carter, P., Presta, L., Eigenbrot, C., Covarrubias, M., Snedecor, B., Speckart, R., Blank, G., Vetterlein, D., and Kotts, C. (1993). Characterization of humanized anti-p185HER2 antibody Fab fragments produced in *E. coli. In* "Protein Folding *in Vivo* and *in Vitro*" (J. L. Cleland, ed.), pp. 218–239. American Chemical Society, Washington, D.C.

Oldfield, E. H., Ram, Z., Culver, K. W., and Blaese, R. M. (1993). Clinical protocol: Gene therapy for the treatment of brain tumors using intra-tumoral transduction with the thymidine kinase gene and intravenous gancyclovir. *Hum. Gene Ther.* **4**, 39–69.

Papahadjopoulos, D., Allen, T. M., Gabizon, A., Mayhew, E., Matthay, K., Huang, S. K., Lee, K. D., Woodle, M. C., Lasic, D. D., Redemann, C., and Martin, F. J. (1991). Sterically stabilized liposomes: Improvements in pharmacokinetics and antitumor therapeutic efficacy. *Proc. Natl. Acad. Sci. U.S.A.* **88**, 11460–11464.

Park, J.-B., Rhim, J. S., Park, S.-C., Kimm, S.-W., and Kraus, M. H. (1989). Amplification, overexpression, and rearrangement of the *erb*B-2 proto-oncogene in primary human stomach carcinomas. *Cancer Res.* **49**, 6605–6609.

Park, J. W., Stagg, R., Lewis, G. D., Carter, P., Maneval, D., Slamon, D. J., Jaffe, H., and Shepard, H. M. (1992). Anti-p185HER2 monoclonal antibodies: Biological properties and potential for immunotherapy. *In* "Genes, Oncogenes, and Hormones: Advances in Cellular and Molecular Biology of Breast Cancer" (R. B. Dickson and M. E. Lippman, eds.), pp. 193–211. Kluwer Academic Publishers, Boston.

Park, J. W., Hong, K., Carter, P., Asgari, H., Guo, L. Y., Keller, G. A., Wirth, C., Shalaby, R., Kotts, C., Wood, W. I., Papahadjopoulos, D., and Benz, C. C. (1995). Development of anti-p185HER2 immunoliposomes for cancer therapy. *Proc. Natl. Acad. Sci. U.S.A.* **92**, 1327–1331.

Park, J. W., Hong, K., Zheng, W., Benz, C. C., and Papahadjopoulos, D. (1997). Development of liposome- and anti-HER2 immunoliposome-plasmid complexes for efficient and selective gene delivery. *Proc. Amer. Assoc. Cancer Res.* **38**, 342.

Parr, M. J., Ansell, S. M., Choi, L. S., and Cullis, P. R. (1994). Factors influencing the retention and chemical stability of poly(ethylene glycol)–lipid conjugates incorporated in large unilamellar vesicles. *Biochim. Biophys. Acta* **1195**, 21–30.

Pegram, M., Lipton, A., Pietras, R., Hayes, D., Weber, B., Baselga, J., Tripathy, D., Twaddell, T., Glaspy, J., and Slamon, D. (1995). Phase II study of intravenous recombinant humanized anti-p185 HER-2 monoclonal antibody (rhuMAb HER-2) plus cisplatin in patients with HER-2/*neu* overexpressing metastatic breast cancer. *Proc. Am. Soc. Clin. Oncol.* **14**, 106.

Phillips, N. C., and Emili, A. (1991). Immunogenicity of immunoliposomes. *Immunol. Lett.* **30**, 291–296.

Phillips, N. G., and Tsoukas, C. (1990). Immunoliposome targeting to CD4+ cells in human blood. *Cancer Detect. Prev.* **14**, 383–390.

Phillips, N. C., Gagne, L., Tsoukas, C., and Dahman, J. (1994). Immunoliposome targeting to murine CD4+ leucocytes is dependent on immune status. *J. Immunol.* **152**, 3168–3174.

Pietras, R. J., Fendly, B. M., Chazin, V. R., Pegram, M. D., Howell, S. B., and Slamon, D. J. (1994). Antibody to HER-2/*neu* receptor blocks DNA repair after cisplatin in human breast and ovarian cancer cells. *Oncogene* **9**, 1829–1838.

Plum, G. E., Arscott, P. G., and Bloomfield, V. A. (1990). Condensation of DNA by trivalent cations. 2. Effects of cation structure. *Biopolymers* **30**, 631–643.

Press, M. F., Cordon-Cardo, C., and Slamon, D. J. (1990). Expression of the HER-2/*neu* proto-oncogene in normal human adult and fetal tissues. *Oncogene* **5**, 953–962.

Ram, Z., Culver, K. W., Walbridge, S., Blaese, R. M., and Oldfield, E. H. (1993). *In situ* retroviral mediated gene transfer for the treatment of brain tumors in rats. *Cancer Res.* **53**, 83–88.

Ranade, V. (1989). Drug delivery systems. 1. Site-specific drug delivery using liposomes as carriers. *J. Clin. Pharmacol.* **29**, 685–694.

Rodrigues, M. L., Presta, L. G., Kotts, C. E., Wirth, C., Mordenti, J., Osaka, G., Wong, W. L., Nuijens, A., Blackburn, B., and Carter, P. (1995). Development of a humanized disulfide-stabilized anti-p185HER2 Fv-β-lactamase fusion protein for activation of a cephalosporin doxorubicin prodrug. *Cancer Res.* **55**, 63–70.

Roth, J. A., and Cristiano, R. J. (1997). Gene therapy for cancer: What have we done and where are we going? *J. Natl. Cancer Inst.* **89**, 21–39.

Roth, J. A., Nguyen, D., Lawrence, D. D., Kemp, B. L., Carrasco, C. H., Ferson, D. Z., Hong, W. K., Komaki, R., Lee, J. J., Nesbitt, J. C., Pisters, K. M. W., Putnam, J. B., Schea, R., Shin, D. M., Walsh, G. L., Dolormente, M. M., Han, C.-I., Martin, F. D., Xu, K., Stephens, L. C., McDonnell, T. J., Mukhopadhyay, T., and Cai, D. (1996). Retrovirus-mediated wild-type *p53* gene transfer to tumors of patients with lung cancer. *Nat. Med.* **2**, 985–991.

Sarup, J.C., Johnson, R. M., King, K. L., Fendly, B. M., Lipari, M. T., Napier, M. A., Ullrich, A., and Shepard, H. M. (1991). Characterization of an anti-p185 HER2 monoclonal antibody that stimulates receptor function and inhibits tumor cell growth. *Growth Regul.* **1**, 72–82.

Shalaby, M. R., Shepard, H. M., Presta, L., Rodrigues, R. L., Beverley, P. L., Feldmann, M., and Carter, P. (1992). Development of humanized bispecific antibodies reactive with cytotoxic lymphocytes and tumor cells overexpressing the *HER2* protooncogene. *J. Exp. Med.* **175**, 217–225.

Shaw, P., Bovey, R., Tardy, S., Sahli, R., Sordat, B., and Costa, J. (1992). Induction of apoptosis by wild-type p53 in a human colon tumor-derived cell line. *Proc. Natl. Acad. Sci. U.S.A.* **89**, 4495–4499.

Shepard, H. M., Lewis, G. D., Sarup, J. C., Fendly, B. M., Maneval, D., Mordenti, J., Figari, I., Kotts, C. E., Palladino, M. A., Ullrich, A., and Slamon, D. (1991). Monoclonal antibody therapy of human cancer: Taking the HER2 protooncogene to the clinic. *J. Clin. Immunol.* **11**, 117–127.

Slamon, D. J., Clark, G. M., Wong, S. G., Levin, W. J., Ullrich, A., and McGuire, W. L. (1987). Human breast cancer: Correlation of relapse and survival with amplification of HER2/*neu* oncogene. *Science* **235**, 177–182.

Slamon, D. J., Godolphin, W., Jones, L. A., Holt, J. A., Wong, S. G., Keith, D. E., Levin, W. J., Stuart, S. G., Udove, J., Ullrich, A., and Press, M. (1989). Studies of the HER2/*neu* proto-oncogene in human breast and ovarian cancer. *Science* **244**, 707–712.

Straubinger, R. M., Lopez, N. G., Debs, R. J., Hong, K., and Papahadjopoulos, D. (1988). Liposome-based therapy of human ovarian cancer: Parameters determining potency of negatively charged and antibody-targeted liposomes. *Cancer Res.* **48**, 5237–5245.

Stribling, R., Brunette, E., Liggitt, D., Gaensler, K., and Debs, R. (1992). Aerosol gene delivery *in vivo. Proc. Natl. Acad. Sci. U.S.A.* **89**, 11277–11281.

Suzuki, S., Watanabe, S., Uno, S., Tanaka, M., Masuko, T., and Hashimoto, Y. (1994). Endocytosis does not necessarily augment the cytotoxicity of Adriamycin encapsulated in immunoliposomes. *Biochim. Biophys. Acta* **1224**, 445–453.

Suzuki, S., Uno, S., Fukuda, Y., Aoki, Y., Masuko, T., and Hashimoto, Y. (1995). Cytotoxicity of anti-c-*erb*B-2 immunoliposomes containing doxorubicin on human cancer cells. *Br. J. Cancer* **72**, 663–668.

Thomas, J. A., Chai, Y. C., and Jung, C. H. (1994). Protein S-thiolation and dethiolation. *In* "Methods in Enzymology" (L. Packer, ed.), Vol. 233, pp. 385–395. Academic Press, San Diego, California.

Torchilin, V. P. (1994). Immunoliposomes and PEGylated immunoliposomes: Possible use for targeted delivery of imaging agents. *Immunomethods* **4**, 244–258.

Valone, F. H., Kaufman, P. A., Guyre, P. M., Lewis, L. D., Memoli, V., Deo, Y., Graziano, R., Fisher, J. L., Meyer, L., Mrozek-Orlowski, M., Wardwell, K., Guyre, V., Morley, T. L., Arvizu, C., and Fanger, M. W. (1995). Phase Ia/Ib trial of bispecific antibody MDX-210 in patients with advanced breast or ovarian cancer that overexpress the proto-oncogene HER2/*neu. J. Clin. Oncol.* **13**, 2281–2292.

VandeVijver, M. J., Peterse, M. L., Mooi, W. J., Wisman, P., Lomans, J., Dalesio, O., and Nusse, R. (1988). *Neu*-protein expression in breast cancer. *N. Engl. J. Med.* **319**, 1239–1245.

Vile, R., and Russell, S. J. (1994). Gene transfer technologies for the gene therapy of cancer. *Gene Ther.* **1,** 88–98.

Waldmann, T. (1991). Monoclonal antibodies in diagnosis and therapy. *Science* **252,** 1657–1662.

Wang, C.-Y., and Huang, L. (1987). pH-sensitive immunoliposomes mediate target-cell-specific delivery and controlled expression of a foreign gene in mouse. *Proc. Natl. Acad. Sci. U.S.A.* **84,** 7851–7855.

Weiner, L. M., Holmes, M., Adams, G. P., LaCreta, F., Watts, P., and Garcia de Palazzo, I. (1993). A human tumor xenograft model of therapy with a bispecific monoclonal antibody targeting c-*erb*B-2 and CD16. *Cancer Res.* **53,** 94–100.

Weissig, V., Lasch, J., Klibanov, A. L., and Torchilin, V. P. (1986). A new hydrophobic anchor for the attachment of proteins to liposomal membranes. *FEBS Lett.* **202,** 86–90.

Wels, W., Harwerth, I.-M., Mueller, M., Groner, B., and Hynes, N. E. (1992). Selective inhibition of tumor cell growth by a recombinant single-chain antibody-toxin specific for the erbB-2 receptor. *Cancer Res.* **52,** 6310–6317.

Wolff, B., and Gregoriadis, G. (1984). The use of monoclonal anti-thy-1 IgG for the targeting of liposomes to AKR-A cells *in vitro* and *in vivo*. *Biochim. Biophys. Acta* **802,** 259–273.

Yanagie, H., Tomita, T., Kobayashi, H., Fujii, Y., Takahashi, T., Hasumi, K., and Nariuchi, H. (1991). Application of boronated anti-CEA immunoliposome to tumour cell growth inhibition in *in vitro* boron neutron capure therapy model. *Br. J. Cancer* **63,** 522–526.

Yokota, J., Yamamoto, T., Miyajima, N., Toyoshima, K., Nomura, N., Sakamoto, H., Yoshida, T., Terada, M., and Sugimura, T. (1988). Genetic alterations of the c-*erb*B-2 oncogene occur frequently in tubular adenocarcinoma of the stomach and are often accompanied by amplification of the v-*erb*A homologue. *Oncogene* **2,** 283–287.

Yonemura, Y., Ninomiya, I., Yamaguchi, A., Fushida, S., Kimura, H., Ohoyama, S., Miyazaki, I., Endou, Y., Tanaka, M., and Sasaki, T. (1991). Evaluation of immunoreactivity for *erb*B-2 protein as a marker of poor short term prognosis in gastric cancer. *Cancer Res.* **51,** 1034–1038.

Zalipsky, S., Hansen, C. B., Menezes, D. L. D., and Allen, T. M. (1996). Long-circulating, polyethylene glycol-grafted immunoliposomes. *J. Controlled Release* **39,** 153–161.

Zhau, H. E., Zhang, X., von Eschenbach, A. C., Scorsone, K., Babaian, R. J., Ro, J. Y., and Hung, M.-C. (1990). Amplification and expression of the c-*erb*B-2/*neu* proto-oncogene in human bladder cancer. *Mol. Carcinog.* **3,** 254–257.

Zhau, H. E., Wan, D. S., Zhou, J., Miller, G. J., and von Eschenbach, A. C. (1992). Expression of c-*erb*B-2/*neu* proto-oncogene in human prostatic cancer tissues and cell lines. *Mol. Carcinog.* **5,** 320–327.

Zhu, N., Liggitt, D., Liu, Y., and Debs, R. (1993). Systemic gene expression after intravenous DNA delivery into adult mice. *Science* **261,** 209–211.

R. E. Kilkuskie
A. K. Field
Hybridon, Inc.
Cambridge, Massachusetts 02139

Antisense Inhibition of Virus Infections

I. Introduction

Antiviral drugs and drug design have traditionally focused on inhibition of key proteins essential for successful viral replication. This approach is seen with the nucleoside and nucleotide analog inhibitors of the herpesvirus DNA polymerases, the human immunodeficiency virus (HIV) reverse transcriptase inhibitors, the HIV protease inhibitors, and the influenza hemagglutinin inhibitors (De Clercq, 1995; Kinchington and Redshaw, 1995; Haffey and Field, 1995; Field, 1994). It is apparent that this approach has resulted in a parade of clinically effective drugs, the best examples in the success story being acyclovir (ACV), which has been used both therapeutically and prophylactically to inhibit acute herpes simplex virus (HSV) lesion formation and duration, and zidovudine (AZT). But as successful as these approaches have been and as promising as the newer antiviral drugs may be, there are limitations to the present approaches.

Advances in Pharmacology, Volume 40

These limitations are best illustrated by the concern for the emergence of clinically important drug-resistant virus mutants, a problem reviewed for the herpesviruses and for HIV (Field and Biron, 1994; De Clercq, 1995; Bowen *et al.,* 1995). For the herpesviruses, the problem has been most apparent in the immunocompromised host treated with suboptimal drug doses. For the HIV antivirals, no one drug treatment has resulted in total viral suppression, and the rapidity of resistance development has varied from drug to drug. As a result of the lack of potent and sustained virus suppression, attention has now turned to combination therapy in the clinic. Thus, the availability of a more potent and diverse group of inhibitors for any one infectious agent should provide a richer armamentarium from which to choose the most effective combinations.

But how does one discover novel antiviral inhibitors? One approach has been the modification of a known inhibitor or an enzyme substrate in a set of structure–activity relationship (SAR) studies, to choose the inhibitor with the most attractive selectivity index (ratio of concentrations for antiviral efficacy compared to toxicity). This approach has produced the array of nucleoside and nucleotide inhibitors of HSV DNA polymerase and HIV reverse transcriptase, and the HIV protease substrate analog inhibitors. In the absence of a defined enzyme substrate, a traditional approach has employed a high-volume screen to evaluate thousands of compounds until a lead with some specificity of action can be identified and lead to the SAR refinement. As fruitful as these approaches have been, and as entrenched as these approaches are in the pharmaceutical *modus operandi,* they miss the great opportunity to design inhibitors against any of the gene targets provided by the virus. Herpes simplex virus has 71 open reading frames, for which the function of many of the resultant proteins is still poorly defined. Human immunodeficiency virus has nine open reading frames, yet only the reverse transcriptase and protease have been effectively addressed as antiviral targets by traditional approaches. Human cytomegalovirus (HCMV) more than 200 open reading frames, which should provide a rich hunting ground in the search for effective antivirals, if only we had the means to identify such inhibitors.

One new approach is just now being recognized as both versatile and practical for identifying and developing antiviral drugs. That approach is the use of antisense nucleic acids complementary to the viral RNAs to block mRNA translation or genome replication. It can take the form of antisense oligonucleotides applied to the infected cell, antisense RNA expressed within the cell, or as ribozymes capable of complementary binding to the target RNA and target cleavage via the inherent catalytic activity. In this chapter we summarize the opportunity to use these approaches to identify novel antiviral drug targets and to develop novel antiviral strategies. These are the approaches embodied in the emerging field of genetic pharmacology: the development of drugs and gene therapy to control gene expression (Field and Goodchild, 1995; Yu *et al.,* 1994; Temsamani and Agrawal, 1996). We review genetic pharmacology as it relates to antiviral antisense research and drug development.

II. Antisense Oligonucleotides as Potential Antiviral Agents

A major research emphasis has focused on the identification of oligonucleotides as antiviral agents. In principle, this suggests that if the sequence of a viral RNA (genome RNA or mRNA) is known, then one should be able to design a complementary oligonucleotide that will hybridize and inhibit its function as a replicating genome or mRNA (Fig. 1). This principle was first documented in the observations of Zamecnik and Stephenson, using a phosphodiester oligodeoxynucleotide to block Rous sarcoma virus replication (Zamecnik and Stephenson, 1978; Stephenson and Zamecnik, 1978). But the road from principle and initial observations in 1978 to practical antiviral utility is a long one, on which we have learned to choose preferred viral RNA target sequences for maximum antisense impact, to

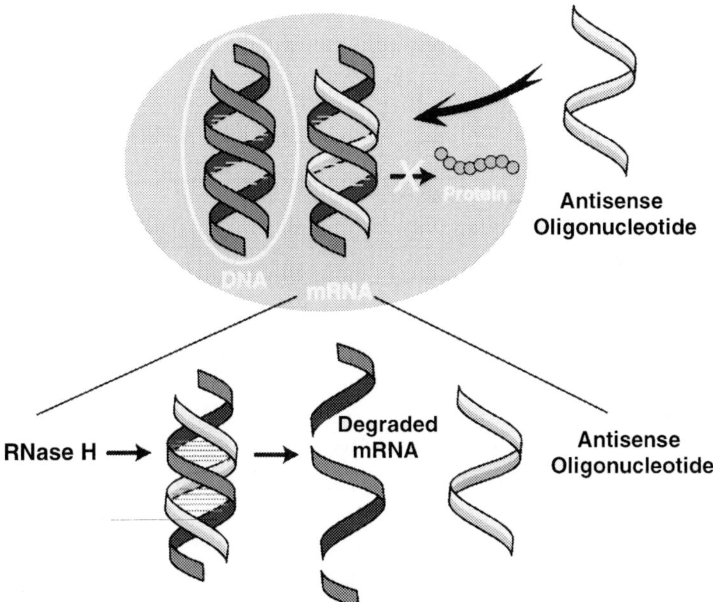

FIGURE I   Mechanisms of antisense oligonucleotide inhibition. Antisense oligonucleotides can bind to target RNA in cells, preventing protein production directly by translation arrest. In addition, DNA–RNA hybrids formed by oligonucleotide binding to RNA are recognized by RNase H, which digests the RNA portion of the hybrid, destroying the mRNA and freeing the oligonucleotide for binding to another mRNA molecule.

chemically alter the antisense oligonucleotides to enhance stability and optimize activities in biological systems, and to recognize the array of biological actions of oligonucleotides that may contribute to antiviral activities and also to toxicities.

An antiviral antisense oligonucleotide should display a high degree of specificity, the reflection of the uniqueness of the nucleotide sequence provided by the genetic code, and the exact complementary base pairing with the target RNA. As a result, introducing mismatch nucleotides to interfere with hybridization of the oligonucleotide with the RNA target, or substituting inappropriate sequence by "scrambling" the order of the nucleotides or reversing the predicted antisense 5'-to-3' orientation, should reduce the antiviral activity. Furthermore, the antisense mechanism predicts selective inhibition of gene expression. Suggested guidelines for acceptance of an oligonucleotide activity as fitting these criteria were presented by Stein and Krieg (1994). Direct demonstration that an antisense oligonucleotide binds to the target RNA and violates its biological role within the cell has generally been difficult to obtain, although numerous indirect lines of evidence have suggested that this is so. For example, studies of antisense oligonucleotides against cytomegalovirus showed selective gene target inhibition resulting in a potent antiviral activity (Smith and Pari, 1995; Pari *et al.*, 1995). In addition, recent studies of antisense mechanism have been fruitful and have demonstrated the direct antisense activities *in situ* (Politz *et al.*, 1995; Giles *et al.*, 1995a,b). The details of these observations as they relate to antiviral activities are described in Section III.

A. Antiviral Target Selection

Antisense oligonucleotides should be highly selective compounds by virtue of their interaction with specific segments of RNA. For potential antivirals, identification of appropriate target RNA sequences for antisense oligonucleotides is performed at two levels: (1) the optimal gene within the virus, and (2) the optimal sequence within the RNA.

"Optimal" genes are those genes that are essential for virus replication and/or are essential for virulence. Thus, a wide variety of targets may be available, and through the antisense approach both validation of the target and discovery of an effective inhibitor can be attained simultaneously. This validation and discovery process are illustrated by studies on HCMV, which are discussed below. To be sure, oligonucleotides have been designed to target expression of structural genes of HIV (Lisziewicz *et al.*, 1993, 1994; Anazodo *et al.*, 1995) and hepatitis B virus (HBV) (Korba and Gerin, 1995), regulatory genes of HIV (Matsukura *et al.*, 1987), HCMV (Azad *et al.*, 1993; Pari *et al.*, 1995), Epstein–Barr virus (EBV) (Roth *et al.*, 1994; Daibata *et al.*, 1996), and human papillomavirus (HPV) (Cowsert *et al.*, 1993), and a variety of functions in HBV (Wu and Wu, 1992; Offensperger *et al.*, 1993; Korba and Gerin, 1995).

But few studies have set out to compare viral targets for vulnerability to antisense inhibition, and fewer have done so while also controlling for nonantisense antiviral effects. For HIV, oligonucleotides targeted against a regulatory gene (*rev*), a viral enzyme (*pol*), and structural gene (*gag*) were compared in a single study (Kinchington *et al.*, 1992). All oligonucleotides were reported to be equally active in an acute infection assay. The *rev* targeted oligonucleotide was the only oligonucleotide active against chronically infected cells, suggesting that this was the "optimal" gene for antisense targeting. However, other sequences against these genes were not evaluated and the inhibition required a 30 μM concentration, so it is not clear if the "optimal" sequences were actually chosen. For example, GEM 91, a 25-mer phosphorothioate oligonucleotide targeted at the *gag* gene, has been found to be active at submicromolar concentrations in acute infection assays and long-term model systems (Agrawal and Tang, 1992; Lisziewicz *et al.*, 1993). For HBV, RNA sequences have been identified that are essential for initiation of reverse transcription and packaging of viral RNA (Pollack and Ganem, 1994). Oligonucleotides against these regions were reported to be potent inhibitors of HBV replication (Korba and Gerin, 1995). For RNA viruses additional genetic targets may also be provided, since both genomic positive (picornaviruses), negative (influenza), and double-stranded (reoviruses), as well as individual transcripts, are available. For instance, we have targeted oligonucleotides to the genomic strand of respiratory syncytial virus to inhibit RNA replication, and it was previously shown that replication of both influenza virus and vesicular stomatitis virus was inhibited by oligonucleotides that were targeted to genomic RNA (Lemaitre *et al.*, 1987; Leiter *et al.*, 1990).

It is clear that there are no defined rules to identify the optimal antisense gene target for a virus, but the versatility of designing antisense inhibitors to heretofore unassailable targets provides great opportunity to eventually clarify those rules.

Much effort has also been made to identify the correct RNA target sequence within a gene. Again, there are no clear rules to easily predict which sequences are most accessible to oligonucleotides. Computer models of RNA secondary structure (Sczakiel *et al.*, 1993), as well as oligonucleotide hybridization efficiency (Stull *et al.*, 1992), and frequency (Han *et al.*, 1994) have been compared to antisense oligonucleotide activity. Antisense activity was shown to correlate well with hybridization strength; however, RNA secondary structure did not predict antisense activity. Often, *in vitro* translation inhibition has been used to screen a series of potential oligonucleotide inhibitors, and then the most active oligonucleotides were evaluated in cellular and/or antiviral assays (Chen *et al.*, 1996).

In our laboratories, oligonucleotide libraries have been screened for binding to target RNA *in vitro* by an assay described by Frank *et al.* (Frank and Goodchild, 1996; Frank *et al.,* 1993) and modified by Ho *et al.* (1996). Binding is measured by RNase H cleavage of RNA–oligonucleotide hybrids. The most sensitive sites are mapped and specific oligonucleotides synthesized and evaluated in cellular assays. An example of the technique is shown in Fig. 2. Correlation exists between accessible RNA regions *in vitro* and in cells, although one would expect more sites to be identified using this *in vitro* technique with the purified RNA target, than would be expected *in situ* where target RNA may carry numerous binding proteins.

B. Oligonucleotide Modifications

As previously mentioned, the fundamental mission of an antisense oligonucleotide is to hybridize to the viral RNA target and inhibit its function. This inhibition may be facilitated by cleaving the target RNA by RNase H, which recognizes the RNA–oligodeoxynucleotide complex, or by inhibiting RNA translation or splicing through hybrid arrest. The end result should

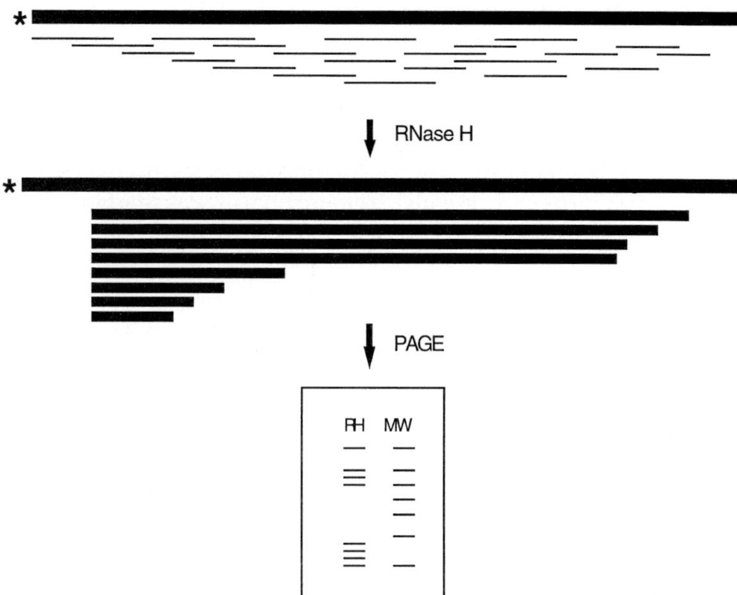

FIGURE 2   Antisense oligonucleotide selection. Oligonucleotide libraries (thin lines) are incubated with 5′ end-labeled RNA (thick line), then treated with RNase H. RNA regions with bound oligonucleotide are digested, producing families of shorter labeled RNAs, which are separated on polyacrylamide gels. Discrete families of RNAs are detected on gels (lane RH) and compared to molecular weight markers (lane MW) to identify regions of maximum oligonucleotide binding.

be either direct inhibition of the targeted RNA genome replication (for an RNA virus), or inhibition of translation of the targeted mRNA. In principle, unmodified oligodeoxynucleotides could satisfy this mission, but in practice chemical modifications of the oligonucleotide are necessary for robust antisense efficacy. Such modifications are designed to enhance stability, while retaining the capacity to hybridize to the target RNA and recruit RNase H. In addition, chemical modifications can be used to alter the hydrophobicity/hydrophilicity, thus altering the presentation of the oligonucleotide to the infected cell. Functional groups, such as phosphate and hydroxyl residues on natural nucleic acids, can be modified as indicated in Fig. 3. In general, those modifications that jeopardize the capacity of the oligonucleotide to participate in Watson–Crick base pairing are avoided, and most modifications have focused on the phosphodiester backbone and/or the sugar moiety.

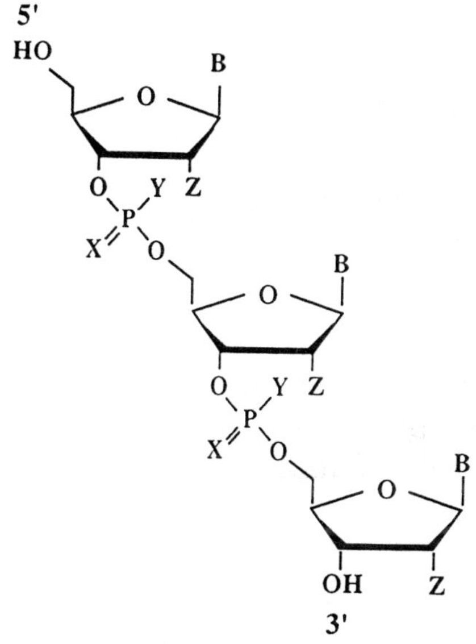

FIGURE 3   Structure of oligonucleotides. B, Any of the nitrogen heterocyclic bases found in DNA (A, C, T, G) or in RNA (where U replaces T). The terminal hydroxyl groups are distinguished as being at either the 5′ or 3′ ends.

	X	Y	Z
Deoxyoligonucleotides			H
Ribooligonucleotides			OH
2′-O-methyl derivatives			OMe
Unmodified	O	O	
Phosphorothioate	O	S	
Methyl phosphonate	O	CH_3	
Phosphoramidate	O	NR^1R^2	

For instance, by replacement of the nonbridging oxygen of the phosphodiester backbone with sulfur, the resulting phosphorothioate (PS) has increased resistance to nuclease degradation. Although the duplex formed with the target RNA has a lower melting temperature (T_m), it is a substrate for RNase H. A similar replacement of the nonester oxygen by a methyl group results in loss of negative charge and greater hydrophobicity, but at the cost of loss of RNase H activation. Alternatively, by replacing a hydrogen at the 2′ position on the deoxyribose with a hydroxymethyl group, the sugar becomes a modified ribose, which hybridizes more strongly with the complementary mRNA ribonucleotide and provides greater stability to the duplex, indicated by an elevated T_m. These and other changes are demonstrated in Fig. 3, and allow one to synthesize tailored oligonucleotides with a balance of characteristics of hybridization affinity, hydrophobicity, and the capacity to recruit RNase H-mediated cleavage of the target RNA.

The importance of these oligonucleotide modifications in designing effective drugs is just now being evaluated, both in animal model systems and in the clinic. The first generation of widely used antisense oligonucleotides has been the PS compounds, and a body of data on biodistribution, pharmacokinetics, and metabolism in animals and in humans is now available. These studies were summarized by Agrawal and Temsamani (1996), Field and Goodchild (1995), and Crooke and Bennett (1996). A second generation of antisense oligonucleotides is now emerging that includes combinations of nucleotide modifications within the oligonucleotide. For instance, *hybrid oligonucleotides* may be defined as substituted at the 3′ and/or 5′ ends with 2′-OCH$_3$ ribonucleosides, while maintaining the phosphorothioate backbone; *chimeric oligonucleotides* may be defined as substituted at the 3′ and/or 5′ ends with nonionic internucleotide linkages; and *self-stabilized oligonucleotides* may be defined as phophorothioates or phosphodiesters that have two domains—a single-stranded antisense sequence and a hairpin loop at the 3′ end. The pharmacological characteristics of these and other modifications are summarized in papers by Agrawal and Temsamani (1996) and by Crooke *et al.* (1996). It is clear from the ongoing studies, that not only can one design an antisense inhibitor with potentially high selectivity in target inhibition, but one can now chemically modify that selected oligonucleotide to tailor the eventual drug for enhanced stability and perhaps tissue-specific uptake and metabolism (Crooke *et al.,* 1996). In addition, the first observation of oral uptake and tissue distribution of a hybrid oligonucleotide (Zhang *et al.,* 1995a) indicates that the opportunity to develop oligonucleotide therapeutics with oral bioavailability may be at hand.

To date, most of the antiviral cell culture studies and all of the animal efficacy and clinical studies have employed PS oligonucleotides. More recently, Hybridon (Cambridge, MA) has introduced the first advanced oligonucleotide, GEM 132, a hybrid oligonucleotide for clinical study (see Section III,B). These studies using PS oligonucleotides have generated useful information concerning antisense activities, but it is apparent that PS compounds may also have additional nonantisense modes of antiviral activity. This can be readily demonstrated for HSV, EBV, and HIV (Gao *et al.*, 1990a; Yao *et al.*, 1993; Wyatt *et al.*, 1994; Ojwang *et al.*, 1994a, 1995; Buckheit *et al.*, 1994). The nonantisense mechanisms include inhibition of virus binding and internalization (Wyatt *et al.*, 1994; Buckheit *et al.*, 1994) and inhibition of virus-specific DNA polymerases (Yao *et al.*, 1993; Gao *et al.*, 1989, 1990a). These effects are related to the general polyanionic nature of oligonucleotides, but are more pronounced for specific sequences. For example, oligonucleotides containing four consecutive G residues formed tetrameric structures that potently inhibited absorption of HIV or HSV (Buckheit *et al.*, 1994). Oligonucleotides containing only G and T residues inhibited absorption and integration of HIV (Ojwang *et al.*, 1994b); this is believed to be due to higher order structure of the oligonucleotides. Certain oligonucleotides containing CpG motifs may also stimulate B cell proliferation and have immunostimulatory characteristics (Krieg *et al.*, 1995). The nonantisense effects may contribute significantly to antiviral effects by oligonucleotides, enhancing their activity in cell culture. Unfortunately, these effects might in certain cases mask antisense effects and make rational selection of gene targets and sequence targets more difficult. These anomalies are more fully discussed below in the context of individual antiviral studies.

By far, most antiviral studies using oligonucleotides have been cell culture evaluations, some of which have neglected to apply a rigorous definition of antisense inhibition. However, as the field has matured and the criteria for an antisense antiviral oligonucleotide have become more precise, antisense mechanisms of antiviral activity have been repeatedly confirmed. In the studies reported below, we have emphasized those studies with a clear definition of the antisense mechanism of antiviral activity. Thus, the literature survey focuses on key studies that illustrate how the field has evolved, and the novelties of those key studies that have helped to build our understanding of the roles of oligonucleotides as antiviral agents.

III. Oligonucleotides and Antiviral Activities

A. The Retroviruses

As was so correctly phrased by John Coffin, "No group of infectious agents has received as much attention from scientists in recent years as the retroviruses" (Coffin, 1996). This group has the capacity to replicate as a productive, cytolytic infection or become latent as genetic information inserted into the host genetic material; they have the capacity to capture and alter host genetic information in the form of oncogenes; they are readily mutable and thus can readily escape the antiviral effects of many initially effective inhibitors; and they are associated with a wide variety of diseases including benign and malignant tumors and acquired immunodeficiency syndrome (AIDS). As a result, they have also been a favorite target for antiviral studies, including antisense studies.

I. Non-HIV Retroviruses

The first detailed observations of inhibition of virus replication by oligonucleotides designed as antisense were published in 1978 (Zamecnik and Stephenson, 1978; Stephenson and Zamecnik, 1978). They used an unmodified 13-mer or a 13-mer blocked at both the 5' and 3' termini as the isourea derivatives. The oligonucleotide sequence was complementary to the reiterated 5'- and 3'-terminal repeats of the virion RNA, and blocked virus replication at about 2 μM, as measured by appearance of Rous sarcoma virus reverse transcriptase. Furthermore, using an *in vitro* translation system, the authors demonstrated selectivity of inhibition of viral RNA translation, and sequence specificity of that inhibition. Since these early observations, an abundance of antisense papers concerning both animal and human retroviruses has appeared.

Bovine leukemia virus encodes a transactivating protein, Tax, which promotes viral transcription and activates cellular genes associated with tumorogenesis. Cantor and Palmer evaluated the capacity of unmodified 15-mers to inhibit Tax translation from Tax message in rabbit reticulocyte lysates (Cantor and Palmer, 1992). Apparent sequence-specific inhibition of translation was observed with oligonucleotides directed against the 5' portion of the Tax RNA, including the AUG. One oligonucleotide, containing a four-G sequence and targeted at a 3' sequence, was actually stimulatory to translation by an undefined mechanism. This research was extended by Kitajima *et al.*. who demonstrated, both in murine cells in culture and in implanted Tax-producing fibrosarcoma cells, that a 20-mer modified by the phosphorothioate substitution at the terminal three nucleotides on the 3' end inhibited Tax protein expression (Kitajima *et al.*, 1992a,b). Greater than 10-fold inhibition of Tax expression in mature tumors occurred with an intraperitoneal injection of 40 μg/g. The similarly modified sense control oligonucleotide was ineffective. Curiously, in Tax-expressing cells the uptake of oligonucleotides was sevenfold higher than in non-Tax-expressing cells, and uptake appeared receptor mediated. The Tax protein causes transcriptional transactivation and is implicated in human T cell leukemia virus type I (HTLV-I)-mediated leukemogenesis, perhaps through activation and elevated expression of NF-κB. Kitajima also reported that antisense oligonucleotides, but not the complementary sense strand, targeted to NFkB mRNA translation start site inhibited Tax-transformed fibroblast growth and HTLV-I-transformed human lymphocyte growth. In mice, the antisense to NF-κB caused a rapid regression of the Tax-transformed tumors (Kitajima *et al.*, 1992a). The studies also suggested that although Tax is necessary to transform HTLV-I-infected cells, it is the maintenance of high levels of NF-κB that is important to sustaining the malignant phenotype.

2. HIV

Since the identification and sequencing of HIV, there has been a strong interest in identifying a potent oligonucleotide inhibitor that would have the potential for development as a therapy for AIDS. Zamecnik *et al.* (1986) and Goodchild *et al.* (1988) described the inhibition of HIV replication by synthetic oligonucleotides, and the phosphorothioate oligonucleotide (GEM 91) designed by Agrawal and Tang to bind to the *gag* region of HIV RNA was identified for further therapeutic development in 1992 (Agrawal and Tang, 1992). GEM 91 was selected because of its potent antiviral effects, and was designed to bind to a well-conserved region of the viral genome of most clinical isolates. This illustrates one of the advantages of antiviral design through antisense—the potential to select among an array of molecular targets for inhibition (Lisziewicz *et al.*, 1994). In these and other studies (Agrawal *et al.*, 1989), oligonucleotides were evaluated for their capacity to inhibit virus replication in long-term infected cell cultures, and selected antisense sequences were compared to unrelated or random sequence oligonucleotides. The results suggest that the selected antisense sequences have a considerable advantage of efficacy. Matsukura and colleagues also evaluated the inhibition of HIV replication by PS oligonucleotides, in this case targeted to the HIV *rev* RNA (Matsukura *et al.*, 1989). They reported sequence specificity of the antiviral effect (lack of activity of the sense, random, or homopolymeric sequences) and the expected effects on the HIV mRNA profile on treatment.

On the basis of the potent antiviral activity for GEM 91 in HIV cell culture infection studies and the well-conserved target sequence, phase I/II clinical evaluations for intravenous therapy of HIV has begun in both the United States and France (Martin and U.S. and French GEM 91 Collaborative Study Groups, 1995; Serini *et al.*, 1994). The US trial is a randomized, double-blind, placebo-controlled dose-escalating study using GEM 91 intravenous continuous infusion for 2 weeks. In the French study, GEM 91 is given as 2-hr infusion every other day for 28 days. Study entrants must have a viral burden of 25,000 copies of viral RNA/ml of plasma. To date, safety has been demonstrated for doses up to 4.4 mg/kg by continuous infusion and 3 mg/kg by repeated intermittent, 2-hr infusion. From the initial pharmacokinetic phase I single dose studies using ^{35}S-labeled GEM 91, plasma disappearance of the radioactivity associated with GEM 91 is the sum of two exponentials with mean half-lives of 0.18 and 26.71 hr. Both intact and degraded materials are found in the plasma. Elimination is mainly by urinary excretion of primarily lower molecular weight metabolites. Maximum tolerated doses have not yet been achieved (Zhang *et al.*, 1995b).

Anazodo *et al.* (1995) demonstrated that a partially phosphorothioated 20-mer targeted to a well-conserved coding region of the *gag* gene inhibited both expression of mRNA for the viral precursor protein p55 and p55 protein and its cleavage product, p24, in COS cells stably transfected with plasmids containing the *gag–pol* region (Fig. 4). In this system, the use of lipofectin with the oligonucleotide enhanced the activity. Inverse sequence and double mismatch control oligonucleotides were less effective. At 1 μM the antisense 20-mer inhibited viral replication, as measured by reverse transcriptase levels, in a sequence-specific manner without inhibiting cell protein biosynthesis (large subunit of ribonucleotide reductase) or cell growth rate.

As with other oligonucleotide antiviral studies mentioned previously, anti-retroviral effects of oligonucleotides by nonantisense mechanisms have been amply demonstrated. Matsukura *et al.* have demonstrated that the phosphorothioate 28-mer homopolymer dC is a potent inhibitor of HIV infection (Matsukura *et al.*, 1988). Direct and potent (K_i values at 6–12 nM) inhibition of the reverse transcriptase (RT), RNase H, and primer extension functions has also been demonstrated for HIV (Hatta *et al.*, 1993; Bordier *et al.*, 1992; Austermann *et al.*, 1992) and avian myeloblastosis virus (Hatta *et al.*, 1993; Boiziau *et al.*, 1992). In addition to inhibition of RT, others have demonstrated the inhibition of HIV adsorption to cells (Zelphati *et al.*, 1994). Most recently an interesting phosphorothioate oligonucleotide sequence ($T_2G_4T_2$) was identified that is a potent inhibitor of HIV viral envelope protein gp120 binding to the CD4 cell receptor. Specifically, $T_2G_4T_2$ as well as other nonantisense oligonucleotide HIV inhibitors such as the phosphorothioate dC bind to the V3 loop of gp120 (Wyatt *et al.*, 1994; Stein *et al.*, 1993). The breadth of anti-HIV mechanisms of oligonucleotides was further expanded with the demonstration that a 17-mer (T3077),

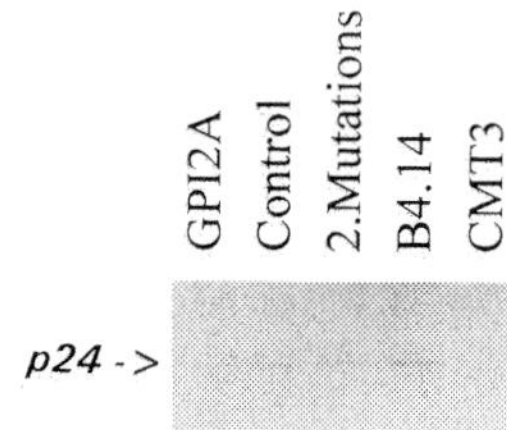

FIGURE 4   Inhibition of HIV p24 production. B4.14 cells, stably transfected to express HIV-1 p24, were treated with an antisense oligonucleotide (GPI2A) spanning bases 1189–1208 of HIV. Two control oligonucleotides containing either two mutations (2.mutations) or the inverse sequence (control), were also tested. p24 production, measured by immunoprecipitation using a rabbit polyclonal antibody, was decreased only in cells treated with the antisense oligonucleotide. Control and 2.mutation oligonucleotides did not alter p24 levels. Cells that were not transfected (CMT3) did not express p24 (Anazodo *et al.*, 1995).

composed of only deoxyguanosine and thymidine and with single phosphorothioate internucleotide linkages at the 5' and 3' ends, is a potent inhibitor of HIV integrase (Ojwang *et al.*, 1995). This same compound and related compounds (Ojwang *et al.*, 1994a) form tetramers owing in part to the motifs of G in the sequence, which appear to enhance the oligonucleotide capacity to block virus adsorption (Bishop *et al.*, 1996).

All of these observations emphasize the importance of defining mechanisms of action of a given oligonucleotide prior to identifying it as an antisense oligonucleotide. However, the fact that oligonucleotides may contribute numerous mechanisms toward the antiviral activity, in addition to the antisense mechanism, may in some cases be an asset in the pursuit of clinically useful antiviral drugs.

B. The Herpesviruses

I. Alphaherpesviruses: Herpes Simplex Viruses and Marek's Disease Virus

Herpes simplex virus types 1 and 2 are responsible for acute infections (cold sores, herpes genitalis, herpes encephalitis) which may result in persistent latent infection of neurons in the sensory ganglia, and may be punctuated by acute recurrences (Roizman and Sears, 1996). Herpes simplex virus replication, like that of other herpesviruses, is the result of a coordinately regulated, sequentially ordered cascade of transcription and translation events that are broadly categorized as immediate early, early, and late events (Roizman and Sears, 1996). The immediate early genes are transcribed without the requirement for previous viral protein synthesis and include transactivators such as the α-4 gene product, ICP 4. The early genes include many required for viral DNA synthesis, including the DNA polymerase, and are not transcribed until immediate early gene translation has been initiated. Finally, the late gene transcription is initiated following viral DNA synthesis. These include the structural proteins such as the virion capsid proteins from genes UL 13 and UL 48 (Vmw 65). Vmw 65 (also known as VP 16 and α-TIF) has been identified as an essential structural element in the virion and also as a transactivator of immediate early gene expression (Roizman and Sears, 1996). Although this is a simplistic description of the events during the replication cycle, it serves to illustrate that multiple antiviral targets are potentially available to truncate replication at various stages.

Draper *et al.* reported the inhibition of transactivation of the ICP 4 promoter and virus replication by unmodified oligodeoxynucleotides targeted to the putative translation initiation site of Vmw 65 (Draper *et al.*, 1990). Sequence specificity of the activity was suggested by the relative activities of two different 18-mers, although rigorous studies to demonstrate that the specificity was an antisense effect were not performed. Further inhibition studies using Vmw 65 as a target were performed by Kmetz et al. using the same 18-mer oligonucleotides, but as PS compounds (Kmetz *et al.*, 1991). The investigators found that a 25-μg/ml (4.3 μM) concentration of oligonucleotide reduced replication of HSV-1 strain KOS by greater than 50%, and that the inhibition correlated well with the reduction of Vmw 65 protein. A random 18-mer served as control and had minimal inhibitory activity. Extension of the 18-mer by 5 nucleotides on the 3′ end resulted in a more potent inhibitor. These studies suggested both a specific target effect (inhibition of Vmw 65 resulted in reduced infectivity), and a sequence specificity (the random 18-mer was less active) and thus potentially an antisense mechanism of action.

A series of studies by the Johns Hopkins group (Kulka and Aurelian, 1995; Kulka *et al.*,1989, 1993, 1994; Kean *et al.*, 1995) have investigated antisense oligonucleotides targeted to the splice donor/acceptor sites of the immediate early pre-mRNAs of IE 4 (ICP 4 protein). For the earlier studies, the investigators used methyl phosphonate oligodeoxynucleotides, and reported a sequence specific inhibition of virus replication. A methyl phosphonate 12-mer targeted to the IE mRNA splice donor site was effective in reducing virus growth by 80% at 100 μM, whereas the same oligonucleotide in which the central two residues were inverted was inactive. The same methyl phosphonate reduced virus yield by localized treatment in the HSV-1 mouse ear infection (Kulka *et al.*, 1993). Although these studies suggest specificity of activity, the effective concentrations make them impractical as potential therapeutic drugs. More recently this group has found that by using a 12-mer methyl phosphonate targeted to the intron/exon junction of the splice acceptor and by substituting 2′-OCH$_3$U for dT, greater affinity for the RNA target and a fivefold reduction in 50% inhibitory concentration IC$_{50}$ (22 to 4 μM versus HSV-1 with no inhibition against HSV-2) was achieved. The sequence specificity of the oligonucleotides was indicated by the relative lack of activity of mismatched oligonucleotides (Kean *et al.* 1995).

UL 13 is a late gene that encodes a protein kinase and appears to be a virion structural protein. A PS 21-mer oligonucleotide targeted to a translation initiation codon is a potent inhibitor of virus replication (IC$_{50}$ values of 0.4 to 1 μM) (Crooke *et al.*, 1992). This activity is quite potent, but demonstration of the characteristic sequence specificity and gene target selectivity must follow to label the antiviral activity as antisense.

Poddevin *et al.* investigated a series of phosphodiester oligonucleotides with 12-mer regions complementary to the target IE 4 pre-mRNA, and 3′ noncomplementary flanking sequences that formed hairpin structures similar to the self-stabilized phosphorothioate oligonucleotides described previously. These studies, which were well controlled for non-sequence-dependent antiviral effects, demonstrated that by adding stability to degradation at the 3′ terminus of an otherwise unaltered phosphodiester oligonucleotide, fairly potent inhibition (IC$_{50}$ value of 1.5 μM) could be achieved (Poddevin *et al.*, 1994).

A further approach to avoid nonspecific effects and yet increase oligonucleotide stability was employed by Peyman *et al.* (1995). They screened an array of 20-mers that were phosphodiester oligonucleotides except for two phosphorothioate nucleotide residues at both the 5' and 3' ends. The antiviral efficacy in cell culture was evaluated by inhibition of virus-induced cytopathic effect (CPE). The most potent compound was targeted at the translation start site of IE 110 mRNA, and had an effective dose of 9 μM. A 2-nucleotide shift in sequence reduced the efficacy by about ninefold, and mismatched oligonucleotides were inactive at 80 μM. Thus, by definition of sequence specificity, the activity appears to be antisense mediated. However, no attempt was made to evaluate the selectivity of inhibition of expression of UL 110 compared to the coexpression of another immediate early gene, which would have provided additional evidence for the antisense mechanism of activity.

Numerous PS oligonucleotides, with no apparent antisense sequence specificity, can have an anti-HSV effect (A. K. Field, personal communication) This observation has been thoroughly investigated by Y.-C. Cheng and colleagues. The most potent antiviral compound was the phosphorothioate dC 28-mer, which had an IC_{90} against HSV-2 of 1 μM, and probably inhibited virus replication by the capacity to block virus adsorption and penetration, and by the additional potent inhibition of viral DNA synthesis (Gao *et al.*,1989, 1990a,b). The dC 28-mer competes as a template for the DNA polymerase and as a competitive inhibitor of viral DNA polymerase exonuclease activity. Most recently, Fennewald and colleagues described PS oligonucleotides that are potent HSV inhibitors (IC_{50} values of 0.02 to 0.2 μM) (Fennewald *et al.*, 1995). These oligonucleotides are composed entirely of dG and dT residues and effectively block virus adsorption and penetration, although these may not be the only antiviral effects.

Marek's disease virus (MDV) is an avian alphaherpesvirus that causes lymphoproliferative disease. Evidence suggests that maintenance of the tumorigenic state of MDV-derived lymphoblastoid cell lines is due to expression of a 1.8-kb gene family. Kawamura *et al.* evaluated a phosphodiester 18-mer oligonucleotide complementary to a splice donor sequence and demonstrated a sequence-specific inhibition of expression of the 1.8-kb mRNA and inhibition of colony growth in soft agar. The multiple copies of the MDV genome were maintained in the lymphoblastoid cells. Appropriate sense and unrelated oligonucleotides were inactive (Kawamura *et al.*, 1991). This relatively early (1991) and well-controlled study provided direct evidence that the transcription of the 1.8-kb region is required for tumorigenicity of MDV-transformed cells.

2. Betaherpesviruses: Human Cytomegalovirus

Infections by HCMV are often silent and result in latency in a high proportion of the population. However, in immunocompromised individuals and in the newborn, HCMV infections may lead to a variety of disease syndromes including retinitis, pharyngitis, esophagitis, systemic disease, and perhaps coronary artery disease (Speir *et al.*, 1994; Haffey and Field, 1995; Falloon and Masur, 1990). As with herpes simplex virus, the replicative cycle is a coordinately regulated cascade with immediate early, early, and late transcription events. However, whereas HSV has 71 open reading frames, HCMV has more than 200, suggesting a more complicated replication pattern and/or a more involved pathogenesis. However, the larger genome also potentially provides more opportunities for antiviral intervention by antisense oligonucleotides.

Two groups of researchers have provided extensive studies on oligonucleotide inhibition of HCMV. Azad *et al.* surveyed a series of oligonucleotides against translation start sites, coding regions, intron/exon regions, and 5′ caps in a variety of genes including the DNA polymerase, and immediate early genes IE 1 and IE 2 (Azad *et al.*, 1993). They reported that the most potent oligonucleotide (ISIS 2922) was a PS 21-mer against the coding region of IE 2. According to the authors, ISIS 2922 reduced both IE 2 and HCMV replication proportionately, with an IC_{50} value of ~0.1 μM (Fig. 5). Although unrelated oligonucleotides were reported to be less active in both reduction of IE 2 and virus replication (suggesting a sequence-specific antiviral affect), mismatches in ISIS 2922 that substantially reduced hybridization did not alter the antiviral effects. The latter observations suggest that antiviral activity may be due in part to a nonantisense mechanism of

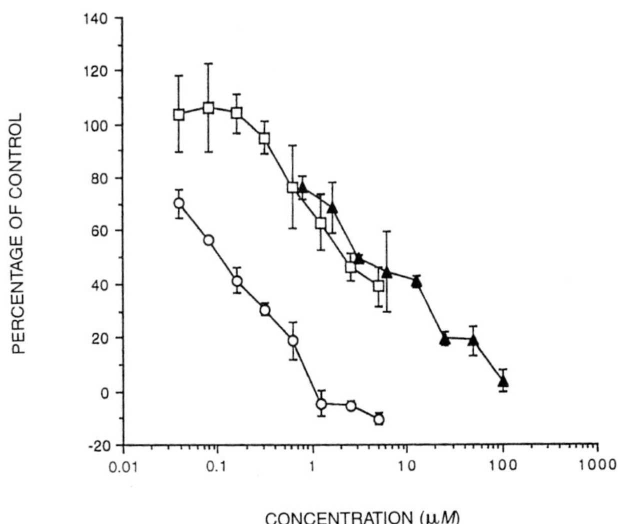

FIGURE 5   Relative antiviral activities of phosphorothioate oligonucleotides and ganciclovir. ISIS 2922 (○, antisense oligonucleotide targeted to the HCMV major immediate early region) was a more potent inhibitor of HCMV antigen expression than ISIS 3383 (□, noncomplementary control oligonucleotide) or ganciclovir (▲) (Azad *et al.*, 1993).

action. When evaluated in combination with ganciclovir or foscarnet, ISIS 2922 was additive in antiviral activity versus HCMV; with AZT it was also mainly additive in inhibiting HIV (Azad *et al.*, 1995). Considering the relative lack of cytotoxicity on uninfected cells, ISIS 2922 was considered by the authors to be an attractive candidate for clinical evaluation. In further support, ISIS 2922 is cleared slowly from vitreous fluid, and intact oligonucleotide accumulates in the retina following intravitreal injection in rabbits. Subacute intravitreal dosing is also well tolerated in monkeys. In a phase I study of repeat intravitreal injections to AIDS patients with refractory HCMV retinitis, ISIS 2922 was well tolerated at doses up to 300 μg—a calculated vitreal concentration of 8 μM. Initial reports state that ISIS 2922 did inhibit the progression of HCMV retinitis in patients who had progressive disease during ganciclovir or foscarnet therapy and ISIS 2922 is now in expanded clinical efficacy studies (Anderson, 1994).

One virtue of the antisense approach to identifying novel antiviral drugs is the ability to target any of a large array of genes. Smith and Pari took this approach in identifying a potent 20-mer PS oligonucleotide (UL36 ANTI) complementary to the splice donor site of UL 36, an immediate early gene that was identified as essential for HCMV DNA origin of replication-dependent synthesis (Smith and Pari, 1995; Pari *et al.*, 1995). Sequence specificity of the antiviral activity, as measured by inhibition of viral DNA replication, was established by comparison of the efficacy of UL 36 ANTI [50% effective concentration (EC$_{50}$) of 0.06 μM], with sense, reverse, and unrelated sequences (EC$_{50}$ values in excess of 0.4 μM) (Fig. 6). Mismatches in the UL 36 sequence reduced activity. The gene target specificity was shown by Northern blots indicating the inhibition of UL 36 transcript, with

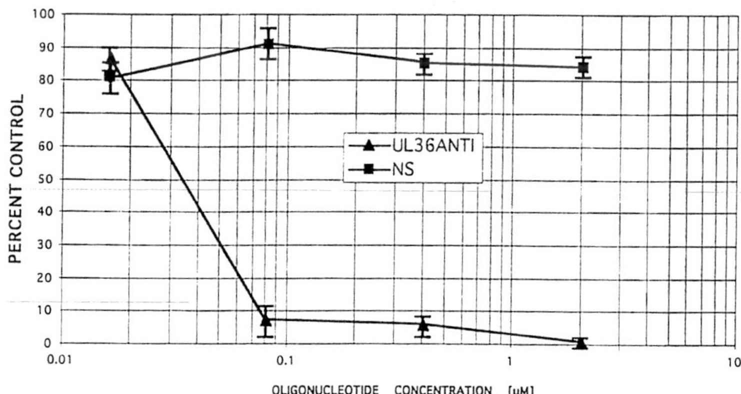

FIGURE 6   UL36 ANTI (▲, antisense oligonucleotide targeted to the UL 36 intron–exon boundary in unspliced RNA) inhibited production of HCMV UL 44 antigen as determined by ELISA of HCMV-infected human foreskin fibroblast cells. A nonspecific control oligonucleotide (■) did not inhibit antigen production (Pari *et al.*, 1995).

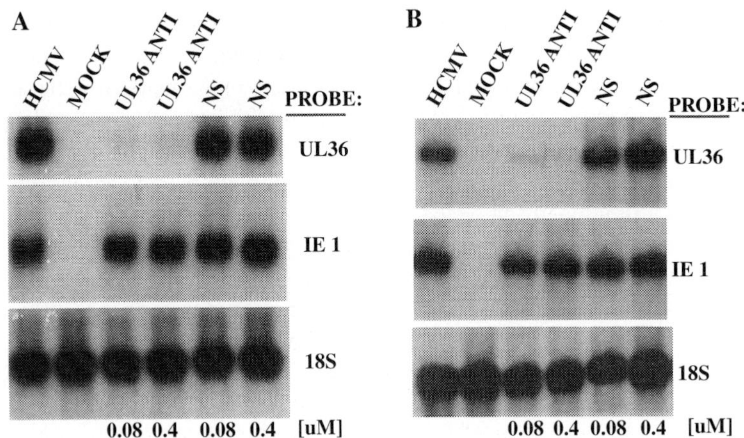

FIGURE 7   UL36 ANTI selectively inhibits UL 36 mRNA. Northern analysis of RNA from oligonucleotide-treated cells shows that UL 36 ANTI decreased UL 36 RNA levels with no effect on either IE 1 RNA (A) or IE 2 RNA (B). Nonspecific control oligonucleotide did not inhibit any RNA production. Ribosomal RNA (18S) levels were constant throughout the experiment (Pari *et al.*, 1995).

no effect on the expression of either IE 1 or IE 2 (Fig. 7). UL 36 ANTI also reduced infectious virus yield by greater than 99% at 0.08 μM. Thus, a novel antiviral target was identified using the antisense oligonucleotide approach, resulting in a potent and selective antiviral agent.

For future clinical evaluations, the UL 36 ANTI sequence has been investigated using chemical modifications described previously in this chapter. Preclinical evaluations are currently ongoing using a hybrid that is phosphorothioate at each nucleotide and that also contains $2'$-OCH$_3$ substitutions at two nucleosides on the $5'$ end and at four nucleosides on the $3'$ end. This compound has been identified as GEM 132 (Fig. 8) and combines the potency embodied in the UL 36 ANTI oligonucleotide sequence with the enhanced stability of a hybrid, providing the potential for systemic and intravitreal use. Its initial entry into clinical studies as a systemic therapy occurred in August 1996. A phase I/II study to assess GEM 132 for intravitreal treatment of HCMV retinitis has been approved and studies are underway.

3. Gammaherpesviruses: Epstein–Barr Virus

Epstein–Barr virus is the causative agent for most cases of infectious mononucleosis, an acute infection associated with active virus replication

$5'$- <u>UG</u>GGGCTTACCTTGCG<u>AACA</u>   -$3'$

FIGURE 8   GEM 132 chemical structure. The underlined bases are $2'$-O-methyl-modified ribonucleotides; others are deoxyribonucleotides. All nucleotides are phosphorothioates.

(Pagano, 1995). Epstein–Barr virus infection can also result in latent infection, which can result in polyclonal B cell lymphoproliferative diseases in immunocompromised individuals and nasopharyngeal carcinoma, which is relatively common in southeast China (Huang, 1991). Among the genes expressed during latency and reactivation to productive infection, two have been the subject for design and evaluation of antisense oligonucleotides. These are the Epstein–Barr nuclear antigen EBNA 1 gene, which is required for maintenance of transformation of latently infected cells, and the BZLF gene, which is required to reactivate latently infected cells to virus-productive infection (Pagano, 1995).

Pagano *et al.* studied the potential for both unmodified and PS 18-mer antisense oligonucleotides as inhibitors of EBNA 1 (Pagano *et al.*, 1992). The oligonucleotides were targeted against the coding region just 3' of the AUG. Prolonged treatment of Raji cells, which carry 60 EBV episomes per cell, with 40 μM unmodified antisense oligonucleotide was reported to result in a progressive reduction of EBNA 1 proteins determined by Western blots and in EBV copy number determined by Southern blots. Similar treatment with the sense oligonucleotide control was reported to be ineffective. Roth *et al.* (1994) also used unmodified oligonucleotides (40 μM) complementary to sequences in the EBNA 1 RNA and observed inhibition of EBNA 1 translation and inhibition of cell proliferation, with no effect on proliferation of EBV-negative cell growth. When PS oligonucleotides were used by Pagano in dose–response evaluations of growth inhibition of EBV-transformed cord blood B cells, control sense and scrambled sequences were reported to be partially effective as inhibitors of cell proliferation, while antisense oligonucleotides were reported to suppress growth totally at 20 μM, and was partially effective at 5 μM (Pagano *et al.*, 1992). Thus, as demonstrated in other virus infections, PS oligonucleotides may have a nonspecific inhibitory effect, and indeed this has been shown in the studies by Yao *et al.* (1993). They described potent inhibition (EC_{90} 0.5 μM) of EBV yield from H1 cells, a chronically infected high-producer cell line, using PS 28-mer without sequence specificity for any EBV target. The mechanism of inhibition may be the result of blocking DNA synthesis, as was also described previously for the anti-HSV nonspecific oligonucleotide efficacy.

Sequence-specific and target-specific antisense inhibition of BZLF 1 expression was described by Daibata *et al.* (1996). Akata cells are latently infected with EBV and are inducible to the lytic viral cycle, a function that requires BZLF 1 expression of the Zebra protein. Both unmodified and PS 25-mer oligonucleotides complementary to the translation initiation codons inhibited the production of Zebra (as determined by Western blots) and replication of virus [shown by lack of production of the replicative linear DNA, viral early antigen (EA-D), and virus capsid antigen (VCA)]. Control sense, reverse sequence, and random oligonucleotides were less effective, and at the effective concentrations of antisense oligonucleotide no inhibition

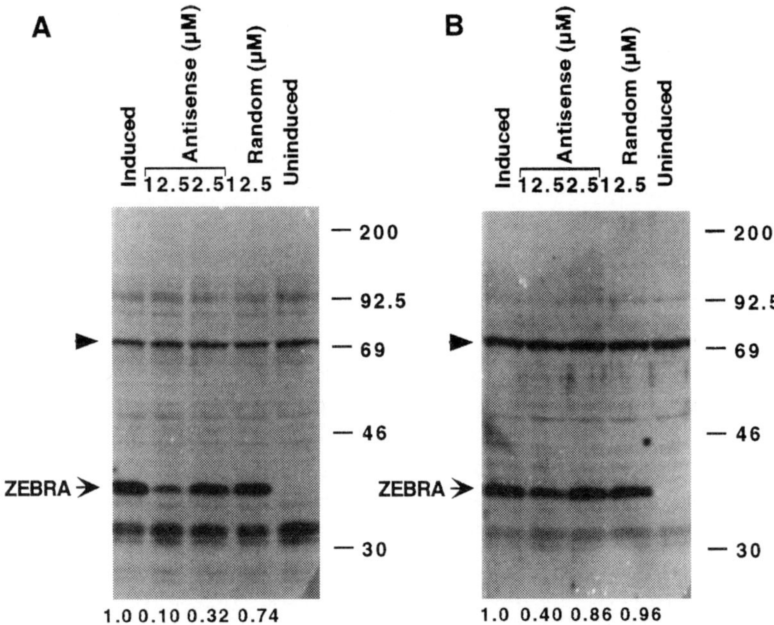

FIGURE 9 Immunoblot analysis of Zebra induction in antisense oligonucleotide-treated Akata cells. Antisense oligonucleotides targeted to the translation start site of Zebra inhibited Zebra production; a random mix of 20-mer oligonucleotides was much less potent. Inhibition was quantitated relative to induced Akata cells induced for EBV expression by stimulation with anti-IgG and normalized to a 72-kDa protein (see arrowhead). Phosphorothioate antisense oligonucleotides (A) and phosphodiester oligonucleotides (B) both inhibited Zebra induction. Reprinted from *Antiviral Res.* **29**, Daibata, M., Enzinger, E. M., Monroe, J. E., Kilkuskie, R. E., Field, A. K., and Mulder, C., Antisense oligodeoxynucleotides against the BZLF1 transcript inhibit induction of produtive Epstein–Barr virus replication, 243–260. Copyright 1996 with kind permission from Elsevier Science-NL, Sara Burgerhartstraat 25, 1055 KV Amsterdam, The Netherlands.

of cellular DNA synthesis or expression of CD19 (a B cell membrane protein) was apparent. Similar inhibition of virus production was observed in P3HR-1 cells, an EBV producer cell line, albeit after a prolonged antisense treatment. Thus, by the criteria of sequence and gene target specificity, these oligonucleotides designed to inhibit expression of the BZLF 1 gene and inhibit virus replication are probably functioning by an antisense mechanism, although nonspecific effects may also occur (Fig. 9 and Table I).

C. Myxoviruses and Paramyxoviruses

1. Influenza Viruses

Influenza viruses are enveloped virions with segmented single-stranded RNA genomes, and are the causative agents for epidemic acute respiratory

TABLE I   Inhibition of Productive Epstein–Barr Virus Replication by Anti-BZLF1 Antisense Oligonucleotides in Anti-IgG-Stimulated Akata Cells[a]

	Mean ± SE (% inhibition)			
Oligonucleotide	0.5 μM	2.5 μM	12.5 μM	25 μM
Phosphorothioate				
Antisense	35 ± 6 (3)[b]	48 ± 13 (3)	68 ± 9 (4)	69 ± 5 (4)
Random	5 ± 5 (3)	21 ± 10 (3)	24 ± 9 (4)	32 ± 10 (4)
Sense[c]			11 (1)	36 (1)
Reverse[d]			17 (1)	36 (1)
Phosphodiester				
Antisense	16 ± 13 (3)	25 ± 13 (3)	46 ± 10 (3)	56 ± 8 (3)
Random	14 ± 14 (3)	14 ± 7 (3)	21 ± 3 (3)	19 ± 6 (3)

[a] Reprinted from *Antiviral Res.* **29,** Daibata, M., Enzinger, E. M., Monroe, J. E., Kilkuskie, R. E., Field, A. K., and Mulder, C., Antisense oligodeoxynucleotides against the BZLF1 transcript inhibit induction of produtive Epstein–Barr virus replication, 243–260. Copyright 1996 with kind permission from Elsevier Science-NL, Sara Burgerhartstraat 25, 1055 KV Amsterdam, The Netherlands. Oligonucleotide-treated Akata cells were stimulated with anti-IgG for 24 hr. Inhibition was measured by the amounts of linear EBV DNA from gel analysis.
[b] The number of experiments (*n*) is listed in parentheses.
[c] Sense is the sequence complementary to the antisense.
[d] Reverse is the sequence with the identical nucleotide sequence as antisense but read in the opposite direction.

illness and severe disease. The negative-stranded RNA genome of influenza A and B viruses contains eight segments, and acts as a template for viral mRNA synthesis as well as synthesis of the antigenomic (positive-stranded) RNA intermediate (Lamb and Webster, 1996). The negative-stranded genome is packaged into virions along with the RNA-dependent RNA polymerase. Viral replication is a complex process; viral mRNA synthesis and replication occurs in the nucleus of infected cells. mRNA synthesis initiates with host cell-derived primers containing 5′-methylated capped RNA fragments from cellular RNA polymerase II transcripts. Four viral proteins are required for mRNA synthesis: NP, which is the major structural protein of the nucleocapsid, and the PB1, PB2, and PA proteins, which are polymerase proteins. Each of these proteins is encoded by a separate segment of the viral genome (Lamb and Webster, 1996), and each is an attractive target for antiviral inhibition.

Current antiviral therapies against influenza include amantidine and rimantidine, which act by blocking the ion channel formed by the M2 protein. This ion channel is pH activated and is required for uncoating of the viral nucleocapsid after virus infection. Amantidine and rimantidine are partially effective, especially when given as a prophylactic drug, but resistant virus variants can readily arise (Belshe *et al.,* 1988). More recently, specific inhibitors of the viral neuraminidase activity (required for virus binding to cells) have been designed. Although this approach appears promising, additional therapies are required. Annual immunization has also been effective for high-risk populations; however, long-term immunization has not been possible due to antigenic variation from epidemic season to season.

Oligonucleotides can be effective anti-influenza agents in cell culture assays. Phosphorothioate oligonucleotides targeted against the viral RNA or mRNA of the PB1 gene inhibited the replication of influenza A and influenza C viruses (Leiter *et al.*, 1990). For influenza C virus, a PS oligonucleotide targeted against the viral RNA of the PB1 gene inhibited plaque formation by >90% at 20 μM and caused a 10^6-fold reduction in infectious virus at 80 μM. A control oligonucleotide, containing one mismatch, was much less active, inhibiting plaque formation by about 50% at 20 μM. Oligonucleotides containing three mismatches or lacking influenza sequence did not inhibit plaque formation at 20 μM. Although this sequence-specific effect was observed, the active oligonucleotide contained a G quartet and mismatches eliminated this motif. It is not known whether the G quartet contributes to additional nonantisense mechanisms. Another study employed phosphodiester oligonucleotides modified at the 5' end with a hydrophobic group, *n*-undecanol (Kabanov *et al.*, 1990). A modified 10-mer targeted at the polymerase III gene was reported to inhibit influenza plaque formation at >10 μM, whereas an unmodified 10-mer and a modified nonsense 10-mer were ineffective. The mechanism of inhibition, although apparently sequence specific, was not elaborated further. This is important since in addition to potential antisense inhibition, oligonucleotides have also been used as inhibitors of influenza virus RNA polymerase *in vitro* (Chung *et al.*, 1994). Short (<9-mer) oligonucleotides containing 5' cap structures (m7pppGm) bind *in vitro* to viral polymerase with high affinity and inhibit cap-dependent transcription. The antiviral effect of these oligonucleotides was not measured.

Clearly, sequence-specific inhibition of influenza virus replication is well documented, but whether this antiviral activity is the result solely of an antisense mechanism of action, or includes an unrelated inhibition of polymerase or adsorption/penetration or a combination of effects, is not entirely clear.

2. Respiratory Syncytial Virus

Respiratory syncytial virus (RSV) causes severe lower respiratory tract disease in infants, young children, and immunocompromised adults (Collins *et al.*,1996). Respiratory syncytial virus is an enveloped virus, containing an unsegmented, negative-stranded genome of approximately 15,000 nucleotides (Collins *et al.*, 1996). The genome encodes 10 viral proteins, which are translated from individual mRNAs. These RNAs are transcribed from the negative-stranded genome by the viral polymerase complex (N, P, and L proteins). In addition, this complex replicates the negative-stranded genome from a positive-stranded intermediate for insertion into progeny virions. Replication occurs in the cytoplasm of the infected cells.

Currently, ribavirin is used therapeutically for RSV disease (Groothuis, 1994; Levin, 1994). Ribavirin is believed to inhibit RSV replication by several potential mechanisms: inhibition of viral polymerase, inhibition of $5'$ cap formation of mRNAs, and inhibition of IMP dehydrogenase, which decreases intracellular GTP levels (Gilbert and Knight, 1986). Which is the dominant mechanism of antiviral activity is uncertain. Unfortunately, clinical benefits from ribavirin are small and occur only in a portion of RSV-infected individuals (Levin, 1994). Clearly, there remains a need for development of effective therapeutics for RSV disease.

Oligodeoxyribonucleotides targeted against the genomic RNA inhibit RSV replication in cell culture by an apparent antisense mechanism. HEp-2 cells were infected with RSV strain A2 in the presence of oligonucleotides, with replication measured by enzyme immunosorbent assay (ELISA) or virus yield assay (Jairath *et al.*, 1997). Using ELISA, EC_{50} values were about $0.5–1$ μM for an antisense oligonucleotide targeted to the start of the NS2 gene. In all assays, the antisense oligonucleotide was more potent than (1) a control oligonucleotide containing the reverse sequence, (2) oligonucleotides targeted at RSV mRNA, (3) a random sequence oligonucleotide, and (4) ribavirin (Table II). Importantly, sequence-specific depletion of the genomic target following treatment of cells with the antisense oligonucleotide was demonstrated by reverse transcriptase-polymerase chain reaction (RT-PCR) (Fig. 10). Specific cleavage of the genomic target RNA has been detected at the antisense oligonucleotide-binding site, suggesting that cellular RNase H participates in the reaction. These observations provide *in situ*

TABLE II   Respiratory Syncytial Virus Inhibition by Oligonucleotides[a]

Compound	$EC_{50} \pm SD$ (μM, F antigen)	$EC_{99} \pm SD$ (μM, infectious virus)
v590 (antisense)	0.64 ± 0.7 (12)[b]	6.6 ± 3.7 (6)
v590s (scrambled)	2.4 ± 2.1 (11)[c]	26 ± 8.9 (5)[d]
r20 (random mix of 20-mers)	17 ± 13 (10)	>30 (5)
Ribavirin	13 ± 5.7 (12)[e]	27 ± 6.2 (6)[f]

[a] Reprinted from *Antiviral Res.* **33**, Jairath, S., Brown Vargas, P., Hamlin, H. A., Field, A. K., and Kilkuskie, R. E., Inhibition of respiratory syncytial virus replication by antisense oligodeoxyribonucleotides, 201–213. Copyright 1997 with kind permission from Elsevier Science-NL, Sara Burgerhartstraat 25, 1055 KV Amsterdam, The Netherlands.

[b] Number of experiments.

[c] $p = 0.02$ vs v590.

[d] $p = 0.006$ vs v590.

[e] $p = 0.001$ vs v590.

[f] $p = 0.0001$ vs v590.

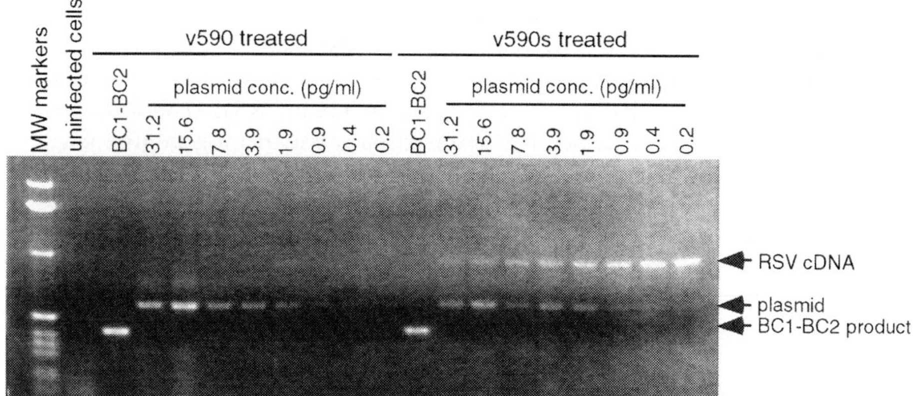

FIGURE 10   Quantitation of genomic RSV RNA. RSV cDNA was prepared by reverse transcription. The PCR was conducted using primers OD1 and BC6 to generate a 940-base pair product from the RSV cDNA. Serial dilutions of a competitive plasmid template that generated a 575-base pair product from the same primers were included with a constant amount of RSV cDNA to determine the concentration of RSV RNA present after oligonucleotide treatment. Cells treated with a scrambled control oligonucleotide (v590s) contained a measurable RSV cDNA PCR product. In cells treated with an antisense oligonucleotide (v590) no RSV cDNA PCR product was detected. Reprinted from *Antiviral Res.* 33, Jairath, S., Brown Vargas, P., Hamlin, H. A., Field, A. K., and Kilkuskie, R. E., Inhibition of respiratory syncytial virus replication by antisense oligodeoxyribonucleotides, 201–213. Copyright 1997 with kind permission from Elsevier Science-NL, Sara Burgerhartstraat 25, 1055 KV Amsterdam, The Netherlands.

evidence that the oligonucleotide, which was designed to bind to the target genome RNA, did trigger cleavage of the heteroduplex. This is precisely what is hoped for from an antisense mechanism of action.

However, the antisense oligonucleotide contains a 4-G quartet, and in some other systems four consecutive guanylic acid residues contribute to nonantisense, sequence-specific inhibition of viruses or other cellular functions (Wyatt *et al.*, 1994; Burgess *et al.*, 1995). The RSV antisense oligonucleotide was about fourfold more active than other oligonucleotides containing 4 Gs, including a control reversed sequence (Table II). This suggests that the four Gs may contribute significantly to the antiviral activity of the oligonucleotide, but are not the predominant antiviral factor. Thus, antisense oligonucleotides targeted against RSV genomic RNA can effectively inhibit RSV replication and may have therapeutic value.

D. Hepadnaviruses

Hepatitis B virus (HBV) is a member of the hepadnavirus family, which also includes hepatitis viruses of the Pekin duck, the heron, the woodchuck, and the ground squirrel. Worldwide, more than 400 million people are infected with HBV, which causes acute (self-limiting) or chronic infection of the liver in the infected individual. Acute infections are characterized by the appearance of neutralizing antibodies, while chronically infected individuals continuously shed virus into the blood. Epidemiologic studies have linked chronic infection with hepatocellular carcinoma. While interferon α has been used to treat chronic hepatitis, it has been only partially successful in suppressing virus shedding and in providing elimination of chronic infection (Locarnini and Cunningham, 1995). While other potential therapeutics such as lamivudine are in clinical evaluation (Dienstag *et al.*, 1995), the need for additional options in therapy has provided impetus to identify novel inhibitors for novel targets.

Hepatitis B virus is a compact DNA virus (3.2 kb) that contains four open reading frames encoding (1) three envelope or surface antigens, (2) two nucleocapsid (HBcAg and HBeAg) proteins, (3) the polymerase (P) gene product, and (4) the X gene product. The genome is transcribed to produce two predominant transcripts (3.5 and 2.1 kb) and two minor transcripts (2.5 and 0.7 kb). In addition to the open reading frames, HBV contains novel genetic targets for oligonucleotides. The encapsidation signal is a short ($\sim$90 base) RNA sequence that has well-defined secondary structure. This sequence is repeated at both ends of the pregenomic RNA and is recognized by viral proteins to package the viral RNA into nucleocapsids. Also, reverse transcription initiates at a sequence within this encapsidation signal. Other HBV RNA regions (DRI and DRII) are recognized by the HBV polymerase and are required for translocation of the enzyme during reverse transcription.

The complex nature of the HBV genome and its replication provides opportunity to identify antisense oligonucleotides with anti-HBV activity. Several groups have reported inhibition by targeting translation of HBV structural proteins(Goodarzi *et al.*, 1990; Blum *et al.*, 1991). A series of oligonucleotides against HBV were evaluated using a hepatocellular carcinoma-derived cell line stably transfected with HBV DNA (Hep2.2.15; Korba and Gerin, 1995). Oligonucleotides were reported to inhibit HBV virion production in a sequence-specific fashion. For oligonucleotides targeted against transcripts encoding structural proteins (S gene and C gene), HBV virion inhibition reportedly correlated well with inhibition of protein production. Antigen inhibition was also reported to be sequence specific; for example, oligonucleotides targeted against the S gene inhibited only HBsAg production, and had no effect on HBeAg or HBcAg levels. Interestingly, oligonucleotides targeted against the encapsidation sequence were reported to be among the most potent inhibitors identified in these experiments. These oligonucleotides inhibited virion production and intracellular HBV DNA replication, which is consistent with the requirement of this region in packaging viral RNA and viral DNA replication. None of the oligonucleotides decreased the level of any HBV RNA species, suggesting that RNase H did not contribute to the antiviral activity in this assay system.

In a different study, a phosphorothioate oligonucleotide targeted against the duck hepatitis B virus pre-S gene inhibited virus replication *in vivo* (Offensperger *et al.*, 1993). This oligonucleotide was the most active of several oligonucleotides that were evaluated in cell culture. *In vivo*, virus replication was measured by the presence of viral DNA replicative intermediates in liver, as well as the presence of viral antigens (surface antigen and core antigen) in livers. *In vivo* inhibition was determined to be dose dependent and sequence specific. A dose of 20 mg/kg for 10 days caused a marked decrease in HBV DNA replication and a reduction in viral antigens in serum and liver, while 5 mg/kg had little effect on HBV DNA or antigen. Sense and random oligonucleotides did not inhibit virus production. These important results represent the first observation of antisense oligonucleotide inhibition of a virus infection *in vivo* (Fig. 11).

HBV X protein (HBx), which is a transactivator, was also reported to be targeted successfully by antisense oligonucleotides *in vivo* (Moriya *et al.*,

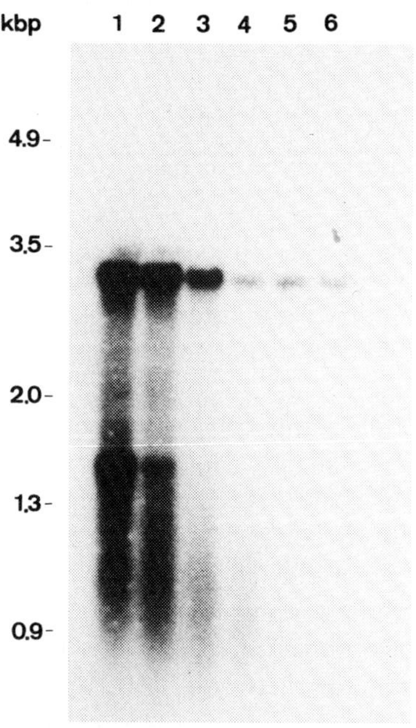

FIGURE 11   Antisense oligonucleotide (AS2, targeted against HBV pre-S mRNA) inhibited DHBV replication *in vivo*. DNA was isolated from livers of DHBV-infected ducks and analyzed by Southern blot. Ducks were treated for 10 days with daily intravenous injections of AS2 at concentrations of 5 μg (lane 2), 10 μg (lane 3), or 20 μg (lanes 4–6) per gram of body weight. Lane 1 is sample from an untreated control duck (Offensperger *et al.*, 1993).

1996). Transgenic mice expressing HBx developed liver lesions at the age of 2 months and liver tumors after 12 months. According to the authors, when treated with PS oligonucleotides targeted at the HBx translation start site intraperitoneally for 8 weeks, HBx production was inhibited as measured by RNA and protein production. In addition, hepatic lesions were reduced. The inhibition was reported to be sequence specific, in that a sense control did not inhibit; also, an oligonucleotide targeted within the HBx coding region was inactive. Subject to future positive studies, HBx therefore, appears potentially to be a useful target in HBV.

Treatment of liver disease may be enhanced by targeting oligonucleotides to hepatocytes, the host cells for hepatitis virus replication. In two studies, oligonucleotides conjugated to asialoorosmucoid were reported to be taken up by cells containing the asialoglycoprotein receptor (Wu and Wu, 1992; Bunnell *et al.*, 1992). The conjugates reportedly attained ~5- to 10-fold greater cell association than did the unconjugated oligonucleotide, and this increase in uptake depended on the presence of the asialoglycoprotein receptor. In both studies, the reported increase in uptake resulted in antisense inhibition of the target gene. Antisense oligonucleotide complexes specifically decreased the production of HBV antigen in the medium of Hep2.2.15 cells by 80–90%; random oligonucleotide had no effect on virus production. In this report, uncomplexed antisense oligonucleotide reduced HBV antigen production by only 30–40%. This is in contrast to the observations of Korba and Gerin, who described uncomplexed phosphorothioates as effective inhibitors of HBV replication (Korba and Gerin, 1995). The differences in these results may be due to the oligonucleotide sequences used by the investigators. These results, especially the *in vivo* studies on duck hepatitis virus, show that antisense oligonucleotides can provide effective therapy against hepatitis B infection. Phosphorothioate oligonucleotides readily attain high concentrations in the liver (Crooke *et al.*, 1996; Agrawal and Temsamani, 1996), and the opportunity to target hepatocytes with the antisense oligonucleotide may provide a unique advantage for therapy.

E. Human Papillomaviruses

Human papillomaviruses (HPVs) include at least 65 types, based on DNA sequence diversity as measured by liquid hybridization. They infect epithelial cells at mucocutaneous surfaces, resulting in lesions from benign warts to cervical carcinoma, with each virus type having a specific anatomical site of replication. Several HPV types infect genital epithelia and represent the most prevalent etiologic agents of sexually transmitted viral disease. The genital HPV types can be further subdivided into "high-risk" types that are associated with the development of neoplasms (most commonly HPV-16 and HPV-18) and "low-risk" types that are rarely associated with malignancy (most commonly HPV-6 and HPV-11) (Shah and Howley, 1996).

The genome of all the HPV types is circular double-stranded DNA of approximately 7900 base pairs. The genome can be divided into three distinct functional domains; the upstream regulatory region, which contains the origin of viral DNA replication plus enhancers and promoters involved in transcription; the E region, which encodes genes required for vegetative functions; and the L region, which encodes the structural proteins L1 and L2. The polypeptides expressed from the E region have several functions: E1 is required for episomal DNA replication; E2 controls transcription and also may be involved in initiation of DNA replication; E4 interacts with cytokeratins; and E5, E6, and E7 are involved in cell transformation. The viral proteins are translated from families of alternatively spliced mRNAs (Howley, 1996).

Current therapies for treating HPV infections involve excision or destruction of infected tissue, rather than inhibition or reversal of the viral directed tumorigenesis. Even though viral replication and transformation are reasonably well understood, the use of that knowledge to identify selective new antiviral drugs has been lacking. This situation provides a rich opportunity for rational design through antisense, and antisense oligonucleotides designed to inhibit viral replication or inhibit cellular transformation have been evaluated. One compound, ISIS 2105, has been targeted at the HPV E2 mRNA translation initiation region of HPV-6 and HPV-11 (Cowsert *et al.,* 1993). They reported that cell culture studies using bovine papillomavirus (BPV) as a model system showed that oligonucleotides targeted at E2 could inhibit both E2-dependent transactivation in a transient assay and BPV-induced transformed cell focus formation, suggesting that the E2 gene is an appropriate target for antiviral therapy. Because HPV does not replicate in cell culture, further studies used a second model system in which E2-dependent transactivation of a reporter gene (chloramphenicol acetyltransferase) was measured. ISIS 2105 was reported to inhibit reporter expression, with an EC_{50} of approximately 5 μM (Fig. 12). The inhibition was said to be sequence specific in that a control oligonucleotide targeted against the BPV E2 translation start site reportedly did not inhibit at concentrations up to 10 μM. These studies led to Phase I and II clinical trials for treatment of genital warts with ISIS 2105; however, these trials were discontinued by the sponsor, ISIS Pharmaceuticals (Carlsbad, CA).

A combined effort of Hybridon (Cambridge, MA) and Roche Research Center (Welwyn-Garden City, UK) has resulted in the identification of sequence-specific and gene target-specific antisense oligonucleotides active in cell culture assays and in a mouse xenograft model of human papillomavirus replication (Roberts *et al.,* 1997; Lewis *et al.,* 1997). The antisense PS oligonucleotides targeted against HPV type 6 and 11 E1 gene selectively reduced E1 mRNA levels in cell culture. A modified version of the same sequence (containing $2'$-OCH_3 nucleosides in addition to the PS backbone) significantly reduced growth of HPV-infected cysts in mice, while a mis-

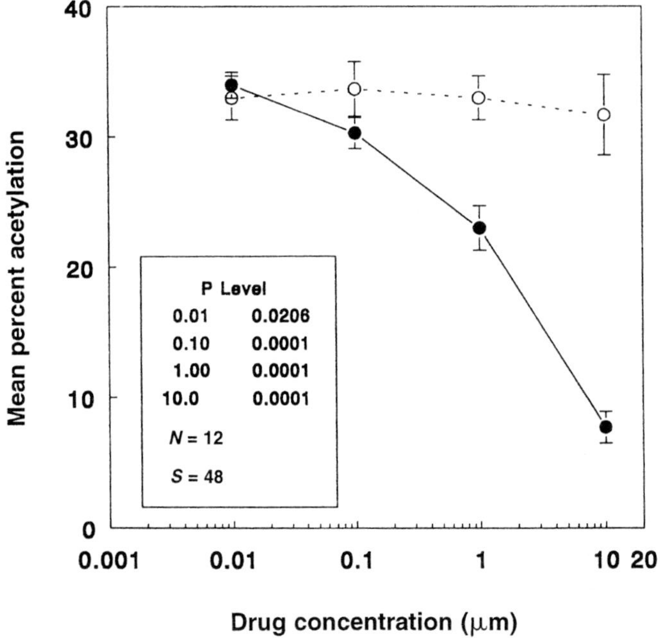

FIGURE 12   Inhibition of HPV-11 E2-dependent transactivation by an antisense oligonucleotide (ISIS 2105, ●). Chloramphenicol acetyltransferase (CAT) expression from a plasmid vector containing E2-dependent sequences was used to measure E2 transactivation. C127 cells were treated with oligonucleotide, then transfected with HPV-11 expression construct, HPV-11 E2-dependent CAT expression reporter. Analysis of variance on acetylated chloramphenicol was performed from four separate experiments. A control oligonucleotide (ISIS 2324, ○) did not inhibit CAT expression (Cowsert *et al.*, 1993).

matched control oligonucleotide did not decrease cyst growth. Histologically, the antisense oligonucleotide-treated cysts carried fewer koilocytes, suggesting that antisense treatment decreased viral load. These findings will be further explored for the potential development of antisense oligonucleotide therapy for genital warts.

In several studies, the genes responsible for cellular transformation, E6 and E7, have been targeted with antisense oligonucleotides. Most studies have used cervical carcinoma-derived cell lines that contain HPV E6 and/or E7 and have measured growth inhibition to assess antisense oligonucleotide activity. In early experiments, plasmids expressing HPV antisense RNA against HPV18 E6 and/or E7 altered growth characteristics of HeLa cells (Steele *et al.*, 1992). Antisense expression was regulated by the inducible mouse mammary tumor virus (MMTV) promoter; induction resulted in a slowing of the growth rate, decrease in growth in soft agar, and an increase in serum requirement. Vero cells, which do not contain HPV sequences, were not effected by transfection with the antisense plasmids. Also, plasmids expressing the sense constructs of E6 and E7 did not change growth characteristics of HeLa cells. All these characteristics suggested that expressed antisense RNA caused a decrease in E6 and/or E7 expression and an alteration of the transformed phenotype of the HeLa cells.

Similarly, HPV-specific PS oligonucleotides inhibited CaSki cell growth, which is dependent on HPV E6/E7 expression, by as much as 80–90% at concentrations between 10 and 20 μM (Storey *et al.*, 1991). This effect was apparently sequence specific since only two oligonucleotides, targeted to the overlapping translation start sites for HPV E6 or HPV E7, were effective. Six HPV-specific oligonucleotides targeting other regions of the E6 or E7 genes, as well as a random sequence oligonucleotide, did not inhibit growth. However, there was no significant decrease in HPV E6 or E7 mRNA or protein, suggesting that the mechanism of inhibition, although selective, was not antisense.

Subsequent studies using PS oligonucleotides targeted at similar regions of E6 and E7 did show apparent antisense effects (Tan and Ting, 1995). In these experiments, oligonucleotides were delivered to cells with a cationic lipid. Cell growth and E7 expression were reportedly specifically inhibited by antisense targeted to E7 RNA. Interestingly, this report also suggested that E7 expression was also inhibited by an E6-specific antisense oligonucleotide, possibly due to the inhibition of the bicistronic E6–E7 transcript. The antisense oligonucleotides also were reported to inhibit tumor growth by SiHa cells in nude mice. In addition, a reported decrease in E7 expression was measured in the tumor, with no effect on a control gene (actin). A control oligonucleotide, containing randomized sequence, was not reported to inhibit tumor growth or E7 expression. These *in vivo* results suggest that E7 inhibition may be an effective treatment for HPV-induced tumors.

F. Picornaviruses

Picornaviruses are nonenveloped viruses containing a single positive-stranded RNA genome. These viruses comprise a large family of human and agricultural pathogens, including poliovirus, hepatitis A virus, and foot and mouth disease virus (FMDV). Translation of picornavirus RNA occurs through a novel cap-independent mechanism involving initiation from internal sites on the viral RNA, specifically recognized by ribosomal and other cellular proteins. Several of these viruses have been targets for antisense oligonucleotide inhibition, and the studies demonstrate the versatility of the antisense approach.

Encephalomyocarditis virus (EMCV) has been used as a model system to study translation initiation by internal ribosomal entry (Jang *et al.*, 1988, 1989). Antisense phosphodiester oligonucleotides (13-mers), targeted at the $5'$ untranslated region and the translation start site of EMCV, inhibited cell-free translation in rabbit reticulocyte lysates (Sankar *et al.*, 1989). Specificity was demonstrated by showing that an oligonucleotide containing a single mismatch did not inhibit translation. The importance of the 5' untranslated region was confirmed; oligonucleotides targeted at the coding region or the 3' end of the genome did not inhibit translation. Although viral replication assays were not performed, the results in the cell-free translation system suggest that inhibition of internal ribosomal entry by oligonucleotides provides a novel target for virus inhibition.

Antisense oligonucleotides inhibited replication of FMDV (Gutiérrez *et al.*, 1993). High concentrations (125–250 μM) of phosphodiester oligonucleotides were microinjected into uninfected cells, which were then infected with FMDV. A modest reduction (35–50%) in the percentage of cells expressing FMDV antigen was detected 5 hr after infection. Sequence specificity was observed, since a scrambled control oligonucleotide was ineffective. Studies following the fate of fluorescein-labeled oligonucleotide indicated localization in the nucleus after microinjection; high concentrations apparently saturated this transport process, and oligonucleotide was then detectable in the cytoplasm, the site of viral replication. Phosphorothioate versions of the same oligonucleotide sequences inhibited infectious virus yield by about 50% when used at high concentrations (50–100 μM). Again, the effect appeared to be sequence specific since the scrambled control was ineffective. However, the modest effects seen in this study using high oligonucleotide concentrations suggest that the utility of antisense oligonucleotides against FMDV is probably quite limited.

G. Coronaviruses

Coronaviruses are large, enveloped virions containing a positive-stranded RNA genome, and the human strains are responsible for acute respiratory disease. Mouse hepatitis virus (MHV), which causes respiratory and gastrointestinal infections as well as hepatitis, has been used as a model system to study coronavirus replication. The viral RNA-dependent RNA polymerase synthesizes a full-length negative-sense copy of the genome prior to copying a series of subgenomic mRNAs from the negative strand. These subgenomic mRNAs contain at their 5' ends a common leader sequence that is derived from the 5' end of the genome. Subgenomic RNAs are derived from discontinuous RNA synthesis from this leader. An antisense oligonucleotide (14-mer) complementary to the leader sequence was reported to inhibit MHV plaque formation when the oligonucleotide was present during infection (Mizutani *et al.*, 1994). In addition, viral RNA was decreased. The effect was reported to be sequence specific and selective; a sense oligonucleotide did not inhibit virus replication, and cellular RNA was not effected by the antisense oligonucleotide.

H. Flaviviruses

The Flaviviridae consist of three groups: flaviviruses, pestiviruses, and hepatitis C. A wide variety of arthropod-borne virus infections of humans is caused by flaviviruses, resulting in severe disease (encephalitis, high fever, or hemorrhagic fever). The virions of all flaviviruses are enveloped and encapsidate a positive-stranded RNA genome that is translated into a single polyprotein. The polyprotein, which is processed by both cellular and viral proteases to produce mature proteins, contains similar functional organization, with structural proteins at the N terminus and nonstructural proteins at the C terminus (Rice, 1995).

The dengue virus group contains four distinct serotypes. Each can cause a debilitating, but nonfatal, fever. More severe disease, dengue hemorrhagic fever, results from reinfection with a heterologous serotype presumably caused by antibody-mediated enhancement of virus replication in monocytic cells. Because of this phenomenon, vaccine approaches have been unsuccessful, and it may be especially difficult to protect against all four serotypes (Monath, 1994).

Antisense oligonucleotides targeted against the translation initiation region and also to the 3' untranslated region were reported to be effective inhibitors of dengue virus type 2 replication (Raviprakash *et al.*, 1995). Specific modification of the C-5 position of uridine and cytosine with propynyl groups was required before sequence-specific inhibition of viral antigen production was observed at concentrations less than or equal to 1 μM. Phosphodiester and PS oligonucleotides were reported not to inhibit in a sequence-specific fashion. Interestingly, inhibition was detected only after microinjection of oligonucleotides, suggesting that the effective presentation of the propynyl-modified compound to the intact cell may be problematic.

Hepatitis C virus (HCV) is a positive-stranded RNA virus that also infects hepatocytes and causes acute and chronic hepatitis. Hepatitis C virus is the major cause of non-A, non-B hepatitis, and like HBV is associated with hepatocellular carcinoma.

The organization of the 9.5-kb HCV genome is similar to that of pestiviruses and flaviviruses, with structural proteins at the 5' end and nonstructural proteins at the 3' end (Houghton, 1996). The virus encodes a single polyprotein that is processed by viral and cellular proteases. Hepatitis C virus also contains short 5' and 3' untranslated regions (UTRs), the 5' UTR representing the most highly conserved region of the virus (Bukh *et al.*, 1992). This 5' UTR region facilitates internal ribosomal entry, so that translation does not occur by ribosomal scanning from the 5' RNA cap. Instead, ribosomes bind to internal secondary structures formed by the 5' UTR (Wang *et al.*, 1994). In addition, separate experiments have shown that HCV 5' UTR sequences can control translation of downstream sequences (Yoo *et al.*, 1992). The conserved nature of the sequence and its requirement for translation suggest that the 5' UTR potentially provides an excellent target for an antisense oligonucleotide antiviral agent.

Hepatitis C virus was only recently suggested to replicate in cell culture (Yoo *et al.*, 1995). Virus replication in a hepatoma cell line was detected only by the presence of both positive- and negative-strand RNA by RT-PCR. The presence of virus antigen was not reported, and although the infection model may have potential for the future it is not available for antiviral studies. However, oligonucleotides have been evaluated as inhibitors of HCV replication in a T lymphocyte cell line (Mizutani *et al.*, 1995). Sequence-specific inhibition was measured by the presence (or absence) of HCV RNA in the cells, although quantitation of RNA and direct evidence of RNA replication were not presented. These limited data do not provide confidence that virus replication was inhibited and that the cell-based assay systems are going to provide model HCV infections.

Because of the limitations of the virus replication systems, novel model systems for evaluation of antisense oligonucleotides have been developed. *In vitro* translation from RNA containing HCV 5′ UTR sequences has been used to screen oligonucleotides (Wakita and Wands, 1994). Sequence-specific oligonucleotides have been identified that were reported to inhibit translation in this artificial system; oligonucleotides near the translation start site were active in all assays. Other domains in the 5′ UTR that contained oligonucleotide-sensitive sequences were also identified. In other assay systems, oligonucleotides have been transfected into cells along with plasmids expressing the HCV 5′ UTR fused to a reporter gene (luciferase) (Alt *et al.*, 1995). As a control, luciferase expression was compared to production of an unrelated gene (hepatitis B surface antigen) transfected on a separate plasmid. Again, sequence-specific inhibition was reported for oligonucleotides targeted near the translation start site and also in the 5′ UTR. Several oligonucleotides targeted to the core protein-coding sequence also inhibited protein production (Alt *et al.*, 1995).

Hepatitis C virus gene expression was reported to be inhibited in stably transfected hepatocytes. The most active oligonucleotides were again targeted to sequences in the 5′ untranslated region, reinforcing the importance of this region for regulating HCV expression (Hanecak *et al.*, 1996). Phosphorothioate oligonucleotides were reported to inhibit RNA and protein production in a sequence-specific, length-dependent fashion. Interestingly, oligonucleotides with all 2′-OCH$_3$ residues were reported to inhibit protein production without an effect on HCV RNA.

In a similar study, sequence-specific interaction of oligonucleotides with the HCV 5′ UTR was reported in cells stably transfected with HCV bases 52–1417 (P. Brown Vargas, H. A. Hamlin, B. Frank, D. Walther, A. K. Field, and R. E. Kilkuskie, unpublished observations). These cells expressed HCV C and E1 proteins; translation was controlled by the 5′ UTR. When treated with individual oligonucleotides in the presence of cationic lipids, HCV RNA cleavage was detected using a ribonuclease protection assay. Cleavage was mapped to the oligonucleotide-binding site. Cleavage required RNase H, since treatment with oligonucleotides containing all 2′-OCH$_3$ residues did not result in RNA cleavage. Importantly, these results suggest that PS oligonucleotides acted by an antisense mechanism through specific interaction with their target ribonucleotide sequence.

I. Rhabdoviruses

Rhabdoviruses are a widespread family that includes rabies virus. Vesicular stomatitis virus (VSV) is a well-studied member of the rhabdovirus family, and causes flulike symptoms in infected humans. Vesicular stomatitis virus is an enveloped virus containing five proteins and a negative-stranded genome. Five individual messenger RNAs are transcribed from the genome. Novel modifications of oligonucleotides have been evaluated as anti-VSV agents. Psoralen-modified methyl phosphonate oligonucleotides were reported to bind specifically to VSV mRNAs *in vitro* (Levis and Miller, 1994). Specific cross-linking occurred for 16-mer oligonucleotides targeted to the N or M mRNAs. A shorter (12-mer) oligonucleotide designed to bind specifically to M mRNA was also reported to bind to N mRNA. The oligonucleotide was partially complementary (9 of 12 residues) to N mRNA; and N mRNA was the predominant RNA species in the mixture. The high degree of homology of different mRNAs in VSV makes specific oligonucleotide targeting more difficult. However, the ability to target more than one RNA with the same oligonucleotide could potentially improve potency of an antiviral compound. Antiviral activity of these psoralen-modified oligonucleotides was not reported.

In a separate study, a 15-mer oligonucleotide, conjugated at the 3′ end to poly-L-lysine, reportedly inhibited VSV antigen synthesis and virus yield (Lemaitre *et al.*, 1987). The conjugated oligonucleotide, complementary to the 5′ end of VSV N mRNA, reportedly inhibited virus production by >95% at 400 nM. Specificity was suggested by the lack of effect on the replication of a control virus, encephalomyocarditis virus, under the same conditions. Also, a second conjugated oligonucleotide, a 13-mer targeted to the coding sequence of N mRNA, had no effect on virus yield. Unconjugated oligonucleotides were not evaluated, so the effect of polylysine on delivery could not be addressed directly.

IV. Antisense RNA, RNA Decoys, and Ribozymes

A second approach to developing an effective antisense antiviral agent is *in situ* transcription of antisense RNA as a selective blockade of virus replication. Antisense RNA has been used to define the activity of a gene product by observing the phenotypic effects of selective inhibition of translation. This approach was used to help define the roles of the human papillomavirus-encoded E6 and E7 transcripts in cervical carcinoma cells. Von Knebel Doeberitz and Gissmann demonstrated that expression of antisense RNA following transfection into HPV-transformed cells reduced the number of surviving cells capable of colony formation (von Knebel Doeberitz and Gissmann, 1987). Similar transfection of the vector expressing the sense RNA did not alter colony formation, and transfection of the antisense RNA vector into control cells that did not carry HPV had no effect on cell growth.

The expression of antisense RNA has been successful for inhibition of mouse hepatitis virus (Mizutani *et al.*, 1994; Thieringer *et al.*, 1995), parvovirus (Ramirez *et al.*, 1995), polyoma virus (Liu *et al.*, 1994), simian immunodeficiency virus (Tung, 1994), human T lymphotropic virus type I (Fujita and Shiku, 1993), and measles virus (Koschel *et al.*, 1995). But not all antisense RNA studies have resulted in virus inhibition, or in readily interpretable results. Leiter *et al.* demonstrated that even with successful inhibition of influenza replication in cells expressing virus-specific RNA but not the vector, the inhibition probably resulted from the induction of interferon following the formation of double-stranded RNAs hybridized from the sense and antisense strands (Leiter *et al.*, 1989). Double-stranded RNA has been repeatedly confirmed as a potent inducer of interferon α, and is a primary mechanism by which interferon is induced during virus infections (Marcus, 1994). In most studies the induction of interferons has not been checked, leaving suspect the direct role of the antisense mechanism in the antiviral activity resulting from vector-expressed antisense RNA; thus further emphasizes the need for extensive controls in such studies.

For HIV, several approaches (Fig. 13) have been taken to inhibit replication by RNA expression (Yu *et al.*, 1994). Transfection of cells to transcribe antisense RNA to the transactivation response (TAR) element (Chuah *et al.*, 1994), the *gag* region (Sczakiel and Pawlita, 1991), the *tat* and *rev* regions (Junker *et al.*, 1994; Sczakiel *et al.*, 1992), and the 5' leader region of HIV (Sun *et al.*, 1995a) conferred various degrees of protection to HIV challenge infection in cells in culture. An additional nonantisense approach has been the overexpression of TAR or the Rev response element (RRE) as decoys. This provides a means to inactivate the Tat and Rev proteins by alternate binding to the antiviral decoy. This approach has had some success as demonstrated by Bevec *et al.* (1994) and by Lori *et al.* (1994). In a single vector, Chang *et al.* combined the use of a TAR decoy and vector-expressed antisense RNA to the Tat mRNA both to block the activity of the Tat protein and block its translation (Chang *et al.*, 1994). Compared to the vector controls, the combined decoy and antisense approach inhibited 94 to 98% of the Tat activity. A combination of RNA decoy and antisense RNA was taken a step further by the construction of a vector that expresses both the polymeric TAR decoy and the antisense to *tat* (referred to as *antitat*). Lisziewicz *et al.* demonstrated that AIDS patient peripheral blood mononuclear cells transfected with the retroviral vectors carrying the *antitat* gene resisted HIV replication and expanded in culture, providing a possible

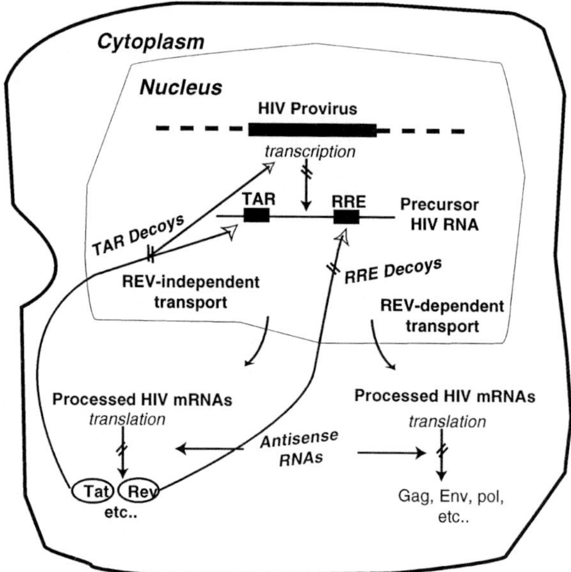

FIGURE 13 HIV inhibition by antisense RNAs and decoys. Antisense RNAs prevent HIV mRNA translation. Alternatively, RNA decoys of TAR and RRE can bind and inactivate the Tat and Rev proteins.

avenue for *ex vivo* gene therapy to provide a competent and expanding pool of CD4$^+$ cells (Lisziewicz *et al.*, 1995).

Ribozymes, which are catalytic RNAs, theoretically combine the specificity of an antisense molecule in binding to the complementary RNA target, with a catalytic core. This combination should provide a potent and selective antiviral agent, and expressed ribozymes are undergoing evaluation and refinement in many laboratories. Sarver *et al.* first demonstrated in 1990 the use of an expressed hammerhead ribozyme targeted to the *gag* sequence of HIV to increase the resistance of HeLa CD4$^+$ cells to HIV infection (Sarver *et al.*, 1990). Since then numerous publications using expressed hammerhead or hairpin ribozymes have appeared and the first clinical trials in AIDS patients should soon be at hand (Leavitt *et al.*, 1996). For a review of the antiviral ribozyme literature see Kijima *et al.* (1995). However, for the purpose of this chapter we highlight only certain key studies. Wong-Staal and colleagues have concentrated on developing potently active hairpin ribozymes that can inhibit HIV replication in peripheral blood lymphocytes and mononuclear cells derived from CD34$^+$ hematopoietic stem cells (Leavitt *et al.*, 1994; Yamada *et al.*, 1994a,b; Yu *et al.*, 1995). They have targeted the ribozymes at either the 5′ leader sequence of HIV-1 or a conserved region of the HIV-1 *pol* gene, and expressed it from a retroviral vector. Hammerhead ribozymes have been utilized by numerous investigators (Homann *et al.*, 1993; Crisell *et al.*, 1993; Sun *et al.*, 1995b; Lo *et al.*, 1992) to target the *tat* gene or the LTR/*gag* region. Both specificity and the requirement of the catalytic activity to obtain optimal anti-HIV activity suggest that the ribozymes are functioning in a catalytic manner, and not merely by an antisense mechanism of action.

V. Future Directions

The aim of this chapter has been to increase awareness of both the opportunities and advances of antisense/gene therapy technology as a means to identify novel antiviral gene targets and to identify molecules with attractive therapeutic potential. Through the use of antisense oligonucleotides and antisense RNA, novel antiviral targets have been reported for HCMV, HIV, the hepatitis viruses and others; in many cases providing the proof of principle needed to launch a concerted effort for serious antiviral drug discovery and development. But in addition, the technology has already yielded antiviral drug candidates, with antisense clinical trials ongoing against HIV and HCMV and others in the preclinical stages of development. In parallel, the antisense, decoy, and ribozyme gene therapy approaches are also gaining credibility as potentially attractive means to enhance cellular resistance to HIV infection.

But where will all this lead? It is our belief that the success of one or more of the antisense therapeutic approaches that are now undergoing evaluation in the clinic will result in accelerated momentum toward developing an array of effective antisense therapeutic oligonucleotides. Viruses that have heretofore been unassailable by conventional therapy should be vulnerable to the antisense approach. And like any evolving field, both specific antisense mechanisms of action and nonspecific activities will be more precisely defined and appreciated for their contribution to the overall antiviral efficacy. In parallel, the identification of novel chemical modifications of the oligonucleotides will provide attractive pharmacological properties as the next generation of oligonucleotide drugs emerges. The end result should be the identification of antisense oligonucleotides that fulfill the promise of antisense—potently active antiviral agents that demonstrate high selectivity for the viral target by virtue of the sequence complementarity of antisense hybridization.

For antisense RNA, RNA decoys, and expressed ribozymes the problems are different. Here the issues concern controlled expression in the appropriate cell. Consequently, emphasis on inducible vector constructs, cell targeting, and RNA stability will be key to practical application.

There should be little doubt that regardless of the outcome of present efforts to show clinical efficacy, the field of genetic pharmacology has scored impressive gains and progress in the field is gaining momentum. This momentum is introducing a new paradigm for antiviral target identification and antiviral drug discovery.

References

Agrawal, S., and Tang, J. Y. (1992). GEM 91—an antisense oligonucleotide phosphorothioate as a therapeutic agent for AIDS. *Antisense Res. Dev.* **2**, 261–266.

Agrawal, S., and Temsamani, J. (1996). *In* "Molecular Medicine: Antisense Therapeutics" (S. Agrawal, ed.), pp. 247–270. Humana Press, Totowa, New Jersey.

Agrawal, S., Ikeuchi, T., Sun, D., Sarin, P. S., Konopka, A., Maizel, J., and Zamecnik, P. C. (1989). Inhibition of human immunodeficiency virus in early infected and chronically infected cells by antisense oligodeoxynucleotides and their phosphorothioate analogues. *Proc. Natl. Acad. Sci. U.S.A.* **86**, 7790–7794.

Alt, M., Renz, R., Hofschneider, P. H., Paumgartner, G., and Caselmann, W. H. (1995). Specific inhibition of hepatitis C viral gene expression by antisense phosphorothioate oligodeoxynucleotides. *Hepatology* **22**, 707–717.

Anazodo, M. I., Wainberg, M. A., Friesen, A. D., and Wright, J. A. (1995). Sequence-specific inhibition of gene expression by a novel antisense oligodeoxynucleotide phosphorothioate directed against a nonregulatory region of the human immunodeficiency virus type 1 genome. *J. Virol.* **69**, 1794–1801.

Anderson, K. P. (1994). Pre-clinical and clinical antiviral activity of an antisense oligonucleotide complementary to cytomegalovirus immediate early RNA. *Antivirals, Second Annual IBC Conference* (Abstr.).

Austermann, S., Kruhoffer, M., and Grosse, F. (1992). Inhibition of human immunodeficiency virus type 1 reverse transcriptase by 3′-blocked oligonucleotide primers. *Biochem. Pharmacol.* **43**, 2581–2589.

Azad, R. F., Driver, V. B., Tanaka, K., Crooke, R. M., and Anderson, K. P. (1993). Antiviral activity of a phosphorothioate oligonucleotide complementary to RNA of the human cytomegalovirus major immediate-early region. *Antimicrob. Agents Chemother.* **37**, 1945–1954.

Azad, R. F., Brown-Driver, V., Buckheit, R. W., Jr., and Anderson, K. P. (1995). Antiviral activity of a phosphorothioate oligonucleotide complementary to human cytomegalovirus RNA when used in combination with antiviral nucleoside analogs. *Antiviral Res.* **28**, 101–111.

Belshe, R. B., Smith, M. H., Hall, C. B., Betts, R., and Hay, A. J. (1988). Genetic basis of resistance to rimantadine emerging during treatment of influenza virus infection. *J. Virol.* **62**, 1508–1512.

Bevec, D., Volc-Platzer, B., Zimmermann, K., Dobrovnik, M., Hauber, J., Veres, G., and Böhnlein, E. (1994). Constitutive expression of chimeric *neo*-Rev response element transcripts suppresses HIV-1 replication in human CD4$^+$ T lymphocytes. *Hum. Gene Ther.* **5**, 193–201.

Bishop, J. S., Guy-Caffey, J. K., Ojwang, J. O., Smith, S. R., Hogan, M. E., Cossum, P. A., Rando, R. F., and Chaudhary, N. (1996). Intramolecular G-quartet motifs confer nuclease resistance to a potent anti-HIV oligonucleotide. *J. Biol. Chem.* **271**, 5698–5703.

Blum, H. E., Galun, E., Weizsacker, F., and Wands, J. R. (1991). Inhibition of hepatitis B virus by antisense oligodeoxynucleotides (letter). *Lancet* **337**, 1230.

Boiziau, C., Thuong, N. T., and Toulmé, J.-J. (1992). Mechanisms of the inhibition of reverse transcription by antisense oligonucleotides. *Proc. Natl. Acad. Sci. U.S.A.* **89**, 768–772.

Bordier, B., Hélène, C., Barr, P. J., Litvak, S., and Sarih-Cottin, L. (1992). *In vitro* effect of antisense oligonucleotides on human immunodeficiency virus type 1 reverse transcription. *Nucleic Acids Res.* **20**, 5999–6006.

Bowen, E. F., Atkins, M., Weller, I. V. D., and Johnson, M. A. (1995). *In* "Antiviral Chemotherapy" (D. J. Jeffries and E. De Clercq, eds.), pp. 65–80. John Wiley & Sons, New York.

Buckheit, R. W., Jr., Roberson, J. L., Lackman-Smith, C., Wyatt, J. R., Vickers, T. A., and Ecker, D. J. (1994). Potent and specific inhibition of HIV envelope-mediated cell fusion and virus binding by G quartet-forming oligonucleotide (ISIS 5320). *AIDS Res. Hum. Retroviruses* **10**, 1497–1506.

Bukh, J., Purcell, R. H., and Miller, R. H. (1992). Sequence analysis of the 5′ noncoding region of hepatitis C virus. *Proc. Natl. Acad. Sci. U.S.A.* **89**, 4942–4946.

Bunnell, B. A., Askari, F. K., and Wilson, J. M. (1992). Targeted delivery of antisense oligonucleotides by molecular conjugates. *Somatic. Cell Mol. Genet.* **18**, 559–569.

Burgess, T. L., Fisher, E. F., Ross, S. L., Bready, J. V., Qian, Y., Bayewitch, L. A., Cohen, A. M., Herrera, C. J., Hu, S. S.-F., Kramer, T. B., Lott, F. D., Martin, F. H., Pierce, G. F., Simonet, L., and Farrell, C. L. (1995). The antiproliferative activity of c-*myb* and c-*myc* antisense oligonucleotides in smooth muscle cells is caused by a nonantisense mechanism. *Proc. Natl. Acad. Sci. U.S.A.* **92**, 4051–4055.

Cantor, G. H., and Palmer, G. H. (1992). Antisense oligonucleotide inhibition of bovine leukemia virus tax expression in a cell-free system. *Antisense Res. Dev.* **2**, 147–152.

Chang, H.-K., Gendelman, R., Lisziewicz, J., Gallo, R. C., and Ensoli, B. (1994). Block of HIV-1 infection by a combination of antisense *tat* RNA and TAR decoys: A strategy for control of HIV-1. *Gene Ther.* **1**, 208–216.

Chen, T. Z., Lin, S. B., Wu, J. C., Choo, K. B., and Au, L. C. (1996). A method for screening antisense oligodeoxyribonucleotides effective for mRNA translation-arrest. *J. Biochem.* (*Tokyo*) **119**, 252–255.

Chuah, M. K. L., Vandendriessche, T., Chang, H.-K., Ensoli, B., and Morgan, R. A. (1994). Inhibition of human immunodeficiency virus type-1 by retroviral vectors expressing antisense-TAR. *Hum. Gene Ther.* **5**, 1467–1475.

Chung, T. D. Y., Cianci, C., Hagen, M., Terry, B., Matthews, J. T., Krystal, M., and Colonno, R. J. (1994). Biochemical studies on capped RNA primers identify a class of oligonucleotide inhibitors of the influenza virus RNA polymerase. *Proc. Natl. Acad. Sci. U.S.A.* **91**, 2372–2376.

Coffin, J. (1996). *In* "Virology" (B. N. Fields, D. M. Knipe, and P. M. Howley, eds.), 3rd Ed., pp. 1767–1847. Lippincott-Raven, Philadelphia.

Collins, P. L., McIntosh, K., and Chanock, R. M. (1996). *In* "Virology" (B. N. Fields, D. M. Knipe, and P. M. Howley, eds.), 3rd Ed., pp. 1313–1351. Lippincott-Raven, Philadelphia.

Cowsert, L. M., Fox, M. C., Zon, G., and Mirabelli, C. K. (1993). *In vitro* evaluation of phosphorothioate oligonucleotides targeted to the E2 mRNA of papillomavirus: Potential treatment for genital warts. *Antimicrob. Agents Chemother.* **37**, 171–177.

Crisell, P., Thompson, S., and James, W. (1993). Inhibition of HIV-1 replication by ribozymes that show poor activity *in vitro*. *Nucleic Acids Res.* **21**, 5251–5255.

Crooke, R. M., Hoke, G. D., and Shoemaker, J. E. (1992). *In vitro* toxicological evaluation of ISIS 1082, a phosphorothioate oligonucleotide inhibitor of herpes simplex virus. *Antimicrob. Agents Chemother.* **36**, 527–532.

Crooke, S. T., and Bennett, C. F. (1996). Progress in antisense oligonucleotide therapeutic. *Annu. Rev. Pharmacol. Toxicol.* **36**, 107–129.

Crooke, S. T., Graham, M. J., Zuckerman, J. E., Brooks, D., Conklin, B. S., Cummins, L. L., Greig, M. J., Guinosso, C. J., Kornblust, D., Manoharan, M., Sasmor, H., Schleich, T., Tivel, K. L., and Griffey, R. H. (1996). Pharmacokinetic properties of several novel oligonucleotide analogs in mice. *J. Pharm. Exp. Ther.* **277**, 923–937.

Daibata, M., Enzinger, E. M., Monroe, J. E., Kilkuskie, R. E., Field, A. K., and Mulder, C. (1996). Antisense oligodeoxynucleotides against the BZLF1 transcript inhibit induction of productive Epstein–Barr virus replication. *Antiviral Res.* **29**, 243–260.

De Clercq, E. (1995). Antiviral therapy for human immunodeficiency virus infections. *Clin. Microbiol. Rev.* **8**, 200–239.

Dienstag, J. L., Perillo, R. P., Schiff, E. R., Bartholomew, M., Vicary, C., and Rubin, M. (1995). A preliminary trial of lamivudine for chronic hepatitis B infection. *N. Engl. J. Med.* **333**, 1657–1661.

Draper, K. G., Ceruzzi, M., Kmetz, M. E., and Sturzenbercker, L. J. (1990). Complementary oligonucleotide sequence inhibits both Vmw65 gene expression and replication of herpes simplex virus. *Antiviral Res.* **13**, 151–164.

Falloon, J., and Masur, H. (1990). *In* "Antiviral Agents and Viral Diseases of Man" (G. J. Galasso, R. J. Whitley, and T. C. Merigan, eds.), 3rd Ed., pp. 669–690. Raven Press, New York.

Fennewald, S. M., Mustain, S., Ojwang, J., and Rando, R. F. (1995). Inhibition of herpes simplex virus in culture by oligonucleotides composed entirely of deoxyguanosine and thymidine. *Antiviral Res.* **26**, 37–54.

Field, A. K. (1994). *In* "Encyclopedia of Virology" (R. G. Webster and A. Granoff, eds.), pp. 42–49. Academic Press, London.

Field, A. K., and Biron, K. K. (1994). "The end of innocence" revisited: Resistance of herpesviruses to antiviral drugs. *Clin. Microbiol. Rev.* **7**, 1–13.

Field, A. K., and Goodchild, J. (1995). Antisense oligonucleotides: Rational drug design for genetic pharmacology. *Exp. Opin. Invest. Drugs* **4**, 799–821.

Frank, B. L., and Goodchild, J. (1997). *In* "Ribozyme Protocols" (P. C. Turner, ed.), pp. 1–7. Human Press, Totowa, New Jersey.

Frank, B. L., Trainor, B. L., Cocuzza, A. J., Hobbs, F. W., and Chidester, D. R. (1993). Selection of mRNA sequences for targeting with antisense reagents. *International Conference on Nucleic Acid Medical Applications* 4–14 (Abstr.).

Fujita, M., and Shiku, H. (1993). A human T lymphotropic virus type I (HTLV-I) long terminal repeat-directed antisense c-*myc* construct with an Epstein–Barr virus replicon vector inhibits cell growth in a HTLV-I-transformed human T cell line. *FEBS Lett.* **322**, 15–20.

Gao, W., Stein, C. A., Cohen, J. S., Dutschman, G. E., and Cheng, Y.-C. (1989). Effect of phosphorothioate homo-oligodeoxynucleotides on herpes simplex virus type 2-induced DNA polymerase. *J. Biol. Chem.* **264**, 11521–11526.

Gao, W., Jaroszewski, J. W., Cohen, J. S., and Cheng, Y.-C. (1990a). Mechanisms of inhibition of herpes virus type 2 growth by 28-mer phosphorothioate oligodeoxycytidine. *J. Biol. Chem.* **265**, 20172–20178.

Gao, W.-Y., Hanes, R. N., Vazquez-Padua, M. A., Stein, C. A., Cohen, J. S., and Cheng, Y.-C. (1990b). Inhibition of herpes simplex virus type 2 growth by phosphorothioate oligodeoxynucleotides. *Antimicrob. Agents Chemother.* **34**, 808–812.

Gilbert, B. E., and Knight, V. (1986). Biochemistry and clinical applications of ribavirin. *Antimicrob. Agents Chemother.* **30**, 201–205.

Giles, R. V., Ruddell, C. J., Spiller, D. G., Green, J. A., and Tidd, D. M. (1995a). Single base discrimination for ribonuclease H-dependent antisense effects within intact human leukaemia cells. *Nucleic Acids Res.* **23**, 954–961.

Giles, R. V., Spiller, D. G., and Tidd, D. M. (1995b). Detection of ribonuclease H-generated mRNA fragments in human leukemia cells following reversible membrane permeabilization in the presence of antisense oligodeoxynucleotides. *Antisense Res. Dev.* **5**, 23–31.

Goodarzi, G., Gross, S. C., Tewari, A., and Watabe, K. (1990). Antisense oligodeoxyribonucleotides inhibit the expression of the gene for hepatitis B virus surface antigen. *J. Gen. Virol.* **71**, 3021–3025.

Goodchild, J., Agrawal, S., Civiera, M. P., Sarin, P. S., Sun, D., and Zamecnik, P. C. (1988). Inhibition of human immunodeficiency virus replication by antisense oligonucleotides. *Proc. Natl. Acad. Sci. U.S.A.* **85**, 5507–5511.

Groothuis, J. R. (1994). Role of antibody and use of respiratory syncytial virus (RSV) immune globulin to prevent severe RSV disease in high-risk children. *J. Pediatr.* **124**(Suppl.), S28–S32.

Gutiérrez, A., Rodríguez, A., Pintado, B., and Sobrino, F. (1993). Transient inhibition of foot-and-mouth disease virus infection of BHK-21 cells by antisense oligonucleotides directed against the second functional initiator AUG. *Antiviral Res.* **22,** 1–13.

Haffey, M. L., and Field, A. K. (1995). *In* "Antiviral Chemotherapy" (D. J. Jeffries and E. De Clercq, eds.), pp. 83–126. John Wiley & Sons, New York.

Han, J., Zhu, Z., Hsu, C., and Finley, W. H. (1994). Selection of antisense oligonucleotides on the basis of genomic frequence of the target sequence. *Antisense Res. Dev.* **4,** 53–65.

Hanecak, R., Brown-Driver, V., Fox, M. C., Azad, R. F., Furusako, S., Nozaki, C., Ford, C., Sasmor, H., and Anderson, K. P. (1996). Antisense oligonucleotide inhibition of hepatitis C virus gene expression in transformed hepatocytes. *J. Virol.* **70,** 5203–5212.

Hatta, T., Kim, S.-G., Nakashima, H., Yamamoto, N., Sakamoto, K., Yokoyama, S., and Takaku, H. (1993). Mechanisms of the inhibition of reverse transcription by unmodified and modified antisense oligonucleotides. *FEBS Lett.* **330,** 161–164.

Ho, S. P., Britton, D. H. O., Stone, B. A., Behrens, D. L., Leffet, L. M., Hobbs, F. W., Miller, J. A., and Trainor, G. L. (1996). Potent antisense oligonucleotides to the human multidrug resistance-1 mRNA are rationally selected by mapping RNA-accessible sites with oligonucleotide libraries. *Nucleic Acids Res.* **24,** 1901–1907.

Homann, M., Tzortzakaki, S., Rittner, K., Sczakiel, G., and Tabler, M. (1993). Incorporation of the catalytic domain of a hammerhead ribozyme into antisense RNA enhances its inhibitory effect on the replication of human immunodeficiency virus type 1. *Nucleic Acids Res.* **21,** 2809–2814.

Houghton, M. (1996). *In* "Virology" (B. N. Fields, D. M. Knipe, and P. M. Howley, eds.), 3rd Ed., pp. 1035–1058. Lippincott-Raven, Philadelphia.

Howley, P. M. (1996). *In* "Virology" (B. N. Fields, D. M. Knipe, and P. M. Howley, eds.), 3rd Ed., pp. 2045–2076. Lippincott Raven, Philadelphia.

Huang, D. P. (1991). *In* "Nasopharyngeal Carcinoma" (C. A. van Hasselt and A. G. Gibb, eds.), pp. 23–36. The Chinese University Press, Hong Kong.

Jairath, S., Brown Vargas, P., Hamlin, H. A., Field, A. K., and Kilkuskie, R. E. (1997). Inhibition of respiratory syncytial virus replication by antisense oligodeoxyribonucleotides. *Antiviral Res.* **33,** 201–213.

Jang, S. K., Krausslich, H. G., Nicklin, M. J. H., Duke, G. M., Palmenberg, A. C., and Wimmer, E. (1988). A segment of the 5′ nontranslated region of encephalomyocarditis virus RNA directs internal entry of ribosomes during *in vitro* translation. *J. Virol.* **62,** 2636–2643.

Jang, S. K., Davies, M. V., Kaufman, R. J., and Wimmer, E. (1989). Initiation of protein synthesis by internal entry of ribosomes into the 5′ nontranslated region of encephalomyocarditis virus RNA *in vivo*. *J. Virol.* **63,** 1651–1660.

Junker, U., Rittner, K., Homann, M., Bevec, D., Bohnlein, E., and Sczakiel, G. (1994). Reduction in replication of the human immunodeficiency virus type 1 in human T cell lines by polymerase III-driven transcription of chimeric tRNA–antisense RNA genes. *Antisense Res. Dev.* **4,** 165–172.

Kabanov, A. V., Vinogradov, S. V., Ovcharenko, A. V., Krivonos, A. V., Melik-Nubarov, N. S., Kiselev, V. I., and Severin, E. S. (1990). A new class of antivirals: Antisense oligonucleotides combined with a hydrophobic substituent effectively inhibit influenza virus reproduction and synthesis of virus-specific proteins in MDCK cells. *FEBS Lett.* **259,** 327–330.

Kawamura, M., Hayashi, M., Furuichi, T., Nonoyama, M., Isogai, E., and Namioka, S. (1991). The inhibitory effects of oligonucleotides, complementary to Marek's disease virus mRNA transcribed from the BamHI–H region, on the proliferation of transformed lymphoblastoid cells, MDCC-MSB1. *J. Gen. Virol.* **72,** 1105–1111.

Kean, J. M., Kipp, S. A., Miller, P. S., Kulka, M., and Aurelian, L. (1995). Inhibition of herpes simplex virus replication by antisense oligo-2′-O-methylribonucleoside methylphosphonates. *Biochemistry* **34,** 14617–14620.

Kijima, H., Ishida, H., Ohkawa, T., Kashani-Sabet, M., and Scanlon, K. J. (1995). Therapeutic applications of ribozymes. *Pharmacol. Ther.* **68,** 247–267.

Kinchington, D., and Redshaw, S. (1995). *In* "Antiviral Chemotherapy" (D. J. Jeffries and E. De Clercq, eds.), pp. 3–40. John Wiley & Sons, New York.

Kinchington, D., Galpin, S., Jaroszewski, J. W., Ghosh, K., Subasinghe, C., and Cohen, J. S. (1992). A comparison of *gag, pol,* and *rev* antisense oligodeoxynucleotides as inhibitors of HIV-1. *Antiviral Res.* **17,** 53–62.

Kitajima, I., Shinohara, T., Bilakovics, J., Brown, D. A., Xu, X., and Nerenberg, M. (1992a). Ablation of transplanted HTLV-I tax-transformed tumors in mice by antisense inhibition of NF-κB. *Science* **258,** 1792–1795.

Kitajima, I., Shinohara, T., Minor, T., Bibbs, L., Bilakovics, J., and Nerenberg, M. (1992b). Human T-cell leukemia virus type 1 tax transformation is associated with increased uptake of oligodeoxynucleotides *in vitro* and *in vivo. J. Biol. Chem.* **267,** 25881–25888.

Kmetz, M. E., Ceruzzi, M., and Schwartz, J. (1991). Vmw65 phosphorothioate oligonucleotides inhibit HSV KOS replication and Vmw65 protein synthesis. *Antiviral Res.* **16,** 173–184.

Korba, B. E., and Gerin, J. L. (1995). Antisense oligonucleotides are effective inhibitors of hepatitis B virus replication *in vitro. Antiviral Res.* **28,** 225–242.

Koschel, K., Brinckmann, U., and Hoyningen-Huene, V. V.(1995). Measles virus antisense sequences specifically cure cells persistently infected with measles virus. *Virology* **207,** 168–178.

Krieg, A. M., Yi, A.-E., Matson, S., Waldschmidt, T. J., Bishop, G. A., Teasdale, R., Koretzky, G. A., and Klinman, D. M. (1995). CpG motifs in bacterial DNA trigger direct B-cell activation. *Nature (London)* **374,** 546–549.

Kulka, M., and Aurelian, L. (1995). Antiviral activity of and oligo(nucleoside methylphosphonate) that targets HSV-1 immediate-early pre-mRNA 4,5 is augmented by cotreatment with replication-defective adenovirus. *Antisense Res. Dev.* **5,** 243–249.

Kulka, M., Smith, C. C., Aurelian, L., Fishelevich, R., Meade, K., Miller, P., and Ts'o, P. O. P. (1989). Site specificity of the inhibitory effects of oligo(nucleoside methylphosphonate)s complementary to the acceptor splice junction of herpes simples virus type 1 immediate early mRNA 4. *Proc. Natl. Acad. Sci. U.S.A.* **86,** 6868–6872.

Kulka, M., Wachsman, M., Miura, S., Fishelevich, R., Miller, P. S., Ts'o, P. O. P., and Aurelian, L. (1993). Antiviral effect of oligo(nucleoside methylphosphonate)s complementary to the herpes simplex virus type 1 immediate early mRNAs 4 and 5. *Antiviral Res.* **20,** 115–130.

Kulka, M., Smith, C. C., Levis, J., Fishelevich, R., Hunter, J. C. R., Cushman, C. D., Miller, P. S., Ts'o, P. O. P., and Aurelian, L. (1994). Synergistic antiviral activities of oligonucleoside methylphosphonates complementary to herpes simplex virus type 1 immediate-early mRNAs 4, 5, and 1. *Antimicrob. Agents Chemother.* **38,** 675–680.

Lamb, R. A., and Webster, R. G. (1996). *In* "Virology" (B. N. Fields, D. M. Knipe, and P. M. Howley, eds.), 3rd Ed., pp. 1353–1396. Lippincott Raven, Philadelphia.

Leavitt, M. C., Yu, M., Yamada, O., Kraus, G., Looney, D., Poeschla, E., and Wong-Staal, F. (1994). Transfer of an anti-HIV-1 ribozyme gene into primary human lymphocytes. *Hum. Gene Ther.* **5,** 1115–1120.

Leavitt, M. C., Yu, M., Wong-Staal, F., and Looney, D. J. (1996). *Ex vivo* transduction and expansion of CD4+ lymphocytes from HIV+ donors: Prelude to a ribozyme gene therapy trial. *Gene Ther.* **3,** 599–606.

Leiter, J. M., Krystal, M., and Palese, P. (1989). Expression of antisense RNA fails to inhibit influenza virus replication. *Virus Res.* **14,** 141–159.

Leiter, J. M. E., Agrawal, S., Palese, P., and Zamecnik, P. C. (1990). Inhibition of influenza virus replication by phosphorothioate oligonucleotides. *Proc. Natl. Acad. Sci. U.S.A.* **87,** 3430–3434.

Lemaitre, M., Bayard, B., and Lebleu, B. (1987). Specific antiviral activity of a poly(L-lysine)-conjugated oligodeoxyribonucleotide sequence complementary to vesicular stomatitis virus N protein mRNA initiation site. *Proc. Natl. Acad. Sci. U.S.A.* **84,** 648–652.

Levin, M. J. (1994). Treatment and prevention options for respiratory syncytial virus infections. *J. Pediatr.* **124**(Suppl.), S22–S27.

Levis, J. T., and Miller, P. S. (1994). Interactions of psoralen-derivatized oligodeoxyribonucleoside methylphosphonates with vesicular stomatitis virus messenger RNA. *Antisense Res. Dev.* **4**, 223–230.

Lewis, E. J., Szymkowski, D. E., Greenfield, I. M., Kreider, J. W., Roberts, P. C., Frank, B. L., Wolfe, J. L., Kilkuskie, R. E., and Mills, J. S. (1997). Identification of an antisense olignucleotide with *in vivo* activity against human papillomavirus. 2. Efficacy against benign human genital condyloma in a mouse xenograft model. *Antiviral Res.* (in press).

Lisziewicz, J., Sun, D., Metelev, V., Zamecnik, P., Gallo, R. C., and Agrawal, S. (1993). Long-term treatment of human immunodeficiency virus-infected cells with antisense oligonucleotide phosphorothioates. *Proc. Natl. Acad. Sci. U.S.A.* **90**, 3860–3864.

Lisziewicz, J., Sun, D., Weichold, F. F., Thierry, A. R., Lusso, P., Tang, J., Gallo, R. C., and Agrawal, S. (1994). Antisense oligodeoxynucleotide phosphorothioate complementary to Gag mRNA blocks replication of human immunodeficiency virus type 1 in human peripheral blood cells. *Proc. Natl. Acad. Sci. U.S.A.* **91**, 7942–7946.

Lisziewicz, J., Sun, D., Lisziewicz, A., and Gallo, R. C. (1995). Antitat gene therapy: A candidate for late-stage AIDS patients. *Gene Ther.* **2**, 218–222.

Liu, Z., Batt, D. B., and Carmichael, G. G. (1994). Targeted nuclear antisense RNA mimics natural antisense-induced degradation of polyoma virus early RNA. *Proc. Natl. Acad. Sci. U.S.A.* **91**, 4258–4262.

Lo, K. M. S., Biasolo, M. A., Dehni, G., Palú, G., and Haseltine, W. A. (1992). Inhibition of replication of HIV-1 by retroviral vectors expressing tat-antisense and anti-tat ribozyme RNA. *Virology* **190**, 176–183.

Locarnini, S. A., and Cunningham, A. L. (1995). *In* "Antiviral Chemotherapy" (D. J. Jeffries and E. De Clercq, eds.), pp. 441–530. John Wiley & Sons, New York.

Lori, F., Lisziewicz, J., Smythe, J., Cara, A., Bunnag, T. A., Curiel, D., and Gallo, R. C. (1994). Rapid protection against human immunodeficiency virus type 1 (HIV-1) replication mediated by high efficiency non-retroviral delivery of genes interfering with HIV-1 tat and gag. *Gene Ther.* **1**, 27–31.

Marcus, P. I. (1994). *In* "Encyclopedia of Virology" (R. G. Webster and A. Granoff, eds.), pp. 733–739. Academic Press, London.

Martin, R. R., and U.S. and French GEM® 91 Collaborative Study Groups (1995). Antisense therapy with GEM 91: Phase I/II clinical trials to evaluate antiretroviral activity of continuous and intermittent IV regimens. *CHI Second Annual HIV Clinical Trials Meeting* (Abstr.).

Matsukura, M., Zon, G., Shinozuka, K., Mitsuya, H., Reitz, M., Cohen, J. S., and Broder, S. (1987). Phosphorothioate analogs of oligodeoxynucleotides: Inhibitors of replication and cytopathic effects of human immunodeficiency virus. *Proc. Natl. Acad. Sci. U.S.A.* **84**, 7706–7710.

Matsukura, M., Zon, G., Shinozuka, K., Stein, C. A., Mitsuya, H., Cohen, J. S., and Broder, S. (1988). Synthesis of phosphorothioate analogues of oligodeoxynucleotides and their antiviral activity against human immunodeficiency virus (HIV). *Gene* **72**, 343–347.

Matsukura, M., Zon, G., Shinozuka, K., Robert-Guroff, M., Shimada, T., Stein, C. A., Mitsuya, H., Wong-Staal, F., Cohen, J. S., and Broder, S. (1989). Regulation of viral expression of human immunodeficiency virus *in vitro* by an antisense phosphorothioate oligodeoxynucleotide against rev (art/trs) in chronically infected cells. *Proc. Natl. Acad. Sci. U.S.A.* **86**, 4244–4248.

Mizutani, T., Hayashi, M., Maeda, A., Sasaki, N., Yamashita, T., Kasai, N., and Namioka, S. (1994). *In* "Coronaviruses" (H. Laude and J. F. Vautherot, eds.), pp. 129–135. Plenum Press, New York.

Mizutani, T., Kato, N., Hirota, M., Sugiyama, K., Murakami, A., and Shimotohno, K. (1995). Inhibition of hepatitis C virus replication by antisense oligonucleotide in culture cells. *Biochem. Biophys. Res. Commun.* **212**, 906–911.

Monath, T. P. (1994). Dengue: The risk to developed and developing countries. *Proc. Natl. Acad. Sci. U.S.A.* **91**, 2395–2400.

Moriya, K., Matsukura, M., Kurokawa, K., and Koike, K. (1996). *In vivo* inhibition of hepatitis B virus gene expression by antisense phosphorothioate oligonucleotides. *Biochem. Biophys. Res. Commun.* **218**, 217–223.

Offensperger, W. B., Offensperger, S., Walter, E., Teubner, K., Igloi, G., Blum, H. E., and Gerok, W. (1993). *In vivo* inhibition of duck hepatitis B virus replication and gene expression by phosphorothioate modified antisense oligodeoxynucleotides. *EMBO J.* **12**, 1257–1262.

Ojwang, J., Elbaggari, A., Marshall, H. B., Jayaraman, K., McGrath, M. S., and Rando, R. F. (1994a). Inhibition of human immunodeficiency virus type 1 activity *in vitro* by oligonucleotides composed entirely of guanosine and thymidine. *J. Acquir. Immune Defic. Syndr.* **7**, 560–570.

Ojwang, J., Elbaggari, A., Marshall, H. B., Jayaraman, K., McGrath, M. S., and Rando, R. F. (1994b). Inhibition of human immunodeficiency virus type 1 activity *in vitro* by oligonucleotides composed entirely of guanosine and thymidine. *J. Acquir. Immune Defic. Syndr.* **7**, 560–570.

Ojwang, J. O., Buckheit, R. W., Pommier, Y., Mazumder, A., De Vreese, K., Esté, J. A., Reymen, D., Pallansch, L. A., Lackman-Smith, C., Wallace, T. L., De Clercq, E., McGrath, M. S., and Rando, R. F. (1995). T30177, an oligonucleotide stabilized by an intramolecular guanosine octet, is a potent inhibitor of laboratory strains and clinical isolates of human immunodeficiency virus type 1. *Antimicrob. Agents Chemother.* **39**, 2426–2435.

Pagano, J. S. (1995). *In* "Antiviral Chemotherapy" (D. J. Jeffries and E. De Clercq, eds.), pp. 155–195. John Wiley & Sons, New York.

Pagano, J. S., Jimenez, G., Sung, N. S., Raab-Traub, N., and Lin, J.-C. (1992). Epstein–Barr viral latency and cell immortalization as targets for antisense oligomers. *Ann. N.Y. Acad. Sci.* **660**, 107–116.

Pari, G. S., Field, A. K., and Smith, J. A. (1995). Potent antiviral activity of an antisense oligonucleotide complementary to the intron–exon boundary of human cytomegalovirus genes UL36 and UL37. *Antimicrob. Agents Chemother.* **39**, 1157–1161.

Peyman, A., Helsberg, M., Kretzschmar, G., Mag, M., Grabley, S., and Uhlmann, E. (1995). Inhibition of viral growth by antisense oligonucleotides directed against the IE110 and the UL30 mRNA of herpes simplex virus type-1. *Biol. Chem. Hoppe-Seyler* **376**, 195–198.

Poddevin, B., Meguenni, S., Elias, I., Vasseur, M., and Blumenfeld, M. (1994). Improved anti-herpes simplex virus type 1 activity of a phosphodiester antisense oligonucleotide containing a 3′-terminal hairpin-like structure. *Antisense Res. Dev.* **4**, 147–154.

Politz, J. C., Taneja, K. L., and Singer, R. H. (1995). Characterization of hybridization between synthetic oligodeoxynucleotides and RNA in living cells. *Nucleic Acids Res.* **23**, 4946–4953.

Pollack, J. R., and Ganem, D. (1994). Site-specific RNA binding by a hepatitis B virus reverse transcriptase initiates two distinct reactions: RNA packaging and DNA synthesis. *J. Virol.* **68**, 5579–5587.

Ramirez, J. C., Santaren, J. F., and Almendral, J. M. (1995). Transcriptional inhibition of the parvovirus minute virus of mice by constitutive expression of an antisense RNA targeted against the NS-1 transactivator protein. *Virology* **206**, 57–68.

Raviprakash, K., Liu, K., Matteucci, M., Wagner, R., Riffenburgh, R., and Carl, M. (1995). Inhibition of dengue virus by novel, modified antisense oligonucleotides. *J. Virol.* **69**, 69–74.

Rice, C. M. (1995). *In* "Virology" (B. N. Fields, D. M. Knipe, and P. M. Howley, eds.), 3rd Ed., pp. 931–960. Lippincott Raven, Philadelphia.

Roberts, P. C., Frank, B. L., Boldt, S. E., Walther, D. M., Wolfe, J. L., Kilkuskie, R. E., Szymkowski, D. D., Greenfield, I. M., Sullivan, V., and Mills, J. S. (1997). Identification of an antisense olignucleotide with *in vivo* activity against human papillomavirus. 1. *In vitro* evaluation of oligonucleotides. *Antiviral Res.* (in press).

Roizman, B., and Sears, A. E. (1996). *In* "Virology" (B. N. Fields, D. M. Knipe, and P. M. Howley, eds.), 3rd Ed., p. 2231. Lippincott-Raven, Philadelphia.

Roth, G., Curiel, T., and Lacy, J. (1994). Epstein–Barr viral nuclear antigen 1 antisense oligodeoxynucleotide inhibits proliferation of Epstein–Barr virus-immortalized B cells. *Blood* **84**, 582–587.

Sankar, S., Cheah, K. C., and Porter, A. G. (1989). Antisense oligonucleotide inhibition of encephalomyocarditis virus RNA translation. *Eur. J. Biochem.* **184**, 39–45.

Sarver, N., Cantin, E. M., Chang, P. S., Zaia, J. A., Ladne, P. A., Stephens, D. A., and Rossi, J. J. (1990). Ribozymes as potential anti-HIV-1 therapeutic agents. *Science* **247**, 1222–1225.

Sczakiel, G., and Pawlita, M. (1991). Inhibition of human immunodeficiency virus type 1 replication in human T cells stably expressing antisense RNA. *J. Virol.* **65**, 468–472.

Sczakiel, G., Oppenlander, M., Rittner, K., and Pawlita, M. (1992). Tat- and rev-directed antisense RNA expression inhibits and abolishes replication of human immunodeficiency virus type 1: A temporal analysis. *J. Virol.* **66**, 5576–5581.

Sczakiel, G., Homann, M., and Rittner, K. (1993). Computer-aided search for effective antisense RNA target sequences of the human immunodeficiency virus type 1. *Antisense Res. Dev.* **3**, 45–52.

Serini, D., Katlama, C., Gouyette, A., Re, M., Lascoux, C., Tubiana, R., and Tournerie, C. (1994). An open-label safety and pharmacokinetic study of single intravenous or subcutaneous ascending doses of GEM 91 in untreated, adult, HIV positive, asymptomatic human volunteer patients. *34th Intersci. Conf. on Antimicrob. Agents and Chemotherapy* Abst. M10. (Abstr.).

Shah, K. V., and Howley, P. M. (1996). *In* "Virology" (B. N. Fields, D. M. Knipe, and P. M. Howley, eds.), 3rd Ed., pp. 2077–2110. Lippicott-Raven, Philadelphia.

Smith, J. A., and Pari, G. S. (1995). Expression of human cytomegalovirus UL36 and UL37 genes is required for viral DNA replication. *J. Virol.* **69**, 1925–1931.

Speir, E., Modali, R., Huang, E.-S., Leon, M. B., Shawl, F., Finkel, T., and Epstein, S. E. (1994). Potential role of human cytomegalovirus and p53 interaction in coronary restenosis. *Science* **265**, 391–394.

Steele, C., Sacks, P. G., Adler-Storthz, K., and Shillitoe, E. J. (1992). Effect on cancer cells of plasmids that express antisense RNA of human papillomavirus type 18. *Cancer Res.* **52**, 4706–4711.

Stein, C. A., and Krieg, A. M. (1994). Problems in interpretation of data derived from *in vitro* and *in vivo* use of antisense oligodeoxynucleotides. *Antisense Res. Dev.* **4**, 67–69.

Stein, C. A., Cleary, A. M., Yakubov, L., and Lederman, S. (1993). Phosphorothioate oligodeoxynucleotides bind to the third variable loop domain (V3) of human immunodeficiency virus type 1 gp120. *Antisense Res. Dev.* **3**, 19–31.

Stephenson, M. L., and Zamecnik, P. C. (1978). Inhibition of Rous sarcoma virus viral RNA translation by a specific oligodeoxyribonucleotide. *Proc. Natl. Acad. Sci. U.S.A.* **75**, 285–288.

Storey, A., Oates, D., Banks, L., Crawford, L., and Crook, T. (1991). Anti-sense phosphorothioate oligonucleotides have both specific and non-specific effects on cells containing human papillomavirus type 16. *Nucleic Acids Res.* **19**, 4109–4114.

Stull, R. A., Taylor, L. A., and Szoka, F. C. (1992). Predicting antisense oligonucleotide inhibitory efficacy: A computational approach using histograms and thermodynamic indices. *Nucleic Acids Res.* **20**, 3501–3508.

Sun, L. Q., Pyati, J., Smythe, J., Wang, L., Macpherson, J., Gerlach, W., and Symonds, G. (1995a). Resistance to human immunodeficiency virus type 1 infection conferred by transduction of human peripheral blood lymphocytes with ribozyme, antisense, or polymeric trans-activation response element constructs. *Proc. Natl. Acad. Sci. U.S.A.* **92**, 7272–7276.

Sun, L. Q., Wang, L., Gerlach, W. L., and Symonds, G. (1995b). Target sequence-specific inhibition of HIV-1 replication by ribozymes directed to tat RNA. *Nucleic Acids Res.* **23**, 2909–2913.

Tan, T. M. C., and Ting, R. C. Y. (1995). *In vitro* and *in vivo* inhibition of human papillomavirus type 16 E6 and E7 genes. *Cancer Res.* **55**, 4599–4605.

Temsamani, J., and Agrawal, S. (1996). Antisense oligonucleotides as antiviral agents. *Adv. Antiviral Drug Design* **2**, 1–39.

Thieringer, H. A., Takayama, K. M., Kang, C., and Inouye, M. (1995). Antisense RNA-mediated inhibition of mouse hepatitis virus replication in L2 cells. *Antisense Res. Dev.* **5**, 289–294.

Tung, F. Y. T. (1994). Suppression of simian immunodeficiency virus replication in primary peripheral mononuclear cells by antisense RNA. *J. Med. Virol.* **42**, 255–258.

von Knebel Doeberitz, M., and Gissmann, L. (1987). Analysis of the biological role of human papilloma virus (HPV)-encoded transcripts in cervical carcinoma cells by antisense RNA. *Hamatologie Bluttransfusion* **31**, 377–379.

Wakita, T., and Wands, J. R. (1994). Specific inhibition of hepatitis C virus expression by antisense oligodeoxynucleotides. *In vitro* model for selection of target sequence. *J. Biol. Chem.* **269**, 14205–14210.

Wang, C., Sarnow, P., and Siddiqui, A. (1994). A conserved helical element is essential for internal initiation of translation of hepatitis C virus RNA. *J. Virol.* **68**, 7301–7307.

Wu, G. Y., and Wu, C. H. (1992). Specific inhibition of hepatitis B viral gene expression *in vitro* by targeted antisense oligonucleotides. *J. Biol. Chem.* **267**, 12436–12439.

Wyatt, J. R., Vickers, T. A., Roberson, J. L., Buckheit, R. W., Jr., Klimkait, T., DeBaets, E., Davis, P. W., Rayner, B., Imbach, J. L., and Ecker, D. J. (1994). Combinatorially selected guanosine-quartet structure is a potent inhibitor of human immunodeficiency virus envelope-mediated cell fusion. *Proc. Natl. Acad. Sci. U.S.A.* **91**, 1356–1360.

Yamada, O., Kraus, G., Leavitt, M. C., Yu, M., and Wong-Staal, F. (1994a). Activity and cleavage site specificity of an anti-HIV-1 hairpin ribozyme in human T cells. *Virology* **205**, 121–126.

Yamada, O., Yu, M., Yee, J.-K., Kraus, G., Looney, D., and Wong-Staal, F. (1994b). Intracellular immunization of human T cells with a hairpin ribozyme against human immunodeficiency virus type 1. *Gene Ther.* **1**, 38–45.

Yao, G.-Q., Grill, S., Egan, W., and Cheng, Y.-C. (1993). Potent inhibition of Epstein–Barr virus by phosphorothioate oligodeoxynucleotides without sequence specification. *Antimicrob. Agents Chemother.* **37**, 1420–1425.

Yoo, B. J., Spaete, R. R., Geballe, A. P., Selby, M., Houghton, M., and Han, J. H. (1992). 5′ End-dependent translation initiation of hepatitis C viral RNA and the presence of putative positive and negative translational control elements within the 5′ untranslated region. *Virology* **191**, 889–899.

Yoo, B. J., Selby, M. J., Choe, J., Suh, B. S., Choi, S. H., Joh, J. S., Nuovo, G. J., Lee, H.-S., Houghton, M., and Han, J. H. (1995). Transfection of a differentiated human hepatoma cell line (Huh7) with *in vitro*-transcribed hepatitis C virus (HCV) RNA and establishment of a long-term culture persistently infected with HCV. *J. Virol.* **69**, 32–38.

Yu, M., Poeschla, E., and Wong-Staal, F. (1994). Progress towards gene therapy for HIV infection. *Gene Ther.* **1**, 13–26.

Yu, M., Poeschla, E., Yamada, O., Degrandis, P., Leavitt, M. C., Heusch, M., Yees, J.-K., Wong-Staal, F., and Hampel, A. (1995). *In vitro* and *in vivo* characterization of a second functional hairpin ribozyme against HIV-1. *Virology* **206**, 381–386.

Zamecnik, P. C., and Stephenson, M. L. (1978). Inhibition of Rous sarcoma virus replication and cell transformation by a specific oligodeoxynucleotide. *Proc. Natl. Acad. Sci. U.S.A.* **75**, 280–284.

Zamecnik, P. C., Goodchild, J., Taguchi, Y., and Sarin, P. S. (1986). Inhibition of replication and expression of human T-cell lymphotropic virus type III in cultured cells by expgenous synthetic oligonucleotides complementary to viral RNA. *Proc. Natl. Acad. Sci. U.S.A.* **83,** 4143–4146.

Zelphati, O., Imbach, J.-L., Signoret, N., Zon, G., Rayner, B., and Leserman, L. (1994). Antisense oligonucleotides in solution or encapsulated in immunoliposomes inhibit replication of HIV-1 by several different mechanisms. *Nucleic Acids Res.* **22,** 4307–4314.

Zhang, R., Lu, Z., Zhao, H., Zhang, X., Diasio, R. B., Habus, I., Jiang, Z., Iyer, R. P., Yu, D., and Agrawal, S. (1995a). *In vivo* stability, disposition and metabolism of a "hybrid" oligonucleotide phosphorothioate in rats. *Biochem. Pharmacol.* **50,** 545–556.

Zhang, R., Yan, J., Shahinian, H., Amin, G., Lu, Z., Saag, M., Jiang, Z., Temsamani, J., Martin, R. R., Schechter, P. J., Agrawal, S., and Diasio, R. B. (1995b). Pharmacokinetics of an anti-HIV antisense oligonucleotide phosphorothioate (GEM 91) in HIV-infected subjects. *Clin. Pharmacol. Therapeut.* **58,** 44–53.

Index

Contents of Previous Volumes

Myocardial Ishemic Preconditioning
Donna M. Van Winkle, Grace L. Chien, and Richard F. Davis

Effects of Hypoxia/Reoxygenation on Intracellular Calcium Ion Homeostasis in Ventricular Myocytes during Halothane Exposure
Paul R. Knight, Mitchell D. Smith, and Bruce A. Davidson

Mechanical Consequences of Calcium Channel Modulation during Volatile Anesthetic-Induced Left Ventricular Systolic and Diastolic Dysfunction
Paul S. Pagel and David C. Warltier

Anesthetic Actions on Calcium Uptake and Calcium-Dependent Adenosine Triphophatase Activity of Cardiac Sarcoplasmic Reticulum
Ning Miao, Martha J. Frazer, and Carl Lynch III

Interaction of Anesthetics and Catecholamines on Conduction in the Canine His-Purkinje System
L. A. Turner, S. Vodanovic, and Z. J. Bosnjak

Anesthetics, Catecholamines, and Ouabain on Automaticity of Primary and Secondary Pacemakers
John L. Atlee III, Martin N. Vincenzi, Harvey J. Woehlck, and Zelijko J. Bosnjak

The Role of L-Type Voltage-Dependent Calcium Channels in Anesthetic Depression of Contractility
T. J. J. Blanck, D. L. Lee, S. Yasukochi, C. Hollmann, and J. Zhang

Effects of Inhibition of Transsarcolemmal Calcium Influx on Content and Releasability of Calcium Stored in Sarcoplasmic Reticulum of Intact Myocardium
Hirochika Komai and Ben F. Rusy

Arrhythmogenic Effect of Inhalation Anesthetics: Biochemical Heterogeneity between Conduction and Contractile Systems and Protein Unfolding
Issaku Ueda and Jan-Shing Chiou

Potassium Channel Current and Coronary Vasodilatation by Volatile Anesthetics
Nediijka Buljubasic, Jure Mariijic, and Zelijko J. Bosnjak

Cerebral Physiology during Cardiopulmonary Bypass: Pulsatile versus Nonpulsatile Flow
Brad Hindman

Anesthetic Actions of Cardiovascular Control Mechanisms in the Central Nervous System
William T. Schmeling and Neil E. Farber

Volume 32

Signal Sorting by G-Protein-Linked Receptors
Graeme Milligan

Regulation of Phospholipase A_2 Enzymes: Selective Inhibitors and Their Pharmacological Potential
Keith B. Glaser

Platelet Activating Factor Antagonists
James B. Summers and Daniel H. Albert

Pharmacological Management of Acute and Chronic Bronchial Asthma
Michael K. Gould and Thomas A. Raffin

Anti-Human Immunodeficiency Virus Immunoconjugates
Seth H. Pincus and Vladimir V. Tolstikov

Recent Advances in the Treatment of Human Immunodeficiency Virus Infections with Interferons and Other Biological Response Modifiers
Orjan Strannegård

Advances in Cancer Gene Therapy
Wei-Wei Zhang, Toshiyoshi Fujiwara, Elizabeth A. Grimm, and Jack A. Roth

Melanoma and Melanocytes; Pigmentation, Tumor Progression, and the Immune Response to Cancer
Setaluri Vijayasaradhi and Alan N. Houghton

High-Density Lipoprotein Cholesterol, Plasma Triglyceride, and Coronary Heart Disease: Pathophysiology and Management
Wolfgang Patsch and Antonio M. Gotto, Jr.

Neurotransmitter-like Actions of ɪ-DOPA
Yoshimi Misu, Hiroshi Ueda, and Yoshio Goshima

New Approaches to the Drug Treatment of Schizophrenia
Gavin P. Reynolds and Carole Czudek

Membrane Trafficking in Nerve Terminals
Flavia Valtorta and Fabio Benfenati

Volume 33

Endothelin Receptor Antagonism
Terry J. Opgenorth

The Ryanodine Receptor Family of Intracellular Calcium
Release Channels
Vincenzo Sorrentino

Design and Pharmacology of Peptide Mimetics
Graham J. Moore, Julian R. Smith, Barry W. Baylis, and John M. Matsoukas

Alternative Approaches for the Application of Ribozymes as Gene
Therapies for Retroviral Infections
Thomas B. Campbell and Bruce A. Sullenger

Inducible Cyclooxygenase and Nitric Oxide Synthase
Kenneth K. Wu

Regulation of Airway Wall Remodeling: Prospects for the
Development of Novel Antiasthma Drugs
Alastair G. Stewart, Paul R. Tomlinson, and John W. Wilson

Advances in Selective Immunosuppression
Luciano Adorini, Jean-Charles Guéry, and Sylvie Trembleau

Monoclonal Antibody Therapy of Leukemia and Lymphoma
Joseph G. Jurcic, Philip C. Caron, and David A. Scheinberg

4-Hydroxyphenylretinamide in the Chemoprevention of Cancer
Harmesh R. Naik, Gregory Kalemkerian, and Kenneth J. Pienta

Immunoconjugates and Immunotoxins for Therapy of Carcinomas
Ingegerd Hellström, Karl Erik Hellström, Clay B. Siegall, and Pamela A. Trail

Nitric Oxide and Peripheral Adrenergic Neuromodulation
Roberto Levi, Kwan Ha Park, Michiaki Imamura, Nahid Seyedi, and Harry M. Lander

A Study on Tumor Necrosis Factor, Tumor Necrosis Factor Receptors, and Nitric Oxide in Human Fetal Glial Cultures
Barbara A. St. Pierre, Douglas A. Granger, Joyce L. Wong, and Jean E. Merrill

Inhaled Nitric Oxide, Clinical Rationale and Applications
Claes G. Frostell and Warren M. Zapol

Inhaled Nitric Oxide Therapy of Pulmonary Hypertension and Respiratory Failure in Premature and Term Neonates
Steven H. Abman and John P. Kinsella

Clinical Applications of Inhaled Nitric Oxide in Children with Pulmonary Hypertension
David L. Wessel and Ian Adatia

Volume 35

Interactions between Drugs and Nutrients
C. Tschanz, W. Wayne Stargel, and J. A. Thomas

Induction of Cyclo-Oxygenase and Nitric Oxide Synthase in Inflammation
Ian Appleton, Annette Tomlinson, and Derek A. Willoughby

Current and Future Therapeutic Approaches to Hyperlipidemia
John A. Farmer and Antonio M. Gotto, Jr.

In Vivo Pharmacological Effects of Ciclosporin and Some Analogues
Jean F. Borel, Götz Baumann, Ian Chapman, Peter Donatsch, Alfred Fahr, Edgar A. Mueller, and Jean-Marie Vigouret

Mono-ADP-ribosylation: A Reversible Posttranslational Modification of Proteins
Ian J. Okazaki and Joel Moss

Activation of Programmed (Apoptotic) Cell Death for the Treatment of Prostate Cancer
Samuel R. Denmeade and John T. Isaacs

Volume 37

Cryptococcosis
Judith A. Aberg and William G. Powderly

Antimalarial Activity of Artemisinin (Qinghaosu) and Related
Trioxanes: Mechanism(s) of Action
Jared N. Cumming, Poonsakdi Ploypradith, and Gary H. Posner

The Role of Endothelin in the Pathogenesis of Atherosclerosis
Mark C. Kowala

The Pharmacology and Molecular Biology of Large-Conductance
Calcium-Activated (BK) Potassium Channels
Valentin K. Gribkoff, John E. Starrett, Jr., and Steven I. Dworetzky

Update on Invasive Candidiasis
Libsen J. Rodriguez, John H. Rex, and Elias J. Anaissie

Volume 38

Antioxidants: The Basics—What They Are and How to
Evaluate Them
Barry Halliwell

Metabolism of Vitamin C in Health and Disease
Ann M. Bode

Regulation of Human Plasma Vitamin E
Maret G. Traber

Glutathione and Glutathione Delivery Compounds
Mary E. Anderson

α-Lipoic Acid: A Metabolic Antioxidant and Potential Redox
Modulator of Transcription
Lester Packer, Sashwati Roy, and Chandan K. Sen

Antioxidant Actions of Melatonin
Russel J. Reiter

Antioxidative and Metal-Chelating Effects of Polyamines
Erik Løvaas

Antioxidant and Chelating Properties of Flavonoids
Ludmila G. Korkina and Igor B. Afanas'ev

Sodium Channels and Therapy of Central Nervous System Diseases
Charles P. Taylor and Lakshmi S. Narasimhan

Anti-adhesion Therapy
Carol J. Cornejo, Robert K. Winn, and John M. Harlan

Use of Azoles for Systemic Antifungal Therapy
Carol A. Kauffman and Peggy L. Carver

Pharmacology of Neuronal Nicotinic Acetylcholine Receptor Subtypes
Lorna M. Colquhoun and James W. Patrick

Structure and Function of Leukocyte Chemoattractant
Richard D. Ye and François Boulay

Pharmacologic Approaches to Reperfusion Injury
James T. Willerson

Restenosis: Is There a Pharmacologic Fix in the Pipeline?
Joan A. Keiser and Andrew C. G. Uprichard

Role of Adenosine as a Modulator of Synaptic Activity in the Central Nervous System
James M. Brundege and Thomas V. Dunwiddie

Combination Vaccines
Ronald W. Ellis and Kenneth R. Brown

Pharmacology of Potassium Channels
Maria L. Garcia, Markus Hanner, Hans-Günther Knaus, Robert Koch, William Schmalhofer, Robert S. Slaughter, and Gregory J. Kaczorowski